D1551128

# PRINCIPLES
# OF
# SOLAR
# ENGINEERING

## SERIES IN THERMAL AND FLUIDS ENGINEERING

JAMES P. HARTNETT and THOMAS F. IRVINE, JR., Editors
JACK P. HOLMAN, Senior Consulting Editor

| | |
|---|---|
| Cebeci and Bradshaw | • Momentum Transfer in Boundary Layers |
| Chang | • Control of Flow Separation: Energy Conservation, Operational Efficiency, and Safety |
| Chi | • Heat Pipe Theory and Practice: A Sourcebook |
| Eckert and Goldstein | • Measurements in Heat Transfer, 2d edition |
| Edwards, Denny, and Mills | • Transfer Processes: An Introduction to Diffusion, Convection, and Radiation |
| Fitch and Surjaatmadja | • Introduction to Fluid Logic |
| Ginoux | • Two-Phase Flows and Heat Transfer with Application to Nuclear Reactor Design Problems |
| Hsu and Graham | • Transport Processes in Boiling and Two-Phase Systems, Including Near-Critical Fluids |
| Kreith and Kreider | • Principles of Solar Engineering |
| Lu | • Introduction to the Mechanics of Viscous Fluids |
| Moore and Sieverding | • Two-Phase Steam Flow in Turbines and Separators: Theory, Instrumentation, Engineering |
| Richards | • Measurement of Unsteady Fluid Dynamic Phenomena |
| Sparrow and Cess | • Radiation Heat Transfer, augmented edition |
| Tien and Lienhard | • Statistical Thermodynamics, revised printing |
| Wirz and Smolderen | • Numerical Methods in Fluid Dynamics |

### PROCEEDINGS

| | |
|---|---|
| Keairns | • Fluidization Technology |
| Spalding and Afgan | • Heat Transfer and Turbulent Buoyant Convection: Studies and Applications for Natural Environment, Buildings, Engineering Systems |
| Zarić | • Thermal Effluent Disposal from Power Generation |

# PRINCIPLES
# OF
# SOLAR
# ENGINEERING

### FRANK KREITH
Solar Energy Research Institute
and
University of Colorado

### JAN F. KREIDER
Consulting Engineer

**HEMISPHERE PUBLISHING CORPORATION**

Washington    London

**McGRAW–HILL BOOK COMPANY**

New York    St. Louis    San Francisco    Auckland    Bogotá    Düsseldorf
Johannesburg    London    Madrid    Mexico    Montreal    New Delhi    Panama
Paris    São Paulo    Singapore    Sydney    Tokyo    Toronto

**PRINCIPLES OF SOLAR ENGINEERING**

Copyright © 1978 by Hemisphere Publishing Corporation. All rights reserved. Printed in the United States of America. No part of this publication may be reproduced, stored in a retrieval system, or transmitted, in any form or by any means, electronic, mechanical, photocopying, recording, or otherwise, without the prior written permission of the publisher.

3 4 5 6 7 8 9 0    K P K P    7 8 3 2 1 0 9

This book was set in Press Roman by Hemisphere Publishing Corporation. The editors were Judith B. Gandy and Mary A. Phillips; the designer was Lilia Guerrero; the production supervisor was Rebekah McKinney; and the compositor was Peggy M. Rote.
The printer and binder was Kingsport Press, Inc.

**Library of Congress Cataloging in Publication Data**

Kreith, Frank.
    Principles of solar engineering.

    (Series in thermal and fluids engineering)
    Bibliography: p.
    Includes index.
    1. Solar energy.    I. Kreider, Jan F., date
joint author.    II. Title.
TJ810.K73      621.47      77-27861
ISBN 0-07-035476-6

# CONTENTS

# PREFACE

There are many books dealing with solar energy utilization. Most of these books have been written during the past few years in response to the great popular appeal that the possibility of using the energy of the sun has generated among people all over the world. A close scrutiny of these books will reveal, however, that they are restricted in scope. Some of them deal with the application of solar energy in the design of buildings, some with direct conversion of solar radiation into electricity, and some with the thermal conversion of solar energy into heat and power. Others describe the utilization of wind energy, or give instructions to the solar enthusiast on how to build a solar energy conversion system. With the exception of Farrington Daniels' classic book "Direct Use of the Sun's Energy" and the collection of articles in "An Introduction to the Utilization of Solar Energy," two books that were written fifteen years ago, none of the recent works attempts by itself to cover all aspects of solar energy utilization. The objective of this book is to update and expand these two classic works and to present all aspects of solar energy utilization in a teachable framework from an engineering point of view.

Oil, coal, and natural gas were originally produced by photosynthetic processes, followed by complex chemical reactions in which decaying vegetation was subjected to very high temperatures and pressures over a long period of time. Strictly speaking, one might therefore argue that the fossil fuel energy sources are also solar in origin and should be included in a comprehensive treatment. This book, however, will not deal with the utilization of fossil forms of solar energy, but will restrict itself to renewable types of solar energy utilization systems. The reason for our philosophic commitment to renewable solar energy is that the total amount of stored fossil fuel is limited, and the educational systems of developed countries have a responsibility to prepare engineers for the inevitable task of switching from nonrenewable to renewable energy sources.

The level of presentation in this book assumes that the reader has had a course in basic thermodynamics, has some background in chemistry and physics, and has a knowledge of calculus and ordinary differential equations. Beyond these requirements, however, the book attempts to present all the information necessary for the design and analysis of solar energy conversion systems. However, as will be shown in

ix

more detail later, an engineering analysis of a nonrenewable energy source requires the simultaneous evaluation of its technical and economic viability. To achieve this end the book combines a basic technical understanding with an appreciation of the economic aspects of utilizing nonrenewable energy sources. It is of course impossible to cover all the specific ideas that have been proposed for the utilization of solar energy. To avoid specialization this book emphasizes basic, fundamental approaches that should make it possible for an engineer to evaluate specific systems in a responsible and professional manner.

Since the audience for this book is expected to have varying amounts of experience in the basic scientific and engineering disciplines needed in the technical analysis of solar energy applications, this book can serve several functions. Senior or graduate courses in engineering can be based on the entire book. Chapter 3, which deals with fundamentals of heat transfer, may be excluded and Chapters 8 and 9 surveyed. Undergraduate engineering courses should use Chapters 1, 2, 3, 4, and 6 as the core of a three-hour course with excerpts on future applications taken from Chapters 7, 8, and 9. Courses in the physical sciences should emphasize Chapters 1, 2, 6, 8, and 9, with selected topics excerpted from Chapter 3 and introductions to flat-plate and concentrating collectors taken from Chapter 4. In addition, the book is suitable as a reference work for the practicing engineer in the field of thermal applications of solar energy; the professional will find Chapter 5 on system analysis and economic analysis particularly useful.

All of the basic material in the book is presented in units used in the Système International d'Unités (SI) because we believe that this system will soon be universally adopted and will facilitate communication among engineers and technologists all over the world. However, we recognize that in the United States the practicing engineer will have to deal with a professional establishment that still requires specifications in the English system of units. To this end some of the problems have been worked out in the English system. We hope that this blend will facilitate communication with the practicing engineers of this country, as well as with the solar technologists of other countries, and that it will prepare the reader to use the literature in the solar field no matter where it may appear. Certain figures and tables taken from other sources have been retained in their original form with the original units.

This book is obviously a new and demanding venture that has stretched our own expertise to the limit—and may at times have exceeded our potential. We beg the indulgence of experts in specialized fields and hope that they will communicate to us constructive criticism on how we might improve the presentation in subsequent editions. A book of this dimension draws of course on many sources and it would be impossible to give credit to all the people who have contributed to its development. We have tried to give credit to original sources wherever possible, and we have included substantial reference lists with each chapter. We recognize, however, that the field of solar energy is old, and that many of the pioneers of the field were engineers who did not publish their results and often do not receive the recognition they deserve. Moreover, at the present time, the field of solar energy is expanding rapidly because of the influx of federal research funds. We would, therefore, encourage those interested in a specific aspect of solar energy utilization to expand and deepen their knowledge by reading the current literature. An annotated list of publications dealing with the field of solar energy is provided, and the reader is encouraged to refer to them for current developments.

At this point we would like to note that since the time we completed our manuscript, some important changes have occurred in the field of solar engineering.

Probably the most important difference concerns the cost of solar systems. On the basis of available data, it appears that it is not the cost of the solar collectors, but rather the cost of the installation that governs the cost of the system. For example, most of the solar-heating systems for buildings constructed with flat-plate collectors costing $12–$15/ft$^2$ have installed costs close to $30/ft$^2$. Thus, to reduce the cost of solar energy, more emphasis will have to be placed on system design and integration and on the reduction of installation costs, rather than on attempts to reduce the cost per square foot of solar collectors. Greater emphasis should also be placed on conservation and passive systems, as well as on bioconversion. The cost of energy from bioconversion processes appears to be much closer to being competitive with other energy sources than any other currently available solar energy utilization technique.

Experience with solar cooling has so far been disappointing. The cost-effectiveness of solar-cooling systems has been poor, and at the present time there does not appear to be much hope of designing and building small-scale solar absorption cooling systems that could compete effectively with conventional vapor-compression cycles or heat pumps. There should therefore be greater consideration of passive approaches and large commercial cooling systems.

In view of the high cost of installed solar systems, current interest should be focused on applications of solar energy to tasks in which the system can be used without interruption throughout the full year. As a result, two applications at the present time appear to be closest to economic viability: the use of solar energy for domestic water heating and, on a larger scale, for industrial process heat. For the first type of application, simple flat-plate collectors appear to be satisfactory, whereas for the latter, single-axis tracking parabolic troughs appear to be the most cost-effective collector systems available. Tracking collectors can reduce appreciably the cosine loss factor experienced by a flat-plate. Moreover, the installed cost of tracking parabolic trough systems is not much greater than that of flat plates per square foot of aperture, according to recent construction costs and contractors' bids. As a corollary to the need for year-round utilization of solar collectors, special emphasis should be placed on all types of energy storage as well as on the integration of storage with collector systems.

From the above discussion it appears much more important to stress that each solar system must be adapted to its task, taking into account the cost factors as well as the thermodynamic limitations. From that point of view we are happy to have chosen a general systems approach that can lead to the design of many types of viable solar energy systems in years to come.

There are a few people who have made some very specific and meaningful contributions to the book. We would like to thank Professor Roland Winston of the

University of Chicago, Dr. Ari Rabl of Argonne National Laboratory, and Dr. G. Steward of the National Bureau of Standards for introducing us to many new concepts in solar concentrators, and Professor G. O. G. Löf for helping us gain a deeper understanding of heating and cooling systems for buildings. We would also like to express appreciation to the students, users, and reviewers of our previous book "Solar Heating and Cooling," who provided many comments that were useful in preparing the present book.

Frank Kreith would like to express his thanks to Dr. Ozer Igra of the Ben Gurion University of the Negev for increasing his understanding of wind generators; to Dr. L. B. Peube and Mr. Bernard Utrey for showing him the operation of the French CNRS Solar Power Plant at Odeillo; to Professor O. D. Hall of King's College for discussions of photosynthetic solar energy generation; and to Dr. S. H. Salter of the University of Edinburgh for introducing him to ocean wave generation systems. Frank Kreith also thanks his wife Marion for helping him organize the book, and Ms. Janet Naumer, Mrs. Lonnie Codding, and Mrs. Aladeen Smith for assisting in typing the manuscript.

Jan Kreider wishes to express his thanks to Charles Berg and George Hotsopolous for insightful comments on the relationship of solar energy to the second law of thermodynamics, especially to Francis deWinter for his very thorough review of the entire book, and to Rosalie Ruegg, S. A. Marton, and Gerard Lameiro for reviewing specific sections of the book. Jan Kreider also thanks Dottie Lang for her unfailing confidence and assistance during the long course of writing and Wendy Klein for her professional assistance in typing and assembling the lengthy manuscript.

*Frank Kreith*
*Jan F. Kreider*

# PRINCIPLES
# OF
# SOLAR
# ENGINEERING

# INTRODUCTION TO SOLAR ENERGY AND ITS CONVERSION FOR USE ON EARTH ① 

*Eyes, though not ours, shall see*
*Sky high a signal flame,*
*The sun returned to power above*
*A world, but not the same.*

**C. D. Lewis**

Ever since Prometheus stole the fire of heaven from the gods, virtually all the energy consumed by humans has come from the sun. Coal, oil, and natural gas are residues of plants and animals, which originally derived all the energy for their development from solar radiation. Solar radiation also drives the earth's rain cycle, which powers modern hydroelectric generators, and large-scale atmospheric circulations provide the winds that have powered windmills for many centuries.

"Solar energy is an essentially inexhaustible source potentially capable of meeting a significant portion of the nation's future energy needs with a minimum of adverse environmental consequences.... The indications are that solar energy is the most promising of the unconventional energy sources...." Despite this encouraging assessment of the potential of solar energy, presented by the National Science Foundation in testimony before the Senate Interior Committee in 1972, considerable technical and economic problems must be solved before large-scale utilization of solar energy can occur. The future of solar power development will depend on how we deal with a number of serious constraints, including scientific and technological problems, marketing and financial limitations, and political and legislative actions favoring conventional and nuclear power. In addition, the education of engineers will have to change its focus from nonrenewable fossil-fuel technology to renewable power sources. There appears to be general agreement that the most significant of the renewable energy sources is solar radiation, and it is the objective of this book to present the basic technical background necessary for the design and economic analysis of solar energy utilization systems.

The design and analysis of solar energy systems are based upon well-established engineering knowledge. However, the conventional approach to engineering education has usually fragmentized the subject matter according to the traditional engineering disciplines. For the design and analysis of solar energy systems, a new viewpoint is necessary that transcends those disciplines and combines many features of the various engineering specializations into one unified approach. To accomplish this goal this book assumes only that the reader is familiar with traditional

1

thermodynamics and basic physics and chemistry, and has a knowledge of calculus and ordinary differential equations. The introductory chapters provide the background in radiation, fluid mechanics, and heat transfer that is necessary for solar engineering. In subsequent chapters the design and analysis of solar utilization schemes are approached from a systems-analysis viewpoint, which combines technical design with economic analysis. As will become clearer later on, there is no single solution to a given task in solar energy utilization, and each problem must be analyzed separately from fundamental principles. Whereas with systems utilizing conventional fuel sources such as coal or gas it is possible for an engineer to specify the heat and power requirements and then expect the manufacturer to meet these requirements with a device guaranteed to meet specifications, in the design of a solar system it is necessary to match the available power source to the task at hand. It is, therefore, not possible to make a general case for or against the utilization of solar energy. Instead, the engineer will have to form a judgment on the basis of the task at hand, the systems available to achieve a technical solution, and the economics involved.

One of the curious aspects of solar technology is that it does not necessarily have to rely on a high level of technology. The requirements for solar systems are on a level which E. Schumacher has termed "intermediate technology." However, this does not mean that the design and analysis of solar energy systems is an easy job. On the contrary, the proper design and optimization of solar energy systems require a higher level of engineering analysis than has heretofore been necessary in many facets of U.S. industry, in particular, those involved in the construction of residential and industrial buildings and in the traditional power industry. This more sophisticated viewpoint requires foremost attention to the problem of matching the system to the goal to be achieved, as well as the abandonment of the traditional brute-force solutions, which have been the practice of conventional engineering when nonrenewable fossil energy sources were still available in large quantities at a reasonable cost.

## THE TIME SCALE OF FOSSIL FUELS AND THE SOLAR ENERGY OPTION

The rate of production of a nonrenewable energy resource increases with time, reaches a maximum, and then decays. Thus, if we plot the rate of production as a function of time, the area under the curve is the total amount of the energy recovered from that resource. Figure 1.1 illustrates this behavior for the case of petroleum liquids in the United States. The production rate increased until about 1968, reached a maximum, and is now decreasing. Foreign sources are currently used to supplement the U.S. requirements, but the rate of production of these supplementary sources will peak in the not-too-distant future.

According to current estimates, about 88 percent of the world's available fossil fuel supply is coal (18). Since this coal reserve represents the only major remaining fossil energy source, it is important to know when it will be exhausted, because this

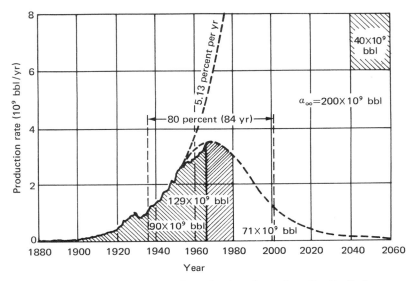

**FIGURE 1.1** Complete cycle of U.S. production of petroleum liquids. Redrawn from (18).

will show how much time is available to develop and deploy other energy sources, which are essential to the survival of industrial society.

Two major studies on the worldwide availability of stored energy are widely used in making projections. Hubbert, a research physicist in the U.S. Department of the Interior, predicted in 1962 that the rate of production of coal would reach a maximum somewhere between the years 2100 and 2150, as shown in Fig. 1.2. Elliott and Turner predicted about 10 years later that production of all fossil fuels would peak possibly as soon as 2030. The reason for this difference in the predicted peaking time may be that the latter projection assumes that gasified and liquified coal would be used for all those energy-consuming functions for which oil and natural gas are currently employed, whereas Hubbert's projection was based on the assumption that the consumption pattern for coal would remain the same as in the past, that is, mostly for electric power production and industrial heat (5).

Of course, the peaking in the production rate of a given energy source does not imply the end of its availability. It does mean, however, that the end of its availability is in sight and that when peaking occurs some other source must become increasingly available to supplement the diminishing source if all the energy requirements are to be met. In case no other source is available at the time the peak in production rate is reached, some energy-consuming activities would obviously have to be curtailed.

Such a forced reduction in energy consumption would have far-reaching effects that could easily threaten modern industrialized culture and society as we know it today. To avoid an energy catastrophe it is therefore imperative that accurate

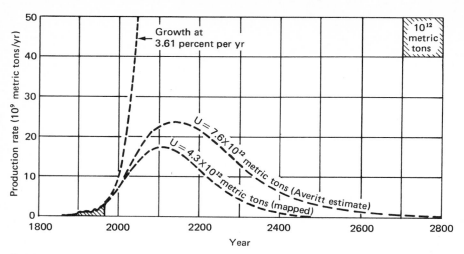

**FIGURE 1.2** Complete cycles of world coal production for two values of $U$. Redrawn from (18).

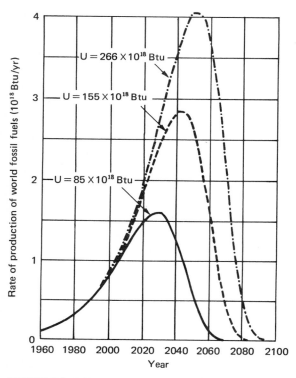

**FIGURE 1.3** Effect of the value of the resource base on the projected rate of world fossil fuel yields according to Elliott and Turner (11).

predictions of future energy supply and demand be available to decision makers. The complete world cycle prediction of Elliott and Turner (11) for all nonrenewable fossil resources is shown in Fig. 1.3 for various assumed values of the total amount of available fossil-fuel energy. An inspection of these curves shows that predicted peaking times depend on the assumed value of the total amount of energy of the source. The following analysis illustrates how the assumed value for the total amount of available energy affects the time scale.

According to Elliott and Turner, the rate of production of a resource is related to the demand and the amount remaining by an expression of the type

$$\frac{dP}{dt} = f(D)f(s) \tag{1.1}$$

where $P$ = cumulative production of the resource
$\quad t$ = time
$\quad f(D)$ = function of demand
$\quad f(s)$ = function of supply

But the supply and demand are functions of the fraction of the resource remaining at any time. If $U$ is the ultimate amount of the resource, then $P/U$ is the fraction that has been produced, and $(1 - P/U)$ is the fraction remaining. Assuming that the demand is linearly related to the fraction produced and the supply, i.e., the fraction remaining, we get

$$\frac{dP}{dt} = C_1 \frac{P}{U}\left(1 - \frac{P}{U}\right) \tag{1.2}$$

where $C_1$ is a constant. The solution of Eq. (1.2) can be written in the form

$$-\ln\left(\frac{U}{P} - 1\right) = \frac{C_1}{U}t + C_2 \tag{1.3}$$

where $C_2$ is a constant of integration. Solving for the cumulative production gives

$$P = \frac{U}{1 + \exp - (C_1 t/U + C_2)} \tag{1.4}$$

To obtain the inflection point in the rate of production of energy, we take the second derivative of $P$ with respect to time and equate it to zero. This yields

$$C_2 = -\frac{C_1}{U}t_m \tag{1.5}$$

where $t_m$ is the time at which the production rate peaks. Substituting Eq. (1.5) for $C_2$ in Eq. (1.3) gives

$$P = \frac{U}{1 + \exp - [(C_1/U)(t - t_m)]} \tag{1.6}$$

Equation (1.6) is similar in form to the relation used by Hubbert as shown by Bockris (5). It illustrates that the time at which the production of a resource will peak depends on the total supply. Quantitatively, however, even for substantial differences in the assumed value of the total remaining supply of energy, the differential in the predicted peaking time is relatively small. This can be seen in Fig. 1.3, where Elliott and Turner's results are plotted for various values of the assumed total energy-source availability.

Table 1.1 gives some estimates of the total energy contents of the world's supply of recoverable fossil fuels according to Hubbert (19). It is apparent that the difference between the optimistic and pessimistic projections places the time at which world production of fossil fuels will peak between the years 2030 and 2050 if coal is used to supply all the energy needs currently obtained from oil and natural gas. Bockris (5) has compared the results of several studies. Table 1.2 summarizes all of these estimates, and it is apparent that reasonable agreement exists among the peaking times predicted by several independent investigations, with the average falling somewhere about the year 2040.

The next question to ask is, "How much time is available to develop other energy sources?" An answer to this question can be given only within the context of the history of science and technology. How long has it taken in the past for a new technology to be introduced into the mainstream of an industrial society? The general estimate is that it takes somewhere between 20 and 40 years for a technology to come to fruition. For example, the Wright brothers made their first flight with a heavier-than-air ship in 1903, and aviation became a major industry

**TABLE 1.1** Energy Contents of the World's Initial Supply of Recoverable Fossil Fuels[a]

| Fuel | Quantity | Energy content | | |
|------|----------|--------------------|-------------------------|---------|
|      |          | $10^{21}$ J(th) | $10^{15}$ kW · hr(th) | Percent |
| Coal and lignite | $7.6 \times 10^{12}$ tons[b] | 201 | 55.9 | 88.8 |
| Petroleum liquids | $2000 \times 10^9$ bbl $(272 \times 10^9$ tons) | 11.7 | 3.25 | 5.2 |
| Natural gas | $10,000 \times 10^{12}$ ft$^3$ $(283 \times 10^{12}$ m$^3$) | 10.6 | 2.94 | 4.7 |
| Tar-sand oil | $300 \times 10^9$ bbl $(41 \times 10^9$ tons) | 1.8 | 0.51 | 0.8 |
| Shale oil | $190 \times 10^9$ bbl $(26 \times 10^9$ tons) | 1.2 | 0.32 | 0.5 |
| Totals |  | 226.3 | 62.9 | 100.0 |

[a]From Hubbert, M. King, Energy Resources for Power Production, in "Environmental Aspects of Nuclear Power Production," International Atomic Energy Agency, Vienna, pp. 13–43, 1971, Table VI.

[b]This is a high estimate because it includes coal beds as thin as 12 in and as deep as 6000 ft. If only beds more than 28 in thick are considered, the estimated quantity is about $2 \times 10^{12}$ tons.

**TABLE 1.2** Estimates of the Year in Which a Maximum Production Rate of Coal Could Occur[a,b]

| Source | Year | Comment |
|---|---|---|
| Chase Manhattan Bank, elucidated (5) | 2033 | Original report states 1500 yr at 1972 rate of consumption; elucidation in text |
| Brennan (6) | 2000 | Assumes energy demand stops increasing at 2000 |
| Elliott and Turner (11) | 2040 | Takes into account use of coal to replace oil |
| Linden (23) | 2010 | Assumes mining technology not radically improved from 1970s |
| Linden (23) | 2052 | Assumes all coal can be mined |
| Linden (23) | 2073 | Highest credible limit for resource base assumed |

[a]From (5).
[b]In all estimates the assumption was made that the logistical difficulty of mining coal *at a sufficient rate* has been solved.

about 40 years later. The first high-speed computer was developed at the end of World War II, and a large-scale computer industry developed about 20 years later. The first nuclear chain reaction was set off in 1942, but despite massive governmental funding nuclear power was supplying less than 2 percent of the total energy consumption of the U.S. economy three decades later. It is therefore prudent to plan for shifting a major share of the total world energy production from fossil fuels to renewable or some other plentiful sources by no later than the end of this century. This leaves less than 25 years for research and development of new energy sources.

The replacement of current energy-producing systems requires careful planning and allocating of current energy resources, as well as time. Of all the currently known available sources, only nuclear and solar energy have the potential of supplying large amounts of power within the available time frame. There are some who believe that in planning for future energy sources a choice between a solar and a nuclear option should be made and that one may be preferable to the other. Any realistic assessment of world energy demand and the technical means of meeting this demand will show, however, that it is not a question of one or another but rather of how much energy each of them can supply and under what conditions one source is more suitable to meet the demand economically and safely than the other. The U.S. Energy Research and Development Administration (ERDA) has made a projection (12) of the total power needs of the United States that will be supplied by solar energy in the years 1985, 2000, and 2020. The rest of the energy needs will be met by nuclear energy, geothermal energy, and the remaining fossil fuel sources. Analyses of the nuclear potential have been presented by Foster and Wright (14) and Häfele (17). The areas in which solar energy is expected to contribute are shown in Table 1.3. The objective of this book is to give an economic and technical analysis of the solar energy option.

**TABLE 1.3**  Estimates of the Heat, Electric Power, and Fuels to be Supplied by Solar Energy in the United States, as Projected by ERDA[a]

| Solar technology | 1985 | | 2000 | | 2020 | |
|---|---|---|---|---|---|---|
| Direct thermal applications (in units of $10^{15}$ Btu = 1 Q per year)[b] | | | | | | |
| Heating and cooling | 0.15 | Q | 2.0 | Q | 15 | Q |
| Agricultural applications | 0.03 | | 0.6 | | 3 | |
| Industrial applications | 0.02 | | 0.4 | | 2 | |
| Total | 0.2 | Q | 3 | Q | 20 | Q |
| Solar electric capacity (in units of $10^9$ W = 1 GWe)[c] | | | | | | |
| Wind | 10 | GWe | 20 | GWe | 60 | GWe |
| Photovoltaic | 0.1 | | 30 | | 80 | |
| Solar thermal | 0.05 | | 20 | | 70 | |
| Ocean thermal | 0.1 | | 10 | | 40 | |
| Total | 1.3 | GWe | 80 | GWe | 250 | GWe |
| Equivalent fuel energy | 0.07 | Q | 5 | Q | 15 | Q |
| Fuels from biomass | 0.5 | Q | 3 | Q | 10 | Q |
| Total solar energy | ~1 | Q | ~10 | Q | ~45 | Q |
| Projected U.S. energy demand | 100 | Q | 150 | Q | 180 | Q |

[a]From (12).
[b]One Q represents the approximate energy consumption of the state of Iowa in 1973.
[c]One GWe is one gigawatt of electrical power.

It is evident from an inspection of Table 1.3 that to speak of a single solar energy option is an oversimplification. The Solar Energy Panel of NSF/NASA suggested a subdivision of the solar energy resource into natural and technological collection systems. Further subdivisions of these are shown in Fig. 1.4. At the present time only the thermal method is sufficiently advanced to compete economically with other sources, including nuclear. It will therefore be instructive to estimate the amount of money and materials that would be required if one-quarter of all the energy requirements of the world were to be provided by thermal conversion of solar energy by 2020.

Assuming that all solar conversion plants could be located in the sunny parts of the world where the average daily insolation is 5 kW · hr/m², about 2.5 million km² of land would have to be covered with solar collectors if the average overall solar system conversion efficiency were 10 percent. Assuming further that these solar systems, including all storage and transmission, could be built with the same capital investment as a current coal-fired thermal power plant (about $500/kW) or a nuclear power plant (about $1,000/kW), the required initial investment would be roughly 25 or 50 trillion dollars, respectively. Such an investment over half a century would require about 15–30 percent of the 1974 gross world product per year, according to a study conducted at the International Institute for Applied

System Analysis (29). If the collection systems weighed 10 kg/m$^2$, which is only about 20 percent of the weight of currently available flat-plate collectors, the mass of material required to build these solar plants would be 50 billion metric tons. Clearly, any plans of converting to a large-scale solar energy industry must include considerations of the availability of suitable materials and work force, as well as of the level of technology required for such an undertaking (22).

The choice of direction for the research and development needed to build a new energy supply system has the dimension of a total commitment. It is laden with uncertainties and value judgments. In practical terms it will depend on the technical feasibility of the proposed schemes and on the cost of the final product. These factors depend on the time scale involved, on the long-term ecological implications of the new industries to be developed, and on the raw material, energy, and manpower resources of the country. Different choices may be made for industrially developed countries with high labor costs than for developing countries with limited resources but low labor costs. Clearly, science, technology, economics, sociology, and certainly politics will play a part. But also wisdom and foresight of decision makers and a national will to conserve energy and make sacrifices for the future will be required if the challenge of energy is to be met.

## OVERVIEW OF SOLAR ENERGY CONVERSION METHODS

In accordance with the suggestions of the Solar Energy Panel of NSF/NASA, the following overview of the principal methods currently under consideration for solar

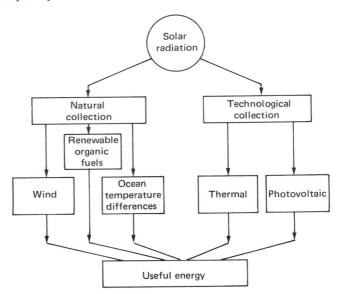

**FIGURE 1.4** A summary of schemes for solar energy conversion. Redrawn from (27).

energy conversion is subdivided into natural and technological collection systems, and these are further subdivided as shown in Fig. 1.4. Each of the methods is described briefly below.

## Thermal Conversion

Thermal conversion is a technological scheme that utilizes a familiar phenomenon. When a dark surface is placed in sunshine, it absorbs solar energy and heats up. Solar energy collectors working on this principle consist of a surface facing the sun, which transfers part of the energy it absorbs to a working fluid in contact with it. To reduce heat losses to the atmosphere one or two sheets of glass are usually placed over the absorber surface to improve its efficiency. These types of thermal collectors suffer from heat losses due to radiation and convection, which increase rapidly as the temperature of the working fluid increases. Improvements such as the use of selective surfaces, evacuation of the collector to reduce heat losses, and special kinds of glass are used to increase the efficiency of these devices.

The simple thermal-conversion devices described above are called flat-plate collectors. They are available today for operation over a range of temperatures up to approximately 365 K (200°F). These collectors are suitable mainly for providing hot service water and space heating and possibly are also able to operate absorption-type air-conditioning systems.

The thermal utilization of solar energy for the purpose of generating low-temperature heat is at the present time technically feasible and economically viable for producing hot water and heating swimming pools. In some parts of the world thermal low-temperature utilization is also economically attractive for heating and cooling buildings. Other applications, such as the production of low-temperature steam, can be expected to become economically feasible soon.

The generation of higher working temperatures as needed, for example, to operate a conventional steam engine requires the use of focusing devices in connection with a basic absorber-receiver. Operating temperatures as high as 4000 K (6740°F) have been achieved in the Odeillo solar furnace in France, and the generation of steam to operate pumps for irrigation purposes has also proved technologically feasible. At the present time a number of focusing devices for the generation of steam to produce electric power are under construction in different regions of the world, and cost estimates suggest that the cost of solar power in favorable locations will be no more than that of nuclear power when the development of these plants has been completed.

## Photovoltaic Conversion

The conversion of solar radiation into electrical energy by means of solar cells has been developed as a part of satellite and space-travel technology. The theoretical efficiency of solar cells is about 24 percent, and in practice efficiencies as high as 15 percent have been achieved with silicon photovoltaic devices. The technology of photovoltaic conversion is well developed, but large-scale application is hampered by the high price of photo cells, which has been estimated by an ad hoc panel on solar cell efficiencies of the National Academy of Sciences at approximately $100 per electric watt in 1972. This is about 100 times more than electric energy from conventional fossil- and nuclear-fuel electric power plants, and a cost reduction of the order of 100 to 1 will be necessary before photovoltaic conversion can be economically viable.

Photovoltaic solar cells, as opposed to conventional collectors, which convert solar radiation into heat, utilize energetic photons of the incident solar radiation directly to produce electricity. Thus, this technique is often referred to as direct solar conversion. Conversion efficiencies of thermal systems are limited by collector temperatures, whereas the conversion efficiency of photo cells is limited by other factors.

Efforts are currently underway to develop new technologies for producing solar cells in the hope that such methods might lead to an appreciable lowering in price. Silicon cells, which have performed well for spacecraft application, have little chance of being produced at a reasonable price. Efforts are directed toward drawing thin monocrystalline foilbands because such a method might lead to appreciable lowering of the production cost. Solar cells using polycrystalline materials can be produced somewhat cheaper, but unfortunately the lower price is counterbalanced by high systems cost resulting from the lower efficiency of these polycrystalline cells. For example, cadmium sulfide-copper sulfide cells, which are one of the candidates in this category, have an efficiency of only about 6 or 7 percent. However, efforts are presently being made to utilize concentrating devices made from less expensive materials to reduce the surface area of solar cells required, and recently highly innovative fabrication techniques have been recommended for reducing the cost of production (1, 28, 31, 32).

One of the most ambitious photovoltaic conversion processes has been suggested by Glaser (16). He proposed placing a solar cell generation station into a geostable orbit at a distance of approximately 36,000 km from earth (see Fig. 1.5). The electric energy collected on an enormous solar cell panel would then be converted to microwaves and radiated by a sharply focused antenna to earth, where it could be received by a correspondingly large antenna and then processed for further use as electricity. In comparison with terrestrial applications, this project offers the advantage that the solar cells receive the full insolation of 1.35 kW/m$^2$ throughout the year. Cost projections for this project vary widely, but it appears that its cost would be high.

## Biological Conversion

Biological conversion of solar energy by means of photosynthesis is a natural process that has been studied by scientists for many decades. This form of solar energy utilization has been of the greatest importance by far to the human race. It provides a small but vital part of our energy consumption in the form of food and for thousands of years has served our ancestors in the form of wood as the only source of heat. Last, but not least, it is this process that in the course of millions of years produced our fossil fuels, which currently provide most of our energy.

**FIGURE 1.5** Design concept for a satellite solar power station. Courtesy of Arthur D. Little, Inc.

In principle, it might be possible to cultivate appropriate plants exclusively for the purpose of generating power either by direct bioconversion or by pyrolytic conversion into liquid or gaseous fuel. Unfortunately, the photosynthetic yield in agricultural application is only of the order of 1 percent. In temperate climates, even under favorable conditions, the average annual harvest is of the order of 20 metric tons of organic dry material per hectare. However, cultivation of special plants such as sugarcane can achieve annual mean conversion efficiencies of the order of 2.5 percent, and recycling processes by which the waste products of foods or animals are used to produce methane, which can also be converted into methanol, are within the realm of possibility. As seen in Table 1.3 bioconversion is expected to provide an increasing amount of our total energy needs in the future.

## Wind Power

The utilization of wind power has been widespread since medieval times. Windmills were used in rural United States to power irrigation pumps and drive small electric generators used to charge batteries that provided electricity during the last century. A windmill or wind turbine converts the kinetic energy of moving air into mechanical motion, usually in the form of a rotating shaft. This mechanical motion can be used to drive a pump or to generate electric power.

The technology of wind-energy conversion is well developed. The energy content of the wind increases with the third power of the wind velocity, and wind-power installations are therefore economical only in regions where winds of sufficient strength and regularity occur. The construction of wind-power installations does not require any new technologies, and cost estimates in favorable regions of the world are fairly close to those of other energy sources. However, in order to produce appreciable amounts of power, installations have to be fairly large. Rotor diameters of 50 m and heights of 100 m are dimensions typically suggested for wind generators, and such giant windmills placed at regular distances might not be aesthetically pleasing. Moreover, wind velocities decrease considerably at night and also, of course, vary with the weather. Consequently, for reliable power from the wind the problem of storing energy must be solved. These are some of the reasons why wind power has not yet found large-scale applications even in favorable regions. However, it is anticipated that in some parts of the world wind power will be able to make an appreciable contribution to the overall energy demand.

## Solar Energy Conversion by Oceans

Almost 71 percent of the world's surface is covered by oceans. Oceans serve as a tremendous storehouse of solar energy because of the temperature differences produced by the sun as well as the kinetic energy stored in the waves. There are a number of places in the ocean where temperature differences of the order of 20–25 K exist at depths of less than 1000 m, and these temperature differences could be used to operate low-pressure heat engines. Although the thermodynamic efficiency of a heat engine operating on such a small temperature difference is low, the available amount of thermal energy is very large. However, putting this energy-conversion method into practice requires the development of efficient and cheap heat exchangers that can withstand the rough marine conditions. Since heat-exchange equipment is the most expensive part of any ocean thermal conversion scheme, the cost of using the temperature gradients in the ocean for practical solar energy utilization depends largely on this development.

The second method of utilizing the storage capacity of the oceans for energy generation is through ocean waves. This approach is being developed in Japan and the United Kingdom, where prototype installations are presently being built. Cost estimates and projections in areas of the world favorable for this type of solar energy utilization method look encouraging. Although the applicability of ocean wave conversion will be limited to relatively few places, it may be one of the more important ways of providing power to some energy-poor nations of northern Europe.

## AVAILABILITY OF SOLAR ENERGY

The amount of solar radiant energy falling on a surface per unit area and per unit time is called *irradiance*. The mean extraterrestrial irradiance normal to the solar beam on the outer fringes of the earth's atmosphere is approximately 1.35 kW/m². Since the earth's orbit is elliptical, the sun-to-earth distance varies slightly with time of year, and the actual extraterrestrial irradiance varies by ±3.4 percent during the year. The angle subtended by the sun when viewed from the earth is only 0.0093 rad (approximately 32 min of arc), and the direct beam radiation reaching the earth is therefore almost parallel. Although the brightness of the solar disc decreases from center to edge, for most engineering calculations the disc can be assumed to be of uniform brightness. The radiant energy from the sun is distributed over a range of wavelengths, and the energy falling on a unit surface area per unit time within a particular spectral band is known as the *spectral irradiance*; its value is usually expressed in watts per square meter per nanometer of bandwidth. The extraterrestrial spectral irradiance is shown in Fig. 1.6; as will be discussed in detail in Chapter 2, it can be approximated by the spectrum of a black body at 5800 K. In the upper part of Fig. 1.6, the wave bands typically useable for different solar applications are shown, and the lowest curve shows the spectral direct-beam irradiance at sea level on earth under clear sky conditions with the sun overhead.

The earth and its atmosphere receive continuously $1.7 \times 10^{17}$ W of radiation from the sun. A world population of 10 billion with a total power need per person of 10 kW would require about $10^{11}$ kW of energy. It is thus apparent that if the irradiance on only 1 percent of the earth's surface could be converted into useful energy with 10 percent efficiency, solar energy could provide the energy needs of all the people on earth. This figure is often quoted by solar energy enthusiasts, but unfortunately the nature of this energy source has technical problems and economic limitations that are not apparent from this macroscopic view of the energy budget. The principal limitations are that the solar energy received on earth is of small flux density, is intermittent, and falls mostly in remote places. The implications of these factors will be discussed below.

## LIMITATIONS OF SOLAR ENERGY

The first problem encountered in the engineering design of equipment for solar energy utilization is the low flux density, which makes necessary large surfaces to collect solar energy for large-scale utilization. Also, the larger the surfaces, the more

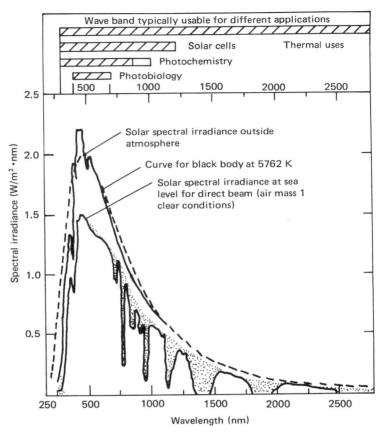

**FIGURE 1.6**  Spectral irradiance curves for direct sunlight extraterrestrially and at sea level with the sun directly overhead. Shaded areas indicate absorption due to atmospheric constituents, mainly $H_2O$, $CO_2$, and $O_3$. Wavelengths potentially utilized in different solar energy applications are indicated at the top.

expensive the delivered energy becomes. When the sun is directly overhead on a cloudless day, $10 m^2$ of surface could theoretically provide energy at 10 percent efficiency of collection at the rate of 1 kW. Several factors reduce this amount in practice, however. One of these is losses in the atmosphere. The solar spectrum is substantially modified in passing through the earth's atmosphere, and approximately 25–50 percent of its energy is lost by scattering and absorption. Even on a cloud-free day with unpolluted skies, about 30 percent of the incident energy is lost; some is scattered back into space by air molecules and some is absorbed by ozone, water vapor, and carbon dioxide in the atmosphere. Even on a clear day, the diffuse energy from the sky that can not be collected efficiently accounts for about 20 percent of the total irradiance on a horizontal surface, and this percentage increases substantially in locations where there are natural clouds and anthropogenic pollution. Moreover, as shown in Fig. 1.6, as a result of almost total absorption of solar energy by ozone at wavelengths below 300 nm and by carbon dioxide at wavelengths beyond 2500 nm the irradiance on the earth's surface is effectively limited to wavelengths between 300 and 2500 nm. In that range only the solar energy falling between 400 and 700 nm can be used by living plants photosynthetically, and this is only about half of the total earth irradiance.

The total solar energy reaching the earth is made up of two parts: the energy in the direct beam and the diffuse energy from the sky. Although plants can use direct and diffuse solar energy, most manmade solar collectors can convert only direct energy efficiently. The amount of direct energy depends on the cloudiness and the position of the sun and is obviously greatest on clear days. Some solar radiation falling on clouds is diffused by scattering, but clouds do not absorb all of the energy. The effect of clouds is mainly to increase the percentage of diffuse energy in the total energy reaching the surface, and diffuse irradiance in summer months with high sun and broken clouds can be as high as 400 W/m$^2$. Thick clouds let less energy pass than thin clouds and scatter proportionally more of the total energy back into space.

The second practical limitation that is not apparent from the macroscopic energy view is that most of the solar energy falls on remote areas and would therefore require some means of transportation to be useful to the industrialized nations. The annual mean surface global irradiance on a horizontal plane as mapped by Budyko (7) is shown in Fig. 1.7. The mean amount of energy available on a horizontal plane is greatest in the continental desert areas around latitudes 25°N and 25°S of the equator and falls off toward both the equator and the poles. The highest annual mean irradiance is 300 W/m$^2$ in the Red Sea area. Clouds reduce the mean global irradiance considerably in equatorial regions, and the annual mean irradiance for Singapore is almost the same as that for Paris. However, whereas in northern climates the mean horizontal surface global irradiance varies from season to season, it remains relatively constant in equatorial regions. Typical values of mean annual horizontal surface irradiance are: Australia, about 200 W/m$^2$; United States, 185 W/m$^2$; United Kingdom, 105 W/m$^2$.

Although the global distribution of solar energy does not favor the industrialized parts of the world, it may be of help to industrially developing countries located in the favorable radiation belt. Table 1.4 shows the total solar electric energy potential for the solar energy–rich areas of the world. It is important to note that many of these parts of the world, for example, Saudi Arabia, central Australia, and parts of South Africa and India, are flat deserts that are practically unuseable for agriculture or any kind of industrial development. At the same time, it should be noted that these are also regions with little or no water, which can create special problems in the generation of thermal electric power.

The third limitation of solar energy as a large-scale source of power and heat is its intermittency. Solar energy has a regular daily cycle due to the turning of the earth around its axis, a regular annual cycle due to the inclination of the earth axis with the plane of the ecliptic and due to the motion of the earth around the sun, and is also unavailable during periods of bad weather. Figure 1.8 shows the variation

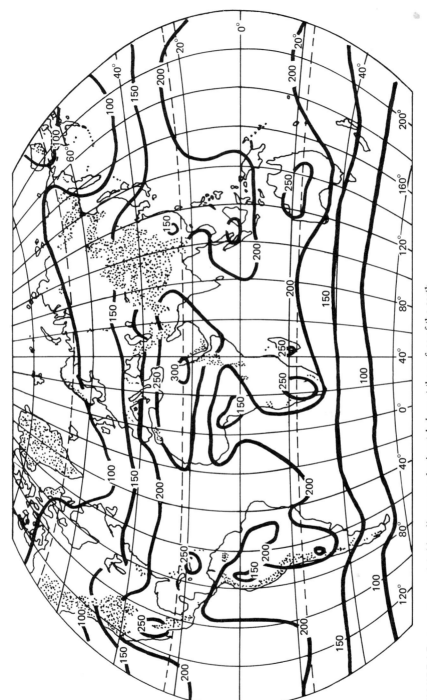

**FIGURE 1.7** Annual mean global irradiance on a horizontal plane at the surface of the earth (W/m² averaged over 24 hr).

**TABLE 1.4**  Solar-Electric Energy from Earth's High-Insolation Areas[a]

| Desert | Nominal areas | | Nominal annual thermal energy flux | | Percentage of area assumed usable | Electric energy extracted at 25% efficiency (GW · hre/yr) |
|---|---|---|---|---|---|---|
| | km² | mi² | GW · hr(th)/km² | GW · hr(th), thousands | | |
| North Africa | 7,770,000 | 3,000,000 | 2,300 | 17,870,000 | 15[b] | 670,000,000 |
| Arabian Peninsula | 1,300,000 | 500,000 | 2,500 | 3,250,000 | 30[c] | 244,000,000 |
| Western and central Australia | 1,550,000 | 600,000 | 2,000 | 3,100,000 | 25 | 194,000,000 |
| Kalahari | 518,000 | 200,000 | 2,000 | 1,036,000 | 50 | 129,000,000 |
| Thar (northwest India) | 259,000 | 100,000 | 2,000 | 518,000 | 50 | 65,000,000 |
| Mojave, southern California | 35,000 | 13,500 | 2,200 | 77,000 | 20 | 3,900,000 |
| Vizcaino, Baja, California (Mexico) | 15,500 | 6,000 | 2,200 | 34,000 | 25 | 2,100,000 |
| Total/Average | 11,447,500 | 4,419,500 | 2,190 ave. | 25,895,000 | 31 ave. | 1,308,000,000 |

[a]From (10).
[b]Parts of Arabian and Libyan deserts.
[c]About 60 percent of Rub' al Khali desert.

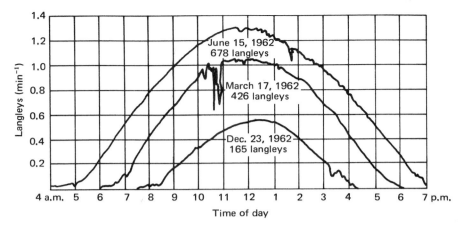

**FIGURE 1.8** Typical solar radiation on clear days at latitude 43°N (1.43 langleys/min equal 1 kW/m²).

in solar irradiance on clear days as measured in 1962 in Madison, Wisconsin, at latitude 43°N. On June 15, 1962, the average irradiance was four times larger than on December 23, 1962. In June the maximum between 10 a.m. and 2 p.m. was about 0.8 kW/m², but when averaged over a 24-hr period this yields only about 0.25 kW/m² under the most favorable conditions and as little as 0.08 kW/m² during the month of December. It is obvious that these daily and seasonal variations in irradiance, exacerbated by variation due to weather, introduce special problems in storage and distribution of this energy, which are entirely different from those problems involved in the utilization of conventional sources such as coal and oil. In assessing the prospects for using solar energy, it therefore becomes particularly important to ascertain for which application the diffuse and cyclic nature of the source will not introduce insurmountable technical and economic problems. It is also necessary to recognize that because of its peculiar characteristics solar energy cannot be relied on as the sole energy source and will need supplementation from other sources.

## Energy Storage

In addition to the technical problems already noted, widespread use of solar energy systems requires satisfactory means of storing the energy once it has been collected. Whereas considerable progress has been made in developing systems for extracting energy from the sun's radiation, the state of development of storage systems is less satisfactory. For low-temperature storage, such as is necessary in heating and cooling buildings, sensible heat storage in water or rocks is a reasonable solution. For electrical energy production systems, however, either thermal storage at elevated temperature or other means of storing energy will be necessary. Table 1.5 presents a comparison of possible energy-storage systems for large-scale electrical generation. The first group of storage possibilities are batteries; but only lead-acid batteries are

**TABLE 1.5** Comparison of Some Energy Storage Systems for Large-scale Electrical Generation[a]

| Type | Storage efficiency (%) | Energy density[b] (kW · hr/ft³) | Power rating of system ($/kW) | Cost of storage capacity of system ($/kW · hr) | State of development |
|---|---|---|---|---|---|
| Lead-acid battery | 60–75 | 1–2 | 80–100 | 20–40 | Ready for commercial use |
| Advanced aqueous battery, such as zinc-chlorine | 60–80 | 1–10 | 60–100 | 10–20 | Feasibility being studied; not available commercially until 1985, although testing will be done in next few years |
| Advanced, compact, alkali metal-sulfur battery, such as lithium or sodium-sulfur units | 60–75 | 0.5–2 | 100–200 | 5–10 | Would not be available commercially until 1986 at the earliest |
| Hydrogen energy system using electrolysis to produce hydrogen, hydrogen storage, and fuel cells to produce electricity from the stored hydrogen | 15–30 | Not applicable | 200–250 | 2–10 | Required technology is available but is expensive and inefficient; large-scale use would require imposing economic adjustments in use of conventional systems; storage of hydrogen requires further research |
| Thermal energy storage using liquid metal or molten salts | ~75 | ? | | | Economic feasibility for large-scale use not certain; earliest commercial development by 1990; technology knowledge is too meager to estimate economics |
| Super flywheels | 85–95 | 0.5–3 | 30–50 | 30–75 | Some flywheel systems now available; advanced systems for large-scale storage may be available by 1980–1985 |
| Hydro storage | 65–70 | 0.001 | 60–90 | 6–12 | The only large-scale energy storage system now in use; energy density increases as water is pumped to greater heights |

| Compressed-air system, using underground storage and gas turbines, the efficiency of which is improved by use of the compressed air | Not applicable | 0.5–2 | 90–110 | 3–20 | Machinery components exist but have not been assembled into integrated system; one system for a 290-mW gas-turbine generator to be completed in West Germany during 1977 |

[a]Information is based on Kalhammer, F. R., "Energy Storage: Incentives and Prospects for Its Development," available from Electric Power Research Institute, 3412 Hillview Avenue, Palo Alto, California 94304. Supplementary information from congressional seminar reported on in this article; from various materials on flywheels by David W. Rabenhorst, Applied Physics Laboratory, the Johns Hopkins University, Laurel, Maryland 20810; and from "The National Power Survey: Energy Conversion Research," Federal Power Commission, Superintendent of Documents, U.S. Government Printing Office, Washington, D.C. 20402, June 1974.

[b]Energy density is a measure of storage potential. The potential varies considerably, from a very low value for stored water to a high value for advanced batteries and flywheels.

currently available at reasonable cost, and considerable research and development will be necessary to produce more advanced battery systems. Hydro storage is another possibility that is currently available and widely used. It could be useful in connection with solar power plants, but would require siting the power plants in areas where hydro storage capability exists. Although there are some places where solar power plants can be combined with hydro storage, the regions with most abundant sunshine are, unfortunately, in those parts of the world that are flat and do not have abundant water. Super flywheels are often mentioned and they appear to be satisfactory, provided the storage requirements are not excessively large. Compressed-air systems using underground storage and gas turbines are technologically possible, and one such system is expected to be completed in West Germany during 1978. Thermal energy storage in liquid metals or molten salts is possible, but its commercialization will require more research and development.

Another possibility is using solar energy to generate hydrogen and storing the energy in the gaseous or liquid phase. There appear to be no technical difficulties to the large-scale production, storage, and subsequent use of hydrogen in either liquid or gaseous form, but the hydrogen production efficiency is relatively low and, consequently, the cost of hydrogen storage and delivery systems is quite high at the present time. A positive aspect of liquid hydrogen as a storage medium is its large amount of energy per unit volume. This method would permit the solar power plants to be located in favorable solar regions of the world and the energy stored in hydrogen to be transported where it is needed. (Figure 1.9 shows a comparison of the mass and volumetric energy densities for various energy-storage media.) There are some experts (5, 8) who believe that a socioeconomic world structure based on combination solar-hydrogen technology is possible within the foreseeable future. This approach certainly has many advantages and is technically feasible, but under current conditions it would increase the cost of energy appreciably.

## THE ECONOMICS OF SOLAR SYSTEMS

There is an important difference between the technological feasibility of utilizing solar energy processes and the prospects for their immediate use. In a free economy the criterion determining whether solar energy or some other energy source will be used is economic competitiveness. Also, in centrally planned economies, decisions on assigning resources are made based on criteria that include cost effectiveness. As shown in more detail in Chapter 5, solar energy systems are economically competitive with other energy sources for some applications, but the economics of using solar energy must be determined for each individual case.

Although solar energy is essentially free, there is a definite cost associated with its utilization. In this respect solar energy is really not very different from gas or oil. For renewable as well as nonrenewable energy sources, the technological cost does not rest with the raw material, but rather with the processes necessary to transform it into a useable form. Whereas for conventional energy sources such as gas and oil the processing cost has traditionally been borne by large industries, which borrow money from a bank and then charge the consumer for each unit of

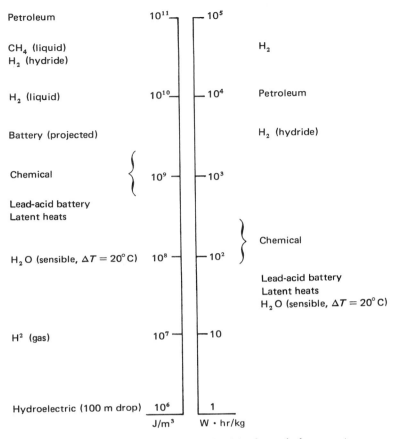

Petroleum $10^{11}$ — $10^5$

CH$_4$ (liquid) H$_2$
H$_2$ (hydride)

H$_2$ (liquid) $10^{10}$ — $10^4$ Petroleum

Battery (projected) H$_2$ (hydride)

Chemical $\{$ $10^9$ — $10^3$

Lead-acid battery
Latent heats

H$_2$O (sensible, $\Delta T = 20°$ C) $10^8$ — $10^2$

$\}$ Chemical

Lead-acid battery
Latent heats
H$_2$O (sensible, $\Delta T = 20°$ C)

H$^2$ (gas) $10^7$ — $10$

Hydroelectric (100 m drop) $10^6$ — $1$

J/m$^3$ W · hr/kg

**FIGURE 1.9** Mass and volumetric energy densities for typical energy-storage media. Redrawn from Offenhartz, P.O.D., Classical Methods of Storing Thermal Energy, in K. W. Böer (ed.), "Sharing the Sun—Solar Technology in the Seventies," vol. 8, Solar Energy Society of Canada, Winnipeg, 1976.

energy used, solar energy installations, especially for low-temperature application in home heating and cooling, are the total responsibility of the user. As a result of this shift in economic risk a solar energy installation appears to the users as a huge additional investment that they personally must make in advance before deriving any benefit from it. If, however, solar energy is viewed as a long-term investment, then its cost can be prorated just like the cost of oil and gas, and acceptance of this view is necessary to assess the cost effectiveness of solar energy.

In its simplest form, neglecting interest charges on capital, one could calculate the cost of solar energy in the following manner. Assume that a solar system will have a life of $T$ years and its initial cost is $C_0$ dollars. If during its life the system will on the average receive every year an amount of energy $Q$, the unit cost of energy, neglecting interest charges, is equal to the cost of the installation divided by the total energy delivery during its lifetime. For example, if a solar energy collector costs $100/m$^2$ surface, has an expected life of 20 yr, and is installed in a part of the country where the mean annual horizontal surface irradiance is 200 W/m$^2$ averaged over 24 hr, the cost of solar energy $C_s$ will be equal to

$$C_s = \frac{C_0}{Q \times T} = \frac{100 \ (\$/m^2) \times 1000 \ (W/kW)}{200 \ (W/m^2) \times 24 \ (hr/day) \times 365 \ (days/yr) \times 20 \ yr}$$

$$= \$0.00285/kW \cdot hr$$

It is obvious, however, that no collector will perform at 100 percent efficiency. Consequently, the collector efficiency enters the economics, because according to thermodynamic laws only a fraction of the incident energy can be converted into useful heat. Assuming that the efficiency of the collector $\eta_c$ is 50 percent, the cost of solar energy will be twice that calculated in the previous example and will be given by

$$C_s = \frac{C_0}{QT\eta_c} = \$0.0057/kW \cdot hr \qquad (1.7)$$

Note that installing the very same collector in another part of the country where the mean irradiance is only 100 W/m$^2$ would again double the cost of the solar energy, although the system cost would be the same.

## Effect of Interest Rate and Inflation

The previous analysis is clearly oversimplified because, as anyone knows, it costs money to borrow money. To install a solar energy system it is necessary to borrow the money to finance the capital investment. Even if one could pay for this capital investment out of savings, the solar investor would immediately recognize that withdrawal of money from a savings account represents loss of the interest that would otherwise accrue. Therefore, one is paying about the same amount of interest whether cash is paid for the initial investment or money is borrowed from a bank.

As shown in Chapter 5, the average *annual cost* of a solar system, neglecting power and maintenance cost, depends on the expected lifetime $T$, the interest rate $i$, and the total initial investment $C_0$, according to the relation

$$C_y = C_0 \ CRF \qquad (1.8)$$

where $CRF = i(1 + i)^T/[(1 + i)^T - 1]$, the capital-recovery factor. Thus, if the solar system receives on the average $Q$ kW $\cdot$ hr/yr, the average cost of solar energy, when all of the money to build the collection system is borrowed, becomes

$$C_s = \frac{C_0 i(1 + i)^T}{[(1 + i)^T - 1]Q\eta_c} \qquad (1.9)$$

It is thus apparent that by paying interest the cost of solar energy increases and the dollar cost of the energy depends on the interest charged to the owner of the installation. For example, if one were to borrow money in a noninflationary economy at an interest rate of 8 percent and repay the money during the 20-yr lifetime of a solar collector, the cost of owning the solar equipment would be approximately 10 percent of the initial investment during each of the 20 yr of ownership, and the cost of energy with a 50 percent efficient collector in a 100 W/m² region would be about $0.0115/kW · hr. This is almost exactly twice the cost calculated by the simple, interest-free approach in Eq. (1.7).

An offsetting feature of borrowing the money from the bank is, of course, the inflation rate. Once a solar system is installed, it is essentially inflation-proof, whereas the cost of the competing energy source will continuously increase with time. As will be shown in more detail in Chapter 5, to assess the economics of solar energy during a period of inflation, one has to guess the average rate of inflation and interest rate during the period of operation of the solar collector $T$. As a result of inflation, the economic viability of solar energy improves, because the real cost to the owner of a solar system per year will be less in an inflationary than in a noninflationary economy. For the special case where the rate of interest is exactly equal to the inflation rate, one finds that economic viability is determined by the simple relation of Eq. (1.7), which can be put in the form

$$\frac{C_0}{C_f \eta_c Q} < T \tag{1.10}$$

if $C_f$ equals the cost in dollars per kilowatt of the conventional energy, which the solar energy is to replace. This relation is known as the *payback condition* and requires that the life of the equipment be sufficiently long to repay the initial capital outlay in the absence of interest and inflation. It is apparent that the economic viability of solar energy improves with increases in collection efficiency, insolation level, alternative fuel cost, and total life of the equipment, but diminishes with increasing initial cost of the conversion equipment.

## Cost Effectiveness of Solar Energy

The question of how competitive solar energy is can be illustrated graphically. Figure 1.10 shows qualitatively a comparison between the predicted annual cost of collecting solar energy in various climates and the cost of an equivalent amount of energy from a competing energy source at three different price levels. Observe that the cost of the competing energy source increases with time. The reason for this is that, because of the scarcity of nonrenewable energy sources and raw materials, the cost of fossil fuels in the world is continuously increasing. Also note that the cost of solar energy is shown to decrease with time and then to level off. This is a result of the expectation that as soon as solar energy systems are mass produced, their cost may decrease and their efficiency will probably increase. It is clear that solar energy is most immediately competitive where existing fuel prices are high and

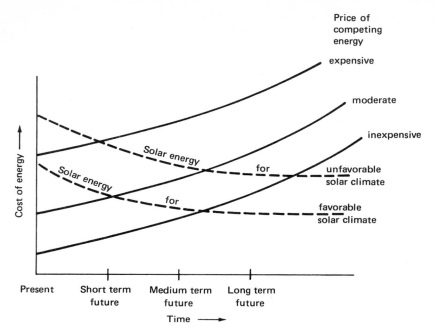

**FIGURE 1.10** Qualitative comparison between the predicted annual real costs of collecting solar energy in a favorable climate and in a less favorable climate, and the cost of an equivalent amount of energy from a competing energy source at three price levels, each assumed to inflate steadily in real terms.

where the insolation level is large. However, a solar energy system may deliver useful energy only during part of the year as, for example, in a system used for heating a home. Under those conditions, the price of solar energy will be higher than if the same system were used throughout the year as, for example, in a hot service water installation, which can utilize all of the collectible solar energy throughout the year. The engineering challenge, which will be discussed in this book, is to discover the most cost-effective solar process for a given application, calculate its true cost, compare it with alternate ways and means of providing the energy, and finally build it to meet its expectations.

## THE SECOND LAW OF THERMODYNAMICS, SOLAR ENERGY, AND ENERGY RESOURCE CONSERVATION

A unique and technically attractive asset of solar energy from a thermodynamic view is the feasibility of matching it to a large range of tasks over widely varying operating temperatures. Although this feature of solar energy has heretofore been largely neglected, this aspect of solar energy utilization may have the greatest long-term impact on the optimum use of the solar resource, as well as on the use of other energy sources that can be supplemented by solar energy. A detailed analysis of solar energy conversion at various temperatures is presented in Chapter 4. For the following discussion it is sufficient to state that solar radiation can be converted to useful heat at temperatures ranging from ambient temperature to at least 3300 K.

## Energy Quantity—The First Law of Thermodynamics

When technical experts, politicians, and users alike discuss the subject of energy, they invariably talk about energy conservation. This approach to providing an adequate energy supply for future use is based on the implicit assumption that energy is a substance that flows or can somehow be transported from place to place by some unspecified means. It also assumes that energy can be saved by not using it, and that the best way of saving energy is not using it. In this pseudoeconometric way of thinking all attention is focused on a system property (internal energy) and not on the interactions among systems. In practice, however, the tasks of transforming energy into work and heat are important and, as shown below, attention to processes and system interactions affords the only effective method of conserving and effectively using energy sources and reserves.

The present widespread application of what Berg (2, 3, 4) has termed "the modern phlogiston theory" of energy accounting completely ignores the quality of energy. It restricts itself to keeping track of the quantity of energy, which according to the first law of thermodynamics will never change in nonrelativistic processes. On the other hand, a system of energy accounting based on second law principles, which assert that work is the highest quality and lowest entropy form of energy, provides a method of using both solar and nonsolar energy sources effectively.

The figure of merit most widely used to evaluate alternative uses of energy is the *first law efficiency* $\eta_1$. It is defined as *the useful energy transfer achieved by a device divided by the energy input required to achieve the effect.* This efficiency is far from adequate, since it may have numerical values greater than one, equal to one, or less than one if constrained by other physical principles. For example, claiming that an oil furnace heats a building with 75 percent efficiency implies that a 100 percent efficient furnace is the best possible use of the available energy. This is an incorrect conclusion, because if the oil for the furnace had been used instead as fuel for a turbine powering a heat pump, which can provide more heat to the building than the total heating value of the raw fuel, the energy of the oil would have been used more effectively than in the furnace.

Other drawbacks of the first law efficiency view include its dependence on the specific system used to achieve a desired effect, its complete neglect of energy quality and the second law, and its lack of generality when applied to complex systems whose output may include a combination of mass transfer, heat production, and work. The fundamental difficulty with the first law efficiency is its reliance on energy as a basic unit of measure. Since energy is a property that cannot be consumed, the first law efficiency is an ambiguous and inadequate measure of energy-use effectiveness. The second law of thermodynamics, on the other hand, affords a method of ranking energy uses by quality-of-energy use and the approach to the thermodynamic limit of a given process.

## Energy Quality—The Second Law of Thermodynamics

Since work (organized energy) is the highest-quality (or lowest-entropy) form of interaction among systems, it is also the most valuable form of energy and should therefore be the index used to rank energy-conversion processes. The second law permits the definition of a property called the *available energy* A—*the maximum amount of work that can be produced from the energy of any system.* Unlike energy, available energy can be consumed in a process. Whereas the first law leads to the conclusion that energy should be hoarded (not used for any purpose) to minimize consumption, the second law provides a quantitative measure that can be used to minimize the consumption of available energy in any process. The goal of energy-process optimization should be to minimize the consumption of available energy, not to minimize the consumption of energy.

The *second law efficiency* $\eta_2$ is defined as *the ratio of the minimum amount of available energy required to perform a task,* $A_{min}$, *to the available energy* A *actually consumed by use of a given system.* Maximizing $\eta_2$ necessarily minimizes consumption of a fuel or other energy source, since availability is an extensive property and is proportional to the mass of fuel. A waste of availability is a waste of fuel mass and hence a waste of energy resources. In the case of solar-thermal systems, no mass of fuel is consumed but rather heat from a thermal collector is used to achieve a desired effect. Note that the maximum value of $\eta_2$ is related to a process or task to be achieved, not to a specific device.

The concept of available energy or maximum work $A$ was defined by Gibbs (15) as

$$A \equiv E - T_0 S + P_0 V - D \tag{1.11}$$

where $E$ = internal energy of fuel
$S$ = entropy of fuel
$V$ = volume of fuel
$T_0, P_0$ = ambient or "dead-state" temperature and pressure
$D$ = maximum useful work from diffusion processes

The diffusion term $D$ is significant only for combustion and chemical processes and need not be considered in solar systems. The availability can be defined in words as the maximum work that can be provided by any system or energy source as it proceeds to a final, equilibrium state with a specified environment.

The second law efficiency in equation form can then be written

$$\eta_2 = \frac{A_{min}}{A} \tag{1.12}$$

Table 1.6 lists the availability and first and second law efficiencies for several common tasks and energy types.

**TABLE 1.6** Availabilities and First and Second Law Efficiencies for Energy Conversion Systems[a]

| Task | Energy input | |
|---|---|---|
| | Input shaft work $W_i$ | $Q_r$ from reservoir at $T_r$ |
| Produce work $W_0$ | $A = W_i$ | $A = Q_r \left(1 - \dfrac{T_0}{T_r}\right)$ |
| | $A_{min} = W_0$ | $A_{min} = W_0$ |
| | $\eta_1 = \dfrac{W_0}{W_i}$ | $\eta_1 = \dfrac{W_0}{Q_r}$ |
| | $\eta_2 = \eta_1$ | $\eta_2 = \eta_1 \left(1 - \dfrac{T_0}{T_r}\right)^{-1}$ |
| | (electric motor, water turbine, aerogenerator) | (solar, Rankine cycle) |
| Add heat $Q_a$ to reservoir at $T_a$ | $A = W_i$ | $A = Q_r \left(1 - \dfrac{T_0}{T_r}\right)$ |
| | $A_{min} = Q_a \left(1 - \dfrac{T_0}{T_a}\right)$ | $A_{min} = Q_a \left(1 - \dfrac{T_0}{T_a}\right)$ |
| | $\eta_1 = \dfrac{Q_a}{W_i}$ | $\eta_1 = \dfrac{Q_a}{Q_r}$ |
| | $\eta_2 = \eta_1 \left(1 - \dfrac{T_0}{T_a}\right)$ | $\eta_2 = \eta_1 \left(\dfrac{1 - T_0/T_a}{1 - T_0/T_r}\right)$ |
| | (heat pump) | (solar water heater) |
| Extract heat $Q_c$ from cool reservoir at $T_c$ (below ambient) | $A = W_i$ | $A = Q_r \left(1 - \dfrac{T_0}{T_r}\right)$ |
| | $A_{min} = Q_c \left(\dfrac{T_0}{T_c} - 1\right)$ | $A_{min} = Q_c \left(\dfrac{T_0}{T_c} - 1\right)$ |
| | $\eta_1 = \dfrac{Q_c}{W_i}$ | $\eta_1 = \dfrac{Q_c}{Q_r}$ |
| | $\eta_2 = \eta_1 \left(\dfrac{T_0}{T_c} - 1\right)$ | $\eta_2 = \eta_1 \left(\dfrac{T_0/T_c - 1}{1 - T_0/T_r}\right)$ |
| | (compression air conditioner) | (absorption air conditioner) |

[a]The heat reservoirs are considered to be isothermal (see Prob. 1.5); $T_0$ is the environmental temperature and processes are reversible; $T_r > T_a > T_0 > T_c$.

The Kelvin-Planck (20) statement of the second law implies that the availability (maximum work) of a process transferring a quantity of heat $Q$ at temperature $T$ is simply the product of the Carnot efficiency and the heat transferred,

$$A_{min} = \left(1 - \frac{T_0}{T}\right) Q \qquad (1.13)$$

The entries in Table 1.6 are derived from the preceding equation.

### EXAMPLE 1.1

Calculate the second law efficiency of a gas-fired furnace providing heat at 50°C (121°F) if the outside temperature is 0°C (32°F). Assume that the available energy $A$ of gas is equal to its lower heating value $\Delta H_L$ (accurate for most hydrocarbons within a few percent) and that the first law efficiency for the furnace is 65 percent.

### SOLUTION

The second law efficiency is given by

$$\eta_2 = \frac{A_{min}}{A}$$

From Table 1.6 the relation for the second law efficiency is

$$\eta_2 = \frac{Q_a(1 - T_0/T_a)}{\Delta H_L}$$

Recognizing that

$$\eta_1 = \frac{Q_a}{\Delta H_L}$$

$\eta_2$ is given by

$$\eta_2 = \eta_1 \left(1 - \frac{T_0}{T_a}\right)$$

Inserting numerical values yields

$$\eta_2 = 0.65 \left(1 - \frac{273}{323}\right) = 0.10$$

The low value of $\eta_2$ suggests a large potential for improvements in space heating. If the task had been defined as delivery of air at 21°C (70°F) to a space—the real purpose of a heater—the second law efficiency would have been 0.046, indicating still greater possibility for improvement. Low second law efficiencies indicate a fundamental mismatch of task and energy source. In this case, low-entropy fossil fuel is used to provide high-entropy hot water for space heating.

## Matching Solar Energy to Its Task

The preceding analysis affords a quantitative methodology for matching solar conversion systems to their tasks. For high-entropy uses such as space heating, water heating, and crop drying, low-temperature, high-entropy solar energy collected by flat-plate collectors provides the best match between energy source and task, whereas the operation of a solar-thermal generating station requires low-entropy (high-temperature) thermal energy to match the task of producing shaft work (low entropy). In addition, capital invested in heat transfer equipment can be reduced if high temperatures are available.

Ford et al. (13) and Reistad (25) have examined many thermal processes, as well as the entire U.S. energy economy, by means of the second law of thermodynamics. Extending their analysis, it is apparent that solar energy is the only major energy source whose entropy level of the form in which it is collected can be manipulated to provide an optimum task-to-entropy-level match. This match can be accomplished by varying degrees of concentration of solar energy or by improved receiver design, for example, with an evacuated cover. An example of secondary solar energy concentration is the damming of rivers to produce hydro-electric power. Rain produced by the solar-powered evaporation-precipitation cycle of the atmosphere falls in small amounts over the earth's surface. This large, high-entropy energy source is converted to a concentrated, low-entropy source by gathering the flow from watersheds into narrow river valleys. The rivers are capable of producing shaft work (low-entropy energy) by concentrating the rainfall. Likewise, in direct solar-to-electric conversion methods, sunlight can be focused by mirrors or lenses onto a high-temperature receiver (see Chapter 4). The high-temperature working fluid can then be used to produce more shaft work in a Rankine-cycle turbine or some other heat engine than could be produced by the same amount of energy if it had been collected by a low-temperature flat-plate collector.

### EXAMPLE 1.2

Using the second law of thermodynamics, evaluate two proposed systems for heating hot water at 50°C with solar energy. The first system is a flat-plate solar collector capable of delivering water to the water heater at 60°C. The second system is a focusing solar collector capable of delivering water to the heater at 200°C. Heat losses in the exchange process between the solar collector fluid and the water heater are 10 percent of the heat collected and the environmental temperature is $0°C$.

## SOLUTION

The second law efficiency of a solar water heater is given by (see Table 1.6)

$$\eta_2 = \eta_1 \left( \frac{1 - T_0/T_a}{1 - T_0/T_r} \right)$$

For the low-temperature system, since losses are 10 percent, $\eta_1 = 0.9$, $T_0 = 273$ K, $T_a = 323$ K, and $T_r = 333$ K. Substituting these values in the preceding equation gives $\eta_2 = 0.77$. Similarly, for the high-temperature system, $\eta_1 = 0.9$, $T_0 = 273$ K, $T_a = 323$ K, and $T_r = 473$ K, from which $\eta_2 = 0.33$. The flat-plate system is more than twice as effective as the concentrator in producing hot water at $50°C$. This phenomenon is reflected in the larger capital investment required to concentrate sunlight. As a result, the second law analysis has reached the same conclusion as an economic analysis based on entirely different principles.

If natural gas (methane) were used to heat water in the preceding analysis with a first law efficiency of 65 percent, the second law efficiency would be significantly lower. For a flame temperature of 1400 K, Table 1.6 shows that

$$\eta_2 = 0.65 \left( \frac{1 - 273/323}{1 - 273/1400} \right) = 0.12$$

The above analysis shows that the thermodynamic effectiveness of a gas water heater is less than one-fifth that of a solar water heater. The poor gas heater performance is a direct consequence of the poor match between task and energy source.

Since diffuse radiation has a higher entropy level than direct radiation, the second law efficiency of converting the latter, at the same intensity level, will be higher. A method for quantifying this effect has been developed by Press (24) who numerically integrated volume integrals for $E$ and $S$ in Eq. (1.11). One of his results indicated that direct radiation conversion allows 25 percent greater work production than does diffuse radiation conversion.

## Summary

The second law of thermodynamics is a powerful tool in analyzing the effectiveness of energy conversion systems, because

1. The second law efficiency of a process is the proper measure of the effectiveness of energy utilization in that process. The first law efficiency is ambiguous,

because it considers only the quantity of energy but does not take the availability of energy into account.

2. The second law provides a method for maximizing the effectiveness of solar energy use by selecting the proper entropy level of solar energy collection.

## PROBLEMS

1.1. Show that the length of time $t$ that a quantity $R$ of a finite resource will last, when the present consumption rate is $C_0$ and the consumption is changing according to the exponential growth curve $C(t) = C_0 \exp (kt)$, is

$$t = \frac{1}{k} \ln \left( \frac{kR}{C_0} + 1 \right)$$

Then show that for the two extreme values of U.S. coal reserves estimated by Hubbert ($R_1 = 0.39 \times 10^{12}$ metric tons and $R_2 = 1.49 \times 10^{12}$ metric tons) the supply will last 74 yr according to the lower estimate and about 100 yr according to the higher estimate, if the annual growth in consumption is 5 percent.

1.2. Show that if the total U.S. coal reserves are $1.49 \times 10^{12}$ metric tons, these reserves would last forever if our consumption rate would decrease at a rate of 3.3 percent per century. Note that this is the ultimate self-sufficiency condition.

1.3. Show that if we want U.S. coal to last through our nation's second 200 yr and the available supplies are $1.49 \times 10^{12}$ metric tons, the maximum rate of increase of coal consumption may not exceed 2 percent per year.

1.4. Evaluate the second law efficiency of solar collectors capable of collecting energy at 100, 500, and $1000°C$ at a first law efficiency of 50 percent.

1.5. Table 1.6 lists second law efficiencies of several processes assuming the existence of isothermal (infinitely large) reservoirs. If the reservoir to which heat is added is finite, show that the minimum availability is

$$A_{\min} = (mc_p) \left[ T_a - T_0 - T_0 \ln \left( \frac{T_a}{T_0} \right) \right]$$

if the initial reservoir temperature is $T_0$; $m$ and $c_p$ are the mass and specific heat of the reservoir.

1.6. Compare a gas-fired and solar-powered absorption air conditioner with a first law efficiency of 0.55. Find the ratio of second law efficiencies if the ambient temperature $T_0 = 35°C$ and the air-conditioner delivery temperature $T_c = 21°C$.

## REFERENCES

1. Anon., "Solar Cells: Outlook for Improved Efficiency," National Academy of Sciences, Washington, D.C., 1973.
2. Berg, C. A. Potential for Energy Conservation in Industry, *Ann. Rev. Energy*, vol. 1, pp. 519–534, 1976.

3. Berg, C. A., A Technical Basis for Energy Conservation, *Technol. Rev.*, vol. 76, pp. 15–23, 1974.

4. Berg, C. A., A Technical Basis for Energy Conservation, *Mech. Eng.*, vol. 96, pp. 30–42, 1974.

5. Bockris, J. O., "Energy: The Solar-Hydrogen Alternative," Architectural Press, Ltd., London, 1976.

6. Brennan, M. H., Energy and Fuels over the Next 50 Years, in "Physics and the Energy Industry," Flinders University, Adelaide, Australia, 1974.

7. Budyko, M. I., "The Heat Balance of the Earth's Surface," English translation by N. Stepanov, U.S. Department of Commerce, Washington, D.C., p. 259, 1958.

8. Commoner, B., "The Poverty of Power," Alfred A. Knopf, New York, 1976.

9. Daniels, F., Direct Use of the Sun's Energy, *Am. Sci.*, vol. 55, no. 1, pp. 15–47, 1967.

10. Ehricke, K. A., The Power Relay Satellite Concept of the Framework of the Overall Energy Picture, *N. Am. Aerospace Rockwell Int. Rept.* E-73-12-1, December 1973.

11. Elliott, M. A., and N. C. Turner, "Estimating the Future Rate of Production of the World's Fossil Fuels," paper presented at the American Chemical Society Meeting, Boston, Massachusetts, April 1972; see also *Proc. Am. Power Conf.*, vol. 24, p. 541, 1962.

12. Energy Research and Development Administration, "National Solar Energy Research, Development, and Demonstration Program, Definition Report," U.S. Government Printing Office, Washington, D.C., 1975.

13. Ford, K. W., et al., Efficient Use of Energy, *Am. Inst. Phys. Rept.*, 25, 1975.

14. Foster, A. R., and R. L. Wright, "Basic Nuclear Engineering," 3rd ed., Allyn and Bacon, Boston, 1977.

15. Gibbs, J. W., "The Collected Works of J. Willard Gibbs," vol. 1, Yale University Press, New Haven, p. 77, 1948.

16. Glaser, P. E., "The Satellite Solar Power Station," paper presented to the Subcommittee on Space Science and Application, U.S. House of Representatives, July 22, 1975.

17. Häfele, W., "Energy Systems: Global Options and Strategies," vol. 1, IIASA Conference, Laxenberg, Austria, May 1976.

18. Hubbert, M. K., Energy Sources, in "Resources and Man," National Academy of Sciences–National Research Council, W. H. Freeman and Co., San Francisco, 1969.

19. Hubbert, M. K., Energy Resources for Power Production, in "Environmental Aspects of Nuclear Power Production," International Atomic Energy Agency, Vienna, pp. 13–43, 1971.

20. Keenan, J. H., "Thermodynamics," John Wiley & Sons, New York, 1948.

21. Kreider, J. F., and F. Kreith, "Solar Heating and Cooling," Hemisphere Publ. Corp., Washington, D.C., 1975.

22. Kreith, F., and J. F. Kreider, "Small Scale Electric Power Generation by Means of SRTA Solar Collector Systems in Developing Countries," paper presented at the 1976 WMO/UNESCO Solar Energy Conference, Geneva, Switzerland, August 1976.

23. Linden, H. R., "Analysis of World Energy Supplies," World Energy Congress, Institute of Gas Technology, Chicago, Illinois, 1974.

24. Press, W. H., Theoretical Maximum for Energy from Direct and Diffuse Sunlight, *Nature*, vol. 264, pp. 734–735, 1976.
25. Reistad, G. M., "Available Energy Conversion and Utilization in the United States," ASME Paper 74-WA/Pwr-1, 1974; see also Reistad, G. M., "Availability: Concepts and Applications," Ph.D. dissertation, University of Wisconsin, Madison, 1970.
26. Reynold, G. T., Energy Research of Physicists, *Phys. Today*, vol. 29, pp. 34–41, 1976.
27. "Solar Energy as a National Resource," NSF/NASA Solar Energy Panel, U.S. Government Printing Office, Washington, D.C., 1972.
28. Spakowski, A. E., and L. Shure, Estimated Cost of Large-Scale Power Generation, in "Solar Energy as a National Resource," NSF/NASA Solar Energy Panel, U.S. Government Printing Office, Washington, D.C., pp. 359–362, 1972.
29. Weingart, J., "Evaluating the Solar Energy Option and the Long-term Major Energy Option for Mankind," IIASA Working Paper WP-74-73, 1974.
30. Winger, J. G., J. D. Emerson, G. D. Gunning, R. C. Sparling, and A. J. Zraly, "Outlook for Energy in the United States," Energy Economics Division, Chase Manhattan Bank, New York, 1972.
31. Winston, R., Dielectric Compound Parabolic Concentrators, *Appl. Opt.*, vol. 15, no. 2, pp. 291–292, 1976.
32. Wolf, M., "Cost Goals for Silicon Solar Arrays for Large Terrestrial Applications," Conference Record 9th IEEE Photovoltaic Specialists Conference, New York, pp. 342–350, 1972.

# FUNDAMENTALS
# OF SOLAR RADIATION

*In houses with a south aspect, the sun's rays penetrate into the porticos
in winter, but in summer the path of the sun is right over our heads and
above the roofs so that there is shade.*

Socrates, ca. 400 B.C.

The quantitative analysis of the sun's motion and its energy flux on earth is the
subject of this chapter. First, the virtual motion of the sun is analyzed on a seasonal
and daily basis. There are quantitative analyses of shading devices and of various
surface orientations of fixed and moving collectors, and the air mass ratio concept is
introduced. The second half of the chapter describes the measurement and
prediction of solar radiation on monthly, daily, and hourly time scales.

In this chapter, the following special terms are used:

**Beam radiation** Solar radiation intercepted by a surface with negligible direction change and scattering in the atmosphere. Beam radiation is also referred to as
*direct radiation.*

**Diffuse radiation** Solar radiation scattered by aerosols, dust, and by the
Rayleigh mechanism; it does not have a unique direction.

**Total radiation** The total of diffuse and beam radiation; total radiation is
sometimes referred to as *global radiation.*

## THE PHYSICS OF THE SUN AND ITS
## ENERGY TRANSPORT

The nature of energy generation in the sun is still the basic unanswered question
about this star. Spectral measurements have confirmed the presence of nearly all the
known elements in the sun. However, 80 percent of the sun is hydrogen and 19
percent helium; therefore, the remaining 100+ observed elements comprise only a
tiny fraction of the composition of the sun. It is generally accepted that a
hydrogen-to-helium thermonuclear reaction is the source of the sun's energy. Since
such a reaction has not been duplicated in the laboratory, it is unclear precisely
what the reaction mechanism is, what role the turbulent flows in the sun play, and
how solar prominences and sunspots are created. One proposed solar structure is
shown in Fig. 2.1.

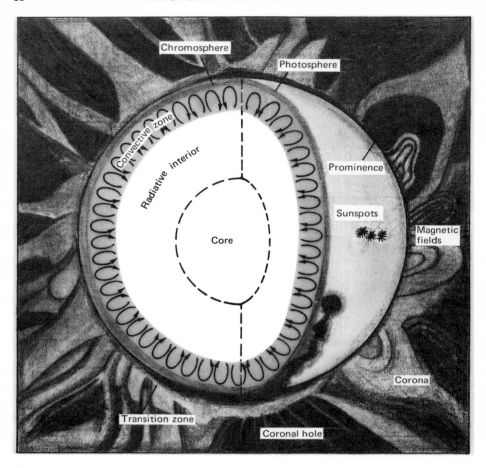

**FIGURE 2.1** Cross section of the sun showing observable exterior features with proposed interior structure.

The nature of the energy-creation process is not of great importance to terrestrial users of the sun's intercepted radiation. Of more interest is the amount of energy, its spectral and temporal distribution, and its reliability. These matters are part of the subject of this chapter. The sun is a $13.9 \times 10^5$ km diameter sphere comprised of many layers of gases, which are progressively hotter toward its center. The outermost layer from which energy is radiated into the solar system is at an equivalent black-body temperature of 5760 K ($10,400°$R) although the center of the sun may be $20 \times 10^6$ K. The rate of energy emission from the sun is $3.8 \times 10^{23}$ kW, which results from the conversion of $4.3 \times 10^9$ kg/sec ($4.7 \times 10^6$ ton/sec) of mass to energy. Of this total, only a tiny fraction, $1.7 \times 10^{14}$ kW, is intercepted by the earth, which is located 150 million km from the sun.

Solar energy is the world's most abundant permanent source of energy. The amount of solar energy intercepted by the planet earth is 5000 times greater than the sum of all other inputs (terrestrial nuclear, geothermal, and gravitational energies and lunar gravitational energy). Of this amount 30 percent is reflected to space, 47 percent is converted to low-temperature heat and reradiated to space, and 23 percent powers the evaporation/precipitation cycle of the biosphere; less than 0.5 percent is represented in the kinetic energy of the wind and waves and in photosynthetic storage in plants.

Total terrestrial radiation is only about one-third of the extraterrestrial total during a year and 70 percent of that falls on the oceans. However, the remaining $1.5 \times 10^{17}$ kW · hr that falls on land is a prodigious amount of energy—about 6000 times the total energy usage of the United States in 1978. However, only a small fraction of this total can be used because of physical and socioeconomic constraints.

## The Solar Constant

The average amount of solar radiation in near earth space is called the *solar constant* $I_0$. Its measured value is 1.353 kW/m$^2$, 429 Btu/hr · ft$^2$, or 1.94 cal/cm$^2$ · min (±1.6 percent). However, since the sun-earth orbit is elliptical, the sun-earth distance varies by ±1.7 percent during a year, and the extraterrestrial radiation also varies slightly by the inverse-square law as shown in Table 2.1 and Fig. 2.2.

## The Spectral Distribution of Extraterrestrial Radiation

The solar constant represents the total energy in the solar spectrum. This quantity, however, is not sufficient for most engineering calculations, and it is necessary to examine the distribution of energy within the spectrum. Figure 2.3 shows the spectral irradiance at the mean sun-earth distance for a solar constant of 1353

**TABLE 2.1**   Annual Variation of Solar Radiation from Orbital Eccentricity

| Date | Radius vector[a] | Ratio of flux to solar constant | Solar radiation (kW/m$^2$) |
|------|-------------|--------------------------------|--------------------------|
| Jan. 1 | 0.9832 | 1.034 | 1.399 |
| Feb. 1 | 0.9853 | 1.030 | 1.394 |
| Mar. 1 | 0.9908 | 1.019 | 1.379 |
| Apr. 1 | 0.9993 | 1.001 | 1.354 |
| May 1 | 1.0076 | 0.985 | 1.333 |
| June 1 | 1.0141 | 0.972 | 1.312 |
| July 1 | 1.0167 | 0.967 | 1.308 |
| Aug. 1 | 1.0149 | 0.971 | 1.312 |
| Sep. 1 | 1.0092 | 0.982 | 1.329 |
| Oct. 1 | 1.0011 | 0.998 | 1.350 |
| Nov. 1 | 0.9925 | 1.015 | 1.373 |
| Dec. 1 | 0.9860 | 1.029 | 1.392 |

[a]Ratio of sun-earth distance to mean sun-earth distance.

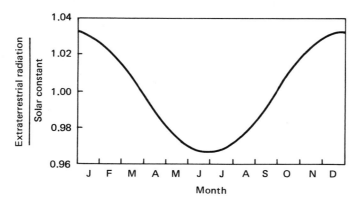

**FIGURE 2.2** Effect of the time of year on the ratio of extraterrestrial radiation to the nominal solar constant.

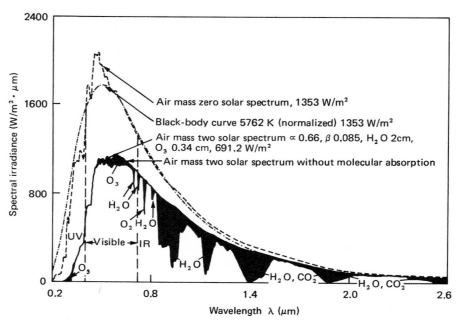

**FIGURE 2.3** Extraterrestrial solar radiation spectral distribution; also shown are equivalent black-body and atmosphere-attenuated spectra.

W/m$^2$ as a function of wavelength according to the standard spectrum data published by NASA in 1971. The same data are also presented in Table 2.2, and their use is illustrated in the following example.

### EXAMPLE 2.1

Calculate the fraction of solar radiation within the visible part of the spectrum, that is, between 0.40 and 0.75 $\mu$m.

### SOLUTION

The first column in Table 2.2 gives the wavelength. The second column gives the averaged solar spectral irradiance in a band centered at the wavelength in the first column. The third column, $D_\lambda$, gives the percentage of solar total radiation at wavelengths shorter than the value of $\lambda$ in the first column. At a value of 0.40 $\mu$m, 8.7 percent of the total radiation occurs at shorter wavelengths. At a wavelength of 0.75 $\mu$m, 51.7 percent of the radiation occurs at shorter wavelengths. Consequently, 43 percent of the total radiation lies within the band between 0.40 and 0.75 $\mu$m, and the total energy received outside the earth's atmosphere within that spectral range is 582 W/m$^2$ (184 Btu/hr $\cdot$ ft$^2$).

## TERRESTRIAL SOLAR RADIATION ESTIMATION— BEAM COMPONENT

Although extraterrestrial radiation can be predicted with certainty,* radiation levels on earth are subject to considerable uncertainty resulting from local climatic interactions. This section describes several ways of estimating radiation under clear skies. Subsequent sections in this chapter will describe prediction methods for conditions when the sky is not completely clear.

### Air Mass Ratio

The simplest method of calculating atmospheric absorption of solar radiation for clear skies is by use of a Bouger's law equation (16):

$$I_b = I_o e^{-km} \qquad (2.1)$$

where $I_b$ and $I_o$ are the terrestrial and extraterrestrial intensities of beam radiation, respectively, $k$ is an absorption constant for the atmosphere, and $m$ is the dimensionless path length of sunlight through the atmosphere and is called the *air mass ratio* (Fig. 2.4). When the sun is overhead, for example, $m = 1$; when the solar-altitude angle is 30°, $m = 2$. By obvious extension air mass zero corresponds to extraterrestrial radiation.

---

*The effect of sunspots, which may cause up to 0.5 percent variation, is neglected.

**TABLE 2.2** Extraterrestrial Solar Irradiance[a,b]

| $\lambda$ (μm) | $E_\lambda{}^c$ (W/m²·μm) | $E_\lambda{}^c$ (Btu/hr·ft²·μm) | $D_\lambda{}^d$ (%) | $\lambda$ (μm) | $E_\lambda$ (W/m²·μm) | $E_\lambda$ (Btu/hr·ft²·μm) | $D_\lambda$ (%) | $\lambda$ (μm) | $E_\lambda$ (W/m²·μm) | $E_\lambda$ (Btu/hr·ft²·μm) | $D_\lambda$ (%) |
|---|---|---|---|---|---|---|---|---|---|---|---|
| 0.115 | 0.007 | 0.002 | $1 \times 10^{-4}$ | 0.43 | 1639 | 520 | 12.47 | 0.90 | 891 | 283 | 63.37 |
| 0.14 | 0.03 | 0.010 | $5 \times 10^{-4}$ | 0.44 | 1810 | 574 | 13.73 | 1.00 | 748 | 237 | 69.49 |
| 0.16 | 0.23 | 0.073 | $6 \times 10^{-4}$ | 0.45 | 2006 | 636 | 15.14 | 1.2 | 485 | 154 | 78.40 |
| 0.18 | 1.25 | 0.397 | $1.6 \times 10^{-3}$ | 0.46 | 2066 | 655 | 16.65 | 1.4 | 337 | 107 | 84.33 |
| 0.20 | 10.7 | 3.39 | $8.1 \times 10^{-3}$ | 0.47 | 2033 | 645 | 18.17 | 1.6 | 245 | 77.7 | 88.61 |
| 0.22 | 57.5 | 18.2 | 0.05 | 0.48 | 2074 | 658 | 19.68 | 1.8 | 159 | 50.4 | 91.59 |
| 0.23 | 66.7 | 21.2 | 0.10 | 0.49 | 1950 | 619 | 21.15 | 2.0 | 103 | 32.7 | 93.49 |
| 0.24 | 63.0 | 20.0 | 0.14 | 0.50 | 1942 | 616 | 22.60 | 2.2 | 79 | 25.1 | 94.83 |
| 0.25 | 70.9 | 22.5 | 0.19 | 0.51 | 1882 | 597 | 24.01 | 2.4 | 62 | 19.7 | 95.86 |
| 0.26 | 130 | 41.2 | 0.27 | 0.52 | 1833 | 581 | 25.38 | 2.6 | 48 | 15.2 | 96.67 |
| 0.27 | 232 | 73.6 | 0.41 | 0.53 | 1842 | 584 | 26.74 | 2.8 | 39 | 12.4 | 97.31 |
| 0.28 | 222 | 70.4 | 0.56 | 0.54 | 1783 | 566 | 28.08 | 3.0 | 31 | 9.83 | 97.83 |
| 0.29 | 482 | 153 | 0.81 | 0.55 | 1725 | 547 | 29.38 | 3.2 | 22.6 | 7.17 | 98.22 |
| 0.30 | 514 | 163 | 1.21 | 0.56 | 1695 | 538 | 30.65 | 3.4 | 16.6 | 5.27 | 98.50 |
| 0.31 | 689 | 219 | 1.66 | 0.57 | 1712 | 543 | 31.91 | 3.6 | 13.5 | 4.28 | 98.72 |
| 0.32 | 830 | 263 | 2.22 | 0.58 | 1715 | 544 | 33.18 | 3.8 | 11.1 | 3.52 | 98.91 |
| 0.33 | 1059 | 336 | 2.93 | 0.59 | 1700 | 539 | 34.44 | 4.0 | 9.5 | 3.01 | 99.06 |
| 0.34 | 1074 | 341 | 3.72 | 0.60 | 1666 | 528 | 35.68 | 4.5 | 5.9 | 1.87 | 99.34 |
| 0.35 | 1093 | 347 | 4.52 | 0.62 | 1602 | 508 | 38.10 | 5.0 | 3.8 | 1.21 | 99.51 |
| 0.36 | 1068 | 339 | 5.32 | 0.64 | 1544 | 490 | 40.42 | 6.0 | 1.8 | 0.57 | 99.72 |
| 0.37 | 1181 | 375 | 6.15 | 0.66 | 1486 | 471 | 42.66 | 7.0 | 1.0 | 0.32 | 99.82 |
| 0.38 | 1120 | 355 | 7.00 | 0.68 | 1427 | 453 | 44.81 | 8.0 | 0.59 | 0.19 | 99.88 |
| 0.39 | 1098 | 348 | 7.82 | 0.70 | 1369 | 434 | 46.88 | 10.0 | 0.24 | 0.076 | 99.94 |
| 0.40 | 1429 | 453 | 8.73 | 0.72 | 1314 | 417 | 48.86 | 15.0 | 0.0048 | 0.015 | 99.98 |
| 0.41 | 1751 | 555 | 9.92 | 0.75 | 1235 | 392 | 51.69 | 20.0 | 0.0015 | 0.005 | 99.99 |
| 0.42 | 1747 | 554 | 11.22 | 0.80 | 1109 | 352 | 56.02 | 50.0 | 0.0004 | 0.0001 | 100.00 |

[a] Adapted from Thekaekara (19).

[b] Solar constant = 429 Btu/hr · ft² = 1353 W/m².

[c] $E_\lambda$ is the solar spectral irradiance averaged over a small bandwidth centered at $\lambda$.

[d] $D_\lambda$ is the percentage of the solar constant associated with wavelengths shorter than $\lambda$.

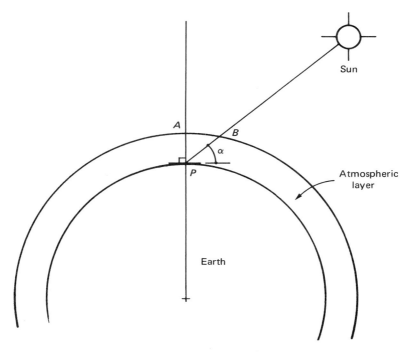

**FIGURE 2.4**  Air mass definition; air mass $m = BP/AP = \csc \alpha$, where $\alpha$ is the altitude angle. The atmosphere is idealized as a constant thickness layer.

In general, atmospheric transmittance $\bar{\tau}_{atm}(\equiv I_b/I_o)$ is given to within 3 percent accuracy for clear skies by the relation

$$\bar{\tau}_{atm} = 0.5(e^{-0.65m(z,\alpha)} + e^{-0.095m(z,\alpha)}) \qquad (2.2)$$

where $\alpha$ is the solar-altitude angle, and the air mass $m$ at an altitude $z$ above sea level is

$$m(z,\alpha) = m(0,\alpha)\frac{p(z)}{p(0)} \qquad (2.3)$$

where $p$ is the atmospheric pressure and the air mass $m(0,\alpha)$ at sea level (altitude 0) is given by

$$m(0,\alpha) = [1229 + (614 \sin \alpha)^2]^{1/2} - 614 \sin \alpha \qquad (2.4)$$

The surface beam radiation $I_b$ is given by

$$I_b = I_o\bar{\tau}_{atm} \qquad (2.5)$$

Equations (2.2) and (2.4) represent an accuracy improvement over Eq. (2.1), since they are based upon measurements in the atmosphere.

## Modified Transmittance Equation

Equations (2.1)–(2.5) apply only to a clear-sky situation in a standard atmosphere in which air pollution is absent. Hottel (5) has made modifications to Eq. (2.2) to account for particulates and water vapor in the air, as follows:

$$\bar{\tau}_{atm} = a_0 + a_1 e^{-k \csc \alpha} \tag{2.6}$$

in which $a_0$, $a_1$, and $k$ are functions only of altitude and visibility as shown in Table 2.3. The coefficient values in the table were calculated for the 1962 standard atmosphere. Other climate types including tropical, subarctic, and temperate show an effect on the values of $a_0$, $a_1$, and $k$ of less than 4 percent in most cases.

## Air Mass Spectral Behavior

In the analysis above, the air mass has been treated as an average quantity to be applied to the solar constant. In practice, it is generally necessary to know the atmospheric attenuation of solar radiation on a spectral basis (13). Tabular values of spectral air mass are given in Table A2.1 in Appendix 2. Detailed spectral calculations are required for photovoltaic devices, reflecting-surface solar reflectance calculations, and selective surface property evaluations (see Chapters 8, 4, and 3, respectively).

In summary, the air mass concept provides a way of estimating the beam or

**TABLE 2.3** Coefficients $a_0$, $a_1$, and $k$ Calculated for the 1962 Standard Atmosphere for Use in Determining Solar Transmittance $\bar{\tau}_{atm}$[a]

| | \multicolumn{6}{c}{Altitude above sea level (km)} | | | | | |
| | 0 | 0.5 | 1 | 1.5 | 2 | (2.5)[b] |
|---|---|---|---|---|---|---|
| | \multicolumn{6}{c}{23 km haze model} | | | | | |
| $a_0$ | 0.1283 | 0.1742 | 0.2195 | 0.2582 | 0.2915 | (0.320) |
| $a_1$ | 0.7559 | 0.7214 | 0.6848 | 0.6532 | 0.6265 | (0.602) |
| $k$ | 0.3878 | 0.3436 | 0.3139 | 0.2910 | 0.2745 | (0.268) |
| | \multicolumn{6}{c}{5 km haze model} | | | | | |
| $a_0$ | 0.0270 | (0.063) | 0.0964 | (0.126) | (0.153) | (0.177) |
| $a_1$ | 0.8101 | (0.804) | 0.7978 | (0.793) | (0.788) | (0.784) |
| $k$ | 0.7552 | (0.573) | 0.4313 | (0.330) | (0.269) | (0.249) |

[a]Adapted from Hottel (5) by permission.
[b]Values in parentheses indicate interpolated or extrapolated values.

direct component of solar irradiation on earth for clear skies. Robinson (16) has described the mechanisms of radiation attenuation in the atmosphere in his thorough treatise on solar radiation. Turner (20) has compiled an extensive bibliography on the subject as well. The estimation of the diffuse component is more difficult and is treated below.

## SOLAR RADIATION GEOMETRY

Many solar collection devices are fixed or move with only limited freedom. As a result the collection apertures do not face directly toward the sun at all times.

The magnitude of the beam component of direct normal radiation, as calculated in the previous section, is therefore insufficient information to estimate the energy intercepted by a solar collector surface. In this section equations describing the virtual motion of the sun are given. The next section describes the view factor of the sun onto fixed and moving terrestrial surfaces. The calculation of shading is also described.

### Solar Motion—Altitude and Azimuth Angles

For purposes of this book, the Ptolemaic view of the sun's motion provides a simplification to the analysis that follows. Since all motion is relative, it is convenient to call the earth fixed and to describe the sun's virtual motion in a coordinate system fixed to the earth with its origin at the site of interest. This approach is used throughout the analysis.

To understand the geometry of the sun's motion, the relationship of the earth's axis of rotation to the plane of its orbit, the ecliptic, must be known. The orbit and rotation occur about axes at an angle of about $23\frac{1}{2}°$ to one another. Figure 2.5 shows this relationship.

In the Ptolemaic sense, the sun is constrained to move with two degrees of freedom on the celestial sphere. As a result the location of the sun can be specified by two angles, as shown in Fig. 2.6. The solar *altitude angle* $\alpha$ is measured from the local horizontal plane upward to the center of the sun. It is measured between a line collinear with the sun's rays and the horizontal plane. The *azimuth angle* $a_s$ is measured in the horizontal plane between a due south line and the projection of the site-to-sun line on the horizontal plane as shown. The sign convention used for azimuth angle is positive east of south and negative west of south. A less convenient angle used by some solar engineers is the zenith angle $z$, which is the complement of the altitude angle $\alpha$.

Solar altitude and azimuth angles are not fundamental angles, however; they must be related to the fundamental angular quantities—hour angle, latitude, and solar declination—all of which will be described in turn. The three angles are shown in Fig. 2.7.

The solar *hour angle* $h_s$ is equal to $15°$ times the number of hours from local solar noon. Again, values east of due south, that is, morning values, are positive; values west, negative. The numerical value of $15°$/hr is based upon the nominal time (24 hr) required for the sun to move around the earth ($360°$) once.

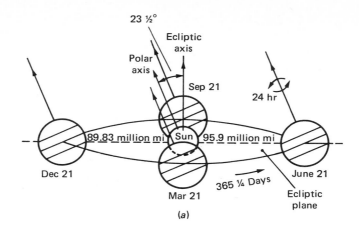

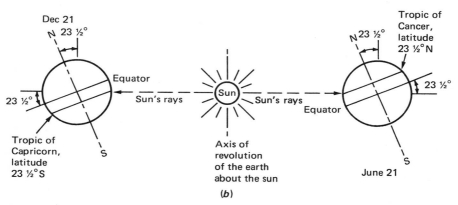

**FIGURE 2.5** (*a*) Motion of the earth about the sun. (*b*) Location of tropics. Note that the sun is so far from the earth that all the rays of the sun may be considered as parallel to one another when they reach the earth.

The *declination* of the sun $\delta_s$ is the angle between the sun's rays and the zenith direction (directly overhead) at noon on the earth's equator as shown in Fig. 2.7. Stated another way, it has the same numerical value as the latitude at which the sun is directly overhead at noon on a given day. The tropics of Cancer ($23\frac{1}{2}°$N) and Capricorn ($23\frac{1}{2}°$S) are at the extreme latitudes where the sun is overhead at least once a year as shown in Fig. 2.5. The Arctic and Antarctic circles are defined as those latitudes above which the sun does not rise above the horizon plane at least once per year. They are located, respectively, at $66\frac{1}{2}°$N and $66\frac{1}{2}°$S. Declinations north of the equator (summer in the northern hemisphere) are positive; those south, negative. Values of solar declination are given in the abbreviated solar ephemeris (Table 2.4). The meaning of the equation of time values given therein will be described shortly. Declination values are also shown in Fig. 2.8 with a nomograph

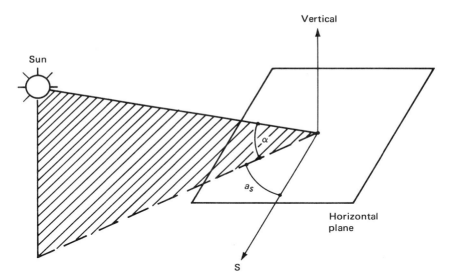

**FIGURE 2.6**  Diagram showing solar-altitude angle α and solar-azimuth angle $a_s$. See also Fig. 2.9a.

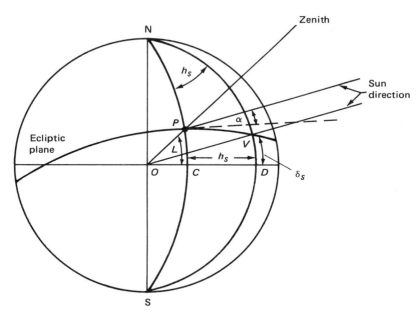

**FIGURE 2.7**  Definition of solar-hour angle $h_s$ (*CND*), solar declination $\delta_s$ (*VOD*), and latitude $L$ (*POC*); $p$, site of interest. Modified from Kreider, J. F., and F. Kreith, "Solar Heating and Cooling," revised 1st ed., Hemisphere Publ. Corp., Washington, D.C., 1977.

**TABLE 2.4**  Summary Solar Ephemeris[a]

| Date | | Declination Deg | Min | Equation of time Min | Sec | Date | | Declination Deg | Min | Equation of time Min | Sec |
|---|---|---|---|---|---|---|---|---|---|---|---|
| Jan. | 1 | −23 | 4 | − 3 | 14 | Feb. | 1 | −17 | 19 | −13 | 34 |
| | 5 | 22 | 42 | 5 | 6 | | 5 | 16 | 10 | 14 | 2 |
| | 9 | 22 | 13 | 6 | 50 | | 9 | 14 | 55 | 14 | 17 |
| | 13 | 21 | 37 | 8 | 27 | | 13 | 13 | 37 | 14 | 20 |
| | 17 | 20 | 54 | 9 | 54 | | 17 | 12 | 15 | 14 | 10 |
| | 21 | 20 | 5 | 11 | 10 | | 21 | 10 | 50 | 13 | 50 |
| | 25 | 19 | 9 | 12 | 14 | | 25 | 9 | 23 | 13 | 19 |
| | 29 | 18 | 8 | 13 | 5 | | | | | | |
| Mar. | 1 | − 7 | 53 | −12 | 38 | Apr. | 1 | + 4 | 14 | − 4 | 12 |
| | 5 | 6 | 21 | 11 | 48 | | 5 | 5 | 46 | 3 | 1 |
| | 9 | 4 | 48 | 10 | 51 | | 9 | 7 | 17 | 1 | 52 |
| | 13 | 3 | 14 | 9 | 49 | | 13 | 8 | 46 | − 0 | 47 |
| | 17 | 1 | 39 | 8 | 42 | | 17 | 10 | 12 | + 0 | 13 |
| | 21 | − 0 | 5 | 7 | 32 | | 21 | 11 | 35 | 1 | 6 |
| | 25 | + 1 | 30 | 6 | 20 | | 25 | 12 | 56 | 1 | 53 |
| | 29 | 3 | 4 | 5 | 7 | | 29 | 14 | 13 | 2 | 33 |
| May | 1 | +14 | 50 | + 2 | 50 | June | 1 | +21 | 57 | + 2 | 27 |
| | 5 | 16 | 2 | 3 | 17 | | 5 | 22 | 28 | 1 | 49 |
| | 9 | 17 | 9 | 3 | 35 | | 9 | 22 | 52 | 1 | 6 |
| | 13 | 18 | 11 | 3 | 44 | | 13 | 23 | 10 | + 0 | 18 |
| | 17 | 19 | 9 | 3 | 44 | | 17 | 23 | 22 | − 0 | 33 |
| | 21 | 20 | 2 | 3 | 34 | | 21 | 23 | 27 | 1 | 25 |
| | 25 | 20 | 49 | 3 | 16 | | 25 | 23 | 25 | 2 | 17 |
| | 29 | 21 | 30 | 2 | 51 | | 29 | 23 | 17 | 3 | 7 |
| July | 1 | +23 | 10 | − 3 | 31 | Aug. | 1 | +18 | 14 | − 6 | 17 |
| | 5 | 22 | 52 | 4 | 16 | | 5 | 17 | 12 | 5 | 59 |
| | 9 | 22 | 28 | 4 | 56 | | 9 | 16 | 6 | 5 | 33 |
| | 13 | 21 | 57 | 5 | 30 | | 13 | 14 | 55 | 4 | 57 |
| | 17 | 21 | 21 | 5 | 57 | | 17 | 13 | 41 | 4 | 12 |
| | 21 | 20 | 38 | 6 | 15 | | 21 | 12 | 23 | 3 | 19 |
| | 25 | 19 | 50 | 6 | 24 | | 25 | 11 | 2 | 2 | 18 |
| | 29 | 18 | 57 | 6 | 23 | | 29 | 9 | 39 | 1 | 10 |
| Sep. | 1 | + 8 | 35 | − 0 | 15 | Oct. | 1 | − 2 | 53 | +10 | 1 |
| | 5 | 7 | 7 | + 1 | 2 | | 5 | 4 | 26 | 11 | 17 |
| | 9 | 5 | 37 | 2 | 22 | | 9 | 5 | 58 | 12 | 27 |
| | 13 | 4 | 6 | 3 | 45 | | 13 | 7 | 29 | 13 | 30 |
| | 17 | 2 | 34 | 5 | 10 | | 17 | 8 | 58 | 14 | 25 |
| | 21 | + 1 | 1 | 6 | 35 | | 21 | 10 | 25 | 15 | 10 |
| | 25 | − 0 | 32 | 8 | 0 | | 25 | 11 | 50 | 15 | 46 |
| | 29 | 2 | 6 | 9 | 22 | | 29 | 13 | 12 | 16 | 10 |
| Nov. | 1 | −14 | 11 | +16 | 21 | Dec. | 1 | −21 | 41 | +11 | 16 |
| | 5 | 15 | 27 | 16 | 23 | | 5 | 22 | 16 | 9 | 43 |
| | 9 | 16 | 38 | 16 | 12 | | 9 | 22 | 45 | 8 | 1 |
| | 13 | 17 | 45 | 15 | 47 | | 13 | 23 | 6 | 6 | 12 |
| | 17 | 18 | 48 | 15 | 10 | | 17 | 23 | 20 | 4 | 17 |
| | 21 | 19 | 45 | 14 | 18 | | 21 | 23 | 26 | 2 | 19 |
| | 25 | 20 | 36 | 13 | 15 | | 25 | 23 | 25 | + 0 | 20 |
| | 29 | 21 | 21 | 11 | 59 | | 29 | 23 | 17 | − 1 | 39 |

[a]Since each year is 365.25 days long, the precise value of declination varies from year to year. *The American Ephemeris and Nautical Almanac* published each year by the U.S. Government Printing Office contains precise values for each day of each year.

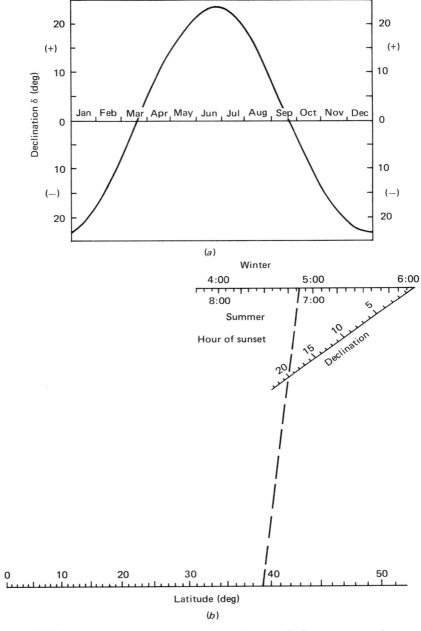

**FIGURE 2.8** (*a*) Graph to determine the solar declination. (*b*) Sunset nomograph example showing determination of sunset time for summer (7:08 p.m.) and winter (4:52 p.m.) when the latitude is 39°N and the solar declination angle is 20°. From Whillier (22). By permission of the publishers.

useful in determining the number of hours of sunlight in a day. For most calculations the declination can be considered constant during any given day.

The final fundamental angle used to calculate the altitude and azimuth angles is site *latitude L*. The latitude may be read from an atlas and is positive north of the equator and negative south.

In order to calculate the solar-altitude $\alpha$, the law of cosines for spherical triangles can be applied to triangle *NPV* in Fig. 2.7 with the result

$$\cos (90 - \alpha) = \cos (90 - L) \cos (90 - \delta_s) + \sin (90 - L) \sin (90 - \delta_s) \cos h_s \quad (2.7)$$

Simplifying,

$$\sin \alpha = \sin L \sin \delta_s + \cos L \cos \delta_s \cos h_s \quad\quad\quad (2.8)$$

Using a similar technique, the solar-azimuth angle can be computed from

$$\sin a_s = \frac{\cos \delta_s \sin h_s}{\cos \alpha} \quad\quad\quad (2.9)$$

### EXAMPLE 2.2

Compute the solar-altitude and -azimuth angles at 900, 1000, 1100, and 1200 hr solar time for Wilmington, Delaware $(L = 40°N)$, on Christmas.

### SOLUTION

From Table 2.4 the declination on December 25 is $-23°25'$. Using Eqs. (2.8) and (2.9) the following table can be constructed:

| Time (hr) | Hour angle (deg) | Sin $\alpha$ | $\alpha$ (deg) | $a_s$ (deg) |
|---|---|---|---|---|
| 900 | +45 | 0.2416 | 14.0 | +42.0 |
| 1000 | +30 | 0.3533 | 20.7 | +29.4 |
| 1100 | +15 | 0.4235 | 25.1 | +15.2 |
| 1200 | 0 | 0.4475 | 26.6 | 0.0 |

Repeat the calculation for Bastille Day (July 14). Note the large seasonal effect on $\alpha$ and $a_s$.

### The Sun-Path Diagram

In any given day the path of the sun is in a plane tilted at an angle from the horizontal plane equal to $(90 - L)$. The isometric sketch of sun paths at the solstices and equinoxes in Fig. 2.9*a* shows the orbital plane. Figure 2.9*b* contains both plan and elevation views of this same sun path showing the horizontal and vertical projections of the sun path for the example site at 40°N latitude.

The projection of the sun's path on the horizontal plane is called a *sun-path diagram.* Such diagrams are very useful in determining shading phenomena associated with solar collectors, windows, and shading devices. As shown earlier, the solar angles ($\alpha$, $a_s$) depend upon the hour angle, declination, and latitude. Since only two of these variables can be plotted on a two-dimensional graph, the usual method is to prepare a different sun-path diagram for each latitude with variations of hour angle and declination shown for a full year. A typical sun-path diagram is shown in Fig. 2.10 for 30°N latitude.

Sun-path diagrams for a given latitude are used by entering them with

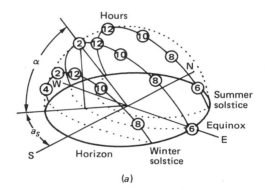

(*a*)

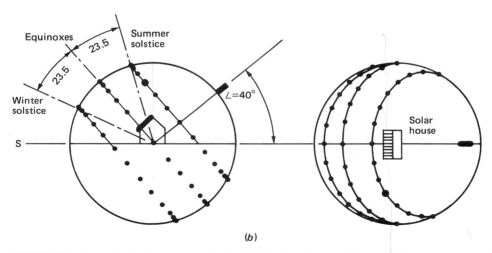

(*b*)

**FIGURE 2.9**  Sun paths for the summer solstice (6/21), the equinoxes (3/21 and 9/21), and the winter solstice (12/21) for a site at 40°N: (*a*) isometric view; (*b*) elevation and plan views.

| Declination | Approx. dates |
|---|---|
| +23°27′ | June 22 |
| +20° | May 21, July 24 |
| +15° | May 1, Aug. 12 |
| +10° | Apr. 16, Aug. 28 |
| + 5° | Apr. 3, Sep. 10 |
| 0° | Mar. 21, Sep. 23 |
| − 5° | Mar. 8, Oct. 6 |
| −10° | Feb. 23, Oct. 20 |
| −15° | Feb. 9, Nov. 3 |
| −20° | Jan. 21, Nov. 22 |
| −23°27′ | Dec. 22 |

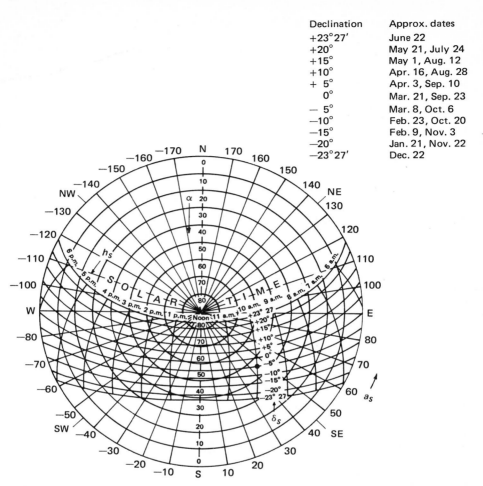

**FIGURE 2.10** Sun-path diagram for 30°N latitude showing altitude and azimuth angles. Modified from Kreider, J. F., and F. Kreith, "Solar Heating and Cooling," revised 1st ed., Hemisphere Publ. Corp., Washington, D.C., 1977.

appropriate values of declination $\delta_s$ and hour angle $h_s$. The point at the intersection of the corresponding $\delta_s$ and $h_s$ lines represents the instantaneous location of the sun. The solar altitude can then be read from the concentric circles in the diagram; the azimuth, from the scale around the circumference of the diagram. A complete set of sun-path diagrams is contained in Appendix 2 (Fig. A2.1).

## EXAMPLE 2.3

Using Fig. 2.10, determine the solar altitude and azimuth for March 8 at 10 a.m. Compare the results to those calculated from the basic Eqs. (2.8) and (2.9).

**SOLUTION**

On March 8 the solar declination is $-5°$; therefore the $-5°$ sun path is used. The intersection of the 10 a.m. line and the $-5°$ declination line in the diagram represents the sun's location; it is marked with a heavy dot in Fig. 2.10. The sun position lies midway between the $40°$ and $50°$ altitude circles, say at $45°$, and midway between the $40°$ and $50°$ azimuth radial lines, say at $45°$. So

$$\alpha \cong 45°$$

$$a_s \cong 45°$$

Equations (2.8) and (2.9) give precise values for $\alpha$ and $a_s$:

$$\sin \alpha = \sin (30°) \sin (-5°) + \cos (30°) \cos (-5°) \cos (30°)$$

$$\alpha = 44.7°$$

$$\sin a_s = \frac{\cos (-5°) \sin (30°)}{\cos (44.7°)}$$

$$a_s = 44.5°$$

Therefore, the calculated values are within $0.5°$ (1 percent) of those read from the sun-path diagram.

## The Shadow-Angle Protractor

The shadow-angle protractor used in shading calculations is a plot of solar-altitude angles, projected onto a given plane, versus solar-azimuth angle. The projected altitude angle is usually called the *profile angle* $\gamma$. It is defined as the angle between the normal to a surface and the projection of the sun's rays on a plane normal to the same surface. The profile angle is shown in Fig. 2.11a with the corresponding solar-altitude angle. The profile angle, which is always used in sizing shading devices, is given by

$$\tan \gamma = \sec a \tan \alpha \qquad (2.10)$$

where $a$ is the wall normal-to-solar azimuth angle.

Figure 2.11b shows the shadow-angle protractor to the same scale as the sun-path diagrams in Fig. 2.10 and Appendix 2. It is used by plotting the limiting values of profile angle $\gamma$ and azimuth angle $a$, which will start to cause shading of a particular point. The shadow-angle protractor is usually traced onto a transparent sheet so that the shadow map constructed on it can be placed over the pertinent sun-path diagram to indicate the times of day and months of the

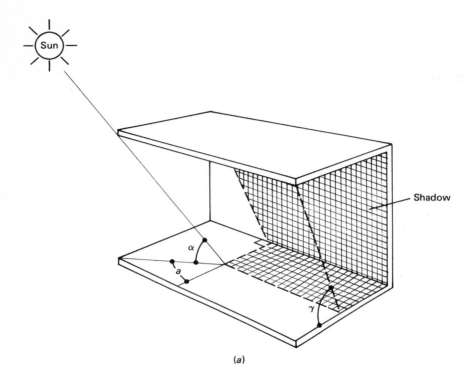

(a)

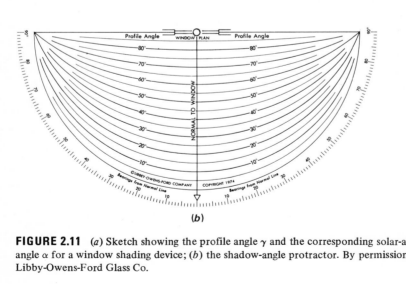

(b)

**FIGURE 2.11** (a) Sketch showing the profile angle γ and the corresponding solar-altitude angle α for a window shading device; (b) the shadow-angle protractor. By permission of the Libby-Owens-Ford Glass Co.

year during which shading will take place. The use of the shadow-angle protractor is best illustrated by an example.

## EXAMPLE 2.4

A solar building with a south-facing collector is sited to the north-northwest of an existing building. Prepare a shadow map showing what months of the year and what part of the day point C at the base of the solar collector will be shaded. Plan and elevation views are shown in Fig. 2.12.

## SOLUTION

The limiting profile angle for shading is 40° and the limiting azimuth angles are 45° and −10° as shown in Fig. 2.12. These values are plotted on the shadow-angle protractor (Fig. 2.13a).

The shadow map, when superimposed on the sun-path diagram (Fig. 2.13b), shows that point C will be shaded during the following times of day for the periods shown:

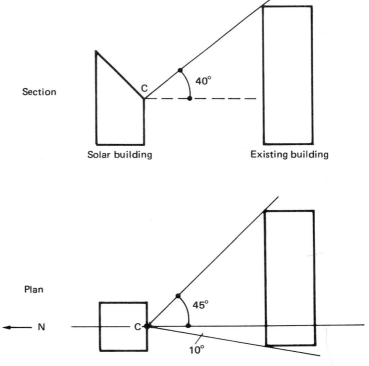

**FIGURE 2.12** Plan and elevation view of proposed solar building and existing building, which may shade solar collector at point C.

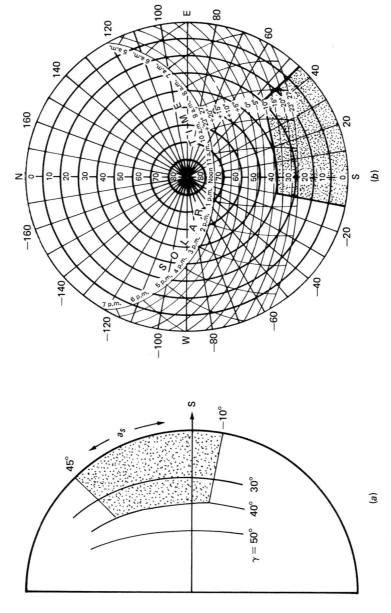

**FIGURE 2.13** (*a*) Shadow map constructed for the example shown in Fig. 2.12; (*b*) shadow map superimposed on sun-path diagram.

| Declination | Date | Time of day |
|---|---|---|
| $-23°27'$ | Dec. 22 | 8:45 a.m.$-$12:40 p.m. |
| $-20°$ | Jan. 21, Nov. 22 | 8:55 a.m.$-$12:35 p.m. |
| $-15°$ | Feb. 9, Nov. 3 | 9:10 a.m.$-$12:30 p.m. |

In summary, during the period November 3 to February 9 point C will be shaded between 3 and 4 hr. It will be shown later that this represents about 50 percent loss in collector performance for point C, which would be unacceptable for a collector to be used for heating a building in winter.

## CALCULATION OF RADIATION INTERCEPTED BY SURFACES—BEAM COMPONENT

The amount of direct radiation intercepted by a surface depends on the incidence angle. The *incidence angle i* is defined as the angle between the normal to a surface and a line collinear with the sun's rays. Figure 2.14 shows the incidence angle for a surface tilted at an angle $\beta$ from the horizontal. By examining the figure, it is easy

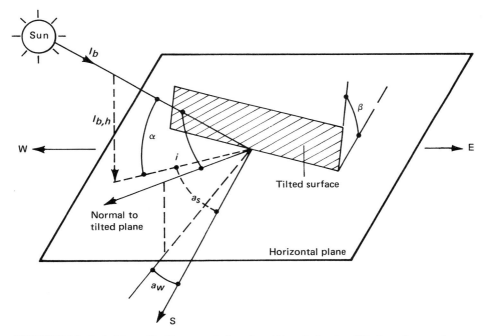

**FIGURE 2.14**  Definition of incidence angle $i$, surface tilt angle $\beta$, solar-altitude angle $\alpha$, wall-azimuth angle $a_w$, and solar-azimuth angle $a_s$ for a non-south-facing, tilted surface. Also shown is the beam component of solar radiation $I_b$ and the component of beam radiation $I_{b,h}$ on a horizontal plane.

to see that the component of beam radiation $I_{b,c}$ intercepted by the surface of a collector is given by

$$I_{b,c} = I_b \cos i \qquad (2.11)$$

where $I_b$ is the beam radiation at the surface in the direction of the sun—the *direct normal* radiation.

### Incidence Angles for Fixed Surfaces

The incidence angle depends on the three basic solar angles—$\delta_s$, $h_s$, $L$—as well as on the two angles that characterize the surface orientation—$\beta$ and $a_w$. The wall-azimuth angle $a_w$ is defined in the same manner as the solar-azimuth angle and is shown in Fig. 2.14. Collector tilt angles $\beta$ are defined as positive for surfaces facing south. Various equations have been developed for the incidence angle for several geometries of interest to the solar designer. The most useful equations are summarized below.

#### South-facing Horizontal and Vertical Surfaces (Fixed)

For vertical surfaces facing due south, that is, $a_w = 0$, $\beta = 90°$, the incidence angle is given by

$$\cos i = -\sin \delta_s \cos L + \cos \delta_s \sin L \cos h_s \qquad (2.12a)$$

For a horizontal surface, that is, $\beta = 0$, the incidence angle is given by

$$\cos i = \sin \delta_s \sin L + \cos \delta_s \cos L \cos h_s \ (= \sin \alpha) \qquad (2.12b)$$

#### South-facing Tilted Surfaces

The incidence angle for a south-facing tilted surface is

$$\cos i = \sin (L - \beta) \sin \delta_s + \cos (L - \beta) \cos \delta_s \cos h_s \qquad (2.13)$$

where $\beta$ is the tilt angle.

#### Non-south-facing Tilted Surfaces

If a tilted wall faces a direction other than due south, Eq. (2.14) is used to calculate the incidence angle $i$:

$$\cos i = \cos (a_s - a_w) \cos \alpha \sin \beta + \sin \alpha \cos \beta \qquad (2.14)$$

#### Generalized Equation for Fixed Planar Surfaces

It is possible for one generalized equation to cover all of the above cases. It is quite cumbersome and should be used only if the simpler equations do not apply. In general, the incidence angle on a planar, fixed surface is

$$\cos i = \sin \delta_s (\sin L \cos \beta - \cos L \sin \beta \cos a_w)$$
$$+ \cos \delta_s \cos h_s (\cos L \cos \beta + \sin L \sin \beta \cos a_w)$$
$$+ \cos \delta_s \sin \beta \sin a_w \sin h_s \quad (2.15)$$

## Fixed, Tilted Cylindrical Surfaces

In the discussion of tubular solar collectors in Chapter 4, the expression for the incidence angle on a tilted tubular collector surface will be required. It is given by

$$\cos i = \{1 - [\sin (\beta - L) \cos \delta_s \cos h_s + \cos (\beta - L) \sin \delta_s]^2\}^{1/2} \quad (2.16)$$

The tube centerline is assumed to be located in a north-south plane.

## Incidence Angles for Moving Surfaces

In Chapter 4 it will be shown that some solar collector designs of the focusing type must move in order to keep the sun's rays sharply focused on a thermal receiver surface. Various tracking regimes are used for various collectors. For example, trough-type collectors can be oriented with their axes east-west and are moved in a north-south manner to maintain a focus. In addition, some collectors move continuously, others periodically. Incidence angle equations for most cases of interest are given below.

## Rotation about an East-West Axis

If a collector tracks the sun continuously and is mounted on a *horizontal* east-west axis, that is, the collector moves north-south, the minimum incidence angle is

$$\cos i = (1 - \cos^2 \delta_s \sin^2 h_s)^{1/2} \quad (2.17)$$

## Rotation about a North-South Axis

A continuously tracking solar collector mounted on a *horizontal* north-south axis will require a significantly greater angular excursion each day to track the sun. For such a collector the minimum incidence angle is

$$\cos i = [(\sin L \sin \delta_s + \cos L \cos \delta_s \cos h_s)^2 + \cos^2 \delta_s \sin^2 h_s]^{1/2} \quad (2.18)$$

To maximize solar energy interception the most favorable axis orientation for an east-west rotating collector is tilted up at an angle equal to the local latitude (that is, parallel to the earth's axis)–the so-called polar mount. In this case, the incidence angle depends only on the solar declination:

$$\cos i = \cos \delta_s \quad (2.19)$$

## Rotation about Two Orthogonal Axes

If a solar collector moves with two degrees of freedom, it can track the sun at zero incidence angle throughout the day to intercept the maximum amount of beam radiation. In this case the minimum incidence angle is zero.

# Sunrise and Sunset Calculation; Integrated, Total Beam Radiation on a Surface

## Sunrise and Sunset Time

The sun is said to rise and set on a surface, either *when the surface-incidence angle is 90°* or *when the altitude angle is zero,* whichever occurs closer to solar noon. In general it is not possible to derive a closed-form equation for the sunset (or sunrise) hour angles $h_{ss}$ (or $h_{sr}$) because of the complexity of the incidence angle equations, for example, Eqs. (2.15)–(2.18). A trial-and-error solution is generally required.

It is quite simple, however, to determine when the sun sets below the horizon, that is, when $\alpha = 0$. In that case from Eq. (2.8)

$$\sin \alpha = 0 = \sin L \sin \delta_s + \cos L \cos \delta_s \cos h_{ss}(\alpha = 0) \qquad (2.20)$$

where $h_{ss}(\alpha = 0)$ is the sunset hour angle for zero altitude. Simplifying and rearranging,

$$|h_{ss}(\alpha = 0)| = |h_{sr}(\alpha = 0)| = \cos^{-1}(-\tan L \tan \delta_s) \qquad (2.21)$$

In winter, when the days are short, Eq. (2.21) is usually the governing equation for sunset on a collector surface. However, in summer the incidence angle will normally reach $90°$ (that is, the sun will pass behind the surface) before the sun sets below the horizon and the condition $i = 90°$ controls. In the simple case of south-facing, tilted surfaces it is possible to write an expression for the sunset/sunrise hour angle directly from Eq. (2.13):

$$\cos i = \cos (90°) = 0 = \sin (L - \beta) \sin \delta_s + \cos (L - \beta) \cos \delta_s \cos |h_{ss}(i = 90)|$$

Solving for $h_{ss}(i = 90)$,

$$|h_{ss}(i = 90)| = |h_{sr}(i = 90)| = \cos^{-1} [-\tan (L - \beta) \tan \delta_s] \qquad (2.22)$$

Then for south-facing surfaces, sunrise and sunset occur at the hour angles $h_{sr}$ and $h_{ss}$ that are equal to the smaller of $h_{ss}(i = 90)$ and $h_{ss}(\alpha = 0)$. Or in equation form:

$$|h_{ss}| = \min [|h_{ss}(\alpha = 0)|, |h_{ss}(i = 90)|] = |h_{sr}| \qquad (2.23)$$

For other surface orientations, a simple closed-form expression for $h_{ss}$ is not available. In addition, for latitudes north or south of the arctic circles, where the sun never sets in summer, the arguments of Eqs. (2.21) and (2.22) will exceed unity. In those cases

$$|h_{sr}| = |h_{ss}| = \pi$$

## Total Beam Radiation on a Surface

Daily, weekly, and hourly totals of beam radiation on a surface are important quantities in the prediction of solar system performance. These totals are given in general by

$$I_{\text{tot}} = \int_{t_0}^{t_0 + \Delta t} I_b(t) \cos i(t)\, dt \tag{2.24}$$

where $\Delta t$ is specified as the time interval of interest. It is generally not possible to use Eq. (2.24) to calculate $I_{\text{tot}}$, however, since the function $I_b(t)$ depends on local weather and microclimate and is, therefore, not known a priori. It is possible to calculate $I_{\text{tot}}$ outside the earth's atmosphere, however, since $I_b(t)$ is simply the solar constant multiplied by the orbital eccentricity factor $e$ from Table 2.1. That is,

$$I_b(t) = e(t)I_0 \tag{2.25}$$

where approximately

$$e(t) = 1 + 0.034 \cos \left[ \frac{2\pi n(t)}{365} \right] \tag{2.26}$$

in which $n$ is the day number counted from January 1.

An important quantity in certain solar radiation predictive correlations is the daily total extraterrestrial radiation on a horizontal surface. It can be calculated from

$$I_{\text{tot}} = \int_{t_0}^{t_0 + \Delta t} I_0 e(t) \sin \alpha(t)\, dt \tag{2.27}$$

Using Eqs. (2.8) and (2.26) for $\alpha(t)$ and $e(t)$, respectively, we have for daily totals $I_{\text{day}}$

$$I_{\text{day}} = \frac{24}{\pi} I_0 \left[ 1 + 0.034 \cos \left( \frac{2\pi n}{365} \right) \right] (\cos L \, \cos \delta_s \, \sin h_{sr} + h_{sr} \sin L \, \sin \delta_s) \tag{2.28}$$

where $h_{sr}$ in the last term of Eq. (2.28) must be in units of radians for dimensional consistency. Daily totals of extraterrestrial radiation are given in Fig. 2.15 and in tabular form in Appendix 2, Table A2.2.

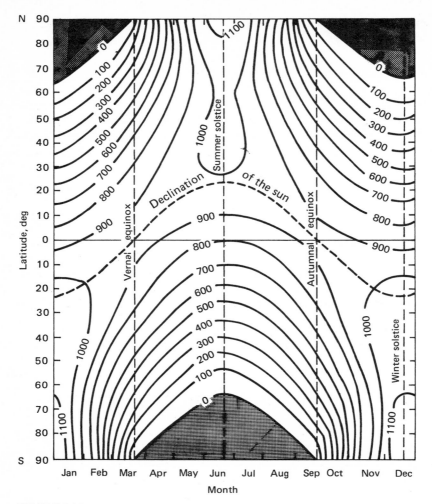

**FIGURE 2.15** The daily variation of the solar radiation at the top of the atmosphere as a function of latitude. The values are measured in langleys per day. (1 langley = 1 cal/cm² = 3.687 Btu/ft² = 11.6 W · hr/m².) From Kreider, J. F., and F. Kreith, "Solar Heating and Cooling," revised 1st ed., Hemisphere Publ. Corp., Washington, D.C., 1977.

## The Equation of Time

In the foregoing analysis it has been assumed that local solar time and local standard time (corrected for longitude) are the same. In actuality, they differ by a small amount called the *equation of time.* That is,

$$\text{True solar time} = \text{corrected standard time} + ET \tag{2.29}$$

where $ET$ is the equation of time and

Corrected standard time = standard time
(mean solar time)

$$+ 4 \text{ (standard time zone meridian} - \text{longitude)} \quad (2.30)$$

The equation of time variation is caused by precession of the earth's axis and orbital and rotational variations. These second-order effects cause a small variation in the time at which the sun is due south (or north) of an observer at true solar noon. A tabular presentation of the equation of time is given in Table 2.4. Figure 2.16 is a graphic representation of the equation called the *analemma*. The standard time zone meridians for use in Eq. (2.30) in the conterminous United States are Pacific–120°, mountain–105°, central–90°, and eastern–75°.

### Magnetic South vs. True South

Solar- and wall-azimuth angles are measured with respect to true south. However, the most common method for locating the south direction is by use of a magnetic compass, which locates magnetic south. In most parts of the world, magnetic south and true south differ by several degrees. This difference, called the magnetic declination, can be read from the isogonic chart for the United States in Fig. 2.17. Magnetic south must be corrected prior to orienting a solar building or solar collector on a building. Another method of locating true south is by taking a siting on the north star in the northern hemisphere.

## MEASUREMENT OF SOLAR RADIATION

In order to predict the performance of any solar conversion system, the diffuse and direct components of solar irradiation must be known on both a temporal and a geographic basis. In addition, spectral information is required to evaluate the performance of photovoltaic devices. The simple beam radiation models and incidence angle equations described above provide insufficient information regarding the amount of solar energy available for the engineering design of solar systems. It is, therefore, necessary to use historical records of measured solar data to determine the magnitude of diffuse and direct radiation at a specific site.

Solar radiation data are available based on several time scales—hourly, daily, and monthly. In Chapters 6 and 7 it will be shown that each of these data sets has its own special utility depending upon the nature of the analysis. The collection of daily and monthly data is more readily accomplished than the collection of hourly data. As a result the monthly and daily data base is more extensive than is the hourly base. Design methods using longer-term data are preferred, therefore.

### Instruments for Measuring Solar Radiation and Sunshine

The energy content in incident solar radiation is usually measured by one of two devices. The more common instrument, called a *pyranometer*, is able to measure

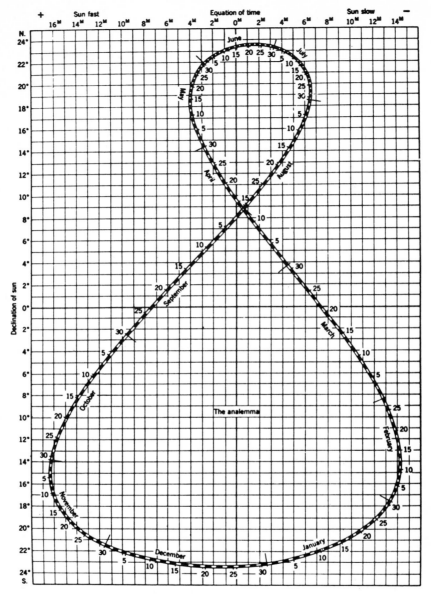

**FIGURE 2.16** The analemma—a graphical presentation of the equation of time. Equation of time values are plotted to the right and left of the vertical centerline.

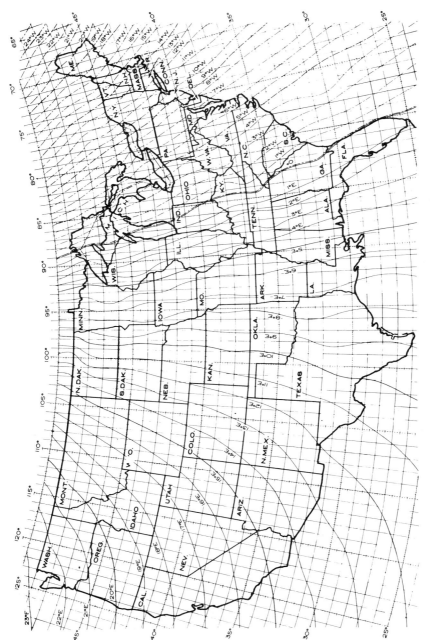

**FIGURE 2.17** The isogonic chart of the United States showing magnetic declination. For example, Los Angeles has a declination of about 15°E, indicating that a compass reads 15°E of true north or 15°W of true south (U.S. Department of Commerce, Coast and Geodetic Survey, 1965).

*total radiation* within its hemispherical field of view. It can be simply modified to measure only diffuse radiation by using an occulting disk or band, which blocks beam radiation from the sensor surface. A typical pyranometer is shown in Fig. 2.18. Pyranometers are calibrated and meant to be used in a horizontal configuration. If tilted or inverted, the free convection regime within the glass dome may change and errors may be introduced. Most manufacturers have measured this effect and can provide corrections at various tilt angles.

The second common solar radiation instrument is the *pyrheliometer*, used to measure *beam radiation*. Diffuse sky radiation is blocked from the sensor surface by mounting the solar sensor at the base of a tube that is pointed directly at the sun during the day. The field of view of a common pyrheliometer is about $5.7°$. If the pyrheliometer is caused to track the sun with a clock drive or other tracker, a continuous record of normal beam radiation can be made. Alternatively, a record of beam radiation can be made at specific values of solar altitude angle. A typical pyrheliometer design is shown in Fig. 2.19.

One difficulty with present pyrheliometric data is that it includes solar radiation beyond the sun's disk, which subtends an arc of only $\frac{1}{2}°$. Most focusing collectors used to generate high temperature (see Chapter 5) focus only the sun's disk ($\frac{1}{2}°$), not the $5+°$ zone seen by a pyrheliometer. As a result the use of current pyrheliometric data to predict concentrator performance can lead to overestimation of the performance because of the viewing-angle disparity between what a focusing device sees ($\sim\frac{1}{2}°$) and what the solar data device sees ($\sim5°$).

For very clear skies this pyrheliometric view-angle effect is small. Some data (21) have indicated that the flux density of the aureole has dropped by $10^3$ for a $1°$ arc beyond the edge of the sun's disk. However, for hazy skies, common over much of the earth, 5 percent of the pyrheliometric data may result from radiation of the sun's aureole, not from the solar disk. This is a significant effect in large-scale solar power plant design.

**FIGURE 2.18**  Solar pyranometer manufactured by the Eppley Laboratory; a typical precision pyranometer. Courtesy of the Eppley Laboratory.

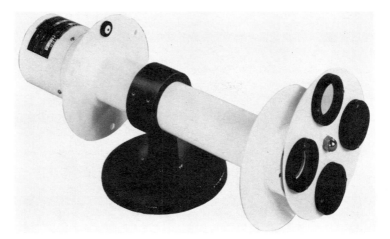

**FIGURE 2.19** Solar pyrheliometer manufactured by the Eppley Laboratory showing typical mounting. Courtesy of the Eppley Laboratory.

Other devices are used to measure sunshine data. Sunshine data are not a measure of solar energy; they are rather a measure of the time during each day when solar radiation is above some minimum, threshold level. The device used by the U.S. National Weather Service (NWS) is called a sunshine switch. It is composed of two photoelectric cells—one shaded, the other not. During daylight a potential difference is created between the two cells, which in turn operates the recorder. The intensity level required to activate the device is that just sufficient to cast a shadow. The main shortcoming of the sunshine switch is that it measures only the duration of sunlight but not its energy content. A simple photocell radiometer accurate to about 5 percent is described in Chapter 8.

## Historical Solar Radiation Data

Various national meteorological services collect solar radiation data on a routine basis. Figure 2.20 shows the stations known to be active in 1977. Most radiation data are collected in the West; only sparse networks exist in central and eastern Asia, Africa, and Australia. Löf et al. (12) and deJong (2) have evaluated the majority of worldwide historical solar radiation and sunshine records to prepare maps of horizontal solar radiation, one of which is shown in Fig. 2.21.

In the United States the NWS presently operates about 60 sites that record hemispherical ("global") solar radiation. Only five sites record beam radiation, however; about two-thirds record daily data only; the remaining one-third report both hourly and daily totals. After many years of less than perfect data collection, the NWS has recently upgraded the solar radiation network by equipping 35 sites with both pyranometers and pyrheliometers.* Instruments used in the new network

*Data available from the U.S. National Climatic Center, Federal Building, Asheville, North Carolina 28801.

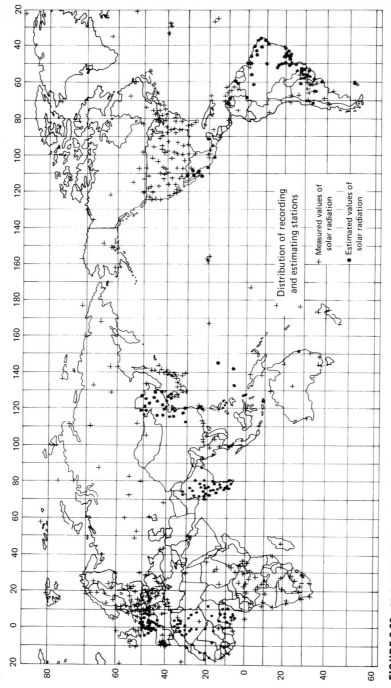

**FIGURE 2.20** Global distributions of stations where solar irradiance data are collected. Crosses indicate pyranometer stations and dots indicate sunshine recorder stations. From Thekaekara (18).

Distribution of recording
and estimating stations

+ Measured values of
solar radiation

● Estimated values of
solar radiation

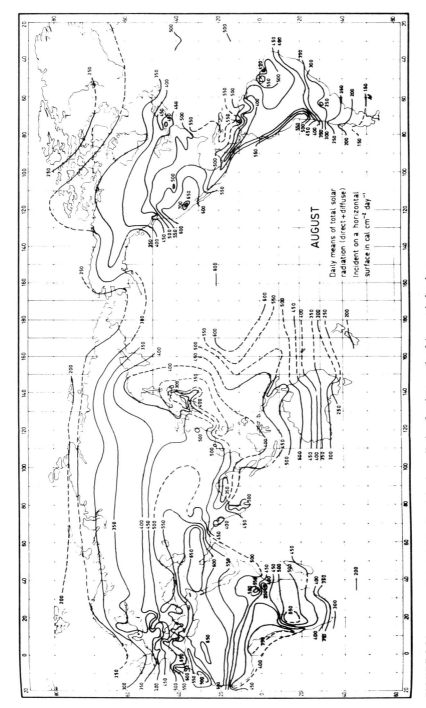

**FIGURE 2.21** Average daily total horizontal solar radiation for the month of August in langleys per day. From deJong (2) with permission.

will be calibrated prior to, during, and after service to eliminate the inaccuracies that have plagued U.S. solar data in the past. It is expected that the accuracy level of the new method will be about ±5 percent (that is, $\sigma/\mu = 0.05$). The master calibration center for the new network is located at the NOAA laboratories in Boulder, Colorado; a map showing the new network is presented in Fig. 2.22.

During the next decade, however, solar designers will be required to use data already recorded. These data, for many sites, are subject to significant systematic errors due to instrument degradation. The NWS has rehabilitated all these past data using a clear-day correlation method. The basic assumption is that clear-day, noon values vary little from year to year. The ratio of the expected value of solar radiation to that observed at true solar noon is applied to all other measurements during the same day and on adjacent cloudy days. All hourly radiation data at 26 sites have been rehabilitated in this manner as well as the 25 nonhourly sites. Any missing data were filled in using regression techniques. The new, corrected solar data along with other meteorological data useful to solar designers are available on magnetic tapes from the National Climatic Center. The new SOLMET tapes have the format shown in Fig. 2.23. A list of all stations for which solar data have been recorded is given in Appendix 2, Table A2.3. A summary of historical solar data on a monthly basis is also contained in Appendix 2, Table A2.4.

## Solar Radiation Maps

A convenient method of visually assessing the solar resource is to plot solar isoflux lines on a geographical map. A typical map for the United States is shown in Fig. 2.24. The units used on this map are langleys (1 langley = 1 cal/cm$^2$), which can be readily converted to kW · hr/m$^2$ or Btu/ft$^2$. Since the maps are plots of radiation on a horizontal surface, they do not completely represent the energy that could be collected on a surface tilted up from the horizontal plane at an angle designed to favor the winter or summer sun. Calculations that convert horizontal radiation to radiation on a tilted surface are described in the next section. A full set of isoflux maps for the United States is contained in Appendix 2. Maps are also presented for monthly and seasonal averages of direct normal radiation and total radiation on tilted surfaces. These maps have been prepared from hourly historical data as described by Boes (1).

## METHODS OF PREDICTING SOLAR RADIATION ON HORIZONTAL AND TILTED SURFACES

Insolation data collection networks around the world do not provide a sufficiently dense data base to ensure data availability for all sites where solar energy conversion is practical. It is therefore necessary to be able to predict solar radiation levels using other meteorological data that are available on a more dense basis. In addition, it is usually necessary to predict incident radiation on surfaces of many orientations and tilts using data for horizontal surfaces.

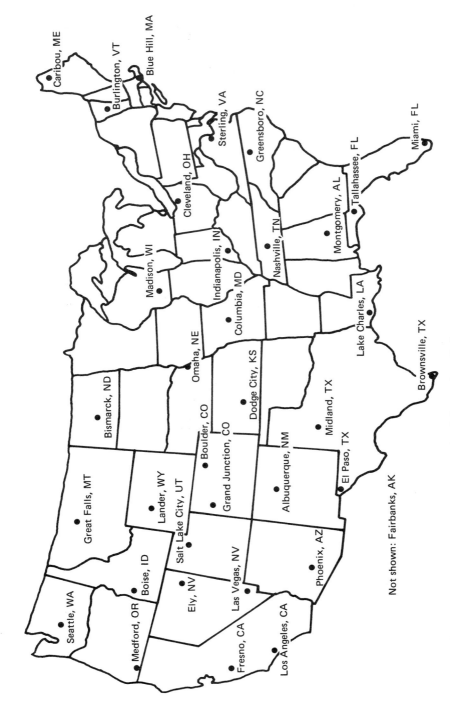

**FIGURE 2.22** New NWS network recording both total and direct solar radiation.

Caribou, ME
Burlington, VT
Blue Hill, MA
Sterling, VA
Greensboro, NC
Miami, FL
Cleveland, OH
Tallahassee, FL
Montgomery, AL
Nashville, TN
Madison, WI
Indianapolis, IN
Columbia, MD
Lake Charles, LA
Omaha, NE
Dodge City, KS
Bismarck, ND
Brownsville, TX
Boulder, CO
Grand Junction, CO
Midland, TX
Albuquerque, NM
Great Falls, MT
Lander, WY
El Paso, TX
Salt Lake City, UT
Seattle, WA
Boise, ID
Ely, NV
Las Vegas, NV
Phoenix, AZ
Medford, OR
Fresno, CA
Los Angeles, CA

Not shown: Fairbanks, AK

## IDENTIFICATION — SOLAR RADIATION OBSERVATION

| TAPE DECK # | WBAN STN # | SOLAR TIME YR MO DY HRMN | LST TIME | ETR KJ/m² | RADIATION VALUES KJ/m² — GLOBAL | | | | | | | | | SUNSHINE MIN |
|---|---|---|---|---|---|---|---|---|---|---|---|---|---|---|
| | | | | | DIRECT | DIFFUSE | NET | TILTED | OBS | ENG COR | STD YR COR | A | B | |
| 9724 | XXXXX | XX XX XX XXXX | XXXX | XXXX | 1XXXX | 1XXXX | 1XXXX | 1XXXX | 1XXXX | 1XXXX | 1XXXX | 1XXXX | 1XXXX | XX |
| 001 | 002 | 003 | 004 | 101 | 102 | 103 | 104 | 105 | 106 | 107 | 108 | 109 | 110 | 111 |

FIELD NUMBER

## SURFACE METEOROLOGICAL OBSERVATION

| OBS CEILING HT | SKY COND | VSBY hm | WEATHER | PRESSURE kPa | | TEMP °C | | WIND | | CLOUDS | | | | | | | | | SNOW COVER X |
|---|---|---|---|---|---|---|---|---|---|---|---|---|---|---|---|---|---|---|---|
| | | | | SEA LEVEL | STA-TION | DRY BULB | DEW- PT. | DIR | SPD | LOWEST | | | SECOND | | | THIRD | | FOURTH | |
| | | | | | | | | deg | m/s | TOTAL | AMOUNT | HEIGHT | TYPE | AMOUNT | HEIGHT | TYPE | AMOUNT | HEIGHT | TYPE |
| dam | | | | | | | | | | | | dam | | | dam | | dam | | dam |
| XX | 1XXXX | XXXX | XXXXXXXX | XXXXX | XXXX | XXXX | XXXX | XXX | XXX | XX | XX | XXXX | XX | XX | XXXX | XX | XX | XXXX | XX |
| 201 | 202 | 203 | 204 | 205 | 206 | 207 | 207 | 208 | 208 | 209 | 209 | 209 | 210 | 210 | 210 | 210 | 210 | 210 | 210 |

**FIGURE 2.23** Revised solar radiation/meteorology data format (SOLMET). Courtesy of F. Quinlan.

72

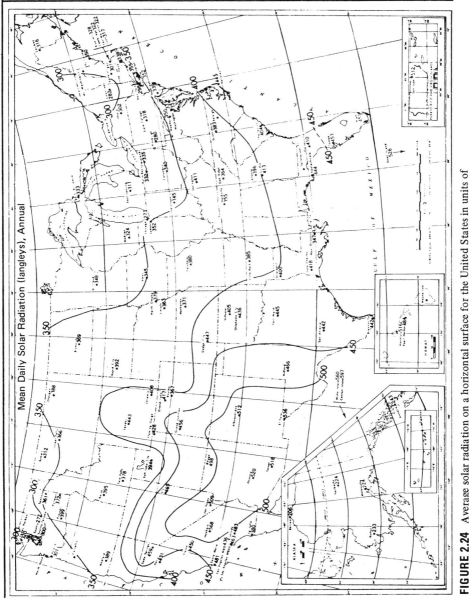

**FIGURE 2.24** Average solar radiation on a horizontal surface for the United States in units of langleys per day. From Kreider, J. F., and F. Kreith, "Solar Heating and Cooling," revised 1st ed., Hemisphere Publ. Corp., Washington, D.C., 1977.

## Monthly Radiation Estimation

One of the earliest methods of estimating solar radiation on a horizontal surface was proposed by the pioneer spectroscopist Angström. It was a simple linear model relating average horizontal radiation to clear-day radiation and to the sunshine level, that is, percent of possible hours of sunshine. Since the definition of a clear day is somewhat nebulous, Page (14) refined the method and based it on extraterrestrial radiation instead of the ill-defined clear day:

$$\bar{H}_h = \bar{H}_o(a + b\overline{PP})$$

**TABLE 2.5** Coefficients $a$ and $b$ in the Angström-Page Horizontal Solar Radiation Regression Equation[a]

| Location | Climate[b] | Sunshine hours in percentage of possible | | $a$ | $b$ |
| | | Range | Avg. | | |
|---|---|---|---|---|---|
| Albuquerque, N.M. | BS–BW | 68–85 | 78 | 0.41 | 0.37 |
| Atlanta, Ga. | Cf | 45–71 | 59 | 0.38 | 0.26 |
| Blue Hill, Mass. | Df | 42–60 | 52 | 0.22 | 0.50 |
| Brownsville, Tex. | BS | 47–80 | 62 | 0.35 | 0.31 |
| Buenos Aires, Arg. | Cf | 47–68 | 59 | 0.26 | 0.50 |
| Charleston, S.C. | Cf | 60–75 | 67 | 0.48 | 0.09 |
| Dairen, Manchuria | Dw | 55–81 | 67 | 0.36 | 0.23 |
| El Paso, Tex. | BW | 78–88 | 84 | 0.54 | 0.20 |
| Ely, Nevada | BW | 61–89 | 77 | 0.54 | 0.18 |
| Hamburg, Germany | Cf | 11–49 | 36 | 0.22 | 0.57 |
| Honolulu, Hawaii | Af | 57–77 | 65 | 0.14 | 0.73 |
| Madison, Wisconsin | Df | 40–72 | 58 | 0.30 | 0.34 |
| Malange, Angola | Aw–BS | 41–84 | 58 | 0.34 | 0.34 |
| Miami, Fla. | Aw | 56–71 | 65 | 0.42 | 0.22 |
| Nice, France | Cs | 49–76 | 61 | 0.17 | 0.63 |
| Poona, India | Am | 25–49 | 37 | 0.30 | 0.51 |
| (monsoon dry) | | 65–89 | 81 | 0.41 | 0.34 |
| Stanleyville, Congo | Af | 34–56 | 48 | 0.28 | 0.39 |
| Tamanrasset, Algeria | BW | 76–88 | 83 | 0.30 | 0.43 |

[a]From Löf et al. (12) with permission.
[b]Af = tropical forest climate, constantly moist, rainfall all through the year
Am = tropical forest climate, monsoon rain, short dry season, but total rainfall sufficient to support rain forest
Aw = tropical forest climate, dry season in winter
BS = Steppe or semiarid climate
BW = desert or arid climate
Cf = mesothermal forest climate, constantly moist, rainfall all through the year
Cs = mesothermal forest climate, dry season in winter
Df = microthermal snow forest climate, constantly moist, rainfall all through the year
Dw = microthermal snow forest climate, dry season in winter

where $\bar{H}_h$ and $\bar{H}_o$ are the horizontal, terrestrial and horizontal, extraterrestrial average radiation levels for a month, $\overline{PP}$ is the monthly averaged percent of possible sunshine [that is, hours of sunshine/maximum possible duration of sunshine from Eq. (2.21)], and $a$ and $b$ are constants for a given site. $\bar{H}_o$ can be calculated using Eq. (2.28) and averaging for the number of days in each month; or data in Appendix 2 (Table A2.2) can be used. Some typical values of $a$ and $b$ are given in Table 2.5, taken from Löf et al. (12). Fritz (3, 4) has also used the Angström-Page method for solar radiation prediction in the United States.

## EXAMPLE 2.5

Using the predictive method, estimate the monthly solar radiation for the North Central Sahara Desert (Tamanrasset, Algeria area) at latitude = 25°N. Percentages of possible sunshine and extraterrestrial radiation for this site are given in the table below.

## SOLUTION

Using the climate data given, the expected monthly average horizontal radiation for the North Sahara is calculated in the following table using $a = 0.30$ and $b = 0.43$ from Table 2.5.

| Month | $\overline{PP}$ | $\bar{H}_o{}^a$ (kJ/m² · day) | $\bar{H}_h$ kJ/m² · day | $\bar{H}_h$ Btu/ft² · day |
|-------|------|------|--------|--------|
| Jan.  | 0.88 | 23,902 | 16,215 | 1428 |
| Feb.  | 0.83 | 28,115 | 18,469 | 1626 |
| Mar.  | 0.87 | 32,848 | 22,143 | 1950 |
| Apr.  | 0.85 | 37,111 | 24,697 | 2174 |
| May   | 0.80 | 39,356 | 25,345 | 2231 |
| June  | 0.76 | 40,046 | 25,101 | 2210 |
| July  | 0.86 | 39,606 | 36,528 | 2336 |
| Aug.  | 0.83 | 37,832 | 24,852 | 2188 |
| Sep.  | 0.77 | 34,238 | 21,608 | 1902 |
| Oct.  | 0.86 | 29,413 | 19,701 | 1735 |
| Nov.  | 0.85 | 24,909 | 16,577 | 1460 |
| Dec.  | 0.80 | 22,669 | 14,599 | 1285 |

[a]Monthly averaged, daily extraterrestrial radiation.

The high levels of radiation predicted above are typical of the North Sahel region and are higher than most U.S. locations except in the Mojave Desert.

Another meteorological variable that could be used for solar radiation prediction is the opaque cloud cover recorded at many weather stations around the world. This quantity is a measure of the percent of the sky dome obscured by opaque clouds. Since this parameter contains even less solar information than sunshine values, it has not been useful in predicting long-term solar radiation values. In a subsequent section, however, it will be shown that cloud cover, when used with solar altitude angle or air mass, is a useful estimator of hourly direct radiation.

## Monthly Solar Radiation on Tilted Surfaces

In a series of papers Liu and Jordan (7–11) have developed an essential simplification in the basically complex computational method required to calculate long-term radiation on tilted surfaces. The fundamental problem in such calculations is the decomposition of total horizontal radiation measured by the NWS into its beam and diffuse components.

If the decomposition can be computed, the trigonometric analysis presented earlier can be used to calculate incident radiation on any surface in a straightforward manner. The key discovery of Liu and Jordan (LJ) is that the diffuse-to-total radiation ratio is correlated by one meteorological parameter, the *monthly clearness index* $\bar{K}_T$, which is defined as

$$\bar{K}_T = \frac{\bar{H}_h}{\bar{H}_o} \tag{2.31}$$

where $\bar{H}_h$ is the monthly averaged, horizontal, terrestrial radiation. $\bar{H}_o$ is the monthly averaged, extraterrestrial, horizontal radiation, which can be calculated from Eq. (2.28) by averaging each daily total for a month. The original LJ method was based upon the extraterrestrial radiation at midmonth, which is not truly an average.

The LJ correlation predicts the monthly diffuse $(\bar{D}_h)$ to monthly total $(\bar{H}_h)$ ratio. It can be expressed (6) by the empirical equation

$$\frac{\bar{D}_h}{\bar{H}_h} = 1.390 - 4.027\bar{K}_T + 5.531\bar{K}_T{}^2 - 3.108\bar{K}_T{}^3 \tag{2.32}$$

Note that the LJ correlations are based upon an early solar constant value of 1394 W/m² (442 Btu/hr · ft²), which was obtained from terrestrial observations, whereas the newer value, based on satellite data, is 1353 W/m² (429 Btu/hr · ft²). The values of $\bar{K}_T$ must be based on this earlier value of the solar constant to use the LJ method. Equation (2.32) is plotted in Fig. 2.25. The monthly average beam component $\bar{B}_h$ on a horizontal surface can be readily calculated by simple subtraction since $\bar{D}_h$ is known:

$$\bar{B}_h = \bar{H}_h - \bar{D}_h \tag{2.33}$$

It will be recalled on an instantaneous basis from Eqs. (2.11) and (2.12) and Fig. 2.14 that

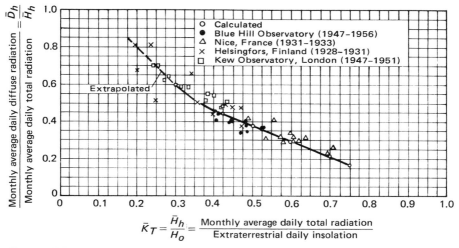

**FIGURE 2.25** The ratio of the monthly average daily diffuse radiation to the monthly average daily total radiation as a function of the clearness index $\bar{K}_T$. From (11).

$$I_b = \frac{I_{b,h}}{\sin \alpha} \tag{2.34}$$

$$I_{b,c} = I_b \cos i \tag{2.35}$$

where $I_{b,h}$ is the instantaneous horizontal beam radiation. Solving for $I_{b,c}$, the beam radiation on a surface,

$$I_{b,c} = I_{b,h}\left(\frac{\cos i}{\sin \alpha}\right) \tag{2.36}$$

The ratio in parentheses is usually called the beam radiation *tilt factor* $R_b$. It is a purely geometric quantity that converts instantaneous horizontal beam radiation to beam radiation intercepted by a tilted surface.

Equation (2.36) cannot be used directly for the long-term beam radiation $\bar{B}_h$, however. To be strictly correct, the instantaneous tilt factor $R_b$ should be integrated over a month with the beam component $I_{b,h}$ used as a weighting factor to calculate the beam tilt factor. However, the LJ method is used precisely when such short-term data as $I_{b,h}$ are not available. The LJ recommendation for the monthly mean tilt factor $\bar{R}_b$ is simply to calculate the monthly average of cos $i$ and divide it by the same average of sin $\alpha$. In equation form for south-facing surfaces this operation yields

$$\bar{R}_b = \frac{\cos (L - \beta) \cos \delta_s \sin h_{sr} + h_{sr} \sin (L - \beta) \sin \delta_s}{\cos L \cos \delta_s \sin h_{sr}(\alpha = 0) + h_{sr}(\alpha = 0) \sin L \sin \delta_s} \tag{2.37}$$

where the sunrise hour angles $h_{sr}(\alpha = 0)$ and $h_{sr}$ are given, in radians, by Eqs. (2.21) and (2.23), respectively, and are evaluated at midmonth. Non-south-facing surfaces require numerical integration or iterative methods to determine $\bar{R}_b$. The long-term beam radiation on a tilted surface $\bar{B}_c$ is then

$$\bar{B}_c = \bar{R}_b \bar{B}_h \qquad (2.38)$$

which is the long-term analog of Eq. (2.36). Values of $\bar{R}_b$ are tabulated in Appendix 2, Table A2.5.

Diffuse radiation intercepted by a tilted surface differs from that on a horizontal surface, since a tilted surface does not view the entire sky dome, which is the source of diffuse radiation. If the sky is assumed to be an isotropic source of diffuse radiation, the instantaneous and long-term tilt factors for diffuse radiation, $R_d$ and $\bar{R}_d$, are equal and are simply the radiation view factor from the plane to the visible portion of a hemisphere. In equation form:

$$R_d = \bar{R}_d = \cos^2 \frac{\beta}{2} \qquad (2.39)$$

In some cases where solar collectors are mounted near the ground, some beam and diffuse radiation reflected from the ground can be intercepted by the collector surface. The tilt factor $\bar{R}_r$ for reflected total radiation $(\bar{D}_h + \bar{B}_h)$ is then calculated to be

$$\bar{R}_r \equiv \frac{\bar{R}}{\bar{D}_h + \bar{B}_h} = \rho \sin^2 \frac{\beta}{2} \qquad (2.40)$$

in which $\rho$ is the diffuse reflectance of the surface south of the collector assumed uniform and of infinite extent.

For snow $\rho \cong 0.75$; for grass and concrete, $\rho \cong 0.2$. A more complete list of reflectances is provided in Table A2.3 of Appendix 2. The total long-term radiation intercepted by a surface $\bar{I}_c$ is then the total of beam, diffuse, and diffusely reflected components

$$\bar{I}_c = \bar{R}_b \bar{B}_h + \bar{R}_d \bar{D}_h + \bar{R}_r (\bar{D}_h + \bar{B}_h) \qquad (2.41)$$

Using Eqs. (2.39) and (2.40), we have

$$\bar{I}_c = \bar{R}_b \bar{B}_h + \bar{D}_h \cos^2 \frac{\beta}{2} + (\bar{D}_h + \bar{B}_h) \rho \sin^2 \frac{\beta}{2} \qquad (2.42)$$

in which $\bar{R}_b$ is calculated from Eq. (2.37).

In summary, monthly averaged, daily solar radiation on a surface is calculated by first decomposing total horizontal radiation into its beam and diffuse components using Eq. (2.32). Various tilt factors are then used to convert these horizontal components to components on the surface of interest.

It is impractical to prepare generalized charts for any latitude for design based upon the preceding equations, since the number of variables—$\bar{K}_T$, $\beta$, $L$—would require preparation of dozens of plots. Instead, it is recommended that designers prepare their own subset of curves for their own locations (fixed $L$ and subrange of $\bar{K}_T$) for several collector tilts.

## Daily Solar Radiation on Tilted Surfaces

Prediction of *daily* horizontal total solar radiation for sites where solar data are not measured can be done using the Angström-Page model. Instead of monthly values, however, daily values are used for percent of possible sunshine $PP$ and extraterrestrial radiation $I_{day}$ from Eq. (2.28). The results of using this simple model would be expected to show more scatter than monthly values, however.

All NWS stations with solar capability report daily horizontal total (beam and diffuse) radiation. Liu and Jordan have extended their monthly method described above to apply to daily data. The equation, analogous to Eq. (2.32), used to calculate the daily diffuse component $\bar{I}_{d,h}$, is (6)

$$\frac{\bar{I}_{d,h}}{\bar{I}_h} = 1.0045 + 0.04349K_T - 3.5227K_T{}^2 + 2.6313K_T{}^3 \qquad K_T \leqslant 0.75 \quad (2.43)$$

where $K_T$ (no overbar) is the *daily* clearness index analogous to the monthly $\bar{K}_T$. (In this section overbars indicate daily radiation totals.) Equation (2.43) is plotted in Fig. 2.26 along with recent Canadian data from Ruth and Chant (17), which are useful for latitudes between 43° and 53°N. For values of $K_T > 0.75$, the diffuse-to-total ratio is constant at a value of 0.166. $K_T$ is given by

$$K_T = \frac{\bar{I}_h}{\bar{I}_{day}} \qquad (2.44)$$

The daily extraterrestrial total radiation $\bar{I}_{day}$ is calculated from Eq. (2.28). Note that Eq. (2.43) is based on the early solar constant value of 1394 W/m² (442 Btu/hr · ft²).

The daily horizontal beam component $\bar{I}_{b,h}$ is given by simple subtraction:

$$\bar{I}_{b,h} = \bar{I}_h - \bar{I}_{d,h} \qquad (2.45)$$

The beam, diffuse, and reflected components of radiation can each be multiplied by their tilt factors $R_b$, $R_d$, and $R_r$ to calculate the total radiation on a tilted surface

$$\bar{I}_c = R_b\bar{I}_{b,h} + R_d\bar{I}_{d,h} + R_r(\bar{I}_{b,h} + \bar{I}_{d,h}) \qquad (2.46)$$

in which

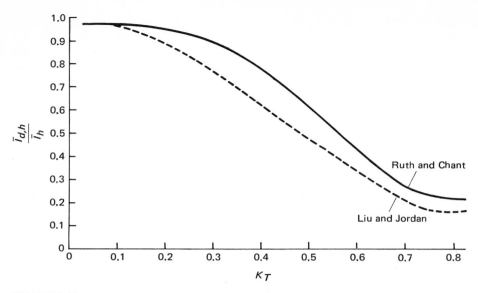

**FIGURE 2.26**  Ratio of daily averaged diffuse to total horizontal radiation as a function of $K_T$, the daily terrestrial to extraterrestrial ratio.[*]

$$R_b = \frac{\cos{(L - \beta)} \cos{\delta_s} \sin{h_{sr}} + h_{sr} \sin{(L - \beta)} \sin{\delta_s}}{\cos{L} \cos{\delta_s} \sin{h_{sr}(\alpha = 0)} + h_{sr}(\alpha = 0) \sin{L} \sin{\delta_s}} \qquad (2.47)$$

$$R_d = \cos^2{\frac{\beta}{2}} \qquad (2.48)$$

$$R_r = \rho \sin^2{\frac{\beta}{2}} \qquad (2.49)$$

by analogy with the previous, monthly analysis.

If daily solar data are available, they can be used for design, the same as monthly data. Daily calculations are necessary when finer time-scale performance is required. In addition, daily data can be decomposed into hourly data, which are useful for calculations made with large, computerized solar system simulation models.

## Hourly Solar Radiation on Tilted Surfaces

Hourly solar radiation can be predicted in several ways. Correlations between hourly total and hourly diffuse (or beam) radiation or meteorological parameters such as

---

[*]Recent analyses of daily diffuse radiation for several new sites by Collares-Pereira and Rabl (private communication, 1978) indicate that the difference between Liu-Jordan and the Ruth-Chant curves is due to the lack of a shading ring correction on the former. A latitude effect does not appear to be present.

cloud cover or air mass may be used. Alternatively, a method proposed by Liu and Jordan based on the disaggregation of daily data into hourly data could be used (7–11). Even if hourly NWS data are available, it is necessary to decompose these total values into beam and diffuse components depending upon the response of the solar conversion device to be used to these two fundamentally different radiation types.

Randall and Leonard (15) have correlated historical data from the NWS stations at Blue Hill, Massachusetts, and Albuquerque, New Mexico to predict *hourly beam radiation $I_b$* and *hourly total horizontal radiation $I_h$*. This method can be used to decompose NWS data into its beam and diffuse components.

The hourly beam radiation was found to be fairly well correlated by hourly *percent of possible insolation $k_t$* defined as

$$k_t = \frac{I_h}{I_{o,h}}$$ (2.50)

in which $I_{o,h}$ is the hourly horizontal extraterrestrial radiation, which can be evaluated from Eq. (2.27) using 1-hr integration periods. Carrying out the integration yields

$$I_{o,h} = I_o \left(1 + 0.034 \cos \frac{2\pi n}{365}\right)(0.9972 \cos L \cos \delta_s \cos h_s + \sin L \sin \delta_s)$$ (2.51)

where the solar hour angle $h_s$ is evaluated at the center of the hour of interest.

The direct normal-beam correlation based on $k_t$ is (1)

$$I_b = -520 + 1800 k_t \text{ (W/m}^2) \qquad 0.85 > k_t \geqslant 0.30$$ (2.52)

$$I_b = 0 \qquad k_t < 0.30$$ (2.53)

This fairly simple correlation gives more accurate $I_b$ values than the more cumbersome Liu and Jordan procedure, at least for Blue Hill and Albuquerque. Vant-Hull and Easton (21) have also devised an accurate predictive method for beam radiation.

Randall and Leonard have made a correlation of *total* horizontal hourly radiation $I_h$ (W/m$^2$) on the basis of *opaque cloud cover CC and air mass m* using data for Riverside, Los Angeles, and Santa Maria, California. Cloud cover is defined as $CC = 1.0$ for full overcast and $CC = 0.0$ for clear skies. A polynomial fit was used (15):

$$I_h = \frac{I_{o,h}}{100}(83.02 - 3.847m - 4.407CC + 1.1013CC^2 - 0.1109CC^3)$$ (2.54)

The average predictive error for this correlation was 2.3 percent of the NWS data; the correlation coefficient of $I_h$ with $CC$ is 0.76 for the data used. Equation (2.54) was used to predict $I_h$ for Inyokern, California, a site not used in the original correlation. Predictions of solar radiation for Inyokern were within 3.2 percent of historical NWS data indicating a good model of $I_h$, at least for the various climatic zones in the Southwest. The diffuse radiation can be calculated from Eqs. (2.53) and (2.54) by subtraction:

$$I_{d,h} = I_h - I_b \sin \alpha \qquad (2.55)$$

The American Society of Heating, Refrigerating, and Air Conditioning Engineers (ASHRAE) has calculated hourly clear-sky values on vertical, horizontal, and tilted surfaces. One table for a full year has been prepared for each of six values of latitude spanning the continental United States. In addition, tables of solar azimuth and altitude angle have been prepared for the same six latitudes. The solar radiation data represent the maximum that could be expected on a clear day and are therefore of limited usefulness in the design of solar systems. They can be used, however, to calculate cooling loads on buildings and the like. The ASHRAE tables are contained in Appendix 2, Table A2.6.

## PROBLEMS

2.1. Show that the hourly averaged, extraterrestrial radiation for a given hour is the same, to within 1 percent, as instantaneous radiation at the hour's midpoint. This is equivalent to deriving Eq. (2.51).

2.2. Prepare shadow maps for point P on the sun-path diagrams for $35°N$ and $40°N$ for the three geometries shown in $a$, $b$, and $c$ below. Determine the hours of shading that occur each month.

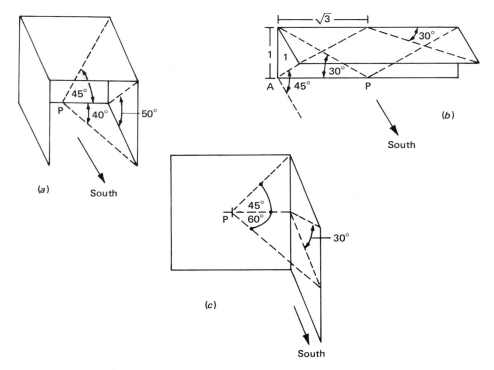

2.3. Repeat Problem 2.2c if the surface containing point P faces due west instead of due south for a 40°N location.

2.4. Calculate the incidence angle at noon and 9 a.m. on a fixed, flat-plate collector located at 40°N latitude and tilted 70° up from the horizontal. Find $i$ for June 21 and December 21.

2.5. (a) If the surface in Problem 2.4 faces S 45°E, what are the incidence angles?

   (b) If the collector in Problem 2.4 has a cylindrical surface, what are the incidence angles on June 21 and December 21?

2.6. Compare the numerical values of air mass when calculated by the equation given in Fig. 2.4 and when calculated by Eq. (2.4). At what value of altitude angle $\alpha$ is the error 1 percent?

2.7. Using a one-term Fourier cosine series, develop an empirical equation for solar declination as a function of day number counted from January 1 (see Table 2.4 and Fig. 2.8).

2.8. Derive an equation for the lines of constant declination in a sun-path diagram, for example, Fig. 2.10. Check your equation by plotting a few declination lines on a piece of polar coordinate graph paper.

2.9. Derive Eq. (2.10) relating profile angle $\gamma$ to azimuth angle $a$ and altitude angle $\alpha$.

2.10. Based upon Eq. (2.13) what value of $\beta$ would result in the annual minimum value of the incidence angle $i$? Note that this tilt angle would result in maximum collection of beam radiation on a fixed, flat, south-facing surface. *Hint*: Use a double integration procedure.

2.11. At what time does the sun set in Calcutta (23°N) on May 1 and December 1?

2.12. What is the mean solar time in Sheridan, Wyoming (107°W) at 10:00 a.m. mountain daylight time on June 10? What is the mean solar time at 10:00 a.m. mountain standard time on January 10?

2.13. If a compass points magnetic north in Cleveland, Ohio, how many degrees east or west of the true north-south line is it pointing?

2.14. Using the Angström-Page method, calculate the average horizontal insolation in Hamburg in May and in October with $\overline{pp}$ 40 percent and 60 percent, respectively.

2.15. Equation (2.32) is based on an early solar constant value of 1394 W/m². Derive a modified form of Eq. (2.32) based on the presently accepted value for the solar constant of 1353 W/m². Replot Fig. 2.25.

2.16. Predict the hourly beam and diffuse radiation on a horizontal surface for Denver (40°N) on September 9 at 9:30 a.m. if the cloud cover ($CC$) is 3.0.

2.17. Derive an expression for the minimum allowable distance between east-west rows of solar collectors that will assure no shading of one row by the row immediately to the south. Use the law of sines and express the result in terms of the collector tilt and face length and the controlling value of the solar profile angle.

# REFERENCES

1. Boes, E. C., et al., Distribution of Direct and Total Solar Radiation Availabilities for the USA, in "Sharing the Sun," vol. 1, pp. 238–263, ISES, Winnipeg, 1976; see also Boes, E. C., et al., Availability of Direct, Total and Diffuse Solar Radiation for Fixed and Tracking Collectors—the USA, *Sandia Laboratory Rept.* SAND 77-0885, 1977.

2. deJong, B., "Net Radiation Received by a Horizontal Surface on Earth," Delft University Press, Delft, 1973.

3. Fritz, S., Solar Radiation During Cloudless Days, *Heat. Vent.*, pp. 69–74, 1949.

4. Fritz, S., Average Solar Radiation in the United States, *Heat. Vent.*, pp. 61–64, July 1949.

5. Hottel, H. C., A Simple Model for Estimating the Transmittance of Direct Solar Radiation Through Clear Atmospheres, *Sol. Energy*, vol. 18, pp. 129–134, 1976.

6. Klein, S. A., "A Design Procedure for Solar Heating Systems," Ph.D. dissertation, University of Wisconsin, Madison, 1976; see also *Sol. Energy*, vol. 19, p. 325, 1977.

7. Liu, B. Y. H., and R. C. Jordan, Availability of Solar Energy for Flat-plate Solar Heat Collectors, in "Low Temperature Engineering Application of Solar Energy," chap. 1, American Society of Heating, Refrigerating, and Air Conditioning Engineers, New York, 1967; see also 1977 revision.

8. Liu, B. Y. H., and R. C. Jordan, A Rational Procedure for Predicting the Long-term Average Performance of Flat-Plate Solar Energy Collectors, *Sol. Energy*, vol. 7, pp. 53–74, 1963.

9. Liu, B. Y. H., and R. Jordan, Daily Insolation on Surfaces Tilted toward the Equator, *Trans. ASHRAE*, vol. 67, pp. 526–541, 1961.

10. Liu, B. Y. H., and R. C. Jordan, Daily Insolation on Surfaces Tilted toward the Equator, *ASHRAE*, vol. 3, no. 10, pp. 53–59, 1961.

11. Liu, B. Y. H., and R. C. Jordan, The Interrelationship and Characteristic Distribution of Direct, Diffuse and Total Solar Radiation, *Sol. Energy*, vol. 4, pp. 1–19, 1960; see also B. Y. H. Liu, "Characteristics of Solar Radiation and the Performance of Flat Plate Solar Energy Collectors," Ph.D. dissertation, University of Minnesota, Minneapolis, 1960.

12. Löf, G. O. G., et al., World Distribution of Solar Energy, *University of Wisconsin, Madison, Eng. Expt. Station Rept. 21*, 1966.

13. Moon, P., Proposed Standard Solar Radiation Curves for Engineering Use, *J. Franklin Inst.*, vol. 230, pp. 583–617, 1940.

14. Page, J. K., The Estimation of Monthly Mean Values of Daily Total Short-wave Radiation on Vertical and Inclined Surfaces from Sunshine Records for Latitudes $40°N–40°S$, *Proc. U.N. Conf. New Sources Energy*, vol. 4, p. 378, 1966.

15. Randall, C. M., and S. L. Leonard, Reference Insolation Data Base: A Case History, with Recommendations, *Rept. Recommendations Solar Energy Data Workshop*, pp. 93–103, 1974.

16. Robinson, N., "Solar Radiation," Elsevier Publ. Co., New York, 1966.

17. Ruth, D. W., and R. E. Chant, The Relationship of Diffuse Radiation to Total Radiation in Canada, *Sol. Energy*, vol. 18, pp. 153–154, 1976.

18. Thekaekara, M. P., Insolation Data for Solar Energy Conversion Derived from Satellite Measurements of Earth Radiance, in "Sharing the Sun," vol. 1, pp. 313–328, ISES, Winnipeg, 1976.

19. Thekaekara, M. P., Solar Energy Outside the Earth's Atmosphere, *Sol. Energy*, vol. 14, pp. 109–127, 1973.

20. Turner, C. P., Bibliography for Solar Energy Workshop, *Solar Energy Data Workshop, NSF Rept.* NSF/RA/N/74/062, 1974 (available from NTIS).

21. Vant-Hull, L., and C. R. Easton, Solar Thermal Power Systems Based on Optical Transmission, *NTIS Rept.* PB253167, 1975.

22. Whillier, A., Solar Radiation Graphs, *Sol. Energy*, vol. 9, pp. 165–166, 1965.

# FUNDAMENTALS
# OF FLUID MECHANICS
# AND HEAT TRANSFER

3

*To climb steep hills requires a slow pace at first.*

Shakespeare, *Henry VIII*, Act 1, Scene 1

This chapter presents the basic principles of fluid mechanics and heat transfer that are required for the design and analysis of solar energy utilization systems. Physical concepts, rather than mathematical derivations, will be emphasized and illustrated by example problems. Also the first law of thermodynamics will be reviewed and units of measurements in the SI and in the English system will be compared.

## PRINCIPLES OF FLUID MECHANICS

The fluid mechanical processes encountered in solar engineering systems require pumping a fluid through the collector and the associated distribution conduits, storing the fluid in a vessel, and controlling its flow rate. In order to cause a fluid to flow through a conduit, one must provide a driving force. This force can sometimes be supplied by gravity when a sufficient difference in elevation between the point of entry and the point of departure of the fluid exists. In the majority of cases, however, this driving force must be supplied by a mechanical device such as a pump or blower driven by an electric motor, which requires power for its operation. The material presented in this section is designed to provide the reader with the information necessary to size the conduits of thermal solar energy conversion systems properly, calculate the pressure drop in flow through conduits, valves, and heat exchangers (including the solar collector), and determine the pumping power and motor size necessary for the operation of the system. The control aspects of the system will be taken up in Chapter 5.

The fluid in a majority of low-temperature solar-thermal systems is either water or air requiring a pump or a fan, respectively, to move it. In most cases the pressure in the system is moderate and the flow rate is relatively low. In all practical situations the fluid can be considered incompressible. Preliminary to a quantitative analysis of flow systems some fundamental concepts will be reviewed: pressure, work, energy, fluid heat, viscosity, and the general energy equation.

## Fundamental Concepts

Pressure is defined as force per unit area. The concept of force is related by Newton's second law of motion to the time rate of change of momentum:

$$\text{Force} \propto \text{time rate of change of momentum}$$

This law can be written in equation form, and for a constant mass one obtains

$$F = \frac{1}{g_c} m \frac{dV}{dt} \tag{3.1}$$

where $m$ = mass

$dV/dt$ = acceleration

$g_c$ = proportionality constant, which serves as the conversion factor necessary for dimensional consistency in some systems of units

Equation (3.1) is used to define a system of units having the dimensions of mass, force, length, and time. In the SI system Eq. (3.1) states that

$$1 \text{ N force will accelerate 1 kg mass at 1 m/sec}^2$$

whereas in the English system it states that

$$1 \text{ lb force will accelerate 1 lb mass at 32.16 ft/sec}^2$$

Consequently, to make Eq. (3.1) dimensionally homogeneous $g_c = 32.16$ $\text{lb}_m \cdot \text{ft/lb}_f \cdot \text{sec}^2$ in the English system. In the SI system $g_c$ is dimensionless ($g_c \equiv 1 \text{ kg} \cdot \text{m/N} \cdot \text{sec}^2$). To facilitate the transition to the SI system of units, most illustrative problems will be developed in the SI system, but corresponding values in the English system will be added in parentheses following their values in the SI system.

Work, as well as energy, has the dimensions of force times distance. In SI units work and energy are expressed in joules (J) or newton-meters (N · m) and

$$1 \text{ N} \cdot \text{m} = 1 \text{ J}$$

In the English system of units work and energy can be expressed in $\text{ft} \cdot \text{lb}_f$, and $1 \text{ ft} \cdot \text{lb}_f = 1.356$ J. However, energy has traditionally also been related to thermal phenomena and measured calorimetrically. The basic unit of energy in this approach is the amount of energy (or work) that will raise the temperature of 1 lb of water at 68°F by 1°F. This unit is called the British thermal unit (Btu), and

$$1 \text{ Btu} = 778.1 \text{ lb}_f \cdot \text{ft} = 1055.06 \text{ J}$$

The weight of a body $W$ is defined as the force exerted on the body as a result of the acceleration of gravity $g$ or

$$W = \frac{g}{g_c} m \tag{3.2}$$

Thus, weight has the dimensions of force, and in the English system 1 $lb_m$ will weigh 1 $lb_f$ at sea level. This simple relation actually has led to the adoption of the English system of units. In the SI system, with $g = 9.806$ m/sec$^2$, 1 kg of mass will exert a force of 9.806 N, and the one-to-one relation between mass and weight does not exist.

Equation (3.2) can be used to determine the static pressure exerted by a fluid at rest. In Fig. 3.1 a column of fluid of height $z$ meters (m) is shown. Since the forces exerted by a stationary fluid are the same in all directions at any given point, the downward force at any cross section also is uniform. If the pressure acting over the upper surface of the fluid column is $p_0$, say atmospheric, then the force $Ap$ exerted by the fluid column at its base is equal to

$$Ap = Az\rho \frac{g}{g_c} + Ap_0 \tag{3.3}$$

where $p$ = pressure in N/m$^2$ (lb$_f$/ft$^2$), also called a Pascal or Pa
$\rho$ = density of the fluid in kg/m$^3$ (lb$_m$/ft$^3$)
$A$ = cross-sectional area of the fluid column in m$^2$ (ft$^2$)
It should be noted (see Fig. 3.1b) that the base pressure (force per unit area) is a function solely of the vertical height, or *fluid head*, and does not depend on the shape of the container. Consequently, the fluid in the vented storage tank shown in Fig. 3.1b will exert the same pressure at the base as the simple cylinder provided $z$ and $\rho$ for the two systems are the same.

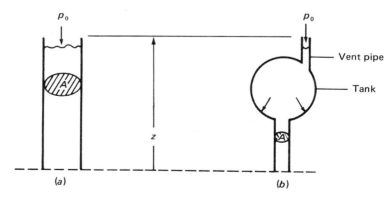

**FIGURE 3.1** Sketch illustrating pressure at the bottom of (a) a simple pipe and (b) a pipe attached to a storage tank.

## EXAMPLE 3.1

Calculate the base pressure available at the outlet pipe of a vented water tank located on a roof as shown in Fig. 3.2. The water temperature is 300 K.

## SOLUTION

The bottom of the tank is 3 m cos $30° = 2.6$ m (8.57 ft) above the ground and $z = 3.6$ m (11.85 ft). With $p_0 = 1$ atm $= 1.01325 \times 10^5$ N/m$^2$ (2120 lb$_f$/ft$^2$) and $\rho_{\text{water}} = 1.0 \times 10^3$ kg/m$^3$ (62.3 lb$_m$/ft$^3$), the base pressure is equal to

$$p = (3.6 \text{ m})(1.0 \times 10^3 \text{ kg/m}^3)\left(\frac{9.806 \text{ m/sec}^2}{1.0 \text{ kg} \cdot \text{m/N} \cdot \text{sec}^2}\right) + 1.013 \times 10^5$$

$$= 3.53 \times 10^4 + 10.133 \times 10^4 \text{ N/m}^2$$

$$= 13.66 \times 10^4 \text{ N/m}^2 \text{ (19.8 psia or 2850 lb}_f/\text{ft}^2)$$

Pressure plays an important role not only in fluid static, but also in fluid dynamic systems. Whenever a fluid flows through a conduit, the viscosity of the fluid produces shear forces, which in turn may cause pressure losses. For Newtonian fluids such as air and water, the viscosity depends mainly on the fluid temperature. The role it plays in determining the shear can be illustrated by considering two parallel plates with a fluid between them. To move the upper plate at a velocity $\Delta V$ relative to the lower plate, the required shear stress $\tau$ is given by (see Fig. 3.3$a$)

$$\tau = \frac{F}{A} = \mu \frac{\Delta V}{\Delta y} \tag{3.4}$$

where $\tau =$ shear in N/m$^2$ (lb/ft$^2$)
  $\Delta V =$ relative velocity in m/sec (ft/sec)
  $\Delta y =$ distance between the plates in m (ft)
  $\mu =$ viscosity in kg/m $\cdot$ sec (lb$_m$/ft $\cdot$ sec)

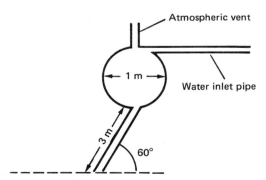

**FIGURE 3.2**  Sketch illustrating system for Example 3.1.

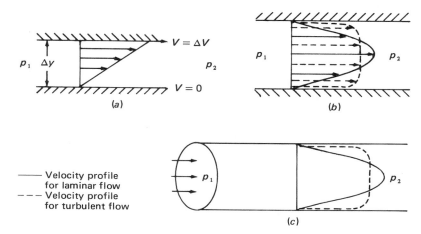

**FIGURE 3.3**  Flow between two parallel flat plates and in a tube. (*a*) One plate stationary, the other moving at velocity $\Delta V$; (*b*) flow between two stationary plates; (*c*) flow through a tube.

In the parallel-plate flow system discussed above, the motive force was in the direction of flow. It was applied to the upper plate and transmitted to the fluid, whose velocity remained zero relative to both the upper and the lower plate. A more common situation is depicted in Fig. 3.3*b*, where a pressure force is applied at the left side of a conduit, also defined by two parallel plates, and in which the fluid is moving from left to right as a result of this pressure difference. As in the previous example, the fluid adjacent to the solid surfaces does not move. But the fluid inside the conduit is flowing with a velocity relative to the plates, which increases with distance from the plates and reaches a maximum at the center plane. The flow pattern in flow through a tube is similar, but axisymmetric, as shown in Fig. 3.3*c*.

In all fluid mechanical systems two different types of flow can exist, laminar or turbulent flow. Figure 3.3 shows the velocity distributions for these two flow types in flow between two parallel plates and in a tube. These velocity distributions are actually based on time-average values, because in turbulent flow there are always fluctuations about the mean value. Only for laminar flow can we plot the velocity distribution as a continuous curve. The situation is quite similar for flow through conduits with other cross-sectional areas, for example, square ducts or annuli.

In practice one is usually more concerned with the volumetric or gravimetric flow rate through a conduit rather than with the velocity distribution. The volumetric flow rate $Q$ can be calculated by integrating the velocity distribution over the flow cross-sectional area $A_f$ or

$$Q = \int_A V \, dA_f = V_{\text{ave}} A_f \tag{3.5}$$

where $V_{ave}$ is the average velocity through the conduit. The ratio between the average velocity and the maximum velocity in the conduit depends on whether the flow is laminar or turbulent. In flow through a circular duct this ratio is 0.5 for laminar flow and about 0.83 for turbulent flow. The meaning of the terms laminar and turbulent will be explained below.

## Laminar and Turbulent Flow

Irrespective of whether a fluid flows through a conduit or over a surface, the flow may be laminar or turbulent. The type of flow is determined primarily by the value of a dimensionless parameter, called the Reynolds number Re, defined for flow through a conduit as

$$\text{Re}_{D_H} = \frac{\rho V_{ave} D_H}{\mu}$$

where $V_{ave}$ = average velocity in m/sec (ft/sec)

$D_H$ = hydraulic diameter in m (ft) $[D_H = 4(\text{cross-sectional area})/(\text{wetted perimeter})]$ (see Table 3.1)

$\rho$ = density in kg/m$^3$ (lb$_m$/ft$^3$)

$\mu$ = viscosity in kg/m $\cdot$ sec (lb$_m$/ft $\cdot$ sec)

The physical basis of the Reynolds number explains its significance. Small instabilities in the flow will be attenuated when the ratio of kinetic or inertial forces to viscous forces in the fluid stream is below a certain critical value. The inertial forces are proportional to the kinetic energy $\rho V^2$ whereas the viscous forces are proportional to the shear stress $\mu V/D_H$. The ratio of these two kinds of forces is the dimensionless Reynolds number, or

$$\frac{\rho V^2}{\mu V/D_H} = \frac{\rho V D_H}{\mu} = \text{Re}_{D_H} \tag{3.6}$$

**TABLE 3.1**   Hydraulic Diameter $D_H$ for Various Cross-sectional Shapes

| Shape | $D_H$ |
|-------|-------|
| Circle of diameter $D$ | $D$ |
| Annulus of inner diameter $D_1$ and outer diameter $D_2$ | $(D_2 - D_1)$ |
| Square of side $D$ | $D$ |
| Rectangle of sides $a$ and $b$ | $\dfrac{2ab}{a+b}$ |
| Ellipse, major axis $2a$, minor axis $2b$ | $\dfrac{4ab}{K(a+b)}$ [a] |

[a] Values of $K$ with $s = (a-b)/(a+b)$:

| $s = 0.2$ | 0.5 | 0.6 | 0.7 | 0.8 | 0.9 | 1.0 |
|-----------|-----|-----|-----|-----|-----|-----|
| $K = 1.01$ | 1.064 | 1.062 | 1.127 | 1.168 | 1.216 | 1.273 |

Thus, the value of the Reynolds number can be used as the criterion to determine whether the flow will be turbulent. For flow through conduits the flow is laminar when $Re_{D_H} < 2100$; it undergoes transition when $2100 < Re_{D_H} < 6000$ and becomes turbulent when $Re_{D_H}$ exceeds 6000, whereas for flow over a flat plate the flow will be laminar when $Re_x (= \rho Vx/\mu)$ is less than about $3 \times 10^5$, if $x$ is the distance from the leading edge and $V$ is the free-stream velocity.

**EXAMPLE 3.2**

Calculate the Reynolds number for air flowing at 350 K and atmospheric pressure through a flat-plate collector. The conduit consists of two flat plates, 0.02 m (0.066 ft) apart, 2 m (6.57 ft) wide. The average air velocity ($V_{ave} = Q/A$) is 0.5 m/sec (1.64 ft/sec).

**SOLUTION**

The hydraulic diameter is from Table 3.1

$$D_H = 2 \frac{(0.02)(2)}{2 + 0.02} = \frac{0.08}{2.02} = 0.04 \text{ m } (0.131 \text{ ft})$$

From Table A3.1 we get

$$\rho = 0.998 \text{ kg/m}^3 \ (0.0623 \text{ lb/ft}^3)$$

$$\mu = 2.075 \times 10^{-5} \text{ kg/m} \cdot \text{sec} \ (1.39 \times 10^{-5} \text{ lb/ft} \cdot \text{sec})$$

$$Re_{D_H} = \frac{(0.5)(0.04)(0.998)}{2.075 \times 10^{-5}} = 962$$

The flow is thus laminar. The reader should verify that the same result obtains by working in English units.

Sometimes it is more convenient to relate the Reynolds number to the mass velocity $G$, defined as

$$G = \rho V_{ave} = \frac{\dot{m}}{A_f} \qquad \text{kg/m}^2 \cdot \text{sec } (\text{lb}_m/\text{ft}^2 \cdot \text{sec}) \tag{3.7}$$

where $\dot{m}$ = mass rate of flow in kg/sec ($\text{lb}_m$/sec)
$\quad A_f$ = flow cross-sectional area in m² (ft²)
$\quad V_{ave}$ = average fluid velocity in m/sec (ft/sec) defined by Eq. (3.5)
$\quad$ The above equation also serves to define the average or mean velocity. It is easily verified that the Reynolds number may be written in the equivalent form

$$\mathrm{Re}_{D_H} = \frac{G D_H}{\mu} \tag{3.8}$$

## Overall Flow System Analysis

In order to determine the power and the size of pump or fan required to move a fluid at a specified flow rate through a system consisting of components such as ducts, valves, tanks, and heat exchangers, an overall energy balance must be made. This energy balance is based on the first law of thermodynamics.

The most general treatment of the first law usually considers an open system in steady state with separate inlet and outlet ducts at different elevations, as shown in Fig. 3.4. For this case, the energy balance can be written in the form

$$\dot{m}\left[z_1\frac{g}{g_c} + \frac{V_1{}^2}{2g_c} + (e+pv)_1\right] + q + W_{\mathrm{sh}} = \dot{m}\left[z_2\frac{g}{g_c} + \frac{V_2{}^2}{2g_c} + (e+pv)_2\right] + \Sigma F_{\mathrm{fr}}$$

$$\tag{3.9}^*$$

where   $\dot{m}$ = mass flow rate through the system in kg/sec (lb$_m$/sec)
       $z$ = specific potential energy of the fluid in m (ft)

*In the remainder of the chapter the conversion factor $g_c$ will be omitted.

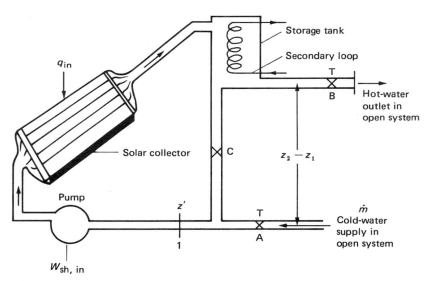

**FIGURE 3.4**   Schematic diagram illustrating open and closed systems. In the open system, cold water enters through valve A and exits through valve B; valve C is shut. In the closed system, valves A and B are shut, valve C is open. Note that a secondary flow loop is required to extract heat from the storage tank when the collector working fluid is in the closed loop.

$g$ = the gravitational constant, 9.806 m/sec$^2$ (32.16 ft/sec$^2$)

$V^2/2g_c$ = specific kinetic energy of the fluid in m/sec$^2$ (ft/sec$^2$)

$(e + pv) = h$ = enthalpy of the fluid (internal energy + $pv$ term) in J/kg (ft $\cdot$ lb$_f$/lb$_m$)

$\Sigma F_{fr}$ = frictional pressure losses in J/sec (ft $\cdot$ lb$_f$/sec)

$q$ = rate of heat transfer to the fluid in J/sec (ft $\cdot$ lb$_f$/sec)

$W_{sh}$ = rate at which work is done on the fluid, that is, the power input in J/sec (ft $\cdot$ lb$_f$/sec)

(Note that in some thermodynamics texts $W_{sh}$ is taken as positive when work is done by the fluid.)

The friction losses have to be calculated from the specific design of the piping and valving to obtain the pressure loss, which must also be made up by work input from a pump or a fan.

The situation depicted above exists in the system shown schematically in Fig. 3.4 when valve C is closed, but A and B are open. However, most solar collector flow loops operate with closed circuits for the working fluid and heat removal from a storage tank through a secondary loop. This condition obtains when valves A and B are closed and valve C is open and no heat is exchanged in the secondary loop.

In this mode of operation, points 1 and 2 coincide,

$$z_1 g + \frac{V_1^{\,2}}{2} + (e + pv)_1 = z_2 g + \frac{V_2^{\,2}}{2} + (e + pv)_2$$

and Eq. (3.9) reduces to

$$q + W_{sh} = \Sigma F_{fr} \tag{3.10}$$

where $\Sigma F_{fr}$ is the sum of all the frictional pressure losses in the loop.

There are two basic modes of operating solar collectors. One mode, encountered in some service hot-water installations, uses the solar heat input $q$ to generate a sufficient change in density to overcome the frictional loss and the difference in fluid head by buoyant forces. In a majority of installations, however, the pressure difference required to move the working fluid through the flow circuit is supplied by a pump or a fan.

A pump is a device for moving liquids, whereas a fan or blower is a device for moving gases. Fans can move fairly large volumetric flows against relatively low pressure losses (of the order of a few centimeters of water), and blowers and compressors can generate higher pressures. In pumps, low-pressure blowers, and fans the density of the fluid does not change appreciably and the fluid can be treated as incompressible. The method of operation and sizing of pumps and blowers is treated in the next section.

## Pumps and Fans

The energy required to move a fluid through a conduit loop such as shown in Fig. 3.4 can be calculated by means of a mechanical energy balance [see Eq. (3.10)]. But since no mechanical pumping device operates at 100 percent efficiency, the power $W$ actually required to operate the fan or the pump is larger than $W_{sh}$. It can be calculated from the relation

$$W = \frac{W_{sh}}{\eta_p} \tag{3.11}$$

where $W_{sh}$ = ideal or shaft work rate calculated from Eq. (3.10)
$\eta_p$ = efficiency of the pump or fan
$W$ = rate at which actual work must be supplied in kW (1 hp = 0.746 kW)

When purchasing a fan or a pump, it is usual practice to specify its performance in terms of the fluid head $z$, rather than the pressure developed. This head is related to the pressure by Eq. (3.3). The mechanical work $W_{sh}$ in J/sec (ft · lb$_f$/sec) can then be calculated from the equation

$$W_{sh} = zg\dot{m} = \Sigma F_{fr} \tag{3.12}$$

where $z$ = fluid head in m (ft)
$\dot{m}$ = mass flow rate in kg/sec (lb$_m$/sec)

In addition to the fluid mechanical friction losses in the pump or fan, the efficiency of the electric motor must be known also in order to specify the size of motor required. Electric motors in the range between 0.1 and 0.4 kW ($\frac{1}{8}$ to $\frac{1}{2}$ hp), the size most commonly required in solar water systems for small buildings, have an efficiency of about 75 percent. Larger motors in the size between 5 and 100 hp are sometimes required for air systems or large buildings, and their efficiency is about 80 percent. The electric power $W_e$ required to drive a pump or a fan is given by

$$W_e = \frac{W_{sh}}{\eta_p \eta_m} \tag{3.13}$$

where $\eta_m$ is the motor efficiency.

## EXAMPLE 3.3

A solar collector system has a pressure drop of 1.27 m (50 in) of $H_2O$ at a flow rate of 0.25 kg/sec (2000 lb$_m$/hr). If the pump efficiency is 60 percent and the electric motor efficiency is 75 percent, calculate the required rate of shaft work and the pumping power in kW and the motor size in hp.

## SOLUTION

$$W = \dot{m}z\frac{g}{g_c} = (0.25 \text{ kg/sec})(1.27 \text{ m})(9.806 \text{ m/sec}^2)$$

$$= 3.1 \text{ kg} \cdot \text{m}^2/\text{sec}^3 = 3.1 \text{ J/sec} = 3.1 \text{ W}$$

The shaft work and electrical power for the motor are

$$W_{sh} = \frac{3.1}{0.60} = 5.2 \text{ W} \qquad W_e = \left(\frac{5.2}{0.75}\right)\left(\frac{1}{746}\right) = 0.0093 \text{ hp}$$

Thus, a 0.01 hp motor would be sufficient.

There are two general classes of industrial pumps: centrifugal and positive displacement (or piston). For solar-thermal system loops centrifugal pumps are used exclusively, because their mode of operation makes it relatively simple to control the flow rate. Such pumps are available in many sizes for flow rates between 0.00003 and 5 $m^3$/sec (about $\frac{1}{2}$ to 100,000 gal/min) and discharge pressures between 1000 and $10^7$ $N/m^2$ (about 0.1 and 1000 psi). Figure 3.5 shows a schematic diagram of a centrifugal pump. Water from the suction line enters the rotating eye of the impeller in the center and is then forced by the centrifugal forces to flow outward through the rotating impeller passages. At the periphery of the impeller the water is collected in the volute chamber and its kinetic energy is converted into pressure.

In engineering practice the efficiency of a pump is obtained experimentally. The performance characteristics of a pump are supplied by the manufacturer by plotting curves of pressure or head developed, pump efficiency, and brake or actual horsepower required to drive the pump at the specified speed versus the flow rate or discharge. In the United States, SI units are rarely used by pump manufacturers, and the performance is usually given in the format shown in Fig. 3.6, a plot of volumetric flow rate versus head.

Fans and low-pressure blowers are used to move air at low pressures. Their mode of operation is similar to that of a pump. They are available for discharge heads up to about 1.5 m $H_2O$ (60 in $H_2O$). The pressure change in a fan is sufficiently small to treat the gas as an incompressible fluid and to use an average density in the energy-balance calculations. It should be noted, however, that the pressure head quoted by fan manufacturers is usually the sum of the velocity head, that is, the kinetic energy converted into pressure, and the static head of the gas at the outlet.

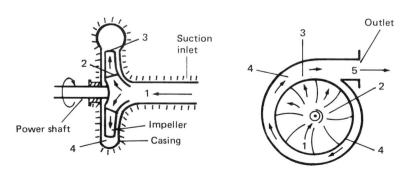

**FIGURE 3.5**  Schematic diagram of a centrifugal pump.

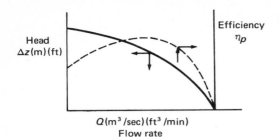

$Q(m^3/sec)(ft^3/min)$
Flow rate

**FIGURE 3.6**  Pump or fan
performance curve.

## EXAMPLE 3.4

A solar collector system uses 2.33 $m^3$/sec (5000 $ft^3$/min) of standard air (1 atm
and 70°F or 294 K). The air enters the fan through a duct at a velocity of 10
m/sec and a pressure of 101.0 kPa or 29.9 in Hg, read on a manometer, and a
temperature of 338.5 K (150°F) read with a thermometer. The fan is to be
sized for an outlet pressure of 101.6 kPa (30.1 in Hg) and a velocity of 50
m/sec. Calculate the size motor required if the fan efficiency is 60 percent and
the electric motor efficiency is 75 percent.

## SOLUTION

The density at the inlet and outlet can be calculated by taking $\rho$ at 350 K and
1 atm (29.9 in Hg) from the Table A3.1 and assuming perfect gas behavior.
Then

$$\rho_1 = (0.998 \text{ kg/m}^3)\left(\frac{29.9}{29.9}\right)\left(\frac{350}{338}\right) = 1.033 \text{ kg/m}^3 \ (0.0620 \text{ lb}_m/\text{ft}^3)$$

$$\rho_2 = (0.998 \text{ kg/m}^3)\left(\frac{30.1}{29.9}\right)\left(\frac{350}{338}\right) = 1.040 \text{ kg/m}^3 \ (0.0644 \text{ lb}_m/\text{ft}^3)$$

$$\rho_{ave} = 1.036 \text{ kg/m}^3 \ (0.0632 \text{ lb}_m/\text{ft}^3)$$

The mass flow rate is

$$\dot{m} = (2.33 \text{ m}^3/\text{sec})(1.036 \text{ kg/m}^3)\left(\frac{338}{294}\right) = 2.79 \text{ kg/sec} \ (6.14 \text{ lb}_m/\text{sec})$$

The pressure head developed by the fan is

$$\frac{\Delta P}{\rho_{ave}} = \frac{600 \text{ Pa}}{1.04 \text{ kg/m}^3} = 577 \ \frac{\text{kg/m} \cdot \text{sec}^2}{\text{kg/m}^3} = 577 \text{ J/kg} \ (203 \text{ ft} \cdot \text{lb}_f/\text{lb}_m)$$

The difference in velocity head is

$$\frac{V_2{}^2 - V_1{}^2}{2g_c} = \frac{50^2 - 10^2}{2g_c} = 1200 \text{ J/kg (421 ft} \cdot \text{lb}_f/\text{lb}_m)$$

Writing the energy equation with $\Delta h = 0$ and $q = 0$ gives

$$W = \frac{P_2 - P_1}{\rho_{\text{ave}}} = \frac{V_2{}^2 - V_1{}^2}{2g_c} = 1200 \text{ J/kg (421 ft} \cdot \text{lb}_f/\text{lb}_m)$$

The motor size required is

$$W_e = \frac{(421 \text{ ft} \cdot \text{lb}_f/\text{lb}_m)(6.14 \text{ lb}_m/\text{sec})}{(0.6)(0.75)(550 \text{ ft} \cdot \text{lb}_f/\text{sec} \cdot \text{hp})} = 10.4 \text{ hp}$$

Observe that the required motor size could be greatly reduced by decreasing the velocity head requirements.

## Frictional Pressure Losses in Piping Systems

The shearing forces at a fluid-solid interface cause frictional pressure losses in flow through conduits. A force balance on a fluid flowing through a pipe of length $L$ and diameter $D$ in steady flow as shown in Fig. 3.7 gives

$$\tau_0 \pi D L = (p_1 - p_2)\frac{\pi D^2}{4} \tag{3.14}$$

The shear $\tau_0$ can be evaluated analytically in laminar flow, but in turbulent flow it must be determined experimentally. If the left-hand side of Eq. (3.14) is

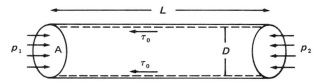

**FIGURE 3.7**  Schematic diagram for force balance in flow through a duct.

evaluated for laminar flow in a long pipe [see, for example, Ref. (33)], one obtains the following solution for the frictional pressure loss of an incompressible fluid:

$$p_1 - p_2 = \frac{32\dot{m}L\mu}{g_c\rho(\pi/4)D^4} = 4f\frac{L}{D}\rho\frac{\bar{V}^2}{2g_c} \tag{3.15}$$

where $\quad \bar{V}$ = average velocity in the duct ($\bar{V} = 4\,\dot{m}/\rho\pi D^2$) in m/sec (ft/sec)
$\quad\quad\quad L$ = length of the duct in m (ft)
$\quad\quad\quad D$ = inside diameter of the duct in m (ft)
$\quad\quad\quad \rho$ = fluid density in kg/m$^3$ (lb$_m$/ft$^3$)
$\quad\quad\quad \mu$ = absolute viscosity of the fluid in Pa $\cdot$ sec (lb$_f$/ft $\cdot$ sec)
$\quad\quad\quad f$ = dimensionless friction factor
$p_1$ and $p_2$ = pressures at points 1 and 2, respectively, in Pa (lb$_f$/ft$^2$)
$\quad$ In laminar flow it can be shown that $f$ is given by the relation

$$f = \frac{16}{\mathrm{Re}_{D_H}} \tag{3.16}$$

### EXAMPLE 3.5

Air at atmospheric pressure and 350 K flows through a 1 m in diameter, 75 m long duct at an average velocity of 0.02 m/sec. Determine the frictional pressure drop in Pascals and in H$_2$O and the volumetric flow rate in m$^3$/sec and ft$^3$/min.

### SOLUTION

The Reynolds number is

$$\mathrm{Re}_D = \frac{\rho V D}{\mu} = \frac{(0.02\ \text{m/sec})(1\ \text{m})(0.998\ \text{kg/m}^3)}{2.07 \times 10^{-5}\ \text{kg/m} \cdot \text{sec}} = 965$$

Since the flow is laminar, Eq. (3.15) applies.

$$\Delta p = 4f\frac{L}{D}\frac{\rho V^2}{2g_c} = \left(\frac{64}{965}\right)\left(\frac{75}{1}\right)\left[\frac{(0.988\ \text{kg/m}^3)(4 \times 10^{-4}\ \text{m}^2/\text{sec}^2)}{2(1\ \text{kg} \cdot \text{m/N} \cdot \text{sec}^2)}\right]$$

$$= (9.9 \times 10^{-4}\ \text{N/m}^2)(0.00406\ \text{in H}_2\text{O/N} \cdot \text{m}^2) = 4 \times 10^{-6}\ \text{in H}_2\text{O}$$

The volumetric flow rate is $V(\pi D^2/4) = 0.0157\ \text{m}^3$/sec, which is equivalent to $(0.0157\ \text{m}^3/\text{sec})(1/0.3048\ \text{ft/m})^3(60\ \text{sec/min}) = 33\ \text{ft}^3$/min. (Note that the density and viscosity for air were taken from Table A3.1.)

$\quad$ Also for turbulent flow through conduits Eq. (3.15) can be used to calculate the frictional pressure loss, but the friction coefficient is more difficult to evaluate. In addition to the Reynolds number, the friction coefficient in turbulent flow also depends on the roughness of the conduit surface, usually defined by the dimensionless ratio $\epsilon/D$, where $\epsilon$ is the averaged height of surface irregularities. The results of many measurements of $f$ have been correlated by plotting the friction factor as a function of the Reynolds number for various values of the roughness parameter $\epsilon/D$ as shown in Fig. 3.8. Observe that this is a log-log plot. In the laminar flow regime roughness has no effect, and a single line representing Eq. (3.16) is shown for $Re_D < 2100$. The lowest line in the turbulent regime is for smooth pipes and ducts of the type usually installed in solar energy systems.

In the hydraulic design of a solar system two types of problems are encountered. The most common one requires the calculation of the pressure drop with the flow rate and the duct or pipe diameter specified. In the other type of problem the available pressure is specified, for example, a storage tank some distance above ground, and the flow rate for a given piping system is asked for. The latter type of problem requires a trial-and-error solution, because both Re and $f$ are functions of velocity and both are unknown. For this type of problem it is often more convenient to use an empirical correlation for $f$ and $Re_D$. One commonly used equation, applicable to smooth ducts in the Reynolds number range between 10,000 and 120,000, is

$$f = 0.079 \ Re_D{}^{-0.25} \qquad\qquad (3.17)$$

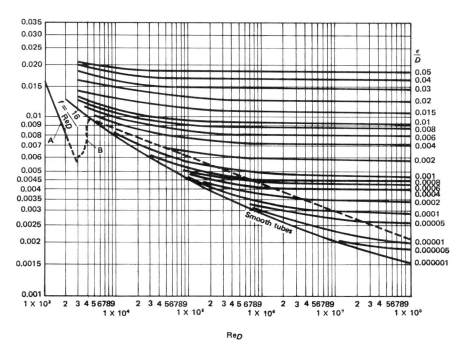

**FIGURE 3.8** Friction coefficient versus Reynolds number for flow through ducts. Based on Moody, L. F., *Trans. Am. Soc. Mech. Eng.*, vol. 66, p. 671, 1944.

For ducts having cross-sectional areas other than circular the same general approach can be used if the hydraulic diameter is used instead of the pipe diameter wherever the latter appears in the governing equations.

In addition to the friction pressure drop in flow through straight ducts, pressure losses in piping systems occur when the velocity or direction of the fluid is changed, as in sudden enlargements of the flow duct or in elbows and in flow through valves and fittings. These losses must be added to the frictional pressure loss in order to obtain the total pressure drop through a complete piping system.

### Sudden Enlargements, Contractions, and Entrance Losses

If the cross section of a duct enlarges very gradually, as in a diffuser, very little extra pressure loss occurs. But if the change in cross section is sudden, eddies form and most of the velocity head is lost by frictional dissipation. For a sudden enlargement, the pressure loss can be calculated from the relation

$$\Delta p = \frac{\rho \alpha (V_1{}^2 - V_2{}^2)}{2g_c} = \alpha \left(1 - \frac{A_1}{A_2}\right)^2 \frac{\rho V_1{}^2}{2g_c} = K_e \frac{\rho V_1{}^2}{2g_c} \tag{3.18}$$

where $V_1$ = average velocity in the smaller duct upstream
     $A_1$ = smaller cross-sectional area
     $A_2$ = large cross-sectional area
     $\rho$ = fluid density
     $\alpha$ = dimensionless velocity correlation coefficient ($\alpha = 1.0$ in turbulent flow; $\alpha = 2.0$ in laminar flow)

For a sudden contraction in size, the following relation applies:

$$\Delta p = 0.55\alpha \left(1 - \frac{A_2}{A_1}\right)^2 \frac{\rho V_2{}^2}{2g_c} = K_c \frac{\rho V_2{}^2}{2g_c} \tag{3.19}$$

where $A_2$ = smaller downstream cross-sectional area
     $A_1$ = larger upstream area
     $V_2$ = average velocity in the smaller duct
     $\alpha$ = velocity correlation coefficient ($\alpha = 1.0$ in turbulent flow; $\alpha = 2.0$ in laminar flow)

For a rounded bell-shaped entrance or contraction the pressure loss is small.

Even though contraction or entrance losses for laminar flow are small or negligible, there is an abnormally high pressure drop for the initial distance along the tube. This distance, known as the entrance length, was found to be $0.057\,\mathrm{Re}_D$ pipe diameters for the case of a circular pipe. The cause of the extra pressure loss is the work required in accelerating fluid particles to set up the parabolic velocity distribution for the laminar flow through a pipe. An increased frictional resistance near the pipe inlet arises from the same cause in turbulent flow also, but the effect is of less importance in the turbulent region than in the laminar region. For a Reynolds number $DG/\mu = 2000$, the entrance effect would persist for about 115 diameters. Since in a straight pipe with a rounded entrance leading from a quiet tank laminar flow may be maintained without great difficulty up to very high Reynolds numbers, the deviation from Eq. (3.16) near the inlet may be important.

For circular tubes the pressure drop for laminar flow in the entrance length following a rounded inlet in a horizontal pipe can be estimated from the curve given in Fig. 3.9.

## Fitting, Bends, and Valves

Pressure losses from fittings and bends in the duct and from manually or automatically operated valves can be evaluated only empirically. In many piping systems, the losses in fittings and valves can be more important than the losses in the straight pipes. The usual approach in practice is to obtain for each bend fitting and valve the appropriate loss factor $K_f$ from a handbook or from the manufacturer and use the relation

$$\Delta p_f = K_f \frac{\rho V_1^2}{2g_c} \tag{3.20}$$

where $K_f$ = appropriate dimensionless pressure loss factor
  $V_1$ = average velocity in the duct upstream of the fitting

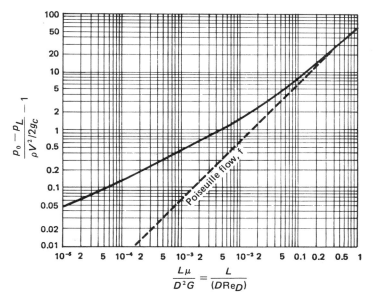

**FIGURE 3.9**  Entrance effect in laminar flow. $P_0$ = pressure; $P_L$ = pressure in duct a distance $L$ from entrance; $D$ = diameter of duct; $\rho$ = fluid density; $\bar{V}$ = average fluid velocity. Based on Kreith, F., and R. Eisenstadt, *Trans. Am. Soc. Mech. Eng.*, vol. 79, p. 1070, 1957; and A. H. Shapiro, R. Siegel, and W. Kline, *Proc. 2d Natl. Congr. Appl. Mech.*, p. 733, ASME, 1954.

**TABLE 3.2**  Additional Frictional Loss for Turbulent Flow through Fittings and Valves[t]

| Type of fitting or valve[k] | Additional friction loss, equivalent no. of velocity heads, $K_f$[s] |
|---|---|
| 45-deg. ell, standard[a,b,e,g,i] | 0.35 |
| 45-deg. ell, long radius[b] | 0.2 |
| 90-deg. ell, standard[a,b,d,g,i,m] | 0.75 |
| Long radius[a,b,e,g] | 0.45 |
| Square or miter[m] | 1.3 |
| 180-deg. bend, close return[a,b,g] | 1.5 |
| Tee, standard, along run, branch blanked off[g] | 0.4 |
| Used as ell, entering run[d,g] | 1.3 |
| Used as ell, entering branch[b,d,g] | 1.5 |
| Branching flow[f,h,l] | 1[o] |
| Coupling[b,g] | 0.04 |
| Union[g] | 0.04 |
| Gate valve,[a,g,i] open | 0.17 |
| $\frac{3}{4}$ open[p] | 0.9 |
| $\frac{1}{2}$ open[p] | 4.5 |
| $\frac{1}{4}$ open[p] | 24.0 |
| Diaphragm valve,[n] open | 2.3 |
| $\frac{3}{4}$ open[p] | 2.6 |
| $\frac{1}{2}$ open[p] | 4.3 |
| $\frac{1}{4}$ open[p] | 21.0 |
| Globe valve,[g,j] bevel seat, open | 6.4 |
| $\frac{1}{2}$ open[p] | 9.5 |
| Composition seat, open | 6.0 |
| $\frac{1}{2}$ open[p] | 8.5 |
| Plug disk, open | 9.0 |
| $\frac{3}{4}$ open[p] | 13.0 |
| $\frac{1}{2}$ open[p] | 36.0 |
| $\frac{1}{4}$ open[p] | 112.0 |
| Angle valve,[a,g] open | 3.0 |
| Y or blowoff valve,[g,j] open | 3.0 |
| Plug cock[c] $\theta =$  5° | 0.05 |
| 10° | 0.29 |
| 20° | 1.56 |
| 40° | 17.3 |
| 60° | 206.0 |
| Butterfly valve[c] $\theta =$  5° | 0.24 |
| 10° | 0.52 |
| 20° | 1.54 |
| 40° | 10.8 |
| 60° | 118.0 |
| Check valve,[a,g,j] swing | 2.0[q] |
| Disk | 10.0[q] |
| Ball | 70.0[q] |

**TABLE 3.2**  Additional Frictional Loss for Turbulent Flow through
Fittings and Valves (*Continued*)

| Type of fitting or valve[k] | Additional friction loss, equivalent no. of velocity heads, $K_f$[s] |
|---|---|
| Foot valve[g] | 15.0 |
| Water meter,[m] disk | 7.0[r] |
| Piston | 15.0[r] |
| Rotary (star-shaped disk) | 10.0[r] |
| Turbine-wheel | 6.0[r] |

[a]Flow of Fluids through Valves, Fittings, and Pipe, *Tech. Paper* 410, Crane Co., 1957.

[b]Freeman, "Experiments upon the Flow of Water in Pipes and Pipe Fittings," American Society of Mechanical Engineers, New York, 1941.

[c]Gibson, "Hydraulics and Its Applications," 5th ed., p. 250, Constable, London, 1952.

[d]Giesecke and Badgett, *Heating, Piping Air Conditioning*, vol. 4, pp. 443–447, 1932.

[e]Giesecke, *J. Am. Soc. Heat. Vent. Eng.*, vol. 32, p. 461, 1926.

[f]Gilman, *Heating, Piping Air Conditioning*, vol. 27, pp. 141–147, 1955.

[g]"Standards of Hydraulic Institute–Tentative Standards, Pipe Friction," Hydraulic Institute, New York, 1948.

[h]Hoopes, Isakoff, Clarke, and Drew, *Chem. Eng. Progress*, vol. 44, pp. 691–696, 1948.

[i]Ito, *Trans. Am. Soc. Mech. Eng.*, vol. 82, Series D, pp. 131–143, 1960.

[j]Lansford, Loss of Head in Flow of Fluids through Various Types of $1\frac{1}{2}$-in. Valves, *Univ. Illinois Eng. Expt. Sta. Bull. Ser.* 340, 1943.

[k]Lapple, *Chem. Eng.*, vol. 56, pp. 96–104, 1949, general survey reference.

[l]McNown, *Proc. Am. Soc. Civil Eng.*, vol. 79, Separate 258, pp. 1–22, 1953; discussion, *ibid.*, vol. 80, Separate 396, pp. 19–45, 1954.

[m]Schoder and Dawson, "Hydraulics," 2d ed., p. 213, McGraw-Hill, New York, 1934.

[n]Streeter, *Prod. Eng.*, vol. 18, pp. 89–91, 1947.

[o]This is pressure drop (including friction loss) between run and branch, based on velocity in the main stream before branching. Actual value depends on the flow split, ranging from 0.5 to 1.3 if main stream enters run and from 0.7 to 1.5 if main stream enters branch.

[p]The fraction open is directly proportional to stem travel or turns of hand wheel. Flow direction through some types of valves has a small effect on pressure drop (see Freeman, *op. cit.*). For practical purposes this effect may be neglected.

[q]Values apply only when check valve is fully open, which is generally the case for velocities more than 3 ft/sec for water.

[r]Values should be regarded as approximate because there is much variation in equipment of the same type from different manufacturers.

[s]Equivalent length in pipe diameter depends upon Reynolds number. For rough estimates, $L_e/D$ can be obtained by multiplying $K$ by 45 for liquids similar to water and by 55 for gases similar to air.

[t]From R. H. Perry, C. H. Chilton, and S. D. Kirkpatrick (eds.), "Chemical Engineers' Handbook," 4th ed., McGraw-Hill Book Co., New York, 1963.

Typical values for $K_f$ are presented in Table 3.2 for various kinds of fittings, bends, and valves in turbulent flow, in Table 3.3 for laminar flow, and in Table 3.4 for miter bends of various angles and straight sections.

Using the approach outlined above, the total pressure drop through a piping system can be calculated as a function of velocity. The pressure drop term in the energy balance of Eq. (3.10), $\Sigma F_{fr}$, can be written

$$\Sigma F_{\text{fr}} = 4f\frac{L}{D_H}\frac{\rho V_1^{\,2}}{2g_c} + K_e\frac{\rho V_1^{\,2}}{2g_c} + K_c\frac{\rho V_2^{\,2}}{2g_c} + \Sigma K_f\frac{\rho V_1^{\,2}}{2g_c} \qquad (3.21)$$

It is always possible to refer *all* the pressure drop terms to the same velocity by using the continuity equation $A_1 V_1 = A_2 V_2$, etc. Then the above relation becomes

$$\Sigma F_{\text{fr}} = \left[4f\frac{L}{D_H} + K_e + K_c\left(\frac{A_1}{A_2}\right)^2 + \Sigma K_f\right]\frac{\rho V_1^{\,2}}{2g_c} \qquad (3.22)$$

For steady-state operation there must be a pump to produce sufficient pressure head to overcome the frictional pressure losses. As shown in Fig. 3.6 a pump or a fan has a performance curve such that a specified head corresponds to a certain rate of flow. The developed pressure is a maximum at zero flow (shut-off pressure) and diminishes with increasing flow rate. The frictional pressure drop in the piping system, on the other hand, exhibits the opposite trend: the larger the desired flow rate, the larger the required pressure. Consequently, for any system design it is necessary to match the pump or the fan to the flow system. The easiest way to match the two is to superimpose the flow-versus-pressure characteristics of the piping network on the flow-versus-pressure performance curve of the pump or fan. The point of intersection will be the operating point. The control of the flow rate can then be accomplished by closing a valve, preferably downstream of the pump or fan. This will change the flow-versus-pressure characteristics of the piping system and produce a different equilibrium operating condition, as shown in Fig. 3.10.

Suppose that in the system shown in Fig. 3.10 the temperature of the fluid leaving the collector is to be maintained. Operating point A corresponds to the control valve wide open (maximum flow rate) while operation point B corresponds to a condition with the valve half closed. In order to maintain a desired outlet temperature, measured by the temperature sensor T.S., the flow rate must be

**TABLE 3.3**  Additional Frictional Loss for Laminar Flow through Fittings and Valves[a]

| Type of fitting or valve | Additional frictional loss expressed as $K_f$ | | | |
| --- | --- | --- | --- | --- |
| | $N_{Re} = 1000$ | 500 | 100 | 50 |
| 90-deg. ell, short radius | 0.9 | 1.0 | 7.5 | 16 |
| Tee, standard along run | 0.4 | 0.5 | 2.5 | |
| Branch to line | 1.5 | 1.8 | 4.9 | 9.3 |
| Gate valve | 1.2 | 1.7 | 9.9 | 24 |
| Globe valve, composition disk | 11 | 12 | 20 | 30 |
| Plug | 12 | 14 | 19 | 27 |
| Angle valve | 8 | 8.5 | 11 | 19 |
| Check valve, swing | 4 | 4.5 | 17 | 55 |

[a]From R. H. Perry, C. H. Chilton, and S. D. Kirkpatrick (eds.), "Chemical Engineers' Handbook," 4th ed., McGraw-Hill Book Co., New York, 1963.

**TABLE 3.4** Additional Frictional Loss for Miter Bends of Various Angles and Straight Sections[a]

**Miter bends (no straight section):**

| Angle | $K_s$ | $K_r$ |
|---|---|---|
| 5° | 0.016 | 0.024[b] |
| 10° | 0.034 | 0.044 |
| 15° | 0.042 | 0.062 |
| 22.5° | 0.066 | 0.154 |
| 30° | 0.130 | 0.165 |
| 45° | 0.236 | 0.320 |
| 60° | 0.471 | 0.684 |
| 90° | 1.129 | 1.265 |

**Miter bends with straight sections:**

| Configuration | $K_s$ | $K_r$ |
|---|---|---|
| 22.5° (1.17D) | 0.112 | 0.284 |
| 30° (1.23D) | 0.150 | 0.268 |
| 30° (2.37D) | 0.143 | 0.227 |
| 22.5° (1.06D) | 0.108 | 0.236 |
| 30° (2.37D / 1.23D) | 0.188 | 0.320 |
| 30° (2.37D / 1.23D) | 0.202 | 0.323 |
| 60° (1.44D) | 0.400 | 0.534 |
| 60° (1.44D) | 0.400 | 0.601 |

**45° bend:**

| $\alpha/D$ | $K_s$ | $K_r$ |
|---|---|---|
| 0.71 | 0.507 | 0.510 |
| 0.943 | 0.350 | 0.415 |
| 1.174 | 0.333 | 0.384 |
| 1.42 | 0.261 | 0.377 |
| 1.50[c] | 0.280 | 0.376 |
| 1.86 | 0.289 | 0.390 |
| 2.56 | 0.356 | 0.429 |
| 3.14 | 0.346 | 0.426 |
| 3.72 | 0.356 | 0.460 |
| 4.89 | 0.389 | 0.455 |
| 5.59 | 0.392 | 0.444 |
| 6.28 | 0.399 | 0.444 |

**22.5° bend:**

| $\alpha/D$ | $K_s$ | $K_r$ |
|---|---|---|
| 1.186 | 0.120 | 0.294 |
| 1.40 | 0.125 | 0.252 |
| 1.50[c] | — | 0.250 |
| 1.63 | 0.124 | 0.266 |
| 1.86 | 0.117 | 0.272 |
| 2.325 | 0.096 | 0.317 |
| 2.40[c] | 0.095 | — |
| 2.91 | 0.108 | 0.317 |
| 3.49 | 0.130 | 0.318 |
| 4.65 | 0.148 | 0.310 |
| 6.05 | 0.142 | 0.313 |

**30° bend:**

| $\alpha/D$ | $K_s$ | $K_r$ |
|---|---|---|
| 1.23 | 0.195 | 0.347 |
| 1.44 | 0.196 | 0.320 |
| 1.67 | 0.150 | 0.300 |
| 1.70[c] | 0.149 | 0.299 |
| 1.91 | 0.154 | 0.312 |
| 2.37 | 0.167 | 0.337 |
| 2.96 | 0.172 | 0.342 |
| 4.11 | 0.190 | 0.354 |
| 4.70 | 0.192 | 0.360 |
| 6.10 | 0.201 | 0.360 |

**30° bend (upper right):**

| $\alpha/D$ | $K_s$ | $K_r$ |
|---|---|---|
| 1.23 | 0.157 | 0.300 |
| 1.67 | 0.156 | 0.378 |
| 2.37 | 0.143 | 0.264 |
| 3.77 | 0.160 | 0.242 |

[a] Reprinted from "Standards of the Hydraulic Institute," 8th ed., Hydraulic Institute, New York, 1947.
[b] $K_s$ = Resistance coefficient for smooth surface; $K_r$ = resistance coefficient for rough surface, $e/D = 0.0022$.
[c] Optimum value of $\alpha$ interpolated.

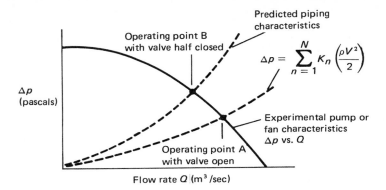

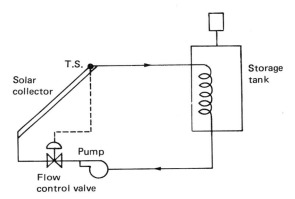

**FIGURE 3.10** Matching a pump or fan to piping system on pressure vs. flow characteristics. Operating point A corresponds to the control valve wide open (maximum flow rate), while operating point B corresponds to a condition with the valve half closed. In order to maintain a desired outlet temperature, sensed by the temperature sensor T.S., the flow rate must be controlled as the amount of heat transferred to the working fluid varies during the day.

controlled as the amount of heat transferred to the working fluid varies during the day. A more detailed analysis of the flow control is presented in Chapter 5.

## PRINCIPLES OF HEAT TRANSFER

The design and analysis of all solar-thermal systems require familiarity with the fundamentals of heat transfer. The remainder of this chapter will therefore be devoted to the principles of heat transfer, their relations to thermodynamics, and on to solar engineering.

a temperature difference exists, energy may be transferred from the er temperature to the region of lower temperature. According to concepts, the energy that is transferred as a result of a temperature difference is called *heat*. Although classical thermodynamics deals with energy and heat transfer, its laws can treat only systems that are in equilibrium. Thermo-dynamic laws can, therefore, predict the amount of energy required to change a system from one equilibrium state to another, but they cannot predict how fast these changes will occur. The engineering science of heat transfer supplements the first and second laws of classical thermodynamics by providing methods of analyses that can be used to predict rates of energy transfer and temperature distributions.

To illustrate the difference in the kind of information that can be obtained from a thermodynamic and heat-transfer analysis, consider the heating of a metal cylinder placed in a tank of hot water. Thermodynamics can be used to predict the final temperature after the two systems have reached equilibrium, but it cannot tell us what the temperature of the cylinder will be after a given time or how long it will take to obtain equilibrium. A heat-transfer analysis, on the other hand, can predict the rate of heat transfer from the water to the cylinder, and from this information we can calculate the temperature of the cylinder as well as the temperature of the water as a function of time.

For a complete heat-transfer analysis it is necessary to deal with three different mechanisms: conduction, convection, and radiation. The design and analysis of solar energy conversion systems require a familiarity with each of the mechanisms of heat transfer, as well as with their interactions. In this chapter we shall consider the basic principles of heat transfer qualitatively and quantitatively.

## Conduction Heat Transfer

Conduction is the only heat-transfer mode in opaque solid media. When a temperature gradient exists in such a body, heat will be transferred from the higher to the lower temperature region. The rate at which heat is transferred by conduction $q_k$ is proportional to the temperature gradient $dT/dx$ times the area through which heat is transferred $A$ or

$$q_k \propto A \frac{dT}{dx} \tag{3.23}$$

where $T(x)$ = temperature
    $x$ = distance in the direction of the heat flow
The actual rate of heat flow depends on the thermal conductivity $k$, a physical property of the medium. For conduction through a homogeneous medium the rate equation can therefore be quantitatively expressed as

$$q_k = -kA \frac{dT}{dx} \tag{3.24}$$

The minus sign is a consequence of the second law of thermodynamics, which requires that heat *must* flow in the direction of lower temperature. The temperature gradient, as shown in Fig. 3.11, will be negative if the temperature decreases with

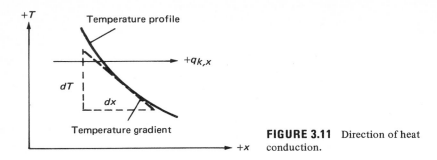

**FIGURE 3.11**  Direction of heat conduction.

increasing values of $x$. Therefore, if heat transferred in the positive direction $x$ is to be a positive quantity, the negative sign must be inserted in the right-hand side of Eq. (3.24), which is called Fourier's law of heat conduction and serves to define the thermal conductivity $k$. If the area is in square meters $(m^2)$, the temperature in Kelvin (K), $x$ in meters (m), and the rate of heat flow in watts (W), $k$ has the units of W/m $\cdot$ K. In the English system the area is in square feet $(ft^2)$, $x$ in feet (ft), the temperature in degrees Fahrenheit $(^\circ F)$, and the rate of heat flow in Btu/hr. Thus, $k$ has the units of Btu/hr $\cdot$ ft $\cdot$ $^\circ$F.

## Plane Walls

A direct application of Fourier's law is the case of heat transfer through a wall (Fig. 3.12). When both surfaces of the wall are at uniform temperatures, the heat flow will be in one direction, perpendicular to the wall surfaces. If the thermal conductivity is uniform, integration of Eq. (3.24) gives

$$q_k = -\frac{kA}{\Delta x}(T_2 - T_1) = \frac{kA}{\Delta x}(T_1 - T_2) \tag{3.25}$$

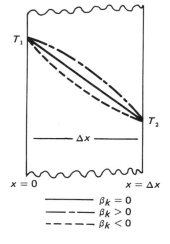

**FIGURE 3.12**  Temperature distribution in conduction through a plane wall with constant and variable thermal conductivity.

where $\Delta x$ = thickness of the wall
$\quad T_1$ = temperature at the left surface where $x = 0$
$\quad T_2$ = temperature at the right surface where $x = L$
For many materials the thermal conductivity is a linear function of temperature

$$k(T) = k_0(1 + \beta_k T) \tag{3.26}$$

In such cases integration of Eq. (3.24) gives

$$q_k = \frac{k_0 A}{\Delta x}\left[(T_1 - T_2) + \frac{\beta_k}{2}(T_1{}^2 - T_2{}^2)\right] \tag{3.27}$$

or

$$q_k = \frac{k_{\text{ave}} A}{\Delta x}(T_1 - T_2) \tag{3.28}$$

where $k_{\text{ave}}$ is the value of $k$ at the average temperature $(T_1 + T_2)/2$. The temperature distributions for a constant value of thermal conductivity $(\beta_k = 0)$, as well as for thermal conductivity increasing $(\beta_k > 0)$ and decreasing $(\beta_k < 0)$ with temperature, are shown in Fig. 3.12.

## EXAMPLE 3.6

The glass cover of a solar collector has an area of 12 m² and a thickness of 1 cm. The thermal conductivity of the glass is 0.8 W/m · K. On a cold day the outside surface temperature of the glass is 272.5 K and the inside surface is 274.5 K. Determine (a) the heat-transfer rate through the glass and (b) the temperature of a plane midway between the inside and outside glass surfaces.

## SOLUTION

(a) The heat-transfer rate through the glass is

$$q = \frac{kA(T_1 - T_2)}{\Delta x} = \frac{(0.8 \text{ W/m} \cdot \text{K})(12 \text{ m}^2)(2 \text{ K})}{0.01 \text{ m}} = 1.92 \times 10^3 \text{ W}$$

(b) The temperature of the midplane is

$$\frac{T - 274.5}{272.5 - 274.5} = \frac{1}{2} \quad \text{and} \quad T = 273.5 \text{ K}$$

This is the mean of the two surface temperatures, because the thermal conductivity is constant and therefore the temperature gradient is linear through the glass.

If heat is conducted through several layers in series, as for example through a multilayer wall as used in the construction of most houses, the analysis is only slightly more difficult. In the steady state the rate of heat flow through all the sections must be the same. However, as shown in Fig. 3.13 for a three-layer system, the gradients are different. The heat-transfer rate can be written for each section and each set equal to the other, or

$$q_k = \left(\frac{kA}{\Delta x}\right)_A (T_1 - T_2) = \left(\frac{kA}{\Delta x}\right)_B (T_2 - T_3) = \left(\frac{kA}{\Delta x}\right)_C (T_3 - T_4) \qquad (3.29)$$

Eliminating the intermediate temperatures $T_2$ and $T_3$ in Eq. (3.29), the rate of heat flow can be written in the form

$$q_k = \frac{T_1 - T_4}{(\Delta x/kA)_A + (\Delta x/kA)_B + (\Delta x/kA)_C} \qquad (3.30)$$

Equation (3.30) is a convenient starting point to introduce a different viewpoint for the analysis of heat transfer, which will be followed up in later chapters. This viewpoint makes use of concepts developed previously in the analysis of electric-circuit theory and is called the electric analogy between the flow of heat and electricity. If the heat-transfer rate $q_k$ is considered to be analogous to the flow of electricity, the combination $(\Delta x/kA)$ is analogous to a resistance $R$, and the temperature difference $\Delta T$ is analogous to a potential difference. Equation (3.24) can thus be written in a form similar to Ohm's law in electric circuit theory, or

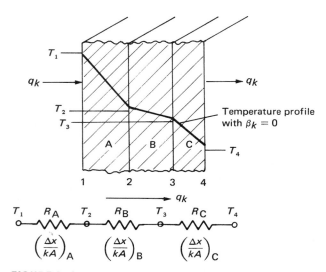

**FIGURE 3.13** One-dimensional heat conduction through a composite wall and corresponding thermal resistance network.

$$q_k = \frac{\Delta T}{R} \qquad (3.31)$$

where $\Delta T = T_1 - T_2$
$R = \Delta x/kA$

Similarly, for heat flow through several sections in series, as shown in Fig. 3.13, Eq. (3.30) can be expressed in the form

$$q_k = \frac{\Delta T}{R_A + R_B + R_C} \qquad (3.32)$$

where $\Delta T = T_1 - T_4$
$R_A = (\Delta x/kA)_A$
$R_B = (\Delta x/kA)_B$
$R_C = (\Delta x/kA)_C$

## EXAMPLE 3.7

The walls of a house consist of a layer of common brick, $\Delta x = 0.1$ m (4.0 in) ($k = 0.70$ W/m $\cdot$ K), and a layer of gypsum plaster board, $\Delta x = 0.0375$ m (1.5 in) ($k = 0.48$ W/m $\cdot$ K). Compare the rate of heat transfer through this uninsulated wall with one that has between the brick and the gypsum plaster a 0.0508 m (2.0 in) thick layer of rock-wool insulation ($k = 0.065$ W/m $\cdot$ K).

## SOLUTION

The rate of heat transfer through the uninsulated wall per square meter area and per Kelvin temperature difference between the surfaces will be given by

$$q_k = \frac{T_1 - T_2}{(\Delta x/k)_A + (\Delta x/k)_B} = \frac{1}{0.10/0.70 + 0.0375/0.48} = \frac{1}{0.221}$$

$$= 4.52 \text{ W/m}^2 \cdot \text{K}$$

The insulation will add a third resistance of $R_C = 0.0508/0.065 = 0.783$ m$^2$ $\cdot$ K/W. Thus, the rate of heat transfer will be

$$q_k = \frac{T_i - T_o}{R_A + R_B + R_C} = \frac{1}{0.145 + 0.783 + 0.079} = 1.00 \text{ W/m}^2 \cdot \text{K}$$

Adding the rock wool reduces the heat loss (or gain) by approximately 78 percent.

In Eq. (3.32) the rate of heat flow $q$ is expressed only in terms of an overall temperature potential and the resistances of the individual sections in the heat-flow path. These values can be combined into what is generally called an *overall transmittance*, or *overall heat-transfer coefficient U*, as shown below:

$$U = \frac{1}{R_A + R_B + R_C} \tag{3.33}$$

Values of $U$ factors for typical, residential wall construction are shown in Table 3.5 and thermal conductivities of building materials in Table 3.6.

The electric analog approach can also be used to solve more complex problems. One application is shown in Fig. 3.14, where heat is transferred through a composite structure involving thermal resistances not only in series, but also in parallel. For this system the resistance of the middle layer, $R_2$ in Fig. 3.14, becomes

$$R_2 = \frac{R_B R_C}{R_B + R_C} \tag{3.34}$$

**TABLE 3.5**  *U*-factor Values for Typical Residential Wall Construction[a]

| Wall no. | Construction component | Component thickness (in) | $U$ value (Btu/hr $\cdot$ ft$^2$ $\cdot$ °F) |
|---|---|---|---|
| 1 | Face brick | 4 | |
| | Block | 8 | |
| | Firring space | $\frac{3}{4}$ | 0.09 |
| | Urethane board | $\frac{1}{2}$ | |
| | Plasterboard | $\frac{1}{2}$ | |
| 2 | Stucco | $\frac{1}{2}$–1 | |
| | Block | 8–12 | |
| | Firring space | $\frac{3}{4}$ | 0.13 |
| | Urethane board | $\frac{1}{2}$ | |
| | Plasterboard | $\frac{1}{2}$ | |
| 3 | Wood siding | $\frac{1}{2}$ $\frac{3}{4}$ | |
| | Building paper | | |
| | Sheathing | $\frac{5}{8}$ | 0.13 |
| | Stud/space (with insulation) | $3\frac{5}{8}$ | |
| | Plasterboard | $\frac{1}{2}$ | |
| 4 | Glass, double-plate | $\frac{1}{4}$ | |
| | Insulated window (separated by $\frac{1}{2}$-in air space) | | 0.60 |

[a]It is assumed the windows are double-glazed, and the glass-to-wall ratio is 20%. From Kreider, J. F., and F. Kreith, "Solar Heating and Cooling," revised 1st ed., Hemisphere Publ. Corp., Washington, D.C., 1977.

**TABLE 3.6**  Thermal Conductivity and Density of
Some Building Materials

|  | $\rho(\text{kg/m}^3)$ | $k[\text{W}/(\text{m} \cdot \text{K})]$ |
|---|---|---|
| Asbestos cement sheet | 1520 | 0.29–0.43 |
| Asbestos felt | 144 | 0.078 |
| Asbestos insulating board | 720–900 | 0.11–0.21 |
| Asphalt, roofing | 1920 | 0.58 |
| Brick, common, dry | 1760 | 0.70–0.81 |
| Brick, wet | 2034 | 1.67 |
| Chipboard | 350–1360 | 0.07–0.21 |
| Concrete, gravel 1:2:4 | 2240–2480 | 1.4 |
|   Vermiculite aggregate | 400–880 | 0.11–0.26 |
|   Cellular | 320–1600 | 0.08–0.65 |
| Cork, granulated, raw | 115 | 0.046 |
|   Slab, raw | 160 | 0.05 |
| Fiberboard | 280–420 | 0.05–0.08 |
| Glass, window | 2500 | 1.05 |
| Glassfiber, mat | 50 | 0.033 |
| Hardboard | 560 | 0.08 |
| Plasterboard, gypsum | 1120 | 0.48–0.50 |
| Polystyrene, expanded board | 15 | 0.037 |
| Polyurethane foam | 30 | 0.026 |
| Polyvinyl chloride, rigid foam | 25–80 | 0.035–0.041 |
| Roofing felt | 960–1120 | 0.19–0.20 |
| Tiles, clay | 1900 | 0.85 |
| Tiles, concrete | 2100 | 1.10 |
| Tiles, PVC asbestos | 2000 | 0.85 |
| Urea formaldehyde foam | 8–30 | 0.032–0.038 |
| Vermiculite granules | 100 | 0.065 |
| Wilton carpet |  | 0.058 |

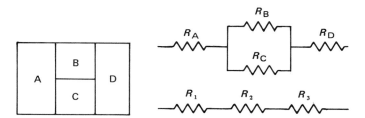

**FIGURE 3.14**  Series and parallel one-dimensional heat conduction
through a composite wall with corresponding thermal resistance
network.

and the rate of heat flow is

$$q_k = \frac{\Delta T_{\text{overall}}}{\displaystyle\sum_{n=1}^{n=N} R_{\text{th},n}} \tag{3.35}$$

where    $N$ = number of layers in series
$R_{\text{th},n}$ = thermal resistance of the $n$th layer
$\Delta T_{\text{overall}}$ = temperature difference between the inner and outer surfaces

## EXAMPLE 3.8

Compute the thermal resistance of a wall of a building constructed by two alternate methods. In method 1 the wall consists of 0.635 cm ($\frac{1}{4}$ in) plasterboard on the interior with 5.08 cm $\times$ 10.16 cm (2 $\times$ 4 in) wood supports spaced 30.48 cm (12 in) center to center and an external layer of shingles 0.635 cm ($\frac{1}{4}$ in) thick. In the alternate method of construction, method 2, the wall consists of the same external and interior material but support consists of 5.08 $\times$ 15.24 cm (2 $\times$ 6 in) wooden studs spaced 40.64 cm (16 in) apart.

## SOLUTION

The thermal circuits for both types of wall construction are identical, consisting of a series resistance element at the interior and exterior, with two parallel thermal resistances between them, as shown in Fig. 3.15.
   The definitions for the example solution are as follows:

| | |
|---|---|
| $A$ | Unit wall surface area |
| $b_1$ | Air space thickness |
| $b_2$ | Wood stud thickness |
| $h$ | Height of wall |
| $k_a$ | Thermal conductivity, air space |
| $k_{\text{in}}$ | Thermal conductivity, inside wall |
| $k_{\text{out}}$ | Thermal conductivity, outside wall |
| $k_w$ | Thermal conductivity, wood studs |
| $L_{\text{in}}$ | Inside wall thickness |
| $L_{\text{out}}$ | Outside wall thickness |
| $R_{k,a}$ | Thermal resistance to conduction, air space |
| $R_{k,\text{in}}$ | Thermal resistance to conduction, inside wall |
| $R_{k,\text{out}}$ | Thermal resistance to conduction, outside wall |
| $R_{k,w}$ | Thermal resistance to conduction, wood studs |
| $R_{k,wa}$ | Combined resistance to conduction, wood studs plus air space |
| $T_{\text{in}}$ | Inside surface temperature |

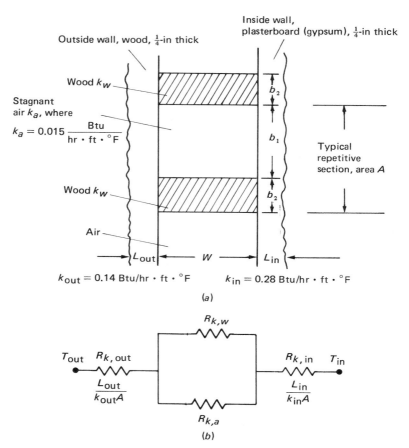

**FIGURE 3.15**  (a) Schematic sketch showing construction of wall for Example 3.8. (b) Thermal circuit for Example 3.8.

$T_{out}$    Outside surface temperature
$W$      Distance between inside and outside walls

If it is assumed the air between the wooden supports is stagnant so that heat is transferred primarily be conduction, the effective resistance $R_{k,wa}$ of the wood-support resistance and the air resistance for a wall section of height 1 m is given by the equation

$$R_{k,wa} = \frac{R_{k,w}R_{k,a}}{R_{kw} + R_{ka}}$$

Method 1, 5.08 × 10.16 cm (2 × 4 in) supports, 30.48 cm apart:

$$R_{k,w} = \frac{W}{k_w b_2/h} = \frac{(10.16\ \text{cm})(1\ \text{m})}{(0.208\ \text{W/m} \cdot \text{K})(5.08\ \text{cm})}$$

$$= 9.62\ \text{m}^2 \cdot \text{K/W}\ (16.7\ \text{hr} \cdot \text{ft}^2 \cdot °\text{F/Btu})$$

$$R_{k,a} = \frac{W}{k_a b_1/h} = \frac{(10.16\ \text{cm})(1\ \text{m})}{(0.026\ \text{W/m} \cdot \text{K})(25.4\ \text{cm})}$$

$$= 15.4\ \text{m}^2 \cdot \text{K/W}\ (26.7\ \text{hr} \cdot \text{ft}^2 \cdot °\text{F/Btu})$$

$$R_{k,wa} = \frac{(9.62)(15.4)}{9.62 + 15.4} = 5.92\ \text{m}^2 \cdot \text{K/W}\ (10.3\ \text{hr} \cdot \text{ft}^2 \cdot °\text{F/Btu})$$

Method 2, 5.08 × 15.24 cm (2 × 6 in) supports, 40.64 cm (16 in) apart:

$$R_{k,w} = \frac{W}{k_w b_2/h} = \frac{(15.24\ \text{cm})(1\ \text{m})}{(0.208\ \text{W/m} \cdot \text{K})(5.08\ \text{cm})}$$

$$= 14.41\ \text{m}^2 \cdot \text{K/W}\ (25.0\ \text{hr} \cdot \text{ft}^2 \cdot °\text{F/Btu})$$

$$R_{k,a} = \frac{W}{k_a b_1/h} = \frac{(15.24\ \text{cm})(1\ \text{m})}{(0.026\ \text{W/m} \cdot \text{K})(35.56\ \text{cm})}$$

$$= 16.5\ \text{m}^2 \cdot \text{K/W}\ (28.6\ \text{hr} \cdot \text{ft}^2 \cdot °\text{F/Btu})$$

$$R_{k,wa} = \left[\frac{(14.4)(16.5)}{14.4 + 16.5}\right]\left(\frac{40.64\ \text{cm}}{30.48\ \text{cm}}\right) = 10.2\ \text{m}^2 \cdot \text{K/W}\ (17.8\ \text{hr} \cdot \text{ft}^2 \cdot °\text{F/Btu})$$

The corresponding $U$ values for methods 1 and 2 are, respectively,

$$U_1 = \frac{1}{5.92} = 0.169\ \text{W/m}^2 \cdot \text{K}\ (0.097\ \text{Btu/hr} \cdot \text{ft}^2 \cdot °\text{F})$$

$$U_2 = \frac{1}{10.2} = 0.098\ \text{W/m}^2 \cdot \text{K}\ (0.056\ \text{Btu/hr} \cdot \text{ft}^2 \cdot °\text{F})$$

which means that construction method 2 reduces the heat loss appreciably. This analysis could be refined by considering convection at the interior and exterior walls, but the essential conclusion would remain the same: a small change in some building codes could reduce the heat loss substantially by reducing the $U$ value without reducing structural strength.

## Cylinders and Spheres

Heat conduction through tubes, pipes, and spherical containers is of importance in many solar engineering systems. Consider first a long, hollow cylinder of inside radius $r_i$, outside radius $r_o$, and length $L$ (see Fig. 3.16). The temperature at the inside surface is $T_i$ and at the outside surface $T_o$. To determine the rate of heat conduction Fourier's law can be used with appropriate coordinates. Assuming that the cylinder is sufficiently long for end effects to be negligible, heat flows only in the radial direction. Then, at a radial distance $r$ the area through which heat is conducted in the cylindrical coordinate system is $A_r = 2\pi r L$ and Fourier's law can be written in the form

$$q_k = -kA_r \frac{dT}{dr} = -2\pi kLr \frac{dT}{dr} \tag{3.36}$$

Separating the variables in Eq. (3.36) gives

$$q_k \frac{dr}{r} = 2\pi kL \, dT$$

which can be integrated subject to the boundary conditions

$$T(r) = T_i \quad \text{at} \quad r = r_i$$

$$T(r) = T_o \quad \text{at} \quad r = r_o$$

This yields the following relation for the rate of heat conduction through the cylinder:

$$q_k = \frac{2\pi kL(T_i - T_o)}{\ln(r_o/r_i)} \tag{3.37}$$

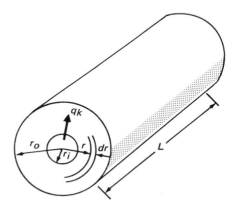

**FIGURE 3.16**  Radial heat
conduction through a hollow
circular cylinder.

In terms of the electric analog Eq. (3.37) can be written as

$$q_k = \frac{T_i - T_o}{R_{th}}$$ (3.38)

where

$$R_{th} = \frac{\ln(r_o/r_i)}{2\pi kL}$$

The thermal resistance concept can be extended easily to multiple-layer cylindrical walls. For example, in a two-layer system (see Fig. 3.17) such as a pipe covered with insulation, the rate of heat conduction can be written in the form

$$q_k = \frac{T_1 - T_3}{R_A + R_B}$$ (3.39)

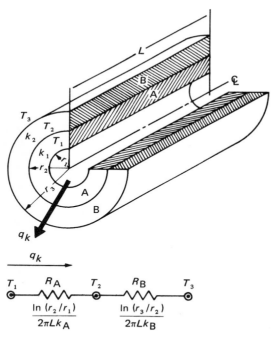

**FIGURE 3.17** Heat conduction through a tube of inner radius $r_1$ and outer radius $r_2$ covered by insulation of inner radius $r_2$ and outer radius $r_3$ with corresponding thermal network.

where

$$R_A = \frac{\ln (r_2/r_1)}{2\pi L k_A}$$

$$R_B = \frac{\ln (r_3/r_2)}{2\pi L k_B}$$

Spherical systems can be treated in a similar manner. When the heat flow is radial, that is, the temperature is only a function of the radial distance, the rate of heat conduction through a shallow spherical container of inner radius $r_i$ and the outer radius $r_o$ with a uniform temperature over the interior surface $T_i$ and over the exterior surface $T_o$ is

$$q_k = \frac{T_i - T_o}{(4\pi k/r_i r_o)(r_o - r_i)} \tag{3.40}$$

For the thermal network, the denominator can be treated as the thermal resistance in a multilayer spherical system.

The heat flow rate from a long rectangular duct of width $w$, height $h$, and insulation thickness $t$ is given by

$$q_{k,\,\text{rect}} = kL\left(\frac{2w + 2h}{t} + 2.16\right)(T_i - T_o) \tag{3.41}$$

## Thermal Conductivity

The thermal conductivity is a material property defined by Eq. (3.24). Except for gases at low temperatures it is not possible to predict this property analytically. Available information about the thermal conductivity of materials is therefore based largely on experimental measurements. In general, the thermal conductivity of a material varies with temperature, but in many practical situations a constant value based on the average temperature of the system will give satisfactory results. Table 3.7 lists typical values of the thermal conductivities for some metals, nonmetallic solids, liquids, and gases to illustrate the order of magnitude to be expected in practice. More detailed information is presented in Tables A3.4, A3.5, and A3.7.

The mechanism of thermal conduction in gases is explained qualitatively by the kinetic theory. All molecules in a gas are in random motion and exchange energy and momentum when they collide with one another. However, since higher temperatures are associated with molecules possessing more kinetic energy, when a molecule from a high-temperature region moves into a region of lower temperature, it transports kinetic energy on a molecular scale to the lower-temperature region. Upon impact with a molecule of lower kinetic energy, an energy transfer occurs, which is seen as a transfer of heat from a macroscopic viewpoint. The physical mechanics of conduction in liquids are qualitatively similar, but since molecules in liquids are more closely spaced, their force fields play a significant role in the

**TABLE 3.7** Thermal Conductivities of Some Metals, Nonmetallic Solids, Liquids, and Gases

| Material | Thermal conductivity at 300 K (W/m · K) |
|---|---|
| Copper | 386 |
| Aluminum | 204 |
| Carbon steel | 54 |
| Glass | 0.75 |
| Plastics | 0.2–0.3 |
| Water | 0.6 |
| Ethylene glycol | 0.25 |
| Engine oil | 0.15 |
| Freon (liquid) | 0.07 |
| Hydrogen | 0.18 |
| Air | 0.026 |

energy transfer during collision, and the picture is even more complex than in gases.

Figure 3.18 shows how the thermal conductivity of some gases varies with temperature. The thermal conductivity of gases is almost independent of pressure, except near the critical point. According to a simplified analysis based on a kinetic exchange model, the thermal conductivity of gases will increase as the square root of the absolute temperature.

Figure 3.19 also shows the thermal conductivity of some liquids as a function of temperature. It can be seen that except for water the thermal conductivity of liquids decreases with increasing temperature, but the change is so small that in most practical situations the thermal conductivity can be assumed constant at some average temperature; there is no appreciable dependence on pressure.

Figure 3.20 shows the thermal conductivities of some metals. In solids thermal energy is transported by free electrons and by vibrations in the lattice structure. In general, the movement of free electrons is the more important transport mode, and since in good electrical conductors a large number of free electrons move within the lattice structure, good electrical conductors are also good heat conductors, for example, copper, silver, and aluminum. On the other hand, good electrical insulations are also good thermal insulators, for example glass and plastics. The best types of thermal insulators, however, rely for their insulating effectiveness on trapping a gas within a porous structure. In those materials the transfer of heat may occur by several modes: conduction through a fibrous or porous solid structure; conduction and/or convection through air trapped in the void spaces; radiation between portions of the solid structure. The last-mentioned mechanism is, as will be shown, especially important at high temperatures or in evacuated enclosures. Special types of superinsulation materials have been developed for cryogenic applications at very low temperatures, down to about 25 K. These kinds of superinsulators consist of several layers of highly reflective materials separated by evacuated spaces to minimize conduction and convection and can achieve effective conductivities as low

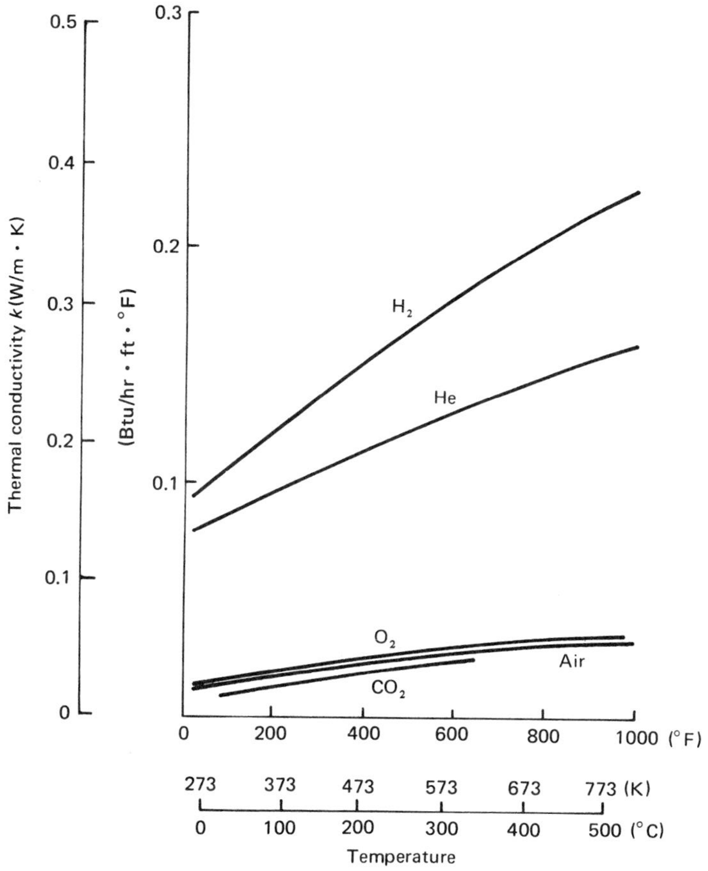

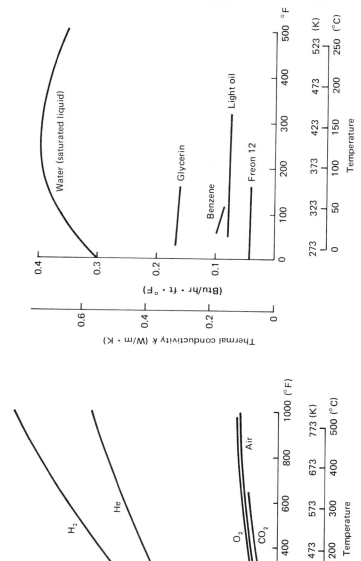

**FIGURE 3.19** Thermal conductivities of some typical liquids.

**FIGURE 3.18** Thermal conductivities of some typical gases. 1 W/m · K = 0.5779 Btu/hr · ft · °F.

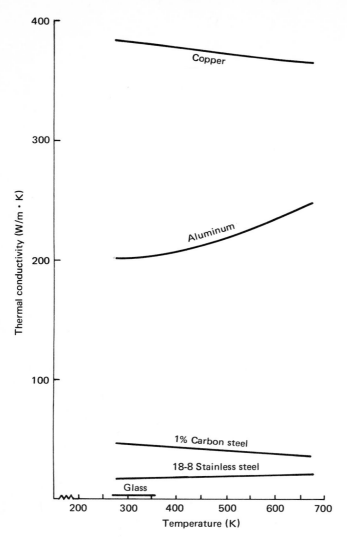

**FIGURE 3.20** Thermal conductivity of some solids.

as $3 \times 10^{-4}$ W/m · K. They may some day become very important if solar energy can be used to produce hydrogen, which would have to be stored as a cryogenic liquid in a hydrogen economy. More complete information on superinsulation is given in Refs. (1, 6).

## Convection Heat Transfer

When a fluid comes in contact with a solid surface at a different temperature, the resulting thermal energy exchange process is called *convection heat transfer*. This process is a common experience, but a detailed description and analysis of the mechanism is complicated. In this book we shall not attempt to treat analytical procedures for predicting convection heat transfer, but rather we shall concentrate first on developing an intuitive grasp of the mechanism and then present correlations of experimental data, which can be used to calculate the rate of heat transfer in those subsystems of complete solar-heating and -cooling systems in which convection occurs.

There are two kinds of convection processes: natural or *free convection* and *forced convection*. In the first type the motive force is the density difference in the fluid, which results from its contact with a surface at a different temperature and gives rise to buoyant forces. Typical examples of such free convection are the heat transfer between the wall or the roof of a house and the surrounding air on a calm day, the convection to the fluid in a tank in which a heating coil is immersed, or the heat loss from the cover surface of a solar collector when there is no wind blowing.

Forced convection occurs when an external force moves a fluid past a surface at a higher or lower temperature than the fluid. Since the fluid velocity in forced convection is higher than in free convection, more heat can be transferred at a given temperature difference. The price to be paid for this increase in the rate of heat transfer is the work required to move the fluid past the surface. But irrespective of whether the convection is free or forced, the rate of heat transfer $q_c$ can be obtained from Newton's law of cooling

$$q_c = \bar{h}_c A(T_s - T_f) \tag{3.42}$$

where $\bar{h}_c$ = average convection heat-transfer coefficient over the surface $A$ in $W/m^2 \cdot K^*$ (Btu/hr $\cdot$ ft$^2$ $\cdot$ °F)

$A$ = surface area in contact with the fluid in m$^2$ (ft$^2$)

$T_s$ = surface temperature in K (°F)

$T_f$ = undisturbed fluid temperature in K (°F)

Table 3.8 lists some approximate values of convection heat-transfer coefficients. Before attempting to calculate a heat-transfer coefficient, we shall examine the energy-transport process in some detail and relate the convection of heat to the

*Note: 5.677 W/m$^2$ $\cdot$ K = 1 Btu/hr $\cdot$ ft$^2$ $\cdot$ °F.

**TABLE 3.8**  Order of Magnitude of Convective Heat-Transfer Coefficients $\bar{h}_c$

|  | W/m² · K | Btu/hr · ft² · °F |
|---|---|---|
| Air, free convection | 6–30 | 1–5 |
| Superheated steam or air, forced convection | 30–300 | 5–50 |
| Oil, forced convection | 60–1,800 | 10–300 |
| Water, forced convection | 300–6,000 | 50–2,000 |
| Water, boiling | 3,000–60,000 | 500–10,000 |
| Steam, condensing | 6,000–120,000 | 1,000–20,000 |

flow of the fluid. Figure 3.21 shows a heated flat plate cooled by a stream of air flowing over it, with the velocity and the temperature distributions in the air stream. The first point to note is that the velocity decreases in the direction toward the surface as a result of viscous forces and the fluid does not move relative to the fluid-solid interface. Since the velocity of the fluid layer adjacent to the wall is zero, the heat transfer at the interface between the surface and the adjacent fluid layer is by conduction, or

$$\frac{q_c}{A} = -k_f \frac{\partial T_f}{\partial y}\bigg|_{y=0} = \bar{h}_c(T_s - T_f)$$

Although this viewpoint suggests that the convection process can be viewed as conduction at the interface, the temperature gradient at the surface $(\partial T_f/\partial y)|_{y=0}$ is determined by the rate at which the fluid farther from the wall can transport the energy into the mainstream. Thus the temperature gradient at the wall depends on the flow field, with higher velocities being able to produce larger temperature gradients and higher rates of heat transfer. At the same time, however, the thermal conductivity of the fluid also plays a role. For example, the value of $k_f$ for the water is an order of magnitude larger than that of air; thus, as shown in Table 3.8 the convection heat-transfer coefficients for water are also larger than for air.

The situation is quite similar in free convection, as shown in Fig. 3.22. The principal difference is that in forced convection the velocity far away from the surface approaches the free-stream value imposed by an external force, whereas in free convection the velocity at first increases with increasing distance from the heat-transfer surface and then decreases. The reason for this behavior is that the action of viscosity diminishes rather rapidly with distance from the surface while the density difference decreases more slowly. Eventually, however, the buoyant

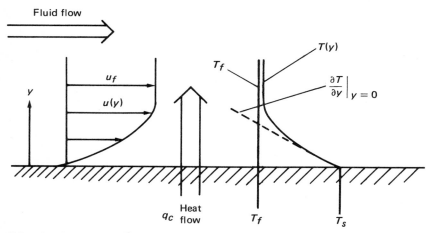

**FIGURE 3.21** Velocity and temperature distributions for forced convection over a heated plate.

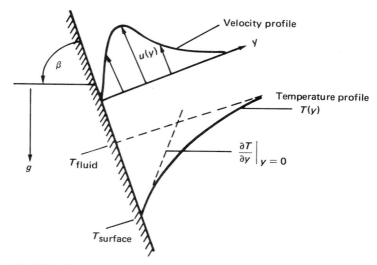

**FIGURE 3.22** Velocity and temperature distributions for free convection over a heated plate inclined on angle $\beta$ from the horizontal.

force decreases also as the fluid density approaches the value of unheated surrounding fluid. This interaction of forces will cause the velocity to reach a maximum and approach zero far away from the heated surface. The temperature fields in free and forced convection have similar shapes and in both cases the heat-transfer mechanism at the fluid-solid interface is conduction.

The preceding discussion indicates that the convection heat-transfer coefficient will depend on the density, viscosity, and velocity of the fluid as well as on its thermal properties (thermal conductivity and specific heat). Whereas in forced convection the velocity is usually imposed on the system by a pump or a fan and can be directly specified, in free convection the velocity will depend on the temperature difference between the surface and the fluid, the coefficient of thermal expansion of the fluid (it determines the density change per unit temperature difference), and the body force field, which in solar systems located on earth is simply the gravitational force.

Also convection heat transfer can be treated within the framework of a thermal resistance network once the heat-transfer coefficient is known. The thermal resistance to convection is given by

$$R_c = \frac{1}{\bar{h}_c A} \qquad (3.43)$$

and this resistance to a surface-to-fluid interface can easily be incorporated into a network. For example, the heat transfer from the interior of a room at $T_i$ through a wall to atmospheric air outside at $T_o$ is shown in Fig. 3.23.

Heat is first transferred by free convection to the interior surface of the wall,

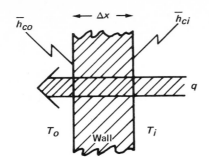

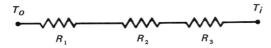

**FIGURE 3.23**   Thermal circuit for heat transfer from house interior at $T_i$ through a wall to exterior environment at $T_o$.

then by conduction through the wall to the exterior surface, and finally from the exterior surface to the air outside. Thus, there are three resistances in series, and the rate of heat transfer is given by

$$q = \frac{T_i - T_o}{\sum\limits_{i=1}^{i=3} R_i} = \frac{T_i - T_o}{R_1 + R_2 + R_3} \tag{3.44}$$

where

$$R_1 = \frac{1}{\overline{h}_{co} A}$$

$$R_2 = \frac{\Delta x}{k_w A}$$

$$R_3 = \frac{1}{\overline{h}_{ci} A}$$

**EXAMPLE 3.9**

Compare the heat loss through a single-plate, double-strength window with the heat loss through a double-plate window. A window consisting of a single plate of glass 0.3175 cm ($\frac{1}{8}$ in) thick is shown with its equivalent thermal circuit in Fig. 3.24,

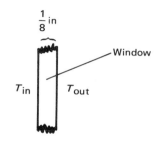

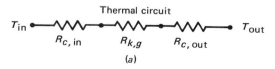

(a)

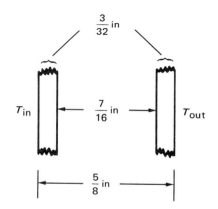

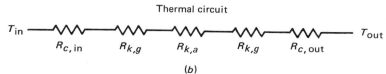

(b)

**FIGURE 3.24** Schematic of window and thermal circuit for (a) single- and (b) double-plate windows.

where $T_{in}$ = inside air temperature

$T_{out}$ = outside air temperature

$R_{c,in}$ = thermal resistance to heat transfer by convection on the inside surface of the glass pane

$R_{c,out}$ = thermal resistance to heat transfer by convection on the outside surface of the glass pane

$h_{c,in}$ = 5.67 W/m$^2$ · K (1 Btu/hr · ft$^2$ · °F)

$h_{c,out}$ = 22.68 W/m$^2$ · K (4 Btu/hr · ft$^2$ · °F)

On the basis of 1 m$^2$ (or 1 ft$^2$) of window the resistances are

$$R_{c,\,in} = \frac{1}{h_{c,\,in}A} = \frac{1}{(5.67 \text{ W/m}^2 \cdot \text{K})(1 \text{ m}^2)} = 0.176 \text{ K/W } (1.0 \text{ hr} \cdot {}^\circ\text{F/Btu})$$

$$R_{k,g} = \frac{W}{k_gA} = \frac{0.003175 \text{ m}}{(0.779 \text{ W/m} \cdot \text{K})(1 \text{ m}^2)} = 4.076 \times 10^{-3} \text{ K/W } (0.023 \text{ hr} \cdot {}^\circ\text{F/Btu})$$

$$R_{c,\,out} = \frac{1}{h_{c,\,out}A} = \frac{1}{(22.68 \text{ W/m}^2 \cdot \text{K})(1 \text{ m}^2)} = 0.0441 \text{ K/W } (0.25 \text{ hr} \cdot {}^\circ\text{F/Btu})$$

where $k_g$ is the thermal conductivity of glass.

If the average inside air temperature $T_{in}$ is 20°C or 293 K (68°F) and the average outside air temperature $T_{out}$ is 6.1°C or 279.1 K (43°F), the heat loss $q$ is

$$q = \frac{T_{in} - T_{out}}{R_{c,\,in} + R_{k,g} + R_{c,\,out}}$$

$$= \frac{13.9 \text{ K}}{(0.176 + 4.076 \times 10^{-3} + 0.0441) \text{ K/W}}$$

$$= 62 \text{ W/m}^2 \ (18.2 \text{ Btu/hr} \cdot \text{ft}^2)$$

A double-plate window consisting of two plates of single-strength glass, each 0.24 cm ($\frac{3}{32}$ in) thick with a 1.11-cm ($\frac{7}{16}$-in) air space between them, is shown with its equivalent thermal circuit in Fig. 3.24,

where $T_{in}$ = inside air temperature

$\qquad T_{out}$ = outside air temperature

$\qquad R_{c,\,in}$ = thermal resistance to heat transfer by convection on the inside surface of the glass pane

$\qquad R_{c,\,out}$ = thermal resistance to heat transfer by convection on the outside surface of the glass pane

$\qquad R_{k,g}$ = thermal resistance to heat transfer by conduction through the glass pane

$\qquad R_{k,a}$ = thermal resistance to heat transfer by conduction in an air gap

$\qquad R_{c,\,in}$ = 0.176 K/W (1 hr $\cdot$ °F/Btu)

$$R_{k,g} = \frac{W}{k_gA} = \frac{(0.24 \text{ cm})(0.01 \text{ m/cm})}{(0.779 \text{ W/m} \cdot \text{K})(1 \text{ m}^2)}$$

$$= 3.08 \times 10^{-3} \text{ K/W } (0.0175 \text{ hr} \cdot {}^\circ\text{F/Btu})$$

$$R_{k,a} = \frac{W}{k_a A} = \frac{0.0111 \text{ m}}{(0.0242 \text{ W/m} \cdot \text{K})(1 \text{ m}^2)} = 0.458 \text{ K/W} \ (2.6 \text{ hr} \cdot {}^\circ\text{F/Btu})$$

$$R_{c,\text{out}} = 0.0441 \text{ K/W} \ (0.25 \text{ hr} \cdot {}^\circ\text{F/Btu})$$

where $k_a$ is the thermal conductivity of heat through the air gap. The heat loss $q$ is calculated as follows:

$$q = \frac{T_{\text{in}} - T_{\text{out}}}{R_{c,\text{in}} + 2R_{k,g} + R_{k,a} + R_{c,\text{out}}}$$

$$= \frac{13.9 \text{ K}}{(0.176 + 2 \times 3.08 \times 10^{-3} + 0.458 + 0.0441) \text{ K/W}}$$

$$= 20.3 \text{ W/m}^2 \ (6.4 \text{ Btu/hr} \cdot \text{ft}^2)$$

The effect of the temperature difference between the air inside and outside on the rate of heat loss is illustrated in Fig. 3.25.

## Correlation of Convection Data

Experimental data for convection heat transfer are usually correlated in terms of dimensionless quantities. One of these, the Reynolds number, has already been

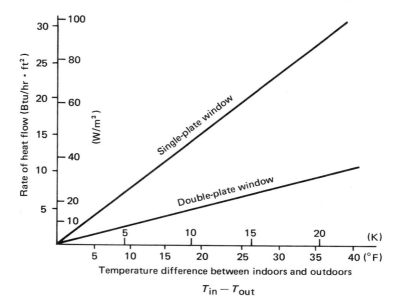

**FIGURE 3.25**  Effect of various temperature differences $T_{\text{in}} - T_{\text{out}}$ for single- and double-plate windows on heat flux ($q/A$).

discussed. The heat-transfer coefficient is nondimensionalized in terms of the Nusselt number $\overline{\text{Nu}}$, defined as

$$\overline{\text{Nu}}_L = \frac{\bar{h}_c L}{k_f} \qquad (3.45)$$

where $L$ is a length dimension characteristic of a system in m (ft) (for example, the diameter for flow in a tube or the distance between the collector surface and the cover in free convection between these surfaces), and $k_f$ is the thermal conductivity of the fluid in W/m · K (Btu/ft · hr · °F), usually evaluated at the arithmetic mean between the surface and the bulk fluid temperature $T_f$.

A third dimensionless parameter of importance is the Prandtl number Pr, defined as

$$\text{Pr} = \frac{c_p \mu_f}{k_f} \qquad (3.46)$$

where $c_p$ is the specific heat of the fluid in J/kg · K (Btu/lb$_m$ · °F) and $\mu_f$ is the viscosity in N · sec/m$^2$ (lb/ft · sec). The Prandtl number is the ratio of momentum diffusivity to thermal diffusivity.

In general, the Nusselt number $\text{Nu}_L$ for forced convection can be related to the Reynolds number $\text{Re}_L$ and the Prandtl number Pr by a relation of the form [25]

$$\overline{\text{Nu}}_L = C \, \text{Re}_L{}^n \, \text{Pr}^m \qquad (3.47)$$

where $C$, $n$, and $m$ are empirically determined constants and $L$ is a characteristic length dimension. Sometimes also a functional relation to account for variations in geometry and physical properties is incorporated into this correlation.

In free convection the velocity is not given explicitly and the Reynolds number, which can only be calculated when the velocity is known, cannot be used. The dimensionless parameter that relates the buoyant inertial and viscous forces and thus determines the velocity indirectly is the Grashof number $\text{Gr}_L$, defined by

$$\text{Gr}_L = \frac{\rho^2 g \beta_T (T_1 - T_2) L^3}{\mu^2} \qquad (3.48)$$

where $\rho$ = fluid density in kg/m$^3$ (lb/ft$^3$)

$g$ = gravitational constant, equal to 9.8 m/sec$^2$ in the SI system (32.2 ft/sec$^2$)

$L$ = characteristic length dimension in m/ft

$\beta_T$ = coefficient of expansion of the fluid in K$^{-1}$ (R$^{-1}$) (for ideal gases, $\beta_T$ equals the reciprocal of the absolute temperature, that is, $\beta_T = 1/T$)

The temperature difference $(T_1 - T_2)$ depends on the system. For a vertical wall, $T_1$ is the surface temperature and $T_2$ is the temperature of the undisturbed fluid. For free convection between two parallel surfaces $T_1$ and $T_2$ are taken as the two surface temperatures.

The gravitational constant must be modified when a flow does not occur in the direction of the force field. Thus, for a plate inclined an angle $\beta$ relative to the horizontal, $g \sin \beta$ is the gravitational force along the surface.

Also in free convection the flow can become turbulent when the Grashof number is large. Normally, for convection from a vertical surface of height $L$, the flow will be laminar as long as $\mathrm{Gr}_L < 10^9$.

In general, the Nusselt number in free convection $\mathrm{Nu}_L$ can be related to the Grashof number and the Prandtl number by a relation of the form

$$\overline{\mathrm{Nu}_L} = C(\mathrm{Gr}_L \, \mathrm{Pr})^n \qquad (3.49)$$

where $C$ and $n$ are constants determined empirically for given system geometries.

In solar energy systems convection heat transfer plays an important role in the transfer of heat from the absorber plate to the working fluid, in the heat loss from the outer cover of a collector, in the transfer of heat between the collector surface and the transparent cover, in the heat transfer in heat exchangers, and in the heat losses from buildings, just to mention a few typical cases. A key problem in constructing the thermal circuit and predicting the rate of heat transfer is to evaluate the heat-transfer coefficients or unit surface conductances. The evaluation of the heat-transfer coefficient in engineering practice is based on the correlation of experimental data by relations such as Eqs. (3.47) and (3.49). Whenever one uses a correlation based on experiments, it is important to make sure that the conditions of the real system are geometrically and dynamically similar to and within the range of the conditions for which the empirical or semiempirical correlation applies. The correlation equations presented below are mainly for geometries and flow conditions encountered in solar energy conversion systems and associated equipment. A more general treatment is presented in Refs. (25, 33) (Fig. 3.26).

## Laminar Flow inside Tubes and Ducts

Because of the small heat flux, the flow rate in the flat-plate collectors is usually low and the flow is laminar. For laminar flow in tubes or ducts the following equation can be used to determine $\overline{h}_c$, as long as $\mathrm{Re} \, \mathrm{Pr} \, D_H/L > 10$:

$$\overline{\mathrm{Nu}_{D_H}} = \frac{\overline{h}_c D_H}{k_b} = 1.86 \left( \mathrm{Re}_{D_H} \mathrm{Pr} \frac{D_H}{L} \right)^{1/3} \left( \frac{\mu_b}{\mu_w} \right)^{0.14} \qquad (3.50)$$

where $D_H$ = hydraulic diameter of the duct
    $L$ = length of the tube or duct
    $k_b$ = fluid thermal conductivity at the bulk temperature
    $\mu_b$ = viscosity of the fluid at the bulk temperature, halfway between inlet and outlet
    $\mu_w$ = viscosity of the fluid at the wall temperature

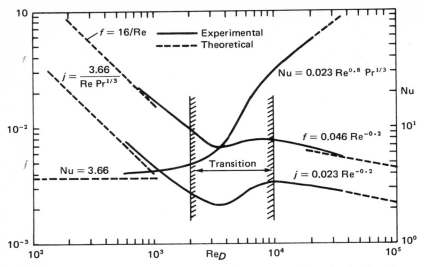

**FIGURE 3.26** Basic heat transfer and flow-friction characteristics of a circular tube: $j = \bar{h}_c/Gc_p \, Pr^{2/3}$.

The other parameters have previously been defined. When the duct is short and $Re_{D_H} \, Pr \, D_H/L > 100$, the following equation applies:

$$\overline{Nu}_{D_H} = \frac{\bar{h}_c D_H}{k_f} = \frac{Re_{D_H} \, Pr \, D_H}{4L} \, \ln \left[ \frac{1}{1 - 2.654/Pr^{0.167} (Re_{D_H} \, Pr \, D_H/L)^{0.5}} \right] \quad (3.51)$$

A summary of Nusselt numbers and friction coefficients for various shapes of long ducts and heat flux conditions in fully developed laminar flow ($L/D_H > 100$) is presented in Table 3.9. In the preceding relations and in Eq. (3.51) all physical properties should be evaluated at the mean temperature of the fluid in the duct, defined by

$$T_{mean} = \frac{T_{bi} + T_{wi} + T_{bo} + T_{wo}}{4} \quad (3.52)$$

where $T_{bi}$ = inlet bulk temperature
$\quad T_{wi}$ = wall temperature at the inlet
$\quad T_{bo}$ = outlet bulk temperature
$\quad T_{wo}$ = outlet wall temperature

In view of the low velocities in many solar convection apparatuses, the velocity produced by buoyant forces, that is, free convection, can also affect the convection heat exchange. In this so-called mixed-flow region when $Gr/Re^2$ is between 1 and 10 and $L/D_H > 50$, the heat-transfer coefficient for flow in horizontal ducts and ducts inclined not too far from the horizontal can be

**TABLE 3.9**  Heat Transfer and Friction for Fully Developed Laminar Flow of a Newtonian Fluid through Specified Ducts[a]

| Geometry $(L/D_H > 100)$ | $Nu_{H1}$ [b] | $Nu_{H2}$ | $Nu_T$ | $fRe$ | $\dfrac{j_{H1}}{f}$ [c] | $\dfrac{Nu_{H1}}{Nu_T}$ |
|---|---|---|---|---|---|---|
| (isosceles triangle) $\dfrac{2b}{2a} = \dfrac{\sqrt{3}}{2}$ | 3.014 | 1.474 | $2.39^d$ | 12.630 | 0.269 | 1.26 |
| (60° equilateral triangle) $\dfrac{2b}{2a} = \dfrac{\sqrt{3}}{2}$ | 3.11 | 1.892 | 2.47 | 13.333 | 0.263 | 1.26 |
| (square) $\dfrac{2b}{2a} = 1$ | 3.608 | 3.091 | 2.976 | 14.227 | 0.286 | 1.21 |
| (hexagon) | 4.002 | 3.862 | $3.34^d$ | 15.054 | 0.299 | 1.20 |
| (rectangle) $\dfrac{2b}{2a} = \dfrac{1}{2}$ | 4.123 | 3.017 | 3.391 | 15.548 | 0.299 | 1.22 |
| (circle) | 4.364 | 4.364 | 3.657 | 16.000 | 0.307 | 1.19 |
| (rectangle) $\dfrac{2b}{2a} = \dfrac{1}{4}$ | 5.331 | 2.930 | 4.439 | 18.233 | 0.329 | 1.20 |
| (rectangle) $\dfrac{2b}{2a} = \dfrac{1}{8}$ | 6.490 | 2.904 | 5.597 | 20.585 | 0.355 | 1.16 |
| (parallel plates) $\dfrac{2b}{2a} = 0$ | 8.235 | 8.235 | 7.541 | 24.000 | 0.386 | 1.09 |
| (parallel plates, insulated) $\dfrac{b}{a} = 0$ | 5.385 | — | 4.861 | 24.000 | 0.253 | 1.11 |

[a]From Shah, R. K., and A. L. London, Thermal Boundary Conditions and Some Solutions for Laminar Duct Flow Forced Convection, *Trans. ASME J. Heat Transfer*, vol. 96, pp. 159–165, 1974.

[b]Subscript $H1$ indicates a uniform heat flux in the flow direction with a uniform wall temperature at a flow cross section; $H2$ indicates a uniform wall heat flux both around the periphery and in the flow direction; $T$ denotes a uniform wall temperature at the boundary.

[c]This heading is the same as $Nu_{H1} Pr^{-1/3}/Re$ with $Pr = 0.7$.

[d]Interpolated values.

evaluated from the relation (25)

$$\overline{\text{Nu}}_{D_H} = \frac{\bar{h}_c D_H}{k} = 1.75 \left(\frac{\mu_b}{\mu_w}\right)^{0.14} \left[\frac{\text{Re}_{D_H} \text{ Pr } D_H}{L} + 0.012 \left(\frac{\text{Gr}_{D_H}^{1/3} \text{ Re}_{D_H} \text{ Pr } D_H}{L}\right)^{4/3}\right]^{1/3}$$

$$(3.53)$$

with properties and symbols the same as for Eq. (3.50) except for $\text{Gr}_D$, which is defined as

$$\text{Gr}_D = \frac{\rho^2 g \beta_T (T_w - T_b) D^3}{\mu^2}$$

## Turbulent Flow inside Tubes and Ducts

When the Reynolds number is above 6000 in a conduit, the flow is turbulent and the rate of heat transfer will be appreciably greater than in the laminar regime. A dimensionless correlation applicable to fluids with Prandtl numbers between 0.7 and 700 and $L/D_H > 60$ is (33)

$$\overline{\text{Nu}}_{D_H} = \frac{\bar{h}_c D_H}{k} = 0.023 \text{ Re}_{D_H}^{0.8} \text{ Pr}^{1/3} \qquad (3.54)$$

where all properties are to be evaluated at the mean temperature. When the duct is so short that $L/D_H < 60$, the heat-transfer coefficient calculated from Eq. (3.54) should be multiplied by $(1 + D_H/L)^{0.7}$.

For turbulent flow between two parallel flat plates with only one surface heated, the relationship

$$\overline{\text{Nu}}_{D_H} = 0.0196 \text{ Re}_{D_H}^{0.8} \text{ Pr}^{1/3} \qquad (3.55)$$

is recommended on the basis of experimental data for air (23).

In turbulent-flow heat transfer the physical properties of the fluid do not undergo as large a variation as in laminar flow. It is therefore sometimes possible to simplify the evaluation of $\bar{h}_c$ for a given fluid by evaluating all the physical properties in Eq. (3.54) for a specific fluid at a given temperature and incorporate them into the coefficient. Note, however, that these simplified equations are only applicable over a limited temperature range, are approximations useful primarily for preliminary design, and are dimensional.

For air at atmospheric pressure flowing through a long duct in turbulent flow at temperatures between 300 K (80°F) and 380 K (223°F), use

$$\bar{h}_c = 3.5 \frac{V^{0.8}}{D_H^{0.2}} \qquad \text{if } D_H \text{ is in m, } V \text{ is in m/sec, and } \bar{h}_c \text{ is in W/m}^2 \cdot \text{K} \quad (3.56)$$

$$\bar{h}_c = 0.50 \frac{V^{0.8}}{D_H{}^{0.2}} \qquad \text{if } D_H \text{ is in in, } V \text{ is in ft/sec, and } \bar{h}_c \text{ is in Btu/hr} \cdot \text{ft}^2 \cdot {}^\circ\text{F} \quad (3.56)$$
$$\text{(cont.)}$$

For atmospheric air at temperatures between 300 K (80°F) and 380 K (223°F) flowing between two flat plates spaced a distance $D$ apart with one side heated, use

$$\bar{h}_c = 2.6 \frac{V^{0.8}}{D^{0.2}} \qquad \text{if } D \text{ is in m, } V \text{ is in m/sec, and } \bar{h}_c \text{ is in W/m}^2 \cdot \text{K}$$

$$(3.57)$$

$$\bar{h}_c = 0.37 \frac{V^{0.8}}{D^{0.2}} \qquad \text{if } D \text{ is in in, } V \text{ is in ft/sec, and } \bar{h}_c \text{ is in Btu/hr} \cdot \text{ft}^2 \cdot {}^\circ\text{F}$$

For water flowing turbulently through a long duct of inside diameter $D$ in the temperature range between 278 and 378 K (40 and 220°F), use

$$\bar{h}_c = \frac{1056(0.02T - 4.06)V^{0.8}}{D^{0.2}} \qquad \text{if } T \text{ is in } K, V \text{ is in m/sec,}$$

$$D \text{ is in m, and } \bar{h}_c \text{ is in W/m}^2 \cdot \text{K}$$

$$(3.58)$$

$$\bar{h}_c = \frac{150(1 + 0.011T)V^{0.8}}{D^{0.2}} \qquad \text{if } T \text{ is in } {}^\circ\text{F}, V \text{ is in ft/sec,}$$

$$D \text{ is in in, and } \bar{h}_c \text{ is in Btu/hr} \cdot \text{ft}^2 \cdot {}^\circ\text{F}$$

For flow in a helical coil the value of the heat-transfer coefficient calculated for flow in a straight tube (33) should be multiplied by $[1 + 3.5(D_{\text{tube}}/D_{\text{coil}})]$.

**EXAMPLE 3.10**

A flat-plate water collector has 12 copper tubes of 0.025 m (1 in) ID soldered to the underside of a copper plate. The collector is 3 m $\times$ 3 m and the average tube temperature is 353 K (176°F). If the entering water is at a temperature of 333 K (140°F) and a velocity of 0.02 m/sec, calculate the exit temperature and the net radiation flux on the surface, assuming a fin efficiency of 100 percent for the flat surface connecting the tubes.

**SOLUTION**

First calculate the Reynolds number at the inlet temperature. Using properties from Table A3.2

$$Re_D = \frac{VD\rho}{\mu} = \frac{(0.02 \text{ m/sec})(0.025 \text{ m})(983.3 \text{ kg/m}^3)}{4.71 \times 10^{-4} \text{ kg/m} \cdot \text{sec}} = 1044$$

Since $Re_D$ is less than 2100, the flow is laminar.

Next, calculate the value of the dimensionless parameter $Re_D \text{ Pr } D/L$. Taking the value of the Prandtl number from Table A3.2, we get

$$Re_D \text{ Pr } D/L = \frac{(1044)(3.01)(0.025)}{3} = 26.5$$

Since $Re_D \text{Pr} D/L$ is larger than 10, Eq. (3.50) is the appropriate relationship. For the first iteration we shall calculate $h_c$ at the inlet temperature. Then we can estimate the outlet temperature and recalculate $\bar{h}_c$ using physical property values at half-way between inlet and outlet. With $\mu$ at 353 K equal to $3.57 \times 10^{-4}$, Eq. (3.50) gives the Nusselt number

$$\overline{Nu_D} = 1.86 \left( \frac{Re_D \text{ Pr } D}{L} \right)^{1/3} \left( \frac{\mu_b}{\mu_w} \right)^{0.14} = (1.86)(26.5)^{1/3} \left( \frac{4.71}{3.55} \right)^{0.14} = 5.8$$

The heat-transfer coefficient is therefore

$$\bar{h}_c = \frac{\overline{Nu_D} k}{D} = (5.8) \frac{0.65}{0.025} = 149 \text{ W/m}^2 \cdot \text{K}$$

and the rate of heat transfer to the fluid is

$$q = \bar{h}_c \pi DL \left( T_w - \frac{T_{b1} + T_{b2}}{2} \right) = \dot{m} c_p (T_{b2} - T_{b1})$$

In the above heat-balance relation $T_{b2}$ is the only unknown. To solve we calculate the mass flow rate

$$\dot{m} = \rho V A_c = \rho V \frac{\pi D^2}{4} = (983 \text{ kg/m}^3)(0.02 \text{ m/sec}) \frac{\pi (0.025 \text{ m})^2}{4} = 0.01 \text{ kg/sec}$$

Substituting numerical values into the heat-balance relation gives

$$(149)\pi(0.025)(3.0) \left( 353 - \frac{T_b - 333}{2} \right) = (0.01)(4179)(T_{b2} - 333)$$

Solving this for the unknown gives $T_{b2} = 345$ K. For the second iteration the preceding calculations are repeated with the physical properties evaluated at $T_{b, \text{mean}} = (345 + 333)/2 = 339$ K. The reader can easily verify that for this problem, where the bulk temperature rise is relatively small, the iteration does not change the results perceptibly. The rate of heat transfer to the fluid is

$$q = \dot{m}c_p(T_{b2} - T_{b1}) = (0.01)(4183)(12) = 502 \text{ J/sec}$$

There are 12 tubes in parallel so that the total rate of heat transfer over the entire 9-m$^2$ surface is 6024 J/sec (or W). Thus, the net radiation flux to the collector is 669 W/m$^2$ (or 212 Btu/hr $\cdot$ ft$^2$).

## EXAMPLE 3.11

Air at 350 K (170°F) and atmospheric pressure is flowing through a flat-plate solar collector at a velocity of 12 ft/sec. The plates are spaced 5.08 cm (2 in) apart and the heat flux is 630 W/m$^2$ (200 Btu/hr $\cdot$ ft$^2$). If the collector is 3.03 m (10 ft) long, calculate the air-outlet temperature and the collector-plate temperature relative to the air temperature.

## SOLUTION

First calculate the Reynolds number

$$\text{Re}_{D_H} = \frac{D_H V \rho}{\mu} = \frac{(0.0508 \text{ m})(3.64 \text{ m/sec})(0.998 \text{ kg/m}^3)}{2.07 \times 10^{-5} \text{ kg/m} \cdot \text{sec}} = 8894$$

The flow will be assumed turbulent and the Nusselt number from Eq. (3.55) is

$$\overline{\text{Nu}} = \frac{\bar{h}_c D_H}{k} = 0.0196 \text{ Re}^{0.8} \text{ Pr}^{1/3} = (0.0196)(8894)^{0.8} (0.697)^{1/3} = 25.1$$

Thus,

$$\bar{h}_c = \frac{(0.030 \text{ W/m} \cdot \text{K})(25.1)}{10.0508 \text{ m}} = 14.8 \text{ W/m}^2 \cdot \text{K} (2.61 \text{ Btu/hr} \cdot \text{ft}^2 \cdot °\text{F})$$

The temperature difference between plate and air is then

$$(T_{\text{wall}} - T_{\text{air}}) = \frac{q}{h_c} = \frac{630 \text{ W/m}^2}{14.8 \text{ W/m}^2 \cdot \text{K}} = 42.6 \text{ K} (76.7°\text{F})$$

The outlet temperature is given by

$$qA_c = A_{\text{cross-sect}} c_p \rho V(T_{\text{out}} - T_{\text{in}})$$

$$T_{out} = T_{in} + \frac{4qL}{Dc_p\rho V} = 350 \text{ K} +$$

$$\frac{(630 \text{ W/m}^2)(3.03 \text{ m})}{(0.0508 \text{ m})(1.0 \times 10^3 \text{ J/kg} \cdot \text{K})(0.998 \text{ kg/m}^3)(3.64 \text{ m/sec})}$$

$$= 350 + 10 = 360 \text{ K}$$

In order to increase the outlet temperature the velocity would have to be reduced.

### Single Cylinder in Cross Flow

The average Nusselt number for air flowing over a single cylinder can be obtained for $Re_D > 500$ from the relation

$$\overline{Nu}_D = 0.46 \ Re_D{}^{1/2} + 0.00128 \ Re_D \qquad (3.59)$$

where the Nusselt and Reynolds numbers are based on the cylinder diameter $D$ and the free-stream velocity, and the fluid properties are to be evaluated at the mean temperature between the surface and the air. This geometry is encountered in the connecting ducts of a solar system using a liquid-to-air heat exchanger.

### Flow across Tube Bundles

Flow across tube bundles is encountered in shell-and-tube heat exchangers, which are used to transfer heat from one fluid to another. Also, in solar water systems this type of exchanger is widely used. The flow and the details of the convection process are quite complex, but for bundles with 10 or more tubes at Reynolds numbers between 100 and 40,000 the average Nusselt number can be estimated from the relation (25)

$$Nu_D = 0.33 \ Re_{max}^{0.6} \ Pr^{0.33} \qquad (3.60)$$

In Eq. (3.60) the Nusselt and Reynolds numbers are based on the tube diameter and the physical properties on the mean temperature. However, the velocity is taken at the maximum value the fluid achieves in the tube bank. This maximum occurs at the smallest free-flow area between adjacent tubes. Equation (3.60) gives only an approximation and for further information the reader is referred to Refs. (14, 33, 40).

### EXAMPLE 3.12

In a shell-and-tube heat exchanger the water on the shell side has a bulk temperature of 333 K and the tubes are at 363 K. The tubes are 0.0254 m OD, spaced at 0.0508 m centers both horizontally and vertically, given a minimum cross-sectional area of 0.0125 m² between baffles. If the water flow rate is 24.6 kg/sec, calculate the heat-transfer coefficient.

**SOLUTION**

Using property values from Table A3.2 at 348 K: $\mu = 3.81 \times 10^{-4}$ kg/m · sec Pr = 2.39, $\rho = 975$ kg/m³, $k = 6.67 \times 10^{-4}$ kW/m · K. The maximum velocity is

$$V_{\max} = \frac{G}{\rho A_{\min}} = \frac{24.6 \text{ kg/sec}}{(975 \text{ kg/m}^3)(0.0125 \text{ m}^2)} = 2.02 \text{ m/sec}$$

Thus,

$$\text{Re}_{\max} = \frac{V_{\max} D \rho}{\mu} = \frac{(2.02 \text{ m/sec})(0.0254 \text{ m})(975 \text{ kg/m}^3)}{3.81 \times 10^{-4} \text{ kg/m} \cdot \text{sec}} = 1.31 \times 10^5$$

and from Eq. (3.60) we get

$$\overline{\text{Nu}} = \frac{\overline{h}_c D}{k} = 0.33(1.31 \times 10^5)^{0.6}(2.39)^{1/3} = 519$$

The heat-transfer coefficient is then found to be

$$\overline{h}_c = \frac{(519)(6.67 \times 10^{-4})}{0.0254} = 13.63 \text{ kW/m}^2 \cdot \text{K}$$

### Air Flow over a Flat Surface

The situation of most interest in flow over a surface occurs when air blows over the roof or the wall of a building or when the outermost cover of a flat-plate collector is exposed to wind. For the heat loss from a flat plate exposed to wind, the heat-transfer coefficient in W/m² · K is related to the wind speed $V$ in m/sec by the approximate dimensional relation

$$\overline{h}_c = 5.7 + 3.8V \qquad (3.61)$$

For a more general discussion of convection in flow over surfaces, the reader is referred to Refs. (25, 33).

### Free Convection from a Flat Surface

Free convection is important to calculate the heat flow to and from buildings on calm days and the heat loss from solar collectors. For heat transfer from a heated plate of length $L$, inclined an angle $\beta$ from the horizontal, to air at atmospheric pressure the relation

$$\bar{h}_c = 1.42 \left( \frac{\Delta T \sin \beta}{L} \right)^{1/4} \tag{3.62}$$

gives the heat-transfer coefficient in laminar flow with $10^4 < \mathrm{Gr}_L < 10^9$ if $L$ is in m and $\Delta T$ in K, while the relation

$$\bar{h}_c = 0.95 (\Delta T \sin \beta)^{1/3} \tag{3.63}$$

applies when the flow is turbulent and $\mathrm{Gr}_L > 10^9$.

### EXAMPLE 3.13

Calculate the heat loss by convection for a flat-plate collector cover 3 m × 3 m and inclined $60°$ if the collector is at 340 K and the ambient air is at 260 K.

### SOLUTION

The Grashof number is

$$\mathrm{Gr} = \frac{\beta_T g \sin 60° \, L^3 \Delta T}{\nu^2}$$

Assuming air to be an ideal gas, the value of $\beta_T$ is taken equal to $1/T_{\text{ave}}$ $= 1/300 \text{ K} = 3.33 \times 10^{-3} \text{ K}^{-1}$ and

$$\mathrm{Gr} = \frac{(3.33 \times 10^{-3} \text{ K}^{-1})(9.806 \text{ m/sec}^2)(\sin 60°)(3 \text{ m})^3 (80 \text{ K})}{(15.68 \times 10^{-6} \text{ m}^2/\text{sec})^2} = 2.5 \times 10^{11}$$

The flow is turbulent and from Eq. (3.63)

$$\bar{h}_c = 0.95 (\Delta T \sin \beta)^{1/3} = 0.95 [(80 \text{ K})(\sin 60°)]^{1/3} = 3.90 \text{ W/m}^2 \cdot \text{K}$$

and the rate of heat transfer is

$$q_c = \bar{h}_c A (T_1 - T_2) = (3.90 \text{ W/m}^2 \cdot \text{K})(9 \text{ m}^2)(80 \text{ K}) = 2809 \text{ W}$$

## *Free Convection between Two Parallel Surfaces*

Free convection between two parallel flat plates has been investigated experimentally by Hollands et al. (17–19) and by Buchberg et al. (14). Their results are correlated for inclination angles from the horizontal $\beta$ between 0 and $75°$ by the relation

$$\overline{Nu}_L = 1 + 1.44 \left(1 - \frac{1708}{Ra_L \cos \beta}\right) \left[1 - \frac{1708 (\sin 1.8\beta)^{1.6}}{Ra_L \cos \beta}\right]^{\bullet}$$

$$+ \left[\left(\frac{Ra_L \cos \beta}{5830}\right)^{1/3} - 1\right]^{\bullet} \quad (3.64)$$

where $L$ is the distance between the plates at temperatures $T_1$ and $T_2$, respectively, $[x]^{\bullet} = (|x| + x)/2$, and the Rayleigh number $Ra_L$ is given by

$$Ra_L = \frac{2g(T_1 - T_2)L^3}{\nu^2(T_1 + T_2)} Pr$$

It should be noted that when $Ra_L < 1708/\cos \beta$ the Nusselt number in Eq. (3.64) is exactly equal to unity. Since by definition

$$q_c = A\overline{h}_c(T_1 - T_2) = A \overline{Nu} \frac{k}{L}(T_1 - T_2)$$

the condition $\overline{Nu} = 1$ implies that the heat transfer is by pure conduction.

For inclinations between $75° < \beta < 90°$ the recommended relation for air is

$$\overline{Nu}_L = [1, 0.288(A \sin \beta \ Ra)^{1/4}, 0.039(\sin \beta \ Ra)^{1/3}]_{MAX} \quad (3.65)$$

where the subscript MAX indicates that, at any given value of Ra, the largest of the three quantities separated by commas should be used. The quantity $A$ in Eq. (3.65) is the aspect ratio of the air layer, defined as the ratio of the thickness $L$ to the length along the layer $H$, measured along either the heated or cooled bounding surface in the upslope direction.

Figure 3.27 is a dimensional plot of Eq. (3.64) for air that illustrates the effect of gap width $L$ and inclination angle $\beta$ on the free-convection coefficient $h$. The quantities $\phi_1$, $\phi_2$, $\phi_3$ are functions of the surface temperatures $T_1$ and $T_2$. For air these functions are

$$\phi_1 = \frac{464}{(\overline{T}_m + 200)^{1/3} \overline{T}_m^{1/2}} \quad (3.66)$$

$$\phi_2 = \left(\frac{T_1 - T_2}{50}\right)^{1/3} \quad (3.67)$$

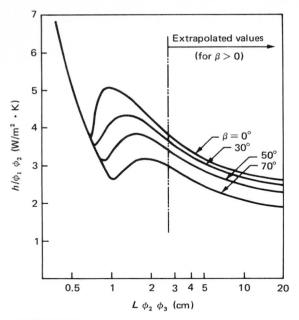

**FIGURE 3.27** Plot of the free convective heat-transfer coefficient across an air gap as a function of the gap spacing $L$. From (19).

$$\phi_3 = \frac{1428(\bar{T}_m + 200)^{2/3}}{\bar{T}_m{}^2}$$ (3.68)

where

$$\bar{T}_m = \frac{T_1 - T_2}{2}$$

To optimize the thermal performance of a flat-plate collector, $\bar{h}_c$ should be as small as possible. A spacing of 5 cm (2 in) appears to be a reasonable compromise, but for more precise design the curves in Fig. 3.27 show that the smallest convective loss corresponds to $L_{\min} = 0.71/(\cos \beta)^{1/3} \phi_2 \phi_3$.

### Free Convection across Inclined Air Layers Constrained by Honeycombs

A review of the state of the art in the application of honeycombs for flat-plate solar collectors is given in Refs. (3, 17), and the theory of the honeycomb collector is presented in Ref. (4). Briefly, the honeycomb is used to suppress free-convection heat transfer across the air gap between a collector plate and its glass cover and to reduce radiation losses from the collector. For an inclined square honeycomb, such as shown in Fig. 3.28, the Nusselt number depends on the Rayleigh number, the inclination, and the aspect ratio of the honeycomb $A_{hc} = L/D$. For the range $0 < \mathrm{Ra} < 6000 \cdot A^4$, $30 < \beta < 90$, and $A = 3$, 4, and 5, the Nusselt number for air is given by

$$\overline{\mathrm{Nu}} = \frac{\overline{h}_c L}{k} = 1 + 0.89 \cos (\beta - 60°) \left(\frac{\mathrm{Ra}}{2420 A^4}\right)^{(2.88 - 1.64 \sin \beta)} \tag{3.69}$$

The above relation may also be used for hexagonal honeycombs if $D$ is replaced by the hydraulic diameter. For engineering design the honeycomb should be chosen to give a Nusselt number of 1.2 according to Hollands (17). For air at atmospheric pressure and moderate temperatures, $370\,\mathrm{K} > T_m > 280\,\mathrm{K}$. The optimum geometry is found from

$$A = C(\beta) \left(1 + \frac{200}{T_m}\right)^{1/2} \left(\frac{100}{T_m}\right)(T_1 - T_2)^{1/4} L^{3/4} \tag{3.70}$$

if $L$ is in cm and $T$ in K. The function $C(\beta)$ is plotted in Fig. 3.29.

### EXAMPLE 3.14

Design a hexagonal honeycomb 5 cm deep to suppress free convection in a solar flat-plate collector operating with an absorber-plate temperature of $T_1 = 375$ K

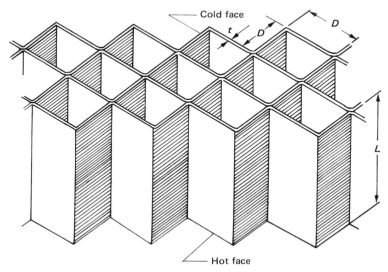

**FIGURE 3.28**  Schematic diagram of honeycomb.

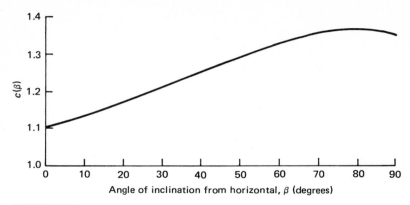

**FIGURE 3.29**  Plot of $C(\beta)$ vs. $\beta$ for use in Eq. (3.71). From (19).

($\sim$200°F) and a glass-cover temperature of 310 K (100°F). The collector is inclined 45°.

**SOLUTION**

From Fig. 3.29, $C(\beta) = 1.27$. Also $T_m = 342$ K, $\Delta T = 65$ K, and $L = 0.05$ m. Thus, the aspect ratio is

$$A = (1.27)(1 + 0.584)^{1/2}(0.292)(65)^{1/4}(5)^{3/4} = 4.44$$

## *Free Convection from Cylinders and Sphere*

The average heat-transfer coefficient by free convection from a horizontal cylinder of diameter $D$ can be calculated from the relation

$$\overline{\mathrm{Nu}_D} = \frac{\bar{h}_c D}{k} = 0.53(\mathrm{Gr}_D\ \mathrm{Pr})^{1/4} \tag{3.71}$$

over a range $10^3 < \mathrm{Gr}_D < 10^9$. This equation also applies to submerged coils.*

The heat-transfer coefficient in free convection to or from a sphere can be obtained from the equation

$$\mathrm{Nu}_D = 2 + 0.45(\mathrm{Gr}_D\ \mathrm{Pr})^{1/4} \tag{3.72}$$

For vertical cylinders Eqs. (3.62) and (3.63) apply.

## *Natural Convection in Annuli*

Very little experimental or theoretical work has been reported applicable to long cylindrical solar collectors other than for annuli formed by concentric circular

---

*For water at 150°F, $\bar{h}_c = 57\ (\Delta T/D)^{1/4}$ in English units.

cylinders. On the basis of flow patterns, temperature profiles, and overall heat-transfer correlations obtained with air, water, and silicon oil contained in annuli of radial gap $L = r_o - r_i$ formed by long isothermal concentric cylinders, the following correlations are recommended (2):

$$Nu = \frac{k_{eff}}{k} = 0.135 \left( \frac{Ra\,Pr}{1.36 + Pr} \right)^{0.278} \tag{3.73}$$

for

$$3.5 \leqslant \log_{10} \left( \frac{Ra\,Pr}{1.36 + Pr} \right) < 8.0 \quad \text{and} \quad 0.25 \leqslant \frac{(r_o - r_i)}{r_i} \leqslant 3.25$$

and

$$Nu = 1 \quad \text{when} \quad \log_{10} \left( \frac{Ra\,Pr}{1.36 + Pr} \right) < 3.0$$

Note that

$$\frac{k_{eff}}{k} = \frac{(dq/dl)\ln(r_o/r_i)}{2\pi k (T_i - T_o)}$$

where $k_{eff} = \bar{h}_c/L$
$dq/dl$ = heat loss per unit length

To properly design cylindrical collectors that have been proposed, additional data must be obtained for nonisothermal concentric and eccentric circular cylinders and nonisothermal noncircular inner cylinders. Until such data are made available, it is recommended that the correlations given by Eq. (3.73) be used. If the inner cylinder is noncircular, one might use the hydraulic radius in the Rayleigh number instead of annulus width $(r_o - r_i)$. An extensive correlation of data for free convection between concentric cylinders is presented in Ref. (31).

## RADIATION HEAT TRANSFER

The material presented in this section has been selected from textbooks on heat transfer or radiation [for example, Refs. (22, 25, 26, 34)] to provide sufficient background in radiation for the thermal design and engineering analysis of solar energy systems. Some material from quantum mechanics and electromagnetic theory is used, but no prior familiarity with these subjects is assumed. However, the treatment of these topics is principally oriented toward the utilization of results to calculate heat transfer by radiation.

All radiation travels at the speed of light, but the actual wavelength depends on the refractive index; the speed of light is equal to the product of the wave~~~~~~

and the frequency of radiation, and this product in turn equals the speed of light in a vacuum divided by the refractive index of the medium through which it travels:

$$c = \lambda \nu = \frac{c_0}{n} \qquad (3.74)$$

where $\lambda$ = wavelength in m (or $\mu$m, 1 $\mu$m = $10^{-6}$ m)
$\quad \nu$ = frequency in $\text{sec}^{-1}$
$\quad c_0$ = speed of light in a vacuum
$\quad n$ = index of refraction of the medium

Thermal radiation is one kind of electromagnetic energy and all bodies emit thermal radiation by virtue of their temperature. When a body is heated, its atoms, molecules, or electrons are raised to higher levels of activity, called excited states. However, they tend to return to lower energy states and in this process energy is emitted in the form of electromagnetic waves. Changes in energy states result from rearrangements in the electronic, rotational, and vibrational states of atoms and molecules. Since these rearrangements involve different amounts of energy changes and these energy changes are related to the frequency, the radiation emitted by a body is distributed over a range of wavelengths. A portion of the electromagnetic spectrum is shown in Fig. 3.30. The wavelengths associated with the various mechanisms are not sharply defined, but thermal radiation is usually considered to fall within the band from about 0.1 to 100 $\mu$m, whereas solar radiation has most of its energy in the range between 0.1 and 3 $\mu$m.

For some problems in solar energy engineering the classical electromagnetic wave theory is not suitable. In such cases, for example, in photovoltaic or photochemical processes, it is necessary to treat the energy transport from the point

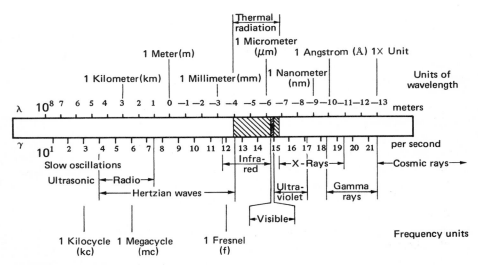

**FIGURE 3.30** Electromagnetic wave spectrum.

of view of quantum mechanics. In this view energy is transported by particles or *photons*, which are treated as energy units or quanta rather than waves. The energy of a photon $E_p$ of frequency $\nu_p$ is

$$E_p = h\nu_p \tag{3.75}$$

where $h$ = Planck's constant ($= 6.625 \times 10^{-34}$ J · sec)

A qualitative physical picture of quantum radiation propagation may be obtained by viewing each quantum as a particle, similar to a molecule in a gas, and the radiation as the transport of a "photon gas" flowing from one place to another. By treating radiation as such a gas, it can be shown from quantum statistical thermodynamics [see, for example, Ref. (20)] that the energy density of radiation per unit volume and per unit wavelength is given by the relation

$$\nu_{b\lambda} = \frac{8\pi h c_0}{(e^{hc_0/\lambda kT} - 1)\lambda^5} \tag{3.76}$$

where $k$ is Boltzmann's constant ($k = 1.38 \times 10^{-23}$ J/molecule · K) and $T$ is the absolute temperature in K (or degrees R). It can also be shown that the energy density at a given wavelength is related to the monochromatic radiation emitted by a perfect radiator, usually called a black body, according to the relation

$$E_{b\lambda} = \frac{\nu_{b\lambda} c_0}{4} = \frac{C_1}{(e^{C_2/\lambda T} - 1)\lambda^5 n^2} \tag{3.77}$$

where $C_1 = 3.74 \times 10^8$ W · $\mu m^4$/$m^2$ ($1.19 \times 10^8$ Btu · $\mu m^4$/hr · $ft^2$)
$C_2 = 1.44 \times 10^4$ $\mu m$ · K ($2.59 \times 10^4$ $\mu m$ · °R)

The quantity $E_{b\lambda}$ has the units of W/$m^2$ · $\mu m$ (Btu/hr · $ft^2$ · $\mu m$) and is called the monochromatic emissive power of a black body, defined as the energy emitted by a perfect radiator per unit wavelength at the specified wavelength per unit area and per unit time at the temperature $T$.

The total energy emitted by a black body can be obtained by integration over all wavelengths. This yields the Stefan-Boltzmann law

$$E_b = \int_0^\infty E_{b\lambda} \, d\lambda = \sigma T^4 \tag{3.78}$$

where $\sigma$ = Stefan-Boltzmann constant
$= 5.67 \times 10^{-8}$ W/$m^2$ · $K^4$ ($0.1714 \times 10^{-8}$ Btu/hr · $ft^2$ · °$R^4$)
$T$ = absolute temperature in K (or °R = 460 + °F)

The concept of a black body, although no such body actually exists in nature, is very convenient in engineering, because its radiation properties can readily be related to those of real bodies. A simple thought experiment will illustrate this in conjunction with definitions of some basic radiation properties.

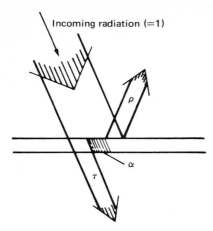

**FIGURE 3.31** Schematic representation of transmittance $\tau$, absorptance $\alpha$, and reflectance $\rho$.

When radiation strikes a body, a part of it is reflected, a part is absorbed, and if the material is transparent, a part is transmitted as shown in Fig. 3.31.

The fraction of the incident radiation reflected is defined as the reflectance $\rho$, the fraction absorbed as the absorptance $\alpha$, and the fraction transmitted as the transmittance $\tau$. According to the first law of thermodynamics these three components must add up to unity, or

$$\alpha + \tau + \rho = 1 \qquad (3.79)$$

Opaque bodies do not transmit any radiation and $\tau = 0$.

The reflection of radiation can be *specular* or *diffuse*. When the angle of incidence is equal to the angle of reflection, the reflection is called specular, whereas when the reflected radiation is uniformly distributed into all directions it is called diffuse (see Fig. 3.32). No real surface is either specular or diffuse, but a highly polished surface approaches specular reflection, whereas a rough surface reflects diffusely.

Using these radiation properties, one can relate the emissive power of a real surface to that of an ideal "black" surface. Suppose an enclosure with perfectly black walls at temperature $T$ contains a body. The enclosure would emit radiation

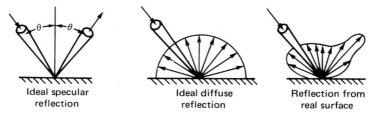

**FIGURE 3.32** Reflections from an ideal specular, an ideal diffuse, and a real surface.

in accordance with Eq. (3.78). After equilibrium is reached, the body inside the enclosure must be at temperature $T$ and the energy emitted by the body must equal the energy it receives. If we let the radiation flux received at some point in the enclosure be $q_i$, equilibrium is established when the radiation emitted is equal to the radiation received, or

$$EA = q_i A \alpha \tag{3.80}$$

If the real body inside the enclosure is now replaced by a black body of the same shape, it too will achieve equilibrium at the temperature of the surrounding walls; since $\alpha_b = 1$, we get

$$E_b A = q_i A (1.0) \tag{3.81}$$

It can now be shown that at a given temperature the absorptance of any body is equal to its emittance. Dividing Eq. (3.80) by Eq. (3.81) one obtains

$$\frac{E(T)}{E_b(T)} = \epsilon(T) = \alpha(T) \tag{3.82}$$

since $E/E_b = \epsilon$ by definition. It should be noted that the absorptance and emittance discussed above are total values for the material. This means that they are representative of the behavior over the entire radiation spectrum.

Not only do real substances emit less than black bodies, but also their emittance varies with wavelength. It is therefore necessary to define monochromatic properties. The monochromatic emittance is defined as the ratio of the monochromatic emissive power of the real body to the monochromatic emissive power of the black body at the same wavelength and temperature, or

$$\epsilon_\lambda = \frac{E_\lambda}{E_{b\lambda}} \tag{3.83}$$

Since the emittance defined by Eq. (3.83) applies to all the radiation emitted into the hemispherical half space above the emitting area, it is, strictly speaking, the hemispherical emittance, but generally the word hemispherical is omitted. The total emittance of a surface $\bar\epsilon$ is related to the monochromatic emittance by the relation

$$\bar\epsilon = \frac{E}{E_b} = \frac{\displaystyle\int_0^\infty \epsilon_\lambda E_{b\lambda}\, d\lambda}{\displaystyle\int_0^\infty E_{b\lambda}\, d\lambda} = \frac{\displaystyle\int_0^\infty \epsilon_\lambda E_{b\lambda}\, d\lambda}{\sigma T^4} \tag{3.84}$$

A special type of surface, called *gray*, is defined as having a monochromatic emittance that does not change with wavelength. For a gray body $\bar{\epsilon} = \epsilon_\lambda = \alpha_\lambda = \bar{\alpha}$. For real surfaces, however, the emittance as well as the absorptance varies with wavelength and to obtain the total values an integration is necessary. The integration is usually performed numerically with the aid of radiation tables discussed below.

## Radiation Function Tables

Engineering calculations of radiative transfer are facilitated by the use of radiation function tables, which present the results of Planck's law in a more convenient form than Eq. (3.77). A plot of the monochromatic emissive power of a black body as a function of wavelength is shown in Fig. 3.33 for temperatures of 400, 1000, and 10,000 K. Notice that the peak of $E_{b\lambda}$ occurs at shorter wavelength as the temperature is increased. These peaks, or inflection points, are uniquely related to the body temperature. By differentiating Planck's distribution law [Eq. (3.77)] and equating to zero, the wavelength corresponding to the maximum value of $E_{b\lambda}$ can be shown to occur when

$$\lambda_{max} T = 2897.8 \ \mu\text{m} \cdot \text{K} \ (5215.6 \ \mu\text{m} \cdot {}^\circ\text{R}) \qquad (3.85)$$

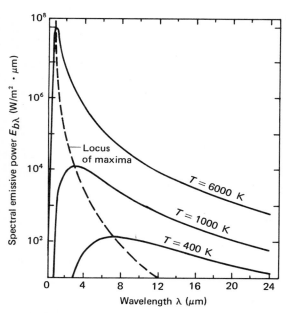

**FIGURE 3.33** Spectral distribution of black-body radiation.

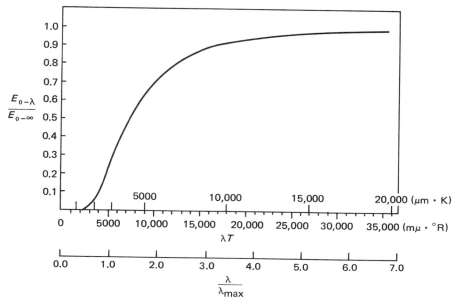

**FIGURE 3.34** Fraction of total emissive power in spectral region between $\lambda = 0$ and $\lambda$ as a function of $\lambda T$ and $\lambda/\lambda_{max}$.

Frequently one needs to know the amount of energy emitted by a black body within a specified range of wavelengths. This type of calculation can be performed easily with the aid of the radiation functions mentioned previously. To construct the appropriate radiation functions in dimensionless form, note that the ratio of the black-body radiation emitted between 0 and $\lambda$ and between 0 and $\infty$ can be made a function of the single variable $(\lambda T)$ by using Eq. (3.77) as shown below (for $n = 1$):

$$\frac{E_{b,0-\lambda}}{E_{b,0-\infty}} = \frac{\int_0^\lambda E_{b\lambda}\, d\lambda}{\sigma T^4} = \int_0^{\lambda T} \frac{C_1 d(\lambda T)}{\sigma(\lambda T)^5 (e^{C_2/\lambda T} - 1)} \qquad (3.86)$$

The above relation is plotted in Fig. 3.34 and the results are also shown in tabular form in Table 3.10. In this table the first column is the ratio of $\lambda$ to $\lambda_{max}$ from Eq. (3.85), the second column the ratio of $E_{b\lambda}$ to $E_{b\lambda,max}$ from Eq. (3.77), and the third column the ratio $E_{b,0-\lambda}$ to $\sigma T^4$ from Eq. (3.86). For use on a digital computer Eq. (3.86) can be approximated by the following polynomials:

$$\nu \geqslant 2 \qquad \frac{E_{b,0-\lambda}}{\sigma T^4} = \frac{15}{\pi^4} \sum_{m = 1, 2, \ldots} \frac{e^{-m\nu}}{m^4} \{[(m\nu + 3)m\nu + 6] m\nu + 6\} \qquad (3.87)$$

**TABLE 3.10**  Thermal Radiation Functions[a]

| $\lambda/\lambda_{max}$ | $\dfrac{E_{b\lambda}}{E_{b\lambda,\,max}}$ | $\dfrac{E_{b\lambda,0-\lambda}}{\sigma T^4}$ | $\lambda/\lambda_{max}$ | $\dfrac{E_{b\lambda}}{E_{b\lambda,\,max}}$ | $\dfrac{E_{b\lambda,0-\lambda}}{\sigma T^4}$ | $\lambda/\lambda_{max}$ | $\dfrac{E_{b\lambda}}{E_{b\lambda,\,max}}$ | $\dfrac{E_{b\lambda,0-\lambda}}{\sigma T^4}$ |
|---|---|---|---|---|---|---|---|---|
| 0.00 | 0.0000 | 0.0000 | 1.50 | 0.7103 | 0.5403 | 2.85 | 0.1607 | 0.8661 |
| 0.20 | 0.0000 | 0.0000 | 1.55 | 0.6737 | 0.5630 | 2.90 | 0.1528 | 0.8713 |
| 0.25 | 0.0003 | 0.0000 | 1.60 | 0.6382 | 0.5846 | 2.95 | 0.1454 | 0.8762 |
| 0.30 | 0.0038 | 0.0001 | 1.65 | 0.6039 | 0.6050 | 3.00 | 0.1384 | 0.8809 |
| 0.35 | 0.0187 | 0.0004 | 1.70 | 0.5710 | 0.6243 | 3.10 | 0.1255 | 0.8895 |
| 0.40 | 0.0565 | 0.0015 | 1.75 | 0.5397 | 0.6426 | 3.20 | 0.1141 | 0.8974 |
| 0.45 | 0.1246 | 0.0044 | 1.80 | 0.5098 | 0.6598 | 3.30 | 0.1038 | 0.9045 |
| 0.50 | 0.2217 | 0.0101 | 1.85 | 0.4815 | 0.6761 | 3.40 | 0.0947 | 0.9111 |
| 0.55 | 0.3396 | 0.0192 | 1.90 | 0.4546 | 0.6915 | 3.50 | 0.0865 | 0.9170 |
| 0.60 | 0.4664 | 0.0325 | 1.95 | 0.4293 | 0.7060 | 3.60 | 0.0792 | 0.9225 |
| 0.65 | 0.5909 | 0.0499 | 2.00 | 0.4054 | 0.7197 | 3.70 | 0.0726 | 0.9275 |
| 0.70 | 0.7042 | 0.0712 | 2.05 | 0.3828 | 0.7327 | 3.80 | 0.0667 | 0.9320 |
| 0.75 | 0.8007 | 0.0960 | 2.10 | 0.3616 | 0.7449 | 3.90 | 0.0613 | 0.9362 |
| 0.80 | 0.8776 | 0.1236 | 2.15 | 0.3416 | 0.7565 | 4.00 | 0.0565 | 0.9401 |
| 0.85 | 0.9345 | 0.1535 | 2.20 | 0.3229 | 0.7674 | 4.20 | 0.0482 | 0.9470 |
| 0.90 | 0.9725 | 0.1849 | 2.25 | 0.3052 | 0.7777 | 4.40 | 0.0413 | 0.9528 |
| 0.95 | 0.9936 | 0.2172 | 2.30 | 0.2887 | 0.7875 | 4.60 | 0.0356 | 0.9579 |
| 1.00 | 1.0000 | 0.2501 | 2.35 | 0.2731 | 0.7967 | 4.80 | 0.0308 | 0.9622 |
| 1.05 | 0.9944 | 0.2829 | 2.40 | 0.2585 | 0.8054 | 5.00 | 0.0268 | 0.9660 |
| 1.10 | 0.9791 | 0.3153 | 2.45 | 0.2447 | 0.8137 | 6.00 | 0.0142 | 0.9790 |
| 1.15 | 0.9562 | 0.3472 | 2.50 | 0.2318 | 0.8215 | 7.00 | 0.0082 | 0.9861 |
| 1.20 | 0.9277 | 0.3782 | 2.55 | 0.2197 | 0.8290 | 8.00 | 0.0050 | 0.9904 |
| 1.25 | 0.8952 | 0.4081 | 2.60 | 0.2083 | 0.8360 | 9.00 | 0.0033 | 0.9930 |
| 1.30 | 0.8600 | 0.4370 | 2.65 | 0.1976 | 0.8427 | 10.00 | 0.0022 | 0.9948 |
| 1.35 | 0.8231 | 0.4647 | 2.70 | 0.1875 | 0.8490 | 20.00 | 0.0002 | 0.9993 |
| 1.40 | 0.7854 | 0.4911 | 2.75 | 0.1780 | 0.8550 | 40.00 | 0.0000 | 0.9999 |
| 1.45 | 0.7477 | 0.5163 | 2.80 | 0.1691 | 0.8607 | 50.00 | 0.0000 | 1.0000 |

[a]
$\lambda$ = wavelength in $\mu$m
$\lambda_{max}$ = wavelength at $E_{b\lambda,\,max}$ in $\mu$m = $2898/T$
$E_{b\lambda}$ = monochromatic emissive power in W/m$^2 \cdot \mu$m
       = $374.15 \times 10^6/\lambda^5$ [exp $(14,387.9/\lambda T) - 1$]
$E_{b\lambda,\,max}$ = maximum monochromatic emissive power in W/m$^2 \cdot \mu$m
       = $12.865 \times 10^{-12}\,T^5$
$E_{b\lambda,0-\lambda}$ = $\int_0^\lambda E_{b\lambda}\,d\lambda$
     $\sigma T^4$ = $E_{b\lambda,0-\infty}$ = $5.670 \times 10^{-8}\,T^4$ W/m$^2$
     $T$ = absolute temperature in K

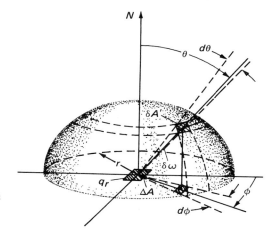

**FIGURE 3.35** Schematic diagram illustrating radiation intensity and flux.

$$v < 2 \quad \frac{E_{b,\,0-\lambda}}{\sigma T^4} = \frac{15}{1-\pi^4}\, v^3 \left( \frac{1}{3} - \frac{v}{8} - \frac{v^2}{60} - \frac{v^4}{5040} + \frac{v^6}{272,160} - \frac{v^8}{13,305,600} \right)$$

$$(3.87)$$
$$\text{(cont.)}$$

where $v = C_2/\lambda T$

## Intensity of Radiation and Shape Factor

The emissive power of a surface gives the total radiation emitted in all directions. To determine the radiation emitted in a given direction we must define another quantity, the radiation intensity $I$. It is defined as the radiant energy passing through an imaginary plane in space per unit area per unit time and per unit solid-angle perpendicular to the plane as shown in Fig. 3.35. $I$ is defined by the relation

$$I = \lim_{\substack{dA' \to 0 \\ d\omega \to 0}} \frac{dE}{dA'\,d\omega} \qquad (3.88)$$

Radiation intensity has both magnitude and direction. It can be related to the radiation flux, defined as the radiant energy passing through an imaginary plane per unit area per unit time in all directions. Note that whereas for the intensity the area $dA'$ is perpendicular to the direction of the radiation, for the flux the area $dA$ is at the base in the center of a hemisphere through which all of the radiation passes. Recalling that the definition for the solid angle between $dA'$ and $dA$ is $d\omega = dA'/r^2$, the radiation flux $q_r$ emanating from $dA$ can be obtained by integrating the intensity over the hemisphere. As shown in Fig. 3.35, the unit projected area for $I$ is $dA \cos \theta$ and the differential area $dA'$ on the hemisphere is $r^2 \sin \theta \, d\theta \, d\phi$; thus

$$q_r = \int_0^{2\pi} \int_0^{\pi/2} I \cos \theta \sin \theta \; d\theta \; d\phi \qquad (3.89)$$

If the area $dA$ is located on a surface, the emissive power $E$ can also be obtained from Eq. (3.89). For the special case of a diffuse surface, for which $I$ is the same in all directions, Eq. (3.89) gives

$$q_r = \pi I \qquad (3.90)$$

Since all black surfaces are diffuse,

$$E_b = \pi I_b \qquad (3.91)$$

Equation (3.91) can, of course, also be written for monochromatic radiation as well as for total radiation, or

$$E_{b\lambda} = \pi I_{b\lambda} \qquad (3.92)$$

In the evaluation of the rate of radiation heat transfer between two surfaces not only their temperatures and their radiation properties but also their geometric configurations and relationships play a part. The influence of geometry in radiation heat transfer can be expressed in terms of the *radiation shape factor* between any two surfaces 1 and 2 defined as follows:

$F_{1-2}$ = fraction of radiation leaving surface 1 that reaches surface 2
$F_{2-1}$ = fraction of radiation leaving surface 2 that reaches surface 1

In general, $F_{m-n}$ = fraction of radiation leaving surface $m$ that reaches surface $n$. If both surfaces are black, the energy leaving surface $m$ and arriving at surface $n$ is $E_{bm}A_m F_{m-n}$ and the energy leaving surface $n$ and arriving at $m$ is $E_{bn}A_n F_{n-m}$. If both surfaces absorb all the incident energy, the net rate of exchange $q_{m \rightleftharpoons n}$ will be

$$q_{m \rightleftharpoons n} = E_{bm}A_m F_{m-n} - E_{bn}A_n F_{n-m} \qquad (3.93)$$

If both surfaces are at the same temperature, $E_{bm} = E_{bn}$ and the net exchange is zero, $q_{m \rightleftharpoons n} \equiv 0$. This shows that the geometric radiation shape factor must obey the reciprocity relation

$$A_m F_{m-n} = A_n F_{n-m} \qquad (3.94)$$

The net rate of heat transfer can therefore be written in two equivalent forms:

$$q_{m \rightleftharpoons n} = A_m F_{m-n}(E_{bm} - E_{bn}) = A_n F_{n-m}(E_{bm} - E_{bn}) \qquad (3.95)$$

The evaluation of geometric shape factors is in general quite involved. For a majority of solar energy applications, however, only a few special cases are of interest. One of these is a small convex object of area $A_1$ surrounded by a large enclosure $A_2$. Since all radiation leaving $A_1$ is intercepted by $A_2$, $F_{1-2} = 1$ and $F_{2-1} = A_1/A_2$.

Another case is the exchange of radiation between two large parallel surfaces. If the two surfaces are near each other, almost all of the radiation leaving $A_1$ reaches $A_2$ and vice versa. Thus, $F_{1-2} = F_{2-1} = 1.0$, according to the definition of the shape factor. A third case of importance is the exchange between a small surface $\Delta A_1$ and a portion of space $A_2$, for example, a flat-plate collector tilted at an angle $\beta$ from the horizontal and the sky it can see. For this situation we refer to the definition of radiation flux (see Fig. 3.35). The portion of the radiation emitted by $\Delta A_1$ that is intercepted by the surrounding hemisphere depends on the angle of tilt. When the surface is horizontal, $F_{1-2} = 1$; when it is vertical, $F_{1-2} = 1/2(\beta = 90°)$. For intermediate values it can be shown that (34)

$$F_{1-2} = \tfrac{1}{2}(1 + \cos \beta) = \cos^2\left(\frac{\beta}{2}\right)$$
(3.96)

If the diffuse sky radiation is uniformly distributed and assumed to be black, then a small black area $A_1$ receives radiation at the rate

$$A_1 F_{1-\text{sky}} E_{\text{sky}} = \frac{A_1}{2}(1 + \cos \beta)\sigma T_{\text{sky}}^4$$
(3.97)

whereas the net radiation heat transfer is given by

$$q_{\text{sky} \rightleftharpoons 1} = A_1 F_{1-\text{sky}} \sigma(T_{\text{sky}}^4 - T_1{}^4)$$
(3.98)

If the receiving area is gray with an absorptance $\bar{\alpha}$ equal to the emittance $\bar{\epsilon}$, the net exchange is given by

$$q_{\text{sky} \rightleftharpoons 1} = A_1 F_{1-\text{sky}} \bar{\alpha}\sigma(T_{\text{sky}}^4 - T_1{}^4)$$
(3.99)

A simple method to evaluate shape factors for two-dimensional surfaces that are infinitely long in one direction and have a uniform cross section normal to the infinite direction has been developed by Hottel and Sarofim (21). The method, called the *crossed-string* method, has applications to trough-type concentrators, which will be discussed in Chapter 4. Two surfaces that satisfy these geometric restrictions are shown in Fig. 3.36. According to this method, the shape factor $F_{1 \rightarrow 2}$ is equal to the sum of the lengths of the cross strings stretched between the ends of the two surfaces minus the sum of the uncrossed strings divided by twice the length of $A_1$. The shape factor $F_{1 \rightarrow 2}$ is therefore

$$F_{1 \rightarrow 2} = \frac{1}{2L_1}[(\overline{ac} + \overline{bd}) - (\overline{ab} + \overline{cd})]$$
(3.100)

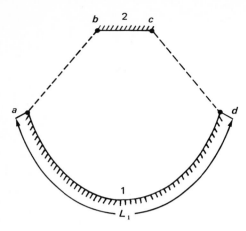

**FIGURE 3.36** Schematic sketch illustrating crossed-string method of evaluating shape factor.

## Radiation Exchange between Two Gray Parallel Flat Plates

The net rate of radiation heat transfer between two gray parallel flat plates $A_1$ and $A_2$ with emittances $\epsilon_1$ and $\epsilon_2$, respectively, can be obtained by tracing the radiation emitted by each of the two surfaces. Surface $A_1$ emits radiation at the rate $\epsilon_1 A_1 \sigma T_1^4$ and since $F_{1-2} = 1$, all of it reaches $A_2$ (see Fig. 3.37), where $\alpha_2 \epsilon_1 A_1 \sigma T_1^4$ of the radiation is absorbed and $(1-\alpha_2)\epsilon_1 A_1 \sigma T_1^4$ is reflected back to $A_1$. Noting that $\epsilon_2 = \alpha_2$ for gray surfaces, a continued tracing of the radiation to obtain the total amount absorbed by $A_2$ gives

$$q_{1\to2} = A_1 \sigma T_1^4 \left[ \epsilon_1 \epsilon_2 + \epsilon_1 \epsilon_2 (1-\epsilon_1)(1-\epsilon_2) + \epsilon_1 \epsilon_2 (1-\epsilon_1)^2(1-\epsilon_2)^2 + \ldots \right]$$

$$(3.101)$$

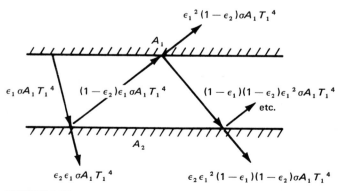

**FIGURE 3.37** Schematic sketch of interreflections between two parallel gray surfaces $A_1$ and $A_2$ at temperatures $T_1$ and $T_2$.

which is a geometric series (9) whose sum is

$$q_{1\to 2} = A_1 \sigma T_1{}^4 \frac{\epsilon_1 \epsilon_2}{1 - (1 - \epsilon_1)(1 - \epsilon_2)} = A_1 \sigma T_1{}^4 \frac{1}{1/\epsilon_1 + 1/\epsilon_2 - 1} \quad (3.102)$$

Repeating the radiation trace for the radiation emanating from $A_2$ and absorbed by $A_1$ gives

$$q_{2\to 1} = A_2 \sigma T_2{}^4 \frac{1}{1/\epsilon_1 + 1/\epsilon_2 - 1}$$

so that the net rate of heat transfer between the two surfaces becomes

$$q_{1\rightleftharpoons 2} = \frac{A\sigma(T_1{}^4 - T_2{}^4)}{1/\epsilon_1 + 1/\epsilon_2 - 1} \quad (3.103)$$

Equation (3.103) applies also to radiation between two concentric long cylinders forming an annulus when the diameter ratio approaches unity and applies generally to radiation across an annulus when the reflection is specular rather than diffuse on both surfaces or on the outer surface alone (22).

It is apparent from Eq. (3.78) and the ensuing discussion that the rate of heat transfer by radiation is not a linear function of the temperature difference, but rather of the difference in the fourth power of the surfaces exchanging radiation. However, if the temperatures of the surfaces exchanging radiation are not very high and the temperature difference between them relatively small, say less than 20 K, Eq. (3.103) can be linearized and radiation heat transfer can be incorporated into the thermal circuit concept. To linearize Eq. (3.103) note that

$$T_1{}^4 - T_2{}^4 = (T_1{}^2 - T_2{}^2)(T_1{}^2 + T_2{}^2) = (T_1 - T_2)(T_1 + T_2)(T_1{}^2 + T_2{}^2)$$

Using this equality, Eq. (3.103) can be expressed in the form

$$q_{1\rightleftharpoons 2} = \frac{T_1 - T_2}{R_r} \quad (3.104)$$

where

$$R_r = \frac{1/\epsilon_1 + 1/\epsilon_2 - 1}{A_1 \sigma(T_1 + T_2)(T_1{}^2 + T_2{}^2)} \quad *$$

---

*A good approximation for conditions when $2(T_1 - T_2)/(T_1 + T_2) < 0.7$ is $T_1{}^4 - T_2{}^4$ $\cong 4[(T_1 + T_2)/2]^3 (T_1 - T_2)$.

### EXAMPLE 3.15

Calculate the rate of heat transfer between the surface of a flat-plate collector ($\epsilon_1 = 0.95$, $T_1 = 360$ K) and the glass cover ($\epsilon_2 = 0.90$, $T_2 = 340$ K) if the convection heat-transfer coefficient is 10 W/m$^2 \cdot$ K (1.76 Btu/hr $\cdot$ ft$^2 \cdot$ °F).

### SOLUTION

Radiation and convection occur in parallel between $T_1$ and $T_2$. We obtain the convection resistance from Eq. (3.43)

$$R_c = \frac{1}{Ah_c} = 0.1 \text{ m}^2 \cdot \text{K/W (0.568 hr} \cdot \text{°F/Btu)}$$

and from Eq. (3.104) the radiation resistance

$$R_r = \frac{1/\epsilon_1 + 1/\epsilon_2 - 1}{\sigma(T_1 + T_2)(T_1{}^2 + T_2{}^2)}$$

$$= \frac{(1.11 + 1.052 - 1) \times 10^8}{5.67(340 + 360)(1.15 + 1.29) \times 10^5}$$

$$= 0.120 \text{ m}^2 \cdot \text{K/W (0.680 hr} \cdot \text{°F/Btu)}$$

Thus

$$\frac{q}{A} = \frac{(T_1 - T_2)/R_1 R_2}{R_1 + R_2} = \frac{20/0.012}{0.1 + 0.120} = 367 \text{ W/m}^2 \text{ (116 Btu/hr} \cdot \text{ft}^2)$$

In the foregoing example the temperature of both surfaces exchanging radiation was specified. In many practical problems one of the temperatures is not given directly, as for example in a flat-plate collector when the collector surface temperature and the environmental air temperature are specified, but the cover glass temperature is unknown. This situation is illustrated in Fig. 3.38.

In this case it is necessary to assume a glass temperature $T_{g1}$, estimate $R_r$ with this assumption, and then recalculate the thermal resistance $R_2$ if it turns out that the value initially assumed was greatly in error. It should be noted, however, that if most of the heat between $T_{g1}$ and $T_{g2}$ is transferred by convection, an error of 5 or 10 percent in $R_r$ will not introduce a serious error in the net rate of heat transfer. In fact, as will be illustrated below, the lack of accuracy of radiation characteristics of real surfaces is often the source of the far greatest uncertainty in the thermal performance prediction by analytical methods. The analytical approach illustrated in this section assumed that all surfaces are gray and diffuse. This is a simplification usually made in engineering design, but cannot be accepted in all

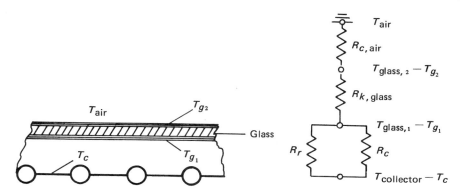

**FIGURE 3.38** Thermal circuit for heat loss from a flat-plate collector with a simple glass cover.

cases. The following discussion of radiation characteristics of real surface will give a more complete picture.

## RADIATION CHARACTERISTICS AND PROPERTIES OF MATERIALS

An understanding of the radiation properties and characteristics of materials is important for the design and analysis of all solar systems. This is a difficult subject and involves concepts that must be carefully defined. The names and symbols used have been selected to facilitate relating a concept to a physical mechanism.

### Opaque Materials

To begin with one must deal with monochromatic characteristics and overall characteristics. Also, one often needs to be aware of the directional effects as compared to total hemispherical characteristics. This can be illustrated by considering the emittance of a surface. The most complete specification is in terms of the monochromatic directional emittance of a surface defined as the ratio of its intensity at a given wavelength in a given direction to the intensity of a black body at the same temperature and wavelength. If the direction is specified by an azimuth angle $\phi$ and a polar angle $\theta$, as shown in Fig. 3.35, the monochromatic directional emittance is

$$\epsilon_\lambda(\theta,\phi) = \frac{I_\lambda(\theta,\phi)}{I_{b\lambda}} \tag{3.105}$$

The overall directional emittance can be defined by the relation

$$\epsilon(\theta,\phi) = \frac{\int_0^\infty \epsilon_\lambda(\theta,\phi) I_{b\lambda}\, d\lambda}{\int_0^\infty I_{b\lambda}\, d\lambda} = \frac{1}{E_b}\int_0^\infty \epsilon_\lambda(\theta,\phi) I_{b\lambda}\, d\lambda \qquad (3.106)$$

The monochromatic hemispherical emittance can be obtained by integration of the beam radiation emittance specified by Eq. (3.105) over the entire hemispherical space into which radiant energy is emitted:

$$\epsilon_\lambda = \frac{\int_0^{2\pi}\int_0^1 \epsilon_\lambda(\mu,\phi) I_{b\lambda}\mu\, d\mu\, d\phi}{\int_0^{2\pi}\int_0^1 I_{b\lambda}\mu\, d\mu\, d\phi} = \frac{1}{\pi}\int_0^{2\pi}\int_0^1 \epsilon_\lambda(\mu,\phi)\mu\, d\mu\, d\phi \qquad (3.107)$$

where for simplicity we let $\cos\theta = \mu$; the overall emittance is then

$$\epsilon = \frac{1}{\sigma T^4}\int_0^\infty \epsilon_\lambda E_{b\lambda}\, d\lambda \qquad (3.108)$$

Observe that both $\epsilon_\lambda$ and $\epsilon$ are properties of the surface.

The next most important surface characteristic is the absorptance. We begin by defining the monochromatic directional absorptance as the fraction of the incident radiation at wavelength $\lambda$ *from* the direction $\theta,\phi$ that is absorbed, or

$$\alpha_\lambda(\theta,\phi) = \frac{I_{\lambda,\,a}(\theta,\phi)}{I_{\lambda,\,i}(\theta,\phi)} \qquad (3.109)$$

where the subscripts $a$ and $i$ denote absorbed and incident, respectively. The monochromatic directional absorptance is also a property of the surface.

More important than $\alpha_\lambda(\theta,\phi)$ is the overall directional absorptance $\alpha(\theta,\phi)$, defined as the fraction of the total radiation from the direction $\theta,\phi$ that is absorbed, or

$$\alpha(\theta,\phi) = \frac{\int_0^\infty \alpha_\lambda(\theta,\phi) I_{\lambda,\,i}(\theta,\phi)\, d\lambda}{\int_0^\infty I_{\lambda,\,i}(\theta,\phi)\, d\lambda} = \frac{1}{I_i(\theta,\phi)}\int_0^\infty \alpha_\lambda(\theta,\phi) I_{\lambda,\,i}(\theta,\phi)\, d\lambda \qquad (3.110)$$

The overall absorptance is a function of the characteristics of the incident radiation and is, therefore, unlike the monochromatic absorptance, not a property of a surface alone. It is this characteristic that makes it possible to have selective surfaces that absorb the radiation from one source at a higher rate than from another. In other words, even though according to Kirchhoff's law the mono-chromatic emittance at $\lambda$ must equal the monochromatic absorptance, i.e.,

$$\alpha_\lambda(\theta, \phi) = \epsilon_\lambda(\theta, \phi) \tag{3.111}$$

the overall emittance is not necessarily equal to the overall absorptance unless thermal equilibrium exists, and the incoming and outgoing radiation have the same spectral characteristics.

The hemispherical monochromatic absorptance is defined by

$$\alpha_\lambda = \frac{\displaystyle\int_0^{2\pi}\int_0^1 \alpha_\lambda(\mu,\theta)I_{\lambda,i}(\mu,\theta)\mu\,d\mu\,d\theta}{\displaystyle\int_0^{2\pi}\int_0^1 I_{\lambda,i}\mu\,d\mu\,d\phi} \tag{3.112}$$

and the hemispherical overall absorptance as

$$\alpha = \frac{\displaystyle\int_0^\infty\int_0^{2\pi}\int_0^1 \alpha_\lambda(\mu,\theta)I_{\lambda,i}(\mu,\theta)\mu\,d\mu\,d\theta\,d\lambda}{\displaystyle\int_0^\infty\int_0^{2\pi}\int_0^1 I_{\lambda,i}\mu\,d\mu\,\bar{d}\phi\,d\lambda} \tag{3.113}$$

The effect of incidence angle on the absorptance is illustrated in Table 3.11

**TABLE 3.11**  Angular Variation of Absorptance of Lampblack Paint[a]

| Incidence angle $i$ | Absorptance $\alpha(i)$ |
| --- | --- |
| 0–30 | 0.96 |
| 30–40 | 0.95 |
| 40–50 | 0.93 |
| 50–60 | 0.91 |
| 60–70 | 0.88 |
| 70–80 | 0.81 |
| 80–90 | 0.66 |

[a]Adapted from Löf and Tybout (1972).

where the angular variation of the absorptance for a nonselective black surface, typical of those used on flat-plate collectors, is shown. The absorptance of this surface for diffuse radiation is approximately 0.90.

The third characteristic to be considered is the reflectance. The reflectance is particularly important for the design of focusing collectors. As mentioned previously, there are two limiting types of reflection: specular and diffuse. As illustrated in Fig. 3.32$a$, when a ray of incident radiation at angle $\theta$ is reflected at the same polar angle and the azimuthal angles differ by $180°$, as for a perfect mirror, the reflection is said to be specular. The reflection is said to be diffuse if the incident radiation is scattered equally in all directions, as shown in Fig. 3.32$b$. For a complete specification of the reflectance it is necessary to specify also its dependence on the wavelength and the spatial distribution of the incident radiation. This characteristic is called the monochromatic biangular reflectance, defined as $\pi$ times the ratio of the intensity reflected in the direction $\theta_r$, $\phi_r$ to the incoming intensity from the direction $\theta_i$, $\phi_i$. As illustrated in Fig. 3.39, $\rho_\lambda(\theta_r, \phi_r, \theta_i, \phi_i)$ is

$$\rho_\lambda(\theta_r, \phi_r, \theta_i, \phi_i) = \pi \frac{I_{\lambda,r}(\theta_r, \phi_r)}{I_{\lambda,i}(\theta_i, \phi_i)} \qquad (3.114)$$

Although conceptually simple, the biangular reflectance is difficult to measure, because the amount of energy within a small solid angle is extremely small. For most practical engineering applications another kind of reflectance is more useful, because it can be related to a widely used experimental technique (13). This is the *monochromatic angular-hemispherical reflectance*, $\rho_\lambda(\theta_i, \phi_i)$, defined as the ratio of the monochromatic radiant energy reflected in all directions to the incident radiant

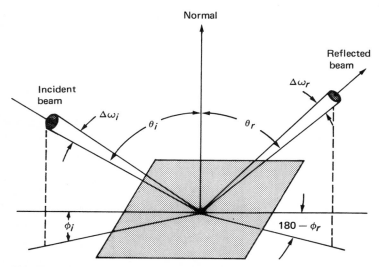

**FIGURE 3.39** Coordinate system for reflectance definitions.

flux within a solid angle $\Delta\omega_i$. From Eq. (3.114) the incident energy at $\lambda$ that is reflected in all directions is

$$q_{\lambda,r} = \frac{1}{\pi} \int_0^{2\pi} \int_0^1 \rho_\lambda(\mu_r, \phi_r, \mu_i, \phi_i) I_{\lambda,i} \mu_i \, \Delta\omega_i \mu_r \, d\mu_r \, d\phi_r \qquad (3.115)$$

Thus, the monochromatic angular-hemispherical reflectance can be expressed as

$$\rho_\lambda(\theta_i, \phi_i) = \frac{q_{\lambda,r}}{I_{\lambda,i}\mu_i \, \Delta\omega_i} = \frac{1}{\pi} \int_0^{2\pi} \int_0^1 \rho_\lambda(\mu_r, \phi_r, \mu_i, \phi_i)\mu_r \, d\mu_r \, d\phi_r \qquad (3.116)$$

Since the integral defining $\rho_\lambda(\theta_i, \phi_i)$ does not contain a specification of the radiation source, the monochromatic angular-hemispherical reflectance is a surface property. It should also be noted that since the direct solar flux is contained within a small solid angle ($\cong 32$ min), $\rho_\lambda(\theta_i, \phi_i)$ is the property that specifies the monochromatic reflectance of solar radiation into space. On the other hand, the total angular-hemispherical reflectance obtained by integrating Eq. (3.116) over all wavelengths is source dependent.

The last kind of reflectance, called the monochromatic hemispherical reflectance $\rho_\lambda(\theta_r, \phi_r)$, is defined as $\pi$ times the ratio of the reflected monochromatic intensity in the direction $\theta_r$, $\phi_r$ to the monochromatic irradiation from all directions, or

$$\rho_\lambda(\theta_r, \phi_r) = \pi \frac{I_{\lambda,r}(\theta_r, \phi_r)}{q_{\lambda,i}} = \pi \frac{I_{\lambda,r}(\theta_r, \phi_r)}{\displaystyle\int_0^{2\pi} \int_0^1 I_{\lambda,i}\mu_i \, d\mu_i \, d\phi_i} \qquad (3.117)$$

It can be shown, however (22), that when the incident radiation is diffuse, the monochromatic hemispherical-angular reflectance is identical to the monochromatic angular-hemispherical reflectance. This equality is of great practical importance, because it is much easier to measure $\rho_\lambda(\theta_r, \phi_r)$ than $\rho_\lambda(\theta_i, \phi_i)$. By placing a sample into an isothermal cavity, its surface is uniformly irradiated. If a small opening is provided, as shown in Fig. 3.40, the radiant energy reflected from the sample can be determined with an instrument capable of measuring radiation over a small wavelength interval within a small, solid angle. By comparing the radiation reflected by the sample with the black-body radiation emanating from the cavity, the reflectance can be obtained with a high degree of precision. The method is described in detail in Refs. (9, 11, 13).

The monochromatic angular-hemispherical reflectance is also significant, because once this property is known, all emittance and absorptance characteristics of opaque surfaces can be calculated. This can be shown easily by considering a body in an

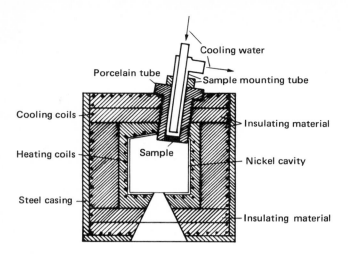

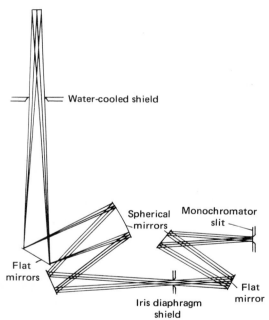

**FIGURE 3.40** Cross section of heated hohlraum and schematic view of optical system.

isothermal enclosure at temperature $T$. Its monochromatic radiation intensity, consisting of the emitted and reflected radiation, must equal the black-body intensity at temperature $T$, or

$$I_{b,\lambda}(T) = \epsilon_\lambda(\theta, \phi)I_{b,\lambda} + \rho(\theta, \phi)I_{b,\lambda} \qquad (3.118)$$

But from Kirchhoff's law $\epsilon_\lambda(\theta, \phi) = \alpha_\lambda(\theta, \phi)$ so that

$$\epsilon_\lambda(\theta, \phi) = 1 - \rho_\lambda(\theta, \phi) = \alpha_\lambda(\theta, \phi) \tag{3.119}$$

Table 3.12 gives total hemispherical emittance at various temperatures and the total normal solar absorptance at room temperature for a few common materials. A more complete list has been compiled by Touloukian et al. (38), who list also angular properties.

For simplified engineering estimates a hemispherical reflectance independent of wavelength (gray-body approximation) and independent of direction (diffuse approximation) is often used. The preceding discussion, however, should make it obvious that appreciable errors may be introduced if this approach is taken without examining the appropriateness of the simplifying assumptions.

## Selective and Reflecting Surfaces

Two types of special surfaces of great importance in solar collector systems are selective and reflecting surfaces. Selective surfaces combine a high absorptance for solar radiation with a low emittance for the temperature range in which the surface emits radiation. This combination of surface characteristics is possible because 98 percent of the energy in incoming solar radiation is contained within wavelengths below 3 $\mu$m, whereas 99 percent of the radiation emitted by black or gray surfaces at 400 K is at wavelengths longer than 3 $\mu$m. As will be shown in Chapter 4, such a combination will give a good collector performance and permit operation at high temperatures. Thus, an ideal selective surface for a collector should have a very high monochromatic reflectance at wavelengths above 3 $\mu$m and a very low reflectance below 3 $\mu$m. The dotted line in Fig. 3.41 illustrates the spectral characteristics of an ideal selective semi-gray surface having a uniform reflectance of 0.05 below 3 $\mu$m,

**TABLE 3.12** Total Hemispherical Emittance and Normal Solar Absorptance of Various Surfaces[a]

| Material | Emittance (temperature in K) | Absorptance |
|---|---|---|
| Aluminum, polished | 0.102 (573), 0.130 (773), 0.113 (873) | 0.09–0.10 |
| Aluminum, anodized | 0.842 (296), 0.720 (484), 0.669 (574) | 0.12–0.16 |
| Aluminum with SiO$_2$ coating | 0.366 (263), 0.384 (293), 0.378 (324) | 0.11 |
| Carbon black in acrylic binder | 0.83 (at 278 K) | 0.94 |
| Copper, polished | 0.041 (338), 0.036 (463), 0.039 (803) | 0.35 |
| Gold | 0.025 (275), 0.040 (468), 0.048 (668) | 0.20–0.23 |
| Iron | 0.071 (199), 0.110 (468), 0.175 (668) | 0.44 |
| Magnesium oxide | 0.73 (380), 0.68 (491), 0.53 (755) | 0.14 |
| Nickel | 0.10 (310), 0.10 (468), 0.12 (668) | 0.36–0.43 |
| Paints | | |
|   Parsons black | 0.981 (240), 0.981 (462) | 0.98 |
|   Acrylic white | 0.90 (298) | 0.26 |
|   White (ZnO) | 0.929 (295), 0.926 (478), 0.889 (646) | 0.12–0.18 |

[a]Abstracted from (38).

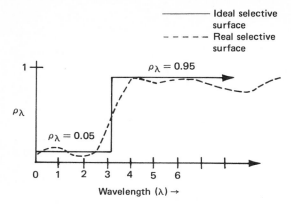

**FIGURE 3.41**  Ideal selective surface characteristics.

but 0.95 above 3 $\mu$m. Real surfaces do not approach this performance. The monochromatic reflectances for two real selective surfaces, typical of those used on commercially available solar collectors, are shown in Fig. 3.42 by the solid and the dotted lines on the same graph. Curve A is for a so-called "Tabor" surface (37) obtained by a special chemical treatment of a galvanized iron and curve B for two layers of black nickel on nickel-coated steel.

For practical applications a good selective surface must not only have the basic spectral characteristics shown in Fig. 3.42, but also be durable, cheap, and replaceable, and it should not chemically decompose if in contact with the working fluid. Also, the spectral selectivity should, if possible, improve with reduced angles

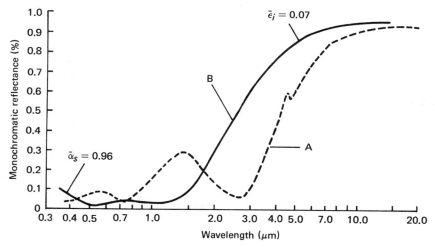

**FIGURE 3.42**  Spectral reflectances of (A) Tabor solar collector surface, chemical treatment of galvanized iron (12) and (B) a two-layer black nickel on nickel-coated steel (28).

of incidence. There is obviously no surface or coating that meets all these requirements, but Table 3.13 gives a summary of the performance, cost, availability, and operating experience for some available selective surfaces. Considerable research effort is being devoted to improve the performance of selective surfaces and the reader is urged to follow developments in this field.

In order to increase the heat flux on a solar collector the incoming solar energy must be concentrated. This requires the use of a reflecting surface with high specular reflectance in the solar spectrum or a refracting device with a high transmittance in the solar spectrum. Reflecting surfaces are usually highly polished metals or metal coatings on suitable substrates. With opaque substrates, the reflective coating must always be front-surfaced, for example, chrome plate on copper or polished aluminum, but if a transparent substrate is used, the coating may be front- or back-surfaced. In any back-surfaced reflector the radiation must pass through the substrate twice and the transmittance of the material becomes very important. It should also be noted that the specular reflectance of most surfaces is wavelength dependent, and to obtain an average value of the specular reflectance $\bar{\rho}_s$, the spectral characteristics should be known and integrated over the solar spectrum, or

$$\bar{\rho}_s = \frac{\displaystyle\int_0^\infty I_{\lambda,rs}\,d\lambda}{\displaystyle\int_0^\infty I_{\lambda,i}\,d\lambda} = \frac{\displaystyle\int_0^\infty \rho_{s,\lambda} I_{\lambda,i}\,d\lambda}{\displaystyle\int_0^\infty I_{\lambda,i}\,d\lambda} \tag{3.120}$$

where $\rho_{s,\lambda}$ is the specular monochromatic reflectance and $I_{\lambda,rs}$ is the solar intensity specularly reflected at $\lambda$.

Manufacturer's data are often restricted to the average reflectance in the visible spectrum and a value for this spectral range may not be the correct average reflectance for the entire solar spectrum. Table 3.14 presents typical values for the normal specular reflectance of new surfaces for beam solar radiation. Some surfaces have smaller reflectances at smaller angles of incidence, but data on the effect of incidence angle are sparse. In practical application, the durability under real conditions must also be known, because maintenance of high specular reflectance can be difficult. Ordinary dust and dirt can generally be removed by washing without difficulty, but front-surfaced reflectors can suffer degradation as a result of oxidation and abrasion. Some protection and better durability can be obtained by surface coatings, but a coating will reduce the initial value of the reflectance.

Most reflecting materials are rather costly and/or structurally weak. They are, therefore, generally attached to some type of substrate. For front-surface mirrors the substrate may be any structural material. For rear- or second-surface mirrors the radiation must pass through the superstrate. It must, therefore, be transparent and also have other desirable optical properties. In some concentrator designs the infrared properties of the transparent medium are important, since appreciable radiation heat

**TABLE 3.13** Properties of Some Selected Plated Coating Systems[a]

| Coating[b] | Substrate | $\bar{\alpha}_s$ | $\bar{\epsilon}_i$ | Durability Breakdown temperature (°F) | Durability Humidity-degradation MIL STD 810B | Estimated manufactured cost per ft² (U.S.) |
|---|---|---|---|---|---|---|
| Black nickel on nickel | Steel | 0.95 | 0.07 | >550 | Variable | 0.30 |
| Black chrome on nickel | Steel | 0.95 | 0.09 | >800 | No effect | 0.35–0.15 |
| Black chrome | Steel | 0.91 | 0.07 | >800 | Completely rusted | 0.10 |
| | Copper | 0.95 | 0.14 | 600 | Little effect | 0.10 |
| | Galvanized steel | 0.95 | 0.16 | >800 | Complete removal | 0.10 |
| Black copper | Copper | 0.88 | 0.15 | 600 | Complete removal | 0.10 |
| Iron oxide | Steel | 0.85 | 0.08 | 600 | Little effect | 0.05 |
| Manganese oxide | Aluminum | 0.70 | 0.08 | 800 | | 0.10 |
| Organic overcoat on iron oxide | Steel | 0.90 | 0.16 | | Little effect | 0.15 |
| Organic overcoat on black chrome | Steel | 0.94 | 0.20 | | Little effect | 0.15 |

[a]From U.S. Dept. of Commerce, "Optical Coatings for Flat Plate Solar Collectors," NTIS No. PB-252-383, Honeywell, Inc., 1975.
[b]Black nickel coating plated over a nickel-steel substrate has the best selective properties ($\bar{\alpha}_s = 0.95$, $\bar{\epsilon}_i = 0.07$) but degraded significantly during humidity tests. Black chrome plated on a nickel-steel substrate also had very good selective properties ($\bar{\alpha}_s = 0.95$, $\bar{\epsilon}_i = 0.09$) and also showed high resistance to humidity.

**TABLE 3.14**  Specular Reflectance Values for
Solar Reflector Materials

| Material | $\rho$ |
|---|---|
| Silver (unstable as front surface mirror) | $0.94 \pm 0.02$ |
| Gold | $0.76 \pm 0.03$ |
| Aluminized acrylic, second surface | 0.86 |
| Anodized aluminum | $0.82 \pm 0.05$ |
| Various aluminum surfaces—range | 0.82 – 0.92 |
| Copper | 0.75 |
| Back-silvered water-white plate glass | 0.88 |
| Aluminized type-C Mylar (from Mylar side) | 0.76 |

transfer can occur between the absorber surface and nearby regions of the reflecting surface. If the reflecting surface is nonmetal or plastic, it may be necessary to use a material that can withstand high enough temperatures. Glass is the only currently available substrate transparent material known to be inert and durable. It can also withstand high temperatures if thermal stresses can be avoided. Other commonly used materials for solar reflecting surfaces are silver and aluminum. Stainless steel, copper, gold, rhodium, or platinum and other exotic metals either have low reflectance values or are too costly to be useful in solar applications.

Specular reflectance varies with wavelength for some materials. For example, anodized aluminum will show spectral absorption bands in the infrared range due to the presence of the anodizing coating. Infrared absorption bands are in fact desirable in some concentrators, since infrared emission can occur at long wavelengths to assist in cooling a mirror exposed to the sun. In other concentrating designs infrared absorption bands are undesirable as described above.

The optical design of specular reflectors is described in Chapter 4.

## Transparent Materials

The last of the radiation parameters to be considered is the transmittance. As mentioned previously, for transparent materials the sum of the absorptance, reflectance, and transmittance must equal unity [see Eq. (3.79)]. However, the mechanism of transmission of radiation through a transparent material, such as glass, is a rather complex process and depends on the wavelength of the radiation, the angle of incidence, the refractive index, and the extinction coefficient.

Transparent covers in various shapes are used to reduce heat losses from the radiation-absorbing surfaces of most solar collectors. Consequently, an understanding of the process and laws that govern the transmission of radiation through a transparent medium is important for the design of solar collectors. Materials for this type of application should be able to transmit solar radiation at all angles of incidence, have dimensional stability, and have long-term durability under exposure to the sun and the weather. Other desirable properties include low weight, low infrared transmittance, and low cost.

## Index of Refraction and Extinction Coefficient

The calculation of transmission of light through a transparent medium is governed by principles laid down long ago by Fresnel, Snell, and Stokes; they are outlined below.

The optical behavior of a substance can be characterized by two wavelength-dependent physical properties—the index of refraction $n$ and the extinction coefficient $K$. The index of refraction, which determines the speed of light in the material [see Eq. (3.74)], also determines the amount of light reflected from a single surface, while the extinction coefficient determines the amount of light absorbed in a substance in a single pass of radiation.

Figure 3.43 defines the angles used in analyzing reflection and transmission of light. The angle $i$ is called the *angle of incidence*. It is also equal to the angle at which a beam is specularly reflected from the surface. Angle $\theta_r$ is the *angle of refraction*, which is defined as shown in the figure. The incidence and refraction angles are related by Snell's law

$$\frac{\sin i}{\sin \theta_r} = \frac{n'_r}{n'_i} = n_r \qquad (3.121)$$

where $n'_i$ and $n'_r$ are the two refractive indices and $n_r$ is the index ratio for the two substances forming the interface. Typical values of refractive indices for various materials are shown in Table 3.15 below. For most materials of interest in solar applications, the values range from 1.3 to 1.6, a fairly narrow range.

The reflectance $\rho'$ from a surface of a transparent substance is related to the refractive index indirectly by the values of $i$ and $\theta_r$ according to Snell's law, Eq. (3.121). The reflectance has two components corresponding to the two components of polarization resolved parallel and perpendicular to the plane of incidence. The perpendicular ($\perp$) and parallel ($\parallel$) components, respectively, are

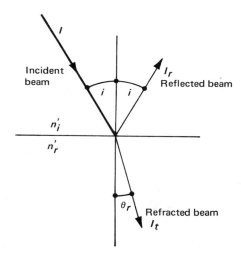

**FIGURE 3.43** Diagram showing incident, reflected, and refracted beams of light and incidence and refraction angles for a transparent medium.

**TABLE 3.15**  Refractive Index for Various Substances in the Visible Range Based on Air

| Material | Index of refraction |
|---|---|
| Air | 1.000 |
| Clean polycarbonate (PCO) | 1.59 |
| Diamond | 2.42 |
| Glass (solar collector type) | 1.50–1.52 |
| Plexiglass[a] (polymethyl methacrylate, PMMA) | 1.49 |
| Mylar[a] (polyethylene terephthalate, PET) | 1.64 |
| Quartz | 1.54 |
| Tedlar[a] (polyvinyl fluoride, PVF) | 1.45 |
| Teflon[a] (polyfluoroethylenepropylene, FEP) | 1.34 |
| Water–liquid | 1.33 |
| solid | 1.31 |

[a]Trademark of the DuPont Company, Wilmington, Delaware.

given by the relations

$$\rho'_\perp = \frac{\sin^2(i - \theta_r)}{\sin^2(i + \theta_r)} \tag{3.122}$$

$$\rho'_\parallel = \frac{\tan^2(i - \theta_r)}{\tan^2(i + \theta_r)} \tag{3.123}$$

For normal incidence the two components are equal, or

$$\rho'_\perp = \rho'_\parallel = \frac{(n_r - 1)^2}{(n_r + 1)^2} \tag{3.124}$$

where $n_r$ is the index of refraction ratio. For grazing incidence $(i = 90°)$ both components are also equal; therefore,

$$\rho'_\perp = \rho'_\parallel = 1$$

The reflectance of a glass-air interface common in solar collectors may be reduced by a factor of four by an etching process. If glass is immersed in a silica-supersaturated fluosilicic acid solution, the acid attacks the glass and leaves a porous silica surface layer. This layer has an index of refraction intermediate between glass and air. By having a gradual change in index of refraction, reflectance losses are reduced significantly.

The efficiency of the process depends upon solution temperature and composition, immersion time, and surface pretreatment. Mar et al. (28) have studied these effects in detail and have devised a repeatable process for producing glass with reflectance of 1 percent per interface (2 percent per pane of glass). They also found that heat treatment at 373 K enhances the durability of the coating significantly. Figure 3.44 shows the spectral reflectance of a pane of glass before and after etching. Attempts to etch plastics have been unsuccessful to date.

When radiation passes through a transparent medium such as glass or the atmosphere, the decrease in intensity can be described by assuming that the attenuation is proportional to the local intensity in the medium. If $I_\lambda(x)$ is monochromatic intensity after radiation has traveled a distance $x$, Bouger's law is expressed by the equation

$$-dI_\lambda(x) = I_\lambda(x)K_\lambda\, dx \tag{3.125}$$

where $K_\lambda$ is the monochromatic extinction coefficient assumed to be a constant of the medium. If the transparent medium is a slab of thickness $L$ and the intensity at $x = 0$ is designated by the symbol $I_{\lambda,0}$, the monochromatic transmittance $\tau_\lambda'$ for the absorption process alone is equal to the ratio of the intensity at $x = L$ to $I_{\lambda,0}$. An expression for $I_\lambda(L)$ can be obtained by integrating Eq. (3.125) between 0 and $L$, which gives

$$\ln\frac{I_\lambda(L)}{I_{\lambda,0}} = -K_\lambda L \quad \text{or} \quad I_\lambda(L) = I_{\lambda,0}\, e^{-K_\lambda L} \tag{3.126}$$

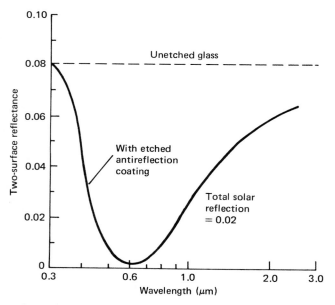

**FIGURE 3.44** Reflection spectra for a sample of glass before and after etching.

Then

$$\tau'_\lambda = \frac{I_\lambda(L)}{I_{\lambda,0}} = e^{-K_\lambda L} \tag{3.127}$$

If the transmittance is sensitive to the direction of the incoming radiation, $I_{\lambda,0}(\theta,\phi)$ must be used in Eq. (3.127) and the transmittance will also be a function of $\theta$ and $\phi$. Fortunately, the absorption part of the transmittance for glass and plastic films is not a strong function of direction. Moreover, as shown below, the spectral transmittance of low-iron glass is almost a constant over the solar spectrum, that is, between 0.3 and 3 $\mu$m. Consequently, it can be treated as a transparent gray medium with $K_\lambda = K$, a wavelength independent property.

Bouger's law (Eq. 3.127) can be used to calculate the absorptance $\alpha'$ of homogeneous substances as follows:

$$\alpha' = 1 - e^{-KL} \tag{3.128}$$

The optical-path length $L$ is the thickness of the material $t$ divided by the cosine of the angle of refraction (see Fig. 3.43), that is,

$$L = \frac{t}{\cos\theta_r} \tag{3.129}$$

It is very difficult to evaluate the extinction coefficient for materials used in solar energy, since they are selected expressly to have very small values of $K$. In order to measure small values of $K$, a large material thickness $L$ is required. However, thick sections made for measurement have properties different from those for thin layers used in solar collectors. Also, if many layers of a thin material are used, the surface reflection from each dominates the optical attenuation of the stack and no increase in accuracy of $K$ measurement is achieved. This problem is particularly acute with plastic films. Consequently, few data exist for transparent materials used in solar collectors. A few values of $K$ are given in Table 3.16 below.

### Stokes' Equations

The total reflectance $\rho$, the flux leaving the top surface divided by the incident flux, in Fig. 3.45 can be calculated by summing the infinite series of components making up the total flux. For simplicity of notation $\tau' = e^{-KL}$ and $\rho'$ denotes both $\rho'_\perp$ and $\rho'_\parallel$. The two series to be summed, which can be written from Fig. 3.45, are

$$\rho = \rho(\rho',\tau') = \rho' + \rho'\tau'^2(1-\rho')^2[1 + \rho'^2\tau'^2 + \rho'^4\tau'^4 + \ldots] \tag{3.130}$$

$$\tau = \tau(\rho',\tau') = (1-\rho')^2\tau'[1 + \rho'^2\tau'^2 + \rho'^4\tau'^4 + \ldots] \tag{3.131}$$

The terms in brackets are geometric series with first-term unity and ratio $\rho'^2\tau'^2$. Using the formula for the sum of an infinite geometric series from elementary algebra, we have

**TABLE 3.16**  Extinction Coefficients for Transparent Materials

| | |
|---|---|
| Polyvinyl fluoride (Tedlar[a]) | 1.4 cm$^{-1}$ |
| Fluorinated ethylene propylene (Teflon[a]) | 0.59 |
| Polyethylene terephthalate (Mylar[a]) | 2.05 |
| Polyethylene | 1.65 |
| Ordinary window glass | ~0.3 |
| White glass (<0.01% $Fe_2O_3$) | ~0.04 |
| Heat-absorbing glass | 1.3–2.7 |

[a]Trademark of the DuPont Company, Wilmington, Delaware; data from (10).

$$\rho(\rho', \tau') = \rho'\left[1 + \frac{\tau'^2(1 - \rho')^2}{1 - \rho'^2\tau'^2}\right] \tag{3.132}$$

$$\tau(\rho', \tau') = \tau'\left[\frac{(1 - \rho')^2}{1 - \rho'^2\tau'^2}\right] \tag{3.133}$$

The total absorptance is simply unity decreased by $(\rho + \tau)$ or

$$\alpha(\rho', \tau') = 1 - \rho(\rho', \tau') - \tau(\rho', \tau') \tag{3.134}$$

Equations (3.132–3.134) were first developed by G. G. Stokes (36) and are called the Stokes' equations.

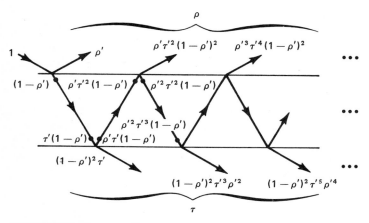

**FIGURE 3.45**  Ray trace diagram for a transparent medium showing the reflected and transmitted fractions accounting for multiple interreflections. Total fraction reflected and transmitted denoted by $\rho$ and $\tau$, respectively; $\tau' \equiv e^{-KL}$.

The quantities $\rho'$ and $\tau'$ apply to single surfaces, single passes, and single reflections. In practice, a layer of glass or other transparent material will have multiple interreflections of radiation, which must all be accounted for in calculating the total reflectance $\rho$, transmittance $\tau$, and absorptance $\alpha$ as shown in Fig. 3.45.

Most materials used for optical transmittance of solar energy have small values of $KL$ as described above. Therefore $\tau' \approx 1$ and Eq. (3.133) can be simplified by separating the absorptance and reflectance effects,

$$\tau(\rho', \tau') \approx \tau \left( \frac{1 - \rho'}{1 + \rho'} \right)$$

which is accurate to a few percent for most materials and incidence angles used in solar applications. This simplification can also be used when more than one absorbing-reflecting layer is involved [see Eq. (3.137 ff)].

### EXAMPLE 3.16

Calculate the total solar transmittance at normal incidence angle for a sheet of glass 0.25 cm thick if the extinction coefficient is $0.1 \text{ cm}^{-1}$ in the solar spectrum, assuming that the attenuation at one air-glass interface $\rho' = 4$ percent.

### SOLUTION

From Eq. (3.127)

$$\tau' = e^{-KL} = e^{-0.025} = 0.975$$

Therefore, from the equations above,

$$\tau(\rho', \tau') = 0.975 \left( \frac{1 - 0.04}{1 + 0.04} \right)$$

$$\tau(\rho', \tau') = 0.90$$

Care must be taken in the use of the Stokes' equations to account for polarization properly. The average reflectance $\bar{\rho}$ and the average transmittance $\bar{\tau}$ for both components of polarization, if they are of equal magnitude, are

$$\bar{\rho} = \tfrac{1}{2} [\rho(\rho'_\perp, \tau') + \rho(\rho'_\parallel, \tau)] \tag{3.135}$$

$$\bar{\tau} = \tfrac{1}{2} [\tau(\rho'_\perp, \tau') + \rho(\rho'_\parallel, \tau')] \tag{3.136}$$

It is incorrect to calculate an average $\rho'$ for both components and use this value in Eqs. (3.132) and (3.133) directly to calculate the average properties. Errors of up to 18 percent in the calculated transmittance of a pane of glass can result from this improper averaging. Figure 3.46 shows the calculated transmittance $\bar{\tau}$ of a layer of glass versus incidence angle $i$ if absorption losses are negligible.

## Multiple, Transparent Layers

Stokes also derived expressions for reflection and transmission of multiple layers of transparent media using the results for one layer. The analysis involves multiple-reflectance ray tracing as in the case of one layer shown above and will not be repeated here. The effective transmittance and reflectance equations are used to generate the properties of multiple layers and are derived by simply increasing the number of layers one by one or by doubling repeatedly. The general equations for $N + M$ layers are

$$\tau_{N+M}(\rho',\tau') = \frac{\tau_N \tau_M}{1 - \rho_N \rho_M} \tag{3.137}$$

$$\rho_{N+M}(\rho',\tau') = \rho_N + \frac{\rho_M \tau_N^2}{1 - \rho_N \rho_M} \tag{3.138}$$

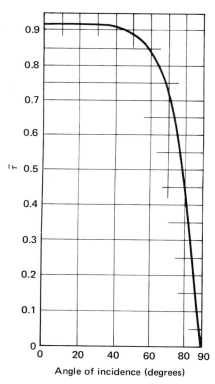

**FIGURE 3.46** Transmission versus angle of incidence for glass, neglecting absorption ($KL = 0$), with refractive index of 1.53.

in which $\rho_M$ and $\rho_N$ are calculated from repeated application of Eq. (3.132) and $\tau_M$ and $\tau_N$ from Eq. (3.133).

For example, for two transparent sheets

$$\tau_2(\rho',\tau') = \frac{\tau_1\tau_1}{1-\rho_1\rho_1} \qquad (3.139)$$

and

$$\rho_2(\rho',\tau') = \rho_1 + \frac{\rho_1\tau_1{}^2}{1-\rho_1\rho_1} \qquad (3.140)$$

where $\rho_1$ and $\tau_1$ are from Eqs. (3.132) and (3.133).

Then for three layers ($M = 2$, $N = 1$) we would have

$$\tau_3(\rho',\tau') = \frac{\tau_2\tau_1}{1-\rho_2\rho_1} \qquad (3.141)$$

and

$$\rho_3(\rho',\tau') = \rho_1 + \frac{\rho_2\tau_1{}^2}{1-\rho_1\rho_2} \qquad (3.142)$$

Values of $\rho_1$ and $\tau_1$ are as above: $\rho_2$ and $\tau_2$ are from Eqs. (3.139) and (3.140). This process can be carried out to calculate the effective transmittance and reflectance of any number of layers from the two basic properties $K$ and $n$. Note that in this analysis each component of polarization must be treated separately as for a single layer. For example, the average transmittance of a three-layer system is

$$\bar{\tau}_3 = \tfrac{1}{2}[\tau_3(\rho'_\perp,\tau') + \tau_3(\rho'_\parallel,\tau')] \qquad (3.143)$$

Equations (3.137) and (3.138) can also be used if the lowest surface is opaque as in a solar collector-absorber plate. In that case $\tau_M = 0$ and $\rho_M = \rho_S$, where $\rho_S$ is one minus the absorber surface solar absorptance. The quantity $1 - \rho_{N+S}$ represents the radiation absorbed by a collector and its cover. The ray trace method can also be used to calculate the effective optical transmittance-absorptance transfer function for the absorber plate of a collector in closed form (see Problem 3.37).

If the layers of material of interest are very thin or have a very small extinction coefficient, the amount absorbed in each layer will be negligible, that is, $\tau' \cong 1$; $\alpha' \cong 0$. In that case the Stokes' equations for single and multiple layers can be simplified. For example, for single layers we have analogs to Eqs. (3.132) and (3.133), which become

$$\rho(\rho') = \frac{\rho'}{1 + \rho'} \qquad (3.144)$$

$$\tau(\rho') = \frac{1 - \rho'}{1 + \rho} \qquad (3.145)$$

$$\alpha(\rho') = 0 \qquad (3.146)$$

For multiple layers Eqs. (3.137) and (3.138) reduce to

$$\tau_{N+M}(\rho') = \frac{1 - \rho'}{1 + (2N + 2M - 1)\rho'} \qquad (3.147)$$

$$\rho_{N+M}(\rho') = 1 - \tau_{N+M}(\rho') \qquad (3.148)$$

$$\alpha_{N+M}(\rho') = 0 \qquad (3.149)$$

These simpler equations can generally be used in solar designs where good, low-iron glass or thin plastic sheets are the material of choice.

### EXAMPLE 3.17

If the reflectance of a single surface of glass in air is 4 percent, calculate the total reflection loss for one through four layers of glass. Neglect absorption of energy in the glass.

### SOLUTION

Equation (3.148) can be used to calculate the effective reflectance of several layers of glass with negligible absorption. For normal incidence the following values are calculated:

| Number of layers $(N + M)$ | Effective reflectance $\rho_{N+M}$ (%) |
|:---:|:---:|
| 1 | 7.7 |
| 2 | 14.2 |
| 3 | 20.0 |
| 4 | 25.0 |

Note that more than two layers reduce the overall transmittance significantly because of reflective losses alone. Radiation lost by reflection is not available for thermal conversion at the absorber of a solar collector and, therefore, represents a nonrecoverable loss.

Note that in the foregoing analysis, the magnitude of the parallel and perpendicular polarization components has been assumed equal. Although not strictly true in all cases, this assumption is sufficiently accurate for engineering calculations. Wehner (39), in an elegant study, has prepared charts and diagrams depicting the degree of polarization of the daytime sky. His measurements indicate that, although the degree of polarization is small by engineering standards, it is large enough to be used by insects such as bees for navigation. He also notes that it is possible to use the small polarization differential, by viewing the sky through the mineral corderite as a polarizing filter, for navigational purposes. The Vikings may have used this navigational aid in the Middle Ages.

## Spectral Transmittance, Absorptance, and Reflectance

The foregoing optical analysis is strictly applicable at one wavelength only or for materials that are "gray" and therefore have wavelength-independent optical properties. In the narrow spectral range of sunlight, approximation of wavelength independence is acceptable. However, when viewed in the broader use required for solar collector performance calculations, both the narrow solar and broader infrared spectral ranges must be considered. Plastics usually have properties different from glass in the infrared range. Most plastics have a higher infrared transmittance than glass, which is nearly opaque beyond 4 $\mu$m. The spectral behavior of transparent materials is an important factor in solar collector performance. Since collector-absorber surfaces radiate in the infrared region, this radiation is intercepted by the infrared-opaque barrier offered by glass, whereas if the absorber has a plastic layer covering it, some infrared radiation is exchanged with the environment directly.

## HEAT EXCHANGERS

Conventional heat exchangers are devices in which two fluid streams, separated from each other by a solid wall, exchange thermal energy: one stream is heated while the other one is cooled. There are a number of arrangements used to transfer heat from one fluid to another. The simplest arrangement is the double-pipe heat exchanger shown in Fig. 3.47. It will be discussed first.

In this system fluid A flows inside a tube of inner radius $r_i$ and outer radius $r_o$. Fluid B flows in the annulus formed between the outer surface of the inner tube, $A_o$, and the inner surface of the outer tube. Individual convection heat-transfer coefficients can be calculated by methods outlined previously, and an overall heat-transfer coefficient $U$ can be calculated from the thermal circuit shown in Fig. 3.47. The overall heat-transfer coefficient may be based on any convenient area of the exchanger, but usually the outside area $A_o = 2\pi r_o L$ of the inner tube is most convenient (25). Then,

$$U_o A_o = \frac{1}{1/h_i A_i + (\ln r_o/r_i)/2\pi kL + 1/h_o A_o}$$ (3.150)

and at any cross section the local rate of heat transfer across the tube is

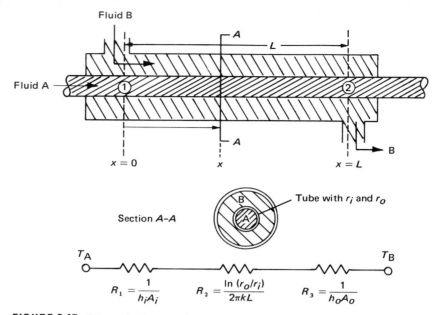

**FIGURE 3.47** Schematic diagram of double-pipe heat exchanger.

$$dq = U_o \, dA_o(T_A - T_B) = U_o 2\pi r_o \, dx(T_A - T_B) \tag{3.151}$$

Simplifications in the evaluation of $U_o$ are possible when one or two of the thermal resistances dominate. For example, the thermal resistance of the tube wall, $R_2$ in Fig. 3.47, is often small compared to the convective resistances and sometimes one of the two convective resistances is negligible compared to the other.

It also happens often that after some period of operation one or both surfaces of a heat exchanger become coated with a deposit or become corroded. Such a coat or scale can offer a substantial thermal resistance over and above those considered in Eq. (3.150) and thereby cause a decrease in the thermal performance. This condition is called *fouling* and a fouling resistance must be added to calculate the overall heat-transfer coefficient under these conditions.

Fouling resistances, often called *fouling factors*, can be measured experimentally by comparing the values of $U$ for a clean and dirty or corroded exchanger. The fouling factors $R_f$ can then be calculated from

$$R_f = \frac{1}{U_{\text{dirty}}} - \frac{1}{U_{\text{clean}}} \tag{3.152}$$

A list of fouling factors for several fluids is given in Table 3.17.

The flow arrangement shown in Fig. 3.47, where both fluids enter from the same end, is called *parallel flow*, and Fig. 3.48a shows the temperature distribution for both fluid streams. If one of the two fluids were to enter at the other end and

**TABLE 3.17** Average Fouling Factors[a]

| Type of fluid | Fouling factor | |
|---|---|---|
| | hr · ft² · °F/Btu | m² · K/W |
| Seawater, below 125°F | 0.0005 | 0.00009 |
| above 125°F | 0.001 | 0.0002 |
| Treated boiler feedwater above 125°F | 0.001 | 0.0002 |
| Fuel oil | 0.005 | 0.0009 |
| Quenching oil | 0.004 | 0.0007 |
| Alcohol vapors | 0.0005 | 0.00009 |
| Steam, non-oil-bearing | 0.0005 | 0.00009 |
| Industrial air | 0.002 | 0.0004 |
| Refrigerating liquid | 0.001 | 0.0002 |

[a]Adapted from (35).

flow in the opposite direction, the flow arrangement would be *counterflow*. The temperature distribution for the latter case is shown in Fig. 3.48b. If a counterflow arrangement is made very long, it approaches the thermodynamically most efficient possible heat-transfer condition (see Example 3.3).

There are two basic methods for calculating the rate of heat transfer in a heat exchanger. One method employs a mean temperature difference between the two

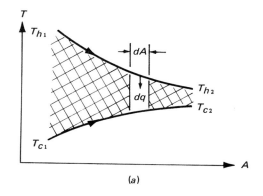

(a)

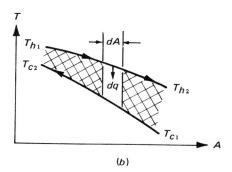

**FIGURE 3.48** (a) Parallel flow; (b) counterflow.

(b)

fluids in the exchanger $\Delta T_{mean}$, and then determines the rate of heat transfer from the relation (25)

$$q = UA \ \Delta T_{mean} \tag{3.153}$$

This mean temperature difference is called the logarithmic mean temperature difference (LMTD). It can be evaluated directly only when the inlet and outlet temperatures of both fluid streams are specified. The use of the LMTD approach is explained in Ref. (25). The other method is called the effectiveness-NTU method (ENTU). The ENTU method offers many advantages over the LMTD approach and will be discussed below.

First, we define the exchanger effectiveness $\epsilon$

$$\epsilon = \frac{\text{actual rate of heat transfer}}{\text{maximum possible rate of heat transfer}} \tag{3.154}$$

The actual rate of heat transfer can be determined by calculating either the rate of internal energy loss of the hot fluid or the rate of internal energy gain of the cold fluid, that is,

$$q = \dot{m}_h c_{p,h}(T_{h,\,in} - T_{h,\,out}) = \dot{m}_c c_{p,c}(T_{c,\,out} - T_{c,\,in}) \tag{3.155}$$

where the subscripts $h$ and $c$ denote the hot and cold fluid, respectively. The maximum rate of heat transfer possible for specified inlet fluid temperatures is attained when one of the two fluids undergoes the maximum temperature difference in the exchanger. This maximum equals the difference in the entering temperatures for the hot and cold fluid. Which of the two fluids can undergo this maximum temperature change depends on the relative value of the product $(\dot{m}c_p)$, the mass flow rate times the specific heat of the fluid at constant pressure, called the heat capacity rate $\dot{C}$. Since a thermodynamic energy balance requires that the energy given up by the one fluid must be received by the other if there are no external heat losses, only the fluid with the smaller value of $\dot{C}$ can undergo the maximum temperature change. Thus, the fluid that may undergo this maximum temperature change is the one that has the minimum value of the heat capacity rate and the maximum rate of heat transfer is

$$q_{max} = \dot{C}_{min}(T_{h,\,in} - T_{c,\,in}) \tag{3.156}$$

where $\dot{C}_{min} = (\dot{m}c_p)_{min}$

The fluid with the minimum $\dot{m}c_p$ or $\dot{C}$ value can be either the hot or the cold fluid. Using subscript $h$ to designate the effectiveness when the hot fluid has the minimum $C$ and the subscript $c$ when the cold fluid has the minimum $\dot{C}$ value, we get for a parallel-flow arrangement

$$\epsilon_h = \frac{\dot{C}_h(T_{h1} - T_{h2})}{\dot{C}_h(T_{h1} - T_{c1})} = \frac{T_{h1} - T_{h2}}{T_{h1} - T_{c1}} \tag{3.157}$$

$$\epsilon_c = \frac{\dot{C}_c(T_{c2} - T_{c1})}{\dot{C}_c(T_{h1} - T_{c1})} = \frac{T_{c2} - T_{c1}}{T_{h1} - T_{c2}} \tag{3.158}$$

where the subscripts 1 and 2 refer to the left- and right-hand side of the heat exchanger as shown in Fig. 3.47. Similarly, for a counterflow arrangement

$$\epsilon_h = \frac{\dot{C}_h(T_{h1} - T_{h2})}{\dot{C}_h(T_{h1} - T_{c2})} = \frac{T_{h1} - T_{h2}}{T_{h1} - T_{c2}} \tag{3.159}$$

$$\epsilon_c = \frac{\dot{C}_c(T_{c1} - T_{c2})}{\dot{C}_c(T_{h1} - T_{c2})} = \frac{T_{c1} - T_{c2}}{T_{h1} - T_{c2}} \tag{3.160}$$

We shall next derive an analytic expression for the effectiveness of a parallel-flow heat exchanger. Using the nomenclature of Fig. 3.47, the rate of heat transfer at any position is given by Eq. (3.151) in the form

$$dq = U_o(T_h - T_c)\, dA_o \tag{3.161}$$

or from Eq. (3.155) in the form

$$dq = \dot{C}_c\, dT_c = -\dot{C}_h\, dT_h \tag{3.162}$$

We can then write the relations $dT_c = dq/\dot{C}_c$ and $dT_h = -dq/\dot{C}_h$ so that

$$d(T_h - T_c) = -dq\left(\frac{1}{\dot{C}_c} + \frac{1}{\dot{C}_h}\right) \tag{3.163}$$

Substitution in Eq. (3.161) for $dq$ gives

$$\frac{d(T_h - T_c)}{T_h - T_c} = -U_o\left(\frac{1}{\dot{C}_h} + \frac{1}{\dot{C}_c}\right) dA_o \tag{3.164}$$

Equation 3.164 can be integrated between positions 1 and 2 in the heat exchanger. The result is

$$\ln\frac{T_{h2} - T_{c2}}{T_{h1} - T_{c1}} = -\frac{U_o A_o}{\dot{C}_c}\left(1 + \frac{\dot{C}_c}{\dot{C}_h}\right) \tag{3.165}$$

or

$$\frac{T_{h2} - T_{c2}}{T_{h1} - T_{c1}} = \exp\left[-\frac{U_o A_o}{\dot{C}_c}\left(1 + \frac{\dot{C}_c}{\dot{C}_h}\right)\right] \tag{3.166}$$

If the cold fluid has the minimum heat capacity rate, Eq. (3.160) gives $\epsilon$. To obtain a convenient expression for $\epsilon$ the temperature ratio in Eq. (3.158) is recast into the form

$$\frac{T_{h2} - T_{c2}}{T_{h1} - T_{c1}} = \frac{T_{h1} + (\dot{C}_c/\dot{C}_h)(T_{c1} - T_{c2}) - T_{c2}}{T_{h1} - T_{c1}} \qquad (3.167)$$

and the outlet temperature of the hot fluid is evaluated from Eq. (3.155)

$$T_{h2} = T_{hi} + \frac{\dot{C}_c}{\dot{C}_h}(T_{c1} - T_{c2}) \qquad (3.168)$$

By substituting $T_{h2}$ from Eq. (3.168) into Eq. (3.167) the latter can be rewritten in the form

$$T_{hi} - T_{ci} + \frac{\dot{C}_c}{\dot{C}_h}(T_{c1} - T_{c2}) + T_{c1} - T_{c2} = 1 - \left(1 + \frac{\dot{C}_c}{\dot{C}_h}\right)\epsilon_c \qquad (3.169)$$

Finally, by reinserting the above relation in Eq. (3.167) we obtain

$$\epsilon_c = \frac{1 - \exp\left[(-U_o A_o/\dot{C}_c)(1 + \dot{C}_c/\dot{C}_h)\right]}{1 + \dot{C}_c/\dot{C}_h} \qquad (3.170)$$

It can easily be verified that the effectiveness for the case of the hot fluid having the minimum heat capacity rate is

$$\epsilon_h = \frac{1 - \exp\left[(-U_o A_o/\dot{C}_h)(1 - \dot{C}_h/\dot{C}_c)\right]}{1 + \dot{C}_h/\dot{C}_c} \qquad (3.171)$$

Equations (3.170) and (3.171) can both be written in terms of $\dot{C}_{max}$ and $\dot{C}_{min}$ in the form

$$\epsilon = \epsilon_c = \epsilon_h = \frac{1 - \exp\left[(-U_o A_o/C_{min})(1 - C_{min}/C_{max})\right]}{1 + C_{min}/C_{max}} \qquad (3.172)$$

The parameter $U_o A_o/C_{min}$, called the number of transfer units ($NTU$), is related to the size of the heat exchanger as shown below.

Kays and London (24) have shown how to calculate the effectiveness for various types of heat exchangers. Important results of this analysis are presented in the charts of Figs. 3.49–3.54 and in Table 3.18, where the analytical expressions from which the charts were prepared are summarized. To use these charts it is necessary to know something about heat-exchanger geometry. The double-pipe parallel and counterflow arrangements (Fig. 3.48) have already been discussed. In gas heating

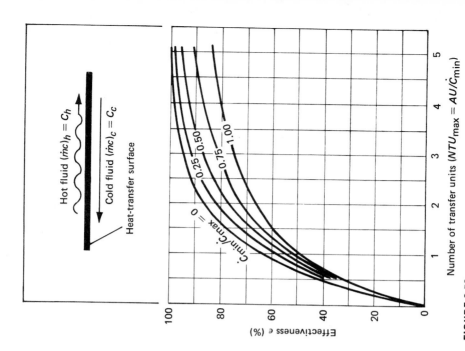

**FIGURE 3.50** Effectiveness for counterflow exchanger performance. From (24).

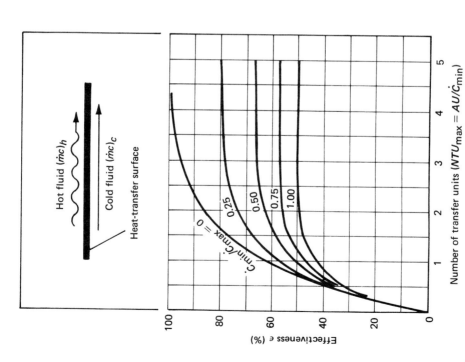

**FIGURE 3.49** Effectiveness for parallel-flow exchanger performance. From (24).

185

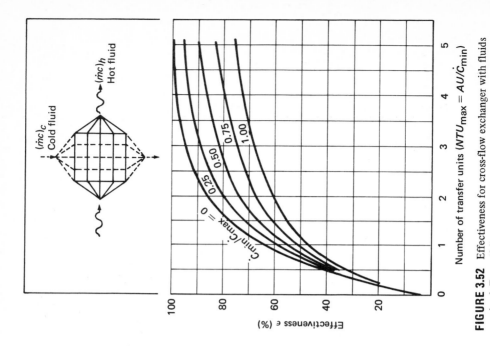

**FIGURE 3.52** Effectiveness for cross-flow exchanger with fluids unmixed. From (24).

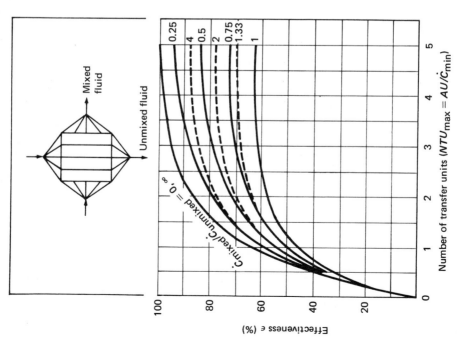

**FIGURE 3.51** Effectiveness for cross-flow exchanger with one fluid mixed. From (24).

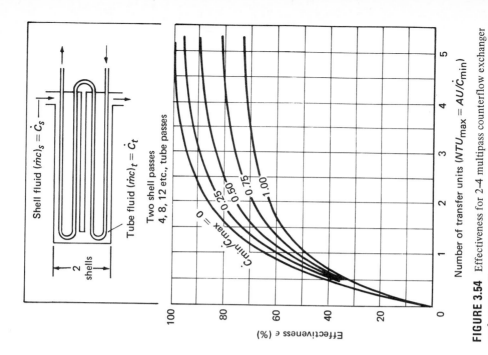

**FIGURE 3.54** Effectiveness for 2-4 multipass counterflow exchanger performance. From (24).

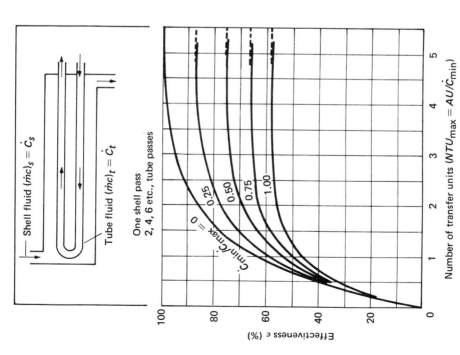

**FIGURE 3.53** Effectiveness for 1-2 parallel-counterflow exchanger performance. From (24).

187

**TABLE 3.18**  Heat-Exchanger Effectiveness Relations[a]

$$N = NTU = \frac{UA}{\dot{C}_{min}}, \quad C = \frac{\dot{C}_{min}}{\dot{C}_{max}}$$

| Flow geometry | Relation |
|---|---|
| Double pipe | |
|   Parallel flow | $\epsilon = \dfrac{1 - \exp\,[-N(1 + C)]}{1 + C}$ |
|   Counterflow | $\epsilon = \dfrac{1 - \exp\,[-N(1 - C)]}{1 - C\,\exp\,[-N(1 - C)]}$ |
| Cross flow | |
|   Both fluids unmixed | $\epsilon = 1 - \exp\left\{\dfrac{C}{n}\,[\exp\,(-NCn) - 1]\right\}$    where $n = N^{-0.22}$ |
|   Both fluids mixed | $\epsilon = \left[\dfrac{1}{1 - \exp\,(-N)} + \dfrac{C}{1 - \exp\,(-NC)} - \dfrac{1}{N}\right]^{-1}$ |
|   $C_{max}$ mixed, $C_{min}$ unmixed | $\epsilon = (1/C)\{1 - \exp\,[C(1 - e^{-N})]\}$ |
|   $C_{max}$ unmixed, $C_{min}$ mixed | $\epsilon = 1 - \exp\,\{(1/C)[1 - \exp\,(-NC)]\}$ |
| Shell and tube | |
|   One shell pass, 2, 4, 6 tube passes | $\epsilon = 2\left\{1 + C + (1 + C^2)^{1/2}\,\dfrac{1 + \exp\,[-N(1 + C^2)^{1/2}]}{1 - \exp\,[-N(1 + C^2)^{1/2}]}\right\}^{-1}$ |

[a]From (24, 29).

and cooling, cross-flow exchangers such as shown in Figs. 3.55 and 3.56 are widely used. In the arrangement shown in Fig. 3.56 one fluid is contained within tubes, whereas the other flows across the tube bundle. The fluid within the tubes is said to be unmixed, whereas the gas flowing across the tubes is not constrained and can mix freely while exchanging heat. The effectiveness for this arrangement is given in the charts of Fig. 3.51. When a fluid is unmixed, there will be a temperature gradient both parallel and normal to the flow direction, whereas when the fluid in the exchanger is mixed, its temperature in the direction normal to the flow tends to equalize as a result of the mixing. This is illustrated in Fig. 3.56b, where the temperature profile for the gas flowing through the heat exchanger configuration of Fig. 3.56a is shown schematically for the gas being heated. The overall rate of heat transfer in an exchanger depends on whether a fluid is mixed or unmixed, because the average temperature difference in an exchanger is a function of the mixing arrangement as illustrated by the difference in effectiveness.

In the arrangement shown in Fig. 3.55 both fluids are restrained from mixing, one fluid by the tubes and the other by the fins attached perpendicular to the tubes. This arrangement is widely used in air conditioning applications.

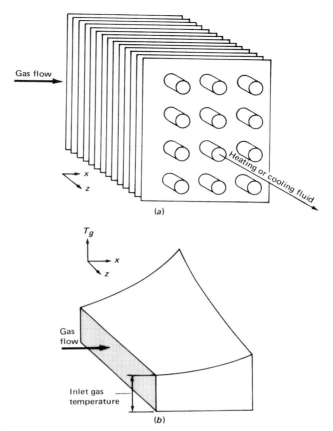

**FIGURE 3.55** (*a*) Cross-flow heat exchanger with both fluids un-mixed; (*b*) typical temperature profile for cross-flow heat exchanger.

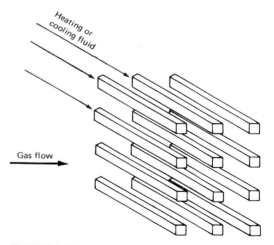

**FIGURE 3.56** Cross-flow heat exchanger, one fluid mixed and one unmixed.

The effectiveness of heat exchangers with both fluids unmixed is shown in Fig. 3.52.

The type of heat exchanger most widely used in commercial practice consists of a shell containing a tube bundle as shown in Fig. 3.57. One fluid flows inside the tubes while the other is forced across the tubes by baffles, as shown. The arrangement at the ends where the tubes emerge, called the header, determines the number of tube passes. If both headers are open, there will be one tube pass. If one header is halved and the other is open, there will be two tube passes and one shell pass. The effectiveness of such a unit with a single shell pass and with two or a multiple of two tube passes is shown in Fig. 3.53.

The use of the effectiveness approach in calculating the size of heat exchangers required in solar-thermal conversion systems is illustrated in the following two examples. The first example shows the sizing of an air-to-water exchanger in which hot water from a solar collector is passing through tubes and the air to be heated is passing over the tubes. The air can be blown over the tubes by means of a fan or can rise naturally upward by free convection. In case of an all-air system, hot air instead of hot water would pass through the tubes and heat the air, but the method of solution would be similar to the water-air case. The second type of problem, illustrated in Example 3.20, is to determine the performance of a given shell-and-tube heat exchanger typical of the kind used in an absorption air conditioning system, in a solar hot-water heating system in which the water from a hot storage tank is used to heat hot service water in a closed loop, in the condenser of a solar power plant, or the economizer in a vapor refrigeration cycle.

**FIGURE 3.57**  Shell-and-tube heat exchanger with one tube pass. BCF exchanger.
*Photograph courtesy of American Standard, Heat Transfer Division.*

**EXAMPLE 3.18**

A 10 m$^2$ (107.2 ft$^2$) flat-plate collector panel with a single cover and selective coating can provide 6.3 kW (21,600 Btu/hr) at a water outlet temperature of 365 K and flow rate of 0.9 kg/sec between 11 a.m. and 1 p.m. on a sunny day. A shell-and-tube heat exchanger is to be placed between the collector and the storage tank to avoid contamination of the water in the tank by the antifreeze in the collector loop. The water from the collector is to heat water entering from the storage tank at 348 to 353 K at a flow rate of 0.3 kg/sec. If the overall heat-transfer coefficient is estimated to be 400 W/m$^2$ · K, determine the area required for a shell-and-tube heat exchanger with a single shell pass and two tube passes. Also, calculate the rate of heat transfer when the solar radiation level decreases and the hot-water inlet temperature drops to 355 K.

**SOLUTION**

A heat balance gives the outlet temperature of the hot stream:

$$q = [\dot{m}c(T_1 - T_2)]_h = [\dot{m}c(T_2 - T_1)]_c$$

$$= (0.9)(4203)(365 - T_2) = (0.3)(4191)(5) = 6286 \text{ W}$$

Thus, $T_2 = 365 - 1.67 = 363.3$ K. The effectiveness based on the cold stream is

$$\epsilon_c = \frac{T_{c2} - T_{c1}}{T_{n1} - T_{c1}} = \frac{5}{15} = 0.294$$

For $\dot{C}_{min}/\dot{C}_{max} = 0.33$ we get from Fig. 3.53

$$\frac{AU}{\dot{C}_{min}} = 0.4$$

and therefore

$$A = \frac{(0.4)(1260)}{400} = 1.2 \text{ m}^2 \text{ (12.9 ft}^2)$$

When the hot-water inlet temperature drops to 355 K, the cold-water outlet temperature is equal to

$$T_{c2} = T_{c1} + \epsilon(T_{h1} - T_{c1}) = 348 + 2.06 = 350.6 \text{ K}$$

and the rate of heat transfer in the exchanger is

$$(\dot{m}c)_c(T_{c2} - T_{c1}) = (0.3)(4191)(2.6) = 3267 \text{ W (11,150 Btu/hr)}$$

## EXAMPLE 3.19

A water solar collector at an average insolation of 64 W/m² (20.3 Btu/hr · ft²) and water outlet temperature of 325 K has an efficiency of 40 percent. This collector is to heat air from the interior of a house at 295 to 302 K at a flow rate of 2.4 m³/sec (5080 ft³/min). A heat exchanger of the type shown in Fig. 3.55a with a total surface area of 20 m² (216 ft²) and overall heat-transfer coefficient of 113 W/m² · K (20 Btu/hr · ft² · °F) is available. Calculate the size of collector required and the outlet temperature of the water, that is, the inlet temperature to the collector.

## SOLUTION

The density of the air at inlet is from Table A3.1

$$\rho = 1.201 \text{ kg/m}^3$$

Thus, $\dot{m}_c = Q\rho = (2.4)(1.201) = 2.88$ kg/sec, and the rate of heat transfer is $q = (\dot{m}c_p)_c(T_2 - T_1)_c = (2.88)(1006)(7) = 20.3$ kW ($6.9 \times 10^4$ Btu/hr). Since it is not known which of the two fluid streams has the smaller heat-capacity rate, it is necessary to assume one and then check it. Since all the information on the air stream is known, it will be used to calculate *NTU*. Once this is known, Fig. 3.52 can be used to determine the water flow rate and a heat balance to calculate the water outlet temperature. If the water were to be the fluid with the minimum heat-capacity rate, a trial-and-error procedure with the effectiveness chart would be necessary. For the air stream $(\dot{m}c_p)$ is 2897 W/K and

$$NTU = \frac{UA}{\dot{C}_{min}} = \frac{(113)(20)}{2897} = 0.78$$

and the effectiveness is

$$\epsilon = \frac{(T_2 - T_1)_{air}}{(T_h - T_c)_{max}} = \frac{7}{30} = 0.234$$

Since no curve on the effectiveness chart in Fig. 3.52 can match these values, the water must be the fluid with $\dot{C}_{min}$. It is therefore necessary to assume a water flow rate and then determine if it fits the performance specified by the analysis represented in Fig. 3.51. From Eq. (3.156) we have

$$(T_2 - T_1)_{water} = \frac{q}{\dot{C}_{min}} = \frac{2.03 \times 10^4}{\dot{C}_{min}}$$

and from Eq. (3.157)

$$\epsilon_h = \frac{(T_2 - T_1)_h}{30}$$

| Assumed $\dot{C}_{min}/\dot{C}_{max}$ | $\dot{C}_{min}$ | $NTU_{max}$ | $(T_2 - T_1)_h$ | $\epsilon_h$ From Fig. 3.51 | $\epsilon_h$ Calculated from Eq. (3.157) |
|---|---|---|---|---|---|
| 0.5 | 1448 | 1.6 | 14.0 | 0.65 | 0.47 |
| 0.7 | 2028 | 1.11 | 10.0 | 0.52 | 0.33 |
| 0.3 | 869 | 2.60 | 23.4 | 0.82 | 0.78 |
| 0.25 | 724 | 3.12 | 28.0 | 0.88 | 0.93 |
| 0.27 | 782 | 2.89 | 26.0 | 0.87 | 0.87 |

Thus $\dot{m}_h = 782/4180$ W/K $= 0.187$ kg/sec and the water temperature entering the solar collector is $325 - 2.03 \times 10^4/782 = 299.0$ K. The area of solar collector required is obtained from a heat balance. At the insolation level of 64 W/m$^2$, 40 percent or 25.6 W/m$^2$ can be delivered as useful energy to the water. The total rate of heat transfer required to heat the air stream was found to be 20,300 W. Thus, the required collector area is $20,300/25.6 = 793$ m$^2$. The trial-and-error solution can be avoided by solving the appropriate equation in Table 3.18.

## FLOW AND CONVECTION HEAT TRANSFER IN FIXED BEDS

In addition to conventional heat exchangers in which the hot and cold fluids are separated by a solid wall, some solar energy systems also make use of another type of heat-exchange device. This device consists of a bed of solid particles, which can be heated by the working fluid when it is hot. However, the particles can transfer the energy stored in them to the same working fluid at a later time when it is cold and must be heated. The most common example of this type of heat exchanger-storage system in solar engineering is the "pebble bed" of a solar collector system using air as the working fluid.

In general, two types of particle heat-exchange systems are encountered in engineering: the fixed bed and the fluidized bed. In the former the particles lie on each other, and their position is not changed by the fluid flowing through the voids. However, if the fluid is flowing upward and its velocity is increased sufficiently, the net drag force will eventually counteract the gravitational force, the fluid will lift the particles in the bed, and the particles will become suspended. This condition is called a fluidized bed, characterized by thorough mixing of particles, which eliminates any temperature gradient through the bed. This type of operation is often sought in the chemical industry, but would be undesirable in a rock storage bed for a solar system.

To analyze the flow and convection heat transfer in a fixed bed of pebbles, certain simplifications are usually made. One of them is to treat the particles as though they were approximately spherical. For such a pebble bed Löf and Hawley (27) investigated the heat transfer and recommended the following empirical relation to evaluate the heat-transfer coefficient:

$$h_v = 650 \left( \frac{\dot{m}_b}{A_b D_s} \right)^{0.7} \tag{3.173}$$

where $h_v$ = *volumetric* heat-transfer coefficient in $W/m^3 \cdot K$
 $\dot{m}_b$ = mass rate of flow in kg/sec
 $A_b$ = cross-sectional area of the pebble bed in $m^2$
 $D_s$ = equivalent spherical diameter of the particles in m given by

$$D_s = \left( \frac{6}{\pi} \frac{\text{net volume of particles}}{\text{number of particles}} \right)^{1/3} \tag{3.174}$$

In English units, an equation similar to Eq. (3.173) is

$$h_c = 0.79 \left( \frac{G}{D_s} \right)^{0.7} \frac{V_b}{A_p} \tag{3.175}$$

where $D_s$ = particle diameter in ft
 $G$ = specific mass flow rate in $lb/hr \cdot ft^2$ (based on the cross-sectional area of the bed)
 $V_b$ = bed volume in $ft^3$
 $A_p$ = total surface area of all the particles in the bed in $ft^2$
 $h_c$ = heat-transfer coefficient based on the surface area of particles in the bed in $Btu/hr \cdot ft^2 \cdot °F$

Experimental verification of the range of variables over which this relation is applicable extends specifically over entering air temperatures between 311 and 367 K (100 and 200°F), gravel sizes between 4.8 and 38 mm (0.19 and 1.5 in) diameter, air-flow rates between 0.66 and 3.6 $m^3/sec \cdot m^2$ (12 and 66 $ft^3/min \cdot ft^2$) of empty bed cross section, and voids obtained by normal filling. For a bed of approximately spherical particles

$$A_p = \frac{6(1 - \epsilon_v) A_b L}{D_s} \tag{3.176}$$

Handley and Heggs (15) measured the heat transfer and frictional pressure drop in flow through packed beds and correlated their results with those of previous investigators. They found that the heat-transfer coefficients for a bed of approximately spherical particles can be correlated by the dimensionless equation

$$j_h = \frac{0.255}{\epsilon_v \, \mathrm{Re}_{D_s}{}^{0.33}} \tag{3.177}$$

where $\epsilon_v$ = void fraction of the bed
$\quad \mathrm{Re}_{D_s}$ = Reynolds number $(GD_s/\mu)$
$\quad D_s$ = sphere diameter
$\quad j_h$ = Stanton number $(\mathrm{Nu}_{D_s}/\mathrm{Re}_{D_s} \, \mathrm{Pr}^{1/3})$
$\quad G$ = superficial mass velocity of the gas)
For air (Pr = 0.71) this relation becomes

$$\mathrm{Nu}_{D_s} = \frac{0.23 \, \mathrm{Re}_{D_s}{}^{2/3}}{\epsilon_v} \tag{3.178}$$

The pressure drop in a bed of length $L$ can be determined from the relation

$$\frac{\Delta p}{\rho V_s{}^2} = \frac{L}{D_s} \frac{1 - \epsilon_v}{\epsilon_v{}^3} \left[ 170 \frac{1 - \epsilon_v}{\mathrm{Re}_{\mathrm{bed}}} + 1.75 \right] \tag{3.179}$$

where

$$\mathrm{Re}_{\mathrm{bed}} = \frac{2}{3} \frac{\rho V_s D_s}{(1 - \epsilon_v)\mu}$$

$$\epsilon_v = \text{void fraction (or porosity) of the bed} \tag{3.180}$$

$$V_s = \frac{\dot{m}}{A_b \rho} \quad \text{the superficial velocity}$$

For pebble bed, $\epsilon_v$ is between 0.35 and 0.45.

More detailed analysis of the transient heat transfer and data on the effect of particle shape on the heat-transfer coefficients are presented in Ref. (16). Additional information on heat storage in pebble beds is presented in Chapter 7.

## PROBLEMS

3.1.  Liquid hydrogen is to be stored at atmospheric pressure in a spherical container having an inside diameter of 10 m. The tank is to be insulated with a super insulation material having a thermal conductivity of $2 \times 10^{-4}$ W/m $\cdot$ K. Estimate the amount of hydrogen vaporized during a 24-hr day for insulation thicknesses of 3, 5, and 10 cm under ambient conditions of 300 K. Assume that the outer temperature of the insulation is approximately at ambient temperature and that the interior surface of the tank is at the evaporation temperature of liquid hydrogen at atmospheric temperature.

3.2. A temperature difference of 50 K is impressed across a layer of fiberglass of 10 cm thickness. If the thermal conductivity of the fiberglass is 0.04 W/m · K, calculate the rate of heat transfer per square meter area per hour.

3.3. Insolation at 700 W/m$^2$ is incident on a metal plate that has a solar absorptance of 0.9 and is perfectly insulated on the back side. If the convection heat-transfer coefficient on the upper side of the plate is 10 W/m$^2$ · K and the ambient temperature is 300 K, calculate the temperature of the plate under equilibrium conditions.

3.4. Repeat Problem 3.3, but assume that the metal plate is treated with a selective surface coating so that its solar absorptance is 0.9 and its emittance at equilibrium is only 0.2.

3.5. If the metal plate in Problems 3.3 and 3.4 was not insulated at the lower surface but formed the upper part of a rectangular metal duct in which air at 330 K is flowing with a unit surface conductance between the metal and the air of 15 W/m$^2$ · K, calculate the rate at which heat can be transferred from the metal plate to the air in W/m$^2$.

3.6. Two large parallel plates having surface properties approximating those of a black body are maintained at 300 and 400 K, respectively. Calculate the rate of heat transfer by radiation between the plates in W/m$^2$.

3.7. Repeat Problem 3.6, but assume that both plates have emittances of 0.1 and both are gray.

3.8. Repeat Problem 3.7, but assume that the two plates are 2 cm apart with atmospheric air between them and that heat is transferred across the gap by conduction through the air and radiation in parallel.

3.9. The wall of a building may be approximated by a 10 cm thick layer of common brick having a thermal conductivity of 0.7 W/m · K followed by a 3.75-cm thick layer of plaster having a thermal conductivity of 0.5 W/m · K. Estimate what thickness of loosely packed rock-wool insulation is necessary to reduce the rate of heat transfer through the wall by 75 percent. Assume that the thermal conductivity of the rock-wool insulation is 0.6 W/m · K.

3.10. A 1-m diameter sheet metal duct is carrying air from a collector to the pebble-bed storage. The duct is covered with a 2.5-cm thick fiberglass insulation ($k = 0.038$ W/m · K). If the overall heat-transfer coefficient at the interior surface is 10 W/m$^2$ · K and at the exterior surface is 5 W/m$^2$ · K, estimate the rate of heat loss if the air temperature inside the duct is 310 K and outside the duct the ambient air is 290 K.

3.11. An electric resistance heater having a total outside surface area of 0.1 m$^2$ is immersed in a water storage tank at 140 K. The average heat-transfer coefficient between the surface and the water is 60 W/m · K. If the electrical energy dissipated within the heater is 1 kW, calculate the average outside surface temperature of the heater under steady-state conditions.

3.12. A 1-cm thick steel plate with a thermal conductivity of 43 W/m · K receives a net heat input by radiation at its upper face of 6000 W/m$^2$. If a gas at 350 K with a heat-transfer coefficient of 50 W/m · K is flowing over the upper and lower faces, calculate the temperatures of the upper and lower faces under steady-state conditions. Neglect their heat losses except the heat transfer to the gas.

3.13. A spherical storage tank of 6 m diameter is exposed to air at 300 K. If the surface temperature of the tank is 50 K and the air is calm, calculate the Nusselt number, the average heat transfer, and the rate of heat loss from the tank to the air in W.

3.14. A 0.1-m diameter steel pipe is exposed to air at 290 K blowing perpendicular to its axis at a velocity of 5 m/sec. If the steam inside the pipe is saturated at 2 atm, calculate the rate of heat loss from the pipe from convection and radiation assuming the emittance of the pipe is 0.9.

3.15. Calculate the spacing between two parallel flat plates inclined 55° from the horizontal that would result in the minimum rate of heat transfer between the two plates.

3.16. Calculate the rate of heat transfer between two flat plates inclined at 45° from the horizontal and spaced 2 cm apart with air between them if the temperature of the lower plate is 340 K and the temperature of the upper plate is 330 K. Assume the surfaces of both plates are black.

3.17. Water flowing at a mass rate of 1 kg/sec is to be heated from 340 to 350 K in a 2.5-cm ID tube whose surface temperature is 360 K. Calculate the heat-transfer coefficient between the tube and the water and estimate the length of tube required to heat the water over the specified temperature difference.

3.18. Calculate the pressure drop of the water for the length of tube calculated in Problem 3.17.

3.19. Water at an average temperature of 340 K flows at a mass rate of 0.7 kg/sec in a 2.5-cm ID copper tube 10 m long. If a constant heat flux is imposed and the average tube wall temperature is 350 K, calculate the exit temperature of the water and the pressure drop.

3.20. Air at 1 atm and 320 K flows through a rectangular duct 3 × 30 cm². A 10-m long section of the duct is maintained at 20 K above the average temperature of the air, and the temperature rise in this length is to be 10 K. Calculate the air flow rate and the net rate of heat transfer. Also estimate the average heat-transfer coefficient and comment whether the value is reasonable.

3.21. A tube bundle using in-line arrangement with $F_n = F_p = 2$ cm uses six rows of tubes of 6 mm diameter in a stack 25 tubes high. If the surface temperature of the tubes in the bundle is constant at 320 K and atmospheric air at 340 K is forced across them at an inlet velocity of 4 m/sec before entering the tube bank, calculate the rate of heat transfer per unit length of tube bank and the pressure drop.

3.22. A surface at 350 K has an emittance of 0.2 at wavelengths larger than 0.3 $\mu m$ and an emittance of 0.8 at short wavelengths. Calculate the rate of radiation emitted from the surface and compare the value with what would obtain if the surface had an emittance of 0.8 over the entire wavelength spectrum.

3.23. Three large flat plates are placed parallel to one another. Plate 1, the one at the top, is maintained at a temperature of 1200 K, while the plate at the bottom is maintained at 60 K. If the emittance of the 1200 K plate is 0.2, the emittance of the plate in the middle is 0.5 on both sides, and the emittance of the 60 K plate is 0.8, calculate the temperature of the intermediate plate.

3.24. A thermocouple is used to measure the temperature in a long air duct. If the duct walls are at a temperature of 370 K and the thermocouple indicates that the air temperature is 330 K, calculate the actual air temperature if the heat-transfer coefficient between the gas and the thermocouple is 100 W/m$^2$ · K and the emittance of the thermocouple material is 0.8.

3.25. Water flowing at a rate of 70 kg/min is to be heated from 340 to 350 K in a counterflow double pipe heat exchanger by a 50 percent glycol solution that enters the heat exchanger at 350 K and exits at 345 K. If the overall heat-transfer coefficient is 300 W/m$^2$ · K, calculate the required heat exchanger area.

3.26. Repeat Problem 3.25, but assume that instead of a counterflow double pipe heat exchanger a shell-and-tube heat exchanger will be used with the water making one shell pass and the glycol solution making two passes.

3.27. A fin-and-tube heat exchanger, such as shown in Fig. 3.55a, is used to heat 3 m$^3$/sec atmospheric air from 15 to 30 C. Hot water enters the tubes at 80 C, and the air flowing across the tubes has an average overall heat-transfer coefficient of 200 W/m$^2$ · K between the water and the air. If the total surface area of the exchanger is 10 m$^2$, calculate the exit water temperature and the rate of heat transfer.

3.28. A shell-and-tube heat exchanger with one shell pass and two tube passes, having an area of 4.5 m$^2$, is to be used to heat high-pressure water initially at 290 K with hot air initially at 400 K. If the exit water temperature is not to exceed 350 K, the air flow rate is 0.5 kg/sec, and the overall heat-transfer coefficient is 300 W/m$^2$ · K, calculate the water flow rate.

3.29. A cross-flow fin-and-tube heat exchanger uses hot water to heat air from 290 to 300 K. Water enters the heat exchanger at 340 K and exits at 310 K. If the total heat-transfer rate is to be 300 kW and the average heat-transfer coefficient between the water and the air is 50 W/m$^2$ · K, calculate the area of the heat exchanger required and the mass rate of air flow through the exchanger.

3.30. A small steam condenser is to be designed to condense 0.8 kg/m steam at 80 kN/m$^2$ with cooling water at 290 K. If the exit temperature of the water is not to exceed 320 K, calculate the area required for a shell-and-tube heat exchanger with the steam making one shell pass and the water making two tube passes. The overall heat-transfer coefficient is 3000 W/m$^2$ · K.

3.31. Determine the ratio of the total hemispheric emittance to the normal emittance for a nondiffused surface if the intensity of emission varies as the cosine of the angle measured from the normal.

3.32. A horizontal uninsulated 0.5-m diameter sheet-metal duct carries air from a solar collector at 233 K (140°F) at a mean velocity of 1 m/sec through a basement at 283 K (40°F). Calculate the heat loss in W/m length and the pressure drop in Pa/m.

3.33. Show that an infinitely long counterflow heat exchanger is thermodynamically the most efficient arrangement for achieving the maximum temperature change (or enthalpy change) in both the hot and the cold fluid. Neglect external losses from the heat exchanger, that is, assume the system is perfectly insulated.

3.34. Estimate the free convection heat-transfer coefficient and the radiation heat-transfer coefficient for a bare horizontal 10-cm OD tube at 320 K in a room at 290 K if the emittance of the surface is equal to 0.9.

3.35. An air collector has a rectangular flow sectional area 0.05 × 1 m$^2$. If the air velocity through the collector is 2 m/sec, calculate the heat-transfer coefficient between the collector assuming one side of the collector is at 330 K, the air is at 320 K, and the other side of the collector is insulated. Heat losses on the side may be neglected.

3.36. The Stokes' equations—Eqs. (3.132) and (3.133)—are based on the assumption that the single-surface reflectance $\rho'$ is the same at both surfaces of the transparent medium (Fig. 3.45). Rederive the equations for the case where $\rho'$ has a different value at the upper and lower surfaces as in the case of a partially mirrored surface.

3.37. Using a ray trace technique show that the effective optical transfer function $(\tau\alpha)_{\text{eff}}$ for a flat-plate collector is given by

$$(\tau\alpha)_{\text{eff}} = \frac{\tau\alpha}{1 - (1 - \alpha)\rho_d}$$

where $\alpha$ is the surface absorptance, $\tau$ is the cover system transmittance [Eq. (3.137)], and $\rho_d$ is the cover system reflectance for radiation diffusely reflected from the absorber surface.

3.38. Calculate the reflectances $\rho'$ and $\rho$ and transmittances $\tau'$ and $\tau$ for a white glass plate 3 mm thick at three values of incidence angle 0, 40, and 75°.

3.39. Using the results of Problems 3.37 and 3.38 calculate the $(\tau\alpha)_{\text{eff}}$ product for a single-cover solar collector with a flat, black surface ($\alpha = 0.96$) if the diffuse reflectance $\rho_d = 0.16$.

3.40. Bouger's law [Eq. (3.125)] assumes a linear decrease in intensity with distance $x$ parameterized by the extinction coefficient $K_\lambda$. In nonhomogeneous diffusion layers, $K_\lambda$ may be a function of $x$. Derive the analog to Eq. (3.127) when $K_\lambda$ is a power law function of $x$, that is, $K_\lambda = K_1 x^n$.

3.41. If a 2 mm thickness of window glass has a transmittance $\tau'$ of 99 percent, how thick would a sheet of polyethylene be if it had the same transmittance?

3.42. Derive Eq. (3.176).

# REFERENCES

1. Barron, R., "Cryogenic Systems," McGraw-Hill Book Co., New York, 1967.
2. Buchberg, H., I. Catton, and D. K. Edwards, "Natural Convection in Enclosed Spaces: A Review of Application to Solar Energy Collection," ASME Paper 74-WA/HT/12, 1974.
3. Buchberg, H., and D. K. Edwards, Design Considerations for Solar Collectors with Cylindrical Glass Honeycombs, *Sol. Energy*, vol. 18, pp. 193–204, 1976.
4. Buchberg, H., O. A. Lalude, and D. K. Edwards, Performance Characteristics of Rectangular Honeycomb Solar-Thermal Converters, *Sol. Energy*, vol. 13, pp. 193–221, 1971.
5. Chi, S. W., "Introduction to Heat Pipe Theory," Hemisphere Publ. Corp., Washington, D. C., 1976.
6. Clark, J. A., Cryogenic Heat Transfer, in T. F. Irvine and J. P. Hartnett (eds.), "Advances in Heat Transfer," vol. 5, Academic Press, New York, 1968.
7. Cotter, T. P., Theory of Heat Pipes, *Los Alamos Scientific Laboratory Rept.* LA-3246-MS, 1965.
8. Dutcher, C. H., and M. R. Burke, Heat Pipes: A Cool Way to Cool Circuits, *Electronics*, vol. 16, pp. 93–100, February 1960.

9. Dwight, H. B., "Tables of Integrals and Other Mathematical Data," The Macmillan Co., New York, 1964.

10. Edlin, F., "Solar Utilization Now, IV," chap. 12, Arizona State University, Tempe, 1976.

11. Edwards, D. K., Measurements of Thermal Radiation Properties, *Proc. Env. Sci.*, pp. 417–424, 1963; see also *J. Opt. Soc. Am.*, vol. 51, pp. 1279–1283, 1961.

12. Edwards, D. K., K. E. Nelson, R. D. Roddick, and J. T. Gier, Basic Studies on the Use and Control of Solar Energy, *University of California, Los Angeles, Department of Engineering Rept.* 60-93, 1960.

13. Gier, J. T., R. V. Dunkle, and J. T. Bevans, Measurement of Absolute Spectral Reflectivity from 1.0 to 15 Microns, *J. Opt. Soc. Am.*, vol. 44, pp. 558–563, 1954.

14. Grimson, E. D., Correlation and Utilization of New Data on Flow Resistance and Heat Transfer for Cross-flow of Gases over Tube Banks, *Trans. ASME*, vol. 59, pp. 583–594, 1937.

15. Handley, D., and P. J. Heggs, Momentum and Heat Transfer Mechanisms in Regularly Shaped Packings, *Trans. Inst. Chem. Eng.*, vol. 46, pp. 251–264, 1968.

16. Handley, D., and P. J. Heggs, The Effect of Thermal Conductivity of the Packing Material on the Transient Heat Transfer in a Fixed Bed, *Int. J. Heat Mass Transfer*, vol. 12, pp. 549–570, 1969.

17. Hollands, K. G. T., Honeycomb Devices in Flat Plate Solar Collectors, *Sol. Energy*, vol. 9, p. 159, 1965.

18. Hollands, K. G. T., L. Konicek, T. E. Unny, and G. D. Raithby, Free Convection Heat Transfer Across Inclined Air Layers, *J. Heat Transfer*, vol. 98, pp. 189–193, 1976.

19. Hollands, K. G. T., G. D. Raithby, and T. E. Unny, "Studies on Methods of Reducing Heat Losses from Flat Plate Collectors," Final Rept. ERDA Contract E (11-1)-2597, University of Waterloo, Ontario, Canada, 1976.

20. Holman, J. P., "Thermodynamics," 2nd ed., McGraw-Hill Book Co., New York, 1975.

21. Hottel, H. C., and A. F. Sarofim, "Radiative Transfer," McGraw-Hill Book Co., New York, 1967.

22. Howell, J. R., and R. Siegel, "Thermal Radiation Heat Transfer," McGraw-Hill Book Co., New York, 1972.

23. Kays, W. M., "Convective Heat and Mass Transfer," McGraw-Hill Book Co., New York, 1966.

24. Kays, W. M., and A. L. London, "Compact Heat Exchangers," 2nd ed., McGraw-Hill Book Co., New York, 1964.

25. Kreith, F., "Principles of Heat Transfer," 3rd ed., Intext Educational Publishers, New York, 1976.

26. Kreith, F., "Radiation Heat Transfer for Spacecraft and Solar Power Plant Design," International Textbook Co., Scranton, Pa., 1962.

27. Löf, G. O. G., and R. W. Hawley, Unsteady State Heat Transfer between Air and Loose Solids, *Ind. Eng. Chem.*, vol. 40, pp. 1061–1066, 1948.

28. Mar, H. Y. B., et al., Optical Coatings for Flat Plate Solar Collectors, *NTIS Rept.* PB-252-383, September 1975.

29. Parker, J. D., J. H. Boggs, and E. F. Blick, "Fluid Mechanics and Heat Transfer," Addison-Wesley Publ. Co., Reading, Mass., 1969.

30. Raithby, G. D., K. G. T. Hollands, and T. E. Unny, "Free Convection Heat Transfer across Fluid Layers of Large Aspect Ratios," ASME Paper 76-HT-45, presented at the 16th National Heat Transfer Conference, St. Louis, August 8–11, 1976.
31. Raithby, G. D., and K. G. T. Hollands, A General Method of Obtaining Approximate Solutions to Laminar and Turbulent Free Convection Problems, in "Advances in Heat Transfer," vol. 11, Academic Press, New York, 1975.
32. Richter, R., "Solar Collector Thermal Power System," vol. I, *Xerox Corp. Rept.* AFAPL-TR-74-89-I, *NTIS Rept.* AD/A-000-940, 1974.
33. Rohsenow, W. M., and J. P. Hartnett (eds.), "Handbook of Heat Transfer," McGraw-Hill Book Co., New York, 1973.
34. Sparrow, E. M., and R. D. Cess, "Radiation Heat Transfer," Wadsworth Publ. Co., Belmont, Calif., 1966.
35. "Standards of Tubular Exchanger Manufacturers Association," 4th ed., Tubular Heat Exchanger Manufacturers Association, New York, 1959.
36. Stokes, G. G., On the Intensity of the Light Reflected from a Transmitter through a Pile of Plates, *Proc. R. Soc. London*, vol. 11, pp. 546–556, 1860–1862.
37. Tabor, H., Radiation, Convection, and Conduction Coefficients in Solar Collectors, *Bull. Res. Counc. Isr., Sect. C*, vol. 6C, pp. 155–176, 1958.
38. Touloukian, Y. S., et al., "Thermophysical Properties of Matter," Plenum Data Corp., vol. 7, "Thermal Radiative Properties—Metallic Elements and Alloys," 1970; vol. 8, "Thermal Radiative Properties—Nonmetallic Solids," 1971; vol. 9, "Thermal Radiative Properties—Coatings," 1972.
39. Wehner, R., Polarized-light Navigation by Insects, *Sci. Am.*, vol. 235, pp. 106–115, 1976.
40. Zhukauskas, A. A., V. Makarevicius, and A. Schlanciankas, "Heat Transfer in Banks of Tubes in Cross-flow of Fluids," Mintis, Vilnius, Lithuania, 1968.

# METHODS OF SOLAR COLLECTION AND THERMAL CONVERSION

*Always take more notice of the corners than you do of a surface to be polished; the middle you cannot very well forget . . . , but the corners and crevices are often neglected.*

**The French Polishers Handbook**

Converting the sun's radiant energy to heat is the most common and well-developed solar conversion technology today. The temperature level and amount of this converted energy are the key parameters that must be known to effectively match a conversion scheme to a specified task.

This chapter analyzes in detail different methods for converting solar energy at temperatures ranging from moderately above ambient to above 3000 K. The thermal and optical performance of several solar-thermal collectors is described. They range from air- and liquid-cooled nonconcentrating, flat-plate types to compound-curvature, continuously tracking types with concentration ratios up to 3000 or more. Applications of the energy produced by solar collectors are described in Chapters 6, 7, and 8.

## ENERGY BALANCE FOR A FLAT-PLATE COLLECTOR

The thermal performance of any type of solar thermal collector can be evaluated by an energy balance that determines the portion of the incoming radiation delivered as useful energy to the working fluid. For a flat-plate collector of area $A_c$ this energy balance is

$$I_c A_c \bar{\tau}_s \alpha_s = q_u + q_{\text{loss}} + \frac{de_c}{dt} \tag{4.1}$$

where $I_c$ = solar irradiation on a collector surface
$\quad \bar{\tau}_s$ = effective solar transmittance of the collector cover(s)
$\quad \alpha_s$ = solar absorptance of the collector-absorber plate surface
$\quad q_u$ = rate of heat transfer from the collector-absorber plate to the working fluid

**203**

$q_{loss}$ = rate of heat transfer (or heat loss) from the collector-absorber plate to the surroundings

$de_c/dt$ = rate of internal energy storage in the collector

The instantaneous efficiency of a collector $\eta_c$ is simply the ratio of the useful energy delivered to the total incoming solar energy, or

$$\eta_c = \frac{q_u}{A_c I_c} \tag{4.2}$$

In practice, the efficiency must be measured over a finite time period. In a standard performance test this period is on the order of 15 or 20 min, whereas for design, the performance over a day or over some longer period $t$ is important. Then we have for the average efficiency

$$\bar{\eta}_c = \frac{\displaystyle\int_0^t q_u \, dt}{\displaystyle\int_0^t A_c I_c \, dt} \tag{4.3}$$

where $t$ is the time period over which the performance is averaged.

A detailed and precise analysis of the efficiency of a solar collector is complicated by the nonlinear behavior of radiation heat transfer. However, a simple linearized analysis is usually sufficiently accurate in practice. In addition, the simplified analytical procedure is very important, because it illustrates the parameters of significance for a solar collector and how these parameters interact. For a proper analysis and interpretation of these test results an understanding of the thermal analysis is imperative, although for design and economic evaluation the results of standardized performance tests are generally used.

## Collector Heat-Loss Conductance

In order to obtain an understanding of the parameters determining the thermal efficiency of a solar collector, it is important to develop the concept of *collector heat-loss conductance*. Once the collector heat-loss conductance $U_c$ is known, and when the collector plate is at an average temperature $T_c$, the second right-hand term in Eq. (4.1) can be written in the simple form

$$q_{loss} = U_c A_c (T_c - T_a) \tag{4.4}$$

The simplicity of this relation is somewhat misleading, because the collector heat-loss conductance cannot be specified without a detailed analysis of all the heat losses. Figure 4.1 shows a schematic diagram of a double-glazed collector, while Fig. 4.2a shows the thermal circuit with all the elements that must be analyzed before

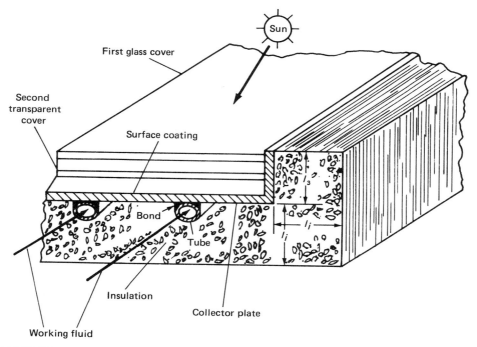

**FIGURE 4.1**  Schematic diagram of solar collector with two covers.

they can be combined into a single conductance element shown in Fig. 4.2b. The analysis below shows how this combination is accomplished.

Figure 4.3 shows the qualitative temperature distributions in a flat-plate collector. Radiation impinges on the top of the plate connecting any two adjacent flow ducts. It is absorbed uniformly by the plate and conducted toward the flow duct, where it is then transferred by convection to the working fluid flowing through the ducts. It is apparent that at any cross section perpendicular to the direction of flow, the temperature is a maximum at the midpoint between two adjacent flow ducts and decreases along the sheet toward the tube. Since heat is transferred to the working fluid, the temperature of the fluid as well as that of the entire collector system will increase in the direction of flow. For example, the increase in temperature at the midpoint between the two tubes is shown qualitatively in Fig. 4.3c. The temperature distribution in both the $x$ and $y$ directions is shown in a three-dimensional view in Fig. 4.3d.

In order to construct a model suitable for a thermal analysis of a flat-plate collector, the following simplifying assumptions will be made:

1. The collector is thermally in steady state.
2. The temperature drop between the top and bottom of the absorber plate is negligible.

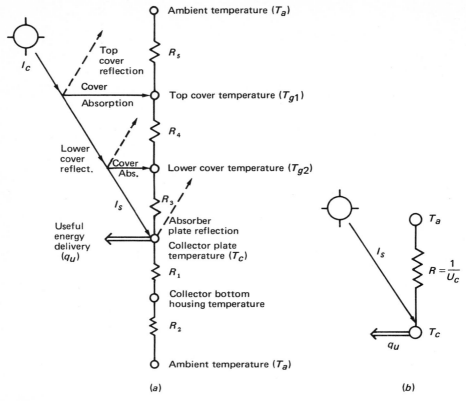

**FIGURE 4.2** Thermal circuits for flat-plate collector shown in Fig. 4.1: (*a*) Detailed circuit; (*b*) approximate, equivalent circuit to (*a*). In both circuits, the absorber plate absorbs incident energy equal to $\alpha_s I_s$, where $I_s = \bar{\tau}_s I_c$. Collector assumed to be at uniform temperature $T_c$.

3. Heat flow is one-dimensional through the covers as well as through the back insulation.
4. The headers connecting the tubes cover only a small area of the collector and provide uniform flow to the tubes.
5. The sky can be treated as though it were a black-body source for infrared radiation at an equivalent sky temperature.
6. The irradiation on the collector plate is uniform.

For a quantitative analysis consider a location at *x,y* on a typical flat-plate collector as shown in Fig. 4.3. Let the plate temperature at this point be $T_c(x,y)$ and assume solar energy is absorbed at the rate $I_s \alpha_s$. Part of this energy is then transferred as heat to the working fluid and if the collector is in the steady state, the other part is lost as heat to the ambient air if $T_c > T_a$. Some of the heat loss occurs through the bottom of the collector. It passes first through the back insulation by conduction $(R_1 = l/k)$ and then by convection $(R_2 = 1/h_{c,\,bottom})$ to the environment. If the collector is part of a sloping roof, the environment for the back losses is the air inside the building; if the collector is installed on a flat roof, the environment is the same air to which the heat is lost from the upper surface of the plate. In a well-insulated collector the external convective resistance is much smaller than the insulation resistance and the back-loss conductance then becomes simply

$$U_b = \frac{1}{R_1} = \frac{k_i}{l_i} \tag{4.5}$$

where $k_i$ and $l_i$ are, respectively, the thermal conductivity and thickness of the insulation.

Evaluation of edge losses is quite difficult for most collectors. However, in a well-designed collector, edge losses are small and need not be predicted with precision. If the insulation around the edges is of the same thickness as the back, the edge losses can be estimated (see Ref. 32, Table 3-1) by assuming one-

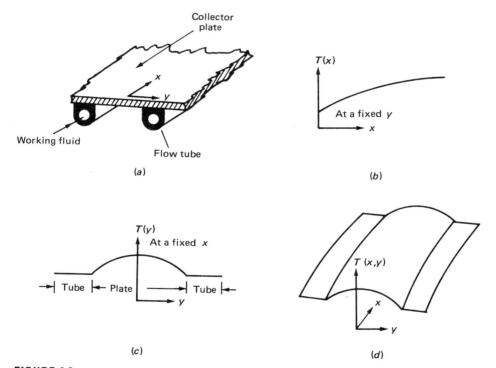

**FIGURE 4.3** Qualitative temperature distribution in the absorber plate of a flat-plate collector. (*a*) Schematic diagram of absorber; (*b*) temperature profile in the direction of the flow of the working fluid; (*c*) temperature profile at given $x$; (*d*) temperature distribution in the absorber plate.

dimensional conduction around the perimeter, but adding a constant term for the edge. Thus, for a collector of area $l_1 l_2$ and thickness $l_3$, and with a layer of insulation $l_i$ thick at the bottom and along the sides (see Fig. 4.1), the effective back loss would be approximately

$$q_{back\ loss} = \frac{A_c k_i}{l_i}(T_c - T_a)\left[1 + \frac{(2.1 + l_i)(l_1 + l_2)}{l_1 l_2}\right] \qquad (4.6)$$

where $k_i$ is the thermal conductivity of the insulation.

The conductance for the upper surface of the collector can be evaluated by determining the thermal resistance $R_3$, $R_4$, and $R_5$ in Fig. 4.2$a$. Heat is transferred between the cover and the second glass plate and between the two glass plates by convection and radiation in parallel. Except for absorptance of solar energy by the second glass plate, the relations for the rate of heat transfer between $T_c$ and $T_{g2}$ and between $T_{g2}$ and $T_{g1}$ are the same. Thus, the rate of heat transfer per unit surface area of collector between the absorber plate and the second glass cover is

$$q_{top\ loss} = A_c \bar{h}_{c2}(T_c - T_{g2}) + \frac{\sigma A_c(T_c{}^4 - T_{g2}{}^4)}{1/\epsilon_{p,i} + 1/\epsilon_{g2,i} - 1} \qquad (4.7)$$

where $h_{c2}$ = convection heat-transfer coefficient between the plate and the second glass cover

$\quad\quad \epsilon_{p,i}$ = infrared emittance of the plate

$\quad\quad \epsilon_{g2,i}$ = infrared emittance of the second cover

As shown in Chapter 3, the radiation term can be linearized; then Eq. (4.7) becomes

$$q_{top\ loss} = (\bar{h}_{c2} + h_{r2})A_c(T_c - T_{g2}) = \frac{T_c - T_{g2}}{R_3} \qquad (4.8)$$

where

$$h_{r2} = \frac{\sigma(T_c + T_{g2})(T_c{}^2 + T_{g2}{}^2)}{(1/\epsilon_{p,i} + 1/\epsilon_{g2,i} - 1)}$$

A similar derivation for the rate of heat transfer between the two cover plates gives

$$q_{top\ loss} = (\bar{h}_{c1} + h_{r1})A_c(T_{g2} - T_{g1}) = \frac{T_{g2} - T_{g1}}{R_4} \qquad (4.9a)$$

where

$$h_{r1} = \frac{\sigma(T_{g1} + T_{g2})(T_{g1}{}^2 + T_{g2}{}^2)}{(1/\epsilon_{g1,i} + 1/\epsilon_{g2,i} - 1)} \qquad (4.9b)$$

and $\bar{h}_{c1}$ = convection heat-transfer coefficient between two transparent covers.

The emittances of the cover will, of course, be the same if they are made of the same material. However, economic advantages can sometimes be achieved by using a plastic cover between an outer cover of glass and the plate and in such a sandwich construction, the radiative properties of the two covers may not be the same.

The equation for the thermal resistance between the upper surface of the outer collector cover and the ambient air has a form similar to the two preceding relations, but the convection heat-transfer coefficient must be evaluated differently. If the air is still, free-convection relations should be used, but when wind is blowing over the collector, forced-convection correlations apply as shown in Chapter 3. Radiation exchange occurs between the top cover and the sky at $T_{sky}$, whereas convection heat exchange occurs between $T_{g1}$ and the ambient air at $T_{air}$. For convenience we shall refer both conductances to the air temperature. This gives

$$q_{\text{top loss}} = (h_{c,\infty} + h_{r,\infty})(T_{g1} - T_{air}) = \frac{T_{g1} - T_{air}}{R_5} \tag{4.10}$$

where

$$h_{r,\infty} = \epsilon_{g1,i}\sigma(T_{g1} + T_{sky})(T_{g1}{}^2 + T_{sky}^2)\left[\frac{(T_{g1} - T_{sky})}{(T_{g1} - T_{air})}\right]$$

The total heat-loss conductance $U_c$ can then be expressed in the form

$$U_c = \frac{1}{R_1 + R_2} + \frac{1}{R_3 + R_4 + R_5} \tag{4.11}$$

for a double-glazed flat-plate collector.

Evaluation of the collector heat-loss conductance defined by Eq. (4.11) requires iterative solution of Eqs. (4.9$a$) and (4.10), because the unit radiation conductances are functions of the cover and plate temperatures, which are not known a priori. A simplified procedure for calculating $U_c$ for collectors with all covers of the same material, which is often sufficiently accurate and more convenient to use, has been suggested by Hottel and Woertz (25) and Klein (27). It is also suitable for application to collectors with selective surfaces. For this approach the collector top loss in $W$ is written in the form

$$q_{\text{top loss}} = \frac{(T_c - T_a)A_c}{N/(C/T_c)[(T_c - T_a)/(N + f)]^{0.33} + 1/h_{c,\infty}}$$

$$+ \frac{\sigma(T_c{}^4 - T_a{}^4)A_c}{1/[\epsilon_{p,i} + 0.05N(1 + \epsilon_{p,i})] + (2N + f - 1)/\epsilon_{g,i} - N} \tag{4.12}$$

where $f = (1 - 0.04h_{c,\infty} + 0.0005h_{c,\infty}{}^2)(1 + 0.091N)$
$C = 365.9\ (1 - 0.00883\beta + 0.00013\beta^2)$

$N$ = number of covers

$h_{c,\infty} = 5.7 + 3.8V$ (m/sec)

$\epsilon_{g,i}$ = infrared emittance of the covers

The values of $q_{\text{top loss}}$ calculated from Eq. (4.12) agreed closely with the values obtained from Eq. (4.11) for 972 different observations encompassing the following conditions:

$320 < T_c < 420$ K
$260 < T_a < 310$ K
$0.1 < \epsilon_{p,i} < 0.95$
$0 \leqslant V \leqslant 10$ m/sec
$1 \leqslant N \leqslant 3$
$0 \leqslant \beta \leqslant 90$

The standard deviation of the differences in $U_c = q_{\text{top loss}}/A_c(T_c - T_a)$ was 0.14 W/m² · K for these comparisons.

## Thermal Analysis of Flat-Plate Collector-Absorber Plate

In order to determine the efficiency of a solar collector the rate of heat transfer to the working fluid must be calculated. If transient effects are neglected (26, 27, 56), the rate of heat transfer to the fluid flowing through a collector depends on only the temperature of the collector surface from which heat is transferred by convection to the fluid, the temperature of the fluid, and the heat-transfer coefficient between the collector and the fluid. To analyze the rate of heat transfer consider first the condition at a cross section of the collector with flow ducts of rectangular cross sections as shown in Fig. 4.4. Solar radiant energy impinges on the upper face of the collector plate. A part of the total solar radiation falls on the upper surface of the flow channels, while another part is incident on the plates connecting any two adjacent flow channels. The latter is conducted in a transverse direction toward the flow channels. The temperature is a maximum at any midpoint between adjacent channels and the collector plate acts as a fin attached to the walls of the flow channel. The thermal performance of a fin can be expressed in terms of its efficiency. The fin efficiency $\eta_f$ is defined as the ratio of the rate of heat flow through the real fin to the rate of heat flow through a fin of infinite thermal conductivity, that is, a fin at a uniform temperature. We shall now derive a relation to evaluate this efficiency for a flat-plate solar collector.

If $U_c$ is the overall unit conductance from the collector-plate surface to the ambient air, the rate of heat loss from a given segment of the collector plate at $x,y$ in Fig. 4.4 is

$$q(x,y) = U_c[T_c(x,y) - T_a] \, dx \, dy \qquad (4.13)$$

where $T_c$ = local collector-plate temperature $(T_c > T_a)$
$T_a$ = ambient air temperature
$U_c$ = overall unit conductance between the plate and the ambient air

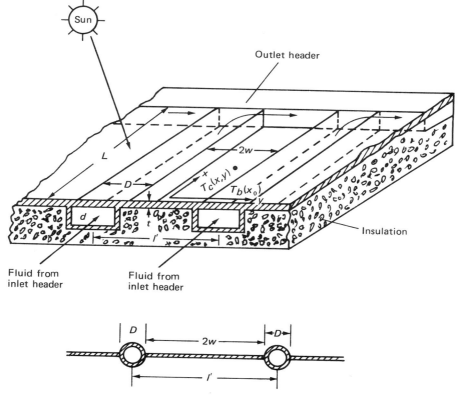

**FIGURE 4.4**  Sketch showing coordinates and dimensions for collector plate and fluid ducts.

$U_c$ includes the effects of radiation and free convection between the plates, the radiative and convective transfer between the top of the cover and the environment, and conduction through the insulation. Its quantitative evaluation has been previously considered.

If conduction in the $x$ direction is negligible, a heat balance at a given distance $x_0$ for a cross section of the flat-plate collector per unit length in the $x$ direction can be written in the form

$$\alpha_s I_s\, dy - U_c(T_c - T_a)\, dy + \left(-kt\,\frac{dT_c}{dy}\bigg|_{y,x_0}\right) - \left(-kt\,\frac{dT_c}{dy}\bigg|_{y+dy,x_0}\right) = 0 \quad (4.14)$$

If the plate thickness $t$ is uniform and the thermal conductivity of the plate is independent of temperature, the last term in Eq. (4.14) is

$$\frac{dT_c}{dy}\bigg|_{y+dy,x_0} = \frac{dT_c}{dy}\bigg|_{y,x_0} + \left(\frac{d^2T_c}{dy^2}\right)_{y,x_0} dy$$

and Eq. (4.14) can be cast into the form of a second-order differential equation:

$$\frac{d^2 T_c}{dy^2} = \frac{U_c}{kt}\left[T_c - \left(T_a + \frac{\alpha_s I_s}{U_c}\right)\right]$$
(4.15)

The boundary conditions for the system described above at a fixed $x_0$, are

1. At the center between any two ducts the heat flow is 0, or at $y = 0$, $dT_c/dy = 0$.
2. At the duct the plate temperature is $T_b(x_0)$, or at $y = w = (1' - D)/2$, $T_c = T_b(x_0)$, where $T_b(x_0)$ is the fin-base temperature.

If we let $m^2 = U_c/kt$ and $\phi = T_c - (T_a + \alpha_s I_s/U_c)$, Eq. (4.4) becomes

$$\frac{d^2 \phi}{dy^2} = m^2 \phi$$
(4.16)

subject to the boundary conditions

$$\frac{d\phi}{dy} = 0 \quad \text{at} \quad y = 0$$

$$\phi = T_b(x_0) - \left(T_a + \frac{\alpha_s I_s}{U_c}\right) \quad \text{at} \quad y = w$$

The general solution of Eq. (4.16) is

$$\phi = C_1 \sinh my + C_2 \cosh my$$
(4.17)

The constants $C_1$ and $C_2$ can be determined by substituting the two boundary conditions and solving the two resulting equations for $C_1$ and $C_2$. This gives

$$\frac{T_c - (T_a + \alpha_s I_s/U_c)}{T_b(x_0) - (T_a + \alpha_s I_s/U_c)} = \frac{\cosh my}{\cosh mw}$$
(4.18)

From the preceding equation the rate of heat transfer to the conduit from the portion of the plate between two conduits can be determined by evaluating the temperature gradient at the base of the fin, or

$$q_{\text{fin}} = -kt \left.\frac{dT_c}{dy}\right|_{y=w} = \frac{1}{m}\{\alpha_s I_s - U_c[T_b(x_0) - T_a]\tanh mw\}$$
(4.19)

Since the conduit is connected to fins on both sides, the total rate of heat transfer is

$$q_{\text{total}}(x_0) = 2w\{\alpha_s I_s - U_c[T_b(x_0) - T_a]\} \frac{\tanh mw}{mw} \qquad (4.20)$$

If the entire fin were at the temperature $T_b(x_0)$, a situation corresponding physically to a plate of infinitely large thermal conductivity, the rate of heat transfer would be a maximum, $q_{\text{total, max}}$. As mentioned previously, the ratio of the rate of heat transfer with a real fin to the maximum rate obtainable is the fin efficiency $\eta_f$. With this definition, Eq. (4.20) can be written in the form

$$q_{\text{total}}(x_0) = 2w\eta_f\{\alpha_s I_s - U_c[T_b(x_0) - T_a]\} \qquad (4.21)$$

where $\eta_f \equiv \tanh mw/mw$.

The fin efficiency $\eta_f$ is plotted as a function of the dimensionless parameter $w(U_c/kt)^{1/2}$ in Fig. 4.5. When the fin efficiency approaches unity, the maximum portion of the radiant energy impinging on the fin becomes available for heating the fluid.

In addition to the heat transferred through the fin, the energy impinging on the portion of the plate above the flow passage provides useful energy. The rate of useful energy from this region available to heat the working fluid is

$$q_{\text{duct}}(x_0) = D\{\alpha_s I_s - U_c[T_b(x_0) - T_a]\} \qquad (4.22)$$

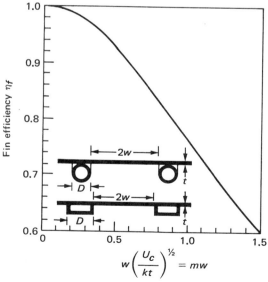

**FIGURE 4.5** Fin efficiency for tube and sheet flat-plate solar collectors.

Thus, the useful energy per unit length in the flow direction becomes

$$q_u(x_0) = (D + 2w\eta)\{\alpha_s I_s - U_c[T_b(x_0) - T_a]\} \qquad (4.23)$$

The energy $q_u(x_0)$ must be transferred as heat to the working fluid. If the thermal resistance of the metal wall of the flow duct is negligibly small and there is no contact resistance between the duct and the plate, the rate of heat transfer to the fluid is

$$q_u(x_0) = 2(D + d)\bar{h}_{c,i}[T_b(x_0) - T_f(x_0)] \qquad (4.24)$$

Contact resistance may become important in poorly manufactured collectors in which the flow duct is clamped or glued to the collector plate. Collectors manufactured by such methods are usually not satisfactory.

### Collector Efficiency Factor

To obtain a relation for the useful energy delivered by a collector in terms of known physical parameters, the fluid temperature, and the ambient temperature, the collector temperature must be eliminated from Eqs. (4.23) and (4.24). Solving for $T_b(x_0)$ in Eq. (4.24) and substituting this relation in Eq. (4.23) gives

$$q_u(x_0) = l'F'\{\alpha_s I_s - U_c[T_f(x_0) - T_a]\} \qquad (4.25)$$

where $F'$ is called the collector efficiency factor (5). It is given by

$$F' = \frac{1/U_c}{l'\left[1/U_c(D + 2w\eta_f) + 1/\bar{h}_{c,i}(2D + 2d)\right]} \qquad (4.26)$$

Physically, the denominator in Eq. (4.26) is the thermal resistance between the fluid and the environment, whereas the numerator is the thermal resistance between the collector surface and the ambient air. The collector-plate efficiency factor $F'$ depends on $U_c$, $\bar{h}_{c,i}$, and $\eta_f$. It is only slightly dependent on temperature and can for all practical purposes be treated as a design parameter. Typical values for the factors determining the value of $F'$ are given in Table 4.1.

The collector efficiency factor increases with increasing plate thickness and plate thermal conductivity, but decreases with increasing distance between flow channels. Also, increasing the heat-transfer coefficient between the walls of the flow channel and the working fluid increases $F'$, but an increase in the overall conductance $U_c$ will cause $F'$ to decrease.

### Collector Heat-Removal Factor

Equation (4.25) yields the rate of heat transfer to the working fluid at a given point $x$ along the plate for specified collector and fluid temperatures. However, in a real collector the fluid temperature increases in the direction of flow as heat is

**TABLE 4.1** Typical Values for the Parameters That Determine the Collector Efficiency Factor $F'$ for a Flat-Plate Collector in Eq. (4.26)

|  |  | $U_c$ |
|---|---|---|
| 2 glass covers | 4 W/m² · K | 0.685 Btu/hr · ft² · °F |
| 1 glass cover | 8 W/m² · K | 1.37 Btu/hr · ft² · °F |
|  |  | $kt$ |
| copper plate, 1 mm thick | 0.4 W/K | 41.5 Btu/hr · °F |
| steel plate, 1 mm thick | 0.005 W/K | 0.52 Btu/hr · °F |
|  |  | $\bar{h}_{c,i}$ |
| water in laminar flow forced convection | 300 W/m² · K | 52 Btu/hr · ft² · °F |
| water in turbulent flow forced convection | 1500 W/m² · K | 254 Btu/hr · ft² · °F |
| air in turbulent forced convection | 100 W/m² · K | 17.6 Btu/hr · ft² · °F |

transferred to it. An energy balance for a section of flow duct $dx$ can be written in the form

$$\dot{m}c_p(T_f|_{x+dx} - T_f|_x) = q_u(x)\,dx \tag{4.27}$$

Substituting Eq. (4.25) for $q_u(x)$ and $T_f(x) + dT_f(x)/dx\,dx$ for $T_f|_{x+dx}$ in Eq. (4.27) gives the differential equation

$$\dot{m}c_p\frac{dT_f(x)}{dx} = l'F'\{\alpha_s I_s - U_c[T_f(x) - T_a]\} \tag{4.28}$$

Separating the variables gives, after some rearranging,

$$\frac{dT_f(x)}{T_f(x) - T_a - \alpha_s I_s/U_c} = \frac{l'F'U_c}{\dot{m}c_p}\,dx \tag{4.29}$$

Equation (4.29) can be integrated and solved for the outlet temperature of the fluid $T_{f,\,out}$ for a duct length $L$, and for the fluid inlet temperature $T_{f,\,in}$ if we assume that $F'$ and $U_c$ are constant, or

$$\frac{T_{f,\,out} - T_a - \alpha_s I_s/U_c}{T_{f,\,in} - T_a - \alpha_s I_s/U_c} = \exp -\left(\frac{U_c l'F'L}{\dot{m}c_p}\right) \tag{4.30}$$

To compare the performance of a real collector with the thermodynamic optimum, it is convenient to define the heat-removal factor $F_R$ as the ratio of the actual rate of heat transfer to the working fluid to the rate of heat transfer at the minimum temperature difference between the absorber and the environment. The thermodynamic limit corresponds to the condition of the working fluid remaining at the inlet temperature throughout the collector. This can be approached when the fluid velocity is very high. From its definition $F_R$ can be expressed as

$$F_R = \frac{Gc_p(T_{f,\text{out}} - T_{f,i})}{\alpha_s I_s - U_c(T_{f,i} - T_a)} \tag{4.31}$$

where $G$ is the flow rate per unit surface area of collector $\dot{m}/A_c$. By regrouping the right-hand side of Eq. (4.31) and combining with Eq. (4.30), it can easily be verified that

$$F_R = \frac{Gc_p}{U_c} \left[ 1 - \frac{\alpha_s I_s/U_c - (T_{f,\text{out}} - T_a)}{\alpha_s I_s/U_c - (T_{f,\text{in}} - T_a)} \right]$$

$$F_R = \frac{Gc_p}{U_c} \left[ 1 - \exp\left(-\frac{U_c F'}{Gc_p}\right) \right] \tag{4.32}$$

Inspection of the above relation shows that $F_R$ increases with increasing flow rate and approaches as an upper limit $F'$, the collector efficiency factor. Since the numerator of the right-hand side of Eq. (4.31) is $q_u$, the rate of useful heat transfer can now be expressed in terms of the fluid inlet temperature, or

$$q_u = A_c F_R \left[ \alpha_s I_s - U_c(T_{f,\text{in}} - T_a) \right] \tag{4.33}$$

This is a convenient form for design, because the fluid inlet temperature to the collector is usually known or can be specified.

## EXAMPLE 4.1

Calculate the averaged hourly and daily efficiency of a water solar collector on January 15, in Boulder, Colorado. The collector is tilted at an angle of $60°$ and has an overall conductance of 8.0 W/m$^2$ · K on the upper surface. It is made of copper tubes, with a 1-cm ID, 0.05 cm thick, which are connected by a 0.05-cm-thick plate at a center-to-center distance of 15 cm. The heat-transfer coefficient for the water in the tubes is 1500 W/m$^2$ · K, the cover transmittance is 0.9, and the solar absorptance of the copper surface is 0.9. The collector is 1 m wide and 2 m long, the water inlet temperature is 330 K, and the water flow rate is 0.02 kg/sec. The horizontal insolation (total) $I_h$ and the environmental temperature are tabulated below. Assume that diffuse radiation accounts for 25 percent of the total insolation.

## SOLUTION

The total radiation received by the collector is calculated from Eq. (2.42):

$$\bar{I}_c = I_{c,\text{diffuse}} + I_{c,\text{beam}} = I_h \times 0.25 \left(\tfrac{1}{2} \cos^2 60\right) + I_h(1 - 0.25)R_b$$

| Time (hr) | $I_h$ (W/m$^2$) | $T_{amb}$ (K) |
|---|---|---|
| 7–8 | 12 | 270 |
| 8–9 | 80 | 280 |
| 9–10 | 192 | 283 |
| 10–11 | 320 | 286 |
| 11–12 | 460 | 290 |
| 12–13 | 474 | 290 |
| 13–14 | 395 | 288 |
| 14–15 | 287 | 288 |
| 15–16 | 141 | 284 |
| 16–17 | 32 | 280 |

The tilt factor $R_b$ is obtained from its definition in Chapter 2 [see Eq. (2.37)]:

$$R_b = \frac{\cos i}{\sin \alpha} = \frac{\sin (L - \beta) \sin \delta_s + \cos (L - \beta) \cos \delta_s \cos h_s}{\sin L \sin \delta_s + \cos L \cos \delta_s \cos h_s}$$

where $L = 40°$  $\delta_s = -21.1$ on January 15 (from Fig. 2.8), and $\beta = 60°$. The hour angle $h_s$ equals $15°$ for each hour away from noon.

The fin efficiency is obtained from Eq. (4.21):

$$\eta_f = \frac{\tanh m(l' - D)/2}{m(l' - D/2)}$$

where

*cu thermal conductivity at 300°K 386 (W/m·K) ρ₁₂₀*

$$m = \left(\frac{U_c}{kt}\right)^{1/2} = \left(\frac{8}{390 \times 5 \times 10^{-4}}\right)^{1/2} = 6.4$$

$$\eta_f = \frac{\tanh 6.4(0.15 - 0.01)/2}{6.4(0.15 - 0.01)/2} = 0.938$$

The collector efficiency factor $F'$ is, from Eq. (4.26),

$$F' = \frac{1/U_c}{l'\,[1/U_c(D + 2w\eta_f) + 1/h_{c,i}\pi D]}$$

$$= \frac{1/8.0}{0.15\,[1/8.0(0.01 + 0.14 \times 0.938) + 1/1500\pi \times 0.01]} = 0.92$$

Then we obtain the heat-removal factor from Eq. (4.32):

$$F_R = \frac{Gc_p}{U_c}\left[1 - \exp - \left(\frac{U_c F'}{Gc_p}\right)\right]$$

| Time (hr) | $I_h$ (W/m²) | $R_b$ | $I_d$ (W/m²) | $I_{b,c}$ (W/m²) | $I_{tot}$ (W/m²) | $q_u$ (W) | $T_{amb}$ (K) | $\eta_c$ |
|---|---|---|---|---|---|---|---|---|
| 7–8 | 12 | 10.9 | 0.4 | 98 | 98 | 0 | 270 | 0 |
| 8–9 | 80 | 3.22 | 2.5 | 193 | 196 | 0 | 280 | 0 |
| 9–10 | 192 | 2.44 | 6.0 | 351 | 357 | 0 | 283 | 0 |
| 10–11 | 320 | 2.18 | 10.0 | 522 | 532 | 133 | 286 | 0.185 |
| 11–12 | 460 | 2.08 | 14.4 | 717 | 732 | 460 | 290 | 0.356 |
| 12–13 | 474 | 2.08 | 14.8 | 739 | 754 | 491 | 290 | 0.366 |
| 13–14 | 395 | 2.18 | 12.3 | 645 | 657 | 331 | 288 | 0.2991 |
| 14–15 | 287 | 2.49 | 9.0 | 525 | 534 | 163 | 288 | 0.201 |
| 15–16 | 141 | 3.22 | 4.4 | 341 | 345 | 0 | 284 | 0 |
| 16–17 | 32 | 10.9 | 1.0 | 261 | 262 | 0 | 280 | 0 |

$$F_R = \frac{0.01 \times 4184}{8.0} \left[ 1 - \exp - \left( \frac{8.0 \times 0.922}{0.01 \times 4184} \right) \right] = 0.845$$

From Eq. (4.33), the useful heat delivery rate is

$$q_u = A_c F_R \left[ \alpha_s I_s - U_c (T_{f,\,in} - T_{amb}) \right]$$

In the relation above $I_s$ is the radiation incident on the collector. If the transmittance of the glass is 0.9,

$$I_s = \tau I_{tot} = 0.9 I_{tot}$$

and

$$q_u = 2 \times 0.0845 \left[ I_{tot} \times 0.81 - 8.0 \, (T_{f,in} - T_{amb}) \right]$$

The efficiency of the collector is $\eta_b = q_u / A I_{tot}$ and the hourly averages are calculated in the table above.

Thus, $\Sigma I_{tot} = 4954$ W/m² and $\Sigma q_u = 2103$ W. The daily average is obtained by summing the useful energy for those hours during which the collector delivers heat and dividing by the total insolation between sunrise and sunset. This yields

$$\bar{\eta}_{c,\,day} = \frac{\Sigma q_u}{\Sigma A_c I_{tot}} = 100 \, \frac{1578}{2 \times 4469} = 17.7 \text{ percent}$$

## Transient Effects

The preceding analysis assumed that steady-state conditions exist during the operation of the collector. Under actual operating conditions the rate of insolation will vary and the ambient temperature and the external wind conditions may change. To determine the effect of changes in these parameters on the performance of a collector it is necessary to make a transient analysis that takes the thermal capacity of the collector into account.

As shown in Ref. (28), the effect of collector thermal capacitance is the sum of two contributions: the *collector storage* effect, resulting from the heat required to bring the collector up to its final operating temperature, and the *transient* effect, resulting from fluctuations in the meteorological conditions. Both effects result in a net loss of energy delivered compared with the predictions from the zero capacity analysis. This loss is particularly important on a cold morning when all of the solar energy absorbed by the collector is used to heat the hardware and the working fluid, thus delaying the delivery of useful energy for some time after the sun has come up.

Transient thermal analyses can be made with a high degree of precision (32), but the analytical predictions are no more accurate than the weather data and the overall collector conductance. For most engineering applications a simpler approach is therefore satisfactory (11). For this approach it will be assumed that the absorber plate, the ducts, the back insulation, and the working fluid are at the same temperature. If back losses are neglected, an energy balance on the collector plate and the working fluid for a single-glazed collector delivering no useful energy can be written in the form

$$(\overline{mc})_p \frac{d\bar{T}_p(t)}{dt} = A_c I_s \alpha_s + A_c U_p [\bar{T}_g(t) - \bar{T}_p(t)] \tag{4.34}$$

where $(\overline{mc})_p$ is the sum of the thermal capacities of the plate, the fluid, and the insulation, and $U_p$ is the conductance between the absorber plate at $\bar{T}_p$ and its cover at $\bar{T}_g$. Similarly, a heat balance on the collector cover gives

$$(mc)_g \frac{d\bar{T}_g(t)}{dt} = A_c U_p [\bar{T}_p(t) - \bar{T}_g(t)] - A_c U_\infty [\bar{T}_g(t) - T_a] \tag{4.35}$$

where $U_\infty = (h_{c,\infty} + h_{r,\infty})$ [see Eq. (4.9b)]

$(mc)_g$ = thermal capacity of the cover plate

Equations (4.34) and (4.35) can be solved simultaneously and the transient heat loss can then be determined by integrating the instantaneous loss over the time during which transient effects are pronounced. A considerable simplification in the solution is possible if one assumes that at any time the collector heat loss and the cover heat loss are proportional, as in a quasi-steady state, so that

$$U_\infty A_c [\bar{T}_g(t) - T_a] = U_c A_c [\bar{T}_p(t) - T_a] \tag{4.36}$$

Then, for a given air temperature, differentiation of Eq. (4.36) gives

$$\frac{d\bar{T}_g(t)}{dt} = \frac{U_c}{U_\infty} \frac{d\bar{T}_p(t)}{dt} \tag{4.37}$$

Adding Eqs. (4.34), (4.35), and (4.37) gives a single differential equation for the plate temperature

$$\left[(\overline{mc})_p + \frac{U_c}{U_\infty}(mc)_g\right]\frac{d\overline{T}_p(t)}{dt} = [\alpha_s I_s - U_c(\overline{T}_p(t) - T_a)]A_c \qquad (4.38)$$

Equation (4.38) can be solved directly for given values of $I_s$ and $T_a$. Since meteorological conditions are usually known only as hourly averages, this restriction is no more limiting than the available input data. The solution to Eq. (4.38) then gives the plate temperature as a function of time, for an initial plate temperature $T_{p,0}$, in the form

$$\overline{T}_p(t) - T_a = \frac{\alpha_s I_s}{U_c} - \left[\frac{\alpha_s I_s}{U_c} - (T_{p,0} - T_a)\right]\exp-\left[\frac{U_c A_c t}{(\overline{mc})_p + (U_c/U_\infty)_c(mc)_g}\right] \qquad (4.39)$$

Collectors with more than one cover can be treated similarly, as shown in Ref. (11).

For a transient analysis the plate temperature $\overline{T}_p$ can be evaluated at the end of a specified time period if the initial value of $\overline{T}_p$ and the values of $\alpha_s$, $I_s$, $U_c$, and $T_a$ during the specified time are known. Repeated application of Eq. (4.39) provides an approximate method of evaluating the transient effects. An estimate of the net decrease in useful energy delivered can be obtained by multiplying the effective heat capacity of the collector, given by $(\overline{mc})_p + (U_c/U_\infty)(mc)_g$, by the temperature rise necessary to bring the collector to its operating temperature. Note that the parameter $[(\overline{mc})_p + (U_c/U_\infty)(mc)_g]/U_c A_c$ is the *time constant* of the collector (11, 32) and small values of this parameter will reduce losses resulting from transient effects.

## EXAMPLE 4.2

Calculate the temperature rise between 8 and 10 a.m. of a 1 m × 2 m single-glazed water collector with a 0.3-cm-thick glass cover if the heat capacities of the plate, water, and back insulation, are 5, 3, and 2 kJ/K, respectively. Assume that the unit surface conductance from the cover to ambient air is 18 W/m$^2$ · K and the unit surface conductance between the collector and the ambient air is $U_c = 6$ W/m$^2$ · K. Assume that the collector is initially at the ambient temperature. The absorbed insolation $\alpha_s I_s$ during the first hour averages 90 W/m$^2$ and between 9 and 10 a.m., 180 W/m$^2$. The air temperature between 8 and 9 a.m. is 273 K and that between 9 and 10 a.m., 278 K.

## SOLUTION

The thermal capacitance of the glass cover is $(mc)_g = (\rho V c_p)_g = (2500$ kg/m$^3$) (1 m × 2 m × 0.03 m) (1 kJ/kg · K) = 15 kJ/K. The combined collector, water, and insulation thermal capacity is equal to

$$(\overline{mc})_p + \frac{U_c}{U_\infty}(mc)_g = 5 + 3 + 2 + 0.3 \times 15 = 15.5 \text{ kJ/K}$$

From Eq. (4.39) the temperature rise of the collector between 8 and 9 a.m. is

$$\overline{T}_p - T_a = \frac{\alpha_s I_s}{U_c} \left\{ 1 - \exp - \left[ \frac{U_c A_c t}{(\overline{mc})_p + (U_c/U_\infty)(mc)_g} \right] \right\}$$

$$= 15 \left[ 1 - \exp - \left( \frac{2 \times 6 \times 3600}{15,500} \right) \right] = 15 \times 0.944 = 14.2 \text{ K}$$

Thus, at 9 a.m. the collector temperature will be 287.2 K. Between 9 and 10 a.m. the collector temperature will rise as shown below:

$$T_p = T_a + \frac{\alpha_s I_s}{U_c} - \left[ \frac{\alpha_s I_s}{U_c} - (T_{p,0} - T_a) \right] \exp - \left[ \frac{U_c A_c t}{(\overline{mc})_p + (U_c/U_\infty)(mc)_g} \right]$$

$$= 278 + \frac{180}{6} - (30 - 9.2)0.056 = 306.3 \text{ K } (91°\text{F})$$

Thus, at 10 a.m. the collector temperature has achieved a value sufficient to deliver useful energy at a temperature level of 306 K.

## Experimental Testing of Collectors

The performance of solar systems for space heating and cooling depends largely on the performance of the solar collectors. Thus, the measurement of the performance of solar collectors is an important and necessary step for an understanding of the total system function. Although analytical design procedures, particularly for flat-plate solar collectors, are in reasonably good agreement with test results, it is preferable to base a system design on actual collector test data whenever available.

There are two basic methods for testing collectors: the *instantaneous* and the *calorimetric* procedures. Each allows determination of fundamental collector characteristics, and each has advantages and disadvantages.

For the instantaneous method it is only necessary to measure simultaneously the mass flow rate of the heat-transfer fluid flowing through the collector, its temperature difference, the collector inlet and outlet, and the insolation incident on the plane of the collector. The instantaneous efficiency can then be calculated from the relation (22, 48)

$$\eta_c = \frac{q_u/A_c}{I_c} = \frac{\dot{m}c_p(T_{f,\text{ out}} - T_{f,\text{in}})}{I_c} \tag{4.40}$$

where $\eta_c$ = solar collector efficiency

   $q_u$ = useful heat output, W (Btu/hr)

   $A_c$ = cross-sectional area, $m^2$ ($ft^2$)

   $I_c$ = total solar energy incident upon the plane of the collector per unit time per unit area, $W/m^2$ (Btu/hr $\cdot$ $ft^2$)

   $\dot{m}$ = mass flow rate of the heat-transfer fluid per unit of collector cross-sectional area, kg/sec $\cdot$ $m^2$ (lb/hr $\cdot$ $ft^2$)

   $c_p$ = specific heat of the heat-transfer fluid, J/kg $\cdot$ K (Btu/lb$_m$ $\cdot$ °F)

   $T_{f,\,out}$ = temperature of the heat-transfer fluid leaving the collector, K (°F)

   $T_{f,\,in}$ = temperature of the heat-transfer fluid entering the collector, K (°F)

The instantaneous method is particularly useful in testing a sample of a collector field on site prior to installation of the entire array, because the measurements are relatively simple and can be carried out with commercially available instrumentation to an accuracy of approximately 10 percent. Moreover, with the instantaneous procedure, one only needs to take measurements on and around the collector, which is an advantage for on-site procedures.

The calorimetric procedure employs a closed system in which the time rate of change of temperature of a constant thermal mass is measured and related to the incident solar energy by the relation

$$\eta_c = \frac{q_u/A_c}{I_c} = \frac{(m/A_c)c_p' \, dT/dt}{I_c} \tag{4.41}$$

where $m$ = the mass of the medium in the calorimeter, kg (lb$_m$)

   $c_p'$ = the average specific heat of the medium in the calorimeter, J/kg $\cdot$ K (Btu/lb$_m$ $\cdot$ °F)

   $T$ = the average temperature of the medium, K (°F)

For the calorimetric procedure, one has to measure the incident solar radiation and the time rate of change of temperature of the mass in the system. However, if the calorimetric procedure is used, a very careful analysis of the calorimeter must be conducted beforehand in order to eliminate errors resulting from temperature gradients inside and to minimize or at least accurately determine the stray thermal losses from the calorimeter. The calorimetric procedure is limited primarily to collectors using a liquid working fluid, because the heat capacity of a gas such as air is too small to be easily measurable and the collector energy cannot readily be stored and determined without transfer to another medium. A complete review of testing procedures was prepared in 1976 by the National Bureau of Standards (NBS) (23).

From Eq. (4.33), the efficiency of a flat-plate collector operating under steady-state conditions can be described by the relationship

$$\eta_c = F_R \bar{\tau}_s \alpha_s - F_R U_c \frac{T_{f,\,in} - T_a}{I_c} \tag{4.42}$$

Alternatively, by means of the collector efficiency factor defined by Eq. (4.25) and an average fluid temperature between inlet and outlet, an approximate relation, satisfactory for liquid collectors only, is

$$\eta_c = F'\bar{\tau}_s\alpha_s - F'U_c \frac{[(T_{f,\text{in}} + T_{f,\text{out}})/2] - T_a}{I_c} \qquad (4.43)$$

Regardless of the relation used, it is apparent that for a constant value of $U_c$, if the efficiency is plotted against the appropriate $\Delta T/I_c$, a straight line will result. In reality, of course, $U_c$ will vary with the operating temperature of the collector and the ambient weather condition causing some deviations from the straight-line relation. For most flat-plate collectors, however, this deviation is not a serious problem over the normal operating conditions, as shown in Fig. 4.6, which presents a typical correlation of test results taken from Ref. (48) by means of Eq. (4.43) for a flat-plate water collector.

### Proposed NBS Standard Test Method

In order to compare the performances of different collectors, the NBS has proposed a standard test method, which was officially adopted in modified form by the American Society of Heating, Refrigerating, and Air Conditioning Engineers (ASHRAE). The tests that comprise this procedure can be conducted outside under real sun conditions as well as indoors with a solar simulator described in Ref. (48). The test procedure is written in a format consistent with other standards adopted within the United States for testing building heating and cooling equipment to facilitate transfer to solar energy systems. Figure 4.7 shows the recommended testing configuration for the solar collector when the heat-transfer fluid is a liquid; Fig. 4.8 shows a comparable configuration to be used when the heat-transfer fluid is air. Reference (33) presents recommended procedures for concentrating collectors in which the fluid may undergo a phase change. Detailed requirements of the apparatus with specifications for instrumentation to be used in standard measurements of incident radiation, temperature, temperature difference, liquid flow rate,

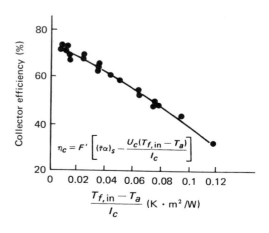

**FIGURE 4.6**  Efficiency curve for a double-glazed flat-plate liquid-heating solar collector with a selective coating on the absorber. $\dot{m} = 0.0136$ kg/sec · m²; $T_a = 29°$C; $T_{f,\text{in}} = 38{-}101°$C; $I_c = 590{-}977$ W/cm²; wind $= 3.1$ m/sec. (The tests were run indoors using a solar simulator.)

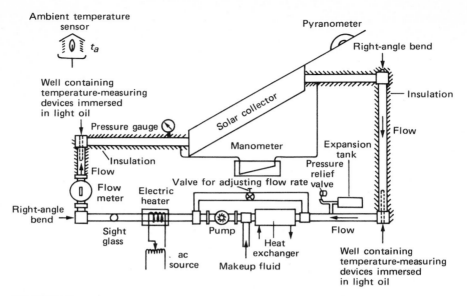

**FIGURE 4.7**  Recommended testing configuration for the solar collector when the heat-transfer fluid is a liquid. From (23).

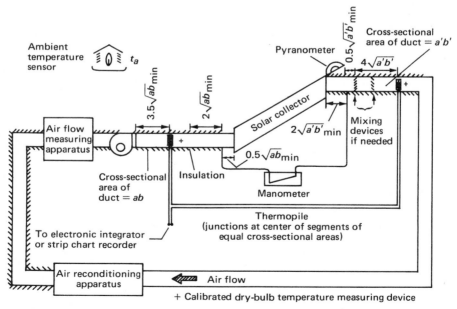

**FIGURE 4.8**  Recommended testing configuration for the solar collector when the heat-transfer fluid is air. From (23).

air flow rate, pressure, pressure drop, time, and weight are given in Ref. (22) and in ASHRAE standard 93-77.

A key requirement for reliable data is the time duration of the test. The NBS recommends that a series of tests should be conducted, each of which determines the average efficiency for 15 min over a range of temperature differences between the average fluid temperature and the ambient air. The efficiency should then be calculated from the relationship

$$\bar{\eta}_c = \frac{\displaystyle\int_0^t \dot{m}c_p(T_{f,\,\text{out}} - T_{f,\text{in}})\,dt}{\displaystyle\int_0^t I_c\,dt} \tag{4.44}$$

The flow rate for the duration of each test must be constant to within ± 1 percent and the heat-transfer fluid must have a known specific heat, which varies by less than 0.5 percent over the temperature range of the fluid during a particular test period. At least 16 data points are required for a complete test series and they must be taken symmetrically with respect to solar noon to prevent biased results resulting from possible transient effects.

During each test period, the incident solar radiation must be *quasi-steady*, as shown in Fig. 4.9. Conditions during a day in which cloud cover causes a time distribution, such as shown in Fig. 4.10, will not be acceptable. Each data point must be taken with an average insolation larger than 630 W/m² (over the 15-min average), and the incident angle between the sun and the normal from the collector surface must be less than 45°. Also, the range of ambient temperatures for the test series must be less than 30°C (54°F).

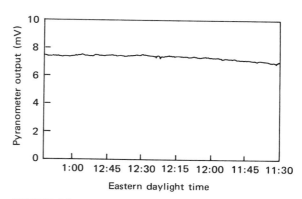

**FIGURE 4.9**  Incident solar radiation on a horizontal surface at the National Bureau of Standards site in Gaithersburg, Maryland. Steady-state conditions. From (23).

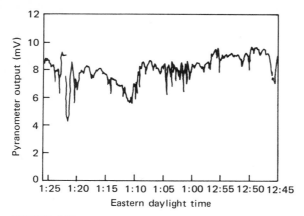

**FIGURE 4.10**  Incident solar radiation on a horizontal
surface at the National Bureau of Standards site in
Gaithersburg, Maryland. Variable conditions. From (23).

The procedure also requires that each of the measurements made and the
calculated efficiency for each data point be reported in tabular as well as graphic
form. This procedure may be adapted for use with concentrating collectors as well
as flat-plate collectors. Some modifications required for application to concentrating
collectors (33) are the following:

1. Measure the direct component as well as the total incident radiation in the plane
   of the collector and include a curve like Fig. 4.6 for each: one in which the
   insolation value in the abscissa is the total radiation and one in which it is only
   the direct component of insolation.
2. Measure the quality as well as the temperature and pressure of the working fluid
   so as to allow for a change of phase of the working fluid as it passes through the
   collector.
3. Measure separately the reflectance of the reflector surface and the absorptance
   and emittance of the selective surface on the absorber if used.

## Comparison of Analysis and Test Results

The results of thermal performance tests for solar energy collectors are generally
presented by plotting the efficiency $\eta_c$ as a function of the difference in
temperature between the inlet to the collector and the ambient air divided by the
solar flux incident on the collector $I_c$. Figure 4.11a shows the effect of temperature
difference on efficiency at given insolation values; Fig. 4.11b shows the effect of
insolation on the efficiency for specified temperature differences between the
collector fluid at the inlet and the ambient air; and Fig. 4.11c shows the correlation
plot of $\eta_c$ versus $\Delta T/I_c$. Note that the $\Delta T/I_c$ plot collapses the several curves of
Figs. 4.11a and 4.11b onto a single curve within a narrow band.

A quantitative comparison of the heat losses from a flat-plate collector for different designs is presented in Fig. 4.12. The three cases illustrated are (1) single glass cover, black absorber (Fig. 4.12a), (2) single glass cover, selective absorber (Fig. 4.12b), and (3) double glass cover, selective absorber (Fig. 4.12c).

The collector heat-loss conductance can be evaluated from Eq. (4.11) or (4.12). For the three configurations that represent a spectrum of commercial practice, the following general conclusions may be drawn:

1. If both surfaces have high emittance ($\epsilon \approx 1$), the radiative heat-transfer loss is the most important part.
2. If the absorber surface has a low emittance ($\epsilon \approx 0.1$), the convective heat loss is more important than the radiative heat loss. The first step toward substantially reducing the heat transfer between two plates is therefore to cover the heat-transfer surface with a low emittance layer.

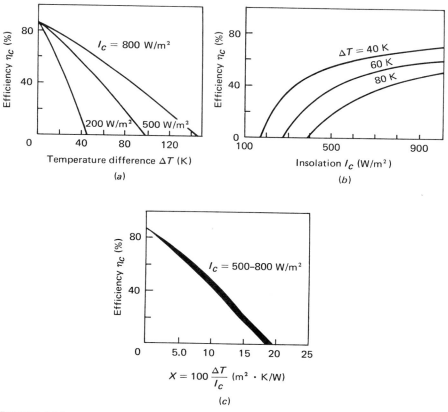

**FIGURE 4.11** Collector efficiency $\eta_c$ versus (a) temperature difference $\Delta T$; (b) total incident radiation $I_c$; (c) $\Delta T/I_c$.

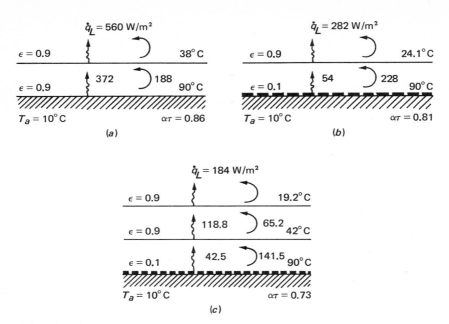

**FIGURE 4.12** Heat losses and temperature profiles for three different flat-plate collectors (ambient temperature $T_a = 10°C$, wind velocity = 3.75 m/sec): (a) single glass, black absorber; (b) single glass, selective absorber; (c) double glass, selective absorber. From (26a).

3. Convective and conductive heat transfer in a flat-plate design can be reduced by introducing an additional cover plate, which normally also reduces the radiative heat transfer. However, this measure will reduce the $\alpha\bar{\tau}$ intercept and decrease the collector performance at small differences in temperature between collector fluid and ambient, that is, at low collector temperatures.

Figure 4.13 shows the performance curve for a flat-plate collector with a selectively coated absorber surface and two glass covers at $0°$ incident flux angle. On this graph $F_R\alpha_s\bar{\tau}_s$ corresponds to the intercept of the best fit through the experimental data points on the ordinate, while $F_R U_c$ is the absolute value of the slope of the line. The values for absorptance and transmittance used in an analysis may differ somewhat from the actual values and one cannot expect perfect correlation between the experimental and analytical values. A comparison of experimental data and analytic predictions for flat-plate collectors (45) indicates that the experimental slope of the performance line is approximately 5 percent steeper than that predicted by analysis. Thus, the analytic prediction tends to underestimate the heat loss from a collector, but the difference between analysis and experiment is well within the limits of practical design information. Table 4.2 shows a comparison of performance coefficients derived from tests and analysis. Inspection of this table shows that the measured slope of the performance curve for good flat-plate collectors varies between 0.52 and 1.3 Btu/hr · ft² · °F. These performance data were obtained under steady-state conditions in a test setup shown in Fig. 4.7. Details of the method used in obtaining these data are presented in Ref. (22).

The efficiency of a collector under real operating conditions may be less than the values obtained in a performance test such as that shown in Fig. 4.7 because of the following (48):

1. The optical efficiency decreases with increasing incidence angles (measured from the normal to the collector surface); that is, the collector curve is displaced toward the abscissa (see Fig. 4.13).
2. The optical efficiency decreases with an increasing ratio of diffuse to beam radiation.
3. The optical efficiency decreases when dust collects on the outer cover or when the outer cover becomes dirty.
4. Transient effects may decrease the amount of useful energy delivered.
5. Selective surface coatings may deteriorate with time.

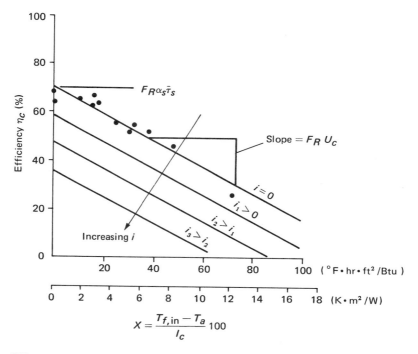

**FIGURE 4.13** Typical collector performance with 0° incident flux angle. Also shown qualitatively is the effect of incidence angle $i$, which may be quantified by $\bar{\tau}_s \alpha_s(i)/\bar{\tau}_s \alpha_s(0) = 1.0 + b_0 (1/\cos i - 1.0)$, where $b_0$ is the incidence angle modifier determined experimentally (ASHRAE 93-77) or from the Stokes and Fresnel equations.

**TABLE 4.2** Comparison of Performance Coefficients Derived from Testing and Analysis[a]

| Collector configuration | Experimental result | | | | | | Analytic prediction | | | | | |
|---|---|---|---|---|---|---|---|---|---|---|---|---|
| | Intercept | $\alpha\tau$ | $\alpha\tau_s$ | $F_R$ | Exp. slope | $U_c$ (Btu/hr · ft² · °F) | Intercept | $\alpha\tau$ | $\alpha\tau_s$ | $F_R$ | Anal. slope | $U_c$ (Btu/hr · ft² · °F) |
| Selective aluminum absorber with two glass covers (baseline collector) | 0.74 | 0.80 | 0.80 | 0.93 | 0.57 | 0.63 | 0.74 | 0.73 | 0.77 | 0.96 | 0.44 | 0.46 |
| Nonselective aluminum absorber with two glass covers | 0.80 | 0.82 | 0.83 | 0.96 | 0.94 | 0.98 | 0.74 | 0.73 | 0.77 | 0.96 | 0.68 | 0.71 |
| Selective aluminum absorber with one glass cover | 0.82 | 0.86 | 0.87 | 0.94 | 0.84 | 0.84 | 0.79 | 0.81 | 0.83 | 0.95 | 0.57 | 0.61 |
| Nonselective aluminum absorber with one glass cover | 0.89 | 0.89 | 0.89 | 1.0 | 1.3 | 1.3 | 0.78 | 0.81 | 0.83 | 0.94 | 0.97 | 1.03 |

[a]From (45).

## Air-Cooled Flat-Plate Collector Thermal Analysis

The basic air-cooled flat-plate collector shown in Fig. 4.14 differs fundamentally from the liquid-based collectors described in preceding sections because of the relatively poor heat-transfer properties of air. For example, in turbulent flow in a given conduit for a fixed value of Reynolds number, the convection heat-transfer coefficient for water is about 50 times greater than that for air. As a result it is essential to provide the largest heat-transfer area possible to remove heat from the absorber surface of an air-heating collector.

The most common way to achieve adequate heat transfer in air collectors is to flow air over the entire rear surface of the absorber as shown. The heat-transfer analysis of such a collector does not involve the fin effect or the tube-to-plate bond conductance problem, which arises in liquid collectors. The heat-transfer process is essentially that of an unsymmetrically heated duct of large aspect ratio (typically 20–40).

Malik and Buelow (38, 39) surveyed the fluid mechanics and heat-transfer phenomena in air collectors. They concluded that a suitable expression for the Nusselt number for a smooth air-heating collector is

$$Nu_{sm} = \frac{0.0192 \, Re^{3/4} \, Pr}{1 + 1.22 \, Re^{-1/8} \, (Pr - 2)} \tag{4.45}$$

If the surface is hydrodynamically rough, they recommended multiplying the smooth-surface Nusselt number by the ratio of the rough-surface friction factor $f$ to the smooth-surface friction factor $f_{sm}$ from the Blasius equation

$$f_{sm} = 0.079 \, Re^{-1/4} \tag{4.46}$$

The convective coefficient $h_c$ in the Nusselt number is based on a unit absorber area

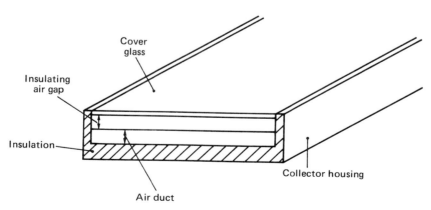

**FIGURE 4.14**  Schematic diagram of a basic air-heating flat-plate collector with a single glass (or plastic) cover.

but the hydraulic diameter $D_H$ in the Nusselt and Reynolds numbers is based on the *entire duct perimeter*.

## Air-Collector Efficiency Factor

The collector efficiency factor $F'$ is defined as the ratio of energy collection to the collection rate if the absorber plate were at the local fluid temperature. $F'$ is particularly simple to calculate, since no fin analysis or bond-conductance term is present for air collectors. For the collector shown in Fig. 4.14, $F'$ is given by (see Problem 4.8)

$$F' = \frac{\bar{h}_c}{\bar{h}_c + U_c} \tag{4.47}$$

where $U_c$ is calculated from Klein's equation, Eq. (4.12), and the duct convection coefficient $\bar{h}_c$ is calculated from Eq. (4.45).

## Air-Collector Heat-Removal Factor

The collector heat-removal factor is a convenient parameter, since it permits useful energy gain to be calculated by knowledge of only the easily determined fluid inlet temperature as shown in Eq. (4.32). The heat-removal factor $F_R$ for a typical air collector such as the one shown in Fig. 4.14 is given by

$$F_R = \frac{\dot{m}_a c_{p,a}}{U_c A_c} \left[ 1 - \exp\left( -\frac{F' U_c A_c}{\dot{m}_a c_{p,a}} \right) \right] \tag{4.48}$$

where $F'$ is given by Eq. (4.47), $\dot{m}_a$ is the air flow rate (kg/sec) and $c_{p,a}$ is its specific heat. The heat-removal factor for air collectors is usually significantly less than that for liquid collectors because $\bar{h}_c$ for air is much smaller than for a liquid such as water.

---

**EXAMPLE 4.3**

Calculate the collector-plate efficiency factor $F'$ and heat-removal factor $F_R$ for a smooth, 1-m-wide, 5-m-long air collector with the following design. The flow rate is 0.7 m³/min · m$_c$² (2.1 ft³/min · ft$_c$²). The air duct height is 1.5 cm (0.6 in), the air density is 1.1 kg/m³ (0.07 lb/ft³), the specific heat is 1 kJ/kg · K (0.24 Btu/lb · °F), and the viscosity is $1.79 \times 10^{-5}$ kg/m · sec ($1.2 \times 10^{-5}$ lb/ft · sec). The collector heat-loss coefficient $U_c$ is 18 kJ/hr · m² · K (5 W/m² · K; 0.88 Btu/hr · ft² · °F).

**SOLUTION**

The first step is to determine the duct heat-transfer coefficient $h_c$ from Eq. (4.45). The Reynolds number is defined as

$$\mathrm{Re} = \frac{\rho \bar{V} D_H}{\mu}$$

in which the average velocity $\bar{V}$ is the volume flow rate divided by the flow area.

$$\bar{V} = \frac{0.7 \times 1 \times 5}{1 \times 0.015} = 233 \text{ m/min} = 3.89 \text{ m/sec}$$

and the hydraulic diameter $D_H$ is

$$D_H = \frac{4(0.015 \times 1)}{1 + 1 + 0.015 + 0.015} = 0.0296 \text{ m}$$

$$\mathrm{Re} = \frac{(1.1)(3.89)(0.0296)}{1.79 \times 10^{-5}} = 7066$$

From Eq. (4.45) the Nusselt number is

$$\mathrm{Nu} = \frac{0.0192(7066)^{3/4}(0.72)}{1 + 1.22(7066)^{-1/8}(0.72 - 2.0)} = 22.0$$

The heat-transfer coefficient is

$$h_c = \mathrm{Nu} \frac{k}{D_H} = \frac{\mathrm{Nu} \, c_p \mu}{\mathrm{Pr} \, D_H}$$

The Prandtl number for air is $\sim 0.72$. Therefore,

$$h_c = \frac{(22.0)(1.0)(1.79 \times 10^{-5})}{(0.72)(0.0296)} = 0.0185 \text{ kJ/m}^2 \cdot \text{sec} \cdot \text{K}$$

$$= 66.5 \text{ kJ/m}^2 \cdot \text{hr} \cdot {}^\circ\text{C} \ (3.26 \text{ Btu/hr} \cdot \text{ft} \cdot {}^\circ\text{F})$$

The plate efficiency is then, from Eq. (4.47),

$$F' = \frac{66.5}{66.5 + 18} = 0.787$$

and the heat-removal factor $F_R$ can be calculated from Eq. (4.48). The mass flow rate per unit area is

$$\frac{\dot{m}}{A_c} = \frac{\rho q}{A_c} = \frac{1.1 \times 0.7}{60} = 0.0128 \text{ kg/sec} \cdot \text{m}_c{}^2$$

Then

$$F_R = \frac{(0.0128)(1)}{(18/3600)} \left\{ 1 - \exp - \left[ \frac{(0.787)(18/3600)}{(0.0128)(1)} \right] \right\} = 0.677$$

That is, the particular collector in question can collect 67.7 percent of the heat it could collect if its surface were at the air-inlet temperature. $F_R$ varies weakly with the fluid temperature through the temperature effect upon $U_c$ [see Eq. (4.12)].

Other air collectors are described in Chapter 6 along with complete heating systems using air as the working fluid. The optical efficiency of liquid and air flat plates is unaffected by the special heat-removal features required in air collectors.

## TUBULAR SOLAR ENERGY COLLECTORS

Two general methods exist for significantly improving the performance of solar collectors above the minimum flat-plate collector level. The first method increases solar flux incident on the receiver; it will be described in the next section on concentrators. The second method involves the reduction of parasitic heat loss from the receiver surface. Tubular collectors, with their inherently high compressive strength and resistance to implosion, afford the only practical means for completely eliminating convection losses by surrounding the receiver with a vacuum on the order of $10^{-4}$ mm Hg. The analysis of evacuated tubular collectors is the principal topic of this section.

Tubular collectors have a second application. They may be used to achieve a small level of concentration—1.5-2.0—by forming a mirror from part of the internal concave surface of a glass tube. This reflector can focus radiation on a receiver inside the tube. Since such a receiver is fully illuminated, it has no parasitic "back" losses. This low-level concentrator can also be warranted for performance improvement. Performance may be improved to levels between those for full evacuation and those for no evacuation without vacuum and its sealing difficulties by filling the envelope with high-molecular-weight noble gases. See Chapter 3 for heat-transfer correlations usable with heavy gases. External concentrators of radiation may also be coupled to an evacuated receiver for improvement of performance over the simple evacuated tube. Collectors of this type are described briefly below.

### Evacuated-Tube Collectors

Evacuated-tube devices have been proposed as efficient solar energy collectors since the early twentieth century. In 1909, Emmett (17) proposed several evacuated-tube concepts for solar energy collection, two of which are being sold commercially today. Speyer (54) also proposed a tubular evacuated flat-plate design for high-temperature operation. With the recent advances in vacuum technology, evacuated-tube collectors can be reliably mass produced. Their high-temperature effectiveness is essential for the efficient operation of solar air-conditioning systems and process heat systems.

Figure 4.15 shows schematic cross sections of several glass evacuated-tube collector concepts. The simplest design is basically a small flat-plate collector housed in an evacuated cylinder (Fig. 4.15a). If the receiver is metal, a glass-to-metal seal is required to maintain a vacuum. In addition, a thermal short may occur from inlet to outlet tube unless special precautions are taken. Alternatively, an all-glass collector can be made from concentric glass tubes as shown in Fig. 4.15b. This collector avoids a glass-to-metal seal but has very limited working fluid pressurization capability. Some investigators have proposed the use of a square absorber circumscribed within the circular region shown. An increased concentration effect would result but the pressurability of the absorber is reduced.

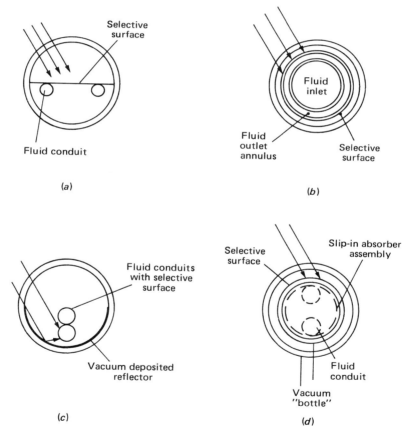

**FIGURE 4.15**  Evacuated-tube solar energy collectors: (a) flat plate; (b) concentric tubular; (c) concentrating; (d) vacuum bottle with slip-in heat exchanger contacting rear surface of receiver.

Mildly concentrating, tubular collectors can be made using the design of Fig. 4.15c. Either a single flow-through receiver with fins or a double U tube as shown can be used. Concentration ratios of from $2/\pi$ to 2.0 can be achieved ideally with this design, but a glass-to-metal seal is required.

One of Emmett's designs is shown by Fig. 4.15d. It consists of an evacuated vacuum bottle much like an unsilvered, wide-mouth Dewar flask into which a metal heat exchanger is inserted. The outer surface of the inner glass tube is the absorber. The heat generated is transferred through the inner glass tube to the metal slip-in heat exchanger. Since this heat transfer is through a glass-to-metal interface that has only intermittent point contacts, significant axial temperature gradients can develop, thereby stressing the glass tube. In addition, a large temperature difference can exist between the inner and outer glass tubes. At the collector ends where the two tubes are joined, a large temperature gradient and consequent thermal stress can exist.

The level of evacuation required for suppression of convection and conduction can be calculated from basic heat-transfer theory. As the tubular collector is evacuated, reduction of heat loss first occurs because of the reduction of the Rayleigh number in Eq. (3.73). The effect is proportional to the square root of density. When the Rayleigh number is further reduced below the lower threshold for convection, the heat-transfer mechanism is by conduction only. For most gases the thermal conductivity is independent of pressure if the mean free path is less than the heat-transfer path length.

For very low pressure, the conduction heat transfer in a narrow gap (12) is given by

$$q_k = \frac{k\Delta T}{g + 2p} \tag{4.49}$$

where $g$ is the gap width and $p$ is the mean free path. For air the mean free path at atmospheric pressure is about $70\ \mu\text{m}$. If 99 percent of the air is removed from a tubular collector, the mean free path increases to 7 mm and conduction heat transfer is affected very little. However, if the pressure is reduced to $10^{-3}$ torr the mean free path is 7 cm, which is substantially greater than the heat-transfer path length, and conduction heat transfer is effectively suppressed. The relative reduction in heat transfer as a function of mean free path can be derived from Eq. (4.49):

$$\frac{q_{\text{vac}}}{q_k} = \frac{1}{1 + 2p/g} \tag{4.50}$$

where $q_k$ is the conduction heat transfer if convection is suppressed and $q_{\text{vac}}$ is the conduction heat transfer under a vacuum. Achieving a vacuum level of $10^{-3}$–$10^{-4}$ torr for a reasonably long period of time is within the grasp of modern vacuum technology.

Thermal and optical analyses of tubular collectors are based on the principles used to analyze other collectors in previous sections. The evacuated-tube flat-plate design (Fig. 4.15a) and the reflecting design (Fig. 4.15c) can be analyzed using exactly the same procedure as that used for flat-plate collectors. The convection term in the receiver and envelope energy balances is simply set to 0. Analysis of the slip-in heat-exchange type is problematic since the glass-to-metal point contact is difficult to quantify so that production tolerances and surface precision and smoothness can be included.

The concentric-tube collector (CTC) illustrates several important principles in collector analysis and will be analyzed in detail in the next section. This collector type is manufactured in the United States by Owens-Illinois, Inc. (OII). Test data on the OII collector will be presented in order to verify the analysis that follows.

## Optical Analysis of the CTC

Since close packing of CTC tubes in an array can result in shading losses at any angle other than normal incidence, it is cost-effective to space the tubes apart and to use a back reflector in order to capture any radiation passing between the tubes. Figure 4.16a shows the geometry of part of a CTC array with tube spacing d; Fig. 4.16b is a cutaway drawing of one tubular assembly. Evacuated-tube flat-plate designs and concentrating designs should be closely packed to optimize solar energy collection. Beekley and Mather (3) have analyzed the CTC in detail and their analysis is the basis of this section.

CTC arrays can collect both direct and diffuse radiation. Each radiation component must be analyzed in turn. The optical efficiency $\eta_o$ may be expressed as

$$\eta_o = \frac{\tau_e \alpha_r I_{\text{eff}}}{I_{b,c} + I_{d,c}} \tag{4.51}$$

where $I_{\text{eff}}$ is the effective solar radiation both directly intercepted and intercepted after reflection from the back reflector and $I_{b,c}$ and $I_{d,c}$ are, respectively, the beam and diffuse radiation components intercepted per unit collector aperture area. $\tau_e$ and $\alpha_r$ are the envelope transmittance and receiver absorptance, respectively.

The beam component of radiation intercepted by the tubes (with no shading loss) is given by

$$I_{b,t} = I_b \cos i_t \tag{4.52}$$

where $\cos i_t$ is calculated from Eq. (2.16).

The reflection of beam radiation from a diffusely reflecting back surface can be calculated by assuming that an illuminated strip $W$ reflects with reflectance $\rho$ onto the rear surfaces of the nearby tubes. The amount of reflected radiation intercepted can be evaluated using radiation shape factors (see Chapter 3). Although the analysis of this radiation problem is quite lengthy, it can be shown that the amount of reflected radiation intercepted by all tubes $I_{b,r}$ is

$$I_{b,r} = I_{b,c}\rho\Delta\frac{W}{D_6} \tag{4.53}$$

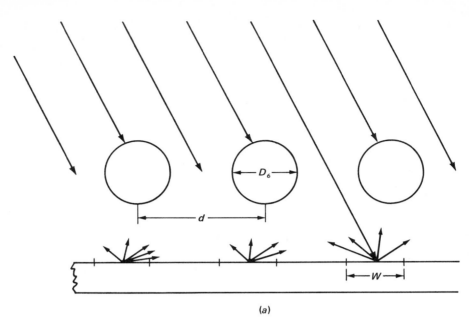

(a)

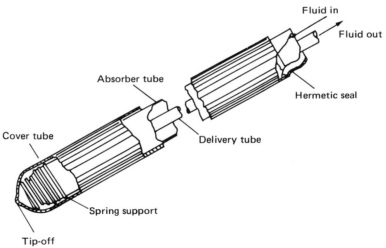

(b)

**FIGURE 4.16** (a) Schematic diagram of concentric-tube collector optics; (b) cutaway view of evacuated-tube solar collector manufactured by Owens-Illinois, Inc. For important dimensions see Table 4.3. From (3).

where $\Delta$ is a sum of shape factors equal to about 0.6–0.7 for tubes spaced one diameter apart, that is, $d = 2D_6$.

Diffuse radiation intercepted directly by the tubes $I_{d,t}$ (per unit collector plane area) is given by

$$I_{d,t} = \pi I_{d,c} F_{TS} \tag{4.54}$$

where $F_{TS}$ is the radiation shape factor from a tube to the sky dome. It depends on tube spacing and lies in the range for closely packed to infinitely spaced tubes, (0.27–0.50, respectively). For tubes spaced one diameter apart $F_{TS} \cong 0.43$.

Diffuse radiation reflected from the array back surface $I_{d,r}$ can be expressed as

$$I_{d,r} = \pi F_{TS} I_{d,c} \rho \bar{F} \tag{4.55}$$

where $\bar{F}$ is a measure of the shape factors of diffusely illuminated strips to collector tubes. For tubes spaced one diameter apart $\bar{F} \cong 0.34$.

The total effective insolation $I_{\text{eff}}$ can be calculated by summing directly intercepted and reflected radiation, that is,

$$I_{\text{eff}} = I_b \left( \cos i_t + \cos i_c \rho \Delta \frac{W}{D_6} \right) + I_{d,c} [\pi F_{TS}(1 + \rho \bar{F})] \tag{4.56}$$

The optical efficiency $\eta_o$ [Eq. (4.51)] is not, therefore, a simple collector property independent of operating conditions, but rather a function of time through the incidence angles $i_t$ and $i_c$, with values increasing away from solar noon for a given solar radiation level.

### Thermal Analysis of the CTC

The heat loss from a CTC occurs primarily through the mechanism of radiation from the absorber surface. The rate of heat loss per unit absorber area $q_L$ then can be expressed as

$$q_L = U_c(T_r - T_a) \tag{4.57}$$

Total thermal resistance $1/U_c$ is the sum of three resistances:

$R_1$—radiative exchange from absorber tube to cover tube
$R_2$—conduction through glass tube
$R_3$—convection and radiation to environment

The overall resistance is then

$$\frac{1}{U_c} = R_1 + R_2 + R_3 \tag{4.58}$$

The conductances $R_i^{-1}$ are given by

$$R_1^{-1} = \frac{1}{1/\epsilon_r + 1/\epsilon_e - 1}\, \sigma(T_r + T_{ei})(T_r^2 + T_{ei}^2) \tag{4.59}$$

$$R_2^{-1} = \frac{2k}{D_r \ln (D_6/D_{ei})} \tag{4.60}$$

$$R_3^{-1} = [h_c + \sigma\epsilon_e(T_{eo} + T_a)(T_{eo}^2 + T_a^2)]\,\frac{D_6}{D_r} \tag{4.61}$$

where the subscript $e$ denotes envelope properties, $T_r$ is the receiver (absorber) temperature, and $h_c$ is the external convection coefficient for the envelope. Test data have shown that the loss coefficient $U_c$ is between 0.5 and 1.0 W/m$^2 \cdot$ °C, thus confirming the analysis.

The CTC energy delivery rate $q_u$ on an aperture area basis can be written as

$$q_u = \tau_e \alpha_r I_{\text{eff}} \frac{A_t}{A_c} - U_c(T_r - T_a)\frac{A_r}{A_c} \tag{4.62}$$

where $A_t$ is the projected area of a tube (its diameter) and $A_r$ is the receiver or absorber area. The receiver-to-collector aperture area ratio is $\pi D_r/d$. Therefore

$$q_u = \frac{D_r}{d}[\tau_e \alpha_r I_{\text{eff}} - \pi U_c(T_r - T_a)] \tag{4.63}$$

Beekley and Mather (3) have shown that a tube spacing one envelope diameter $D_6$ apart maximizes daily energy gain. Test data on such a configuration with properties shown in Table 4.3 are shown in Fig. 4.17 for both a diffuse reflector and a specularly reflecting, cylindrical back surface. The specular reflector improves performance by 10 percent or more.

## INTRODUCTION TO CONCENTRATING SOLAR ENERGY COLLECTORS

Solar collection devices that increase the receiver surface flux intensity over that incident on the collector aperture are called concentrators. Concentration is achieved by the use of reflecting or refracting elements positioned to concentrate the incident flux onto a receiver.

All solar collectors can be divided into three classes. Flat plates, which operate without concentration, have been described in the preceding section. This is the largest group, and includes collectors for heating, solar ponds, and solar stills. Another group of collectors is designed to produce very high energy density and

**TABLE 4.3** Physical and Thermal Characteristics of the Owens-Illinois Evacuated-Tube Solar Collector[a]

| Component | Size or value |
|---|---|
| Glass cover tube | |
| OD | 2.09 in |
| ID | 1.93 in |
| Glass absorber tube | |
| OD | 1.69 in |
| ID | 1.54 in |
| Delivery tube | |
| OD | 0.47 in |
| ID | 0.35 in |
| Tube active length | 42.0 in |
| Overall collector module size | 24 tubes |
| | 4 × 8 ft |
| | 27.4 ft² net area |
| Tube spacing in module | 4.18 in on centers |
| Weight | 3.7 lb/ft² dry |
| | 6.7 lb/ft² water-filled |
| Vacuum level | $10^{-3}$ to $10^{-4}$ mm Hg |
| Absorber solar absorptance | 0.9 (approximate) |
| Absorber solar emittance | 0.1 or less |
| Cover transmittance | 0.92 |
| Specific heat | 2.5 Btu/ft$_c$² · °F (approximately five times typical flat-plate value) |

[a]From (3).

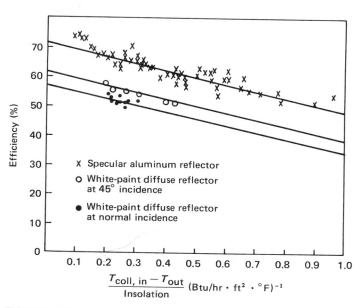

**FIGURE 4.17** Owens-Illinois, Inc., evacuated-tube solar collector efficiency. From Owens-Illinois brochure, 1977.

temperature at the receiver by means of accurate focusing devices and continuous precise tracking of the sun's virtual motion.

Another group of collectors, intermediate between the extremes described above, increases energy density by a factor of 2-10. They usually do not require accurate tracking but only occasional turning, and they do not produce a sharp focus. In addition, they need not be fabricated from precise optical components. These collectors have possible wide application and are described on p. 251; a later section describes the tracking devices.

Concentrating or focusing devices are much older than flat-plate collectors. The ancient Chaldeans are known to have used burning glasses or lenses made from quartz, which were discovered in the Middle East along with tablets containing very small characters readable only with these early lenses. In 1695 a diamond was melted for the first time in Florence by an early solar practitioner. Lavoisier used his famous double-lens concentrator to carry out a number of high-temperature experiments in the eighteenth century. Flat-plate collectors were first introduced in the eighteenth century. In the eighteenth and nineteenth centuries many concentrators were built to power heat engines and to produce steam.

Several terms are used to describe concentrating collectors. The *concentration ratio*, denoted by *CR*, is the ratio of the net collecting aperture area to the area of the receiver or absorber. The *aperture area* is the net area of the collector that intercepts radiation. The *absorber area* is the total area that receives the concentrated radiation. It is the area from which useful energy is removed and from which some heat may be lost to the environment. Since the absorber in some concentration devices is not fully or uniformly illuminated, some authors distinguish between brightness (or flux) and geometric (or area) concentration ratios. In this text the geometric concentration ratio based on total absorber area—illuminated or not—is used exclusively since it is the more useful in thermal balance calculations and economic analyses.

In this section the rationale for concentrating solar energy is introduced along with a description of generic types of concentration methods. The limits of concentration imposed by the second law of thermodynamics are also developed. In subsequent sections, several common concentration types are described in detail both thermally and optically and performance comparisons are made.

## Reasons for Using Concentrating Collectors

Three reasons are commonly cited for using concentrators:

1. To increase energy delivery temperatures in order to achieve a thermodynamic match between temperature level and task. The task may be to operate thermionic, magnetohydrodynamic, thermodynamic, or other higher temperature devices.
2. To improve thermal efficiency by reducing the heat-loss area relative to the receiver area. There would also be a reduction in transient effects, since the thermal mass is usually much smaller than for a flat-plate collector.

3. To reduce cost by replacing an expensive receiver by a less expensive reflecting or refracting area.

Since the upper limit of energy delivery temperature for standard flat-plate collectors is on the order of 380 K, the Carnot efficiency of a heat engine powered by such a collector is limited to about 20 percent. In order to realize a significant economic return on the high capital investment in a heat engine, it is necessary to achieve higher efficiencies by using higher fluid temperatures. For example, if a solar collector is capable of delivering energy at 580 K instead of 380 K, the Carnot efficiency and energy transformation to work can be doubled. In Chapter 8 solar-powered heat-engine systems of this type are described. Figure 4.18 shows that less than 5 percent of the industrial process heat needs of the United States exist at

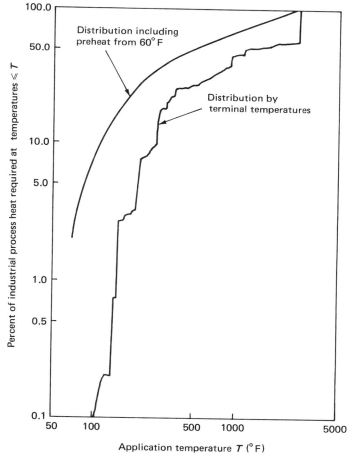

**FIGURE 4.18**   Percentage distribution of industrial process heat energy delivery at various temperatures. From (19).

temperatures below 100°C. For any significant impact of solar energy in this sector, concentrators are essential.

Concentrators are inherently more efficient at a given temperature than are flat-plate collectors, since the area from which heat is lost is smaller than the aperture area. In the flat-plate device both areas are equal in size. A simple energy balance illustrates this principle. The useful energy delivered by a collector $q_u$ is given by

$$q_u = \eta_o I_c A_a - U_c (T_c - T_a) A_r \qquad (4.64)$$

in which $\eta_o$ is the optical efficiency, $A_a$ is the aperture area, $A_r$ is the receiver (or absorber) area, and other terms are as defined previously. The instantaneous collector efficiency is given by

$$\eta_c = \frac{q_u}{I_c A_a} \qquad (4.65)$$

from which, using Eq. (4.64),

$$\eta_c = \eta_o - \frac{U_c (T_c - T_a)}{I_c} \frac{1}{CR} \qquad (4.66)$$

where the concentration ratio $CR$ is defined as

$$CR \equiv \frac{A_a}{A_r} \qquad (4.67)$$

For the flat plate $CR \cong 1$ and for concentrators $CR > 1$. As a result, the loss term (second term) in Eq. (4.66) is smaller for a concentrator and the efficiency is higher. This analysis is necessarily simplified and does not reflect the reduction in optical efficiency that frequently but not always occurs because of the use of imperfect mirrors or lenses in concentrators. The evaluation of $U_c$ in Eq. (4.66) in closed form is quite difficult for high-temperature concentrators, since radiation heat loss is usually quite important and introduces nonlinearities ($\propto T^4$). One disadvantage of concentrators is that they can collect only a small fraction of the diffuse energy incident at their aperture. This property is an important criterion in defining the geographic limits to the successful use of concentrators and is described shortly.

## The Thermodynamic Limits to Concentration

A simple criterion is developed below for the upper limit of concentration (43) of a solar collector. Figure 4.19 is a schematic diagram of any concentrating device in which the source, aperture, and receiver are shown. The source represents a diffuse source or a diffuselike source that could be formed by a moving point source, that

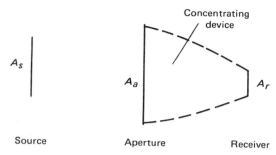

**FIGURE 4.19** Generalized schematic diagram of any two-dimensional solar energy concentrating device showing source of radiation $A_s$, aperture $A_a$, and receiver $A_r$ areas.

is, the sun. The evaluation of the maximum achievable concentration ratio $CR_{max}$ uses the concept of radiation exchange factors described in Chapter 3.

The factor $\mathfrak{F}_{12}$ is defined as the fraction of radiation emitted from surface 1 that reaches surface 2 by whatever means—direct exchange, reflection, or refraction. It is shown in Chapter 3 that reciprocity relations exist for area exchange factors; for this analysis the second law of thermodynamics requires, in addition, that

$$A_s \mathfrak{F}_{sa} = A_a \mathfrak{F}_{as} \qquad (4.68)$$

and

$$A_s \mathfrak{F}_{sr} = A_r \mathfrak{F}_{rs} \qquad (4.69)$$

where $a$ denotes aperture area, $r$ denotes receiver area, and $s$ denotes the source. By means of these expressions, the concentration ratio $CR$ can be expressed as

$$CR \equiv \frac{A_a}{A_r} = \frac{\mathfrak{F}_{sa}\,\mathfrak{F}_{rs}}{\mathfrak{F}_{as}\,\mathfrak{F}_{sr}} \qquad (4.70)$$

For the best concentrator possible, all radiation entering the aperture $A_a$ reaches the receiver $A_r$, that is,

$$\mathfrak{F}_{sa} = \mathfrak{F}_{sr} \qquad (4.71)$$

In addition, if the source is representable as a black body,

$$\mathfrak{F}_{as} = F_{as} \qquad (4.72)$$

where $F_{as}$ is the radiation shape factor between two black surfaces. Using Eqs. (4.70) and (4.72), we have

$$CR = \frac{\mathfrak{F}_{rs}}{F_{as}} \tag{4.73}$$

Since $\mathfrak{F}_{rs} \leqslant 1$ by the second law,

$$CR \leqslant CR_{\max} = \frac{1}{F_{as}} \tag{4.74}$$

Equation (4.74) states the simple and powerful result that the maximum concentration permitted by the second law is simply the reciprocal of the radiation shape factor $F_{as}$.

The shape factor $F_{as}$ for a solar concentrator in two dimensions can be calculated from the diagram in Fig. 4.20. This sketch represents a troughlike or single-curvature concentrator formed from mirrors or a linear (cylindrical) lens. It is illuminated by a line source of light of length $2r$ representing a portion of the sun's virtual trajectory. By reciprocity, we have

$$F_{as} = F_{sa} \frac{A_s}{A_a} \tag{4.75}$$

If an angle $2\theta_{\max}$ is defined as the maximum angle within which light is to be collected, we have

$$F_{as} = \sin \theta_{\max} \tag{4.76}$$

from Hottel's crossed-string method (24) for $L \ll r$. The angle $\theta_{\max}$ is called the *acceptance half-angle*. From Eq. (4.74) the maximum concentration is then

$$CR_{\max,2D} = \frac{1}{\sin \theta_{\max}} \tag{4.77}$$

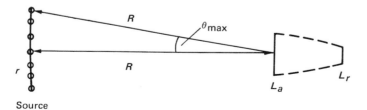

Source

**FIGURE 4.20** Schematic diagram of two-dimensional sun-concentrator geometry used to calculate shape factor $F_{as}$ and showing acceptance half-angle $\theta_{\max}$ and several positions of the sun.

The name acceptance half-angle denotes coverage of one-half of the angular zone within which radiation is accepted (that is, "seen") by the receiver of a concentrator. Radiation is said to be accepted over an acceptance angle $2\theta_{max}$ because radiation incident within this angle reaches the absorber after passing through the aperture. Practical acceptance angles range from a minimum subtending the sun's disk (about $\frac{1}{2}°$) to 180°, a value characterizing a flat-plate collector accepting radiation from a full hemisphere.

Double-curvature or dish-type concentrators have an upper limit of concentration that can be evaluated by extending the method used above to three dimensions. The result of such a calculation for a compound curvature device is given by

$$CR_{max,3D} = \frac{1}{\sin^2 \theta_{max}} \tag{4.78}$$

The second law prescribes not only the geometric limits of concentration as shown above, but also the operating temperature limits of a concentrator. The radiation emitted by the sun and absorbed by the receiver of a concentrator $q_{abs}$ is

$$q_{abs} = \tau\alpha_s A_s F_{sa} \sigma T_s^4 \tag{4.79}$$

where $T_s$ is the effective temperature of the sun and $\tau$ is the overall transmittance function for the concentrator, including the effects of any lenses, mirrors, or glass covers. If the acceptance half-angle $\theta_{max}$ is selected to just accept the sun's disk of angular measure $\theta_s(\theta_s \sim \frac{1}{4}°)$, we have, by reciprocity [Eq. (4.75)], for a concentrator with compound curvature,

$$q_{abs} = \tau\alpha_s A_a \sin^2 \theta_s \sigma T_s^4 \tag{4.80}$$

If convection and conduction could be eliminated, all heat loss $q_L$ is by radiation, and

$$q_L = \epsilon_{ir} A_r \sigma T_r^4 \tag{4.81}$$

where $\epsilon_{ir}$ is the infrared emittance of the receiver surface. Radiation inputs to the receiver from a glass cover or the environment can be ignored for this upper limit analysis.

An energy balance on the receiver is then

$$q_{abs} = q_L + \eta_c q_{abs} \tag{4.82}$$

where $\eta_c$ is the fraction of energy absorbed at the receiver that is delivered to the working fluid. Substituting for Eqs. (4.80) and (4.81) in Eq. (4.82) we have

$$(1 - \eta_c)\tau\alpha_s A_a \sin^2 \theta_s \sigma T_s^4 = \epsilon_{ir} A_r \sigma T_r^4 \tag{4.83}$$

Since $CR = A_a/A_r$ and $CR_{max} = 1/\sin^2 \theta_s$,

$$T_r = T_s \left[(1 - \eta_c)\tau \frac{\alpha_s}{\epsilon_{ir}} \frac{CR}{CR_{max}}\right]^{1/4} \tag{4.84}$$

In the limit as $\eta \to 0$ (no energy delivery) and $\tau \to 1$ (perfect optics), we have

$$\lim T_r \to T_s \left(\frac{CR}{CR_{max}}\right)^{1/4} \tag{4.85}$$

Since $\epsilon_{ir} \to \alpha_s$ as $T_r \to T_s$, Eq. (4.85) shows that

$$T_r \leqslant T_s \tag{4.86}$$

as expected for an optically and thermally idealized concentrator. Equation (4.86) is equivalent to the Clausius statement of the second law for a solar concentrator.

## Optical Limits to Concentration

Equations (4.77) and (4.78) define the upper limit of concentration that may be achieved for a given concentration viewing angle. Of interest are the upper and lower limits of concentration defined by practical viewing angle limits—the maximum $CR$ limited only by the size of the sun's disk and achieved by *continuous* tracking; and the minimum $CR$, based on a specific number of hours of collection with *no* tracking.

The upper limit of concentration for two- and three-dimensional concentrators is on the order of

$$CR_{max,2D} = \frac{1}{\sin \frac{1}{4}^\circ} = O(200) \tag{4.87}$$

$$CR_{max,3D} = \frac{1}{\sin^2 \frac{1}{4}^\circ} = O(40,000) \tag{4.88}$$

In practice, these levels of concentration are not achievable because of the effects of tracking errors and imperfections in the reflecting- or refracting-element surface, as described later.

For a fixed single-curvature (troughlike) collector the maximum achievable concentration can be determined from the motion of the sun as described in Chapter 2. If a troughlike concentrator is oriented north-south and tilted up from the horizontal so that the plane of the sun's virtual motion is normal to the aperture, the acceptance angle and hence concentration ratio are related to the range of the hour-angle over which collection is required. If 8 hr of collection is required, the acceptance half-angle is $60°$ ($15°/\text{hr} \times 8 \text{ hr}/2$) and the maximum concentration is

$$CR_{max,NS} = \frac{1}{\sin 60°} = 1.15 \tag{4.89}$$

If the fixed trough collector is oriented east-west, the acceptance angle is limited by the profile angle excursion from solstice to solstice. From Eq. (2.10) the profile angle $\gamma$ at 40°N latitude at 4 p.m. on the winter solstice is

$$\tan \gamma \ (h = 60°; \text{winter}) = \sec (53°) \tan (5.5°) \tag{4.90}$$

or

$$\gamma \ (h = 60°; \text{winter}) = 9.1° \tag{4.91}$$

At 4 p.m. or 8 a.m. the summer solstice profile angle is

$$\tan \gamma \ (h = 60°; \text{summer}) = \sec (90.7°) \tan (37.4°) \tag{4.92}$$

or

$$\gamma \ (h = 60°; \text{summer}) = 91° \tag{4.93}$$

Hence, the total profile angle excursion is 81.9° (91 − 9.1°). In order to accept solar radiation over this range, the maximum $CR$ is

$$CR_{max,EW} = \frac{1}{\sin 81.9°/2} = 1.53 \tag{4.94}$$

which is a little better than that for a flat-plate collector. The values above are for 8 hr of collection, all year long, at 40°N latitude. For other combinations of latitudes, collector tilt, and periods of collection the limiting $CR$ values will vary. Figure 4.21 shows the acceptance angle range on a plot of solar profile angles for the solstices. This type of plot can be used to evaluate graphically hours of collection for various tilts of troughlike concentrators.

## Acceptance of Diffuse Radiation

Diffuse or scattered radiation is not associated with a specific direction as is beam radiation. It is expected, therefore, that some portion of the diffuse component will fall beyond the acceptance angle of a concentrator and not be collectable. The minimum amount of diffuse radiation that is collectable can be estimated by assuming that the diffuse component is isotropic at the aperture. The exchange factor reciprocity relation shows that

$$A_a \mathfrak{F}_{as} = A_r \mathfrak{F}_{rs} \tag{4.95}$$

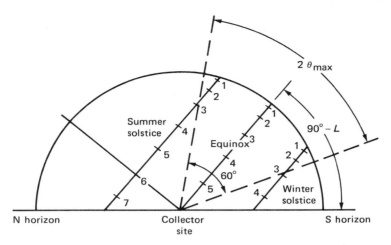

**FIGURE 4.21** Profile angles of the sun for equinoxes and solstices showing hours of collection for a concentrator with acceptance angle of 60° ($CR \approx 2$); collection period of approximately $5\frac{1}{2}$ hr at the solstices for a collector tilted at an angle equal to the latitude 40°N is shown. The numbers attached to the sun positions give the time in hours from local solar noon.

For most practical concentrating devices that can accept a significant fraction of diffuse radiation, $\mathfrak{F}_{rs} = 1$, that is, all radiation leaving the receiver reaches the aperture and the environment eventually. Then

$$\mathfrak{F}_{as} = \frac{A_r}{A_a} = \frac{1}{CR} \tag{4.96}$$

for any concentrator. Equation (4.96) indicates that at least $1/CR$ of the incident diffuse radiation reaches the receiver. In actual practice, the collectable diffuse portion will be greater than $1/CR$, since diffuse radiation is usually concentrated near the solar disk except during the cloudiest of days. (See Chapter 2.)

## Concentrator Types

Concentrators can be classified according to

1. Operating temperature range
2. Amount of tracking required to maintain the sun within the acceptance angle
3. Type of tracking—single or double axis

Figure 4.22 is a qualitative map (not to be used for design) of the operating temperature ranges of various concentrator types. The highest temperatures to date were achieved by the French solar furnace at Odeillo. This furnace system is described in Chapter 8. Single-curvature concentrators (troughs) usually have a concentration ratio of up to 50 and double-curvature units, from above 30 up to several hundred. Delivery temperatures up to 300°C can be achieved by single-curvature devices. The Odeillo furnace has achieved temperatures above 3000°C.

As shown in the preceding sections, tracking requirements depend on the acceptance half-angle $\theta_{max}$: the larger the $\theta_{max}$, the less frequently and less accurately the tracker must operate. Two tracking levels may be identified:

1. Intermittent tilt change or completely fixed.
2. Continuously tracking reflector, refractor, or receiver. If oriented east-west, it requires an approximate ± 30°/day motion; if north-south, a ~ 15°/hr motion. Both must accommodate to a ± 23°/yr declination excursion.

Single-curvature devices may be of either type but double-curvature devices are usually of the continuously tracking, high-*CR* type. Both concentrator types are described in detail by example analyses of some common archetypes of each class in the next sections.

## FIXED OR INTERMITTENTLY TURNED CONCENTRATORS

The least complex concentrators are those not requiring continuous accurate tracking. These are necessarily of large acceptance angle, moderate

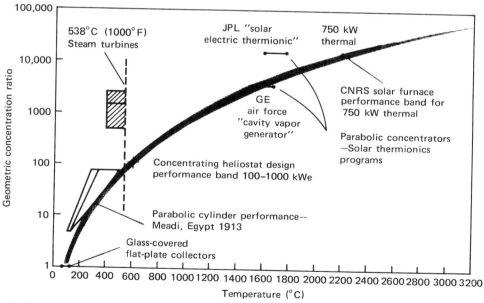

**FIGURE 4.22** Typical temperatures achievable by concentrating solar collectors. Courtesy of *Popular Science Magazine.*

concentration ratio, and usually single-curvature design. Since the smallest diurnal, angular excursion of the sun is in a north-south plane, the fixed or intermittently turned concentrators must be oriented with their axis of rotation perpendicular to this plane, that is, in an east-west direction, in order to capitalize on the large acceptance angles. Table 4.4 shows the motion required to ensure 7 hr of collection for a range of concentration ratios up to 10. Beyond $CR = 10.5$, 7 hr of collection cannot be achieved without readjustment during the day.

## Fixed Concentrators

The general discussion of concentrators in the previous section showed that little concentration is achievable for collectors with a yearlong, 8-hr collection period. If the collection period requirement is reduced to 6 or 7 hr, concentration ratios up to 2.0 can be achieved with fixed concentrators. Concentrations of this level by lens or mirror may be justified in some cases by the resultant efficiency improvements over flat plates. However, this method of performance enhancement and other proposed methods for flat-plate collectors must be evaluated by a complete systems and economic analysis.

A fixed-concentrator concept that does have some merit is the boosting of flat-plate collector performance on a *seasonal basis* by the use of simple planar mirrors (56a) as shown in Fig. 4.23. McDaniels et al. (40) have evaluated the optical boost of beam radiation that can be realized by this method. The optical enhancement $P$ is defined as the ratio of radiation intercepted by a collector in the collector-reflector combination to that intercepted by a collector alone tilted to provide normal solar incidence at solar noon.

Figure 4.24 shows the enhancement factor for solar noon and 3 hr from noon for a horizontal reflector at 45°N latitude for a near-vertical collector. During the summer the enhancement is less than unity, since the reflector-collector combination is compared with a collector with normally incident radiation.

Figure 4.25 shows the boost to summer solar radiation pickup that can be achieved by the use of tilted reflectors. By proper tilt, a seasonal peak of collected energy can be matched to a seasonal peak of energy demand as shown. The curves in Fig. 4.25 are calculated from optical analyses; data collected on this system (located in Atlanta, Georgia) indicated that the performance of the system as built was somewhat better than that calculated.

## Intermittently Turned Concentrators—The Compound Parabolic Concentrator Family

As an example of the class of moderate concentration, turning collectors, the compound parabolic concentrator (CPC) family developed by Winston (60–62) and Rabl (41–44), and described by Baranov and Melnikov (2), will be described in detail. The CPC family can achieve in practice the upper bound of concentration ratio as set by the laws of physics. Most other concentrators fall short of the ideal limit by a factor of 2 or more. These CPC collectors are of either single or compound curvature and are characterized by the highest concentration permitted

**TABLE 4.4** Frequency of Tilt Adjustment Required to Ensure Seven Hours of Solar Collection for Single-Curvature Concentrators[a]

| Acceptance half-angle $\theta_{max}$ (ideal concentration for perfect mirrors and point sun) | Collection time avg. over year (hr/day) | No. of adjustments per year | Shortest period without adjustment (days) | Avg. collection time if tilt adjusted every day (hr/day) |
|---|---|---|---|---|
| 19.5° (3.0) | 9.22 | 2 | 180 | 10.72 |
| 14° (4.13) | 8.76 | 4 | 35 | 10.04 |
| 11° (5.24) | 8.60 | 6 | 35 | 9.52 |
| 9° (6.39) | 8.38 | 10 | 24 | 9.08 |
| 8° (7.19) | 8.22 | 14 | 16 | 8.82 |
| 7° (8.21) | 8.04 | 20 | 13 | 8.54 |
| 6.5° (8.83) | 7.96 | 26 | 9 | 8.36 |
| 6° (9.57) | 7.78 | 80 | 1 | 8.18 |
| 5.5°[b] (10.43) | 7.60 | 84 | 1 | 8.00 |

[a]Concentrators aligned east-west. From (41).
[b]For $\theta_{max} = 5.5°$, minimum collection time is 6.78 hr/day.

253

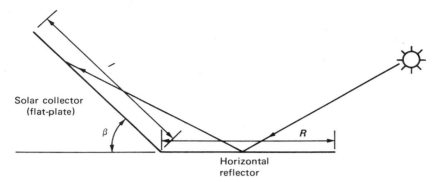

**FIGURE 4.23** Horizontal reflecting surface used to boost flat-plate collector performance.

for a given acceptance angle. Since the compound-curvature (dishlike) CPCs require almost continuous tracking at useful values of the concentration ratio, the single-curvature devices requiring only periodical movement are described herein.

Figure 4.26 shows a schematic cross section of the original CPC concept. It is seen to be formed from two distinct parabolic segments, the foci of which are located at the opposing receiver surface end points. The axes of the parabolic segments are oriented away from the CPC axis by the acceptance angle $\theta_{max}$. The slope of the reflector surfaces at the aperture is parallel to the CPC optical axis. Figure 4.27 is a photograph of a CPC collector.

Since the CPC achieves the maximum concentration ratio permitted by the second law, the *CR* is

$$CR_{CPC} = \frac{1}{\sin \theta_{max}} \tag{4.97}$$

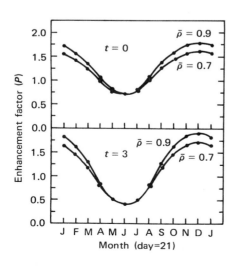

**FIGURE 4.24** Enhancement $P$ of beam radiation intercepted by a flat-plate collector by virtue of horizontal mirror boosters. Collector tilt is 80°, latitude 45°N, and reflectances 0.7 and 0.9 as shown. From (40).

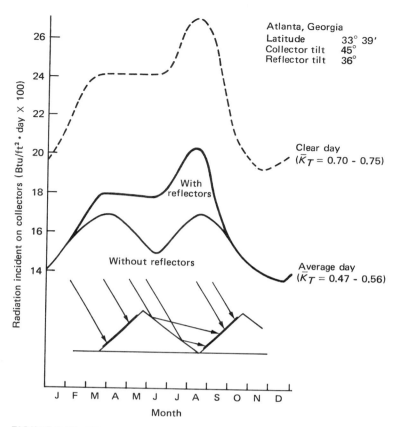

**FIGURE 4.25** Effect of tilted reflector booster for flat-plate collector array in Atlanta, Georgia, for both clear-sky and average conditions (1 Btu/ft² = 11.4 kJ/m²). From (59).

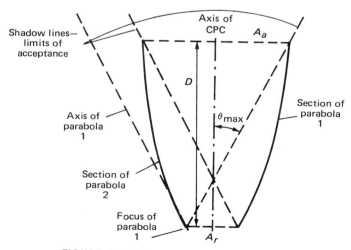

**FIGURE 4.26** Schematic cross section of a CPC showing parabolic segments, aperture, and receiver.

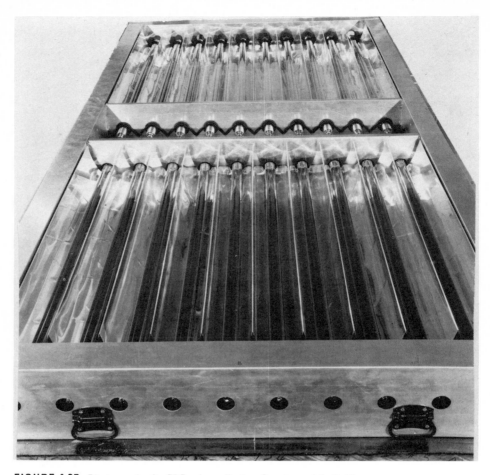

**FIGURE 4.27** Photograph of a CPC solar collector. Courtesy of A. Rabl.

By analyzing the geometry of a CPC it can be shown that the depth $D$ of a full CPC trough is given by

$$\frac{D}{A_r} = \frac{CR + 1}{2} \sqrt{CR^2 - 1} \qquad (4.98)$$

**EXAMPLE 4.4**

Calculate the depth and aperture dimensions of a CPC collector with $CR = 5$ if the receiver is 20 cm wide.

**SOLUTION**

From the definition of concentration ratio we have

$$A_a = CR \times A_r = 5 \times 20 \text{ cm} = 100 \text{ cm}$$

The collector depth is given by Eq. (4.98). Inserting values for this example gives

$$\frac{D}{A_r} = \frac{5+1}{2}\sqrt{5^2 - 1} = 14.7$$

Therefore

$$D = 294 \text{ cm}$$

As described below, practical CPCs can be truncated by 50 percent or more to reduce cost with little performance effect. For the example collector, the depth of a practical unit would be 1.5 m or less.

### Optical Analysis of the CPC Collector

An optical analysis of a CPC is required in order to determine its optical efficiency, which is found by calculating *the average number of reflections ñ that radiation undergoes between the aperture and the receiver.* Consideration of Fig. 4.26 shows that normally incident radiation in the central region of the aperture undergoes no reflection between the aperture and receiver, whereas radiation near the edges of the aperture must undergo one or more reflections. The average number of reflections is an important parameter and is defined as the average taken over all radiation entering the aperture. The attenuation or loss of radiation, to the lowest order in $1 - \rho_m$, is then (41)

$$\text{Reflectance loss} = 1 - \rho_m{}^{\bar{n}} \tag{4.99}$$

where $\rho_m$ is the specular mirror reflectance.

The average number of reflections depends on incidence angle, collector depth, and concentration ratio (41). Since CPC collectors are quite deep, most practical designs are usually truncated in height by one-half or more to reduce reflecting surface costs and to reduce physical size to practical levels. Since the CPC trough has steep sides, little concentration is lost by truncation, whereas the average number of reflections is reduced, thereby improving the optical efficiency.

Table 4.5 shows the number of reflections averaged over all incidence angles (in the transverse plane) $i < \theta_{max}$ and over the full aperture area for several values of CPC acceptance angles, for both full and truncated varieties. The data show that a 50 percent truncation, that is, 50 percent reduction in height/aperture ratio, causes

**TABLE 4.5** Optical and Geometric Properties of Full and Truncated Compound Parabolic Concentrator Solar Collectors[a]

| Acceptance half-angle of CPC (degrees) | Concentration ratio CR | Height/aperture ratio | Reflector/aperture ratio | Average number of reflections $\bar{n}$ |
|---|---|---|---|---|
|     | 1.40 | 0.38 | 0.80 | 0.25 |
|     | 1.50 | 0.50 | 1.06 | 0.34 |
| 36  | 1.60 | 0.65 | 1.36 | 0.42 |
|     | 1.70[b] | 1.09 | 2.24 | 0.61 |
|     | 3.65 | 1.04 | 2.22 | 0.60 |
|     | 4.51 | 1.62 | 3.38 | 0.79 |
|     | 4.90 | 2.25 | 4.62 | 0.91 |
| 11.5 | 5.0[b] | 2.94 | 6.00 | 0.99 |
|     | 7.28 | 1.86 | 3.86 | 0.87 |
|     | 9.08 | 3.03 | 6.17 | 1.06 |
| 5.7 | 9.80 | 4.17 | 8.44 | 1.17 |
|     | 10.0[b] | 5.47 | 11.05 | 1.25 |

[a]From (41).
[b]Untruncated.

a CR reduction of only about 10 percent while reducing the average number of reflections by 20 percent or more. The average number of reflections $\bar{n}$ increases with CR roughly as $\ln \sqrt{CR}$. The effect of truncation on mirror area is also shown to be significant. Reflector surface area increases linearly with CR, for example, $1 + a \, CR$, where $0.5 < a < 1.0$ depending on truncation.

Figures 4.28 and 4.29 show more complete data describing the truncation effects. The former depicts the effect of truncation on CR for a given acceptance angle (second column, Table 4.5). The latter shows the optical effect of truncation on the number of reflections. Both figures show that appreciable truncation and cost saving can be realized simultaneously without significant performance degradation.

For detailed optical analyses of CPC or other trough collectors, the number of reflections must be known as a function of aperture incidence angle in the transverse plane $\theta_{in}$ (this is the absolute value of profile angle $\gamma$ minus the aperture tilt angle $\beta$, that is, $|\gamma - \beta|$—see Fig. 2.11). Since no analytical method has been developed for this problem, a computerized Monte-Carlo approach can be used. Table 4.6 shows typical values of $\bar{n}$ for an attenuated CPC trough of acceptance angle equal to 5°. However, Rabl (43) has shown that for practical CPC devices the variation of $\bar{n}$ with $\theta_{in}$ can be neglected in most engineering designs.

The preceding optical analysis of CPC concentrators was based on idealized optics in that the sun was assumed to be a point source and the mirror was assumed to conform exactly to the specified curve. In practice, surface imperfections and the nonuniform energy distribution across the finite angular width of the sun's disk

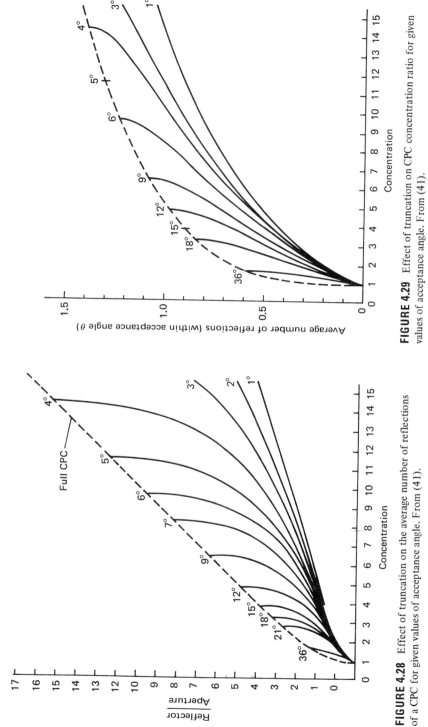

**FIGURE 4.29** Effect of truncation on CPC concentration ratio for given values of acceptance angle. From (41).

**FIGURE 4.28** Effect of truncation on the average number of reflections of a CPC for given values of acceptance angle. From (41).

**TABLE 4.6** Average Number of Reflections for an Attenuated Compound Parabolic Concentrator Collector[a]

| Incidence angle $\theta_{in}$ (degrees) | Number of reflections averaged over aperture $\bar{n}(\theta_{in})$ |
|:---:|:---:|
| 0.0 | 1.29 |
| 0.25 | 1.29 |
| 0.50 | 1.29 |
| 0.75 | 1.28 |
| 1.00 | 1.28 |
| 1.25 | 1.29 |
| 1.50 | 1.29 |
| 1.75 | 1.29 |
| 2.00 | 1.26 |
| 2.25 | 1.23 |
| 2.50 | 1.20 |
| 2.75 | 1.16 |
| 3.00 | 1.12 |
| 3.25 | 1.08 |
| 3.50 | 1.07 |
| 3.75 | 1.04 |
| 4.00 | 1.01 |
| 4.25 | 0.99 |
| 4.50 | 0.97 |
| 4.75 | 0.93 |
| Average $\bar{n}$ | 1.17 |

[a]Acceptance angle = 5°; truncated by 58%. From R. Winston, private communication, 1974.

must also be considered in determining what $CR$ is achievable by the CPC for a given acceptance angle.

Figure 4.30 shows ray traces of reflection from a nominal surface (perfect optics) and an imperfect surface (random surface angle). If the measures of surface error $\psi_1$ and solar beam displacement $\psi_2$ are taken to be their standard deviations $\sigma_{\psi_1}$ and $\sigma_{\psi_2}$, the maximum CPC concentration achievable for a design to capture a given percentage of all rays is reduced from the ideal maximum by

$$CR_{CPC,\,max} = \frac{1}{\sin\left(\theta_{max} + k\sqrt{\sigma_{\psi_2}{}^2 + 4\sigma_{\psi_1}{}^2}\right)} \qquad (4.100)$$

The parameter $k$ may be chosen to include a specific fraction of the randomly distributed rays. For a normal probability distribution and for $k = 2$, about 95 percent of the rays will be captured. The distribution of the sun's energy is probably more rectangular than normal; therefore Eq. (4.100) is a conservative estimate. However, Eq. (4.100) is most significant near cutoff, because when the sun is in the center of the acceptance angle most rays reach the absorber directly.

Typical values of $\sigma_{\psi_1}$ and $\sigma_{\psi_2}$ are $\frac{1}{4}°$ ($\sim 4$ mrad). The former corresponds to a carefully fabricated surface, the latter to the angle subtended by the sun. The effect of errors of this size is to increase the effective acceptance angle by about $\frac{1}{2}°$ and to reduce the $CR$ value correspondingly.

Instead of decreasing $CR$ (by increasing $\theta_{max}$) to compensate for optical uncertainties associated with a desired acceptance angle, no adjustment need be made. In that case some rays, otherwise collectable, would be lost. For transverse plane incidence angles $\theta_{in} < (\theta_{max} - k\sqrt{4\sigma_\psi{}^2 + \sigma_\phi{}^2})$ all rays will be collected. For $\theta_{in} > (\theta_{max} + k\sqrt{4\sigma_\psi{}^2 + \sigma_\phi{}^2})$, all rays are rejected. In the transition band some are captured and some are lost. Since the rays rejected are incident only near the cutoff angle, which has a value of 6° or more in practical designs, overall energy collection is not greatly affected by optical imperfections on the order of $\frac{1}{4}°$, an order of magnitude smaller than $\theta_{max}$.

If a CPC collector is to be used all year long, tilt adjustments are required periodically to locate the sun within the acceptance angle. The frequency of tilt adjustment depends on the number of hours of collection per day $n_h$. This important quantity can be calculated from the result of Rabl (43) who showed that

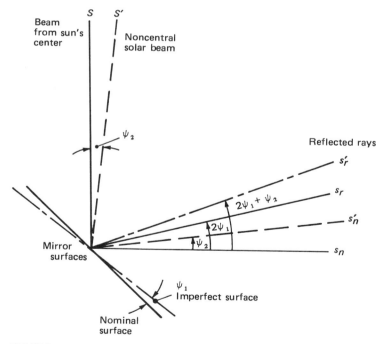

**FIGURE 4.30**  Ray traces of reflection of beam radiation from perfect and imperfect (random) mirror surfaces showing nominally reflected central ray $s_n$, nominally reflected noncentral ray $s'_n$, randomly reflected central ray $s_r$, and randomly reflected noncentral ray $s'_r$.

$$\tan\left(\theta_{max} + \beta - L\right) = \frac{\tan \delta_s}{\cos \omega t_c} \qquad (4.101)$$

where

$$\omega = \frac{\pi}{12}\, \text{hr}^{-1} \qquad (4.102)$$

$$n_h = 2t_c \qquad (4.103)$$

and where all other quantities are as defined previously.

## Thermal Analysis of the CPC Collector

In order to determine the energy balance on a CPC collector all important heat fluxes must be calculated from basic heat-transfer principles (29, 30). The fluxes are then combined in heat-balance equations for the absorber, the aperture cover (assumed glass), and the transport fluid. Since the various flux terms are nonlinear, a simultaneous iterative solution for the several temperatures is required.

The following nine macroscale (lumped) heat fluxes, all based on a unit absorber area $A_r$, are considered in the thermal analysis of a CPC:

1. Direct solar radiation absorbed both directly and indirectly by the receiver after reflection from the envelope—$q_{b,r}$ (solar wavelength region)
2. Direct solar radiation absorbed both directly and indirectly by the aperture cover after reflection from the absorber—$q_{b,a}$ (solar wavelength region)
3. Diffuse solar radiation absorbed by the receiver—$q_{d,r}$ (solar wavelength band)
4. Diffuse solar radiation absorbed by the aperture cover—$q_{d,a}$ (solar wavelength band)
5. Radiative exchange between the receiver and cover—$q_{ir}$ (infrared wavelength band)
6. Radiative exchange between the cover and the environment—$q_{sky}$ (infrared wavelength band)
7. Convective exchange between the receiver and the cover—$q_{c,ra}$
8. Convective loss from cover to the environment—$q_{c,e}$
9. Useful heat extraction—$q_u$

The heat loss through the outer walls of the collector and absorber can be made very small by proper insulation, and is therefore of higher order than the above nine heat fluxes and can be ignored in this first-order analysis.

The heat-flux terms listed above are defined in Eqs. (4.104)-(4.111), to the first order:

$$q_{b,r} = I_{b,c}\tau_a(i)\rho_m{}^{\bar{n}}\alpha_r(1 + \rho_m{}^{2\bar{n}}\rho_r\rho_a)\frac{A_a}{A_r} \qquad (4.104)$$

$$q_{b,a} = I_{b,c}[\alpha_a(i) + \tau_a(i)\rho_m{}^{2\bar{n}}\rho_r\bar{\alpha}_a]\frac{A_a}{A_r} \tag{4.105}$$

$$q_{d,r} = I_{d,c}\bar{\tau}_a\rho_m{}^{\bar{n}}\alpha_r{}^* \tag{4.106}$$

$$q_{d,a} = I_{d,c}\bar{\alpha}_a\frac{A_a}{A_r} \tag{4.107}$$

$$q_{ir} = \epsilon_{\text{eff}}\sigma(T_r{}^4 - T_a{}^4) \tag{4.108}$$

$$q_{\text{sky}} = \epsilon_{a,ir}\sigma(T_a{}^4 - T_{\text{sky}}^4)\frac{A_a}{A_r} \tag{4.109}$$

$$q_{c,ra} = h_{c,ra}(T_r - T_a) \tag{4.110}$$

$$q_{c,e} = h_{c,e}(T_a - T_\infty)\frac{A_a}{A_r} \tag{4.111}$$

where  $I_{b,c}$ = beam radiation incident on collector plane
$I_{d,c}$ = diffuse radiation incident on collector plane
$i$ = incidence angle (see Chapter 2)
$T_{\text{sky}}$ = effective sky temperature for radiation (see Chapter 7)
$h_{c,e}, h_{c,ra}$ = convection coefficients
$\alpha_r, \alpha_a$ = receiver and cover absorptances
$\rho_r, \rho_a$ = receiver and cover reflectances
$\epsilon_{\text{eff}} = (\epsilon_r{}^{-1} + \epsilon_a{}^{-1} - 1)^{-1}$

The angle-dependent cover transmittance $\tau_a(i)$ and absorptance $\alpha_a(i)$ for direct solar radiation are calculated from the equations of Stokes (Chapter 3). Transmittance and absorptance values for diffuse radiation $\bar{\tau}_a$ and $\bar{\alpha}_a$ are assumed independent of the angle. The convection coefficient $h_{c,ra}$ within the CPC enclosure has not been measured. It can be estimated from free convection correlations for flat surfaces in Chapter 3 or from Ref. (36).

Steady-state energy equations relate input energy terms to losses and to the useful output of the collector. The unknown quantities in the energy equations are $q_u$, $T_r$, and $T_a$, for which there are three equations to be solved simultaneously. The absorber conservation of energy equation is

$$q_{b,r} + q_{d,r} = q_u + q_{c,ra} + q_{ir} \tag{4.112}$$

The cover conservation of energy equation is

$$q_{b,a} + q_{d,a} + q_{ir} + q_{c,ra} = q_{\text{sky}} + q_{c,e} \tag{4.113}$$

---

*The assumption here is that the accepted diffuse radiation is equal to the incident diffuse radiation divided by $CR$; see Eq. (4.96).

The transport fluid energy equation is ($h_o$, $h_i$ = enthalpy)

$$h_o - h_i = \frac{q_u A_r}{\dot{m}} \tag{4.114}$$

where $\dot{m}$ is the fluid flow rate.

An order of magnitude analysis showed (29) that the resistance offered to heat transfer for steam or water at the inner surface and in the wall of the absorber was of higher order than the external surface resistance. Consequently, the fluid and absorber temperature are equal to lowest order. The energy balance equations are solved in a simultaneous iterative manner by computing $T_a$ from Eq. (4.113), $q_u$ from Eq. (4.112), and $h_o$ (or $T_{f,\text{out}}$) from Eq. (4.114). This iterative technique is continued until $T_a$, $q_u$, and $h_o$ (or $T_{f,\text{out}}$) are known to a required level of accuracy.

The instantaneous collector efficiency $\eta_c$ is defined as the system output divided by the incident radiation:

$$\eta_c = \frac{q_u A_r}{A_a(I_{b,c} + I_{d,c})} \tag{4.115}$$

The above detailed analysis can be expressed more simply by analogy with the Hottel-Whillier-Woertz-Bliss (HWWB) formulation (5, 25, 26) used for flat-plate collectors in an earlier section. Useful energy gain per unit absorber area may be expressed as

$$q_u = [CR\, \tau_a(i)I_{b,c} + \bar{\tau}_a I_{d,c}]\rho_m{}^{\bar{n}}\alpha_r - U_c(T_r - T_\infty) \tag{4.116}$$

This simplified expression assumes that the receiver absorptance for diffuse and beam radiation is independent of the angle and that all heat-loss terms can be parameterized by one quantity $U_c(T_r - T_\infty)$.

The instantaneous efficiency of a CPC may be expressed in further simplified form by assuming angle-independent cover transmittance as

$$\eta_c \equiv \frac{q_u A_r}{I_{\text{tot},c} A_a} = \rho_m{}^{\bar{n}}\tau\alpha\,\delta - \frac{U_c(T_r - T_\infty)}{CR\, I_{\text{tot},c}} \tag{4.117}$$

where the beam-to-diffuse properties of local solar radiation are contained in the intercepted radiation parameter $\delta$ defined as

$$\delta \equiv \frac{I_{b,c}}{I_{\text{tot},c}} + \frac{1}{CR}\frac{I_{d,c}}{I_{\text{tot},c}} \tag{4.118}$$

Although Eq. (4.117) is similar in appearance to the HWWB formulation, the optical efficiency $\rho_m{}^{\bar{n}}\tau\alpha\,\delta$ in this case is not simply a parameter of the collector independent of its operating conditions. It depends on the diffuse-to-beam radiation ratio. The order of magnitude of the optical efficiency $\eta_o$ of a typical CPC collector is

$$\eta_o \approx 0.85^{0.9}(0.9 \times 0.9)0.95$$
$$\eta_o \approx 0.6\text{-}0.7 \tag{4.119}$$

This efficiency value is less than that for flat-plate collectors, so the CPC device is less effective at low values of $\Delta T/I_c$, where optical efficiency is important. At elevated temperatures the CPC will usually have the advantage because of the reduced thermal losses.

Equation (4.117) cannot be used to calculate thermal performance of a CPC, since $U_c$ is not known a priori. The loss coefficient $U_c$ can be determined by experimentation, however, or by performance mapping using a computer model based on the detailed analysis.

Sample useful results of the detailed thermal analysis above [Eqs. (4.104)–(4.117)] are presented in Figs. 4.31 and 4.32 (29). The former figure shows the effect of flow rate and selective surface on performance for a collector with $CR = 3$. The effect of cover removal is also shown to be significant. The latter figure shows the effect of insolation to be important, as it is with all solar collectors.

The theoretical analysis above has been verified by experimental data collected at Argonne National Laboratory for a collector with $CR = 10$. The analysis above shows good correlation with the data and can be used for extrapolation purposes. Figure 4.33 depicts a comparison of test data and computer predictions.

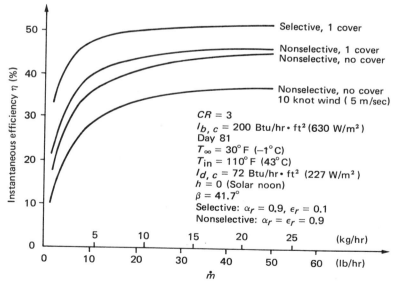

**FIGURE 4.31** Midday CPC efficiency versus flow rate. $CR = 3, A_a = 9.23$ ft$^2$ (0.86 m$^2$). From (29).

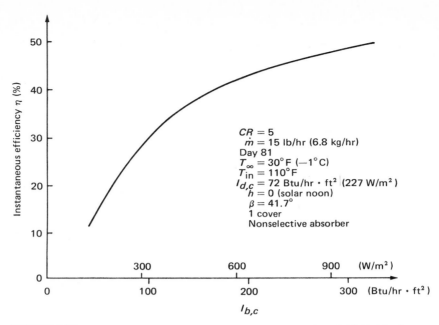

**FIGURE 4.32** Midday CPC efficiency versus direct insolation. $A_a = 6$ ft² (0.56 m²). From (29).

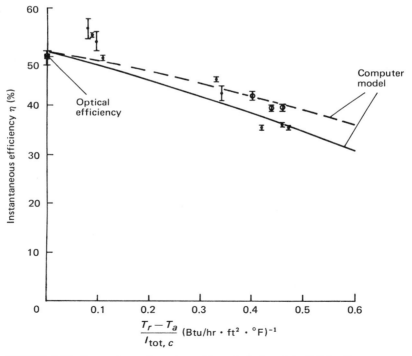

**FIGURE 4.33** Comparison of test data with analytical model of CPC collector. $CR = 10$, single cover, nonselective absorber surface. From (41).

## Other CPC Designs

The basic CPC shown in Fig. 4.26 can be improved in several ways. Since the rear surface of the receiver is exposed to the environment, it is a source of heat loss from the collector. In Fig. 4.34 three ways of eliminating the parasitic rear surface loss are shown. The "fin" receiver of Fig. 4.34a is illuminated on both sides thereby reducing the receiver material by one-half as well as eliminating rear losses.

Figure 4.34c shows a circular pipe as a receiver. This device has no parasitic back loss and is structurally desirable for use as a collector support or to confine a high-pressure working fluid. Any absorber shape may be used with a CPC-type concentrator. Each absorber shape requires a slightly different reflector surface as shown in Fig. 4.34 and as described by Winston and Hinterberger (62) wherein a formula for reflector construction is given. They showed that radiation incident within the acceptance zone of $2\theta_{max}$ can be concentrated onto any convex receiver of area $A_r = A_a \sin \theta_{max}$ for perfect optics.

The reflector section bounded between the shadow lines extended is the convolute of the receiver surface below the tangency point of the shadow line. The upper reflector portion is specified by simply requiring that the reflector normal bisect angle $\psi$ (see Fig. 4.34a). This angle is formed by the absorber tangent line $PT$ and a ray through $P$ incident at $\theta_{max}$ on the aperture $A$. This prescription is sufficient to construct any of the CPC family of concentrators for any absorber shape. Another useful CPC-like collector is called the angle transformer (43), which accepts radiation over $2\theta_{max}$ and restricts receiver incidence angles to $2\theta_2$. The concentration of the angle transformer is $\sin \theta_2 / \sin \theta_{max}$; the basic CPC (Fig. 4.26) is a special case of the angle transformer with $\theta_2 = 90°$. Restricted absorber plane incidence angles are useful in photovoltaic and other applications where absorber reflectance losses are very high at large absorber incidence angles.

## Optical Boosters for Tubular Collectors

An inexpensive method of improving the performance level of tubular collectors is to couple them to a single-curvature concentrating reflector of the CPC family. Although a normally fixed tubular array may require some tracking if coupled to a concentrator, a substantial performance improvement can be realized. A simplified analysis illustrates the magnitude of the effect.

The instantaneous efficiency of a single tubular collector with a *flat* absorber (Fig. 4.15a) can be expressed as

$$\eta_t = \eta_{ot} - \frac{2U_c(T_r - T_a)}{I_c} \tag{4.120}$$

where $U_c$ is based on the projected area of the glass envelope. For a *circular* receiver (CTC-type) the efficiency is

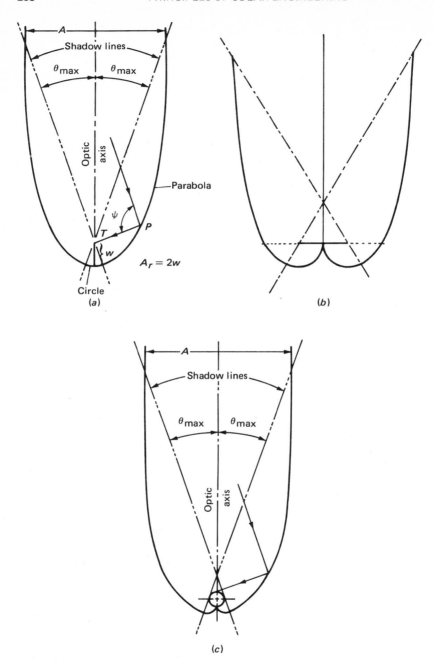

**FIGURE 4.34** Alternative CPC designs for reduced parasitic rear losses: (*a*) fin-type receiver; (*b*) fully illuminated plate-type receiver; (*c*) circular pipe receiver for improved structural strength.

$$\eta_t = \eta_{ot} - \frac{\pi U_c(T_r - T_a)}{I_c} \tag{4.121}$$

Loss coefficients $U_c$ on the order of 0.5–1.0 W/m² · °C have been measured for evacuated tubes with selective absorbers (3).

If a reflector is coupled to a glass tube as shown in Fig. 4.35, two opposite effects occur. First, thermal losses per unit aperture area decrease approximately as $1/CR$. Second, optical efficiency is reduced because of imperfect mirror optics and because some light is lost through the gap between the receiver and the mirror cusp or ends. It will be recalled that in all CPC-like collectors, the receiver surface must intersect the mirror to capture all light funneled to the receiver.

The instantaneous efficiency of a reflector-enhanced, uninsulated tubular collector with *flat* absorber is

$$\eta_{t,CPC} = \eta_{ot}\, \delta \rho_m{}^{\bar n} - \frac{2U_c(T_r - T_a)}{CR_{CPC}I_c} \tag{4.122}$$

where $\delta$ is the optical capture factor and $\rho_m$ the mirror reflectance. The factor 2 enters since heat is lost from both the front and back of the absorber. Goodman et al. (20) have shown that $\delta$ for collectors shown in Fig. 4.35b and c is given by

$$\delta = \left(1 - \frac{g}{p_r}\right)\frac{I_{b,c} + (1/CR_{CPC})I_{d,c}}{I_{\text{tot},c}} \tag{4.123}$$

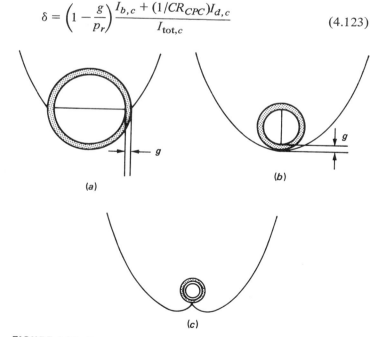

(a)

(b)

(c)

**FIGURE 4.35**  Examples of CPC-type concentrators coupled to receivers within evacuated glass tubes, with absorber-to-mirror gap $g$ shown.

where $p_r$ is the absorber perimeter and $g$ is the gap size between the absorber and reflector surfaces. For the basic CPC in Fig. 4.35a,

$$\delta = \left\{ 1 - \frac{g}{p_r} \left[ 1 - \sin \left( \frac{\pi}{4} - \frac{\theta_{max}}{2} \right) \right] \right\} \frac{I_{b,c} + (1/CR_{CPC})I_{d,c}}{I_{tot,c}} \qquad (4.124)$$

where $\theta_{max}$ is the acceptance angle. In Eq. (4.120) note that the concentration ratio of the receiver is taken to be 1/2, since the area from which heat can be lost is twice the aperture area. This convention is used in most tubular analyses.

Figure 4.36 shows the instantaneous efficiency $\eta$ for basic tubular collectors and mirror-enhanced, tubular collectors. The data indicate that efficiency can be

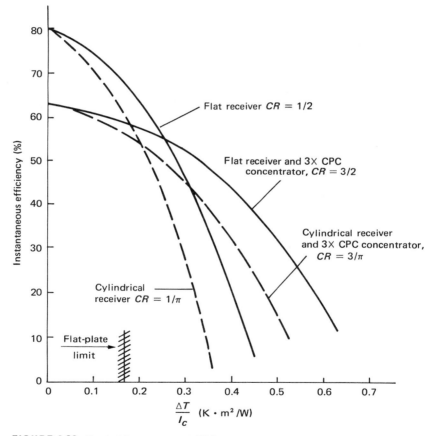

**FIGURE 4.36** Typical instantaneous efficiency curves for evacuated tubes with and without reflector boosters. Flat and cylindrical geometries are shown. $I_{tot,c} = 500$ W/m², $CR_{CPC} = 3.0$, $\delta = 0.9$, $\eta_o = 0.8$, $\rho_m = 0.85$, $\bar{n} = 0.6$. Concentration ratios are geometric ratios of aperture to *total* absorber surface area.

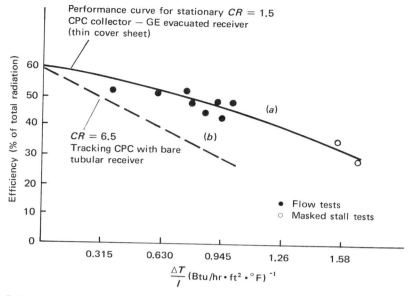

**FIGURE 4.37**  Thermal performance data of (*a*) a CPC-boosted evacuated tubular receiver and (*b*) an uninsulated tubular receiver. From A. Rabl, private communication, 1977.

doubled for an evacuated flat absorber at 200°C above ambient by the use of a 3X CPC. At 150°C above ambient, a tubular-receiver collector will realize a 75 percent increase if equipped with a 3X CPC booster. Figure 4.37 contains test data for a mirror-boosted tubular collector.

The question as to whether reflector enhancement is viable for a specific application depends not only on the probable increase in efficiency but also on the probable decrease in collector cost because of the replacement of evacuated tube area with reflector area. For example, if $250/m² evacuated tubes can be replaced with a $40/m² reflector in a 3X concentrator, the resultant cost savings will amount to $140/m².

The use of tubular receivers also affords the opportunity to use selectively transmitting glass (infrared reflectant such as $SnO_2$ or $In_2O_3$) instead of a selective absorber surface. The choice between the two options must again be based on a combined study of thermal performance and economics.

## V-Trough Concentrators

The most versatile turning concentrator, the CPC, has been described in detail. A less versatile but simpler collector with straight reflectors has also been used. This V-groove device cannot achieve the levels of concentration of a CPC but its simplicity may offer advantages in some applications.

According to Rabl (43), the V-trough device may be analyzed by referring to the diagram in Fig. 4.38. If $\theta_{max}$ is the acceptance half-angle (within which all rays are accepted) and $\theta_c$ is the cutoff angle beyond which no ray reaches the absorber, it can be shown that

$$\theta_c = \theta_{max} + 2\phi \qquad (4.125)$$

where $\phi$ is the trough half-angle as shown. A simplification allows the angles to be found to lowest order by drawing tangents to the reference circle through the aperture end points. The concentration ratio is thus found to be

$$CR = \frac{1}{\sin(\theta_{max} + \phi)} \qquad (4.126)$$

indicating a lesser concentration than the CPC for all $\phi$.

The acceptance of a V-trough is not as sharp a cutoff at $\theta_{max}$ as is the case with the CPC. All rays within $\theta_{max}$ are accepted; all beyond $\theta_c$ are rejected. In the intermediate band between complete acceptance or rejection, some rays are rejected and some accepted, depending on the point of incidence upon $A_a$. The average

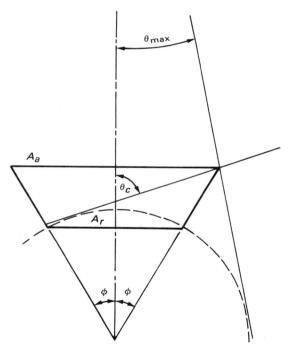

**FIGURE 4.38** V-trough solar concentrator schematic showing trough half-angle $\phi$, acceptance half-angle $\theta_{max}$, and cutoff angle $\theta_c$.

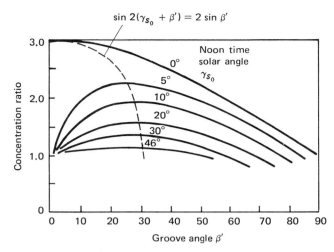

**FIGURE 4.39**  Concentration ratios for optimal groove geometries. From (1).

number of reflections $\bar{n}$ is roughly the same as that for comparable CPCs but the concentration is significantly less.

For example (43), if the acceptance half-angle is $5.7°$, a truncated CPC achieves $CR_{CPC} = 7.8$ and $\bar{n} = 1.0$. An analogous V-trough can achieve $\bar{n} = 1.0$ with a trough half-angle $\phi = 10°$. For this configuration the concentration is less than half that of a CPC, that is, $CR_V = 3.7$. Above $CR = 3$, V-troughs seem impractical. Below this level V-troughs can compete with CPCs for comparable levels of $\bar{n}$ and $CR$, but the broad, ill-defined acceptance transition zone will always be disadvantageous for the V-trough.

Bannerot and Howell (1) have shown that concentration ratios up to 3.0 can be achieved with grooved concentrators under the restriction that, at most, radiation will undergo one reflection and all incident beam radiation will be fully captured at noon. Figure 4.39 shows the concentration ratio achievable for various included groove-angles as a function of solar-noon incidence angle $\gamma_{s_0}$.

## Summary

Table 4.7 summarizes thermal applications for various fixed and turning concentrators. Evacuated glass jackets surrounding receivers can boost temperatures by $50-100°C$ in some cases permitting efficient operation of solar cooling machinery (see Chapter 7). As expected, fixed concentrators have little performance gain over basic flat-plate collectors unless used with an evacuated receiver where they offer some performance gain over nonconcentrating collectors at high temperature. They can be more cost-effective, however, since $1\ m^2$ of mirror is usually less expensive than $1\ m^2$ of collector assembly.

**TABLE 4.7** Comparison of Performance of Fixed and Turning Solar Concentrators[a]

| Tracking requirement | Examples | Approximate range of CR | Approximate maximum operating temperature above ambient |
|---|---|---|---|
| None—tilt fixed | CPC | 1.5–2.0 | Up to 100°C no vacuum |
| | | | Up to 140°C vacuum |
| None—two tilt | V-trough | 1.5–2.0 | Up to 140°C vacuum |
| adjustments per year | CPC | 3.0 | Up to 180°C vacuum |
| per year | | | |
| None—tilt adjust | V-trough | 2–3 | Up to 180°C vacuum |
| seasonal to daily | CPC | 3–10 | 100–150°C no vacuum |
| | | | 150–250°C vacuum |

[a]From (44).

## CONTINUOUSLY TRACKING SOLAR CONCENTRATORS

In order to achieve concentration ratios above 20 for year-round periods of solar collection of 6 hr or more, collectors that track the sun continuously are required. Two types of continuously tracking collectors can be identified:

1. Single-curvature (or cylindrical) concentrators characterized by one plane of symmetry; frequently called *trough* concentrators.
2. Compound-curvature concentrators characterized by two planes (or a line) of symmetry; frequently called *dish* concentrators.

   Single-curvature concentrators generally track with mechanisms having one degree of freedom, whereas compound-curvature concentrators track with two degrees of freedom. Many absorber shapes are commonly used, including tubular or planar for trough types, and tubular, circular-planar, or spherical for dish types. Reflectors are usually parabolic, paraboloidal, circular, or spherical in shape. Frequently, Fresnel-like surfaces, made from planar segments that approximate the geometric shapes listed, are used.

   Tracking of the sun by solar collectors can be done approximately by using a simple clock device. However, this method is suitable only for low-concentration devices that can accommodate some error and still maintain the sun within the acceptance angle. For more precise tracking, a feedback control device with sun sensors should be used. A discussion of these devices is beyond the purview of this book, but the electronic and mechanical design of sun trackers has been well developed by the aerospace industry. Usually these devices move at a base rate to keep the sun, even if obscured by clouds, within the acceptance angle. A fine correction feedback device is used to compensate for small errors, windy conditions, or irregularities in the mechanical drive.

   Many realizations of tracking concentrators have been described in the technical literature—each with specific advantages and disadvantages. In subsequent sections one single-curvature and two double-curvature concentrators are described in detail as examples of each generic type. Other practical members of each family are described only in general. A complete and detailed analysis of each member is repetitive since the basic features of all are similar.

## Single-Curvature Tracking Concentrators—
## Parabolic Trough Example

Figure 4.40 shows a number of troughlike concentrators that have been built and tested. Each has a continuously tracking component that maintains a sharp focus of the sun on the receiver. The parabolic trough and Fresnel lens types move the entire collector assembly continuously as a unit; however, the Fresnel mirror designs require independent movement of mirror segments, thereby reducing structural demands but increasing tracking complexity. The fixed mirror trough uses a moving absorber instead of a moving reflector or refractor in order to reduce structural requirements. As an example of the single-curvature concentrator, the parabolic trough concentrator (PTC) (Fig. 4.40a), as developed by the University of Minnesota and Honeywell, Inc., is analyzed in detail (14, 15, 49–52).

### Optical Analysis of the PTC Collector

Figure 4.41a is an isometric view of the PTC showing the fixed absorber tube and tracking mirror assembly. The optical performance of the PTC depends on optical properties of the various surfaces involved and the shading by both trough ends ($A_s$) and the receiver tube.

The optical efficiency $\eta_o$ of the collector based on direct normal insolation can be expressed (46) as

$$\eta_o = \rho_m \tau_c \alpha_r f_t \delta \left[(1 - A \tan i) \cos i\right] F(\theta_t) \tag{4.127}$$

where interreflections of radiation between the cover (receiver envelope) are ignored for simplicity. In Eq. (4.127),

$\rho_m$ = mirror reflectance
$\tau_c$ = receiver tube cover transmittance
$\alpha_r$ = receiver tube absorptance for solar radiation
$i$ = solar incidence angle
$f_t$ = fraction of mirror area not shaded by receiver brackets
$\delta$ = fraction of reflected radiation intercepted by receiver for perfect tracking and imperfect mirror and mirror surfaces but with specular reflection
$F(\theta_t)$ = fraction of reflected radiation intercepted by receiver for perfect optics and imperfect tracking
$\theta_t$ = tracking angle error measured in plane of rotation

The direct normal radiation $I_b$ is used as the efficiency basis in this analysis. Other investigators have used the component of direct solar radiation normal to the collector aperture $I_{b,c} (= I_b \cos i)$. In studying the literature the reader must

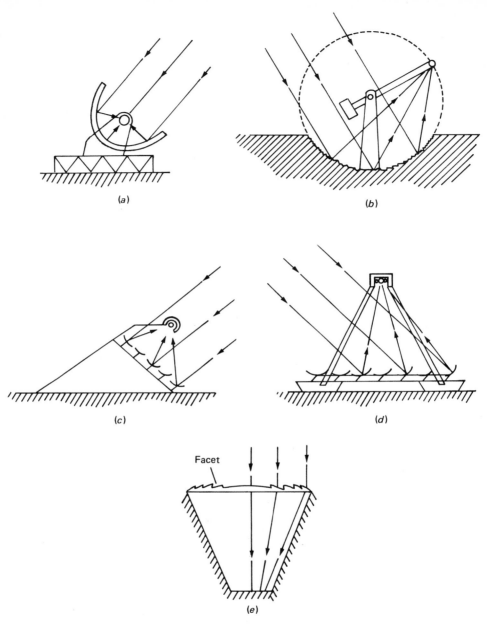

**FIGURE 4.40** Single-curvature solar concentrators: (*a*) parabolic trough; (*b*) fixed circular trough with tracking absorber; (*c*) and (*d*) Fresnel mirror designs; and (*e*) Fresnel lens.

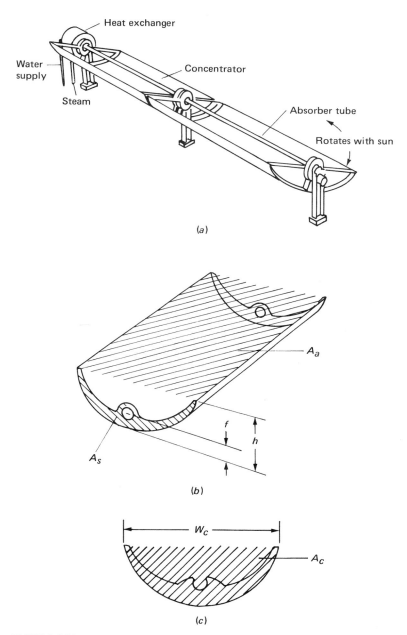

**FIGURE 4.41**  PTC assembly drawing and schematic diagram defining parameters used in its optical analysis. From (46).

carefully determine the efficiency basis. The confusion can be easily avoided by using the integrated energy delivery instead of instantaneous efficiency as the figure of merit.

The term $A \tan i$ in Eq. (4.127) is a measure of the effective reduction in aperture area because of shading, blocking, and loss of radiation reflected beyond the receiver tube end. For the configuration shown, $A$ is given by (46)

$$ A = \frac{W_c(f + h) + A_s - A_c}{A_a} \tag{4.128} $$

Typical values for $A$ are between 0.2 and 0.3. Other optical parameters are all on the order of 0.90–0.95 with the exception of $\rho_m (\approx 0.8)$. The tracking accuracy function $F(\theta_t)$ measured by Ramsey et al. (46) on a PTC prototype shown in Fig. 4.42 increases with tracking accuracy. Tracking errors of $1°$ reduce optical efficiency by 20 percent, a result of significantly greater magnitude than that for turning collectors described in the previous section where only a negligible effect is experienced.

Optical performance data measured on a carefully fabricated PTC are shown in Fig. 4.43 for two collector orientations. The east-west orientation shows a large hour-angle effect, since the incidence angle for this tracking mode varies directly with the hour-angle [see Eq. (2.17)]. The tilted north-south position is much less sensitive to the hour-angle, as expressed in Eq. (2.19). The data for both orientations are predicted very accurately by Eq. (4.127), which is represented by the solid lines in Fig. 4.43. Equation (4.127) therefore is adequate for the optical design of the PTC collector.

Duff and Lameiro (9) have shown that the optical property $\delta$, sometimes called the intercept function, can be expressed analytically for several concentrator types. It depends on the uncertainty in reflector surface slope $\psi_1$ and light ray orientation

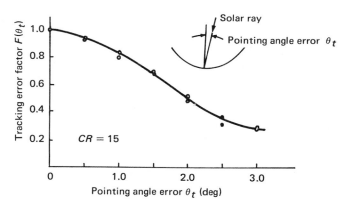

**FIGURE 4.42**  Tracking accuracy correction factor versus pointing angle error. From (46).

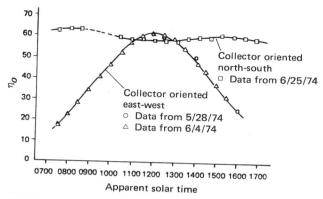

**FIGURE 4.43**  Optical efficiency $\eta_o$ versus time of day (40°C operating temperature). From (46).

$\psi_2$ as shown in Fig. 4.30. From Fig. 4.44 it is easy to see that

$$y = r \sin (2\psi_1 + \psi_2) \approx r(2\psi_1 + \psi_2) \tag{4.129}$$

for small values of $\psi_1$ and $\psi_2$. The distance $r$ from mirror to receiver center for a PTC is given by (9)

$$r = \frac{A_a (1 + \cos \phi)}{2 \sin \phi (1 + \cos \theta)} \tag{4.130}$$

where $\phi$ is the mirror-rim half-angle and $\theta$ is the angle measured between the optical axis and the nominal reflected ray direction as shown in the figure.
   Substituting for $r$ in Eq. (4.129) gives

$$y = \frac{A_a (1 + \cos \phi)(2\psi_1 + \psi_2)}{2 \sin \phi (1 + \cos \theta)} \tag{4.131}$$

If the random variables $\psi_1$ and $\psi_2$ are normally distributed with mean zero, the variance $\sigma_y{}^2$ of the intercept point $y$ is given by

$$\sigma_y{}^2 = \frac{A_a{}^2(4\sigma_{\psi_1}{}^2 + \sigma_{\psi_2}{}^2)(1 + \cos \phi)^2}{4 \sin^2 \phi} \frac{1}{2\phi} \int_{-\phi}^{\phi} \frac{d\theta}{(1 + \cos \theta)^2} \tag{4.132}$$

where the average is taken over all values of $\theta$ between $\phi$ and $-\phi$. Evaluation of the integral gives

$$\sigma_y{}^2 = \frac{A_a{}^2(4\sigma_{\psi_1}{}^2 + \sigma_{\psi_2}{}^2)(2 + \cos \phi)}{12\phi \sin \phi} \tag{4.133}$$

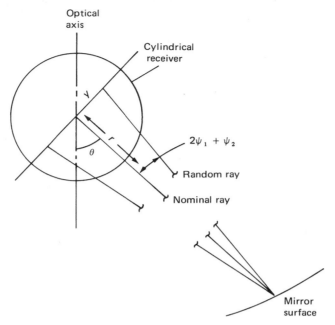

**FIGURE 4.44** Schematic diagram of parabolic reflector-circular receiver optics showing a nominal ray for perfect optics with a point sun and a random ray for imperfect optics and a distributed sun.

The intercept factor $\delta$ is the amount of energy contained in a normally distributed flux pattern of mean zero and standard deviation $\sigma_y$. In the closed interval $[-D_r/2, D_r/2]$, where $D_r$ is the receiver diameter, the intercept factor $\delta$ is given by the expression

$$\delta(\sigma_{\psi_1}, \sigma_{\psi_2}, \phi, A_a) = \frac{1}{\sigma_y\sqrt{2\pi}} \int_{-D_r/2}^{D_r/2} \exp\left[-\frac{1}{2}\left(\frac{y}{\sigma_y}\right)^2\right] dz \qquad (4.134)$$

If measures of mirror quality $\psi_1$ and solar distribution $\psi_2$ are used with the standard PTC design variables $A_a$ and $\phi$, Eq. (4.134) can be used directly to evaluate the percentage of radiation intercepted by the absorber. A change of variable in Eq. (4.134) will permit evaluation of the integral from tables of the standard normal distribution. Let $z \equiv y/\sigma_y$ and Eq. (4.134) becomes

$$\delta(\sigma_{\psi_1}, \sigma_{\psi_2}, \phi, A_a) = \frac{1}{\sqrt{2\pi}} \int_{-d_r/2}^{d_r/2} e^{-z^2/2}\, dz \qquad (4.135)$$

where $d_r \equiv D_r/\sigma_y$. The right-hand side of Eq. (4.135) is the standard normal probability distribution tabulated in Appendix 4.

Another method of quantifying mirror, tracking, and solar-disk flux spreads is to define an effective standard deviation $\Sigma$ for all of these factors

$$\Sigma^2 = 4\sigma_{\psi_1}{}^2 + \sigma_{\psi_2}{}^2 + \sigma_{track}^2 + \sigma_{mirror}^2$$

The first and last terms represent mirror slope and diffuseness errors, respectively. The factor $F(\theta_t)$ in Eq. (4.127) is now represented by the $\sigma_{track}^2$ term. Both approaches, properly formulated, should provide the same conclusion.

## EXAMPLE 4.5

Calculate the maximum concentration ratio and minimum receiver diameter for a PTC that captures 90 percent of the incident radiation with a cylindrical receiver if the rim half-angle $\phi$ is $45°$ and the surface and solar standard deviations $\sigma_{\psi_1}$ and $\sigma_{\psi_2}$ are each $\frac{1}{4}°$. The collector aperture is 1 m and tracking errors can be ignored.

## SOLUTION

From Eq. (4.133) the variance $\sigma_y{}^2$ is

$$\sigma_y{}^2 = \frac{A_a{}^2(4\sigma_{\psi_1}{}^2 + \sigma_{\psi_2}{}^2)(2 + \cos \phi)}{12\phi \sin \phi}$$

$$= \frac{(1.0)(\pi/180)^2 (4 \times 1/4 \times 1/4 + 1/4 \times 1/4)(2 + \cos 45°)}{(12)(\pi/180)(45°) \sin 45°}$$

$$= 38.7 \times 10^{-6} \text{ m}^2$$

or

$$\sigma_y = 6.2 \text{ mm}$$

The optical intercept factor $\delta = 0.9$. Therefore

$$d_r \left( = \frac{D_r}{\sigma_y} \right) = 1.65$$

from the probability table in Appendix 4. Therefore

$$D_r = 1.65 \, \sigma_y$$

$$D_r = 10.23 \text{ mm}$$

The concentration ratio $CR$ is given by

$$CR = \frac{A_a}{\pi D_R} = \frac{1000 \text{ mm}}{\pi \times 10.23}$$

$$CR = 31.1$$

Repeat for a concentrator that captures 95 percent of the incident radiation. Answer: $D_r = 12.2$ mm and $CR = 26.2$.

### Thermal Analysis of the PTC Collector

A simplified, macroscale thermal analysis of the PTC can be made by considering only radiative losses from the absorber surface and conductive losses through the absorber tube ends, assuming that the receiver is surrounded by an evacuated jacket. The simplified heat loss $q_L$ is given by

$$q_L = A_{\text{cond}} \frac{T_r - T_a}{R_k} + \epsilon_r \sigma (T_r^4 - T_c^4) A_r \qquad (4.136)$$

where $R_k$ is the composite thermal resistance for conductivity from the absorber to ambient via absorber supports and vacuum seals. $T_c$ is the receiver cover temperature, and $A_{\text{cond}}$ is the effective area for conduction heat transfer.

Of the two components of heat loss, the second is expected to be the larger at high temperatures ($\sim 300°C$) useful for process heat production or electric power generation. Figure 4.45 is a plot of typical heat losses for a 5-m$^2$ prototype PTC tested in Arizona. At temperatures up to 200°C the radiative and conductive heat losses are of the same magnitude for a selectively coated absorber ($\alpha_s = 0.91$, $\epsilon_r \sim 0.2$). The nonselective absorber ($\alpha_s = \epsilon_r = 0.96$) has radiative losses much greater than those for the comparable selective absorber. For the data shown, the conductive resistance $R_k/A_{\text{cond}}$ was found to be 1.8 °C/W. The magnitude of the conductive loss is seen to be significant even for a carefully built prototype.

Löf et al. (37) have also measured the performance of a PTC with $CR = 22$. Figure 4.46 summarizes their results for an exposed receiver (no glass cover) with a nonselective coating. The convection losses in this configuration are of the same order as conduction losses. Both taken together exceed radiative losses from the nonselective receiver, thereby illustrating the importance of convection suppression by either evacuation or other means.

Instantaneous collector efficiency $\eta_b$ is based on beam radiation only, since most PTC collectors collect little diffuse radiation because of their concentration levels of 25 or more. Equation (4.137) gives the efficiency as

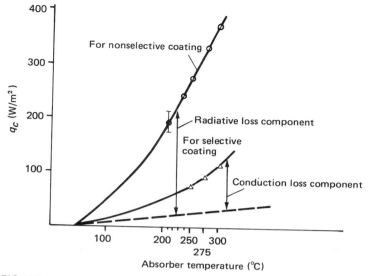

**FIGURE 4.45**  Heat loss as a function of absorber temperature. From (46).

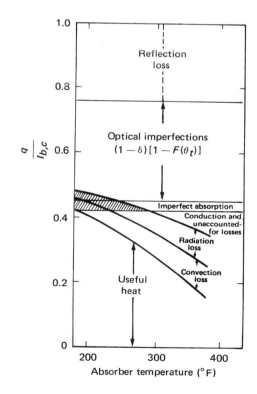

**FIGURE 4.46**  Useful heat delivery, and reflection, optical, and heat losses for 6.19-ft aperture parabolic reflector with 1.05-in diameter, bare absorber vs. absorber surface temperature. From (37).

$$\eta_b \equiv \frac{q_u}{A_a I_b} = \eta_o - \frac{1}{I_b} \left[ \frac{T_r - T_a}{R_k} \frac{A_{\text{cond}}}{A_a} + \frac{\epsilon_r \sigma (T_r^4 - T_c^4)}{CR} \right] \qquad (4.137)$$

Ramsey et al. (46) measured thermal efficiency for the prototype described above. Some results are shown in Fig. 4.47 where direct, normal-incidence beam radiation $I_b$ and energy delivery are plotted for an entire clear day. Energy delivery at 300°C at noon exhibits a peak *thermal* efficiency of about 50 percent, only 10 percent below the prototype *optical* efficiency of 60 percent. A north-south polar orientation could be expected to deliver energy at the 50 percent level for a full day because of the continuous small incidence angle for the polar mount. Data recorded in the same series of tests on a nonselective receiver showed a peak noontime efficiency of only 20 percent, thereby illustrating the importance of low-surface emittance for high-temperature applications. Edenburn (16) has also analyzed and tested PTC collectors for a total energy project in the U.S. Southwest.

## Other Single-Curvature Concentrators

Thermal and optical analyses of other cylindrical concentrators are quite similar to those described above. For each, optical properties of all surfaces as well as the

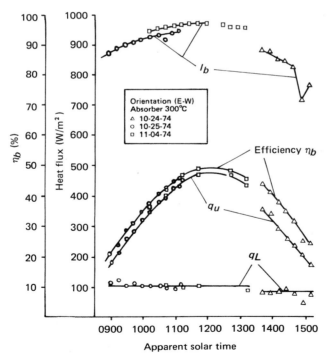

**FIGURE 4.47** Performance data for selective coated absorber tube (east-west orientation; 300°C absorber temperature). From (46).

**TABLE 4.8** Comparison of Single-Curvature Solar Concentrators[a]

| Generic type | Figure number | Approximate range of CR | Approximate maximum operating temperature (°C) |
|---|---|---|---|
| Fresnel lens | 4.40e | 6–30 | 100–200 |
| Parabolic trough and fresnel mirror | 4.40a 4.40c and d | 15–50 | 200–300 |
| Fixed mirror moving receiver | 4.40b | 20–50 | 300 (with second-stage concentrator, vacuum, and selective absorber, up to 400) |

[a]From (44).

effects of mirror and tracking inaccuracies must be known. Shading variation with incidence angle must also be calculated. This calculation is quite simple for reflecting devices, but requires a more complex analysis for Fresnel lenses.

In order to mold a Fresnel lens in one piece, the vertical segments between facets (Fig. 4.40e) must have a nonzero draft angle to permit removal from the die. These steps will not refract light to the receiver, but instead cause losses of incident radiation. This effect varies with facet size, facet angle, and incidence angle. Extruded Fresnel lenses do not require a draft angle but are usually less precise optically because of residual stresses introduced by the extrusion process. The greater tolerance of Fresnel lenses to surface errors vis-à-vis reflectors tends to offset this disadvantage. The beam displacement error resulting from a reflector inaccuracy $\Delta\psi$ is $2\Delta\psi$ whereas the error in a lens is only $(n_r - 1)\,\Delta\theta$ where $n_r$ is the refractive index ($\sim 1.5$).

Thermal balance analyses of cylindrical collectors are all quite similar. Radiation, conduction, and convection loss mechanisms must be evaluated for each collector component operating above ambient temperature—receiver, receiver cover, supports, mirrors (in designs where the mirror and receiver are adjacent and radiative transfer can occur), and fluid conduits. For preliminary design a simplified macroscale thermal analysis of the type used above may be adequate. More complete microscale thermal analyses, for example, those shown in the next section, are needed for detailed collector design.

Table 4.8 summarizes the expected range of operating temperature for the various types of single-curvature concentrators discussed. Although operating ranges overlap, each collector type has a distinct subrange of CR from 6 to 50 at which it operates best. The type used in a specific application depends on its economic viability. These matters are discussed in detail in Chapter 5.

## Compound-Curvature Tracking Concentrators—
## SRTA and Fresnel Mirror Examples

In order to achieve concentration ratios above 50 (one-fourth the theoretical limit for trough types) without inordinately precise tracking and optical features, compound-curvature concentrators must be used. This group of concentrators includes paraboloidal and spherical reflectors of both continuous and Fresnel designs as well as circular Fresnel lenses. The focal pattern of most dish-type concentrators is called, ideally, a point focus, which is the theoretical focal pattern for perfect optics and a point sun. Spherical mirrors and lenses have a focal line instead of a focal point and for these concentrators, linear absorbers are required to capture all reflected rays.

Other dish-type concentrators have been demonstrated, including straight-sided, conical light funnels and compound-curvature versions of the CPC. These two collectors are simply continuous-tracking, compound-curvature analogs of the single-curvature V-trough and CPC. They can be analyzed by extension of the principles presented in the section on intermittently turning concentrators.

The efficiency of solar collectors capable of tracking with two degrees of freedom is defined somewhat differently from that for fixed, turning, or single degree-of-freedom tracking concentrators. Fully tracking collectors can be operated ideally at normal solar incidence for the entire period of collection, since the sun's virtual motion is of two degrees of freedom. Therefore, the upper limit of collection of beam radiation is that radiation $I_b$ intercepted on a plane normal to the sun's rays. Even though the SRTA (spherical, stationary reflector, tracking absorber) collector described below is a unique device with two degrees of freedom in that it cannot operate continuously at normal incidence, its efficiency is likewise based on the direct normal solar radiation $I_b$ and not the projection $I_{b,c}$ normal to the aperture. In further study of the technical literature on concentrating collectors, the reader must determine for each analysis whether $I_b$ or $I_{b,c}$ is the basis for efficiency calculation, since no broadly accepted convention exists.

Two compound-curvature collectors are analyzed in detail in this section. The first is the SRTA concentrator first described analytically by Steward (55). The SRTA is of particular interest from a practical viewpoint since the mirror need not track the sun; only the linear absorber moves to maintain a focus. The second concentrator, analyzed in less detail, uses a Fresnel mirror consisting of a large horizontal array of tracking mirrors (heliostats) and a centrally located receiver positioned above the surface on a support tower. This collector concept is commonly called the *power tower*, since it has been considered for use in electric power production. It was first proposed by Baum (2a).

## SRTA Optical Analysis

Figure 4.48 is an isometric drawing of the SRTA collector concept. The major components of the collector include a fixed spherical mirror, a tracking linear absorber directed parallel to the sun's rays, a support structure, and a tracking device. Because of the spherical symmetry of the collector assembly, the optical behavior of this concentrator is independent of solar-incidence angle unlike any other concentrator. Figure 4.49 shows the reflection pattern for three values of incidence angle. All ray trace patterns, as viewed by the absorber, are identical except for their partial truncation resulting from the varying amount of absorber illuminated.

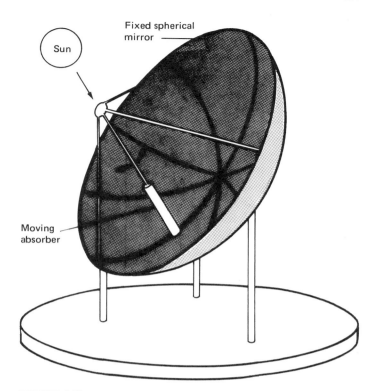

**FIGURE 4.48** Stationary reflector, tracking absorber solar collector. From (34).

Figure 4.49 shows an SRTA collector with a 60° rim half-angle. If only a 15° half-angle mirror were used instead, little spherical aberration would exist and a linear receiver would not be required to intercept all radiation. In that case an approximate point focus could be achieved, albeit with a relatively low concentration ratio.

The optical efficiency $\eta_o$ of the SRTA can be expressed to the lowest order *for perfect optics and tracking* as

$$\eta_o = \rho_m \tau_e(i_r) \alpha_r f_t \cos i \tag{4.138}$$

where $i_r$ is the *receiver cover (envelope) incidence angle* (see Fig. 4.50), $\tau_e(i_r)$ is the receiver envelope transmittance, $\rho_m$ is the mirror reflectance, and $\alpha_r$ is the receiver absorptance. The parameter $f_t$ is the fraction of aperture area not shaded by absorber supports or other blocking devices. Unlike other concentrators, the SRTA *absorber* instantaneously receives radiation over an incidence angle $i_r$ range from 0 to 90°. The effect of $i_r$ on envelope transmittance must therefore be included in the analysis of this concentrator. Stokes' equations are used to quantify this effect.

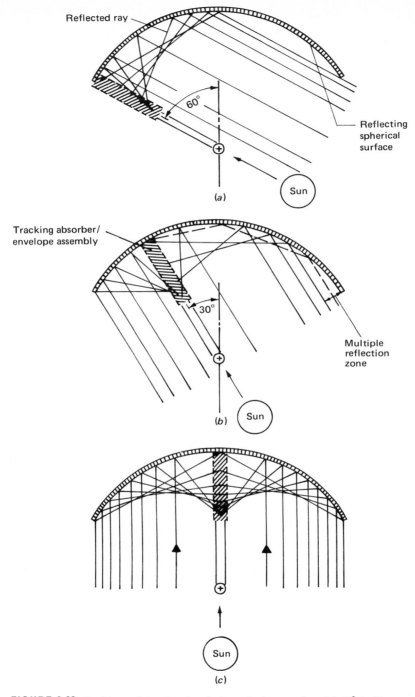

**FIGURE 4.49** Positions of the absorber during a day's operation: (*a*) 60° incidence angle; (*b*) 30° incidence angle; (*c*) normal incidence. Courtesy of W. Gene Steward.

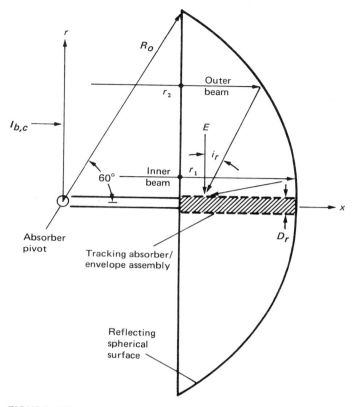

**FIGURE 4.50** Schematic diagram of the SRTA solar collector showing incoming ray traces and coordinate system for analysis for a 60° half-angle mirror.

The receiver incidence angle $i_r$ can be calculated by referring to the diagram in Fig. 4.50. For specular reflection it is easy to show that

$$i_r = \frac{\pi}{2} - 2 \sin^{-1} \frac{r}{R_o} \tag{4.139}$$

The reflected insolation can be resolved into components parallel and perpendicular to the receiver surface. The useful perpendicular component $E$ (see Fig. 4.50) can be calculated from an energy balance on a differential area of aperture and receiver:

$$\rho_m I_{b,c} 2\pi r \, dr = E\pi D_r \, dx \tag{4.140}$$

from which

$$E = \rho_m \frac{2r I_{b,c}}{D_r} \frac{dr}{dx} \tag{4.141}$$

where $D_r$ is the receiver diameter.

The derivative $dr/dx$ can be calculated from the function that maps aperture points onto receiver points. The mapping function, as shown by Kreider (31), is

$$\frac{x}{R_o} = \frac{1}{2\sqrt{1 - (r/R_o)^2}} \left[ 1 + \frac{D_r}{2R_o} \left( \frac{R_o}{r} - \frac{2r}{R_o} \right) \right] \tag{4.142}$$

The mapping function is seen to have a pair of $r$ values for a single receiver coordinate value $x$. The two rays are called inner and outer beams as shown in Fig. 4.50. Since the inner ray always intercepts a cylindrical receiver surface at large incidence angles, reflection losses from the transparent receiver cover can be quite large unless special precautions are taken or a cover is not used.

The geometric concentration ratio of an SRTA collector with a cylindrical receiver is given by

$$CR \equiv \frac{A_a}{A_r} \tag{4.143}$$

$$= \frac{\pi (R_o \sin \phi)^2}{\pi D_r L_r} \tag{4.144}$$

where $\phi$ is the rim half-angle and $L_r$ is the receiver length. Since no reflected radiation intercepts the focal line for $x > R_o/2$, the receiver length is usually confined to its illuminated area from which

$$L_r = \frac{R_o}{2} \tag{4.145}$$

The concentration ratio can then be expressed as

$$CR = \frac{2R_o}{D_r} \sin^2 \phi \tag{4.146}$$

Clausing (6) has considered the second-order, multiple-reflection zone of an SRTA as shown in Fig. 4.49b. Although this zone usually represents less than 5 percent of the aperture area, all the multiply reflected radiation intercepts the receiver in a small area adjacent to the mirror. The number of reflections $n(r)$ is given by (6)

$$n(r) = \left( \frac{\sin^{-1} r/R_o}{\pi - 2\sin^{-1} r/R_o} \right) \tag{4.147}$$

where the parentheses denote the greatest integer.

The point of intersection $x_o$ of these rays on a small absorber is given by (6)

$$x_o = \frac{r}{(-1)^{(n-1)} \sin 2n(\sin^{-1} r/R_o)} \tag{4.148}$$

Equation (4.148) shows that $n \to \infty$ as $r \to R_o$. The optical consequence of multiple reflections is that the sun's image at the focal line is larger than would be calculated for a single reflection. The multiple-reflection effect reduces the maximum possible concentration ratio for perfect optics (no multiple reflections) by about 6 percent. The optical consequences of multiple reflections are not great, but the aggregation of all multiple-reflected rays near the receiver end causes a zone of increased heat flux, which must be considered in the thermal design of the receiver.

For nonideal optics in which mirror imperfections and a finite solar disk are considered, the intercept factor $\delta$ analogous to that in Eq. (4.135) must be calculated. Because of the optical complexities and nonuniform flux distribution along the SRTA receiver, no closed-form expression for $\delta$ analogous to Eq. (4.135) for the PTC has yet been obtained. A Monte-Carlo technique or other numerical method is the only technique presently available for calculation of $\delta$ for arbitrary surface errors $\psi_1$ and beam radiation direction errors $\psi_2$.

An estimate of the optical losses can be made by considering the optical efficiency shown in Eq. (4.138). The optical efficiency $\eta_o$ will fall somewhere between 0.4 and 0.85. The blocking factor $f_t$ should be on the order of 0.95–0.99; the reflectivity $\rho_m$ between 0.7 and 0.9; the intercept factor $\delta$ between 0.90 and 0.98; the transmittance of the envelope around the receiver between 0.75 and 1.0; and the receiver absorptance between 0.85 and 0.95.

### Simplified SRTA Thermal Analysis

The main advantage of an SRTA system is that it uses a fixed reflector, which makes possible the construction of large units at reasonable cost. A spherical radio telescope with a surface of $7.5 \times 10^4$ m$^2$ (18.5 acres) located near Arecibo in Puerto Rico has demonstrated the technical feasibility of constructing large spherical shapes. Large collecting areas drastically reduce the losses that occur in transporting thermal energy, as well as the costs of plumbing and pumping the working fluid. With the exception of the central receiver, to be discussed in the last section, arrays of other concentrators consist of many small individual collectors that suffer thermal and mechanical losses in transporting the working fluid.

The major disadvantage of the SRTA collector system is the so-called cosine or incidence angle loss, which is common to all collection systems whose aperture does not track the sun. Thus, the beam component of solar radiation incident on a collector with fixed aperture is proportional to the cosine of the angle between the sun's ray and the normal to the plane of the aperture—the angle of incidence $i$. This angle depends on the latitude and the orientation of the reflector, and varies with the time of day and time of year as shown in Chapter 2. Degradation in collectors' performance occurs during times of the day that are appreciably removed from solar noon and during times of year at which the aperture is not at a favorable inclination. Since the convection and radiation losses from the absorber are constant at a given collection temperature, the thermal efficiency of the collector system decreases with increasing $i$ at a rate faster than the rate of decrease of the cosine of $i$, similar to the performance of a flat-plate collector.

The effective concentration ratio $CR_e$ for a fixed reflector of the SRTA type is given by $CR_e = CR \cos i$. For angles of incidence greater than $60°$, the incident radiant energy flux and hence $CR_e$ is reduced to less than half that for a completely tracking reflector.

Approximate calculations can be made to determine the thermal efficiency of an SRTA, if one assumes that it operates at a uniform absorber temperature. This would be a reasonable assumption if a heat pipe were to be used as the receiver or if the receiver were a boiler. For the simplified performance analysis of the SRTA collector system, Clausing (6) suggests that the absorber be assumed to have a selective gray surface with a solar absorptance $\alpha_r$, infrared emittance $\epsilon_r$, and with absorptance independent of incident angle. The absorber is assumed to be uncovered in order to avoid reflective losses at the large incident angle (see Fig. 4.50) always present on an SRTA absorber surface. Under these assumptions, the heat balance on the absorber at $T_r$ is given by the relation

$$q_u = I_b \rho_m \alpha_r \delta f_t \cos i - \frac{1}{CR}[h_c(T_r - T_a) + \sigma\epsilon_r(T_r{}^4 - T_a{}^4)] \qquad (4.149)$$

where $h_c$ is the average convective heat coefficient from the receiver and $I_b$ is the direct normal insolation.

An idealized picture of the daily output available from an SRTA can be obtained by integrating the thermal output $q_u$ as shown below. The total energy $E$ per unit aperture produced from time $t$ before solar noon to time $t$ after solar noon is then given by

$$E = \int_{-t}^{t} q_u{}^+ dt$$

The overall daily efficiency $\bar{\eta}$, which is probably the best indicator of the performance, can then be calculated from the expression

$$\bar{\eta} = \frac{E}{\displaystyle\int_{t_{sr}}^{t_{ss}} I_b \, dt} \qquad (4.150)$$

where $t_{sr}$ and $t_{ss}$ are the times from solar noon to sunrise and sunset, respectively. Some of the results that can be obtained from this simplified approach are discussed below.

Assuming that there is no cover so that the transmittance $\tau_e(i_r)$ is unity, and that the emittance $\epsilon_r$ is 0.22, the absorptance $\alpha_r$ is then 0.9, the concentration ratio is 100 for a 90° rim angle, $f_t$ is 0.99, $\delta$ is 0.96, and the reflectance $\rho_m$ is 0.9. Therefore, the optical efficiency is 0.76. Figure 4.51 shows the normalized insolation based on a simple air mass attenuation model (see Chapter 2) and instantaneous system output at the vernal equinox for a receiver temperature of 500°C as a function of time before or after solar noon. An overall heat-loss coefficient of 8 W/m² · K (1.4 Btu/hr · ft² · °F) from the absorber and a latitude and tilt of 34° have been assumed. In this figure both the insolation and the collector output are normalized with respect to their values at solar noon. It should be noted that the daily collector output peaks at noon and decreases by more than 50 percent $3\frac{1}{2}$ hr before or after noon, similar to the effect shown for the north-south tracking PTC in Fig. 4.47.

The magnitude of SRTA losses are shown in Fig. 4.52 for a range of operating temperatures. Optical and view factor losses account for about a 50 percent performance penalty. Thermal losses are relatively smaller and account for a loss of from 2 to 15 percent of the total incident radiation over a day. The remaining 35–50 percent of the incident radiation can ultimately be converted to heat for operation of a power plant or other thermal conversion device.

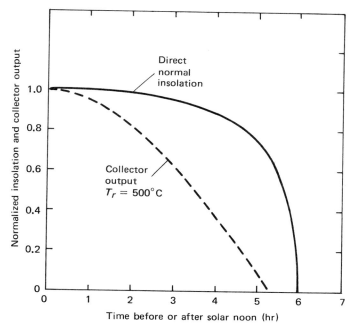

**FIGURE 4.51** Diurnal variations in direct normal insolation and collector output at vernal equinox. From (6).

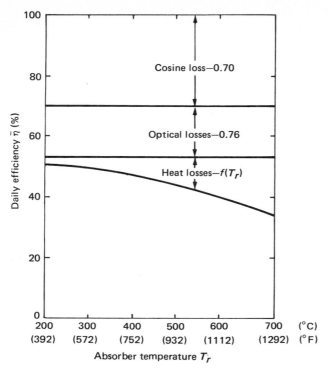

**FIGURE 4.52**  Comparison of cosine, optical, and thermal losses
for an SRTA.

A more detailed numerical model developed to predict the SRTA thermal
performance has been verified at low temperatures (31) by tests with a 3-m
diameter SRTA prototype built by G. W. Steward. This unit was tested by the
National Aeronautics and Space Administration (NASA) and showed good correlation between experimental data and model predictions at temperatures up to 100°C
(55). Complete model verification will require high-temperature testing.

### Central Receiver Solar Collector

A central receiver collector or power tower is shown schematically in Fig. 4.53 and
Fig. 4.54 shows an artist's conception of a full-scale power plant based on the Fresnel
mirror concept. Beam radiation is reflected from a number of independently steered
mirrors to a central receiver-collector. The arrangement overcomes some problems
associated with other concentration systems, namely, the need to insulate plumbing
carrying the working fluid over large distances, rotating joints, and heat losses from
large collectors. It can achieve concentration ratios as large as 3000 and thus heat a
working fluid to extremely high temperatures. Francia (18) built a pilot model of the
solar power tower plant in 1967 at the University of Genoa, Italy. The mirrors were
driven kinematically to reflect the incoming solar radiation from the field to a central
boiler where useful steam at temperatures up to 950 K could be generated.

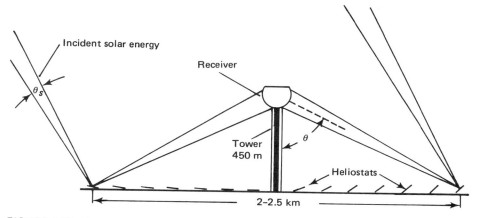

**FIGURE 4.53**  Cross-sectional view of solar power tower system. (The heliostats and receiver are not to scale.)

**FIGURE 4.54**  Artist's conception of Fresnel mirror, central receiver solar collector. Courtesy of J. V. Otts and Sandia Laboratories, Albuquerque, New Mexico.

A preliminary engineering design of a power tower system has been presented by Vant-Hull and Hildebrandt (57). They divided the overall design into several independent steps and concluded that the following four items represent the major problems in the thermal design:

1. Heliostats—The large number of heliostats required to reflect and track solar radiation during the day represents the major cost item in the installation. Consequently, they must be able to be mass produced, able to withstand extreme weather conditions, easily installed, and give long life at low maintenance.
2. Receiver—The receiver must be able to handle very large heat-flux densities (about 1–2 MW/m$^2$) and transfer the radiant energy with minimum heat loss to a working fluid. At temperatures below 850 K the authors suggested using an unprotected absorbing surface, but at higher temperatures they believe that vacuum jacketing will be necessary.
3. Optics—In the first design the parabolic heliostat surfaces were approximated by an array of nominally flat mirrors of good quality glass. The flatness requirement was that the angular deviation of the mirror surface be less than 16' of arc. For tracking, each mirror must be aimed by an electronically controlled hydraulic system with an accuracy that must be on the order of a few minutes of arc. The actual optical performance can be checked experimentally by reflecting sunlight on a shaded target at a specified distance from the mirror, where it can be photographed. The photograph of the pattern can then be transformed to intensity contours by means of a photometer. The technique has been described by Blake (4) and Fig. 4.55 shows the optical image from a 4 ft$^2$ mirror with a concentration of 4.3:1 at a distance 108 ft from the target.
4. Mirror field—A detailed design by individual ray tracing is possible, but was deemed too complicated. Instead, simplifying assumptions, such as perfect reflection from aberration-free mirrors, were made, and the effects of mirror aberration and optical losses were considered under items 1 and 2. The procedure is shown below.

**Design of mirror field** The procedure for designing the mirror field consisted of choosing a specific season (midwinter in this example) and then laying out a linear north-south array of mirrors that would experience no shading but would still intercept the largest possible fraction $\phi_s$ of the incident solar radiation. The heliostats are permanently located, but steered to direct the solar image to the receiver on the central tower, as shown in Fig. 4.54. If $\gamma$ is the fixed angle from the center of a mirror to the receiver and $\omega$ is the varying angle from the mirror to the sun, then the law of reflection requires that the mirror be lifted relative to the horizontal to the angle $(\gamma + \omega)/2$. This will raise one edge of the mirror above its center and change the area of the ground shaded from the sun by that mirror. Of course, any part of any other mirror whose projection on the ground falls into the shaded area will not produce "redirected" solar energy.

Once the mirror arrangement for the selected part of the year is laid out, the

**FIGURE 4.55** Photograph of images of flat and warped heliostat mirrors 4 ft$^2$ made from silvered glass; concentration shown is 4.3:1. Courtesy of Martin Marietta Corp.

fraction redirected during any other season can be obtained by geometric construction. More closely spaced mirrors would redirect more energy for a given size field during the summer when the solar elevation is high, but less in midwinter because of increased shading.

In a similar fashion, an east-west array that should give no shading at 9 a.m. and 3 p.m. can be established. This requires a symmetrical mirror arrangement with respect to east and west of the tower, but increases the separation distance between heliostats on the side nearer the sun at 9 a.m. With this arrangement a somewhat larger fraction of the solar irradiation is bypassed, that is, does not hit the target, but further from noon nearly all of the solar radiation incident on this linear array will be redirected toward the receiver. Once this east-west array is laid out, the net radiation redirected toward the receiver can be determined throughout the day. If $\phi_n$ represents the ratio of mirror to ground area in the east-west array and $\phi_s$ represents the fraction of the incident solar radiation intercepted by the mirrors in the north-south array, the total area of mirror $A_m$ for a given ground area $A_g$ is $\phi A_g$, where $\phi$ is $\phi_s\phi_n$. This analysis gives $\phi_s = 0.617$ and $\phi_n = 0.725$ for midwinter, so that $\phi = 0.447$. To determine the value of the rim angle of the receiver field that would avoid shading the following empirical relation was derived (57). It is valid for rim angles $\theta$ between 0 and 70°.

$$\phi = 1.06 - 0.23 \tan \theta \qquad (4.151)$$

Thus, once a value for $\phi$ is found, the relation between the tower height and maximum field diameter is given by Eq. (4.151) (see Fig. 4.53).

The final parameter to be determined is the mirror utilization factor $\tilde{\rho}$, defined as the ratio of the energy redirected toward the receiver by a mirror to the solar energy incident on an equal area of level ground. If the mirrors are spaced too closely, $\tilde{\rho}$ can be less than unity because of mutual shading. For an isolated mirror $\tilde{\rho}$ can be more or less than unity. Its value, just as for a flat-plate collector, will depend on the solar-altitude angle $\alpha$, mirror azimuth, and tilt. The average value of $\tilde{\rho}$ for the entire array must be determined by computer. In a design of the full system the competing effects of shading and orientation must be combined with economic factors to determine the optimum value of $\phi$ for a given geographic location. For the system optimized by Vant-Hull and Hildebrandt (57) under winter conditions in the United States, $\tilde{\rho}$ was found to vary from 0.78 in midsummer to 2.0 on a winter afternoon with the approximate relation

$$\tilde{\rho} = 0.78 + 1.8 \times 10^{-4} \, (90 - \omega)^2 \tag{4.152}$$

The power redirected to the receiver per square meter of mirror versus time of day for the solstices and equinoxes can be calculated by combining the foregoing analysis with solar radiation levels and altitude angles. The resulting curves are shown in Fig. 4.56 along with a sketch of the collector field configurations and

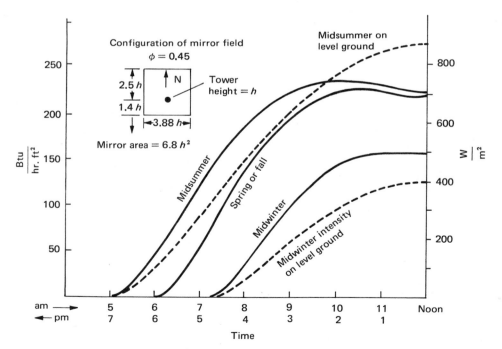

**FIGURE 4.56** Solar energy redirected by mirrors, 35°N latitude. From (57).

**TABLE 4.9** Solar Power Redirected by a Field of Ideal[a] Mirrors[b]

|  | Midwinter | Spring-fall | Midsummer |
|---|---|---|---|
| 95 percent of maximum power | 470 W[c] | 680 W[c] | 710 W[c] |
| Hours 95 percent exceeded | 3.6 | 4.6 | 5.8 |
| Hours 300 W[c] exceeded | 6.1 | 9.0 | 10.6 |
| Total power available | 3200 W[d] | 5900 W[d] | 7200 W[d] |
| Total at > 300 W[c] | 2700 W[d] | 5600 W[d] | 6800 W[d] |
| Mean at < 300 W[c] |  | 5200 W[d] |  |

[a]About 0.75 of this power will be absorbed by the receiver, assuming reasonable absorption and reflection losses.

[b]From (57). Values of $\rho E_o$ and $\int \rho E_o \, dt$ are averaged over the field.

[c]Thermal watts per square meter of mirror.

[d]Thermal watt hours per square meter of mirror per day.

curves of the intensity of solar radiation on level ground in midwinter and midsummer. It is obvious that winter operation can be improved only at the expense of peak summer performance. Similarly, the curves are relatively flat because of optimizing performance 3 hr from noon. Both of these effects reduce the peak energy handling requirements on the receiver. Some interesting characteristics of these curves are summarized in Table 4.9.

**Thermal analysis of energy collector and receiver** The total power absorbed by the receiver can be expressed as

$$q = I_b A_g \phi \tilde{\rho} \eta_o \alpha_r \qquad (4.153)$$

where $I_b$ is the direct solar radiation incident on $A_g$. (Appropriate values for the actual solar angle can be obtained by the method described in Chapter 2, for example, 300 W/m² corresponds to the sun's being about 26° above the horizon on a clear winter day at 2 p.m. at 35°N latitude.)

$A_g$ is the ground area of the concentrator, which is related to the tower height $h$ and the rim half-angle $\theta$ for a square concentrator by

$$A_g = 4h^2 \tan^2 \theta$$

$\phi$ is the fraction of the concentrator area covered by mirrors (about 0.45 for this example).

$\tilde{\rho}$ is the mirror utilization factor, that is, the effective tilt factor, described above. To reduce variations in the available energy, optimize $\tilde{\rho}$ at about 3 p.m. on a midwinter day by judicious placing of mirrors. Then $\tilde{\rho}$ varies with the hour and season from 0.78 at midsummer noon to 1.74 at the time of optimization.

$\eta_o$ represents the fraction of light striking the mirrors that actually reaches the receiver, and includes the mirror reflection ($\sim 0.85$), the transmittance of a possible glass receiver jacket ($\sim 0.95$), and the transmission of the support structure (estimated at $\sim 0.94$). Thus $\eta_o \sim 0.75$.

$\alpha_r$ is the absorptance of the receiver surface for the incident radiation. If the receiver is coated with iron oxide or titanium carbide, use $\alpha_r = 0.90$ for solar radiation incident on the receiver at angles less than about 75°.

Combining these factors, we have finally

$$q = 2.67 I_b \tilde{\rho}\phi(h \tan \theta)^2 \tag{4.154}$$

for the total power absorbed by the receiver. The power absorbed normalized by the square of the tower height $q/h^2$ is tabulated in Table 4.10 for several conditions of operation. A fraction $\beta$ of this power is re-radiated. The power reradiated is given by

$$\beta q = \epsilon_r \sigma(T_r^4 - T_\infty^4)A_r \tag{4.155}$$

where $T_r =$ the receiver temperature in degrees Kelvin ($T_\infty \sim 300$ K is ambient temperature)

$A_r =$ the radiating surface area of the receiver

$\epsilon_r =$ the emittance of the surface at the radiating temperature

In order to evaluate $\beta$, we need to estimate the minimum receiver area that will efficiently intercept the energy incident on the concentrator. Assuming the mirrors are ideal and form a minimum image of the sun,

$$\frac{\text{Diameter of optimum image}}{\text{Distance to image}} = \frac{\text{diameter of sun}}{\text{distance to sun}} = \theta_s = 0.0093$$

or 32' of arc. This implies that those mirrors at the maximum radial distance ($2h$ for a rim angle of 63°) from the tower are dished by $\frac{1}{2}$ mm (for a mirror diameter

**TABLE 4.10** Concentrator and Receiver Performance[a]

| Season | Midwinter | | Equinox | | Midsummer | |
|---|---|---|---|---|---|---|
| Rim half-angle $\theta$ (degrees) | 45[b] | 63 | 45 | 63 | 45 | 63 |
| $\rho\phi E_O$ (W/m²) | 208 | 220 | 294 | 317 | 328 | 329 |
| $q/h^2$ (W/m²) | 555 | 2260 | 785 | 3260 | 876 | 3380 |
| $\beta$(1000 K) | 0.033 | 0.024 | 0.023 | 0.017 | 0.021 | 0.016 |
| $\beta$(1500 K) | 0.168 | 0.124 | 0.120 | 0.085 | 0.106 | 0.083 |
| $q(h = 450$ m) Megawatt-T | 112 | 458 | 159 | 660 | 177 | 684 |

[a]From (57).
[b]Concentration $\sim 2300$ for $\theta = 45°$, $\sim 3000$ for $\theta = 63°$.

$D_m$ of 3 m) to optimally focus the sun on the collector. Deviations from the ideal shape of about one-half this amount over the mirror would increase the size of the image by about one-half the diameter of the mirror. Hence the effective diameter of the image $D_i$ is

$$D_i = \frac{h}{\cos \theta} \theta_s + \frac{D_m}{2} \qquad (4.156)$$

For configurations in the range considered in this design a good approximation is to replace $D_i$ by a nominal value. For example, choose $D_m = 3$ m, $h \sim 450$ m, and $\cos \theta \sim 0.5$. Then

$$D_i = \frac{h}{\cos \theta} \theta_s \left( 1 + \frac{D_m \cos \theta}{2 \theta_s h} \right) = 1.2 \frac{h}{\cos \theta} \theta_s = 8.4 \text{ m} \qquad (4.157)$$

(For a flat mirror 6 m in diameter the numerical factor would be 1.8.) If a reasonable allowance for mirror aberrations and steering errors is made as previously discussed, essentially all the reflected energy will be contained in a circle of diameter about $1.2\, D_i$. In order to intercept essentially all of the image on the receiver, while minimizing reflection and reradiation losses, one would choose a receiver having a spherical bottom with a conical section appended as shown in Fig. 4.57. The apparent "height" of the receiver would thus be equal to $D_r$ when viewed from a mirror on the rim of the concentrator, that is, at a radius $h \tan \theta$.

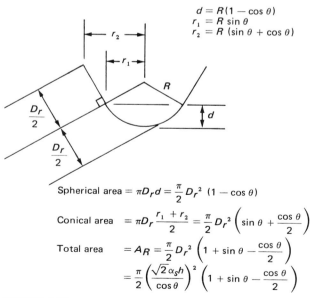

$$d = R(1 - \cos \theta)$$
$$r_1 = R \sin \theta$$
$$r_2 = R (\sin \theta + \cos \theta)$$

$$\text{Spherical area} = \pi D_r d = \frac{\pi}{2} D_r^2 (1 - \cos \theta)$$

$$\text{Conical area} = \pi D_r \frac{r_1 + r_2}{2} = \frac{\pi}{2} D_r^2 \left( \sin \theta + \frac{\cos \theta}{2} \right)$$

$$\text{Total area} = A_R = \frac{\pi}{2} D_r^2 \left( 1 + \sin \theta - \frac{\cos \theta}{2} \right)$$

$$= \frac{\pi}{2} \left( \frac{\sqrt{2} \alpha_s h}{\cos \theta} \right)^2 \left( 1 + \sin \theta - \frac{\cos \theta}{2} \right)$$

**FIGURE 4.57** Schematic diagram of receiver as seen from mirror at the rim.

Table 4.10 summarizes the performance for each season of the year, with the concentrator optimized for midwinter operation. Data are given for rim angles of 45° (diameter equal to tower height) and 63° (diameter approximately twice the tower height). The entry $\rho\phi E_o$ in the table gives the mean energy redirected to the receiver per square meter for a field of ideal mirrors during midday. For a given tower height, $q$ can then be obtained from the entry $q/h^2$. Finally, under the assumptions quoted above, $\beta$ is evaluated for each condition for two operating temperatures. For a given configuration, $\beta$ is nominally independent of the tower height as are the combinations $\rho\phi E_o$ and $q/h^2$. If one had to operate with the entire receiver surface at the temperature $T_r$, the thermal loss fraction $\beta$ would double.

## Summary

Table 4.11 summarizes the common types of compound-curvature solar collectors. The SRTA is limited in achievable concentration since it cannot form a true point focus. The paraboloidal dish and power tower are two different realizations of the same concept and are capable of achieving the highest temperatures. The great solar furnace at Odeillo, France, is of this type and has achieved temperatures above 3000°C. The decision as to which concentrator type to use for a specific task must be made by using systems analysis with economic constraints as described in Chapters 5 and 8.

## Solar Concentrator Summary

Table 4.12 contains a summary of all important optical properties of the common reflecting solar concentrators:

Concentration ratio $CR$           Capture parameter $\delta$

Average number of reflections $\bar{n}$       Beam spread variance $\sigma_y{}^2$

The $CR$ standard of comparison is the thermodynamic limit derived in the first section describing concentrators. It is seen that all concentrators except the CPC

**TABLE 4.11** Comparison of Compound-Curvature Solar Concentrators[a]

| Concentration type | Approximate range of $CR$ | Approximate maximum operating temperature (°C) |
|---|---|---|
| SRTA | 50–150 | 300–500 |
| Fresnel lens | 100–1000 | 300–1000 |
| Paraboloidal dish | 500–3000 | 500–2000 |
| Power tower | 1000–3000 | 500–2000 |

[a]From (44).

**TABLE 4.12** Comparison of Design Parameters of Solar Concentrators[a]

| Collector parameter | Single curvature | | | Compound curvature | | |
| --- | --- | --- | --- | --- | --- | --- |
| | Nonimaging CPC family: | Parabolic mirror | | Paraboloidal mirror | | Spherical mirror cylindrical receiver |
| | | Cylindrical receiver | Flat receiver | Spherical receiver | Flat receiver | |
| $CR/CR_{max}$[b] for perfect optics | 1.0 | $\dfrac{\sin\phi^c}{\pi}$ | $\dfrac{\sin\phi\cos(\phi+\theta_{max})^c}{-\sin\theta_{max}}$ | $\dfrac{\sin^2\phi^c}{4}$ | $\dfrac{\sin^2\phi\cos^2(\phi+\theta_{max})^c}{-\sin^2\theta_{max}}$ | $2\sin^2\phi\sin\theta_{max}$ |
| $\bar{n}$ | 0.7–1.2 | 1.0 | | 1.0 | | 1.0 |
| $\delta$ | $\dfrac{1}{1+\sigma_\theta\cot\theta_{max}}$[d] | $\dfrac{1}{\sqrt{2\pi}\,\sigma_y}\displaystyle\int_{-L_c/2}^{L_c/2}\exp\left[-\dfrac{1}{2}\left(\dfrac{y}{\sigma_y}\right)^2\right]dy$ | | $1-e^{-(L_c/\sigma_y^2)}$ | | (No closed-form expression) |
| $\sigma_y^2$ | — | $\dfrac{A_a^2\sigma_\theta^2(2+\cos\phi)}{12\phi\sin\phi}$ | $e$ | $\dfrac{2A_a\sigma_\theta^2(2+\cos\phi)}{3\phi\sin\phi}$ | $\dfrac{2A_a\sigma_\theta^2}{\sin^2\phi}$ | (No closed-form expression) |
| $L_c$ | $\dfrac{D_r}{2}$ | $\dfrac{D_r}{2}$ | $\dfrac{D_r}{2}$ | $\dfrac{\pi D_r^2}{4}$ | $\dfrac{\pi D_r^2}{4}$ | $\dfrac{D_r}{2}$ |
| $A_a$ | Aperture width | Aperture width | | Aperture area | | |

Note: See page 304 for footnotes.

**TABLE 4.12** Comparison of Design Parameters of Solar Concentrators[a] (*Continued*)

[a]Adapted from (9).

[b]Symbols in table:

$CR_{max}$ = thermodynamic limit to concentration; see Eqs. (4.77) and (4.78)

$L_c$ = characteristic dimension of receiver

$D_r$ = receiver diameter or width

$\theta_{max}$ = acceptance angle

$\phi$ = mirror rim half-angle

$\sigma_\theta{}^2 = 4\sigma_{\psi_1}{}^2 + \sigma_{\psi_2}{}^2$

$\delta$ = optical capture factor

$\bar{n}$ = average number of reflections

Other parameters defined in the text

[c]For Fresnel mirror arrangements or power towers, multiply expressions given by the ground cover fraction. It is usually on the order of $\frac{1}{2}$.

[d]Approximate expression for radiation incident near the limits of acceptance only; otherwise $\delta = 1.0$.

[e]

$$\left\{\frac{A_a{}^2\left(4\sigma_{\psi_1}{}^2 + \sigma_{\psi_2}{}^2\right)}{4\phi\,\tan^2\left(\phi/2\right)}\left[\frac{-1}{3\sin^3\phi\cos\phi} + \frac{2}{3\sin^3\phi} + \frac{2}{\sin\phi} - \frac{\cos\phi}{3\sin\phi} - \frac{2\phi}{\sin\phi} + \frac{4\sin\phi}{3\cos\phi} - \ln\tan\left(\frac{\pi}{4}+\frac{\phi}{2}\right) + \ln\tan\left(\frac{\pi}{4}-\frac{\phi}{2}\right)\right]\right\}^{1/2}$$

Duff (10) has given an alternative expression for $\sigma_y{}^2$ in which a different integration procedure was used.

family fall short of the ideal limit by a factor of 3 or more. For any meaningful concentration and rim half-angles $\phi > 45°$, the SRTA is seen to fall short of the ideal limit by the greatest amount because of its focal aberration.

The optical parameter $\delta$ and its independent variable $\sigma_y$ have been developed using analyses analogous to that used above for the PTC collector. Duff and Lameiro (9) have also derived expressions for $\sigma_y$ for reflector systems with flat mirror segments instead of curved ones. They are slightly different in form from those in Table 4.12 but are developed by using the same analysis method.

Thermal properties of concentrators cannot be summarized conveniently since they are complex functions of surface properties, geometry, and conduction-convection suppression techniques. Each collector concept must be analyzed using the approach described in detail for the SRTA above. Alternatively, the approximate method illustrated by the preceding PTC thermal analysis can be used for preliminary performance estimation. The type of receiver used is also a principal design variable. Planar and tubular receivers are described herein. Sparrow and Cess (53) have evaluated the efficacy of cavity devices, which may have advantages in some applications for solar receivers.

Table 4.13 summarizes the equations most often used for calculation of the incidence angles of beam radiation on the aperture of all common types of solar collectors. These equations are used to calculate daylong energy delivery of both concentrators and flat-plate collectors. A complete discussion of incidence angles is contained in Chapter 2.

**TABLE 4.13**  Fundamental Orientations and Motions of a Plane Unit of Surface Serving as a Collector of Solar Radiation[a]

| Orientation of collector | Incidence factor $\cos i$ |
|---|---|
| A. Fixed, horizontal, plane surface. | $\sin L \sin \delta_s + \cos \delta_s \cos h \cos L$ |
| B. Fixed plane surface tilted so that it is normal to the solar beam at noon on the equinoxes. | $\cos \delta_s \cos h$ |
| C. Rotation of a plane surface about a horizontal east-west axis with a single daily adjustment permitted so that its surface normal coincides with the solar beam at noon every day of the year. | $\sin^2 \delta_s + \cos^2 \delta_s \cos h$ |
| D. Rotation of a plane surface about a horizontal east-west axis with continuous adjustment to obtain maximum energy incidence. | $\sqrt{1 - \cos^2 \delta_s \sin^2 h}$ |
| E. Rotation of a plane surface about a horizontal north-south axis with continuous adjustment to obtain maximum energy incidence. | $[(\sin L \sin \delta_s + \cos L \cos \delta_s \cos h)^2 + \cos^2 \delta_s \sin^2 h]^{1/2}$ |
| F. Rotation of a plane surface about an axis parallel to the earth's axis with continuous adjustment to obtain maximum energy incidence. | $\cos \delta_s$ |
| G. Rotation about two perpendicular axes with continuous adjustment to allow the surface normal to coincide with the solar beam at all times. | $1$ |

[a]The incidence factor denotes the cosine of the angle between the surface normal and the solar beam.

## SUMMARY

This chapter summarizes the thermal and optical properties of many common solar energy thermal collectors, such as air- and liquid-cooled flat-plate collectors, high efficiency evacuated-tube collectors with and without mirror boosters, as well as a number of single- and compound-curvature concentrating devices. Each collector has been analyzed for instantaneous energy delivery under any operating conditions. Other concentrators not specifically considered in this chapter can be analyzed by the methods put forth herein with appropriate modifications for the specific design under consideration. However, the analysis of collector performance is not an end in itself. Only when a collector is coupled with all other components of a thermal conversion system can its long-term energy delivery be calculated. The thermal efficiency is a useful figure of merit, but it is not sufficient to enable an engineer to select a specific collector for a specific task. The integration of solar collectors into several practical thermal systems is the subject of Chapters 5 through 8.

## PROBLEMS

4.1.  Calculate the heat-removal factor and the duct flow factor for a collector having an overall heat-loss coefficient of 6 $W/m^2 \cdot K$ constructed of aluminum fins and tubes. Tube-to-tube centered distance is 15 cm; fin thickness is 0.05 cm; tube diameter is 1.2 cm; fluid tube heat-transfer coefficient is 1200 $W/m^2 \cdot K$; and the cover transmittance to solar radiation is 0.9 and is independent of direction. The solar absorptance of the absorber plate is 0.9; the collector is 1 m wide and 3 m long and the water flow rate is 0.02 kg/sec. The water temperature is 330 K.

4.2.  Calculate the efficiency of the collector described in Problem 4.1 on March 1 at a latitude of $40°N$ between 11 and 12 a.m. Assume that the total insolation is 450 $W/m^2$, the ambient temperature is 280 K, and the collector is facing south.

4.3.  Calculate the plate temperature in Example 4.2 at 11 a.m., if the insolation during the third hour is 270 $W/m^2$ and the air temperature is 285 K.

4.4.  Calculate the overall heat-transfer coefficient, neglecting edge losses, for a collector with a double glass cover, with the following specifications:

| | |
|---|---|
| Plate-to-cover spacing | 3 cm |
| Plate emittance | 0.9 |
| Ambient temperature | 275 K |
| Wind speed | 3 m/sec |
| Glass-to-glass spacing | 3 cm |
| Glass emittance | 0.88 |
| Back insulation thickness | 5 cm |
| Back insulation thermal conductivity | 0.04 $W/m \cdot K$ |
| Mean plate temperature | 340 K |
| Collector tilt | $45°$ |

4.5.  The graph below gives the results of an NBS standard performance test for a single-glazed flat-plate collector. If the transmittance for the glass is 0.90 and the absorptance of the surface of the collector plate is 0.92, determine

1. The collector heat-removal factor $F_R$.
2. The overall heat-loss conductance of the collector $U_c$ in Btu/hr $\cdot$ ft$^2$ $\cdot$ °F.
3. The rate at which the collector can deliver useful energy in Btu/hr $\cdot$ ft$^2$ when the insolation incident on the collector per unit area is 200 Btu/ft$^2$ $\cdot$ hr, the ambient temperature is 30°F, and inlet water is at 60°F.
4. The collector efficiency factor if water at a flow rate of 150 lb/hr is flowing through the collector, which has a total surface area of 10 ft$^2$.
5. The maximum flow rate through the collector that will give an outlet temperature of at least 140°F, if this collector is to be used to supply heat to a hot-water tank, with cold water entering at a temperature of 60°F.

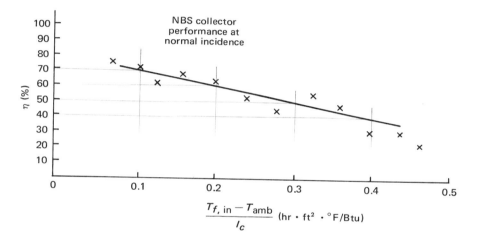

4.6.  Calculate the overall heat-loss coefficient for a solar collector with a single glass cover having the following specifications:

| | |
|---|---|
| Spacing between plate and glass cover | 5 cm |
| Plate emittance | 0.2 |
| Plate absorptance at equilibrium temperature | 0.2 |
| Ambient air temperature | 283 K |
| Wind speed | 3 m/sec |
| Back insulation thickness | 3 cm |
| Conductivity of back insulation material | 0.04 W/m $\cdot$ K |
| Mean plate temperature | 340 K |
| Collector tilt | 45° |

Note that the solution to this problem requires trial and error. Start by assuming a glass temperature and then determine whether the heat gained by the glass equals the heat loss from the glass at that temperature. If the heat gain is larger than the heat loss, repeat the calculations with a slightly higher temperature. After you have calculated the heat-loss coefficient, compare your answer with that obtained from Eq. (4.12).

4.7.   What is the second law efficiency of a flat-plate collector operating at $70°C$ if the environmental temperature is $10°C$ and the first law efficiency is 50 percent? Compare with a single-curvature concentrator operating at $200°C$ and with a double-curvature concentrator operating at $2500°C$, all with first law efficiencies of 50 percent.

4.8.   Show that the plate efficiency $F'$ for an air-cooled flat-plate collector is given by Eq. (4.47).

4.9.   The heat-removal factor $F_R$ permits solar collector delivery to be written as a function of collector fluid *inlet* temperature $T_f$, in Eq. (4.33). Derive the expression for a factor analogous to $F_R$ relating collector energy delivery to fluid *outlet* temperature.

4.10.   In nearly all practical situations the argument of the exponential term in Eq. (4.32) for $F_R$ is quite small. Use this fact along with a Taylor's series expansion to derive an alternate equation for $F_R$. Determine the range where the alternate equation and Eq. (4.32) agree to within 1 percent.

4.11.   The stagnation temperature $T_{c,\,max}$ of a solar collector corresponds to the temperature at the zero efficiency point, i.e., the no net energy delivery point. Using Eq. (4.64) calculate the stagnation temperature of a flat-plate collector with 75 percent optical efficiency and a $U_c$ value of 4.5 $W/m^2 \cdot K$ if the insolation is 900 $W/m^2$ and the ambient temperature is $20°C$.

4.12.   Calculate the stagnation temperature of the collector in Problem 4.11 if it were used with an optical booster of concentration ratio 2.0 having an optical efficiency of 70 percent.

4.13.   What is the operating temperature for an evacuated-tube collector operating at 50 percent efficiency if the insolation is 290 $Btu/hr \cdot ft^2$? Use data from Fig. 4.17.

4.14.   A method of reducing heat loss from a flat-plate collector is to pull a partial vacuum in the dead air spaces between cover plates. What vacuum level is required to completely eliminate convection in a single cover flat-plate collector operating at $85°C$ and tilted at $45°$ if the cover plates are at $40°C$ and are spaced 2 cm apart? [See Eq. (3.64).]

4.15.   Derive an expression for the heat-loss conductance $U_c$ for a flat-plate collector in which convection and conduction are completely eliminated in the air layers by use of a hard vacuum.

4.16.   The effect of air flow rate in an air-cooled flat-plate collector appears in the heat-removal factor $F_R$ [Eq. (4.48)]. Calculate the effect of doubling the flow rate on the heat-removal factor for the collector analyzed in Example 4.3. What percentage increase in energy delivery would be achieved by doubling the fan size?

4.17.   Calculate the optical efficiency of an evacuated-tube collector array (Fig. 4.16) at noon and 2 p.m. if the direct normal insolation is 600 $W/m^2$ and the diffuse insolation is 100 $W/m^2$. The effective optical transfer function $\tau_e \alpha_r$ is 70 percent and the tubes are spaced one diameter apart in front of a white painted surface with reflectance $\rho$ of 60 percent.

4.18.   Estimate the heat-loss conductance of an evacuated-tube collector at $200°C$ by assuming that the envelope temperature is the same as the ambient temperature—$30°C$. Use dimensions and physical properties in Table 4.3 and assume the glass envelope to be opaque to infrared radiation. Evaluate for a wind of 10 m/sec. Which thermal resistances are negligible?

4.19. What is the maximum concentration ratio for a trough concentrator with acceptance angle of $10°$; $25°$; $36°$?

4.20. What is the maximum achievable temperature of a double-curvature concentrator with a concentration ratio of 5000, a nonselective surface, and an 80 percent transmittance function?

4.21. Calculate the depth of a full CPC collector if the aperture is 1 m and $CR = 5$. Repeat for an aperture of 1 cm.

4.22. What is the reflectance loss in a 50 percent truncated CPC collector using silver mirrors if it has a $36°$ acceptance half-angle? What is the loss for an anodized aluminum reflector?

4.23. Compare the average number of reflections for a full and 50 percent truncated CPC if the acceptance half-angle is $9°$.

4.24. How much concentration effect is lost by truncating a $7°$ half-angle CPC by $\frac{1}{4}$, $\frac{1}{3}$, and $\frac{1}{2}$?

4.25. Explain how Figs. 4.28 and 4.29 could be used to prepare a map showing the effect of truncation on average number of reflectors of a CPC collector. Sketch qualitatively what such a map would look like.

4.26. How would Eqs. (4.104)–(4.111) be modified if no transparent cover were used over the CPC aperture? How many heat-flux terms are present and what are the equations for each?

4.27. What optical gap losses would occur in a CPC of the type shown in Fig. 4.35c if an evacuated-tube receiver (Table 4.3) were placed in contact with the reflector cusp?

4.28. A parabolic trough 1 m wide and 10 m long with no end support plates is 30 cm deep and has a focal length of 20.83 cm. Calculate its optical efficiency at a $30°$ incidence angle for perfect optics and tracking if its reflector reflectance is 80 percent, $\tau\alpha$ product is 75 percent, and if receiver supports shade 5 percent of the aperture.

4.29. Calculate the maximum concentration ratio for a parabolic trough collector with a 50 cm aperture, which captures 92 percent of the incident radiation if the rim half-angle is $50°$ and the surface is accurate to 2 mrad in slope. What is the receiver diameter?

4.30. Derive Eq. (4.142) relating aperture points to receiver points in an SRTA concentrator.

# REFERENCES

1. Bannerot, R. B., and J. R. Howell, "Modulating Concentrating Flat-Plate Solar Energy Collectors," *ASME Paper* 75-HT/54, 1975; see also Howell, J. R., and R. B. Bannerot, The Evaluation of Surface Geometrical Modification to Improve the Directional Selectivity of Solar Energy Collectors, *NSF Rept.* NSF/AER/ 73-003357 AO1/FR/76/1, 1976 (available from NTIS).

2. Baranov, V. K., and G. K. Melnikov, Study of the Illumination Characteristics of Hollow Focons, *Sov. J. Opt. Technol.*, vol. 33, pp. 408–411, 1966; Parabolocylindric Reflecting Unit and Its Properties, *Geliotekhnika*, vol. 11, pp. 45–52, 1975.

2a. Baum, V. A., *Proc. World Symp. Solar Energy, Phoenix*, pp. 289–298, 1955.

3. Beekley, D. C., and G. R. Mather, Analysis and Experimental Tests of a High Performance, Evacuated Tube Collector, Owens-Illinois, Toledo, Ohio, 1975.
4. Blake, F., Solar Thermal Power, *Proc. Energy Technol. Update Colorado Decision Makers, Colorado School of Mines*, Golden, 1975.
5. Bliss, R. W., The Derivations of Several "Plate Efficiency Factors" Useful in the Design of Flat-Plate Solar-Heat Collectors, *Sol. Energy*, vol. 3, p. 55, 1959.
6. Clausing, A. M., The Performance of a Stationary Reflector/Tracking Absorber Solar Concentrator, in "Sharing the Sun," vol. 2, pp. 304–326, ISES, Winnipeg, 1976; see also, by the same author, *ERDA Rept.* SAND 76-8039, 1976.
7. Colorado State University and Westinghouse Electric Corporation, Solar Thermal Electric Power Systems, Final Report, 3 vols., *Colorado State University Rept.* NSF/RANN/SE/GI-37815/FR/74/3, 1974.
8. deWinter, F., Solar Energy and the Flat Plate Collector, *ASHRAE Rept.* S-101, 1975.
9. Duff, W. S., and G. F. Lameiro, "A Performance Comparison Method for Solar Concentrators," ASME Paper 74-WA/Sol-4, 1974.
10. Duff, W. S., "Optical and Thermal Performance of Three Line Focus Collectors," ASME Paper 76-WA/HT-15, 1976.
11. Duffie, J. A., and W. A. Beckman, "Solar Energy Thermal Processes," John Wiley & Sons, New York, 1974.
12. Dushman, S., in "Scientific Foundations of Vacuum Technology," 2nd ed., ed. by J. M. Lafferty, John Wiley & Sons, New York, 1962.
13. Duwez, P., et al., The Operation and Use of a Lens-Type Solar Furnace, *Trans. Int. Conf. Use Solar Energy–Sci. Basis*, pp. 213–221, 1958.
14. Eckert, E. R. G., et al., Research Applied to Solar-Thermal Power Systems, *University of Minnesota Rept.* NSF/RANN/SE/GI-34871/PR/72/4, 1973.
15. Eckert, E. R. G., Research Applied to Solar-Thermal Power Systems, *University of Minnesota Rept.* NSF/RANN/SE/GF-34871/PR/73/2, 1973.
16. Edenburn, M. W., Performance Analysis of a Cylindrical Parabolic Focusing Collector and Comparison with Experimental Results, *Sol. Energy*, vol. 18, pp. 437–444, 1976.
17. Emmett, W. L. R., Apparatus for Utilizing Solar Heat, U.S. Patent 980,505, 1911.
18. Francia, G., Pilot Plants of Solar Steam Generating Stations, *Sol. Energy*, vol. 12, pp. 51–64, 1968.
19. Fraser, M. D., Survey of the Applications of Solar Thermal Energy to Industrial Process Heat, *Proc. Solar Industrial Process Heat Workshop*, University of Maryland, 1976; see also, InterTechnology Corporation, "Analysis of the Economic Potential of Solar Thermal Energy to Provide Industrial Process Heat," *InterTechnology Rept.* 00028-1, Rept. for U.S. ERDA Contract EY-76-C-02-2829, p. 53, 1977.
20. Goodman, N. B., A. Rabl, and R. Winston, Optical and Thermal Design Considerations for Ideal Light Collectors, in "Sharing the Sun," vol. 2, pp. 336–350, ISES, Winnipeg, 1976; see also Rabl, A., N. B. Goodman, and R. Winston, Optical and Thermal Design Considerations for Ideal Light Collectors, *Sol. Energy*, vol. 19, 1977.
21. Grimmer, D. P., and K. C. Herr, Solar Process Heat from Concentrating Flat-Plate Collectors, in "Sharing the Sun," vol. 2, pp. 351–373, ISES, Winnipeg, 1976.

22. Hill, J. E., and E. R. Streed, A Method of Testing for Rating Solar Collectors Based on Thermal Performance, *Sol. Energy*, vol. 18, pp. 421–431, 1976.

23. Hill, J. E., E. R. Streed, G. E. Kelly, J. C. Geist, and T. Kusuda, Development of Proposed Standards for Testing Solar Collectors and Thermal Storage Devices, *NBS Tech. Note* 899, 1976.

24. Hottel, H. C., and A. F. Sarofim, "Radiative Transfer," McGraw-Hill Book Co., New York, 1967.

25. Hottel, H. C., and B. B. Woertz, Performance of Flat-Plate Solar-Heat Collectors, *Trans. Am. Soc. Mech. Eng.*, vol. 64, p. 91, 1942.

26. Hottel, H. C., and A. Whillier, Evaluation of Flat-Plate Collector Performance, *Trans. Conf. Use Sol. Energy*, vol. 2, part 1, p. 74, 1958.

26a. Kauer, E., R. Kersten, and F. Madrdjuri, Photothermal Conversion, *Acta Electron.*, vol. 18, pp. 297–304, 1975.

27. Klein, S. A., Calculation of Flat-Plate Collector Loss Coefficients, *Sol. Energy*, vol. 17, pp. 79–80, 1975.

28. Klein, S. A., J. A. Duffie, and W. A. Beckman, Transient Considerations of Flat-Plate Solar Collectors, *J. Eng. Power*, vol. 96A, pp. 109–114, 1974.

29. Kreider, J. F., "Performance Study of the Compound Parabolic Concentrator Solar Collector," Environmental Consulting Services, Inc., Boulder, Colorado, 1974.

30. Kreider, J. F., "Thermal Performance of the Compound Parabolic Concentrator Solar Collector—Phase II," Environmental Consulting Services, Inc., Boulder, Colorado, 1975.

31. Kreider, J. F., Thermal Performance Analysis of the Stationary Reflector/ Tracking Absorber (SRTA) Solar Concentrator, *J. Heat Transfer*, vol. 97, pp. 451–456, 1975.

32. Kreith, F., "Principles of Heat Transfer," 3rd ed., Intext Educational Publishers, New York, 1976.

33. Kreith, F., Evaluation of Focusing Solar Energy Collectors, *ASTM Stand News*, vol. 3, pp. 30–36, 1975.

34. Kreider, J. F., and F. Kreith, "Solar Heating and Cooling," Hemisphere Publ. Corp., Washington, D.C., 1975.

35. Kumar, S., and H. S. Liers, "Performance of Flat-Plate Collectors with Planar Reflectors," ASME Paper 76-WA/HT-27, 1976.

36. Kutateladze, S. S., and V. M. Borishanskii, "A Concise Encyclopedia of Heat Transfer," Pergamon Press, New York, 1966.

37. Löf, G. O. G., Fester and Duffie, Energy Balances on a Parabolic Cylinder Solar Collector, *J. Eng. Power*, vol. 84A, pp. 24–32, 1962.

38. Malik, M. A. S., and F. H. Buelow, Hydrodynamic and Heat Transfer Characteristics of a Heated Air Duct, in "Heliotechnique and Development," (COMPLES 1975), vol. 2, pp. 3–30, Development Analysis Associates, Cambridge, Massachusetts, 1976.

39. Malik, M. A. S., and F. H. Buelow, Heat Transfer in a Solar Heated Air Duct—A Simplified Analysis, in "Heliotechnique and Development" (COMPLES 1975), vol. 2, pp. 31–37, Development Analysis Associates, Cambridge, Massachusetts, 1976.

40. McDaniels, D. K., et al., Enhanced Solar Energy Collectors Using Reflector-Solar Thermal Collector Combinations, *Sol. Energy*, vol. 17, pp. 277–283, 1975.

41. Rabl, A., Optical and Thermal Properties of Compound Parabolic Concentrators, *Sol. Energy*, vol. 18, p. 497, 1976; see also A. Rabl, "Radiation through Specular Passages," *Int. J. Heat Mass Trans.*, vol. 20, p. 323, 1977.

42. Rabl, Ideal Two Dimensional Concentrators for Cylindrical Absorbers, *Appl. Opt.*, vol. 15, p. 871, 1976.

43. Rabl, A., Comparison of Solar Concentrators, *Sol. Energy*, vol. 18, pp. 93–111, 1975; see also Rabl, A. and R. Winston, Ideal Concentrators for Finite Sources and Restricted Exit Angles, *Appl. Opt.*, vol. 15, p. 2886, 1976.

44. Rabl, A., The Technical Design of Solar Thermal Systems for Buildings, in "Seminar Notes," ed. by J. F. Kreider, University of Colorado, 1977.

45. Ramsey, J. W., J. T. Borzoni, and T. H. Hollands, Development of Flat Plate Solar Collectors for Heating and Cooling of Buildings, *NASA Rept.* CR-134804, 1975.

46. Ramsey, J. W., et al., "Experimental Evaluation of a Cylindrical Parabolic Solar Collector," ASME Paper 76-WA/HT-13, 1976; see also, same title, *J. Heat Transfer*, 1977.

47. Seitel, S. C., Collector Performance Enhancement with Flat Reflectors, *Sol. Energy*, vol. 17, pp. 291–295, 1975.

48. Simon, F. F., Flat Plate Solar Collector Performance Evaluation with a Solar Simulator as a Basis for Collector Selection and Performance Prediction, *NASA Rept.* TM X-71793, 1975; see also *Sol. Energy*, vol. 18, pp. 451–466, 1976.

49. Sparrow, E. M., et al., Research Applied to Solar Thermal Power Systems, *University of Minnesota Rept.* NSF/RANN/SF/GI-34871/PR/73/4, 1974.

50. Sparrow, E. M., et al., Research Applied to Solar-Thermal Power Systems, *University of Minnesota Rept.* NSF/RANN/SE/GI-34871/PR/74/2, 1974.

51. Sparrow, E. M., et al., Research Applied to Solar Thermal Power Systems, *University of Minnesota Rept.* NSF/RANN/SE/GI-37871/PR/74/4, 1975.

52. Sparrow, E. M., et al., Research Applied to Solar Thermal Power Systems, *University of Minnesota Rept.* NSF-RANN-75-219, 1975.

53. Sparrow, E. M., and R. D. Cess, *Radiation Heat Transfer*, Augm. Ed., Hemisphere Publ. Corp., Washington, D.C., 1977.

54. Speyer, F., Solar Energy Collection with Evacuated Tubes, *J. Eng. Power*, vol. 87, p. 270, 1965.

55. Steward, W. G., "A Concentrating Solar Energy System Employing a Stationary Spherical Mirror and a Movable Collector," *Proc. Solar Heating Cooling Buildings Workshop*, pp. 24–25, 1973; Steward, W. G., et al., "Experimental Evaluation of a Stationary Spherical Reflector Tracking Absorber Solar Energy Collector," ASME Paper 76-WA/HT-10, 1976.

56. Tabor, H., Radiation, Convection, and Conduction Coefficients in Solar Collectors, *Bull. Res. Coun. Isr.*, vol. 6C, p. 155, 1958.

56a. Tabor, H. Z., Stationary Mirror Systems for Solar Collectors, *Sol. Energy*, vol. 2, p. 27; 1958; see also, by the same author, Mirror Boosters for Solar Collectors, *Sol. Energy*, vol. 10, p. 111, 1966.

57. Vant-Hull, L. L., and A. F. Hildebrandt, A Solar Thermal Power System Based on Optical Transmission, *Proc. Solar Thermal Conversion Workshop, NSF Rept.* NSF/RANN/GI32488, 1974; see also, same title, *Sol. Energy*, vol. 18, pp. 31–39, 1976.

58. Weinstein, A., et al. Lessons Learned from Atlanta (Towns) Solar Experiments, in "Sharing the Sun," vol. 3, pp. 153–167, ISES, Winnipeg, 1976; see also, same title, *Sol. Energy*, vol. 19, pp. 421–427, 1977.

59. Westinghouse Electric Corporation, Solar Heating and Cooling Experiment for a School in Atlanta, *NTIS Rept.* PB240611, 1974.
60. Winston, R., Solar Concentrators of a Novel Design, *Sol. Energy*, vol. 16, pp. 89–95, 1974; includes references of early CPL work from 1966.
61. Winston, R., Radiant Energy Collection, U.S. Patent 3,923,381, 1975.
62. Winston, R., and H. Hinterberger, Principles of Cylindrical Concentrators for Solar Energy, *Sol. Energy*, vol. 17, pp. 255–258, 1975; see also, *J. Opt. Soc. Am.*, vol. 60, pp. 265–270, 1970.

# SYSTEM ANALYSIS, COMPONENTS, AND ECONOMICS OF SOLAR SYSTEMS  5

*The Buddha ... resides quite as comfortably in the circuits of a digital computer or the gears of a cycle transmission as he does at the top of a mountain or in the petals of a flower. To think otherwise is to demean the Buddha—which is to demean oneself.*

**Robert Pirsig**

This chapter describes the performance of the mechanical components, other than solar collectors, of low- and medium-temperature solar systems for heating, cooling, or agricultural applications. The features and components common to these and other systems are described, whereas details unique to specific applications are discussed in Chapters 6, 7, and 8. A summary of system mathematical modeling is included.

The methodology of the economic analysis and of the optimization of solar systems is presented. Principles of discounted cash flows, production functions, and marginal cost-benefit analysis are developed in detail. The importance of the study of the complete system, not its separate components, is emphasized throughout.

## INTRODUCTION TO SOLAR SYSTEM ANALYSIS

The interaction of solar system components in response to climate and energy demands determines the useful energy delivery of the system. The characteristics of a collector alone, for example, cannot be used directly to calculate energy delivery, since collector delivery depends on its operating temperature, which, in turn, is determined by the temperature of thermal storage. Storage temperature is, in turn, determined by the net effect of energy withdrawals to supply demands and energy inputs for the collection system. It is the behavior of the integrated system, not its components, that is important. Likewise, peak ratings of equipment such as solar collectors, heat exchangers, etc., do not provide sufficient information for design. It is the long-term, integrated behavior at both peak and nonideal or nonpeak conditions that must be known.

This chapter describes the system analysis of *low- and medium-temperature* solar-thermal systems. High-temperature and direct conversion systems are described in Chapter 8. The generic system considered in this chapter includes configurations for

Solar space heating
Solar service water heating
Solar space cooling
Solar dehumidification
Low-temperature process heat
Crop drying
Distillation
Swimming pool heating
Stock-tank heating, etc.

Important common features of these systems are described in this chapter. Differences among the basic systems when used for specific applications are described in Chapters 6 and 7.

## The Generic Solar-Thermal System

All solar systems listed above have five basic components that determine their performance. These five subsystems are

Solar collection
Thermal storage
Fluid circulation and energy distribution
Controls
Auxiliary, nonsolar energy source

The subsystems are interconnected as shown in Fig. 5.1. The basic system thus formed responds to local climate—sunshine, solar radiation level, wind speed, and ambient temperature—and to energy-demand characteristics as governed by a control subsystem.

### Solar Collection and Conversion Subsystem

This subsystem was described in detail in Chapter 4. Energy delivery of a collector was shown to depend on its physical configuration, its operating temperature, and the climatic parameters of solar radiation level, ambient temperature, and wind speed. Collector performance is also determined by secondary variables, including fluid flow rate, collector orientation, geographic location, and system control strategy. Table 5.1 provides a collector-task classification showing the best combination of collectors and tasks.

### Thermal Storage

Thermal storage is used in most solar systems to provide a thermal inertia or capacitance effect. Thermal capacitance is required to damp out fluctuations of collector energy delivery that occur in response to short-term fluctuations in sunshine level and diurnal variations related to the 24-hr day-night cycle. The amount of thermal storage is usually subject to economic constraints, since very

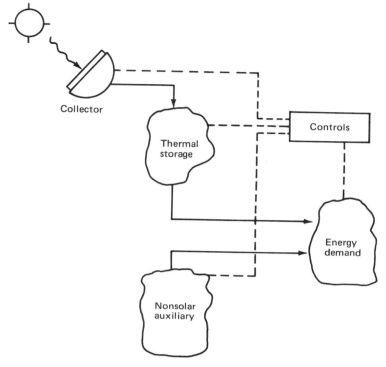

**FIGURE 5.1** The generic solar-thermal conversion system. Solid lines denote energy flows; the dashed lines, information flows.

**TABLE 5.1** Classification of Collector Type and Task[a]

| Collector type | Task | | | | | | |
|---|---|---|---|---|---|---|---|
| | Space heating, small bldg. | Space heating, large bldg. | Water heating | Space cooling | Process heat[c] | Crop drying | Distillation |
| Flat-plate (liquid) | S | S | S | $S^b/U^c$ | S | S | S |
| Flat-plate (air) | S | U | U | U | S | S | U |
| Evacuated-tube | S | S | – | S | S | – | U |
| Intermediate concentration | – | – | – | S | S | – | – |
| High concentration | – | – | – | – | S | – | – |

[a]U, unsatisfactory; S, satisfactory; –, uneconomic or task mismatch.
[b]Depends on local dry-bulb temperatures and solar radiation level.
[c]Temperature level dictates collector type.

large storage, which could produce a nearly uniform system output, is generally prohibitively expensive. In the network analysis of a solar system, storage acts as a summing junction for energy flows.

### Energy Flows

In a solar system energy flows usually occur when fluids convey internal energy because there are temperature or enthalpy elevations above a base state. Gases and liquids are widely used. An ideal working fluid would have a high heat capacitance and a low circulation-energy requirement, and would be chemically inert.

### Controls

In solar systems controls are designed to make decisions on energy flows based on preprogrammed criteria. For example, a cooling system controller will extract cooling effect from storage or solar-operated cooling machinery when the temperature level in a building (the load) rises. Likewise, the controller will cause fluid to flow through a solar collector when the collector is at a temperature sufficiently above storage temperature to make energy collection worthwhile. Control systems also call upon the nonsolar auxiliary energy source when solar capacity is below the demand level.

### Load

The load or demand component of a solar system is peculiar to each application. Therefore, this component is not described here but rather in Chapters 6, 7, and 8 by application type. The load generally contains some thermal inertia, which is important in determining its transient and time-dependent behavior. This thermal mass may be introduced by design or may be an integral part of the load characteristic. For example, a solar-heated building has a thermal mass by its very nature. The thermal mass of load must be considered in all careful solar system designs.

The major solar system components other than collectors (see Chapter 4) are discussed in this chapter. Components specific to a given application (for example, a solar cooling system) are discussed in subsequent chapters.

## THERMAL STORAGE

Although thermal storage represents extra cost and additional complexity in solar-thermal systems, it is almost always required to buffer the fluctuations of energy collection and demand in a solar system. This thermal capacitance effect is needed to match fluctuating energy collection to energy demands, which tend to be more uniform on a temporal basis. Solar stills and some agricultural crop-drying applications do not use separate storage, since the time variation of energy delivery to load is not important. The effects of storage thermal mass are shown in the following example.

## EXAMPLE 5.1

On a clear day the energy delivered by a solar system is approximately a sinusoidal function of time with a maximum near noon and a minimum at sunrise and sunset. If the energy demand on a solar-thermal system is a constant $L_0$, calculate the temperature history of storage for a day for the block system in Fig. 5.2.

## SOLUTION

Let the useful energy delivery be given by

$$Q_u = Q_{u0} \sin \frac{\pi t}{\tau} \tag{5.1}$$

where $Q_{u0}$ is the peak delivery rate, $\tau$ is the length of a day, and $t$ is measured from sunrise for convenience.

The energy balance on the storage tank, which acts as a summing junction for energy flows, is

$$(mc_p)_s \frac{dT_s}{dt} = Q_u - L_0 \tag{5.2}$$

where $(mc_p)_s$ is the heat capacitance of storage kJ/K and $T_s$ is the storage temperature. Equation (5.2) ignores thermal losses from storage, which can be made very small by proper insulation.

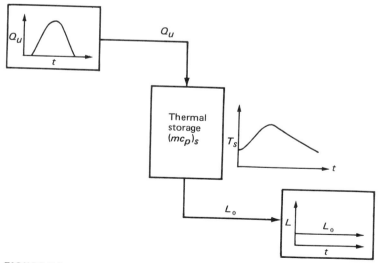

**FIGURE 5.2** Energy flow in Example 5.1 showing solar input to storage and load withdrawals from storage.

Equation (5.2) can be solved for $T_s(t)$ after substituting Eq. (5.1) for $Q_u$.

$$\frac{dT_s}{dt} = \frac{1}{(mc_p)_s}\left(Q_{uo}\, \sin \frac{\pi t}{\tau} - L_0\right) \tag{5.3}$$

Integrating gives

$$T_s = T_{so} - \frac{1}{(mc_p)_s}\left(\frac{\tau}{\pi}\, Q_{uo}\, \cos \frac{\pi t}{\tau}\bigg|_0^t + L_0 t\right) \tag{5.4}$$

$$T_s = T_{so} - \frac{1}{(mc_p)_s}\left(L_0 t - \frac{2\tau Q_{uo}}{\pi}\, \sin^2 \frac{\pi t}{2\tau}\right) \tag{5.5}$$

where $T_{so} \equiv T_s\,(t = 0)$ and $t \leqslant \tau$.

Equation (5.5) illustrates that the larger the storage capacitance $(mc_p)_s$, the smaller the fluctuations of storage temperature represented by the second term of the equation.

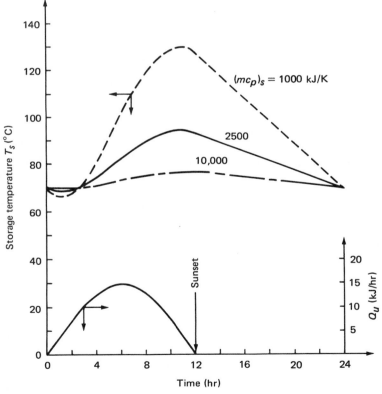

**FIGURE 5.3** Storage temperature history for Example 5.1 showing the effect of storage size. Useful energy delivery to storage shown in lower curve $Q_u$.

It is frequently important to control storage temperature swings. For example, if water is used it is necessary to maintain $T_s$ below the boiling point and yet above a minimum useful temperature $T_{s,\,min}$. Below $T_{s,\,min}$ fluid temperature levels are too low to permit efficient operation of the load device—tube-and-fin heat exchanger, absorption air conditioner, etc. Figure 5.3 shows Eq. (5.5) plotted for three sizes of storage and for other parameter values of $L_0 = 4775$ kJ/hr, $Q_{u0} = 15,000$ kJ/hr, $\tau = 12$ hr, and $T_{s0} = 70°C$. Note that the thermal mass of storage results in a phase shift between the energy input peak and the storage temperature peak. Also, the larger the storage mass, the smaller the storage temperature excursion as predicted by Eq. (5.5).

This chapter describes only thermal storage for low- and medium-temperature storage systems. In other types of solar systems, other storage types—electrical, chemical, kinetic—may be more suitable. The judgment as to which storage type is best suited to a given task requires consideration of six parameters:

Second law match of storage to task—entropy level match
Suitability of storage configuration to task
Longevity of stored energy—does storage "leak"?
Cyclic durability of storage
Storage energy density
Transportability of stored energy to task

## Storage Materials

Heat may be stored in a material by virtue of sensible heat or latent heat. The former requires a temperature rise; the latter, a phase change or chemical reaction. For example, energy may be stored in water by raising its temperature or in paraffin by melting it. The best sensible heat storage materials have high specific heat and density but low cost, toxicity, flammability, and chemical activity. Materials storing latent heat should have similar properties with additional requirements of high energy of phase change or chemical reaction, negligible corrosion, negligible expansion, and good repeatability of the freeze-thaw temperature. All storage materials must be resistant to chemical or physical change resulting from thermal cycling.

The required match between the solar energy source and the task, or load, is an important criterion in the selection of storage size and type. If large excursions in storage temperature are permitted, sensible-heat storage may be used. If delivery to load must be in a narrow temperature range, isothermal phase-change storage or a secondary heat exchanger with flow control may be required. The type, size, and cost of storage-associated heat exchanges, to be discussed shortly, are intimately related to storage thermal properties. For example, if high-temperature (low-entropy) storage can be used to supply a low-temperature demand, a relatively small heat exchanger can be used. If storage and demand subsystems must operate at approximately the same temperature, larger more expensive heat exchangers will be required.

### Sensible-Heat Storage—Liquids

The most common medium for storing sensible heat for use with low- and medium-temperature solar systems is water. Water is cheap and abundant and has a number of particularly desirable properties. Table 5.2 lists advantages and disadvantages of aqueous storage of thermal energy.

Water is the standard storage medium for solar-heating and -cooling systems for buildings today. For these systems, useful energy can be stored below the boiling point of water (without pressurization). The storage of water at temperatures above its normal boiling point requires expensive pressure vessels and is rarely cost-effective.

Other liquids can be used for thermal storage above 100°C without a pressure vessel. Such liquids are usually organic chemicals with lower density and specific heat than water and with higher flammability. It is possible, however, to store substantial amounts of energy per unit volume, because higher temperatures may be utilized. The thermal properties of a number of common organic liquids with particularly high specific heats are given in Table 5.3. Since these liquids are flammable, special fire-prevention methods must be used. Although high-molecular-weight oils have a slightly higher density than the liquids in Table 5.3, their specific heats are lower than those for the liquids shown. In addition, their viscosity is prohibitively high and pump and pipe sizes must be increased.

### Sensible-Heat Storage—Solids

Sensible-heat storage in solids is advantageous for some applications. For example, thermal energy from an air collector can be stored in a bed of solid material for later heating of a building, or steam can be condensed in a particle bed and the heat removed later by a counterflow of air. Storage beds formed from uniformly sized

**TABLE 5.2** Advantages and Disadvantages of Water as a Thermal Storage Medium

| Advantages | Disadvantages |
|---|---|
| Abundant | High vapor pressure |
| Low cost | Difficult to stratify[a] |
| Not toxic | Low surface tension— |
| Not combustible | leaks easily |
| Excellent transport properties | Corrosive medium |
| High specific heat | Freezing and consequent |
| High density | destructive expansion |
| Good combined storage medium | Nonisothermal energy |
| and working fluid | delivery |
| Well-known corrosion control | |
| methodology | |

[a]See Chapter 6.

**TABLE 5.3** Physical Properties of Liquid Storage Media[a]

| Material | Specific heat[b] | | Density[b] | | Boiling point[b] (°C) |
|---|---|---|---|---|---|
| | kJ/kg · K | Btu/lb · °F | kg/m³ | lb/ft³ | |
| Ethanol | 2.4 | 0.58 | 790 | 49 | 78 |
| Propanol | 2.5 | 0.59 | 800 | 50 | 97 |
| Butanol | 2.4 | 0.56 | 809 | 51 | 118 |
| Isobutanol | 3.0 | 0.72 | 808 | 50 | 100 |
| Isopentanol | 2.2 | 0.54 | 831 | 52 | 148 |
| Octane | 2.4 | 0.58 | 704 | 44 | 126 |
| Water | 4.2 | 1.00 | 1000 | 62.4 | 100 |

[a]Abstracted from (18).
[b]Properties at 20–25°C.

particles of solid storage medium can act as both a storage medium and heat exchanger, thereby saving the cost of a separate heat exchanger. In addition, solid-phase storage in bed form lends itself to thermal stratification, which is quite desirable as described in Chapter 6. This special stratification capability arises from the poor particle-to-particle thermal contact in a bed of rock or mineral storage particles of the type commonly used in solar air-heating systems, for example.

It is rarely economical to use a single large mass of a solid material for sensible-heat storage. A large mass has a very low surface-to-volume ratio, and the rapid transfer of heat to or from the storage medium is difficult for small temperature differences. If a volume $V$ of storage material is subdivided into $n$ subvolumes of size $v$, the surface area increases as the cube root of $n$.

Solid-phase storage has an advantage over liquid storage in the size of the maximum allowable storage temperature excursion. Since most solid storage media do not melt readily, they are suited for use with high-temperature, concentrating collectors. Care must be taken to avoid thermal fracturing, however, since this could induce an overall particle size reduction and result in bed constriction.

Table 5.4 contains density and specific heat values for several inexpensive, solid-phase, thermal storage materials. The density-specific heat product is an important parameter in evaluating storage energy density. The highest product in the table is that for cast iron, which nearly approaches the energy-density level of water storage. However, cast iron is more than 10 times as expensive as river rocks, the most commonly used solid storage material.

## Latent-Heat Storage

Latent-heat exchanges are involved in all changes of phase—solid-solid, solid-liquid, solid-gas, liquid-gas, etc. Phase changes involving gases should generally be avoided in solar systems, since the large volume of gas involved precludes compact storage. Of the many substances undergoing phase changes at useful temperatures, all corrosive, combustible, expensive, and toxic materials are usually eliminated from consideration.

**TABLE 5.4** Density and Specific Heat of Solid-Phase Storage Materials[a]

| Material | Density[b] | | Specific heat[b] | |
|---|---|---|---|---|
| | kg/m³ | lb/ft³ | kJ/K · kg | Btu/lb · °F |
| Aluminum | 2700 | 168 | 0.88 | 0.21 |
| Aluminum sulfate | 2710 | 169 | 0.75 | 0.18 |
| Aluminum oxide | 3900 | 240 | 0.84 | 0.20 |
| Brick[c] | 1698 | 106 | 0.84 | 0.20 |
| Calcium chloride | 2510 | 157 | 0.67 | 0.16 |
| Earth–dry[c] | 1698 | 106 | 0.84 | 0.20 |
| Magnesium oxide | 3570 | 223 | 0.96 | 0.23 |
| Potassium chloride | 1980 | 124 | 0.67 | 0.16 |
| Potassium sulfate | 2660 | 166 | 0.92 | 0.22 |
| Sodium carbonate | 2510 | 157 | 1.09 | 0.26 |
| Sodium chloride | 2170 | 135 | 0.92 | 0.22 |
| Sodium sulfate | 2700 | 168 | 0.92 | 0.22 |
| Cast iron[c] | 7754 | 484 | 0.46 | 0.11 |
| River rocks | 2245–2566 | 140–160 | 0.71–0.92 | 0.17–0.22 |

[a]From (18) (unless otherwise specified).
[b]At 50–100°C.
[c]From (25).

**TABLE 5.5** Melting Points and Heats of Fusion
for Solid-to-Liquid Phase Changes[a]

| Material | Melting point | | Heat of fusion | |
|---|---|---|---|---|
| | °C | °F | kJ/kg | Btu/lb |
| Aluminum bromide | 97 | 207 | 42 | 18 |
| Aluminum iodide | 191 | 376 | 81 | 35 |
| Ammonium bisulfate | 144 | 291 | 125 | 54 |
| Ammonium nitrate | 169 | 336 | 77 | 33 |
| Ammonium thiocyanate | 146 | 295 | 260 | 112 |
| Anthracine | 96 | 203 | 105 | 45 |
| Arsenic tribromide | 32 | 89 | 37 | 16 |
| Beeswax | 62 | 143 | 177 | 76 |
| Boron hydride | 99 | 211 | 267 | 115 |
| Metaphosphoric acid | 43 | 109 | 107 | 46 |
| Naphthalene | 80 | 176 | 149 | 64 |
| Naphthol | 95 | 203 | 163 | 70 |
| Paraffin | 74 | 166 | 230 | 99 |
| Phosphoric acid | 70 | 158 | 156 | 67 |
| Potassium | 63 | 146 | 63 | 27 |
| Potassium thiocyanate | 179 | 355 | 98 | 42 |
| Sodium | 98 | 208 | 114 | 49 |
| Sodium hydroxide[b] | 318 | 604 | 167 | 72 |
| Sulfur[b] | 110 | 246 | 56 | 24 |
| Tallow | 76 | 169 | 198 | 85 |
| Water (ice)[b] | 0 | 32 | 335 | 144 |

[a]From (15) (unless otherwise specified).
[b]From (18).

Solid-liquid phase changes are of two types—the first, melting-freezing; the second, a chemical reaction. Table 5.5 is a partial list of melting points and heats of fusion for organic and inorganic compounds that could be used for energy storage in solar systems. Phase-change materials have specific heats in the range 1-2 kJ/kg · K (0.25-0.5 Btu/lb · °F). They are, therefore, also useful for sensible-heat storage. When sensible- and latent-heat storage capabilities are combined over a temperature range useful for space heating, for example, these compounds can store from two to four times the energy per unit volume of water storage. However, the space required for working fluid circulation within the storage mass reduces the storage density advantage by a significant amount.

A second phase-change process involves a hydration-dehydration chemical reaction. Heats of hydration in the range 30-50 kJ/kg are obtained in substances useful in solar systems. At a temperature below the hydration point, the anhydrate becomes hydrated and crystalline with the evolution of heat. Upon heating, the crystal dissolves in its water of hydration (or crystallization), thereby absorbing heat. Table 5.6 shows the physical properties of several salt hydrates useful for storage of internal energy at temperature levels above or below ambient. However, these materials tend to be more corrosive than most simple melting substances described earlier. A typical chemical hydration reaction is

$$Na_2SO_4 + 10H_2O \rightarrow Na_2SO_4 \cdot 10H_2O + 241 \text{ kJ/kg} \qquad (5.6)$$

**TABLE 5.6**  Melting Points and Heats of Fusions for Hydration-Dehydration Materials for Low-Temperature Storage

| Material | Melting point | | Heat of fusion | | Specific heat (kJ/kg · K) | |
|---|---|---|---|---|---|---|
| | °C | °F | kJ/kg | Btu/lb | Solid | Liquid |
| $NH_4Cl \cdot Na_2SO_4 \cdot 10H_2O^a$ | 11 | 52 | 163 | 70 | 1.33 | 2.51 |
| $NaCl \cdot NH_4Cl \cdot 2Na_2SO_4 \cdot 20H_2O^a$ | 13 | 55 | 181 | 78 | 1.46 | 2.72 |
| $NaCl \cdot Na_2SO_4 \cdot 10H_2O^a$ | 18 | 65 | 186 | 80 | 1.76 | 3.26 |
| $CaCl_2 \cdot 6H_2O^c$ | 30 | 86 | 168 | 72 | — | — |
| $Na_2SO_4 \cdot 10H_2O^a$ | 32 | 90 | 241 | 108 | 1.76 | 3.30 |
| $Na_2CO_3 \cdot 10H_2O^b$ | 33 | 92 | 267 | 115 | — | — |
| $Na_2HPO_4 \cdot 12H_2O^c$ | 40 | 104 | 279 | 120 | — | — |
| $Ca(NO_3)_2 \cdot 4H_2O^c$ | 47 | 117 | 153 | 66 | — | — |
| $Na_2S_2O_3 \cdot 5H_2O^c$ | 45 | 113 | 95 | 41 | 1.46 | 2.38 |
| $NaC_2H_3O_2 \cdot 3H_2O^a$ | 58 | 136 | 265 | 114 | 1.97 | 3.22 |
| $Cd(NO_3)_2 \cdot 4H_2O^b$ | 59 | 139 | 107 | 46 | — | — |
| $Mg(NO_3)_2 \cdot 6H_2O^c$ | 90 | 194 | 167 | 72 | — | — |
| $N_2H_4HCl^b$ | 93 | 199 | 223 | 96 | — | — |
| $MgCl_2 \cdot 6H_2O^c$ | 115 | 239 | 165 | 71 | — | — |

[a]From (1).
[b]From (30).
[c]From (15).

Phase-change storage has promise because of higher energy density $(kJ/m^3)$ and resultant economic advantage, but despite many years of research, it has not yet been demonstrated to be reliable. The fundamental problem with salt hydrates is the deterioration of the freeze-thaw cycle repeatability. After many cycles, the rehydration reaction requires progressively more subcooling to reach completion, thereby cancelling the beneficial effects of isothermal phase change. Since some hydrated salts are heavier than their water of hydration, a phase separation resulting from gravity occurs. This separation exacerbates the subcooling rehydration problem. If the cycle repeatability problem can be solved, economic costs may negate the energy-density advantage of phase-change materials. Hydrated salts, for example, are more expensive than water or rock used for storage up to 130°C or so. Nucleating agents commonly used to avoid subcooling also add to the cost of phase-change materials.

Melting phase-change materials generally have poor heat-transfer properties (low thermal conductance). As a result, the solid-liquid interface moves slowly (0.9 mm/hr · °C) and energy flows into and out of storage are slow.* It is possible to disperse some phase-change storage substances in a liquid with which the storage material is immiscible. If the fluid has good heat-transfer properties, its capability for energy storage can be enhanced by the dispersal of phase-change material. In order to be widely used, phase-change storage must possess the following properties: (1) high latent heat, (2) an appropriate and well-defined melting point, (3) high thermal diffusivity and thermal conductivity, (4) low coefficient of thermal expansion, (5) low phase-change volume change, and (6) low cost.

## Storage Material Containers

### Liquid Storage

Storage tanks for aqueous storage can be fabricated from steel, fiberglass (support required), aluminum, or concrete. They must be leakproof, capable of withstanding long exposure to temperature cycles, and corrosion resistant (inert liners are required in metal tanks). Solid-phase storage media can be contained in tanks made from the above materials or from wood. These storage vessels must be sealed airtight with a high-temperature sealant.

Ideally, liquid storage tanks should be spherical to minimize heat loss and tank material requirements. However, spherical tanks are difficult to fabricate, consume a volume in a building about twice that of the sphere, and require special supports. A compromise shape, easier to fabricate, is a right circular cylinder. If its height is equal to its diameter, the surface area of such a tank is only about 15 percent greater than that of a sphere.

The thickness and quality of insulation on storage tanks depend on the economic value of stored energy. For a given heat loss, high-temperature storage requires greater insulation than low-temperature storage. The optimal amount of insulation is that which saves energy equal in value to the cost of the extra

*See Appendix 5 for a tabulation of interface velocities for some freeze-thaw materials.

insulation required to provide that saving. Marginal-cost analysis used to evaluate the best insulation configuration is described in detail later in this chapter. For liquid storage tanks for space heating and/or cooling, a typical $U$ value of about 0.25 $W/m^2 \cdot K$ (0.04 Btu/hr $\cdot$ ft$^2 \cdot$ °F) should be used. For particle-bed storage for space heating, $U$ values of about twice those for liquid storage are recommended. Instead of a tank's being insulated directly, it can be enclosed in an insulated box or small room. This approach has the advantage of permitting ventilation of the storage space during periods when storage temperature may be high and may cause unnecessary additional cooling load in a building.

### Particle-Bed Storage

The geometric shape of particle storage beds is a compromise of heat-transfer and pressure-drop requirements. For a given volume of storage, the shorter the flow path (that is, the flatter the bed), the lower the pressure drop. However, for flow paths that are very short, the fluid residence time in the bed is too short to permit effective heat transfer. In addition, fluid velocity must be sufficiently high to ensure a good convective heat-transfer coefficient between the fluid of the particles. A convenient way to determine the length of a rock bed for air as the working fluid is to require that the length be greater than that required to transfer more than 90 percent of the energy contained in the working fluid to storage (5). The analysis of the thermal performance of particle beds is contained in Chapters 3 and 6 and Ref. (27).

### Phase-Change Storage

Phase-change storage containers must be carefully designed to account for poor heat-transfer properties of the media, possible corrosion, and leakage. If a volume change is associated with the phase change (30 percent is typical), the container must be capable of accommodating the change. No general rules can be stated for shaping these storage containers other than that they must have a small cross section and large area normal to the direction of motion of the solid-liquid interface. Water-of-hydration heat storage containers must be well sealed to avoid evaporation. Since most water-of-hydration materials are electrolytes in solution, great care must be taken to avoid use of dissimilar metals in their containers. Electrochemical replacement or gas evolution could result from mixing metals with various oxidation potentials in a water-of-hydration energy-storage container. Use of hermetically sealed plastic containers having high-temperature stability can avoid corrosion problems.

### Storage Container Costs

The optimum size of storage containers is an economic question, which is analyzed in Chapters 6, 7, and 8. It depends on the required operating temperature and planned use of solar energy. The cost of metal tanks for energy storage depends on the shape, material of construction, and size. Chase (10) and Guthrie (16) have collected cost data on tank sizes and types. They found that cost is correlated by a power law in the tank volume and by a material multiplier $F_m$ as follows:

$$\text{Cost} = \text{base cost} \times F_m \times \left(\frac{\text{volume}}{\text{base volume}}\right)^n$$

The values of material multiplier, based on carbon steel cylindrical tanks as unity, are as follows:

| Material | Multiplier $F_m$ |
|---|---|
| Carbon steel | 1.00 |
| Aluminum | 1.40 |
| Rubber-lined | 1.48 |
| Stainless steel | 3.20 |
| Glass-lined | 4.25 |

The scale-up exponents in the cost power law are shown in Table 5.7. The table shows that carbon steel tanks are the least expensive, although the required rubber lining adds significant cost.

Expressions for calculating the heat loss from storage tanks are given in the next section.

## ENERGY TRANSPORT IN SOLAR SYSTEMS

The practical use of solar energy requires a means of energy distribution and transport from the solar collector to the storage subsystems and in turn to the load or demand points. In a solar-thermal system, energy transports are ordinarily carried out by the internal energy of a gaseous (usually air) or liquid transport fluid. The technology of energy transport in fluids is well developed and described in several basic texts (25, 33).

Energy transport is subject to two parasitic loss modes, which should be minimized subject to economic constraints. Any fluid flow is subject to viscous and turbulent resistance to flow that must be overcome by a motive force, which requires mechanical energy. The motive force may be supplied by shaft work for

**TABLE 5.7** Scale-up Exponents for Storage Tanks

| Tank type | Scale-up exponent |
|---|---|
| Atmospheric–steel | 0.46 |
| Steel | |
|     100–250 $kP_a$ (15–35 psi) | 0.47 |
|     250–850 $kP_a$ (35–120 psi) | 0.49 |
| Rubber-lined | 0.57 |
| Aluminum | 0.61 |
| Atmospheric–stainless steel | 0.50–0.54 |
| Glass-lined[a] | 0.43 |

[a]Limited to 38 m³ (10,000 gal); adapted from Chase (10).

pumps, blowers, compressors, or fans; by gravity resulting from buoyancy effects; or by a magnetic or electrical field resulting from the flow of an electric current. The most common motive force is pressure generated by a pump or blower.

Since fluid energy transport involves temperatures above the environmental level, heat loss from ducts or pipes within which transport fluids are contained occurs. This internal energy loss is the second parasitic energy sink in any solar-thermal system. Heat losses can be controlled, but not eliminated, by careful design.

Flowing fluids require additional precaution because of their tendency to corrode, erode, freeze, expand upon freezing, or change chemical composition. Specific load point apparatuses—baseboard convection heaters, air-conditioning coils, etc.—are described in Chapters 6 and 7. This chapter describes considerations related to energy flow common to all solar-thermal systems except high-temperature systems, which are analyzed in Chapter 8.

Figure 5.4 shows a typical solar-heating system for a building. Energy flows between one subsystem and another—storage-to-load, for example—occur by fluids flowing in closed loops. The system in Fig. 5.4 has five distinct loops, each requiring a motive force (pump) and pipes for energy distribution. Valves for air or liquid flows are used to divert flows within loops or to control flow rates in loops.

## Thermal Losses in Energy Distribution Systems

Since most solar-thermal applications require fluids at temperatures above the environmental level, pipes and ducts conveying those fluids are a source of energy loss. This loss can be made arbitrarily small by insulating. However, economic constraints place an upper limit on the amount of insulation actually used. The last increment of insulation added must save enough energy to pay for itself. This type of marginal-cost analysis is described later in the chapter.

Heat losses occur from all components of a solar system, including collectors, pipes, tanks, valves, pumps, heat exchangers, and blowers. The losses from collectors are treated in Chapter 4 since they are an inherent part of the collector. Losses from storage tanks are treated in previous sections of this chapter. Valves and pumps constitute a small fraction of the surface area of a distribution system and therefore have a correspondingly small heat loss. Pipes and ducts are the major source of parasitic loss in the distribution system and are analyzed below.

Fluid conduits in solar systems are usually circular or rectangular in cross section, as shown in Fig. 5.5. Rectangular or circular ducts are used for air; round pipes, for liquids. Insulation may be placed inside or outside the air ducts. As shown in Chapter 3, the heat loss for circular conduits is given by

$$Q_{\text{cyl}} = \left[ \frac{2\pi L\ k}{\ln(D + 2t)/D} \right](T_f - T_a) \tag{5.7}$$

where $L$ is the conduit length, $k$ is the conductivity of insulation, $T_f$ is the fluid temperature, and $T_a$ is the ambient temperature. The parameters $D$ and $t$ are

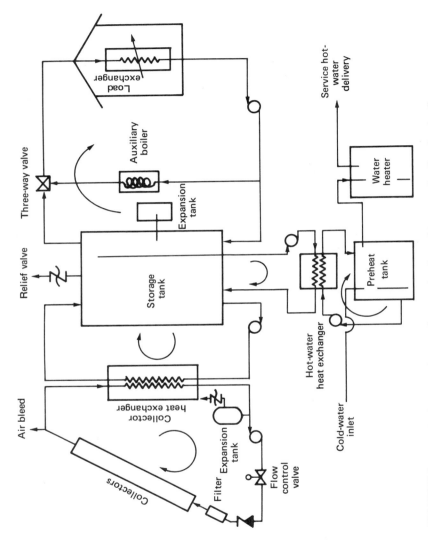

**FIGURE 5.4** Typical solar-thermal system for space heating and hot-water heating showing five fluid transport loops and pumps.

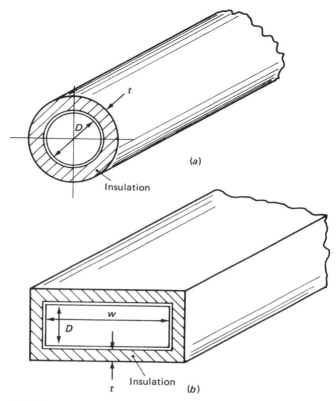

**FIGURE 5.5** Typical fluid conduits for fluids used in solar-thermal systems: (*a*) pipes or circular ducts for liquids or air; (*b*) rectangular ducts for air.

defined in Fig. 5.5. The heat flow is dominated by the insulation resistance; therefore, the small convection resistances are neglected in Eq. (5.7).

Heat loss from rectangular ducts is given (25) by

$$Q_{\text{rect}} = kL\left(\frac{2w + 2D}{t} + 2.16\right)(T_f - T_a) \tag{5.8}$$

In a solar system a range of pipe sizes is usually encountered for flow balancing purposes. In order to determine the optimum amount (volume) of insulation and its optimum distribution among all components of all sizes from which heat loss takes place, the following analysis is used.

The heat loss $Q_L$ from a system operating at a given flow rate and temperature level depends on the amount of insulation $V_i$ on each component. The total heat loss from the several components is given by

$$Q_L = Q_1(V_1) + Q_2(V_2) + \ldots \tag{5.9}$$

$Q_L$ should be minimized subject to the constraint of a fixed total volume $V_T$ of insulation:

$$V_T = V_1 + V_2 + V_3 + \ldots \tag{5.10}$$

Equation (5.10) may be unconstrained by use of the Lagrange multiplier $\lambda$:

$$Q_L = Q_1(V_1) + Q_2(V_2) + \ldots + \lambda(V_1 + V_2 + \ldots - V_T) \tag{5.11}$$

In order to find a stationary point of Eq. (5.11), the differential of Eq. (5.11) is taken, giving

$$dQ_L = \frac{dQ_1}{dV_1} dV_1 + \frac{dQ_2}{dV_2} dV_2 + \ldots + \lambda(dV_1 + dV_2 + \ldots) = 0 \tag{5.12}$$

Since each differential is unconstrained, each may be set equal to 0 in order to get

$$\frac{dQ_1}{dV_1} = \frac{dQ_2}{dV_2} = \ldots = -\lambda \tag{5.13}$$

The expression $Q(V_i)$ relating heat loss to insulation volume can be derived for three common shapes encountered in solar systems as shown in Table 5.8. It is easy to show (see Problem 5.5) that the derivatives in Eq. (5.13) for the three basic shapes are given (14) by

$$\frac{dQ_i}{dV_i} = -\frac{q_i^2}{k\Delta T} \tag{5.14}$$

where $q_i$ is the heat loss per unit of external area for component $i$. Substituting Eq. (5.14) into Eq. (5.13), simplifying, and extracting the square root for fixed insulation conductivity and fixed $\Delta T$, give

$$q_1 = q_2 = q_3 = \ldots \tag{5.15}$$

**TABLE 5.8**  Rate of Heat Loss and Insulation Volume for Basic Insulation Shapes

| Shape | Rate of heat loss | Volume |
|-------|-------------------|--------|
| Rectangular prism | $\dfrac{2kL(w+D)}{t}\Delta T$ | $DwtL$ |
| Cylinder | $\dfrac{2\pi kL}{\ln(D+2t)/D}\Delta T$ | $\dfrac{\pi}{4}[(D+2t)^2 - D^2]L$ |
| Sphere | $\dfrac{k\pi D(D+2t)}{t}\Delta T$ | $\dfrac{\pi}{6}[(D+2t)^3 - D^3]$ |

This equation states that the *optimal amount of insulation results in the same heat flux* for all distribution system components. Although it is not possible to calculate the $V_i$ directly, in general the analysis above can easily be programmed for iterative solution on a hand calculator.

Thermal conductivity values for common pipe and tank insulations are given in tabular form in Appendix 3.

## Mechanical Losses in Solar Systems

The flow of fluids within conduits requires a motive force, usually a pressure difference, to overcome viscous and turbulent resistance to flow. In liquid systems pumps are used; for air systems, blowers. Pressure drops to be overcome by a pump or blower include losses in piping, collectors, heat exchangers, valves, pipe fittings, size changes, filters, and hydrostatic head development.

The methods for calculating the total pressure drop through all components are described in Chapter 3. Nomographs for these calculations are provided in Figs. A5.3–A5.5.

The design of liquid and air pumping systems requires care to avoid vapor lock, stalling, or cycling. These matters are treated in Refs. (2, 3, 33).

### Flow Distribution in Networks

In nearly all solar-thermal conversion systems a number of collectors must be connected to provide sufficient area to capture energy adequate for the solar energy end use. Series connection of collectors results in high pressure drops at low flow and inefficient pump or blower operation. For that reason a parallel or series-parallel configuration is generally used in large arrays of solar collectors.

In a parallel configuration like that shown in Fig. 5.6, it is desirable to have equal flow through each fluid branch. If the flows in all branches are equal, the branches are spaced equally at distance $L$, and the manifold diameter $D$ is uniform,

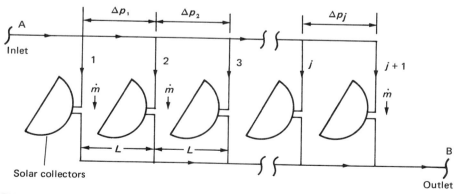

**FIGURE 5.6** Typical manifold connections for solar collector array. Branches are numbered and assumed to be spaced equally.

a value for the pressure drop in the manifold can be calculated in closed form. If the pressure drop through a collector is added to that in the manifold, the pressure drop $\Delta p_{AB}$ can be computed. The pressure drop through a collector is dependent on its design and is a unique function of fluid type and flow rate. Because of the complexity of flow passages in collectors, it is more efficacious to measure the collector pressure drop experimentally than to calculate it. The calculation of the manifold pressure drop follows.

The pressure drop in the manifold $\Delta p_{1j}$ is the sum of each incremental pressure drop $\Delta p_i$

$$\Delta p_{1j} = \sum_{L=1}^{j} \Delta p_i \tag{5.16}$$

If the velocity in the pressure-drop equation [see Eq. (3.15)] is expressed in terms of the mass flow rate $\dot{m}$ (kg/sec, lb/sec),

$$V = \frac{\dot{m}}{\rho_f A} = \frac{\dot{m}}{\rho_f \pi D^2/4} \tag{5.17}$$

where $A$ is the flow area and $D$ is its diameter. Substituting in Eq. (3.15) gives,

$$\Delta p_i = \bar{f} \frac{L\dot{m}^2}{\rho_f D^5} \frac{8}{\pi^2} \tag{5.18}$$

where $\bar{f}$ is the average friction factor in the manifold evaluated at one-half the total flow rate in the manifold $(j+1)\dot{m}/2$. Since $\bar{f}$, $L$, $\rho_f$, and $D$ are constants, the pressure drop is given by

$$\Delta p_i = K\dot{m}^2 \tag{5.19}$$

The sum in Eq. (5.16) can now be formed:

$$\Delta p_{1j} = \Delta p_1 + \Delta p_2 + \Delta p_3 + \ldots + \Delta p_j \tag{5.20}$$

$$= K(j\dot{m})^2 + \ldots + K(3\dot{m})^2 + K(2\dot{m})^2 + K\dot{m}^2 \tag{5.21}$$

Simplifying gives

$$\Delta p_{1j} = K\dot{m}^2 (1 + 2^2 + 3^2 + \ldots + j^2) \tag{5.22}$$

The series in parentheses can be summed in closed form (37) from which

$$\Delta p_{1j} = K\dot{m}^2 \left[ \frac{j(2j+1)(j+1)}{6} \right] \tag{5.23}$$

The total pressure drop from inlet $A$ to outlet $B$ of the solar collector array is given by

$$\Delta p_{AB} = \frac{K\dot{m}^2}{6} \left[ j(2j+1)(j+1) \right] + \Delta p_{coll} \tag{5.24}$$

where $\Delta p_{coll}$ is the average collector pressure drop at flow $W$. This pressure drop is then added to that for pipes and fittings in the collector loop to compute the total dynamic pressure drop in the collector loop.[*]

The preceding analysis of collector array pressure drop assumes balanced or equal flow through individual collectors. This is not realized automatically by simply requiring equal flow path lengths for fluid passing through all collectors. Since the pressure distribution in the manifold changes because of momentum withdrawals or additions at each header connection to collector inlet or outlet, special measures to ensure equal collector flows are required. Löf et al. (29) have shown that, if the pressure drop in one manifold $\Delta p_{1j}$ is 10 percent of the entire array pressure drop $\Delta p_{AB}$, the individual collector pressure drops $\Delta p_{coll}$ will vary by 8 percent at most for turbulent flow; if the manifold drop is 4 percent of the total, $\Delta p_{coll}$ will vary by 3 percent at most throughout the array.

Equation (5.24) also applies to individual collectors as well as collector arrays if the absorber-plate conduits are connected in parallel to a header within the collector. Löf et al. (29) recommended that the head-to-collector pressure-drop ratio be 0.04 or less in order to have good flow balance in a collector. In collectors it is ill-advised to use large headers to achieve this end, since the liquid volume contained in the collector would cause the collector to have a large time constant—an undesirable feature of any collector. In a flat-plate development project, Honeywell (19) calculated collector-plate flow distributions by a numerical method for various header and riser sizes.

Pressure drops or differences between points in a fluid circuit are important design parameters. However, the magnitude of the pressure itself is equally important in circuits using liquid heat-transfer fluids. In order to avoid vapor lock, boiling, or pump cavitation, the pressure at all points in a loop must be maintained above the local boiling point corresponding to the local fluid temperature. For example, the lowest pressure point in closed, residential collector loops is usually at the top of the collector. This is also the point of highest fluid temperature. Therefore, this location controls the operating pressure of the entire loop. Pressure should not be allowed to approach the local boiling point by less than 15 kPa ($\sim$ 2 psi). Another low pressure point in many liquid solar transport systems is located at the fluid pump inlet if the storage tank is located below the storage pump inlet.

## Corrosion Control and Freeze Protection in Aqueous Working Fluids

If dissimilar metals are used in distribution systems with aqueous working fluids, corrosion may occur because of differences in electrochemical potential. This corrosion will result in the eventual disappearance into solution of the material with

---

[*]The pressure drop and flow distribution in complex networks may be calculated using the Hardy-Cross method (T. Hicks, "Standard Handbook of Engineering Calculations," McGraw-Hill Book Co., 1976).

the higher oxidation potential. Electrochemical corrosion is generally not a problem in systems using dry air or organic working fluids.

Table 5.9 presents a partial electromotive series of elements that may be present in solar systems. The potential difference between any two elements is a measure of the tendency for the reaction to occur. For example, aluminum metal may be oxidized by the reduction of copper ions in solution. The potential tending to promote the reaction is 1.66 V − (−0.34 V) = 2.00 V. To slow such reactions, metals that differ in the series must be electrically isolated by dielectric couplings or the like. Alternatively, in systems not containing aluminum a sacrificial magnesium anode can be used. In systems containing aluminum a sacrificial getter of aluminum could be used upstream of the aluminum component. In addition, filters should be used to capture metal particles in all solar systems in which dissimilar metals are present.

Corrosion in solar systems can also be caused by the presence of acidic, basic, or other ions conducive to electrochemical corrosion. Table 5.10 contains a list of common beneficial and detrimental water-quality parameters. A pH between 7.0 and 8.5 (slightly basic) seems best for corrosion control. A pH value below 7.0 in the presence of chloride or sulfate ions will result in particularly rapid corrosion. Dissolved oxygen, also an active corrosive agent, can be removed from liquid solar systems by boiling and then sealing the systems. Use of deionized water in liquid solar systems can avoid later water treatment to remove dissolved corrosive substances. External corrosion of collector housings and absorber plates may result from air pollutants such as $SO_2$, $NH_3$, $CO_2$, or $NO_2$.

Erosion, a nonchemical phenomenon with damage similar to corrosion, results from excessive fluid velocity in direction-change sections of conduits. Erosion can be eliminated by keeping fluid velocities below 1 m/sec (3 ft/sec) in aluminum

**TABLE 5.9** Partial Electromotive Series for Solar System Components[a]

| Element | Oxidation potential in 1-molal solution (V) |
|---|---|
| Sodium | +2.71 |
| Magnesium | 2.37 |
| Aluminum | 1.66 |
| Zinc | 0.76 |
| Iron (to feric ion) | 0.44 |
| Cobalt | 0.28 |
| Nickel | 0.25 |
| Tin | 0.14 |
| Lead | 0.13 |
| Hydrogen (reference) | 0.00 |
| Copper (to cupric ion) | −0.34 |
| Bromide ion (to $Br_2$ gas) | 1.09 |
| Chloride ion (to $Cl_2$ gas) | 1.36 |

[a]Adapted from (18).

**TABLE 5.10**  Water Quality Parameters and Corrosion Control[a]

| Detrimental parameters | Beneficial parameters |
|---|---|
| Dissolved oxygen | Calcium ion |
| Sulfides | Bicarbonate ion |
| Tin | Metaphosphate ion |
| Magnesium ion (with Al) | Monohydrogen phosphate ion |
| Chloride ion | Silica ($SiO_2$) |
| Sulfate ion | Organic color |
| Nitrate ion | Borax (sodium borate decahydrate) |
| Carbonate ion | |
| Hydroxyl ion (high pH enhances corrosion) | |
| Hydronium ion (low pH enhances corrosion) | |
| Cupric ion | |
| Carbonyl ion | |
| Nickel ion | |
| Lead ion | |

[a]From (6, 35).

conduits and below 2 m/sec (6 ft/sec) in copper conduits. Popplewell (32) has prepared a very thorough study of corrosion and erosion in aluminum, copper, and steel flat-plate collectors.

## HEAT EXCHANGERS IN SOLAR-THERMAL SYSTEMS

Heat exchangers are used in solar-thermal systems to isolate one fluid loop from another. This isolation may be necessary to avoid contamination of one fluid loop by another (possibly toxic) or purchase of a large amount of an expensive fluid (such as antifreeze) to fill thermal storage (see Fig. 5.4). Exchangers may also act as boilers for vapor generation in heat engine cycles (see Chapter 8). Although heat exchangers represent additional expense and complexity in a solar system, they are an economic, legal, or thermodynamic necessity in many cases.

Heat exchangers can ideally execute a no-loss energy exchange between fluid streams, but the exchange is always accompanied by a temperature decrement and resultant loss in available energy. This temperature penalty will require a solar collector to operate at a higher temperature than a system lacking a heat exchanger, in order to deliver the same fluid temperature to the load. Since higher collector temperatures result in lower collection efficiency (see Chapter 4), the use of a heat exchanger results in lower solar system energy delivery. This phenomenon is illustrated in the  following example.

### EXAMPLE 5.2

Calculate the efficiency at which the solar collector in Fig. 5.7 operates in order to deliver energy to a working fluid at 65°C for several values of the approach temperature difference. The approach temperature is the difference between the incoming cool fluid and the exiting warm fluid at the heat exchanger (see Fig. 5.7). The temperature rise through the heat exchanger is 10°C and is equal to the fluid temperature rise through the solar collector.

The collector has a $\tau\alpha$ product of 0.80 and loss coefficient $U_c$ of 5 W/m² · K. If the solar radiation normal to the collector surface is 500 W/m² and the ambient temperature $T_a$ is 20°C, calculate the collector efficiency for the following five values of approach $\Delta T$:

| Case | Approach $\Delta T$ (°C) |
|------|--------------------------|
| 1 | 0 (thermodynamic limit) |
| 2 | 5 |
| 3 | 10 |
| 4 | 15 |
| 5 | 20 |

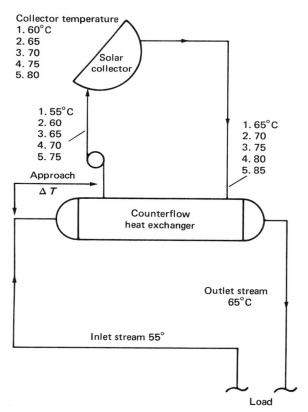

**FIGURE 5.7** Fluid stream and solar collector temperatures for Example 5.2 (heat-exchanger problem).

**TABLE 5.11**  Collector Efficiency Calculation for
Example 5.2

| Case | $\bar{T}_c$ (°C) | $\Delta T$ (°C) | $\Delta T/I_c$ (K · m²/W) | $\eta$ (%) |
|---|---|---|---|---|
| 1 | 60 | 40 | 0.08 | 40 |
| 2 | 65 | 45 | 0.09 | 35 |
| 3 | 70 | 50 | 0.10 | 30 |
| 4 | 75 | 55 | 0.11 | 25 |
| 5 | 80 | 60 | 0.12 | 20 |

## SOLUTION

Fluid stream temperatures are shown at several points for each case in Fig. 5.7. The collector temperature $T_c$ can be taken as the average of inlet and outlet fluid temperatures for purposes of the example as shown in the figure.

Collector efficiency can be calculated from the following equation (see Chapter 4):

$$\eta = \tau\alpha - U_c\left(\frac{\Delta T}{I_c}\right)$$

(5.25)

where $\Delta T = \bar{T}_c - T_a$

(5.26)

For the collector specified here

$$\eta = \left(0.8 - 5\,\frac{\Delta T}{I_c}\right) \times 100\%$$

(5.27)

The collector efficiency can be calculated in tabular form as shown in Table 5.11.

Note that the heat-exchanger design has a very strong effect on collector efficiency; for relatively small changes of approach temperature difference, efficiency and, therefore, energy delivery change significantly.

A general analytical discussion of heat exchangers is contained in Chapter 3. In this section the integration of heat exchangers into solar-thermal systems is discussed. DeWinter (13) has developed a closed-form expression for the energy delivery penalty resultant from the use of a collector-to-storage or collector-to-load heat exchanger. The heat-exchanger penalty $F_{hx}$ is defined as

$$F_{hx} = \frac{q_u \text{ (with exchanger)}}{q_u \text{ (with no exchanger and collector fluid inlet equal to the storage fluid outlet temperature)}}$$

(5.28)

The analytical expression for $F_{hx}$ (13) is

$$F_{hx} = \frac{1}{1 + [F_R U_c A_c / (\dot{m} c_p)_c] [(\dot{m} c_p)_c / (\dot{m} c_p)_{min} \, \varepsilon_{hx} - 1]}$$ (5.29)

where  $F_R$ is the collector heat-removal factor (see Chapter 4)
$U_c$ is the collector loss coefficient (see Chapter 4)
$A_c$ is the collector area
$(\dot{m} c_p)_c$ is the fluid capacitance rate through the collector array
$(\dot{m} c_p)_s$ is the fluid capacitance rate from load or storage
$(\dot{m} c_p)_{min}$ is the minimum $[(\dot{m} c_p)_c, (\dot{m} c_p)_s]$
$\varepsilon_{hx}$ is the heat-exchanger effectiveness

The collector thermal properties are contained in the group $F_R U_c A_c$, while the information for operating conditions (flow rate) is contained in the two groups $(\dot{m} c_p)_c$ and $(\dot{m} c_p)_s$. The heat-exchanger effectiveness is defined as the ratio of heat exchanged to the maximum possible heat exchange amount limited by the second law of thermodynamics. Figure 5.8a is a plot of heat-exchanger effectiveness for several generic heat-exchanger types and Fig. 5.8b, of the heat-exchanger penalty factor $F_{hx}$.

Equation (5.29) shows that the heat-exchanger penalty $F_{hx}$ decreases as heat-exchanger effectiveness increases. It is therefore desirable to use a heat exchanger with largest possible effectiveness. Figure 5.8a indicates that the counter-flow exchanger has the highest heat-transfer effectiveness and is therefore the

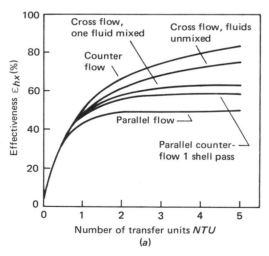

**FIGURE 5.8a** Heat-exchanger effectiveness as a function of $NTU$; effect of flow type for $(\dot{m} c_p)_{min} = (\dot{m} c_p)_{max}$. From "Compact Heat Exchangers" by W. M. Kays and A. L. London. Copyright 1964 by McGraw-Hill Book Co. Used with permission of McGraw-Hill Book Co.

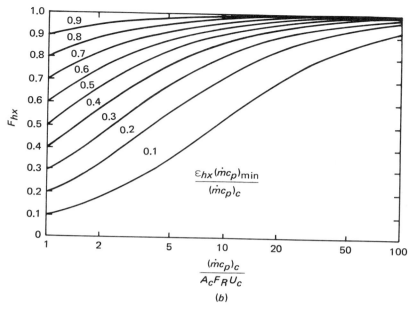

**FIGURE 5.8b**  Heat-exchanger penalty factor $F_{hx}$. In most solar-heating systems $F_{hx} > 0.90$.

preferred design for solar systems. Exchangers with parallel flow are to be avoided, since their effectiveness is low. For the special case of $(\dot{m}c_p)_{min} = (\dot{m}c_p)_{max}$ a special effectiveness equation applies for counterflow exchangers

$$\varepsilon_{hx} = \frac{NTU}{NTU + 1} \tag{5.30}$$

where the number of transfer units $NTU$ as defined in Chapter 3 is

$$NTU = \frac{(UA)_{hx}}{(\dot{m}c_p)_{min}} \tag{5.31}$$

Table 5.12 contains counterflow heat-exchanger effectiveness values for a range of $NTU$ and fluid capacitance values.

**EXAMPLE 5.3**

Calculate the energy-delivery penalty in a solar system caused by a required heat exchanger for values of effectiveness of from 0.2 to 0.99. Compare the economic value of this energy penalty to the heat-exchanger cost if the annual amortized cost (see following sections on economics in this chapter) of a heat

**TABLE 5.12** Counterflow Heat-Exchanger Performance[a]

| | $\varepsilon_{hx}$ for indicated capacity rate ratios $(\dot{m}c_p)_{min}/(\dot{m}c_p)_{max}$ | | | | | | | |
|---|---|---|---|---|---|---|---|---|
| NTU | 0 | 0.25 | 0.50 | 0.70 | 0.75 | 0.80 | 0.90 | 1.00 |
| 0 | 0 | 0 | 0 | 0 | 0 | 0 | 0 | 0 |
| 0.25 | 0.221 | 0.216 | 0.210 | 0.206 | 0.205 | 0.204 | 0.202 | 0.200 |
| 0.50 | 0.393 | 0.378 | 0.362 | 0.350 | 0.348 | 0.345 | 0.339 | 0.333 |
| 0.75 | 0.528 | 0.502 | 0.477 | 0.457 | 0.452 | 0.447 | 0.438 | 0.429 |
| 1.00 | 0.632 | 0.598 | 0.565 | 0.538 | 0.532 | 0.525 | 0.513 | 0.500 |
| 1.25 | 0.713 | 0.675 | 0.635 | 0.603 | 0.595 | 0.587 | 0.571 | 0.556 |
| 1.50 | 0.777 | 0.735 | 0.691 | 0.655 | 0.645 | 0.636 | 0.618 | 0.600 |
| 1.75 | 0.826 | 0.784 | 0.737 | 0.697 | 0.687 | 0.677 | 0.657 | 0.636 |
| 2.00 | 0.865 | 0.823 | 0.775 | 0.733 | 0.722 | 0.711 | 0.689 | 0.667 |
| 2.50 | 0.918 | 0.880 | 0.833 | 0.788 | 0.777 | 0.764 | 0.740 | 0.714 |
| 3.00 | 0.950 | 0.919 | 0.875 | 0.829 | 0.817 | 0.804 | 0.778 | 0.750 |
| 3.50 | 0.970 | 0.945 | 0.905 | 0.861 | 0.848 | 0.835 | 0.807 | 0.778 |
| 4.00 | 0.982 | 0.962 | 0.928 | 0.886 | 0.873 | 0.860 | 0.831 | 0.800 |
| 4.50 | 0.989 | 0.974 | 0.944 | 0.905 | 0.893 | 0.880 | 0.850 | 0.818 |
| 5.00 | 0.993 | 0.982 | 0.957 | 0.921 | 0.909 | 0.896 | 0.866 | 0.833 |
| 5.50 | 0.996 | 0.998 | 0.968 | 0.933 | 0.922 | 0.909 | 0.880 | 0.846 |
| 6.00 | | | 0.975 | 0.944 | | 0.921 | 0.892 | 0.857 |
| 6.50 | | | 0.980 | 0.953 | | 0.930 | 0.902 | 0.867 |
| 7.00 | | | 0.985 | 0.960 | | 0.939 | 0.910 | 0.875 |
| 7.50 | | | 0.988 | 0.966 | | 0.946 | 0.918 | 0.882 |
| 8.00 | | | 0.991 | 0.971 | | 0.952 | 0.925 | 0.889 |
| 8.50 | | | 0.993 | 0.975 | | 0.957 | 0.931 | 0.895 |
| 9.00 | | | 0.994 | 0.979 | | 0.962 | 0.936 | 0.900 |
| 9.50 | | | 0.996 | 0.982 | | 0.966 | 0.941 | 0.905 |
| 10.00 | | | 0.997 | 0.985 | | 0.970 | 0.945 | 0.909 |
| $\infty$ | 1.000 | 1.000 | 1.000 | 1.000 | 1.000 | 1.000 | 1.000 | 1.000 |

[a]From "Compact Heat Exchangers" by W. M. Kays and A. L. London. Copyright 1964 by McGraw-Hill Book Co. Used with permission of McGraw-Hill Book Co.

exchanger per year is $6/m_{hx}^2$. The economic value of the solar energy collection penalty is the value of nonsolar energy required to make up for the penalty resulting from the heat exchanger. The value of nonsolar, makeup energy is $10/GJ for this example. System specifications are listed below (water is the working fluid).

$$A_c = 100 \text{ m}^2$$

$$(\dot{m}c_p)_{min} = (\dot{m}c_{p_{max}}) = 50 \text{ kg/hr} \cdot m_c^2 \times 4182 \text{ J/kg} \cdot \text{K}$$

Energy delivery with
no heat exchanger     $= 200 \text{ MJ/m}_c^2 \cdot \text{yr}$

$$F_R U_c = 5 \text{ W/m}_c{}^2 \cdot {}^\circ\text{C} \ (= 5 \times 3600 \text{ J/hr} \cdot \text{m}_c{}^2 \cdot {}^\circ\text{C})$$

$$U_{hx} = 1400 \text{ W/m}_c{}^2 \cdot {}^\circ\text{C}$$

## SOLUTION

The heat-exchanger penalty factor $F_{hx}$ equation with the above values substituted in consistent units is

$$F_{hx} = \frac{1}{1 + 0.0861 \, (1/\varepsilon_{hx} - 1)} \tag{5.32}$$

Table 5.13 summarizes the thermal penalty from heat-exchanger use. Also shown is the heat-exchanger area required to provide a given effectiveness. The heat-exchanger area $A_{hx}$ is evaluated by the use of Eq. (5.31) and values of $NTU$ required to provide a given effectiveness for counterflow heat exchangers from Table 5.12. The annual energy delivery $Q_u$ in Table 5.13 is given by

$$Q_u = F_{hx} \times A_c \times 200 \text{ MJ/m}_c{}^2 \cdot \text{yr} \tag{5.33}$$

The solar energy penalty, which must be made up by other fuels, is

$$Q_a = (1 - F_{hx}) \times A_c \times 200 \text{ MJ/m}_c{}^2 \cdot \text{yr} \tag{5.34}$$

The economic value of the heat-exchanger energy penalty is

$$C_a = Q_a c_a$$

**TABLE 5.13**  Thermal Performance Summary for Example 5.3

| Effectiveness $\varepsilon_{hx}$ | Exchanger penalty $F_{hx}$ | Energy collection $Q_u$ (GJ/yr) | $NTU^a$ | Exchanger area $A_{hx}$ (m$_{hx}{}^2$) |
|---|---|---|---|---|
| 0.2 | 0.744 | 14.88 | 0.25 | 1.04 |
| 0.333 | 0.853 | 17.06 | 0.50 | 2.07 |
| 0.5 | 0.921 | 18.42 | 1.00 | 4.15 |
| 0.6 | 0.946 | 18.92 | 1.50 | 6.22 |
| 0.7 | 0.964 | 19.28 | 2.33 | 9.75 |
| 0.8 | 0.979 | 19.58 | 4.00 | 16.60 |
| 0.9 | 0.991 | 19.82 | 9.00 | 37.34 |
| 0.99 | 0.999 | 19.98 | 99.00 | 411.00 |
| 1.0$^b$ | 1.000$^b$ | 20.00$^b$ | $c$ | 0.00$^b$ |

$^a$Table 5.12 or Eq. (5.30).
$^b$Baseline.
$^c$No exchanger baseline.

where $c_a$ is the cost of auxiliary energy in dollars per gigajoule. The cost of heat-exchanger area $C_{hx}$ is

$$C_{hx} = A_{hx} c_{hx} \qquad (5.35)$$

where $c_{hx}$ is the amortized cost per year of heat exchanger per unit area in dollars per square meter per year.

The results of this example show that the extra exchanger area to provide extra effectiveness $\varepsilon_{hx}$ rises rapidly with $\varepsilon_{hx}$. Since the heat-exchanger cost to provide the last few percent of solar energy delivery is high, the value of this energy is relatively small when compared with the additional exchanger cost. It is, therefore, not cost-effective to provide this energy. Likewise, small heat exchangers have a large energy penalty, which has a correspondingly larger dollar value in replacement fuel. It is therefore cost-effective to add heat-exchanger area in order to recover a portion but not all of the solar energy required.

Figure 5.9 is a plot of the total extra cost required to provide a total demand of 20 GJ/yr. This extra cost is required by the design stipulation that a heat exchanger be used. It is a cost above the basic solar and backup system costs. The least cost heat-exchanger configuration is for an effectiveness value of about 0.46. It is to be noted that this example considers only one component of the solar system. A complete system optimization requires making simultaneous trade-offs of many system components. This methodology is described later in this chapter. Note that the direct application of the $F_{hx}$ factor to *annual* energy delivery instead of *instantaneous* delivery is correct only for the case of negligible loss from storage and no heat rejection by storage boiling or other means.

**TABLE 5.14**  Economic Summary for Example 5.3

| Effectiveness $\varepsilon_{hx}$ | Extra nonsolar $Q_a$ (GJ/yr) | Economic value of energy penalty $C_a$ ($/yr) | Cost of exchanger area $C_{hx}$ ($/yr) | Total cost[a] ($/yr) |
|---|---|---|---|---|
| 0.2   | 5.12 | 51.20 | 6.24    | 57.44   |
| 0.333 | 2.94 | 29.40 | 12.42   | 41.82   |
| 0.5   | 1.58 | 15.80 | 24.90   | 40.70   |
| 0.6   | 1.08 | 10.80 | 37.32   | 48.12   |
| 0.7   | 0.72 | 7.20  | 58.50   | 65.70   |
| 0.8   | 0.42 | 4.20  | 99.60   | 103.80  |
| 0.9   | 0.18 | 1.80  | 224.04  | 225.85  |
| 0.99  | 0.02 | 0.20  | 2466.00 | 2466.20 |

[a] $C_a + C_{hx}$.

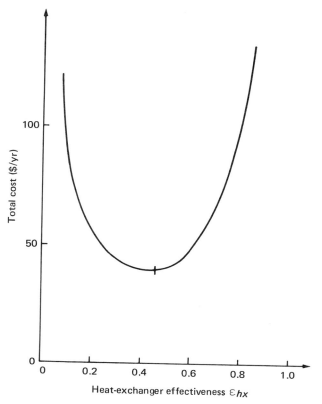

**FIGURE 5.9** Total cost of heat exchanger and nonsolar energy penalty for Example 5.3. Least-cost configuration has a heat-exchanger effectiveness of 0.46.

## ANCILLARY AND NONSOLAR MECHANICAL SYSTEM COMPONENTS

The primary components of a solar system—collector, storage, distribution, pumps and blowers, and heat exchangers—have been described in detail above. There remain a number of small but important components of all systems, including valves, dampers, expansion tanks, relief valves, and flow-balancing valves. Although not technically sophisticated, these ancillary components are practical necessities.

### Expansion Tanks

In many liquid systems, the working fluids in closed loops are subject to large temperature excursions. The fluid-specific volume can change by 10 percent in some cases over the common temperature ranges. It is necessary, therefore, to have an expansion tank to accommodate expected fluid volume changes, since the fluid conduits must remain full at all temperatures. Standard procedures are available for expansion tank design. An analytical expression has been developed for tank volume as shown below. The factors governing the size of an expansion tank are

1. Fluid volume
2. Maximum operating temperature
3. Fill pressure and temperature of the system
4. Tank location (that is, hydrostatic head)
5. Maximum permitted operating temperature

The minimum required volume of a tank in an aqueous system is given by (3) in English units over a restricted range of fill temperatures

$$V_{\exp} = \frac{(0.00041T - 0.0466)}{(P_a/P_0 - P_a/P_{\max})} \times V_f \tag{5.36}$$

where $V_f$ = the volume of the fluid loop including, for example, the solar collector, piping, and heat exchangers; or the storage tank and distribution loop

    $T$ = maximum liquid temperature (°F)

    $P_a$ = atmospheric pressure

    $P_0$ = absolute pressure created by initial fill; that is, the pressure to force water to the highest point in the system

    $P_{\max}$ = maximum operating absolute pressure

    $V_{\exp}$ = minimum expansion tank volume

For tanks in nonaqueous systems, $V_{\exp}$ can be increased or decreased in size by the relative expansivity of the nonaqueous fluid.

Vented expansion tanks are generally not recommended, since they introduce corrosive oxygen into the fluid system. Most systems should be closed to ensure long-term reliability, and expansion tanks are best located near the inlet side of the loop circulating pump. This location is preferred to ensure a positive head at the pump inlet at all times. In addition, an expansion tank is one point for collection of air entrained in a liquid system. Reference (3) contains detailed design information for expansion tank specifications.

## Filters

Filters are generally used in solar systems to prevent plugging of solar collectors by foreign matter or sludge suspended in the collector working fluid. Collectors are particularly susceptible to plugging, since the smallest flow areas in a solar system occur in the collector fluid passages. A filter passing particles no larger than 150 $\mu$m seems appropriate in most heating and cooling systems.

## Nonsolar Backup Systems

These systems are used to provide the fraction of energy demand not generated by solar-thermal conversion. As shown later in this chapter, it is rarely cost-effective to provide 100 percent of the annual energy demand by solar energy. Backup systems are designed by conventional means and are identical in function to those in nonsolar systems. Control strategies for solar-plus-nonsolar systems are generally different from those in nonsolar systems; controls are described in the next section. It is beyond the scope of this book to describe nonsolar energy systems. The reader is referred to Refs. (2, 3).

## CONTROLS FOR SOLAR-THERMAL SYSTEMS

An information-processing subsystem must be used in all but the simplest solar systems (solar-heated stock tanks, for example) to control energy flows in response to instantaneous values of temperature (or solar radiation) and energy demands. Solar control systems process data received from strategically placed sensors to control pumps, diverting valves, load devices, the backup system, or collector tracking devices. Controls are generally tuned to the solar-thermal process to be controlled so that a comprehensive treatment is not possible. Two simple examples will be described as illustrations of methodology. More detailed examples will be described by use type in Chapters 6 and 8.

### Solar Water-Heating Controls—Example System

Figure 5.10 is a simplified schematic diagram of a solar water-heating system with only major components shown. The controller for the collector pump is particularly simple and fulfills two functions only:

1. Operate collector pump when sufficient energy can be collected to make pump operation worthwhile
2. Control the maximum temperature of the hot-water tank

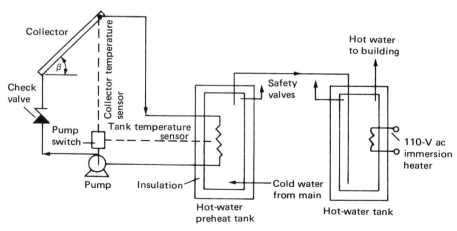

**FIGURE 5.10**   Forced-circulation, service water-heating system.

The decision to operate the collector pump is based on the temperature difference between the collector and the storage. In water-heating systems the turn-on difference is usually 10°C (18°F). It is usually profitable to continue to operate the pump until the collector and storage temperatures differ by an amount less than 10°C. Usually a turn-off difference of one-fifth to one-seventh the turn-on difference, that is, 1.5-2.0°C, works well. The difference between the turn-on and turn-off criteria is called hysteresis.

Figure 5.11 shows the sequence of operation of a controller on a plot of the difference between collector outlet and storage temperature $\Delta T_{cs}$ versus pump flow rate. For simplicity the collector inlet temperature has been assumed to be the same as the storage temperature. The sequence of operation of the controller can be understood by following the dashed lines in Fig. 5.11. At point a the collector begins to warm as the sun first strikes it, a process that continues to point b, at which time the pump turn-on condition is met and fluid begins to flow at rate $Q_2$. As the fluid continues to remove heat from the collector, its temperature drops toward the steady-state point 2. However, at point d the turn-off criterion is met and flow ceases. From point e the collector temperature must again rise to point b before circulation begins.

This cycling mode is undesirable, since it causes excessive wear and reduced efficiency; therefore, a lower flow rate $Q_1$ must be used to ensure stable operation

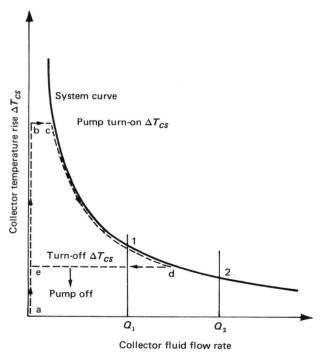

**FIGURE 5.11**  Solar hot-water system performance curve showing stable and unstable control modes for flow rates $Q_1$ and $Q_2$, respectively.

at point 1 above the turn-off temperature set point. This simplified analysis illustrates the intimate relation between flow rate and controller set points.

An alternative method of preventing cycling is to cause the pump to remain on for a minimum period of time—15 or 20 min. This approach has the disadvantage that the pump may operate when climatic conditions have changed and no energy is collectable. In that case the collector would be heated by storage—a clearly undesirable operating condition.

To prevent scalding and boiling, it is necessary to set an upper limit on the storage temperature. The controller also performs this function. The simplest method for achieving this result is to turn off the pump when $T_s$ exceeds some set-point. This simple approach is rarely used, since exposure to high-temperature, no-flow conditions will cause most fluids to boil or thermally degrade in the collector. A better heat-rejection method is to cause the controller to draw off hot water from the preheat tank. This water is replaced by cold supply water and tank temperature and fluid flow are maintained.

The backup for the hot-water system is a conventional nonsolar water heater. No backup actuator is required in the solar controller, since the thermostat in the conventional heater serves this purpose. In general, however, the solar controller will actuate the backup system as well as the solar system.

Figure 5.12 shows a typical solid-state controller used for low-temperature applications.

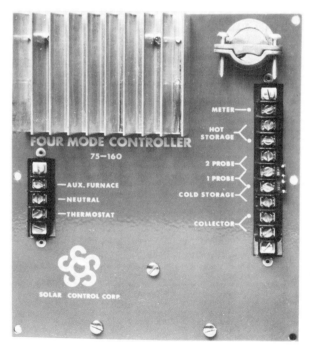

**FIGURE 5.12** Typical solid-state controller for solar hot-water heating systems. Courtesy of Solar Control Corp.

## Solar Swimming Pool Heater Controls—Example System

An example of a different control philosophy is embodied in the solar swimming pool heater shown in Fig. 5.13. Instead of sensing the temperature difference in order to control flows, a sun sensor, which senses solar energy above a threshold level is used. Above that threshold the diverter valve causes flow through the collector if the pool temperature can be increased by so doing. The on-off decision is programmed into the controller. A special pump is not required for this system, since most pools contain a pump-operated filtration system, which is retained in the solar system. The diverter valve will bypass the collector if the sun sensor-pool temperature signals so dictate.

A high-temperature cutoff is not required for pools with a properly sized solar collector. As pool temperature rises, convection and evaporation increase, offsetting the temperature rise. It is also possible to use night heat losses from unglazed pool collectors to cool the pool if required; this feature is easily programmed into the controller. No backup is generally used with pool heaters. Swimming pool controls

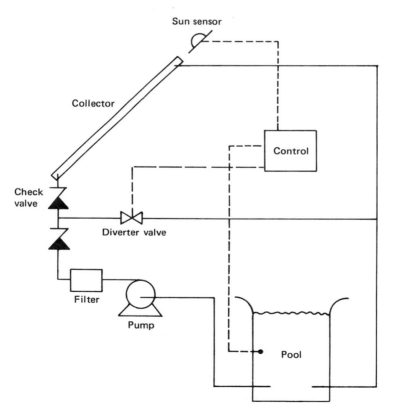

**FIGURE 5.13** Schematic diagram of solar swimming pool heater showing controller signal sources—sun sensor and pool temperature sensor.

are similar in their level of sophistication to crop-drying controls. Essentially, these simple controls use solar energy when it is available and cause operation to cease when not available.

## General Features of Solar Control Systems

The preceding examples of simple solar systems have illustrated the major features of solar controls except for those used in industry. Industrial process heat systems require more accurate temperature and flow control than do most other low- and medium-temperature applications. Special controls with feedback and feedforward capability are therefore required. Information inputs to the controller can include temperatures, temperature differences, load information, and sunshine level. Outputs from the controller include actuation of valves, pumps, or flow-control valves. Mode controls—solar delivery to load, backup system actuation, heat rejection, or null state—are also exercised.

Information inputs for controllers in buildings are usually in the form of an analog voltage signal from temperature and sun sensors and in the form of an off-on signal from the thermostat. In most solar controllers, output signals are on-off, with the exception of an analog voltage or compressed-air signal for flow control valves and solar trackers. The on-off signal generally actuates relays or power transistors, which in turn operate motors, valves, dampers, etc.

Well-designed control systems contain fail-safe modes to avoid equipment damage during power outages, pump failures, or failures of solar concentrator trackers. It is essential that solenoid valves fail in a safe mode, that backup power be provided automatically if solar equipment can be damaged by long exposure to no-flow conditions in bright sun, and that solar concentrators defocus automatically if fluid ceases to flow. If fail-safe controls are not utilized, damage far exceeding their small cost can result.

## MATHEMATICAL MODELS OF SOLAR SYSTEMS

The rational design of a solar-thermal system requires a knowledge of the dynamic interaction of all system components described in this chapter. Although essential and valuable experience can be gained by testing solar systems in the field, the generalization of experimental results and their application to other systems in other locations can best be handled by a modeling approach.

Models are tools to assist in learning, solving problems, and communicating. One of the earliest models was the spoken language. In the seventeenth century coordinate systems were devised. Graphs plotted in coordinates in turn led to mathematical functions, which are the basis for modern science and technology. Various levels of models are used every day; each represents something in the physical world. A model is simply a symbolic or abstract representation of a phenomenon in the real world.

Mathematical modeling is a calculation approach by which mathematical representations of individual components are integrated into a dynamic numerical system. This system of algebraic, differential, and integral equations can be used to calculate the response of a physical solar system to arbitrary forcing functions, which represent time series of energy demands, solar radiation levels, and meteorologic data. Numerical experiments may readily be conducted to evaluate control strategies, component size and type effects, or geographic variations of system behavior, for example.

A simple block diagram of a computer model of a solar system using hourly weather data is shown in Fig. 5.14. The basic steps in any model are shown therein:

1. Solar radiation data processor
2. Mathematical model of solar-thermal system
3. Mathematical model of thermal demand
4. Comparison of solar delivery to demand
5. Economic optimization routines

Computers are generally required for calculation of numerical solar models because many component models are nonlinear and a closed-form, simultaneous

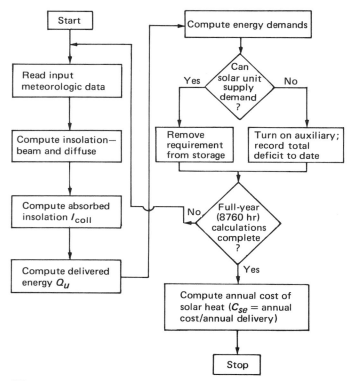

**FIGURE 5.14** Simplified block diagram of computer-simulation model of a solar-thermal system. Adapted from Kreider, J. F., and F. Kreith, "Solar Heating and Cooling," revised 1st ed., Hemisphere Publ. Corp., 1977.

solution of the equations describing a system is not possible in all but the simplest cases. In addition, most solar systems operate in a continuously changing, transient manner subject to short time-scale changes in all forcing functions. Since meaningful modeling results require energy delivery totals for fairly long time periods (months or years), and since solar systems respond on a fairly short time scale (minutes or hours), a great many calculations are required. For these reasons computerized models are the method of choice for analysis of most solar-thermal systems.

A second type of numerical model has also been used successfully for solar system analysis. Stochastic solar models regard solar processes as sequences of random events that can be considered in a statistical manner. Stochastic models have an important advantage over simulation models in that the calculation effort can be reduced by an order of magnitude. In developing general solar system design guidelines requiring hundreds of model runs, this is a major improvement in methodology.

It is beyond the purview of this book to describe modeling methodology in detail. The salient features of a typical simulation model and a stochastic model are described, however, in order to provide the reader with a summary of these methods.

The first widely used solar simulation model TRNSYS was developed by Klein et al. (22-24). This quasi-steady-state model consists of detailed FORTRAN models of about two dozen components including collectors, storage devices, valves and other pipe fittings, heat exchangers, controllers, heat pumps, and thermal loads. Component models are mathematical representations of energy and mass conservation laws applied to the component. The components may be connected in many ways to permit simulation of nearly all practical solar-thermal systems. Also included is a solar radiation data processor, which transforms historical solar data to radiation on the tilted surface of a collector.

Many component models have optional representations in TRNSYS (22). For example, storage may be treated as well mixed or stratified; the solar collector can be modeled by empirical test data or by algebraic equations. Accuracy criteria and integration time steps can be specified depending on the level of detail and stability criteria required. An input data checker is included to avoid spurious model runs using erroneous or incomplete data.

Component connections in TRNSYS are of two types—energy and mass transports in fluid conduits and information flows in electrical wires. Information flows are controller inputs and outputs, while fluid flows represent energy inputs and outputs of solar system components. Since components are interconnected, a set of simultaneous algebraic and differential equations results. For example, storage fluid exit temperature is the load device inlet temperature; collector fluid outlet temperature is the same as storage inlet temperature (if no heat exchanger is used). In TRNSYS algebraic equations are solved by iterative methods subject to user-specified accuracy criteria. Differential equations are solved by a first-order predictor-corrector method.

Numerical outputs from TRNSYS include solar energy delivery, energy demands, nonsolar energy requirements, and many other component performance indices selected by the user. These model data can be used to examine the performance of effects of the myriad design variables in solar systems. Model results can also be used for economic analyses and feasibility studies for specific systems in specific locations.

A particularly efficacious use of detailed models such as TRNSYS is the generation of generalized performance correlations that are independent of system size or location, or thermal-demand profile (24). The generalized correlation can be used directly for prediction of overall system behavior for a nominal system configuration. This approach represents an essential simplification in solar system design, and reduces the computer requirements by two orders of magnitude. Three generalized correlations for design of this type are described in Chapter 6.

Löf and Tybout (28), Buchberg and Roulet (8), and Butz et al. (9) have also developed quasi-steady-state models of solar systems, which were predecessors of TRNSYS. They were capable of identifying important parameter trends in solar-heating systems of specific designs.

A modeling approach that shows considerable promise uses statistical properties of solar and weather phenomena instead of a step-by-step calculation for each hour's meteorologic data as used in the simulation approach. The discipline of stochastic (or time-varying, random) processes (4) can be used to devise a stochastic model of solar-thermal systems. Lameiro (26) has prepared the first stochastic model of a solar-thermal system. It models the behavior of a solar-heated residence using the statistical properties of hourly solar and weather data.

## INTRODUCTION TO PRINCIPLES OF ECONOMIC ANALYSIS

Although solar radiant energy is free, the equipment required to convert it to a useful form—thermal or electrical—is not free. Therefore, a cost must be assigned to solar-thermal or solar-electrical energy that reflects the conversion equipment cost prorated on the number of kilowatt-hours or Btu delivered by the solar system. If the solar cost is less than the cost of other energy sources that can perform the same task, there is an economic incentive to use solar energy. The purpose of an economic optimization is to maximize the savings resulting from use of solar energy. In this and following sections the method of calculating the cost of solar energy in various economic scenarios is described. The methods for calculating the amount of solar energy delivered by a given system for a given cost are given in detail in Chapters 6 and 7.

### Overview of Solar Economics

It is rarely cost-effective to provide all the energy requirements of a thermal or mechanical system by means of solar energy. If this were to be done, the solar system would be required to be capable of providing 100 percent of the energy demand for the worst set of operating conditions ever expected—inclement weather, maximum demand, and no sunshine. A solar system capable of providing the peak demand for a lengthy period would be oversized for all less severe operating conditions; it would thus be greatly oversized most of the time. An oversized system delivers more energy than can be used, thereby requiring energy rejection in some manner. However, the oversized system must be paid for even though it delivers below its capability nearly continuously. A solar system with such a low load factor is uneconomical and impractical.

The best use of solar energy is in conjunction with conventional fuels, which are used as an auxiliary source for special high-demand situations. It is then possible to design a base-load solar system that operates at nearly full capacity (high load factor) most of the time although it provides less than 100 percent of the annual demands. Such a system is more economical, since it delivers more energy per unit system cost than an oversized full-capacity solar system. Peak loads, for which it is uneconomical to size a solar system, are carried by the auxiliary system. It will be shown shortly that the solar-plus-auxiliary method is nearly always the least-cost method. The determination of the fraction of the annual demand to be provided by solar energy and what solar system size is required for that energy delivery is the key problem addressed in this portion of the chapter.

Several indices or figures of merit can be used to select the best solar-auxiliary mix depending on what national economic system applies. In the Western economies it is the annual cost—dollars, marks, pounds, francs, etc.—of providing energy to a given load that is the figure of merit. The lower the cost, the more desirable the configuration. The least-cost approach is used in this book. Other approaches could include minimization of entropy production or minimization of fossil energy usage.

## Life-Cycle Costing

A solar-thermal system differs fundamentally from a fossil fuel or electrical energy system in a manner that requires an economic analysis reflecting the benefits accrued by solar usage throughout the lifetime of the system. Nonsolar systems usually have relatively small initial costs and relatively large annual operating costs reflecting raw energy purchases. Solar systems, however, are relatively expensive initially but have negligible nonsolar energy cost during their lifetime. If the selection of a system (solar versus nonsolar) were made on the basis of initial cost alone, the solar system would rarely be selected. However, the naive initial-cost judgment cannot account for the principal reason a solar system is considered—to reduce the usage of expensive and scarce fuels.

The concept of life-cycle costing includes both the initial capital cost and the year-to-year operating cost in making economic decisions. The life-cycle cost of any energy system is defined as the total of the following cost components over the life of a system:

1. Capital equipment cost
2. Acquisition costs
3. Operating costs—fuels, etc.
4. Interest charges if money for capital is borrowed
5. Maintenance, insurance, and miscellaneous charges

6. Taxes—sales, local, state, federal
7. Other recurring or one-time costs associated with the system
8. Salvage value (usually a negative cost)

For example, the life-cycle cost of a conventional forced-air, gas-heating system would include the cost of the furnace, distribution system and controls, the total cost of gas burned throughout the period of analysis, the portion of property tax associated with the heating system, any repair charges and replacement parts (filters, motors, dampers), and the mortgage interest charge on the heating system. It is clear that the life-cycle cost is a much more useful cost index than is the initial cost of the gas furnace.

In the following sections of this chapter the parameters affecting the life-cycle cost of solar systems will be described in analytical detail. The important parameters include

Initial cost, including hardware, transportation, installation, system designer's fee, special building features, and value of space used by system
Down payment and mortgage interest rate
Life-cycle time
Depreciation rate and final salvage value
Repair and replacement costs
Maintenance costs
Nonsolar fuel costs
Inflation rate of cost of nonsolar fuel
General inflation rate
Local, state, and federal taxes

It will be shown that residential and commercial (income-producing) life-cycle cost analyses differ somewhat in the treatment of taxes. Inflation, mortgage, and tax parameters can be significant for the homeowner, whereas repair and maintenance costs for a well-designed solar system can be made quite small.

## PRINCIPLES OF DISCOUNTED CASH FLOW ANALYSIS

### Annualized Present Worth

Since it is convenient to conduct economic analyses in a unit of fixed monetary value, the concept of present worth is used. The *present* worth of a *future* cash flow or payment is the future value of that flow with the time value of money factored out. The time value of money represents the opportunity for investment yielding a future return; it is an effect separate from inflation. For example, the future value $X$ of a sum of money $P$ invested at an annual interest rate $i$ dollars per dollars per year is

$$X = P(1 + i)^t \tag{5.37}$$

in which $t$ is the future time expressed in years (or other time unit corresponding to the time basis of the interest rate $i$). Equation (5.37) states that an initial sum of money $P$ appreciates by a multipler factor $(1 + i)$ each year (disregarding general inflation). Stated alternatively, a future sum of money $X$ has a present worth $P$ given by

$$P = \frac{X}{(1 + i)^t} \tag{5.38}$$

Equation (5.38) indicates that the present worth of a given amount of money in the future is discounted, in constant dollars, by a factor $(1 + i)^{-1}$ for each year in the future. In this context, $i$ is generally called the discount rate. The factor $(1 + i)^{-t}$ is called the present-worth factor $PWF(i, t)$ given by

$$PWF(i, t) \equiv \frac{1}{(1 + i)^t} \tag{5.39}$$

The present-worth factor can be multiplied by any cash flow in the future at time $t$ to give its present value. Table 5.15 is a tabulation of present-worth factors.

**TABLE 5.15**  Present-Worth Factors[a]

| $n$ \ $i$ | 0% | 2% | 4% | 6% | 8% | 10% | 12% | 15% | 20% | 25% |
|---|---|---|---|---|---|---|---|---|---|---|
| 1 | 1.0000 | 0.9804 | 0.9615 | 0.9434 | 0.9259 | 0.9091 | 0.8929 | 0.8696 | 0.8333 | 0.8000 |
| 2 | 1.0000 | 0.9612 | 0.9246 | 0.8900 | 0.8173 | 0.8264 | 0.7972 | 0.7561 | 0.6944 | 0.6400 |
| 3 | 1.0000 | 0.9423 | 0.8890 | 0.8396 | 0.7938 | 0.7513 | 0.7118 | 0.6575 | 0.5787 | 0.5120 |
| 4 | 1.0000 | 0.9238 | 0.8548 | 0.7921 | 0.7350 | 0.6830 | 0.6355 | 0.5718 | 0.4823 | 0.4096 |
| 5 | 1.0000 | 0.9057 | 0.8219 | 0.7473 | 0.6806 | 0.6209 | 0.5674 | 0.4972 | 0.4019 | 0.3277 |
| 6 | 1.0000 | 0.8880 | 0.7903 | 0.7050 | 0.6302 | 0.5645 | 0.5066 | 0.4323 | 0.3349 | 0.2621 |
| 7 | 1.0000 | 0.8706 | 0.7599 | 0.6651 | 0.5835 | 0.5132 | 0.4523 | 0.3759 | 0.2791 | 0.2097 |
| 8 | 1.0000 | 0.8535 | 0.7307 | 0.6274 | 0.5403 | 0.4665 | 0.4039 | 0.3269 | 0.2326 | 0.1678 |
| 9 | 1.0000 | 0.8368 | 0.7026 | 0.5919 | 0.5002 | 0.4241 | 0.3606 | 0.2843 | 0.1938 | 0.1342 |
| 10 | 1.0000 | 0.8203 | 0.6756 | 0.5584 | 0.4632 | 0.3855 | 0.3220 | 0.2472 | 0.1615 | 0.1074 |
| 11 | 1.0000 | 0.8043 | 0.6496 | 0.5268 | 0.4289 | 0.3505 | 0.2875 | 0.2149 | 0.1346 | 0.0859 |
| 12 | 1.0000 | 0.7885 | 0.6246 | 0.4970 | 0.3971 | 0.3186 | 0.2567 | 0.1869 | 0.1122 | 0.0687 |
| 13 | 1.0000 | 0.7730 | 0.6006 | 0.4688 | 0.3677 | 0.2897 | 0.2292 | 0.1625 | 0.0935 | 0.0550 |
| 14 | 1.0000 | 0.7579 | 0.5775 | 0.4423 | 0.3405 | 0.2633 | 0.2046 | 0.1413 | 0.0779 | 0.0440 |
| 15 | 1.0000 | 0.7430 | 0.5553 | 0.4173 | 0.3152 | 0.2394 | 0.1827 | 0.1229 | 0.0649 | 0.0352 |
| 16 | 1.0000 | 0.7284 | 0.5339 | 0.3936 | 0.2919 | 0.2176 | 0.1631 | 0.1069 | 0.0541 | 0.0281 |
| 17 | 1.0000 | 0.7142 | 0.5134 | 0.3714 | 0.2703 | 0.1978 | 0.1456 | 0.0929 | 0.0451 | 0.0225 |
| 18 | 1.0000 | 0.7002 | 0.4936 | 0.3503 | 0.2502 | 0.1799 | 0.1300 | 0.0808 | 0.0376 | 0.0180 |
| 19 | 1.0000 | 0.6864 | 0.4746 | 0.3305 | 0.2317 | 0.1635 | 0.1161 | 0.0703 | 0.0313 | 0.0144 |
| 20 | 1.0000 | 0.6730 | 0.4564 | 0.3118 | 0.2145 | 0.1486 | 0.1037 | 0.0611 | 0.0261 | 0.0115 |
| 25 | 1.0000 | 0.6095 | 0.2751 | 0.2330 | 0.1460 | 0.0923 | 0.0588 | 0.0304 | 0.0105 | 0.0038 |
| 30 | 1.0000 | 0.5521 | 0.3083 | 0.1741 | 0.0994 | 0.0573 | 0.0334 | 0.0151 | 0.0042 | 0.0012 |
| 40 | 1.0000 | 0.4529 | 0.2083 | 0.0972 | 0.0460 | 0.0221 | 0.0107 | 0.0037 | 0.0007 | 0.0001 |
| 50 | 1.0000 | 0.3715 | 0.1407 | 0.0543 | 0.0213 | 0.0085 | 0.0035 | 0.0009 | 0.0001 | — |
| 100 | 1.0000 | 0.1380 | 0.0198 | 0.0029 | 0.0005 | 0.0601 | — | — | — | — |

[a]For interest rates $i$ of from 0 to 25 percent and for periods of analysis $n$ of from 1 to 100 yr.

The present-worth concept is the key idea in the use of discounted cash flow analysis. Its use permits all calculations to be made in present, discounted monies. In life-cycle cost analyses in this book, the present-worth approach is used instead of forming the total of all life-cycle costs. This approach seems more useful, since the sums of money are more intuitively familiar when expressed on an annual, not total, basis.

## EXAMPLE 5.4

Assume that opportunity $A$ generates benefits equal to $100, $150, and $200 at the end of years 1, 2, and 3, respectively. Assume that opportunity $B$ yields benefits of $225 in year 2 and $225 in year 3. Therefore over 3 yr, both opportunity $A$ and $B$ yield benefits of $450. However, the timing of the benefits received is different in each case. By using the present-worth technique, their two benefit flows can be viewed in terms of today's dollar value.

## SOLUTION

### Step 1

Compute the present-worth factor $PWF$ using Eq. (5.39). With an interest rate of 10 percent, the calculation of present-worth factors is as follows:

$$\text{Year } 1 = \frac{1}{(1 + 0.10)^1} = \frac{1}{(1.10)^1} = 0.9091$$

$$\text{Year } 2 = \frac{1}{(1 + 0.10)^2} = \frac{1}{(1.10)^2} = 0.8264$$

$$\text{Year } 3 = \frac{1}{(1 + 0.10)^3} = \frac{1}{(1.10)^3} = 0.7513$$

### Step 2

Compute the present worth of each opportunity benefit flow by multiplying the present-worth factor by the annual benefit amount.

| Year | Present-worth factor | Annual benefit ($) A | B | Present worth ($) A | B |
|---|---|---|---|---|---|
| 1 | 0.9091 | 100 | 0 | 90.91 | 0 |
| 2 | 0.8264 | 150 | 225 | 123.96 | 185.94 |
| 3 | 0.7513 | 200 | 225 | 150.26 | 169.04 |
| Total | | 450 | 450 | 365.13 | 354.98 |

The effect of the time value of money on the investment opportunity is seen by comparing the total annual benefit with the total present worth for the two investment opportunities.

The discount rate is a measure of an investor's preference for the time value of money. A discount rate of 8 percent means that an investor regards a sum of money as worth 8 percent less next year than it is worth this year. Alternatively, the investor is indifferent when offered the option of $100 this year or $108 next year (apart from inflation).

Large businesses have well-established values of discount rate or required rates of return on investments. Homeowners do not. Most analysts agree that the lower bound of a homeowner's discount rate is his return on personal savings or other investments. Likewise, the upper bound is the interest rate on personal loans or mortgages. An individual would not borrow a sum of money at a higher interest rate than that expected for the investment of the borrowed sum in an income-producing venture. Since the discount rates of homeowners and small business owners are somewhat subjective, both high and low limits should be considered in a life-cycle cost analysis.

The effect of inflation on the future value of an invested sum is to reduce the future value by a factor $(1 + j)$ per year, where $j$ is the inflation rate per year (for example, dollars per dollars per year). The future value $X$ from Eq. (5.37), taking inflation into account, is

$$X = P \left( \frac{1 + i}{1 + j} \right)^t \tag{5.40}$$

Alternatively, the present worth $P$ of a future sum $X$ under inflation is

$$P = \frac{X}{[(1 + i)/(1 + j)]^t} \tag{5.41}$$

If an *effective* interest rate $i'$ is defined to include both interest and inflation effects such that

$$P = \frac{X}{(1 + i')^t} = X\, PWF(i', t) \tag{5.42}$$

it is easy to show that the effective rate is

$$i' = \frac{1 + i}{1 + j} - 1 = \frac{i - j}{1 + j} \tag{5.43}$$

The concept of the present-worth factor can be modified to account for inflation by replacing the interest rate $i$ with the effective or real interest rate $i'$. For small inflation rates $j$ the effective rate $i'$ is approximately the difference between the interest rate and the inflation rate:

$$i' \simeq i - j \qquad \text{for small } j \qquad (5.44)$$

The general inflation rate is usually measured by the consumer price index in the United States. See Hirst (17) for a survey of the usefulness of this index.

## Series of Payments

In many economic analyses a *series* of annual or monthly sums is invested or used to pay off a loan. The method above can easily be extended from a single payment to a series of payments. If a sum $P_{ann}$ is invested each year at interest rate $i$, the present worth of the *sum* of these payments $S$ is

$$S = \frac{P_{ann}}{1+i} + \frac{P_{ann}}{(1+i)^2} + \frac{P_{ann}}{(1+i)^3} + \cdots \qquad (5.45)$$

or

$$S = P_{ann} \left[ (1+i)^{-1} + (1+i)^{-2} + (1+i)^{-3} + \ldots \right] \qquad (5.46)$$

The expression in brackets is a geometric series with first term $P_{ann}/(1+i)$ and ratio $(1+i)^{-1}$. It follows from the expression for the sum of such a series (37) that

$$S = P_{ann} \left[ \frac{1 - (1+i)^{-t}}{i} \right] \qquad (5.47)$$

Alternatively, the annual payment $P_{ann}$ required to form a total amount $S$, for example, to purchase a solar system, is

$$P_{ann} = S \left[ \frac{i}{1 - (1+i)^{-t}} \right] \qquad (5.48)$$

The term in brackets in Eq. (5.48) is called the capital-recovery factor $CRF(i, t)$:

$$CRF(i, t) \equiv \frac{i}{1 - (1+i)^{-t}} \qquad (5.49)$$

Table 5.16 is a summary tabulation of capital-recovery factors; Appendix 5 contains detailed tables of capital-recovery factors. The term $P_{ann}$ in Eq. (5.48) is the annual payment on a self-amortizing loan of value $S$. Each payment is a mix of interest and principal repayment. Early payments are mostly interest because of the large outstanding balance; later payments are primarily principal repayment. Most solar systems owned by private firms or individuals are purchased with self-amortizing loans.

**TABLE 5.16** Capital-Recovery Factors[a]

| n | 0% | 2% | 4% | 6% | 8% | 10% | 12% | 15% | 20% | 25% |
|---|---|---|---|---|---|---|---|---|---|---|
| 1 | 1.00000 | 1.02000 | 1.04000 | 1.06000 | 1.08000 | 1.10000 | 1.12000 | 1.15000 | 1.20000 | 1.25000 |
| 2 | 0.50000 | 0.51505 | 0.53020 | 0.54544 | 0.56077 | 0.57619 | 0.59170 | 0.61512 | 0.65455 | 0.69444 |
| 3 | 0.33333 | 0.34675 | 0.36035 | 0.37411 | 0.38803 | 0.40211 | 0.41635 | 0.43798 | 0.47473 | 0.51230 |
| 4 | 0.25000 | 0.26262 | 0.27549 | 0.28859 | 0.30192 | 0.31547 | 0.32923 | 0.35027 | 0.38629 | 0.42344 |
| 5 | 0.20000 | 0.21216 | 0.22463 | 0.23740 | 0.25046 | 0.26380 | 0.27741 | 0.29832 | 0.33438 | 0.37184 |
| 6 | 0.16667 | 0.17853 | 0.19076 | 0.20336 | 0.21632 | 0.22961 | 0.24323 | 0.26424 | 0.30071 | 0.33882 |
| 7 | 0.14286 | 0.15451 | 0.16661 | 0.17914 | 0.19207 | 0.20541 | 0.21912 | 0.24036 | 0.27742 | 0.31634 |
| 8 | 0.12500 | 0.13651 | 0.14853 | 0.16101 | 0.17401 | 0.18744 | 0.20130 | 0.22285 | 0.26061 | 0.30040 |
| 9 | 0.11111 | 0.12252 | 0.13449 | 0.14702 | 0.16008 | 0.17364 | 0.18768 | 0.20957 | 0.24808 | 0.28876 |
| 10 | 0.10000 | 0.11133 | 0.12329 | 0.13587 | 0.14903 | 0.16275 | 0.17698 | 0.19925 | 0.23852 | 0.28007 |
| 11 | 0.09091 | 0.10218 | 0.11415 | 0.12679 | 0.14008 | 0.15396 | 0.16842 | 0.19107 | 0.23110 | 0.27349 |
| 12 | 0.08333 | 0.09156 | 0.10655 | 0.11928 | 0.13270 | 0.14676 | 0.16144 | 0.18148 | 0.22526 | 0.26845 |
| 13 | 0.07692 | 0.08812 | 0.10014 | 0.11296 | 0.12652 | 0.14078 | 0.15568 | 0.17911 | 0.22062 | 0.26454 |
| 14 | 0.07143 | 0.08260 | 0.09467 | 0.10758 | 0.12130 | 0.13575 | 0.15087 | 0.17469 | 0.21689 | 0.26150 |
| 15 | 0.06667 | 0.07783 | 0.08994 | 0.10296 | 0.11683 | 0.13147 | 0.14682 | 0.17102 | 0.21388 | 0.25912 |
| 16 | 0.06250 | 0.07365 | 0.08582 | 0.09895 | 0.11298 | 0.12782 | 0.14339 | 0.16795 | 0.21144 | 0.25724 |
| 17 | 0.05882 | 0.06997 | 0.08220 | 0.09544 | 0.10963 | 0.12466 | 0.14046 | 0.16537 | 0.20944 | 0.25576 |
| 18 | 0.05556 | 0.06670 | 0.07899 | 0.09236 | 0.10670 | 0.12193 | 0.13794 | 0.16319 | 0.20781 | 0.25459 |
| 19 | 0.05263 | 0.06378 | 0.07614 | 0.08962 | 0.10413 | 0.11955 | 0.13576 | 0.16134 | 0.20646 | 0.25366 |
| 20 | 0.05000 | 0.06116 | 0.07358 | 0.08718 | 0.10185 | 0.11746 | 0.13388 | 0.15976 | 0.20536 | 0.25292 |
| 25 | 0.04000 | 0.05122 | 0.06401 | 0.07823 | 0.09368 | 0.11017 | 0.12750 | 0.15470 | 0.20212 | 0.25095 |
| 30 | 0.03333 | 0.04465 | 0.05783 | 0.07265 | 0.08883 | 0.10608 | 0.12414 | 0.15230 | 0.20085 | 0.25031 |
| 40 | 0.02500 | 0.03656 | 0.05052 | 0.06646 | 0.08386 | 0.10226 | 0.12130 | 0.15056 | 0.20014 | 0.25003 |
| 50 | 0.02000 | 0.03182 | 0.04655 | 0.06344 | 0.08174 | 0.10086 | 0.12042 | 0.15014 | 0.20002 | 0.25000 |
| 100 | 0.01000 | 0.02320 | 0.04081 | 0.06018 | 0.08004 | 0.10001 | 0.12000 | 0.15000 | 0.20000 | 0.25000 |
| ∞ | | 0.02000 | 0.04000 | 0.06000 | 0.08000 | 0.10000 | 0.12000 | 0.15000 | 0.20000 | 0.25000 |

[a]For interest rates $i$ of from 0 to 25 percent and for periods of analysis $n$ of from 1 to 100 yr.

## EXAMPLE 5.5

Compute the annual cost of a solar energy system with the characteristics tabulated below.

| Factor | Specification |
|---|---|
| Expected system lifetime $t$ (yr) | 20 |
| Discount rate (%) | 8 |
| Collector area $A_c$ (m²) | 20 |
| Collector cost ($/m²) | 100 |
| Storage cost ($/m²) | 6.25 |
| Cost of control system ($) | 100 |
| Miscellaneous costs (for example, pipes, pumps, motors, heat exchangers) ($) | $200 + (5A_c)$ |

## SOLUTION

To obtain the total cost of this solar energy system, add the costs for the controls, the collector, the storage, and miscellaneous items, or,

$$S = \$100 + (\$100 \times 20) + (\$6.25 \times 20) + [\$200 + (5.00 \times 20)]$$

$$S = \$2525$$

From Table 5.16 a capital-recovery factor $CRF$ for 20 yr at 8 percent is

$$CRF(20, 0.08) = 0.102$$

The annual cost $P_{ann}$ is then

$$P_{ann} = \$2525 \times 0.102 = \$257/\text{yr}$$

The effect of inflation on a series of payments is analyzed in the same manner as for a single payment [see Eqs. (5.40 to 5.55)]. Inflation effects are included by replacing the basic interest rate $i$ with the effective or real rate $i'$.

## EXAMPLE 5.6

The owner of a four-unit apartment building of masonry construction is considering the retrofit installation of a solar heater. Under the existing climatic conditions it is expected that the solar unit can supply $87.5 \times 10^6$ kJ/yr and will thus save 5530 liters of fuel oil (with an effective heating value of 15,800 kJ/liter and an efficiency of 43 percent in the existing heating plant). In present dollars, this savings resulting from the solar unit represents a sum of $510 (9.2¢/liter of fuel oil saved). If the owner can borrow money for the solar retrofit at 10 percent per year and wants to repay the loan in 7.8 yr, how much should the owner invest in the solar-heating system?

**SOLUTION**

If

$P_{ann}$ = $510 (savings resulting from of solar energy for fuel oil)
$j = 0.12$ (expected increase in fuel cost)
$i = 0.10$ (interest rate for loan)
$t = 7.8$ (time to repay loan, in years)

then

$$i = \frac{1 + 0.10}{1 + 0.12} - 1 = -0.0179 \text{ (effective interest rate)}$$

and

$$S = 510 \times \frac{0.982^{7.8} - 1}{-0.0179 \,(0.982)^{7.8}} = \$4,300$$

The owner can pay $4,300 or less for the retrofit installations.

Repeat the problem, assuming the repayment time can be extended to 15 yr and government-backed, FHA solar-improvement loans of 5 percent interest are available. Observe the large effect of interest rate on the calculations.

# SOLAR SYSTEM LIFE–CYCLE COSTS

The principal costs of a solar system are the capital and interest required to pay for equipment. In the life-cycle cost analysis of a solar system, it is the *extra cost* of the solar system in excess of the basic auxiliary system that is considered. It is assumed that a nonsolar system is present whether solar energy is used or not. The savings in conventional fuels pay for the extra investment in solar equipment, along with the recurring expenses of maintenance, taxes, and insurance.

Although a life-cycle cost analysis normally deals with totals of annual costs during the life-cycle period, it is more convenient to deal with life-cycle costs on an annual basis. Both approaches are equivalent, as shown by the duality of annual costs (or payments $P_{ann}$) and the total costs $S$ in the preceding analysis.

## Solar System Initial Costs

The initial extra cost of a solar system consists of a number of components. For a solar-heating system for a building, for example, the following initial costs will usually be present:

Solar collector—cost and delivery; special siting requirements
Thermal storage and insulation; special siting requirements
Pumps or fans, piping, valves, and insulation; working fluids
Heat exchangers
Tanks—expansion, etc.
Controls
Wiring
Labor
Value of any building floor space used by solar components
Extra collector support structure
Special distribution system features
Testing and checkout
Profit and overhead of installer

Solar system costs are of two types. If the costs are dependent on system size, they are called *variable*; examples are costs of collector, storage, heat exchangers, pumps, etc. Costs independent of system size are called *fixed*; examples are controls, some labor, building floor space used for solar components, etc. Numerous experimental and theoretical analyses of solar systems for buildings (described later) have shown that the sizes of most components are related to collector area by specific rules; these rules will be described in Chapters 6 and 7. Consequently, *most variable costs in a solar system can be related to collector size or area, which, therefore, becomes the obvious index of size of a solar system* over a limited range.

Since solar collector area is a measure of system size and therefore of each component's size, variable costs are usually expressed as cost per unit area of collector. For example, storage, while measured in physical units of kg or $m^3$, is usually priced as $/m_c^2$, the subscript c denoting collector. Of course, fixed costs do not depend on system size and are usually expressed in units such as dollars. Fixed costs can be considered as those required to purchase and install a solar system with an infinitesimally small collector area, that is, small energy delivery.

In most cases the initial costs listed above are paid off by the owner in monthly or annual installments. The size of the periodic payments is determined by applying the appropriate value of *CRF* to the initial sum. These periodic costs, combined with recurring future costs described below, comprise the total annual cost of solar energy.

## Solar System Future Costs

The components of life-cycle costs that recur throughout the useful life of a solar system are called future costs and include

Maintenance—personnel, materials
Repair—personnel, materials
Replacement—personnel, materials
Power costs for operating pumps, fans, and controls
Taxes—local, state, and federal
Insurance

These costs generally vary throughout the life-cycle of a system subject to inflationary pressures and political decisions. They must each be converted to their present value with appropriate values of the present-worth factor.

Maintenance and repair costs vary with the solar application. High-temperature systems, using tracking concentrating collectors, will require more maintenance and repair than low-temperature systems for building heating. Likewise, power costs will vary. For example, a heating system for a building using air as the working fluid will require more motive power than a system using water because of the relatively low heat capacitance of air vis-à-vis water and the consequent higher air flow rates.

Taxes and insurance are widely variable. Local or state decisions on solar system tax policy have resulted in situations where solar systems are tax-free in one location but are taxed at the prevailing rate in another. Assessments of solar systems can range from 0 to 50 percent of their market value. Likewise, insurance rates for solar systems vary widely depending on the insurer's guidelines, location of the solar building, and location of the collector on the building. Since no general rules apply, a determination of these costs is required for each solar project in each location.

## Annualized Solar Costs—Residential Applications

The total cost of solar energy on an annual basis $C_y$ is the total of amortized initial costs and distributed costs expressed on a present-worth basis. Since most residences are not income producing, the only federal and state tax credits against these costs are for local taxes and interest. Income-producing installations will have more extensive tax credits and are described in the next section. Annualized solar costs $C_y$ can be expressed in constant dollars as

$$C_y = C_{s,\text{tot}}\, CRF(i', t) \qquad\qquad\qquad \text{Initial investment} \qquad (5.50)$$

$$- C_{s,\text{salv}}\, PWF(i', t)\, CRF(i', t) \qquad\qquad \text{Salvage value}$$

$$+ \left[ \sum_{k=1}^{t} R_k\, PWF(i', t_k) \right] CRF(i', t) \qquad \text{Replacements}$$

$$+ C_e \frac{CRF(i', t)}{CRF(i'', t)} \qquad \text{Energy}$$

$$+ T_{\text{prop}} C_{s,\text{ass}} \qquad \text{Property tax}$$

$$- T_{\text{inc}} T_{\text{prop}} C_{s,\text{ass}} \qquad \text{Property tax, tax deduction}$$

$$- T_{\text{inc}} i_m \left[ \sum_k \frac{P_k}{(1 + i')^k} \right] CRF(i', t) \qquad \text{Interest, tax deduction}$$

$$+ M \qquad \text{Maintenance}$$

$$+ I \qquad \text{Insurance}$$

where $C_{s,\text{tot}}$ = the total initial solar investment (p. 364) including sales tax
$\quad C_{s,\text{salv}}$ = solar system salvage value at end of period of analysis
$\quad C_{s,\text{ass}}$ = assessed value of solar system
$\quad C_e$ = energy cost to operate solar system in year one
$\quad i' = i - j/1 + j$ = effective discount rate
$\quad i'' = i - j_e/1 + j_e$ = effective discount rate for energy
$\quad i$ = discount rate, that is, bounded below by opportunity cost to home-owner or the rate of return foregone on the next best alternative investment in market value; bounded above by the cost of borrowing. The rate $i$, strictly speaking, is the marginal discount rate, that is, the rate applicable to the solar unit purchaser's next set of investment decisions. It can be different from the average discount rate on previous investments
$\quad M$ = maintenance (\$/yr)
$\quad j$ = general inflation rate
$\quad i_m$ = market mortgage rate (real mortgage rate + general inflation rate)
$\quad j_e$ = energy inflation rate (\$/\$ · yr)
$\quad k$ = years at which replacements or repairs are made ($k$ can denote any time increments if $i, j$, etc. correspond)
$\quad I$ = insurance charges (\$/yr)
$\quad R_k$ = replacement costs in year $k$ (some $R_k = 0$) (\$/yr) in constant \$
$\quad P_k$ = outstanding principal (unpaid balance) of $C_{s,\text{tot}}$ in year $k$
$\quad t$ = life-cycle time or period of analysis (yr)
$\quad T_{\text{prop}}$ = property tax rate (\$/\$ assessed · year)
$\quad T_{\text{inc}}$ = state tax rate + federal tax rate − state tax rate × federal tax rate, where the rates are based on the last dollar earned

The summation in the seventh term of Eq. (5.50) can be evaluated by noting that the remaining principal $P_k$ during the year $k$ is

$$P_k = C_{s,\text{tot}} \left[ (1 + i_m)^{k-1} + \frac{(1 + i_m)^{k-1} - 1}{(1 + i_m)^{-t} - 1} \right] \qquad (5.51)$$

The summation is then given by

$$i_m \sum_k \frac{P_k}{(1+i')^k} = C_{s,\text{tot}} \left\{ \frac{CRF(i_m, t)}{CRF(i', t)} + \frac{1}{1+i_m} \frac{1}{CRF(i' - i_m, t)} \right.$$
$$\left. [i_m - CRF(i_m, t)] \right\} \quad (5.52)$$

The terms in Eq. (5.50) represent two types of real payments—those that remain constant in time and those that vary. For example, maintenance, insurance, property tax, and property tax credits are usually the same, in constant dollars, for the period of analysis $t$. If a real escalation is expected, these terms can readily be treated as the energy (fourth) term of Eq. (5.50). The initial investment (first) term of Eq. (5.50) represents the amortized cost of the solar system on an annual basis. Tending to offset this cost is the salvage value (second) term, which acts as a credit when reduced to its present worth by use of the *PWF* value.

Repair and replacement $R_k$ are considered to be a series of charges, small and infrequent for a well-designed system, which are reduced to their respective present values by the use of the *PWF* for year $k$. Energy requirements (usually only electricity) for solar system operation may escalate in price at a rate $j_e$ differing from the general inflation rate $j$. Therefore, the ratio of capital-recovery factors is required to determine the present value of operating energy.

Interest tax credits diminish during a life-cycle as the principal of the solar loan is reduced year by year. This is reflected by the expression in Eq. (5.51), which, when multiplied by the market mortgage rate $i_m$, is the interest payment for year $k$. A fraction of the interest cost is deductible from state and federal returns depending on the tax bracket of the building owner. Property taxes based on the extra assessment of property value resulting from the solar system $C_{s,\text{ass}}$ are, likewise, partially deductible. During a period of general inflation, a price index must be applied to annual tax deductions to reduce them to constant dollars. Tending to offset this inflation effect, however, is the increase in tax rate $T_{\text{inc}}$ (that is, higher tax bracket) because of income increases that reflect cost-of-living increases frequently applied to many homeowners' wages.

Local and federal governments can provide tax incentives for the adoption of solar systems by eliminating property taxes, initiating special tax credits, subsidizing solar equipment manufacturers, offering low-interest loans, or causing grants to be made for the purchase and installation of solar systems. The efficacy of these measures can be evaluated by the use of Eq. (5.50). Although an annual time scale has been used in the analysis, any time scale (months, days, continuous cash flow) compatible with the time scale of the interest and discount rates can be used.

## EXAMPLE 5.7

Determine the annualized cost of a solar system with the following specifications over a 15-yr period ($t = 15$).

| | |
|---|---|
| Solar system cost | $8,000 |
| Salvage value | $0 |
| Assessed value | $4,000 |
| Effective discount rate | $i' = 0.02$ |
| Effective discount rate for energy | $i'' = -0.02$ (4 percent differential energy inflation) |
| Market mortgage rate | $i_m = 0.10$ |
| Repairs: one of $200 at year 10 | $R_{10} = \$200; R_{k \neq 10} = \$0$ |
| Maintenance and insurance | neglect |
| Property tax rate (local law exempts solar systems) | $T_{prop} = 0$ percent |
| Income tax bracket | $T_{inc} = 25$ percent |
| Electric power | $C_e = \$20/yr$ |

## SOLUTION

For clarity each term in Eq. (5.50) is evaluated in the table below. The required capital-recovery factors and present-worth factors are calculated first.

$$CRF(i', t) = CRF(0.02, 15) = 0.0778$$

$$CRF(i'', t) = CRF(-0.02, 15) = 0.0565$$

$$PWF(i', t_k) = PWF(0.02, 10) = 0.820$$

| Term type | Analytical expression | Value ($) |
|---|---|---|
| Annualized extra cost | $C_{s,tot}\, CRF(i', t)$ | 622.40 |
| Salvage value | $C_{s,salv}\, (= 0)$ | 0 |
| Repair and replacement | $R_{10}\, PWF(i', 10)\, CRF(i', t)$ | 12.76 |
| Electric energy | $C_e\, \dfrac{CRF(i', t)}{CRF(i'', t)}$ | 27.53 |
| Property tax | (Tax exempt) | 0 |
| Property tax credit | (Tax exempt) | 0 |
| Interest credit [use Eq. (5.52)] | $T_{inc}\, i_m\, CRF(i', t) \sum \dfrac{P_k}{(1 + i')^k}$ | $-124.16$ |
| Maintenance and insurance | $I = M = 0$ | 0 |
| | | 538.53/yr |

The annualized solar system cost $C_y$ is $538.53. Note that the interest tax credit is quite significant in reducing the annualized cost.

In many cases a down payment $D_s$ is made on a solar system to reduce the mortgage amount. This payment reflects the increase in down payment for a solar-equipped building required by a banking institution above that required for a nonsolar building. The annual solar cost equation can be modified to include the down payment $D_s$ by replacing the first term with two terms, that is,

$$(C_{s,\text{tot}} - D_s)\, CRF(i', t) + \frac{D_s i'^*}{(1 + i')^t - 1} \rightarrow C_{s,\text{tot}}\, CRF(i', t) \qquad (5.53)$$

Likewise, the tax deduction (seventh term) in Eq. (5.50) replaces $C_{s,\text{tot}}$ with $(C_{s,\text{tot}} - D_s)$. An investment tax credit is treated as a negative cost occurring in year one.

## EXAMPLE 5.8

If the homeowner in the preceding example makes a $1,000 down payment, how much is the annualized solar cost reduced because of interest savings on the mortgage?

## SOLUTION

The difference between the two terms in Eq. (5.53) represents the annual cost difference. From the preceding example

$$C_{s,\text{tot}}\, CRF(i', t) = \$622.40$$

If a $1,000 down payment is made, this sum is replaced by

$$(C_{s,\text{tot}} - D_s)\, CRF(i', t) + \frac{D_s i'}{(1 + i')^t - 1} = \$544.60 + \$57.83 = \$602.43$$

The difference in the two values is $19.97. In addition, the annualized tax credit is reduced to $108.64. The net annual payment is then $534.08.

Equation (5.50) is based on the effective discount rate $i'$, net of inflation. An equivalent method of analysis could be based on simple discount rates by inflating costs at the general inflation rate $j$. The results of the two analytical modes are identical.

The complete annual cost equation [Eq. (5.50)] does not consider certain indirect costs and benefits. For example, the use of solar energy to displace fossil fuels or electricity for heating would be expected to have the effect of reducing air pollution, conserving scarce fossil fuels for use as petrochemical feedstocks, and increasing industrial energy usage in the sectors providing materials for solar collectors. It is not anticipated that those factors would be considered in most microeconomic analyses. However, on the national macroeconomic scale, such factors are quite important.

---

*It is assumed here that the time value of the down payment amount is $i'$, the same as the real discount rate.

## Annualized Solar Costs—Commercial Applications

The calculation of annualized solar costs for an income-producing application is similar in most respects to that for noncommercial usages. However, in addition to the deductions described above, additional deductions apply for solar buildings. These include depreciation, operating costs, maintenance costs, and insurance. There will be an offsetting effect in these deductions because of the reduced deduction for fuels, since solar-plus-fossil fuel systems consume less fuel than nonsolar systems. The method of calculating all additional deductions except depreciation for commercial systems is identical to that used in Eq. (5.50) for property tax deductions and need not be described in detail.

There are three widely used methods for depreciating investments—straight line, sum-of-years digits, and declining balance. Each method results in a different depreciation schedule, as shown in Fig. 5.15. The basic annualized solar cost equation [Eq. (5.50)] can be modified to account for depreciation by subtracting the term

$$\text{Annualized depreciation tax deduction} = T_{\text{inc}} CRF(i', t) \sum_{k} \frac{D_k}{(1 + i')^k} \qquad (5.54)$$

where $D_k$ is the annual depreciation amount in current dollars.

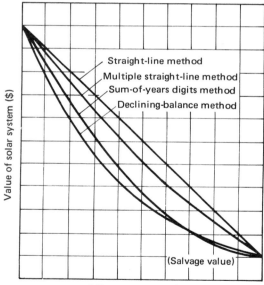

FIGURE 5.15 Comparison of straight-line, multiple straight-line, sum-of-years digits, and declining-balance methods for calculating solar system depreciation for income-producing property. From "Plant Design and Economics for Chemical Engineers," by M. S. Peters and R. D. Timmerhaus. Copyright 1968 by McGraw-Hill Book Co. Used with permission of McGraw-Hill Book Co.

The annual depreciation amount $D_k$ is given below for the three common write-off procedures.

Straight line (uniform):

$$D_k = \frac{C_{s,\text{tot}} - C_{s,\text{salv}}}{t} \qquad\qquad (5.55)$$

Sum-of-years digits (based on remaining service life):

$$D_k = \frac{2(t - k + 1)}{t(t + 1)} \; (C_{s,\text{tot}} - C_{s,\text{salv}}) \qquad\qquad (5.56)$$

Declining balance (based on fixed percent of value in each year):

$$D_k = C_{s,\text{tot}} \left(\frac{C_{s,\text{salv}}}{C_{s,\text{tot}}}\right)^{k/t} \left[1 - \left(\frac{C_{s,\text{salv}}}{C_{s,\text{tot}}}\right)^{1/t}\right] \qquad\qquad (5.57)$$

Of the three methods, the straight-line approach is easiest to use and is recommended for preliminary studies. The declining-balance method is particularly sensitive to the salvage value, a parameter difficult to evaluate a priori. In addition, this method cannot be used for an assumed zero salvage value. The sum-of-years digits method has the advantage of accelerated early depreciation and simplicity of use. As salvage value estimates improve in accuracy through the life of a system, a multiple straight-line method can be used to refine the depreciation schedule.

A special tax situation exists for commercial rental property. Rents on solar buildings are expected to be higher than on nonsolar buildings, since the owner must recover solar system capital costs. This is not a disincentive for renting solar buildings, since conventional fuel savings would normally exceed the extra rental costs. However, the owner of a solar building realizes a larger income on the property because of the increased rentals. This increased benefit after taxes acts as an incentive to use solar energy; it is a negative cost to be subtracted from the annualized solar cost, Eq. (5.50).

Sales taxes on solar equipment can be partially recovered by an income tax deduction. To calculate the deduction the credit on the one-time tax payment during the purchase in the first year can be annualized as a negative cost in the same way as the one-time down payment is annualized in Eq. (5.53).

## Continuous Cash Flows

In some economic analyses, cash flows occur with such regularity and frequency that they may be assumed to occur continuously. Although this situation is uncommon in small-scale solar applications, it is common in industrial practice. The capital-recovery factors and present-worth factors for continuous cash flows differ from those for discrete cash flows and they are given in Eqs. (5.58) and (5.59).

$$CRF(i', t) = \frac{i'}{1 - e^{-i't}} \tag{5.58}$$

$$PWF(i', t) = e^{-i't} \tag{5.59}$$

where $i' \equiv i - j$

The annualized solar cost equation can be used directly by substituting the above factors for their discrete-cash-flow counterparts in Eq. (5.50).

## COST–BENEFIT ANALYSIS AND OPTIMIZATION

The optimization of a solar energy system requires that all components be sized to provide the *least-cost mix of solar energy and conventional fuel* for the application and site in question. In the preceding section the methods of calculating the annual solar cost for a given system have been detailed. In this section the concept of a production function is used to evaluate the energy delivery or *benefit* resulting from a solar system investment. The values of costs and benefits are those used to determine the optimal system configuration that maximizes benefits for a given level of solar-plus-nonsolar energy demand—the constraint on the optimization process.

### Production Functions

A production function is the technical relationship specifying the maximum amount of output capable of being produced by each and every set of specified inputs (34). It is defined for a given state of technical knowledge. For example, suppose a solar energy system consisting of a 100-m$^2$ collector, 2 metric tons of aqueous storage and associated pumps, heat exchangers, and controls can deliver 200,000 MJ/yr of thermal energy. The *technical production function* relates the physical solar components' sizes and characteristics to the output of 200,000 MJ. The *economic production function* relates the costs of the physical inputs to the value of the energy output. The value of energy output is equal to the value of nonsolar energy displaced by solar energy. In this book the technical production function is used rather than the economic production function. There exists a duality between the two based on energy prices and solar equipment prices so either could be used in practice. In most economic analyses the production function is taken to be a long-term average applicable for the entire period of analysis.

The calculation of a production function is not an exact science; rather, an empirical method is used to relate energy output to endogenous solar system parameters such as collector size, efficiency and orientation, storage size, control strategy, and many other factors. The exogenous variables of climate and energy demand type and profile are also system output determinants. The empirical methods used differ with each solar application—heating, cooling, steam production, direct conversion—and are discussed in turn in Chapters 6, 7, and 8. Although a thermal demand may not be thought of as a production function input, it behaves as one, since a given solar system will behave differently when coupled to different demands (Chapter 6). In this chapter several types of production functions are described and their use in system optimization is derived in detail.

Nearly all inputs to a solar system production function are subject to the law of diminishing returns. This basic technological law states that extra output from a system diminishes relatively when successive equal units of an input are added to fixed amounts of other inputs (34). For example, doubling storage capacity in a solar cooling system will not double the solar cooling effect. Figure 5.16 shows qualitatively this law applied to a solar-heating system as an example. It is seen that adding extra storage, collector area, heat-exchanger area, or pump capacity gains relatively little beyond a certain point. This fundamental law necessitates the optimization method of sizing solar systems. If equal or greater returns-to-scale were possible, 100 percent solar systems could be used with no auxiliary.

## Polynomial Production Functions

A particularly simple production function is a multiple-variable polynomial with coefficients based on regression analysis of system performance—either data from computer models or actual equipment. It is expressed in the form

$$Q_s = \sum_{i=1}^{N} P_i(X_i) \tag{5.60}$$

where $P_i$ is a polynomial in the parameter $X_i$. The $X_i$'s include system component sizes, solar-radiation levels, weather parameters, energy-demand parameters, and the

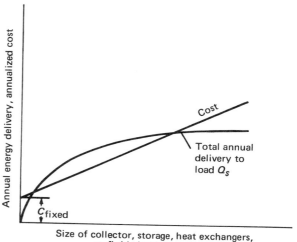

**FIGURE 5.16** Examples of the law of diminishing returns for a solar-heating system. $C_{fixed}$ is the cost component independent of the size of the solar system component.

like; $Q_s$ is the annual solar energy *delivery-to-load*. Instead of a summation of polynomials, a product could be used

$$Q_s = \prod_{i=1}^{N} P_i(X_i) \tag{5.61}$$

Polynomial production functions can be selected to have any number of terms required to ensure the best accuracy as adjudged by an *F*-test. However, the basic information contained in such a production function is limited to calculating $Q_s$. The coefficients and exponents do not provide any additional information as is the case in other production functions below. The principal advantage of the polynomial is its simplicity.

## Logarithmic and Exponential Production Functions

A solar energy production function can be formed from logarithms (or their inverse, exponents) as follows:

$$Q_s = \ln \prod_{i=1}^{N} a_i X_i^{n_i} \tag{5.62}$$

where $n_i$ are powers (integral or nonintegral) and $a_i$ are coefficients. This form is equivalent to

$$Q_s = (a_1' + n_1 \ln X_1) + (a_2' + n_2 \ln X_2) + \ldots$$
$$= A + n_1 \ln X_1 + n_2 \ln X_2 + \ldots \tag{5.63}$$

The coefficients $n_i$ can be evaluated by regression methods using data relating $Q_s$ to $X_i$.

Although the logarithmic form has a limited number of coefficients and possibly lower regression accuracy than the polynomial form, the coefficients $n_i$ contain additional information. If Eq. (5.63) is differentiated, we have

$$dQ_s = n_1 \frac{dX_1}{X_1} + n_2 \frac{dX_2}{X_2} + \ldots \tag{5.64}$$

The coefficients are seen to be the change in $Q_s$ resulting from a unit change in the independent variables $X_i$. The coefficients, therefore, indicate where a change in physical inputs can have the largest impact.

### EXAMPLE 5.9

A production function for a particular solar space-heating system in Chicago has been determined to be of the form

$$Q_s = L \left\{ 0.8 + \ln \left[ \left( \frac{A_c}{L} \right)^{1/3} \left( \frac{S}{L} \right)^{1/20} \right] \right\} \tag{5.65}$$

where $A_c$ is the collector area, $L$ is the annual energy demand of 100 GJ, and $S$ is the storage size. Determine the expected increase in delivery $Q_s$ for a 1 percent increase in both storage size and collector size if the load remains unchanged.

**SOLUTION**

Expand Eq. (5.65) to get

$$Q_s = L\left(0.8 + \frac{1}{3}\ln A_c + \frac{1}{20}\ln S - \frac{23}{60}\ln L\right)$$

For small changes in $A_c$ or $S$, the differential approximation can be used. Differentiating we have

$$dQ_s = \frac{L}{3}\frac{dA_c}{A_c} + \frac{L}{20}\frac{dS}{S}$$

Therefore, for a 1 percent increase in collector area $A_c$ $(dA_c/A_c = 0.01)$,

$$dQ_s = \frac{100,000 \text{ MJ}}{3} \times 0.01 = 333 \text{ MJ}$$

For a 1 percent increase in storage $S$

$$dQ_s = \frac{100,000 \text{ MJ}}{20} \times 0.01 = 50 \text{ MJ}$$

In this example a unit increase in collector area $A_c$ delivers $6\frac{2}{3}$ times more energy than the same percentage increase in storage delivers.

## Power-Law Production Functions

A production function can be formed from the product of powers of physical inputs:

$$Q_s = \nu_o \prod_{1=i}^{N} X_i^{\nu_i} \tag{5.66}$$

As in previous cases the $\nu_i$'s are determined by a regression analysis of performance data. The exponents in the power-law production function are similar to those in the logarithmic expression and represent the percentage change in output for a percentage change in input. It is easy to show that

$$\frac{dQ_s}{Q_s} = \sum_{i=1}^{} \nu_i \frac{dX_i}{X_i} \tag{5.67}$$

In addition, *the sum of the $v_i$'s represents the return-to-scale* of a solar system. Returns-to-scale are the change in output $Q_s$ resulting from a uniform unit change in all inputs $X_i$. If $\Sigma_{i=1}^{N} v_i$ is greater than one, returns-to-scale are increasing; if less than one, decreasing; and if equal to one, proportionate.

**Numerical production functions** Instead of evaluating a closed-form empirical production function from data by regression methods, the data itself can be used as a tabular production function. Such a function does not lend itself to ready manipulation by hand, but can be used by a digital computer to perform optimization calculations. No information regarding returns-to-scale can be determined directly from a numerical production function, however.

## Economic Optimization Methodology

### Marginal Cost Analysis

*The optimal size of a solar system is that which minimizes the annual cost of solar-plus-auxiliary energy.* The total annual cost $C_T$ can be expressed as

$$C_T = C_y(Q_s) + C_a(Q_a) \tag{5.68}$$

where $C_a(Q_a)$ is the annual cost of nonsolar energy and $C_y(Q_s)$ is the annualized solar cost—Eq. (5.50). Both costs are assumed to depend on the quantity of energy delivered. This is always true for solar systems and for nonsolar sources. Three typical total cost curves are shown in Fig. 5.17. The three generic types of cost curves can result from the following sets of conditions:

1. High nonsolar energy costs and/or high sun-high load factor location (curve 1); largest system possible is optimal.
2. Average or typical mix of energy cost, sun amount, and demand (curve 2).
3. Low nonsolar energy costs and/or low sun-low load factor location (curve 3); the no-solar system is optimal.

A discussion of nonsolar energy prices is presented at the end of this section.

The value of the nonsolar energy cost $C_a(Q_a)$ in Eq. (5.68) represents the *annualized life-cycle fuel cost*. As such it is not the cost for any single year during the life cycle. It is rather the total life-cycle cost in present dollars annualized by the capital-recovery factor $CRF\,(i',t)$. It is given analytically by an expression of the form of the fourth (nonsolar) term in the annualized solar cost equation [Eq. (5.50)]. The quantity $C_e$ in Eq. (5.50) can be thought of, in the present analysis, as representing the backup fuel cost in current dollars for the first year of the period of analysis. In most economic analyses it is assumed that the amount of fuel displaced by solar energy each year is the same; only its value increases year by year because of the differential inflation of fuel price.

The optimization of total cost is subject to the constraint

$$Q_T = Q_s + Q_a \tag{5.69}$$

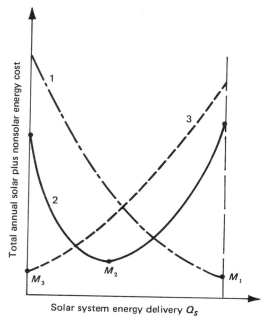

**FIGURE 5.17**  Solar-plus-nonsolar total annual cost curves; minima shown by $M$. Curves 1 and 3 represent boundary minima; boundaries represent maximum and minimum solar delivery for a specific application ($M_3$ normally represents zero collector area or the null system).

where $Q_T$ is the total annual energy demand to be met, $Q_s$ is the solar portion of load and $Q_a$ is the nonsolar portion of load. Equation (5.68) can be unconstrained by use of the Lagrange multipler $\lambda$:

$$C_T = C_y(Q_s) + C_a(Q_a) + \lambda(Q_T - Q_s - Q_a) \tag{5.70}$$

It is desired to find a stationary point of Eq. (5.70), that is, a point where $dC_T = 0$:

$$dC_T = \frac{dC_y}{dQ_s}\,dQ_s + \frac{dC_a}{dQ_a}\,dQ_a + \lambda(-dQ_a - dQ_s) = 0 \tag{5.71}$$

Since the differentials $dQ_s$ and $dQ_a$ are unconstrained, they may be independently set to 0 to get

$$\frac{dC_y}{dQ_s} = \frac{dC_a}{dQ_a} = \lambda \tag{5.72}$$

Equation (5.72) states that the minimum annual cost occurs at the point where *both marginal energy costs are equal.* This condition means that investment in any solar system should continue up to the point where the last dollar invested generates precisely $1 in nonsolar energy savings. Figure 5.18 shows the above analysis graphically. At the point where total cost curves for solar and nonsolar energy have equal slope, marginal costs are equal as required for the optimal system.[*]

At any point between A and B in Fig. 5.18, the solar source represents a less expensive alternative than the nonsolar source. That is, at any point between A and B the solar system owner saves both money and nonsolar energy vis-à-vis the full nonsolar option. Point M is the least-cost optimum. If the goal of economic policy

[*]If there is a budget upper limit, the marginal cost criterion may not be achievable. In that case the full budget amount should be spent.

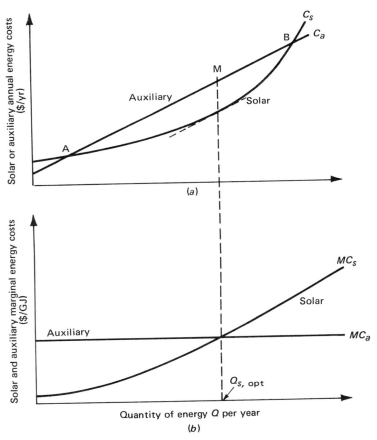

**FIGURE 5.18** Optimum solar size determination: (*a*) total annualized costs of solar and nonsolar energy; (*b*) marginal-cost curves for solar and nonsolar energy showing optimal system size at intersection of marginal-cost curves.

were to minimize nonsolar energy use while paying the same total annual cost for energy, the design point will be B in Fig. 5.18. In this book the least-cost point M, representing rational consumer behavior in most Western economies, is adjudged the optimum. If the solar curve does not intercept or is not tangent to the nonsolar curve in Fig. 5.18a, there is no microeconomic incentive to consider solar energy.

## EXAMPLE 5.10

A single-family dwelling in St. Joseph, Missouri, has an annual energy demand of 70 GJ for heating. Computer model simulations of a solar system for this building have indicated that the annualized cost function for the solar system is

$$C_y = 400 - 20Q_s + 0.4\,Q_s{}^2 \qquad (20 < Q_s < 70) \qquad (5.73)$$

in which $Q_s$ is measured in gigajoules and $C_y$ in dollars. Fuel oil is the nonsolar energy source for this building and its cost function is

$$C_a = 5Q_a \qquad (5.74)$$

where $Q_a$ is measured in gigajoules. Determine the optimum mix of solar energy and fuel oil.

## SOLUTION

The optimal solar fraction of load is determined by equating marginal costs of solar energy $MC_s$ and fuel oil $MC_a$. From Eqs. (5.73) and (5.74),

$$MC_s = \frac{dC_y}{dQ_s} = -20 + 0.8Q_s \qquad (5.75)$$

$$MC_a = \frac{dC_a}{dQ_a} = 5$$

Equating and solving for the optimal $Q_s$ we have

$$-20 + 0.8Q_s = 5$$

$$Q_s = 31 \text{ GJ}$$

The fraction of load carried by solar energy $f_s$ is

$$f_s = \frac{31}{70} = 0.44 \text{ or } 44 \text{ percent}$$

## Simplified Analysis

In many solar analyses it is possible to simplify the above method. As mentioned in a preceding section, collector area is a reliable index by which to determine the size and cost of all ancillary components in most solar building systems. The cost function Eq. (5.68) can then be determined directly by inverting the technical production function as shown below. If the production function is given by

$$Q_s = f(X_1, X_2, \ldots) \tag{5.76}$$

then it can be simplified to

$$Q_s = f(A_c) \tag{5.77}$$

using rules referred to earlier. The cost of a solar system, likewise, can be keyed to collector area by

$$C_y = c_y A_c + C_f \tag{5.78}$$

where $c_y$ is the solar cost per unit area and $C_f$ is the fixed cost. The nonsolar portion of the cost function can also be keyed to collector area. If

$$C_a = p(Q_a) \tag{5.79}$$

then

$$C_a = p(Q_T - Q_s) = q(Q_s) \tag{5.80}$$

where $Q_T$ is the total annual demand. By use of Eq. (5.77) we have

$$C_a = q[f(A_c)] = h(A_c) \tag{5.81}$$

Combining $C_y$ and $C_a$ to form total cost $C_T$

$$C_T = g(A_c) + h(A_c) \tag{5.82}$$

or

$$C_T = F(A_c) \tag{5.83}$$

Determination of the optimal system size can be done directly by differentiating Eq. (5.83) with respect to $A_c$ to get the optimization criterion

$$F'(A_{c,\text{opt}}) = 0 \tag{5.84}$$

subject to

$$F''(A_{c,\text{opt}}) > 0 \tag{5.85}$$

and subject to

$$C_T(A_{c,\text{opt}}) < C_T(A_{c,\text{max}}) \tag{5.86}$$

$$C_T(A_{c,\text{opt}}) < C_T(A_{c,\text{min}}) \tag{5.87}$$

where the $A_{c,\text{min}}$ and $A_{c,\text{max}}$ are the extreme values of the collector area domain of interest. It is necessary to check end points, since in some cases the cost curve may have a boundary minimum as shown in Fig. 5.17. The following example shows how a technical production function can be inverted and used directly to find the optimum collector area.

### EXAMPLE 5.11

The simplified solar energy production function for a building in Denver, Colorado, is given by

$$Q_s = 35\sqrt{A_c} \tag{5.88}$$

where $A_c$ is the collector area in square meters $(350 > A_c > 50$ m$^2)$. If the annualized cost of the solar system $c_y$, per square meter, is \$10, find the optimum collector size. The total annual energy demand is 650 GJ and the cost of the nonsolar energy source $c_a$ is \$8/GJ.

### SOLUTION

The solar cost function $C_y$ is

$$C_y = c_y A_c = 10 A_c$$

Note that the annualized cost $c_y = 10$ is based on an initial cost of \$150/m$^2$ and an annual multiplier [Eq. (5.50)] of 0.0667.

The nonsolar cost function is calculated from Eqs. (5.80) and (5.81):

$$C_a = c_a (Q_T - Q_s)$$
$$= 8(650 - 35\sqrt{A_c})$$

The total cost function $C_T$ is

$$C_T = C_y + C_a$$
$$= 10 A_c - 280\sqrt{A_c} + 5200 \tag{5.89}$$

Equation (5.89) must be differentiated and equated to 0 to find the optimal area:

$$\frac{dC_T}{dA_c} = 10 - \frac{280}{2} \frac{1}{\sqrt{A_{c,\text{opt}}}} = 0$$

Solving for $A_{c,\text{opt}}$,

$$A_{c,\text{opt}} = 196 m^2$$

Equation (5.88) permits calculation of the solar energy produced:

$$Q_s = 35\sqrt{196} = 490 \text{ GJ}$$

This represents the following portion of load $f_s$

$$f_s = \frac{490}{650} = 0.75 \text{ or } 75 \text{ percent}$$

The total cost curve for the example is shown in Fig. 5.19 with the least-cost point illustrated.

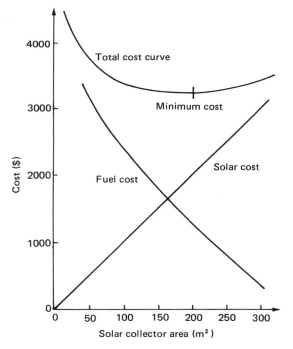

**FIGURE 5.19**  Cost curves for solar system used in Example 5.11. Total cost curve is the sum of the solar and fuel curves.

It is important to recognize that the production function and cost functions used in the two preceding examples are for a specific system in a specific building in a specific location. *It is not possible to use a production function derived for one system in the analysis of another system.* Since no a priori or theoretical method exists to calculate the technical production functions, they can only be derived from simulation studies of system thermal performance as described in Chapters 6 and 7.

## Other Economic Indices

Although the proper method for choosing the optimal solar configuration is the method outlined above, other indices are sometimes used. The payback-period and return-on-investment approaches are commonly used, since they are easy to calculate and have an intuitive appeal because of their simplicity.

However, these indices are incomplete and therefore incorrect for two reasons. First, they do not consider the time value of money. Secondly, they are not a quantitative measure of cash flow, that is, they do not have a dollar magnitude associated with them for either costs or benefits. In addition, the payback method ignores any cash flows beyond the payback period, thereby giving only a partial view of the entire time series of cash flows.

Since the investment in a solar system does require a fixed number of years for repayment, it is a long-term investment. It may not be as attractive as a shorter, more liquid investment at the same rate of return. Therefore, a somewhat larger return, that is, larger discount rate, may be needed to induce investment in a solar system. This effect is usually small because of compensating considerations such as the effect of inflation on the investment market.

If a solar system owner sells a facility before the term of the mortgage, it could be argued that a loss is experienced, because the lifetime of the system has not been realized. This is an incomplete view, since a solarized building would command a higher price in the marketplace, making it easier for the owner to recover the unamortized portion of the solar investment.

A year-by-year calculation of costs and benefits can be made to determine the profitability of a solar system for periods of time shorter than the mortgage term. The annual payment for the solar system represents a constant number of current dollars (the tax credit decreases during the life cycle, however) whereas the payment for fuel would represent a constantly increasing dollar amount in a period of inflation. Since the dollar value of fuel savings will be largest near the end of the life cycle, it may be expected that during the initial years the solar system is not cost-effective. Tending to offset this trend is the tax deduction for the solar system, which is largest in the earlier years because of large mortgage interest payments. The following example shows annual cash flows and a calculation of the cash benefit of the solar system.

## EXAMPLE 5.12

Calculate the annual cash flows for a solar-heating system on a residence providing an average of 100 GJ/yr. The initial cost of the system is $6,500, and the owner pays no property taxes and is in the 25 percent federal tax bracket. The mortgage interest rate $i_m$ is 10 percent for 15 yr; for simplicity, ignore maintenance and insurance costs. Fuel costs are \$5/GJ and are expected to increase by 12 percent/yr. Perform the calculation in current dollars not constant dollars. Compare the cost year by year of both solar and nonsolar systems capable of delivering 100 GJ/yr.

**SOLUTION**

Cash flows will be tabulated below. The annual mortgage payment is

$$CRF(10 \text{ percent, } 15 \text{ yr}) \times 6500 = \$854.58$$

The charge for fuel to supply the full load in the first year is

$$100 \text{ GJ} \times \$5/\text{GJ} = \$500/\text{yr}$$

This sum inflates at a rate of 12 percent/yr. Table 5.17 summarizes cash flows for the first 10 yr of the mortgage.

The table shows that the homeowner has no positive benefit (savings for nonsolar versus solar) until the fifth year, when fuel inflation overtakes the fixed annual solar system cost. Within another 1-2 yr the total early diseconomy of the solar system has been more than eliminated because of fuel inflation. A calculation such as this can be made to show the homeowner's cash flows in current dollars. Unless extensive simplifying assumptions are made, it is not possible to carry out the calculation above in a general analytical form to determine the break-even point or the payout period.

**TABLE 5.17** Annual Cash Flows for Example 5.12[a]

| Annual solar cost $CRF \times C_{s,\text{tot}}$ (1) | Remaining principal $P_k$ (2) | Interest $i_m P_k$ (3) | Principal payment [(1) − (3)] (4) | Tax credit $i_m T_{\text{inc}} P_k$ (5) | Solar cost [(1) − (5)] (6) | Fuel cost for no solar usage[b] |
|---|---|---|---|---|---|---|
| 854.58 | 6500.00 | 650.00 | 204.58 | 162.50 | 692.08 | 500.00 |
| 854.58 | 6295.42 | 629.54 | 225.04 | 157.39 | 697.19 | 560.00 |
| 854.58 | 6070.38 | 607.04 | 247.54 | 151.76 | 702.82 | 627.20 |
| 854.58 | 5822.84 | 582.28 | 272.30 | 145.57 | 709.01 | 702.46 |
| 854.58 | 5550.54 | 555.05 | 239.53 | 138.76 | 715.82 | 786.76 |
| 854.58 | 5251.01 | 525.10 | 329.48 | 131.28 | 723.30 | 881.17 |
| 854.58 | 4921.53 | 492.15 | 362.43 | 123.04 | 731.54 | 989.91 |
| 854.58 | 4559.10 | 455.91 | 398.67 | 113.98 | 740.60 | 1105.34 |
| 854.58 | 4160.43 | 416.04 | 438.54 | 104.01 | 750.57 | 1237.98 |
| 854.58 | 3721.89 | 372.19 | 482.39 | 93.05 | 761.53 | 1386.54 |

[a]In current, not constant dollars.
[b]This column represents the cost of providing the solar energy portion of load (100 GJ/yr) from a nonsolar source. It does not represent the total cost of heating the subject building.

It should be noted that savings in fuel costs by use of a solar system are a kind of nontaxable return on an investment (the solar system). As such they are equivalent to about twice their dollar value in earnings from a conventional investment subject to capital gains taxes (maximum of 50 percent in the United States).

In summary, economic indices such as payout period, return-on-investment, or break-even point are incomplete and incorrect indices to use for selection of the optimal solar system. They are, at best, intuitively appealing methods because of their simplicity.

### Nonsolar Energy Costs—Utility Pricing Strategies

The cost of the auxiliary energy used to provide that fraction of an energy demand for which it is uneconomical to use solar energy is equal in importance to the cost of the solar system itself. Utilities that provide nonsolar energies can have a significant effect on the future viability of solar energy.

Two pricing strategies have been widely used in the United States. The first, the declining-block rate, results in charges that, on a unit-energy basis, decline with increasing energy usage rate. Part of the rationale for use of this rate by electric utilities is the higher generator load factor and increased return to the utility for a fixed capital investment.

The second strategy—used mainly by electric utilities—is called the peak-demand rate. It is frequently a declining-block rate with an additional charge related to the peak monthly demand. Often, a fraction of the peak annual demand rate (50–75 percent) is charged as the minimum utility bill for a given month even if no utility-based energy is used. This fraction is called the *billing demand* and is distinct from the *actual demand*. Part of the rationale for the demand rate is that those who use peaking power provided by extra peaking capacity should pay for the capital investment in that extra capacity.

Electric utilities have large investments in capital and therefore use the demand pricing strategy in parts of the United States to recover that investment. For solar systems in small and large buildings using electricity as the backup, the type of rate structure in force can make the difference between investment in a solar system or investment in an energy conservation system (to reduce peak demand) costing the same amount. The following example illustrates the trade-offs.

**EXAMPLE 5.13**

The owner of a building with a peak heat demand in January of 15.8 kW (54,000 Btu/hr) and a total demand of 5600 kW · hr ($19 \times 10^6$ Btu) wishes to reduce the utility bill. Compare the economies of a solar system that can provide half the January heating energy demand for $50/month (including amortization of the solar system loan, taxes, and maintenance) with an energy conservation system (insulation, inflation control, etc.) that can reduce the building heat loss by one-third. The extra mortgage payment for this conservation system is the same as that for the solar system—$50/month.

The utility uses a peak-demand pricing strategy wherein each hour of peak demand costs $2.70/month and energy costs $0.008/kW · hr.

**SOLUTION**

The *solar system* cost for January will be

$$\text{Demand charge} = 15.8 \times \$2.70 = \$42.66$$

$$\text{Electric energy charge} = \frac{5600}{2} \times \$.008 = \$22.40$$

$$\text{Solar energy mortgage charge} = \$50.00$$

$$\text{Total charge with the solar system} = \$115.06$$

The *energy conservation system* cost for January will be

$$\text{Demand charge} = \left(\frac{2}{3} \times 15.8\right) \times \$2.70 = \$28.44$$

$$\text{Electric energy charge} = \left(\frac{2}{3} \times 5600\right) \times \$.008 = \$29.87$$

$$\text{Extra mortgage charge} = \$50.00$$

$$\text{Total charge with the energy conservation system} = \$108.30$$

For this example, in January, the energy conservation system costs less than the solar system even though more electricity is purchased from the utility with the conservation scheme than with the solar scheme. A complete economic analysis requires that the above calculations be made and totals determined for an entire year. In addition, peak-shaving strategies should be analyzed.

As an exercise, compare the economies of the energy conservation and solar systems described above for a scenario in which the auxiliary electricity is priced on a declining-block basis instead of a peak-demand basis (see Problem 5.20).

The example illustrates that solar systems reduce consumption, not demand, in general. It is the relative cost of reduced consumption versus reduced demands that dictates the investment size and direction.

Other utility pricing strategies have been considered. For example, as an inducement to flatten diurnal peaks, a utility could price its "valley" energy at a lower cost than its peak energy. This could be accomplished by a time-of-day pricing strategy. Presumably this would lead to a redistribution of diurnal loads in the utility grid to flatten peaks and use generating capacity at a higher load factor. It must be recognized that a solar system simply displaces the lowest-cost energy available from a utility if diurnal pricing is used. If solar energy were not used, a simple storage system that could store offpeak power for use during peak periods would be the first investment made, not the investment in a solar system. Therefore, the cost-effectiveness of a solar system must be weighed against the off-peak energy rate, not the peak rate.

## Summary of Optimization Methodology

The optimization and sizing methodology described in this chapter can be summarized in a series of steps that must be followed to ensure the proper sizing of the cost-optimal solar system. These steps are

1. Establish the technical production function for a range of all important solar system component sizes. Determine the annualized cost for each configuration using discounted cash flow analysis. The energy delivery $Q_s$ of the production function is usually established by modeling techniques or empirical data (see Chapters 6, 7, and 8).
2. Establish the technical production function for the nonsolar auxiliary energy alternative.
3. Calculate marginal costs from the solar and nonsolar production functions. Equate to determine the optimal amount of solar energy to be supplied. The portion of load supplied by the nonsolar source is the difference between the total demand and the optimal solar fraction. Marginal costs can be calculated numerically or algebraically.

# PROBLEMS

5.1. Compare the energy storage capability of sodium sulfate decahydrate in a $30°C$ range from 30 to $60°C$ with that of water in the same range. What is the ratio of volumes of storage for the two media?

5.2. A $100\text{-}m^2$ liquid solar collector (flat-plate type) is located 40 m from the building it serves. If the liquid pressure drop through the collector is 7 kPa, what is the total pressure drop in the collector fluid loop? What pipe size would you recommend for the collector loop? Why?

5.3. A cylindrical storage tank for the collector in Problem 5.2 is located in the basement of the dwelling it serves. If the basement temperature is $16°C$ and the storage tank liquid is $70°C$ what is the heat-loss rate from the tank? The tank is a cylinder with diameter equal to its height; it is insulated with 15-cm-thick fiberglass batt.

5.4. How large should a water storage tank be if it is to supply the total daily heat load (1.5 GJ) of a building if storage tank temperature may vary by $20°C$. Neglect parasitic transmission and conversion losses. If a $4\text{-}m^3$ carbon steel tank costs $1,000, what will this tank cost?

5.5. Derive Eq. (5.14) for rectangular and cylindrical geometries.

5.6. What is the pressure drop in 50 m of 1-in pipe for fluid flowing at 0.4 m$^3$/min?

5.7. If the fluid velocity in a pipe is to be kept below 4 ft/sec, what pipe size should be used to flow 500 gal/min? What is the pressure drop in 500 ft of this pipe?

5.8. An array of twenty 40-ft$^2$ solar panels is connected in parallel. What is the pressure drop through the manifold if water is used at the rate of 0.03 gal/min $\cdot$ ft$_c$$^2$ and the collectors are spaced 5 ft apart. Pipe with 1-in diameter is used.

5.9. A heat exchanger is used to isolate a collector loop from a storage loop in a solar-cooling system. Collector and storage fluid capacitance rates are 300 kJ/m$_c$$^2$ $\cdot$ hr $\cdot$ $^\circ$C and the collector heat loss conductance is 5 kJ/m$^2$ $\cdot$ hr $\cdot$ $^\circ$C with a heat-removal factor $F_R = 0.9$. What is the energy-delivery penalty from heat-exchanger use if the heat-exchanger effectiveness is 0.25, 0.5, 0.9?

5.10. Size an expansion tank for a closed solar collector loop containing 0.15 m$^3$ of water if the initial fill pressure is 0.3 atm and if the maximum operating pressure and temperature are 1.0 atm and 100$^\circ$C, respectively.

5.11. Prepare a block diagram of a computer model for predicting the performance of a solar crop dryer. Show all inputs required, loops needed, and output listing.

5.12. If the initial extra cost of a solar system is $5,000, what is the annual payment if the interest rate is 8 percent and the mortgage term is 15 yr?

5.13. A solar system in China Lake, California, delivers the amounts of energy each year shown in the table. If the solar system costs $195/m$^2$ and is amortized (paid off) over 25 yr at 10 percent interest, prepare the total cost curve and specify the cost-optimal system. The backup fuel is fuel oil at a cost of $7.00/GJ; fuel price is inflating 8 percent/yr.

| Collector area (m$^2$) | Energy delivered (GJ) |
| --- | --- |
| 100 | 336 |
| 150 | 444 |
| 200 | 531 |
| 250 | 612 |
| 300 | 673 |
| 400 | 791 |
| 500 | 856 |
| 600 | 915 |
| (no solar) | 0 |

The total annual energy demand is 1000 GJ/yr.

5.14. What is the present value of a $1,000 payment in 1987 if the discount rate is 7 percent?

5.15. What is the present value in constant dollars of the $1,000 payment in Problem 5.14 if the inflation rate is 4 percent/yr?

5.16. Calculate and tabulate the annual cash flows associated with a solar system for a 10-yr period of analysis if the extra initial solar cost is $6,000 and if

Interest rate          = 9 percent
Power cost            = $30/yr escalating at 10 percent/yr
Property tax          = 0
Income tax bracket = 32 percent
Maintenance         = $\frac{1}{2}$ percent/yr
Scrap value           = 50 percent of initial cost

If the solar system saves $550/yr in conventional fuel (escalating at 10 percent/yr), is it cost effective? Work the problem in current dollars.

5.17. Capital-recovery factors based on monthly and yearly payments differ slightly. What is the percentage difference in these two payment schedules for an 8 percent loan over 20 yr?

5.18. What is the one-term logarithmic production function for the system in Problem 5.13? Use a regression method or a plot on semi-log graph paper.

5.19. The production function of a solar water-heating system is $Q_s = 12\ A_c^{0.6}$ GJ/yr. The annual solar system cost is $20/m_c^2 \cdot$ yr. If the conventional energy is valued at $9/GJ and the annual energy demand is 140 GJ, what is cost-optimal collector area?

5.20. Rework Example 5.13 on page 385 if the utility uses a declining-block rate instead of a demand rate. The declining-block rate is

| 0–200 kW $\cdot$ hr | $7.41 |
| 200–1000 kW $\cdot$ hr | $0.03117/kW $\cdot$ hr |
| over 1000 kW $\cdot$ hr | $0.02130/kW $\cdot$ hr |

Is the solar system or the energy conservation system cheaper for January?

# REFERENCES

1. Altman, M., Conservation and Better Utilization of Electric Power by Means of Thermal Energy Storage and Solar Heating, *NTIS Rept.* PB210359, 1971.
2. ASHRAE, "Handbook of Fundamentals," American Society of Heating, Refrigerating, and Air Conditioning Engineers, New York, 1972.
3. ASHRAE, "Handbook and Product Directory, Systems," American Society of Heating, Refrigerating, and Air Conditioning Engineers, New York, 1973.
4. Bailey, N. T. J., "The Elements of Stochastic Processes with Application to the Natural Sciences," John Wiley & Sons, New York, 1964.
5. Balcomb, J. D., J. C. Hedstrom, and B. T. Rogers, Design Considerations of Air Cooled Collector/Rock-Bin Storage Solar Heating Systems, *Los Alamos Scientific Laboratory Rept.* LA-UR-75-1335, 1975.
6. Betz Laboratories, "Betz Handbook of Industrial Water Conditioning," Betz Laboratory, Philadelphia, 1962.
7. Bird, R. B., W. E. Stewart, and E. L. Lightfoot, "Transport Phenomena," John Wiley & Sons, New York, 1960.
8. Buchberg, H., and J. R. Roulet, Simulation and Optimization of Solar Collection and Storage for House Heating, *Sol. Energy,* vol. 12, pp. 31–50, 1968.
9. Butz, L. W., W. A. Beckman, and J. A. Duffie, Simulation of a Solar Heating and Cooling System, *Sol. Energy,* vol. 16, pp. 129–136, 1974.

10. Chase, J. D., "Plant Cost vs. Capacity: New Way to Use Exponents," Chemical Engineering Reprint, McGraw-Hill Book Co., New York, 1970.

11. Crane Company, "Flow of Fluids," The Crane Co. Technical Paper 410, 1957.

12. DeKold, D., Diode Pair Senses Differential Temperature, in "Designer's Casebook," pp. 44–45, McGraw-Hill Book Co., New York, 1976.

13. deWinter, F., Heat Exchanger Penalties in Double Loop Solar Water Heating Systems, *Sol. Energy*, vol. 17, pp. 335–337, 1975; see also Ref. (24).

14. Eckert, E. R. G., et al., Research Applied to Solar Thermal Power Systems, *University of Minnesota Rept.* NSF/RANN/SE/GI-34871/PR/73/2, 1973.

15. Edlin, F., Thermal Energy Storage, in "Solar Utilization Now," Arizona State University, Tempe, 1976.

16. Guthrie, K. M., "Capital Cost Estimating," Chemical Engineering Reprint, McGraw-Hill Book Co., New York, 1969.

17. Hirst, E., Changes in Retail Energy Prices and the Consumer Price Index, *Energy*, vol. 1, pp. 33–43, 1976.

18. Hodgman, C. D., R. C. Weast, and S. M. Selby, "Handbook of Chemistry and Physics," 40th ed., Chemical Rubber Co., Cleveland, 1959.

19. Honeywell, Inc., Design and Test Report for a Transportable Solar Laboratory, *NTIS Rept.* PB240609, 1974.

20. Johnson, G. R., "A Data Acquisition, Performance Evaluation and Monitoring System for Solar Heated/Cooled Residential Dwellings," ASME Paper 76-WA/Sol-13, 1976.

21. Kays, W. M., and A. L. London, "Compact Heat Exchangers," McGraw-Hill Book Co., New York, 1964.

22. Klein, S. A., TRNSYS–A Transient Simulation Program, *Solar Energy Laboratory, University of Wisconsin, Madison, Rept.* 38, 1973.

23. Klein, S. A., W. A. Beckman, and J. A. Duffie, A Method of Simulation of Solar Processes and Its Applications, *Sol. Energy*, vol. 17, pp. 29–37, 1975.

24. Klein, S. A., "A Design Procedure for Solar Heating Systems," Ph.D. dissertation, University of Wisconsin, Madison, 1976.

25. Kreith, F., "Principles of Heat Transfer," Intext Educational Publishers, New York, 1973.

26. Lameiro, G. F., "Stochastic Models of Solar Energy Systems," Ph.D. disseration, Colorado State University, Fort Collins, 1977; see also Lameiro, G. F. and W. S. Duff, "The Development and Utilization of a Stochastic Model for Solar Energy Space Heating Systems," American Section ISES Conf., Orlando, Florida, June 1977; Lameiro, G. F. and M. C. Bryson, The Effect of Markovian and Discrete-Climate Assumptions on the Performance Analysis of Solar Heating Systems, submitted to *J. Appl. Meteorol.*, July 1977; by the same authors, The Solution of a Large System of Equations by Means of Successive Approximation-Exponential Curve Fit Method, submitted to *ACM Trans. on Math. Software*; Lameiro, G. F., W. S. Duff, and M. C. Bryson, A General Stochastic Model for Solar Energy Space Heating Systems, submitted to *Sol. Energy*.

27. Löf, G. O. G., and R. W. Hawley, "Unsteady-State Heat Transfer between Air and Loose Solids," *Industrial and Engineering Chemistry*, vol. 40, pp. 1061–1070, 1948.

28. Löf, G. O. G., and R. A. Tybout, "A Model for Optimizing Solar Heating Design," ASME Paper 72-WA/Sol-8, 1972.

29. Löf, G. O. G., et al., Design and Construction of a Residential Solar Heating Cooling System, *NSF Rept.* NSF/RANN/SE/GI-40457/PR/74/2, 1974.
30. Lorsch, H., Conservation and Better Utilization of Electric Power by Means of Thermal Energy Storage and Solar Heating, University of Pennsylvania, *NSF Rept.* NSF/RANN/SE/GI-27976/PR/73/5, 1973.
31. Peters, M. S., and R. D. Timmerhaus, "Plant Design and Economics for Chemical Engineers," McGraw-Hill Book Co., New York, 1968.
32. Popplewell, J. M., "Corrosion Considerations in the Use of Aluminum, Copper and Steel Flat Plate Solar Collectors," Olin Metals Research Laboratory, New Haven, Conn., 1976.
33. Potter, P. J., "Power Plant Theory and Design," Ronald Press, New York, 1959.
34. Samuelson, P., "Economics," 10th ed., McGraw-Hill Book Co., New York, 1976.
35. Sawyer, C. N., and P. L. McCarty, "Chemistry for Sanitary Engineers," McGraw-Hill Book Co., New York, 1967.
36. Union Carbide Corporation, "Glycols," Union Carbide Corp., New York, 1971.
37. Weast, R. C., "Handbook of Chemistry and Physics," Chemical Rubber Co., Cleveland, 1968.

# SOLAR-HEATING SYSTEMS

*For the well being and health ... the homesteads should be airy in
summer and sunny in winter. A homestead promising these qualities
would be longer than it is deep and the main front would face south.*

**Aristotle**

The use of solar energy for heat production dates from antiquity. This chapter
describes in detail the function and design of systems for heating buildings and
service water. Other applications including the heating of swimming pools (1, 11)
and greenhouses (9) or the use of low-temperature solar heat for agriculture,
agricultural drying, or aquaculture can be analyzed with the principles set forth in
this chapter and Chapters 4 and 5. Solar stills and solar ponds are described in
Chapter 8.

Energy for heating buildings and hot water consumes about one-fourth of the
annual energy production in the United States. In many areas of the United States
and the world, solar heating can compete economically with other types of fuel for
heating. There has been recent widespread interest in the use of solar energy for
these applications as manifested by the rapid increase in construction of solar-
conditioned buildings. In 1975, more solar buildings were built worldwide than the
cumulative total for all prior years.

## CALCULATION OF HEATING AND HOT-WATER LOADS IN BUILDINGS

Energy requirements for space heating or service water heating can be calculated
from basic conservation of energy principles. For example, the heat required to
maintain the interior of a building at a specific temperature is the total of all losses
from the structure including

Heat transmission through the opaque building skin—roof, doors, ceilings, walls,
floors
Heat transmission through windows and skylights
Heat required to warm air exchanged with the environment
Heat required to maintain building humidity levels

Comfort in buildings has been a subject of long investigation by the American Society of Heating, Refrigerating, and Air Conditioning Engineers (ASHRAE). This professional organization has developed extensive heat load calculation procedures embodied in the ASHRAE "Handbook of Fundamentals" (2). The most frequently used load calculation procedures will be summarized in this section; the reader is referred to the ASHRAE handbook for details.

Figure 6.1 shows the combinations of temperature and humidity that are required for human comfort. Figure 6.2 contains more detailed comfort information, including wet-bulb temperatures. The shaded area is the standard U.S. comfort level for sedentary persons. Many European countries have human comfort levels from 3 to 7°C below U.S. levels. If activity of a continuous nature is anticipated, the comfort zone lies to the left of the shaded area; if extra clothing is worn, the comfort zone is displaced similarly. For temperature standards in solar-heated agricultural buildings housing animals or plants, ASHRAE and the U.S. Department of Agriculture guidelines are recommended.

## Calculation of Heat Loss

In this section the common heat-loss mechanisms in buildings at steady state are evaluated analytically. In a subsequent section, water-heating energy demands are described.

### Transmission Losses

Heat losses by transmission occur through walls, windows, floors, and ceilings, that is, the skin of a building. Such losses occur by the combined mechanisms of

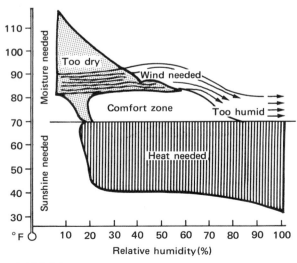

**FIGURE 6.1** Bioclimatic comfort chart showing temperature and humidity regions for comfort.

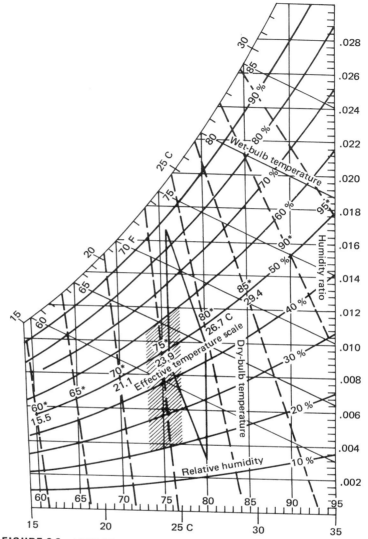

**FIGURE 6.2** ASHRAE comfort chart for sedentary humans; crosshatched region of humidity and dry-bulb temperature is comfortable for the majority of humans. From (2).

heat transfer (see Chapter 3). For example, heat is transferred through a single-pane window in three steps—by convection to the inner surface of the window, by conduction through the window glass, and by convection from the window. The rate of heat transfer in each step is the same ($q_{loss}$) and the several temperature differences involved can be calculated as described in Chapter 3 using an effective, overall heat-transfer coefficient $U_{trans}$. Thus, the rate of heat loss

$$q_{loss} = U_{trans}A_{trans}(T_{inside} - T_{outside}) \qquad (6.1)$$

Heat losses through walls and doors likewise follow the heat-transfer principles developed in Chapter 3. Complete tables of thermal properties of building materials are given in Appendix 6. Appendix 6 also contains average wind and design temperature data for many U.S. cities.

Transmission heat losses through attics, unheated basements, and the like are buffered by the thermal resistance of the unheated space. For example, the temperature of an unheated attic lies between that of the heated space and that of the environment. As a result the ceiling of a room below an attic is exposed to a smaller temperature difference and consequent lower heat loss than would the same ceiling without the attic. The effective conductance of thermal buffer spaces can easily be calculated by forming an energy balance on such spaces.

**EXAMPLE 6.1**

Consider an unheated attic of a building with its upper surface $A_o$ exposed to the environment and lower surface $A_i$ exposed to a heated space. If the respective $U$ values are $U_o$ and $U_i$, calculate the effective $U$ value for heat loss from the building through the attic to the environment.

**SOLUTION**

Under steady-state conditions, the net heat flow into the attic is 0. Therefore

$$U_o A_o(T_{at} - T_a) = U_i A_i(T_i - T_{at}) \qquad (6.2)$$

where $T_{at}$ is the attic temperature, $T_a$ is the ambient temperature, and $T_i$ is the building interior temperature. The effective $U$ value $U_{eff}$ is defined as

$$q_{at} \equiv U_{eff}A_i(T_i - T_o) \qquad (6.3)$$

where $q_{at}$ is the rate of heat loss to the environment via the attic. This heat loss is also equal to

$$q_{at} = U_i A_i(T_i - T_{at}) \qquad (6.4)$$

If Eq. (6.2) is solved for $T_{at}$ and that value is inserted into Eq. (6.3), it is easy to show that

$$U_{eff} = \frac{U_i U_o A_o}{U_o A_o + U_i A_i} \qquad (6.5)$$

The definition of thermal resistance $R$ is

$$R \equiv \frac{1}{UA} \qquad (6.6)$$

and Eq. (6.5) is seen to be simply the effective conductance of two resistances in series. The same method of analysis applies to other thermal buffer spaces.

Heat loss through basement floors and walls is generally at a fixed rate independent of the ambient air temperature. The temperature difference causing heat flow is that between the interior space and the surrounding ground temperature—usually nearly constant during the heating season. Therefore, the rate of heat loss through the floor $q_f$ can be expressed as

$$q_f = U_f^* A_f \qquad (6.7)$$

in which $U_f^*$ has special units of W/m² or Btu/hr · ft². Values of $U_f^*$ for various groundwater temperatures are given in Table 6.1.

If concrete floors exposed to the environment at *ground level* are used in a building, the heat loss from the *edge* of the slab is much larger than the heat loss downward through the slab. In that case, the heat loss from a slab depends on the perimeter length of the exposed edge $P_e$. The rate of heat loss at the floor edge $q_{fe}$ (2) is given by

$$q_{fe} = F_e P_e (T_i - T_a) \qquad (6.8)$$

where $P_e$ is the perimeter (m or ft) and $F_e$ is the heat-loss coefficient (W/m · °C or Btu/hr · ft · °F). Typical values of $F_e$ are shown in Table 6.2. Note that insulation on slab edges should both cover the vertical edges and extend about 50–70 cm under the slab and inward toward the slab interior.

**TABLE 6.1** Heat-Loss Rates for Basement Floors and Walls[a]

| Groundwater temperature | | Basement floors | | Basement walls | |
|---|---|---|---|---|---|
| °C | °F | W/m² | Btu/hr · ft² | W/m² | Btu/hr · ft² |
| 5 | 40 | 9.5 | 3.0 | 18.9 | 6.0 |
| 10 | 50 | 6.3 | 2.0 | 12.6 | 4.0 |
| 15 | 60 | 3.2 | 1.0 | 6.3 | 2.0 |

[a]Adapted from (2).

**TABLE 6.2** Values of Heat-Loss Coefficient $F_e$ at Floor Edge for Various Insulation $U$ Values[a]

| Insulation $U$ value | | $F_e$ | |
|---|---|---|---|
| W/m² · °C | Btu/hr · ft² · °F | W/m · °C | Btu/hr · ft · °F |
| 0.85 | 0.15 | 0.43 | 0.25 |
| 1.13 | 0.20 | 0.57 | 0.33 |
| 1.42 | 0.25 | 0.73 | 0.42 |
| 1.70 | 0.30 | 0.87 | 0.50 |
| 1.98 | 0.35 | 1.02 | 0.59 |
| 2.27 | 0.40 | 1.16 | 0.67 |

[a]Loss coefficients shown are for unheated slabs with 60 cm of total insulation length. Adapted from (2).

## Heat Losses Resulting from Air Change

If air is exchanged between the environment and a building interior, either actively by a ventilation system or passively because of air leaks and infiltration, the temperature of the inflowing air must be raised to the design comfort level. The calculation of energy requirements for air-change heating $q_{chg}$ is given by

$$q_{chg} = Q\rho_a c_{pa}(T_i - T_o) \tag{6.9}$$

where $Q$ is the volumetric exchange rate, $\rho_a$ is the air density, and $c_{pa}$ is the specific heat of air.

The evaluation of $Q$ in Eq. (6.9) for forced ventilation is quite simple. It is the total exchange of outside and inside air. If air changes occur passively by infiltration through pores and cracks, however, it is much more difficult to make reliable estimates of the rate of air change. The recommended method is based on empirical data for air leakage through cracks around windows and doors. This leakage rate depends on crack size and the pressure difference across the crack. The pressure difference exists because of wind stagnation on the windward sides of buildings. However, in the absence of wind a pressure difference can also exist because of the stack effect in a building. The upper part of a building will be warmer and the lower density air in such areas will give rise to a pressure difference causing infiltration and exfiltration. These matters are treated in detail in Ref. (2).

It is beyond the purview of this book to consider infiltration losses in detail. Therefore, only a general summary is presented. The stagnation pressure $p_v$ exerted on a building resulting from wind is $\frac{1}{2}\rho_a V_w^2$ where $V_w$ is the wind speed in m/sec. Since some pressure buildup occurs within a building subject to wind load, the pressure differences across cracks $\Delta p_c$ is not the full value of $p_v$ but about one-half that value (2)

$$\Delta p_c \cong \frac{1}{2}p_v = \frac{1}{4}\rho_a V_w^2 \tag{6.10}$$

Three levels of building skill can be distinguished with corresponding differences in window and door fit. The quality of fit is reflected in infiltration rates as shown in Table 6.3. The infiltration functions in Table 6.3 are based on numerical data contained in the ASHRAE handbook (2).

## EXAMPLE 6.2

Calculate the leakage rate through a double-hung window measuring 1 m $\times$ 1.6 m with a wind of 10 m/sec. The window is weather stripped and of average fit.

## SOLUTION

The total crack length $L_c$ is

$$L_c = 2 \times 1.6 + 3 \times 1 = 6.2 \text{ m}_c$$

The wind stagnation pressure $p_v$ at sea level is

$$p_v = \tfrac{1}{2}(1.2)(10)^2 = 60 \text{ N/m}^2$$

From Eq. (6.10)

$$\Delta p_c = \tfrac{1}{2}p_v = 30 \text{ N/m}^2$$

From Table 6.3 the infiltration function is

$$Q_i = 0.65 + 0.026 \, \Delta p_c$$

**TABLE 6.3**  Infiltration Functions $Q_i$ for Windows and Doors[a]

| Quality of skill | $Q_i$ | |
|---|---|---|
| | m³/hr · m$_c$[b] | ft³/hr · ft$_c$[c] |
| First level | | |
| Loose fit without weather strip | $3.7 + 0.14 \, \Delta p_c$ | $40 + 370 \, \Delta p_c$ |
| Second level | | |
| Average fit without weather strip | $1.3 + 0.049 \, \Delta p_c$ | $14 + 132.5 \, \Delta p_c$ |
| Loose fit with weather strip | | |
| Third level | | |
| Average fit with weather strip | $0.65 + 0.026 \, \Delta p_c$ | $7 + 70 \, \Delta p_c$ |

[a]Infiltration through doors is about twice that through windows of the same quality. At low pressures the infiltration rate is proportional to $\sqrt{\Delta p_c}$. Prepared from (2).
[b]$Q_i$ = m³/hr · m$_c$; $\Delta p_c$ = N/m².
[c]$Q_i$ = ft³/hr · ft$_c$; $\Delta p_c$ = in H$_2$O.

so

$$Q_i = 0.65 + 0.026(30) = 1.43 \text{ m}^3/\text{hr} \cdot m_c$$

For a crack length $L_c$ of 6.2 m, the total infiltration is

$$L_c Q_i = 6.2 \times 1.43 = 8.9 \text{ m}^3/\text{hr}$$

If wind blows from a given direction, roughly one-half the total crack length in a building has an air inflow. The remaining half is the area through which warm air leaves the building. In addition, the use of storm doors and windows can reduce infiltration losses significantly. The above treatment of infiltration losses is necessarily simplified; detailed procedures are contained in the ASHRAE handbook.

The example below is an illustration of the heat-loss calculation method described in this section. The total energy demand from a building is the total of transmission and infiltration losses. In many older buildings the uncontrolled infiltration loss is the dominant heat-loss mechanism. The least-cost mix of transmission and infiltration loss controls can be calculated using economic analysis methods described in Chapter 5. Some results of calculations of this type are described in the next section.

## EXAMPLE 6.3

Calculate the heat load on a house for which the wall area is 2000 ft$^2$, the roof area is 3000 ft$^2$, and the window area totals 1000 ft$^2$. The construction of the wall is shown in Fig. 6.3; the calculation of the $U$ factor for the walls ($U_{wa}$) follows [heat-loss estimates for these calculations are based on property values presented in Ref. (7)].

## SOLUTION

$$U_{wa} = \cfrac{1}{\underset{\substack{\text{Outside} \\ \text{convection}}}{(1/6)} + \underset{\text{Wood}}{(1/1.0)} + \underset{\text{Drywall}}{(0.5/0.25)} + \underset{\substack{\text{Glass} \\ \text{wool}}}{(3.5/0.27)} + \underset{\substack{\text{Air} \\ \text{space}}}{(1/1.1)} + \underset{\substack{\text{Gypsum} \\ \text{board}}}{(0.5/1.41)} + \underset{\substack{\text{Inside} \\ \text{convection}}}{(1/1.65)}}$$

$$U_{wa} = \frac{1}{0.166 + 1.0 + 2.0 + 12.96 + 0.909 + 0.35 + 0.606}$$

$$= 0.0556 \text{ Btu/hr} \cdot \text{ft}^2 \cdot °\text{F}$$

The heat loss through the windows depends on whether they are single- or double-glazed. In this example, single-glazed windows are installed, and a $U$

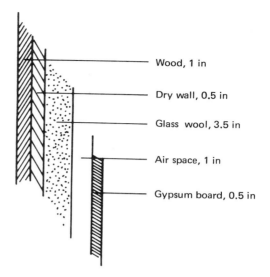

Wood, 1 in

Dry wall, 0.5 in

Glass wool, 3.5 in

Air space, 1 in

Gypsum board, 0.5 in

**FIGURE 6.3** Wall construction for calculation of $U$ factor. From (23).

factor equal to 1.13 Btu/hr · ft² · °F is used. (If double-glazed windows were installed, the $U$ factor would be 0.55 Btu/hr · ft² · °F.)

The roof is constructed of 1-in-thick foam and $\frac{1}{2}$-in siding. The calculation of the $U$ factor for the roof $U_{rf}$ is shown below.

$$U_{rf} = \frac{1}{\underbrace{(1/6) + (1/0.1)}_{\text{Foam}} + \underbrace{(0.5/1.0)}_{\text{Siding}} + (1/1.65)}$$

$$= \frac{1}{0.166 + 10 + 0.5 + 0.606} = 0.0887 \text{ Btu/hr} \cdot \text{ft}^2 \cdot \text{°F}$$

If the respective areas and $U$ factors are known, the rate of heat loss per hour for the walls, windows, and roof $q_{tr}$ can be calculated.

$$q_{wa} = (2000 \text{ ft}^2) \times 0.0556 \text{ Btu/hr} \cdot \text{ft}^2 \cdot \text{°F} = \quad 111 \text{ Btu/hr} \cdot \text{°F}$$

$$q_{wi} = (1000 \text{ ft}^2) \times 1.13 \text{ Btu/hr} \cdot \text{ft}^2 \cdot \text{°F} \quad = 1130 \text{ Btu/hr} \cdot \text{°F}$$

$$q_{rf} = (3000 \text{ ft}^2) \times 0.0887 \text{ Btu/hr} \cdot \text{ft}^2 \cdot \text{°F} = \quad 266 \text{ Btu/hr} \cdot \text{°F}$$

$$\text{Total } q_{tr} = 1507 \text{ Btu/hr} \cdot \text{°F}$$

If double-glazed windows were used, the heat loss would be reduced to 927 Btu/hr · °F.

The infiltration and ventilation rate $Q$ for this building is 100 ft³/min; thus, from Eq. (6.9),

$$q_{chg} = 100 \times (0.070 \text{ lb/ft}^3) \times (0.24 \text{ Btu/lb} \cdot {}^\circ\text{F}) \times (60 \text{ min/hr})$$

$$= 101 \text{ Btu/hr} \cdot {}^\circ\text{F}$$

The total rate of heat loss $q_{tot}$ is the sum of $q_{chg}$ and $q_{tr}$:

$$q_{tot} = (1507 + 101) = 1608 \text{ Btu/hr} \cdot {}^\circ\text{F}$$

This calculation is simplified for purposes of illustration. Heat losses through the slab surface and edges have been neglected, for example. The procedure to be used for heat-loss calculation is described in detail in the ASHRAE "Handbook of Fundamentals" (2).

More refined methods of calculating energy requirements on buildings do not require the steady-state assumption used above (24). The thermal inertia of buildings may be expressly used as a load-leveling device. If so, the steady-state assumption is not met and the energy capacitance of the structure must be considered for accurate results. Many adobe structures in the U.S. Southwest are built intentionally to use daytime sun absorbed by 1-ft-thick walls for nighttime heating, for example.

One method of including transient effects of building components is the time-response method (15). This technique calculates detailed transient performance of a building for several generic types of days. The results of these detailed calculations are used to generate a transfer function for the building, which transforms time series of weather data into time series of building responses. The method has been shown to be quite accurate and is recommended if steady-state calculations are not sufficiently accurate.

### Controlling Transmission and Infiltration Losses

Jacobs and Peterson (16) have considered the economics of heat-loss control from buildings. Their economic analysis is identical to that used in determining the cost-effectiveness of various solar system sizes described in Chapter 5. For this analysis six winter heating zones are identified as shown on the map in Fig. 6.4.

Table 6.4 is used to evaluate a *heating index* to be used to judge the efficacy of certain heat-loss control strategies. The heating index depends on the heating zone and the cost of energy used for heating. Although solar energy is not shown in Table 6.4, it can be added by evaluating the cost of solar energy as described in Chapter 5 and using the "Other" heading in the table. For high energy prices and high heating demands, the index is higher, indicating that more investment can be made in heat-loss control. Table 6.4 can be interpolated for costs not shown. Costs of energy used in the table are marginal, that is, the cost of the last unit of energy used in the time period of analysis including taxes and surcharges. In many cases the marginal cost is the same as the average cost of fuel.

For a given value of heating index, Tables 6.5A–6.5C list the cost-optimal

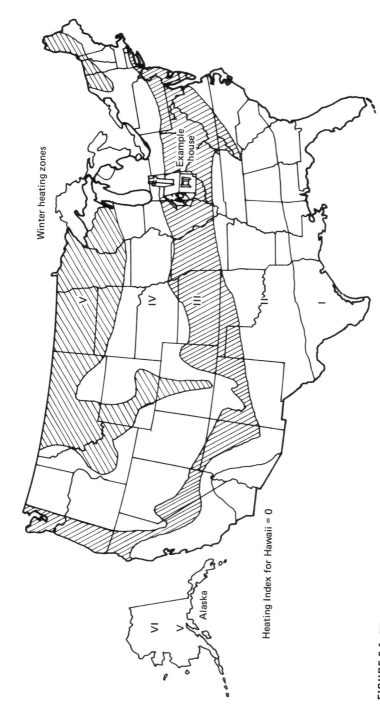

**FIGURE 6.4** Winter heating zones for assessment of energy conservation in buildings. From (16).

403

**TABLE 6.4** Heating Indices for Various Heating Zones in the United States for Various Fuel Prices[a]

| Type of fuel | Marginal fuel cost per unit[b] ($) | | | | | | | | | |
|---|---|---|---|---|---|---|---|---|---|---|
| Gas (therm) | 0.09 | 0.12 | 0.15 | 0.18 | 0.24 | 0.30 | 0.36 | 0.54 | 0.72 | 0.90 |
| Oil (gal) | 0.13 | 0.17 | 0.21 | 0.25 | 0.34 | 0.42 | 0.50 | 0.75 | 1.00 | 1.25 |
| Electric (kW · hr) | | | 0.01 | 0.013 | 0.016 | 0.02 | 0.03 | 0.04 | 0.05 | |
| Heat pump (kW · hr) | 0.009 | 0.011 | 0.015 | 0.018 | 0.023 | 0.029 | 0.035 | 0.053 | 0.07 | 0.088 |
| Other (MMBtu) | 1.46 | 1.79 | 2.44 | 2.93 | 3.80 | 4.69 | 5.86 | 8.79 | 11.72 | 14.65 |
| Heating zone | Heating indices | | | | | | | | | |
| I | 2 | 2 | 3 | 3 | 4 | 5 | 6 | 9 | 12 | 15 |
| II | 5 | 6 | 8 | 9 | 12 | 15 | 18 | 27 | 36 | 45 |
| III | 8 | 10 | 13 | 15 | 20 | 25 | 30 | 45 | 60 | 75 |
| IV | 11 | 14 | 18 | 21 | 28 | 35 | 42 | 63 | 84 | 105 |
| V | 14 | 18 | 23 | 27 | 36 | 45 | 54 | 81 | 108 | 135 |
| VI | 22 | 28 | 36 | 42 | 56 | 70 | 84 | 126 | 168 | 210 |

[a]From (16).
[b]One therm is 100,000 Btu or 105.5 GJ. One gallon of oil is equivalent to 140,000 Btu or 147.7 GJ. Cost of other energy sources must include the efficiency of energy delivery from basic fuel to delivered heat energy. Heat pump seasonal performance factor (SPF) is assumed to be 1.8; see p. 471.

**TABLE 6.5A**  Cost-Optimal Heat-Loss Control Configurations for Buildings with
Attic Floor Insulation and Attic Duct Insulation[a]

| Heating index for attics | Attic insulation | | | | Duct insulation[b] | |
|---|---|---|---|---|---|---|
| | | Approximate thickness | | | | |
| | R value | Mineral fiber Batt/blanket (in) | Mineral fiber Loose fill[c] (in) | Cellulose Loose fill[c] (in) | R value | Approximate thickness (in) |
| 1–3 | 0 | 0 | 0 | 0 | 8 | 2 |
| 4–9 | 11 | 4 | 4–6 | 2–4 | 8 | 2 |
| 10–15 | 19 | 6 | 8–10 | 4–6 | 8 | 2 |
| 16–27 | 30[d] | 10 | 13–15 | 7–9 | 16 | 4 |
| 28–35 | 33 | 11 | 14–16 | 8–10 | 16 | 4 |
| 36–45 | 38 | 12 | 17–19 | 9–11 | 24 | 6 |
| 46–60 | 44 | 14 | 19–21 | 11–13 | 24 | 6 |
| 61–85 | 49 | 16 | 22–24 | 12–14 | 32 | 8 |
| 86–105 | 57 | 18 | 25–27 | 14–16 | 32 | 8 |
| 106–130 | 60 | 19 | 27–29 | 15–17 | 32 | 8 |
| 131– | 66 | 21 | 29–31 | 17–19 | 40 | 10 |

[a]R values are given in (Btu/hr · ft² · °F)⁻¹. To convert to (W/m² · °C)⁻¹, multiply by 0.176.
1 in = 2.54 cm; 1 ft² = 0.093 m². Tables 6.5A–6.5C from (16).
[b]Use heating index only if ducts are not used for air conditioning.
[c]High levels of loose-fill insulation may not be feasible in many attics.
[d]Assumes that joists are covered; otherwise use R-22.

**TABLE 6.5B**  Cost-Optimal Heat-Loss Control Configurations for
Buildings with Insulation under Floors and Storm Doors

| Heating index | Insulation under floors[a] | | Storm doors |
|---|---|---|---|
| | R value | Mineral fiber Batt thickness (in) | |
| 0–7 | 0[b] | 0[b] | None |
| 8–15 | 11[b] | 4[b] | None |
| 16–30 | 19 | 6 | Optional |
| 31–65 | 22 | 7 | Optional |
| 66– | 22 | 7 | On all doors |

[a]If the furnace and hot-water heater are located in an otherwise unheated
basement, cut the heating index in half to find the level of floor insulation.
[b]In zones I and II R-11 insulation is usually economical under floors, over open
crawl spaces, and over garages; in zone I insulation is not usually economical if crawl
space is closed off.

**TABLE 6.5C** Cost-Optimal Heat-Loss Control Configurations for Buildings with Wall Insulation, Duct Insulation, and Storm Windows

| Heating index for walls | Wall insulation (Blown in) | Insulation around ducts in crawl spaces and in other unheated areas (except attics)[a] Resistance and approximate thickness (in) | Storm windows (Triple-track) minimum economical window size (ft²) |
|---|---|---|---|
| 0–10 | None | R-8 (2) | None |
| 11–12 | | R-8 (2) | 20 |
| 13–15 | | R-8 (2) | 15 |
| 16–19 | Full- | R-16(4) | 12 |
| 20–28 | wall | R-16(4) | 9 |
| 29–35 | insulation | R-16(4) | 6 |
| 36–45 | approximately | R-24(6) | 4 |
| 46–65 | R-14 | R-24(6) | All windows[b] |
| 66– | | R-32(8) | All windows[b] |

[a]Use heating index only if ducts are not used for air conditioning.

[b]Windows too small for triple-track windows can be fitted with one-piece windows.

heat-loss control investment for attic, floor, and wall insulation; storm windows; and heating duct insulation. The cost-optimal configurations are based on a specific set of component costs. *If other costs are applicable, an adjusted heating index is used.* The adjusted index is the heating index from Table 6.4 multiplied by the ratio of assumed cost (Table 6.6) to applicable cost.

The ASHRAE has prepared an energy conservation standard for new buildings (3). Although it is primarily prescriptive and not oriented toward design guidance, it is a substantial beginning for dealing with the conservation problem in buildings.

### Heat Gains in Buildings

Heat supplied to a building to offset energy losses is derived from both the heating system and internal heat sources. Table 6.7 lists the common sources of internal heat generation for residences. Commercial buildings such as hospitals, computer facilities, or supermarkets will have large internal gains specific to their function. Heat gains tend to offset heat losses from a building but will add to the cooling load of an air conditioning system. The magnitude of the reduction in heating system operation will be described in the next section. Heat gains from sunlight are covered in more detail in Chapter 7.

## The Degree-day Method

The preceding analysis of heat loss from buildings expresses that loss on a per unit temperature difference basis (except for unexposed floor slabs). In order to calculate the peak load and total annual load for a building, appropriate temperature differences ($\Delta T$) must be defined for each. The peak load $\Delta T$ is usually defined statistically as that temperature difference expected to exist no more than 1 percent of the time over a long period. It is usually expressed in terms of a 99 percent design temperature as shown in Table 6.8. The design temperature difference is then the interior building temperature minus the 99 percent design temperature. The design $\Delta T$ is used for rating nonsolar heating systems but is not useful for selection of solar systems, since solar systems rarely provide 100 percent of the energy demand of a building (see Chapter 5).

A more useful index of heating energy demand is the total annual energy requirement for a building. This quantity is somewhat more difficult to calculate than the peak load. It requires a knowledge of day-to-day variations in ambient

**TABLE 6.6**  Assumed Heat-Loss Control Component Costs[a]

| Component | R value | Installed cost |
|---|---|---|
| Attic insulation (all materials) | 11 | 15[b] |
| | 19 | 25 |
| | 22 | 29 |
| | 30 | 39 |
| | 33 | 43 |
| | 44 | 57 |
| | 49 | 64 |
| | 57 | 74 |
| | 60 | 78 |
| | 66 | 86 |
| Floor insulation (mineral fiber batt) | 11 | 20[c] |
| | 19 | 30 |
| | 22 | 34 |
| Wall insulation (all materials) | | 60[d] |
| Duct insulation (mineral fiber blanket) | 8 | 30[e] |
| | 16 | 50 |
| | 24 | 70 |
| | 32 | 90 |
| | 40 | 110 |
| Storm windows (triple-track, custom made, installed)[f] | | |
| Up to 100 united in (height + width) | | $30.00 |
| Greater than 100 united in | | $30.00 + $0.60 per united in greater than 100 in |
| Storm doors (custom fitted, installed) | | |
| All sizes | | $75.00 |
| Weather stripping and caulking | | Varies according to material Use most durable materials available |

[a]Costs assumed in preparation of Table 6.6. See footnote a in Table 6.5A for definition of R values. From (16).
[b]Installed cost in cents per square foot of attic.
[c]Installed cost in cents.
[d]Installed cost in cents per square foot of net wall area. Price includes allowance for painting inside surface of exterior walls with water vapor-resistant paint.
[e]Installed cost in cents per square foot of material.
[f]Prices may be considerably less for stock sizes, installed by the homeowner.

**TABLE 6.7** Some Common Internal Sensible Heat Gains That
Tend to Offset the Heating Requirements of Buildings[a]

| Type | Magnitude (Btu/hr) |
|---|---|
| Incandescent lights | 3.4 × total W |
| Fluorescent lights | 4.1 × total W |
| Electric motors | 2544 × (hp/efficiency) |
| Natural gas stove | 1000 × ft$^3$/hr |
| Appliances | 3413 × total kW |
| A dog | 150–300 |
| People | |
|   Sitting | 230 |
|   Walking | 250 |
|   Dancing | 300 |
|   Working hard | 580 |
| Sunlight | Solar heat gain × fenestration transmittance × shading factor[b] |

[a]For more data see Ref. (2).
[b]Shading factor is the amount of a window not in a shadow expressed
as a decimal between 1.0 and 0.0.

temperature during the heating season and the corresponding building heat load for
each day. Experimental measurements of building heat loads have exhibited a daily
heat load profile like that shown in Fig. 6.5. The environmental temperature $T_{nl}$
above which no heat need be supplied to the building is a few degrees below the
required interior temperature $T_i$ because of internal heat-generation effects.

The no-load temperature at which internal source generation $q_i$ just balances
transmission and infiltration losses can be determined from the energy balance (see
Chapter 3)

$$q_i = \overline{UA}(T_i - T_{nl}) \tag{6.11}$$

where $\overline{UA}$ is the overall loss coefficient for the building (W/°C). Then

$$T_{nl} = T_i - \frac{q_i}{\overline{UA}} \tag{6.12}$$

The total annual heat load on the building $Q_T$ (kJ) can be expressed as

$$Q_T = \int_{365 \text{ days}} 86.4\overline{UA}(T_{nl} - T_a)^+ \, dt \tag{6.13}$$

in which all arguments of the integral are functions of time. In practice it is difficult to evaluate this integral; therefore, three simplifying assumptions are made:

1. $\overline{UA}$ is independent of time.
2. $T_{nl}$ is independent of time.
3. The integral can be expressed by the sum

$$\overline{UA} \sum_{n=1}^{365} (T_{nl} - \bar{T}_a)_n^+ \tag{6.14}$$

where $n$ is the day number, and the daily average temperature $\bar{T}_a$ can be approximated by $\frac{1}{2}(T_{a,\,max} + T_{a,\,min})$, in which $T_{a,\,max}$ and $T_{a,\,min}$ are the daily maximum and minimum temperatures, respectively.

**TABLE 6.8**  Normal Degree F-days and Design Outside Temperatures[a, b]

| State | City | Degree F-days, Sep. 1– May 31 | Design outside temp. (°F) | State | City | Degree F-days, Sep. 1– May 31 | Design outside temp. (°F) |
|---|---|---|---|---|---|---|---|
| Ala. | Birmingham | 2780 | 12 | Mont. | Helena | 8250[c] | −39 |
| Alaska | Juneau | 8088 | −5 | Neb. | Omaha | 6160 | −17 |
| Ariz. | Phoenix | 1698 | 36 | Nev. | Reno | 6036[c] | 3 |
| Ark. | Little Rock | 2982 | 8 | N.H. | Concord | 7612[c] | −11 |
| Calif. | San Francisco | 3421[c] | 37 | N.J. | Trenton | 5068[d] | 2 |
| Colo. | Denver | 6132[c] | −12 | N. Mex. | Albuquerque | 4389 | 8 |
| Conn. | Hartford | 6139 | −2 | N.Y. | New York | 5050[d] | 5 |
| D.C. | Washington | 4333 | 10 | N.C. | Raleigh | 3369 | 14 |
| Fla. | Jacksonville | 1243 | 28 | N. Dak. | Bismarck | 9033[c] | −31 |
| Ga. | Atlanta | 2826 | 11 | Ohio | Cleveland | 5950 | 0 |
| Hawaii | Honolulu | | | Okla. | Oklahoma City | 3647 | −1 |
| Idaho | Boise | 5890[c] | −10 | Oreg. | Portland | 4632[c] | 10 |
| Ill. | Chicago | 6310 | −11 | Pa. | Harrisburg | 5258 | 4 |
| Ind. | Indianapolis | 5611 | −8 | R.I. | Providence | 6125 | 1 |
| Iowa | Des Moines | 6446[c] | −13 | S.C. | Columbia | 2435 | 19 |
| Kan. | Topeka | 5209 | −8 | S. Dak. | Rapid City | 7535[c] | −22 |
| Ky. | Louisville | 4434 | −2 | Tenn. | Nashville | 3513 | 3 |
| La. | New Orleans | 1317 | 26 | Tex. | Fort Worth | 2361 | 8 |
| Maine | Portland | 7681[c] | −9 | Utah | Salt Lake City | 5866 | −1 |
| Md. | Baltimore | 4787 | 8 | Vt. | Burlington | 7865[c] | −17 |
| Mass. | Boston | 5791 | 0 | Va. | Richmond | 3955 | 11 |
| Mich. | Detroit | 6404[c] | −4 | Wash. | Spokane | 6852[c] | −16 |
| Minn. | Minneapolis | 7853[c] | −23 | W. Va. | Elkins | 5733 | −4 |
| Miss. | Vicksburg | 2000[d] | 15 | Wis. | Milwaukee | 7206[c] | −15 |
| Mo. | St. Louis | 4699 | −5 | Wyo. | Cheyenne | 7652[c] | −19 |

[a]Adapted from (7).
[b]All readings except those followed by $d$ were taken at the airport of the city listed.
[c]Degree-days for the entire year.
[d]Readings were taken within the city listed.

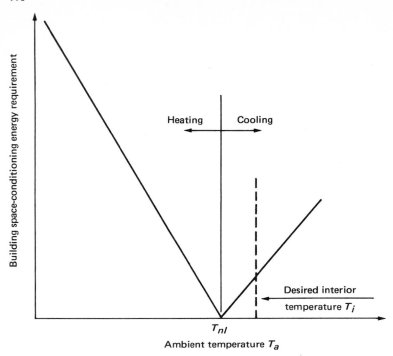

**FIGURE 6.5**  Building load profile versus ambient temperature showing no-load temperature $T_{nl}$ and desired interior temperature $T_i$.

The quantity $(T_{nl} - \bar{T}_a)^+$ is called the degree-day unit. For example, if the average ambient temperature for a day is 5°C and the no-load temperature is 20°C, 15 degree C-days are said to exist for that day. However, if the ambient temperature is 25°C, 0 degree-days exist, indicating 0 demand for heating that day. Degree-day totals for monthly and annual periods can be used directly in Eq. (6.13) to calculate the monthly and annual heating energy requirements.

In the past a single value of $T_{nl}$ has been used throughout the United States as a universal degree-day base—65.0°F (18.3°C). This practice is now outdated, since many homeowners and commercial building operators have lowered their thermostat settings in response to increased heating fuel costs, thereby lowering $T_{nl}$. Likewise, warehouses and factories operate well below the 18.3°C level. Therefore, a more generalized data base of degree-days to several bases (values of $T_{nl}$) has been created by the U.S. National Weather Service (NWS).*

Thom (31–34), in a series of papers, developed a statistically rigorous method of calculating degree-days to any base from values of ambient temperature and monthly standard deviations of ambient temperature. The details of this method are

---

*The degree-day base in SI units is defined as 19.0°C, not 18.3°C, which corresponds to 65.0°F. Therefore, precise conversion between the two systems is not possible by a simple multiplication by 5/9.

too lengthy to present here. However, the NWS has used this method to prepare tabulations of degree F-days for many U.S. locations. An example tabulation for Denver, Colorado, is shown in Table 6.9. In addition, maps of degree F-days to the standard base, like the map of Fig. 6.6, are available on a monthly basis (37). A more useful map for solar designers is provided in Fig. 6.7, wherein zones of heating degree-days and solar radiation are shown. High heat load, high insolation areas are the best regions for solar heating for given auxiliary energy and solar system prices.

## EXAMPLE 6.4

A building located in Denver, Colorado, has a heat-loss coefficient $\overline{UA}$ of 1000 kJ/hr · °C and internal heat sources of 4440 kJ/hr. If the interior temperature is 20°C (68°F), what are the monthly and annual heating energy requirements? A gas furnace with 65 percent efficiency is used to heat the building.

## SOLUTION

In order to determine the monthly degree-day totals, the no-load temperature (degree-day basis) must be evaluated from Eq. (6.12).

$$T_{nl} = 20 - \frac{4440}{1000} = 15.6°C \ (60°F)$$

From Table 6.9, degree F-days to base 60°F are read, entered in Table 6.10, and converted to degree C-days for this example.

The annual energy demand of 62.9 GJ is delivered by a 65 percent efficient device. Therefore

$$\frac{62.9}{0.65} GJ = 96.8 \ GJ$$

of gas must be purchased on the average per year.

## Service Hot-Water Load Calculation

Service hot-water loads can be calculated precisely with the knowledge of only a few variables. This accuracy is in contradistinction to building heat load calculations, which are subject to uncertainties in construction quality and materials, variability in occupant use pattern, and inaccuracy in air change calculation methods. The data required for calculation of hot-water demand are

Water source temperature      $(T_s)$
Water delivery temperature      $(T_d)$
Volumetric demand rate      $(Q)$

**TABLE 6.9** Heating Degree F-days to Selected Bases for Stapleton International Airport, Denver, Colorado (1941–1970)[a]

| Base | Jan | Feb | Mar | Apr | May | June | July | Aug | Sep | Oct | Nov | Dec | Annual |
|---|---|---|---|---|---|---|---|---|---|---|---|---|---|
| Below 70 | 1243 | 1042 | 1023 | 675 | 403 | 162 | 19 | 33 | 230 | 558 | 918 | 1159 | 7465 |
| 65 | 1088 | 902 | 868 | 525 | 253 | 80 | 0 | 0 | 120 | 408 | 768 | 1004 | 6016 |
| 60 | 933 | 762 | 713 | 386 | 123 | 26 | 0 | 0 | 47 | 266 | 618 | 849 | 4723 |
| 55 | 778 | 622 | 565 | 256 | 41 | 7 | 0 | 0 | 12 | 153 | 473 | 694 | 3601 |
| 50 | 630 | 493 | 421 | 156 | 7 | 0 | 0 | 0 | 0 | 73 | 334 | 539 | 2653 |
| 45 | 485 | 367 | 292 | 81 | 0 | 0 | 0 | 0 | 0 | 24 | 214 | 389 | 1852 |
| 43 | 430 | 321 | 248 | 57 | 0 | 0 | 0 | 0 | 0 | 14 | 175 | 330 | 1575 |
| 40 | 352 | 258 | 191 | 31 | 0 | 0 | 0 | 0 | 0 | 7 | 125 | 247 | 1211 |
| 35 | 242 | 174 | 112 | 11 | 0 | 0 | 0 | 0 | 0 | 0 | 58 | 135 | 732 |
| 32 | 189 | 131 | 73 | 0 | 0 | 0 | 0 | 0 | 0 | 0 | 32 | 87 | 512 |
| 30 | 156 | 104 | 52 | 0 | 0 | 0 | 0 | 0 | 0 | 0 | 21 | 58 | 391 |
| 25 | 86 | 52 | 22 | 0 | 0 | 0 | 0 | 0 | 0 | 0 | 7 | 17 | 184 |
| Mean temperature | 29.9 | 32.8 | 37.0 | 47.5 | 57.0 | 66.0 | 73.0 | 71.6 | 62.8 | 52.0 | 39.4 | 32.6 | 50.1 |
| Standard deviation | 4.70 | 5.00 | 4.30 | 3.70 | 2.10 | 3.00 | 1.70 | 1.70 | 2.70 | 3.10 | 3.70 | 2.90 | — |

[a]From unpublished data of the U.S. National Weather Service. Courtesy of F. Quinlan, National Climatic Center.

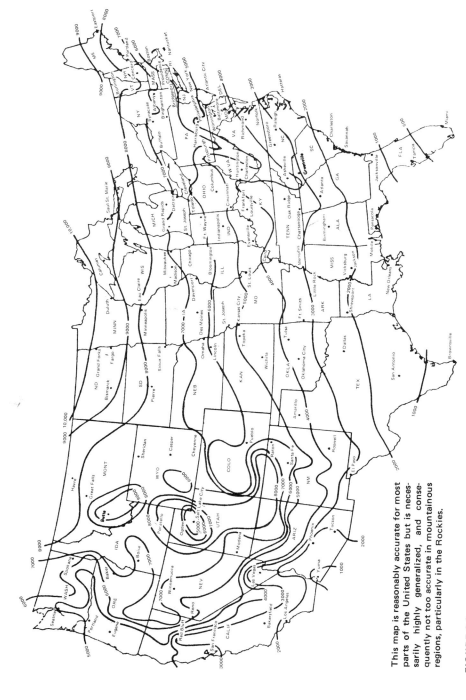

This map is reasonably accurate for most parts of the United States but is necessarily highly generalized, and consequently not too accurate in mountainous regions, particularly in the Rockies.

**FIGURE 6.6** Normal number of degree F-days per year for the continental United States. Based on data from (37).

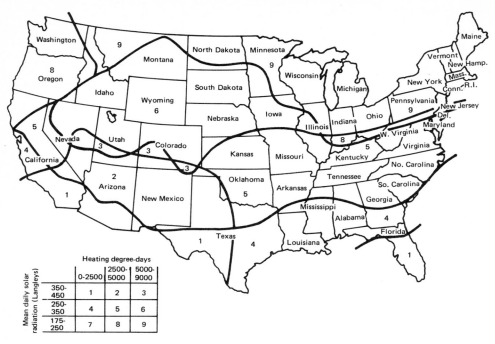

**FIGURE 6.7** Composite heating degree F-day and solar-radiation map for the heating season (November to April) for the United States. Adapted from (36).

**TABLE 6.10** Monthly and Annual Energy Demands for Example 6.4

| Month | Degree F-days | Degree C-days[a] | Energy demand[b] (GJ) |
|---|---|---|---|
| Jan. | 933 | 518 | 12.4 |
| Feb. | 762 | 423 | 10.2 |
| Mar. | 713 | 396 | 9.5 |
| Apr. | 386 | 214 | 5.2 |
| May | 123 | 68 | 1.6 |
| June | 26 | 14 | 0.3 |
| July | 0 | 0 | 0 |
| Aug. | 0 | 0 | 0 |
| Sep. | 47 | 26 | 0.6 |
| Oct. | 266 | 148 | 3.6 |
| Nov. | 618 | 343 | 8.2 |
| Dec. | 849 | 472 | 11.3 |
| Total | 4723 | 2622 | 62.9 |

[a]The approximate conversion to degree C-days has been made by multiplying by 5/9.
[b]Energy demand equals $\overline{UA}$ × 24 × degree C-days.

**TABLE 6.11** Approximate Service Hot-Water
Demand Rates

| | Demand per person | |
|---|---|---|
| Usage type | liters/day | gal/day |
| Retail store | 2.8 | 0.75 |
| Elementary school | 5.7 | 1.5 |
| Multifamily residence | 76.0 | 20.0 |
| Single-family residence | 76.0 | 20.0 |
| Office building | 11.0 | 3.0 |

The energy requirement for service water heating $q_{hw}$ is given by

$$q_{hw}(t) = \rho_w Q(t) c_{pw} [T_d - T_s(t)] \tag{6.15}$$

where $\rho_w$ is the water density and $c_{pw}$ is its specific heat. The demand rate $Q(t)$ varies in general with time of day and time of year; likewise, the source temperature varies seasonally. Source temperature data are not compiled in a single reference; local water authorities are the source of such temperature data.

Few generalized data exist with which to predict the demand rate $Q$. Table 6.11 indicates some typical usage rates for several common building types. Process water heating rates are peculiar to each process and can be ascertained by reference to process specifications.

Residential and commercial water usage is highly variable during a given day depending on habits of residents and duration of business hours. Industrial process heat requirements tend to be more uniform during a day—a factor that may be either advantageous or disadvantageous in solar systems. Mutch (28) has generated an average, typical hot-water use profile for residences. These data are shown in Table 6.12.

## EXAMPLE 6.5

Calculate the monthly energy required to heat water for a family of four in Nashville, Tennessee. Monthly source temperatures for Nashville are shown in Table 6.13 and the water delivery temperature is 60°C (140°F).

## SOLUTION

For a family of four the demand rate $Q$ is

$$Q = 4 \times 76 \text{ liters/day} = 0.30 \text{ m}^3/\text{day}$$

The density of water can be taken as 1000 kg/m$^3$ and the specific heat as 4.18 kJ/kg · °C.

**TABLE 6.12** Average Residential Hot-Water Usage by Time of Day[a]

| End of time period (hr) | Percentage of daily total usage |
|---|---|
| 100 | 2.25 |
| 600 | 0.0 |
| 700 | 1.50 |
| 800 | 4.60 |
| 900 | 7.25 |
| 1000 | 8.49 |
| 1100 | 6.90 |
| 1200 | 4.50 |
| 1300 | 3.60 |
| 1400 | 5.10 |
| 1500 | 2.70 |
| 1600 | 2.40 |
| 1700 | 2.10 |
| 1800 | 3.70 |
| 1900 | 6.75 |
| 2000 | 11.60 |
| 2100 | 9.60 |
| 2200 | 6.90 |
| 2300 | 5.46 |
| 2400 | 4.60 |

[a] Adapted from (28).

**TABLE 6.13** Water Heating Energy Demands for Example 6.4

| Month | Days/month | Demand (m³/month) | Source temperature °C | K | Energy requirement (GJ/month) |
|---|---|---|---|---|---|
| Jan. | 31 | 9.3 | 8 | 281 | 2.0 |
| Feb. | 28 | 8.4 | 8 | 281 | 1.8 |
| Mar. | 31 | 9.3 | 12 | 285 | 1.9 |
| Apr. | 30 | 9.0 | 19 | 292 | 1.5 |
| May | 31 | 9.3 | 17 | 290 | 1.7 |
| June | 30 | 9.0 | 21 | 294 | 1.5 |
| July | 31 | 9.3 | 22 | 295 | 1.5 |
| Aug. | 31 | 9.3 | 24 | 297 | 1.4 |
| Sep. | 30 | 9.0 | 24 | 297 | 1.4 |
| Oct. | 31 | 9.3 | 22 | 295 | 1.5 |
| Nov. | 30 | 9.0 | 14 | 287 | 1.7 |
| Dec. | 31 | 9.3 | 12 | 285 | 1.9 |

Monthly demands are given by

$$q_m = (Q \times \text{days/month})(\rho_w c_{pw})[T_d - T_s(t)]$$

$$= (0.30 \times \text{days/month})(1000 \times 4.18)[60 - T_s(t)]$$

The monthly energy demands calculated from the equation above with these data are tabulated in Table 6.13.

# SOLAR WATER HEATING

Solar water heating represents a low-temperature use of solar energy in a mechanically simple system. Since most hot-water loads are quite uniform on a monthly basis, this solar application has a particularly high load factor and therefore can be quite cost-effective. From a thermodynamic viewpoint the best match of collection temperature to use temperature dictates use of a nonconcentrating collection device. Flat-plate liquid collectors are normally used in conjunction with aqueous storage (26).

In this section, two common solar hot-water systems are described in detail. A design method for water heating is also described.

## Thermosiphon Systems

The natural tendency of a less dense fluid to rise above a more dense fluid can be used in a simple solar water heater to cause fluid motion through a collector (10, 29). The density difference is created within the solar collector where heat is added to the liquid. This collection concept is called a thermosiphon. Figure 6.8 shows schematically the major components of such a system.

Since the driving force in a thermosiphon system is only a small density difference and not a pump, larger-than-normal plumbing fixtures must be used to reduce pipe friction losses (30). In general, one pipe size larger than would be used with a pump system is satisfactory. Under no conditions should piping smaller than $\frac{1}{2}$-in national pipe thread (NPT) be used. Most commercial thermosiphons use 1-in NPT pipe. The flow rate through a thermosiphon system is about 1 gal/ft² · hr (40 liters/m² · hr) in bright sun. Isometric views of Israeli and Australian thermosiphon heaters are shown in Fig. 6.9.

Since the hot-water system loads vary little during a year, the best angle of tilt is that equal to the latitude, that is, $\beta = L$. The temperature difference between the collector inlet water and the collector outlet water is usually 15–20°F during the middle of a sunny day (10). After sunset, a thermosiphon system can reverse its flow direction and lose heat to the environment during the night. To avoid reverse flow, the top header of the absorber should be at least 1 ft (30 cm) below the cold leg fitting on the storage tank, as shown.

To provide heat during long cloudy periods, an electrical immersion heater can be used as a backup for the solar system. The immersion heater is located near the

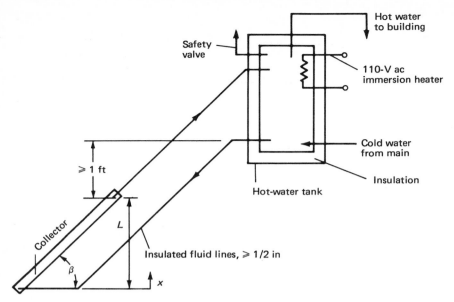

**FIGURE 6.8** Schematic diagram of thermosiphon loop used in a natural circulation, service water-heating system. The flow pressure drop in the fluid loop must equal the buoyant force "pressure" $[\int_0^L g\rho(x)\,dx - \rho_{\text{stor}}gL]$ where $\rho(x)$ is the local collector fluid density and $\rho_{\text{stor}}$ is the tank fluid density, assumed uniform.

top of the tank to enhance stratification and so that the heated fluid is at the required delivery temperature. Tank stratification is desirable in a thermosiphon to maintain flow rates as high as possible. Insulation must be applied over the entire tank surface to control heat loss.

Several features inherent in the thermosiphon design limit its utility. If it is to be operated in a freezing climate, a nonfreezing fluid must be used, which in turn requires a heat exchanger between collector and potable water storage. (If potable water is not required, the collector can be drained during cold periods instead.) Heat exchangers of either the shell-and-tube type or the immersion-coil type require higher flow rates for efficient operation than a thermosiphon can provide. Therefore, the thermosiphon is limited to nonfreezing climates unless a heat exchanger with a very large area—for example, a jacketed tank—is used.

A further restriction on thermosiphon use is the requirement for an elevated tank. In many cases structural or architectural constraints prohibit raised-tank locations. In residences, collectors are normally mounted on the roof; tanks mounted above the high point of the collector can easily become the highest point in a building. Practical considerations usually do not permit this application.

## Forced-Circulation Systems

If a thermosiphon system cannot be used for climatic, structural, or architectural reasons, a forced-circulation system is required. Figure 6.10 shows such a system

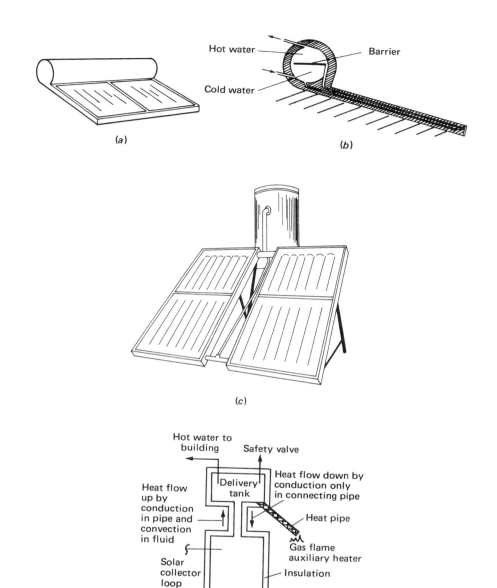

**FIGURE 6.9** Thermosiphon solar water heaters: (*a*) Australian compact model, isometric view; (*b*) Australian model, section view showing fluid compartments; (*c*) Israeli tank-and-collector assembly; (*d*) deWinter system using gravity return heat pipe for auxiliary heating and thermal diode between preheat tank and delivery tank. Adapted from (23).

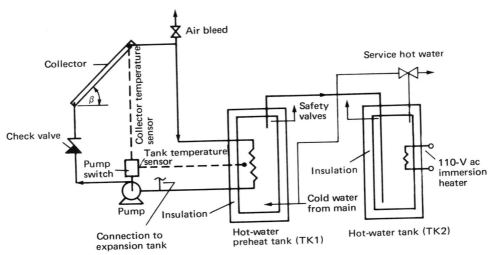

**FIGURE 6.10**    Forced-circulation, service water-heating system with optional mixing valve.

designed for use in a freezing climate. For this system one pump, a heat exchanger, a controller, piping, an expansion tank, and a check valve have been added to the basic system in Fig. 6.9. Although an immersed-coil heat exchanger is shown, an external shell-and-tube exchanger with a second pump provides much better heat transfer per unit exchanger area (see Fig. 6.14).

The preferred control strategy for a forced-circulation water-heating system is described in Chapter 5, but other procedures can be used. For example, the conventional backup heater (TK2) can be connected in parallel with the solar preheat tank (TK1). In this case no solar preheat is used. Solar-heated water is used only when it has been heated to the delivery temperature solely by the sun. This parallel-tank method results in less solar delivery per year per unit collector area, since the collector must operate at higher temperature (lower efficiency) more of the time than the system shown in Fig. 6.10 (14).

A second alternative backup interface is to add nonsolar heat directly to the tank (TK2) to maintain the required water temperature. This approach has two disadvantages—a tank maintained at the delivery temperature causes a collector to operate less efficiently (that is, at higher temperature), and a tank filled with nonsolar energy cannot be filled with solar energy.

Most of the features of water-heating systems that are common to all other solar-thermal systems are described in Chapter 5 (see sections on working fluids, heat-exchanger sizing, pump and pipe sizing, storage design, and distribution systems).

## Performance Prediction of Solar Water-Heating Systems—The f-Chart

The annual average insolation on a surface tilted at an angle equal to the latitude in the mid-latitudes of the United States is about 23,000 kJ/m$^2$ · day (2030

Btu/ft$^2$ · day). If the average collection efficiency is 30 percent, 6900 kJ/m$^2$ · day (610 Btu/ft$^2$ · day) is delivered by the collector system. The service hot-water demand can be taken as 76 liters/day (20 gal/day) per person. If the tap water has a temperature of 283 K (50°F) and the hot water, 333 K (140°F), about 210 kJ/liters (750 Btu/gal) is required for water heating. For a delivery of 230 liters/day (60 gal/day) for a family of three, 48.3 MJ/day (46,000 Btu/day) is needed. If the collector delivers 6900 KJ/m$^2$, a collector with a size of approximately 7 m$^2$ is needed. This collector is about one-tenth the size of the collector required for heating a typical building in the mid-latitudes. A rule of thumb is that 1 m$^2$ of collector is required for each 30 liters of hot water to be delivered. The storage tank should be large enough to hold about a 2-day supply of hot water, or 500 liters for a family of three.

The preceding qualitative estimation of water delivery is too crude for accurate system design. As discussed in Chapter 5, the proper method of calculation of system delivery is by means of a mathematical model of the system. However, many architects and engineers do not routinely use computers in design; therefore, a method of calculation amenable to hand calculation is necessary. Klein and his co-workers (8a, 19–22) have developed such a method. It is described in this section in summary form for liquid solar water-heating systems. *The method applies only to the system shown in Fig. 6.10 (without the mixing valve).*

Since hourly system calculations are necessary for accurate performance prediction (8), Klein's approach involved performing many hundreds of hourly simulations for various system configurations in various locations in the United States. These results were then correlated in a dimensionless fashion on charts that are general in form and usable anywhere. The charts are called f-*charts*, denoting their parameter $f_s$, the percentage of monthly load carried by solar energy. Dimensionless groups used in the f-charts are derived from a nondimensionalization of the equations governing system energy flows; the groups therefore have a physical significance as described below. The f-chart for hot-water systems is shown in Fig. 6.11.

The percentage of monthly load $f_s$ is seen to depend on two dimensionless numbers called the loss parameter $P_L$ and the solar parameter $P_s$. For a pair of values, $P_L$ and $P_s$, the f-chart will give directly a unique value for $f_s$. $P_L$ and $P_s$ are defined as follows:

$$P_L = \frac{F_{hx}F_R A_c U_c \, \Delta t(11.6 + 1.18 \, T_{w,o} + 3.86 \, T_{w,i} - 2.32 \, \bar{T}_a)}{L} \tag{6.16}$$

$$P_s = \frac{F_{hx}F_R A_c \bar{I}_c (\tau\alpha)}{L} \tag{6.17}$$

$P_s$ and $P_L$ are measures, respectively, of the long-term insolation gain by the collector-absorber surface and the long-term thermal loss from the absorber per unit load. The parameters in Eqs. (6.16) and (6.17) and in the f-chart are defined in Table 6.14. The f-chart is based on nominal values of storage size and fluid flow

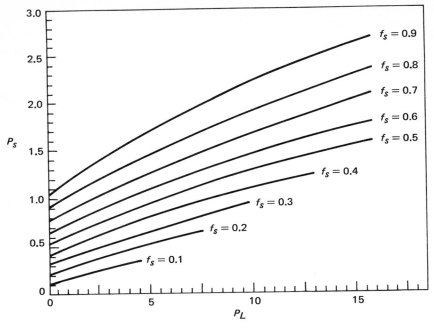

**FIGURE 6.11**  The *f*-chart for solar water-heating systems. From (19).

**TABLE 6.14**  Definition of Parameters in *f*-Chart

| Parameter | Definition |
|---|---|
| $A_c$ | Net collector aperture area ($m^2$). |
| $f_s$ | The solar load fraction: percentage of monthly load carried by solar system. |
| $F_R$ | Collector heat-removal factor [see Chapter 4, Eq. (4.32)]. |
| $F_{hx}$ | Collector loop heat-exchanger factor [see Chapter 5, Eq. (5.29)]. $F_{hx} = 1$ if no collector heat exchanger is used. |
| $\bar{I}_c$ | Total monthly collector-plane insolation (see Chapter 2) (J/month). |
| $L$ | Total monthly water-heating load (see Example 6.4) (J/month). |
| $\bar{T}_a$ | Monthly average ambient temperature (°C). |
| $T_{w,i}$ | Water supply (inlet) temperature (°C). |
| $T_{w,o}$ | Water delivery (outlet) temperature (°C). |
| $U_c$ | Collector heat-loss coefficient (see Chapter 4) ($W/m^2 \cdot °C$). |
| $\Delta t$ | Number of seconds per month (sec/month). |
| $(\overline{\tau\alpha})$ | Monthly averaged collector transmittance-absorptance product (see below). |

rate; these values are given in Table 6.16 together with factors that permit any other storage or flow rate to be used.

As shown in Chapter 4, $F_R U_c$ and $F_R(\overline{\tau\alpha})$ in the equations above can be determined directly from the slope and intercept of a collector efficiency curve plotted as shown in Fig. 6.12. In addition, Klein suggests that the time-averaged value $F_R(\overline{\tau\alpha})$ is related to the normal incidence value $F_R(\tau\alpha)_n$:

$$\frac{F_R(\overline{\tau\alpha})}{F_R(\tau\alpha)_n} = 0.95 \tag{6.18}$$

for a surface tilted within ±20° of the local latitude.

Note that Fig. 6.12 is a plot of efficiency versus collector *inlet* fluid temperature, as distinguished from *average* fluid temperature. Such curves are usually available from the majority of U.S. collector manufacturers. The *f*-chart is meant to be used to determine the average *annual* delivery of a given solar system. Figures for any given month may be less precise.

The results of the computer model used to generate the $f_s$ curves in Fig. 6.11 can also be expressed in the form of an empirical equation. One such equation (8a) is

$$f_s = 1.029\,P_s - 0.065\,P_L - 0.245\,P_s^{\,2} + 0.0018\,P_L^{\,2} + 0.0215\,P_s^{\,3} \tag{6.19}$$

valid for the range

$$0 \leqslant P_s \leqslant 3.0$$

$$0 \leqslant P_L \leqslant 18.0$$

$$0 \leqslant f_s \leqslant 1.0$$

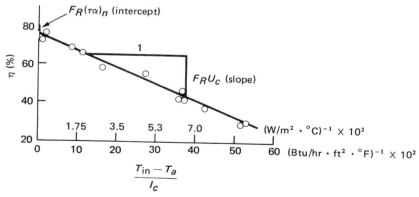

**FIGURE 6.12** Typical flat-plate collector efficiency curve showing method of evaluating $F_R(\tau\alpha)_n$ and $F_R U_c$. From (23).

It is further required that $P_s > P_L/12$ to ensure that the monthly solar radiation is above the useful threshold and that thermal losses $U_c$ are below an upper bound at which energy absorbed is equal to energy lost from the absorber plate.

A detailed example of the step-by-step application of the $f$-chart is given in the next section dealing with liquid-based space-heating systems.

The $f$-chart applies to only one system configuration, and performance predictions for this system are satisfactory for engineering purposes. The method is not to be used for any other system type. Design methods for other configurations are under development.

## LIQUID-BASED SOLAR-HEATING SYSTEMS
## FOR BUILDINGS

The earliest active solar space-heating systems were constructed from enlarged water-heating components. Experiments beginning in 1938 at the Massachusetts Institute of Technology (MIT) showed that solar heating with liquid working fluids could be done without any major technical problems. The early MIT work formed the basis of many of the design techniques used today. Other experiments after World War II provided additional fundamental information on collector designs and storage operation for liquid-based heating systems. Figure 6.13 shows a modern solar-heated, multifamily building in Colorado.

In this section configurations, design methods, and control strategies for solar-heating systems are described. Components common to all liquid-based solar-thermal systems—water heating, space heating, cooling, etc., are described in Chapter

**FIGURE 6.13** Multi-unit residence located in Boulder, Colorado, heated by an active solar-heating system. Courtesy of Joint Venture, Inc., Boulder, Colorado.

5. The specific application of these components to space heating is addressed below. Applications for heat pumps are described in a subsequent section.

Since space heating is a relatively low-temperature use of solar energy, a thermodynamic match of collector to task indicates that an efficient flat-plate collector or low-concentration collector is the thermal device of choice. The application of medium-concentration collectors to space heating is warranted, however, in large systems where the large energy density of steam dictates its use for the distribution fluid. Steam is readily produced by efficient concentrators with concentration ratios on the order of 10.

## Physical Configurations of Heating Systems

Figure 6.14 is a schematic diagram of a typical space-heating system. The system consists of three fluid loops—collector, storage, and load. In addition, most space-heating systems are integrated with a domestic, water-heating system to improve the yearlong solar load factor (see additional fluid loops in Fig. 6.14).

The collector fluid loop contains fluid manifolds, the collectors, the collector pump and heat exchanger, an expansion tank, and other subsidiary components. In many systems a device for heat rejection is required in order to purge excess energy collected during periods of low load. The details of this device are described later. A collector heat exchanger and antifreeze in the collector loop are normally used in all solar space-heating systems, since the existence of a significant heating demand implies the existence of some subfreezing weather.

The storage loop contains the storage tank and pump as well as the tube side of the collector heat exchanger. To capitalize on whatever stratification may exist in the storage tank, fluid entering the collector heat exchanger is generally removed from the bottom of storage. This strategy ensures that the lowest temperature fluid available in the collector loop is introduced at the collector inlet for high efficiency.

The energy delivery-to-load loop contains the load device—baseboard heaters or fin-and-tube coils—and the backup system with a flow control (mode selector) valve. The usual complement of valves, a safety valve, and an expansion tank are generally employed. The water heating loops are of a design described in the preceding section and are not discussed further herein.

### Solar Collector Orientation

The best solar collector orientation is such that the average solar incidence angle is smallest during the heating season. For tracking collectors this objective is automatically realized. For fixed collectors in the northern hemisphere the best orientation is due south, tilted up from the horizon at an angle of about 15° greater than the local latitude.

Although due south is the optimum azimuthal orientation for collectors in the northern hemisphere, variations of 20° east or west have little effect on annual energy delivery (23). Off-south orientations greater than 20° may be required in some cases because of obstacles in the path of the sun. These effects may be analyzed using sun-path diagrams and shadow-angle protractors as described in Chapter 2.

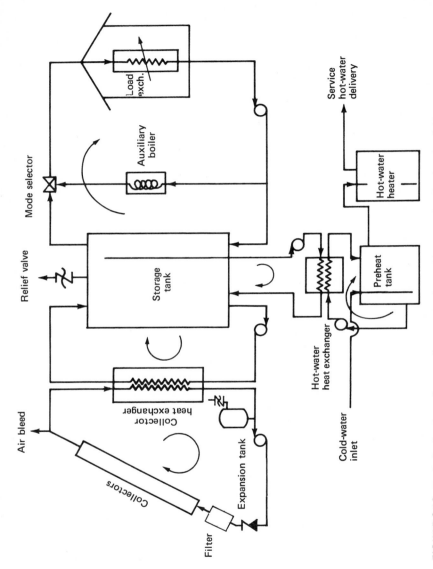

**FIGURE 6.14** Typical solar-thermal system for space heating and hot-water heating showing fluid transport loops (five) and pumps.

## Number of Glass Covers

Löf and Tybout (25) found that the optimum number of collector covers varies with climate, as shown in Table 6.15. The results in Table 6.15 were based on a nonselective absorber surface coating. In some locations a selective surface can be traded for the second collector cover. This assumption must be checked for each application, however, since selective surface properties vary from one manufacturer to another.

## Fluid Flow Rates

For the maximum energy collection in a solar collector, it is necessary that it operate as closely as possible to the lowest available temperature—the collector inlet temperature. Very high fluid flow rates are needed to maintain a collector-absorber surface nearly isothermal at the inlet temperature. Although high flows maximize energy collection, practical and economic constraints put an upper limit on useful flow rates. Very high flows require large pumps and excessive power consumption and lead to fluid conduit erosion.

Figure 6.15 shows the effect of mass flow rate on annual energy delivery from a solar system. It is seen that the law of diminishing returns applies and that flows beyond about 50 kg/hr $\cdot$ m$^2$ ($\sim$ 10 lb/hr $\cdot$ ft$^2$) have little marginal benefit for collectors with loss coefficients on the order of 6 W/m$^2$ $\cdot$ °C ($\sim$ 1 Btu/hr $\cdot$ ft$^2$ $\cdot$ °F). In practice, liquid flows in the range of 50–75 kg/m$_c^2$ $\cdot$ hr (10–15 lb/hr $\cdot$ ft$_c^2$) of water equivalent are the best compromise among collector heat-transfer coefficient, fluid pressure drop, and energy delivery, although an infinitely large flow rate will deliver the most energy if pumping power is ignored for a nonstratified storage. If storage stratification is desired, lower flow rates must be used, since high flow destroys stratification.

In freezing climates, an antifreeze working fluid is recommended for collectors. Attempts to drain collectors fully for freeze protection have usually been unsuccessful unless collector fluid conduits are very large and smooth and unless all piping is sloped to assure drainage. The potential damage risk from incomplete draining will usually dictate that the additional investment in antifreeze be made.

**TABLE 6.15** Optimal Number of Collector Covers for Space Heating[a]

| Climate type | Example location | No. of covers |
|---|---|---|
| Tropical and subtropical desert | Phoenix | 1 |
| Tropical savannah | Miami | 1 |
| Tropical or subtropical steppe | Albuquerque | 1 or 2 |
| Mediterranean or dry summer subtropical | Santa Barbara | 1 or 2 |
| Humid subtropical | Charleston | 2 |
| Marine West Coast | Seattle | 2 |
| Humid continental, warm summer | Omaha | 2 |
| Humid continental, cool summer | Boston | 2 |

[a]Adapted from (25).

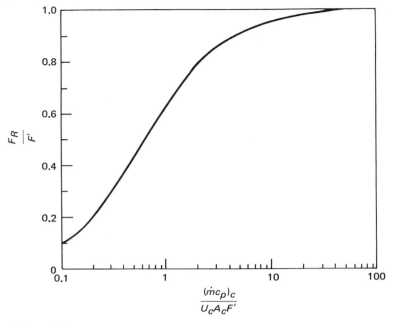

**FIGURE 6.15** Effect of fluid flow rate on collector performance as measured by the heat-removal factor $F_R$; $F'$ is the plate efficiency factor (see Chapter 4).

## Thermal Storage

Thermal storage tanks must be insulated to control heat loss, as described in Chapter 5. If a storage tank is located within a structure, any losses from the tank tend to offset the active heating demands of the building. However, such storage loss is uncontrolled and may cause overheating in seasons with low heat loads. Some solar-heating systems have used a ventilatable structure surrounding storage. This enclosure may be vented to the building interior in winter and to the environment in summer.

Safety concerns may cause storage to be located external to a building in some cases. Large volumes of hot water could be released to a building interior if a storage tank were to fail. Potential personal injury or property damage, which could result from an accident, must be assessed in siting storage tanks. Tank burial would seem to be the safest approach in some cases. Buried storage tanks must be sealed from ground moisture, insulated with waterproof insulation, and galvanically protected.

The amount of thermal storage used in a solar-heating system is limited by the law of diminishing returns. Although larger storage results in larger annual energy delivery, the increase at the margin is small and hence not cost-effective. Seasonal storage is, therefore, usually uneconomic, although it can be realized in a technical sense. Experience has shown that liquid storage amounts of 50–75 kg/m$^2$ (10–15 lb/ft$^2$) are the best compromise between storage tank cost and useful energy delivery. Klein (19) has calculated the effect of storage size on annual energy delivery. His results, shown in Fig. 6.16, exhibit the expected diminishing returns to scale.

Since solar energy heating systems operate at temperatures relatively close to the temperatures of the spaces to be heated, storage must be capable of delivering and receiving thermal energy at relatively small temperature differences. The designer must consider the magnitude of these driving forces in sizing heat exchangers, pumps, and air blowers. The designer must also consider the nonrecoverable heat losses from storage—even though storage temperatures are relatively low, surface areas of storage units may be large and heat losses therefore appreciable.

Some investigators have proposed heating the storage medium with conventional fuels to maintain its temperature at useful levels during sunless periods. This approach has two major flaws:

1. If storage is heated with conventional fuels, it cannot be heated with solar energy when available; therefore, some collected solar energy cannot be used.
2. If storage is partially heated with conventional fuels, the collector inlet temperatures will be higher and efficiency lower than it would be if storage were not boosted. Therefore, the useful return on the solar system investment would be diminished.

In conclusion it should be emphasized that storage heating with conventional fuels is uneconomic in any practical solar-thermal system designed to date.

## Other Mechanical Components

Heat exchangers in solar systems are selected based on economic criteria described in Chapter 5. The best trade-off of energy delivery increase with increasing heat-exchanger size usually results from use of an exchanger with effectiveness in the range of 0.6–0.8. Counterflow heat exchangers are required for this level of effectiveness. A detailed example of heat-exchanger selection is contained in Chapter 5.

Achievement of the required effectiveness level may dictate fairly high flow rates in the storage tank side of the collector heat exchanger. Flows up to twice that in the collector side can improve exchanger performance significantly in many cases. Since the storage side loop is physically short and has a small pressure drop, increased flow in this loop increases pump energy requirements by a negligible amount. Typical solar heat-exchanger sizes range from 0.05 to 0.10 m$^2$ of heat-exchanger surface per square meter of net collector area.

In hydronic heating systems it is essential that all air be pumped from the system. To facilitate this process, air bleed valves located at the high points in a system are used. These are opened during system fill and later if air should collect. Air bleeds are also required at points of low velocity in a piping system where air may collect because the local fluid velocity is too low for entrainment.

Pressure-relief valves sized to flow the maximum possible blowoff rate are used. In the following section, devices for heat rejection are described.

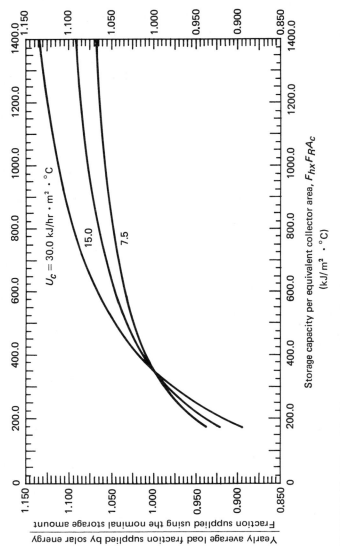

**FIGURE 6.16** Effect of liquid storage capacity on liquid-based solar-heating system energy delivery. From (19).

## Subsystems for Heat Rejection

A solar-heating system that carries an appreciable fraction of the annual heating load of a building will generate more heat than is required in spring and fall—the lower load periods of the heating season. High collector incidence angles in summer usually result in less excess heat generation during that season than during spring or fall. Some means of rejecting this heat must be employed. Several approaches can be used:

1. Drain collectors. This has the disadvantages of possible thermal damage to collector and complete cutoff of off-season water heating.
2. Boil storage. This has the disadvantage of mineral buildup from hardness salts in the storage tank.
3. Active rejection by heat exchanger. This is reliable but additional investment is needed.

The first method accomplishes heat rejection by allowing collector temperature to rise to its no-net-output thermal equilibrium point with the environment. For a well-designed flat-plate collector this temperature will approach 200°C (about 400°F) on a sunny day. Elevated temperature over long periods can degrade collector components such as the insulation and the dust-rain seals. In addition, the thermal expansion of large arrays can cause extra expense in structural members.

Heat rejection by boiling of storage can be used in some areas if water hardness is low. The system controller activates the boiling mode by causing the collector and storage pumps to operate whenever storage is near its boiling point. This is generally an extra feature of controllers independent of the normal differential temperature operation. Water boiled from storage can automatically be replaced by using a float valve located within the tank or by a differential pressure water valve, which senses the drop in storage loop pressure at night resulting from water volume loss from boiling during the day. To control the mineral buildup in boiling storage, the entire storage tank must be drained or blown down periodically.

Active devices for heat rejection are heat exchangers in the collector loop connected in parallel or in series with the storage heat exchanger. Figure 6.17 shows both a liquid-to-air heat rejector and a liquid-to-liquid subsystem. All liquid systems require a liquid heat sink such as a cooling tower, river, or cooling pond. Heat sinks of this type are generally used on large solar installations whereas other methods are used on single-family dwellings. Active heat rejectors are activated by the solar system controller above a preset storage tank temperature. The mode selector shown in Fig. 6.17 cuts off flow to the collector heat exchanger and diverts it to the purge device instead. Auxiliary fans and pumps required by the rejector subsystem are activated at the same time.

## Controls in Liquid Systems

Control strategies and hardware used in current solar system designs are quite simple and are similar in several respects to those used in conventional systems. The single fundamental difference lies in the requirement for one differential measurement

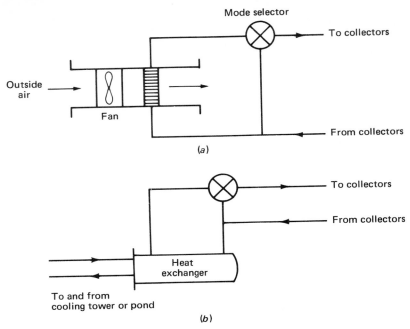

**FIGURE 6.17** Schematic diagrams of typical subsystems for heat rejection for use with liquid systems: (*a*) liquid-to-air; (*b*) liquid-to-liquid. The subsystem is connected in parallel with the collector heat exchanger.

instead of simple temperature sensing. In the space heating system shown in Fig. 6.14, two temperature signals determine which of three modes is used. The signals used are the collector-storage differential and room temperature. The collector-storage difference is sensed by two thermistors, diodes, or transistors, the difference being determined by a solid-state comparator, which is a part of the control device. Room temperature is sensed by a conventional dual-contact thermostat.

The control system operates as follows. If the first room thermostat contact closes, the mode selector valve and distribution pump are activated in an attempt to deliver the thermal demand from solar-thermal storage. If room temperature continues to drop, indicating inadequate solar availability, the mode selector diverts flow through the backup system instead of the solar system and the backup is activated until load is satisfied.

The collector-storage control subsystem operates independently of the heating subsystem described above. If collector temperature, usually sensed by a thermistor thermally bonded to the absorber plate, exceeds the temperature in the bottom of the storage tank by 5-10°C (9-18°F), the collector pump and heat-exchanger pump (if present) are activated and continue to run until the collector and storage temperature are within about 1-2°C (2-4°F) of each other. At this point it is no longer worthwhile to attempt to collect energy and the pump(s) are turned off. The collector-storage subsystem also has a high temperature cutout that activates a heat rejector if storage is above the maximum design temperature—usually 93–96°C (200–205°F) for nonpressurized storage at sea level. The heat-rejection method used can be a cooling tower, an industrial liquid-to-air heat exchanger, or a collector draindown system as described above.

Controls are usually of solid-state design, and several commercial manufacturers are producing such devices. Controls for air systems are similar to those for liquid systems except that more mode selector valves (dampers) are generally used. In addition, if only one blower is used, it may need to be a two-speed air handler.

In summary, the design of controls for solar systems employs principles long used in the heating industry. The methods described above are simple and adequate to control systems properly. However, new controls utilizing microprocessors and other advanced concepts offer the possibility of improving overall operation of not only the solar system but also that of the entire building.

## Load Devices in Liquid Solar-Heating Systems

A heating load device converts heat contained in the working liquid in a solar-heating system to heat usable for warming a space. It is, therefore, a liquid-to-air heat exchanger that must be sized based on the energy demand of a building. Several generic types of load devices are in common use.

1. Forced-air systems—tube-and-fin coil located in the main distribution duct of a building or zone of a building (see Fig. 6.18).
2. Baseboard convection systems—tube-and-fin coils located near the floor on external walls. These operate by natural convection from the convectors to the room air.
3. Heated floors or ceilings—water coils. These transfer heat to large thermal masses that in turn radiate or convect into the space. Usually called radiant heating.

Each load device requires fluid at a different temperature in order to operate under design load conditions as shown in Fig. 6.19. Since baseboard heaters are

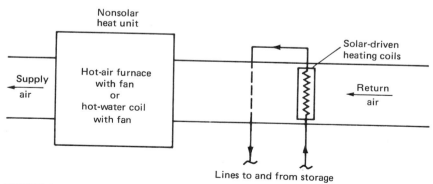

**FIGURE 6.18** Forced-air heating system load device location upstream of nonsolar heat exchanger or furnace.

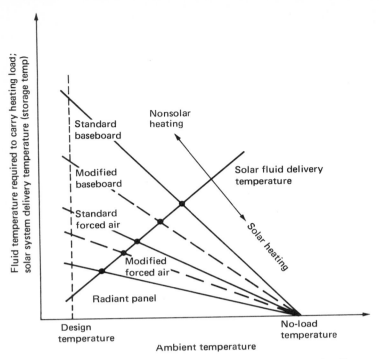

**FIGURE 6.19** Heating load diagram for baseboard, forced-air, and radiant systems. Modified baseboard and forced-air systems are oversized in order to carry heating demands at lower temperature. Balance points are indicated by large dots at intersections.

small in heat-transfer area and rely on the relatively ineffectual mechanism of natural convection, they require the highest fluid temperature. Forced-air systems involve the more efficient forced-convection heat-transfer mode and hence are operable at lower fluid temperature (see Fig. 6.19). Radiant heating can use very large heat-transfer areas and is, therefore, operable at relatively low fluid temperatures.

In Fig. 6.19 the intersection of the solar fluid temperature line and the load line for a specific configuration is called the balance point. At ambient temperatures below the balance point solar energy cannot provide the entire demand, and some backup is required; above the balance point solar capacity is sufficient to carry the entire load. Note that the load lines are specific to a given building. The solar fluid temperature line is not fixed for a building but depends on solar collector and storage size as well as local solar radiation levels. The line shown in Fig. 6.19 is, therefore, an average line. The instantaneous solar line changes continuously in response to load and climatic forcing functions as described in Chapter 5.

It is possible to modify load devices to lower the balance point, as shown in Fig. 6.19. For example, a forced-air tube-and-fin exchanger can be enlarged by adding one

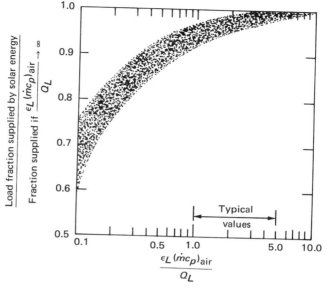

**FIGURE 6.20** Effect of load device capacity for forced-air heating systems; $Q_L$ is the building heat load expressed in units consistent with $(\dot{m}c_p)_{air}$. From (19).

or more additional rows of tubes. This increased heat-transfer area will permit the same energy delivery at a lower fluid temperature. Figure 6.20 depicts the effect of forced-air load device size (heat-transfer effectiveness $\epsilon_L$) on annual energy delivery. The law of diminishing returns is evident as increasing effectiveness returns progressively less energy. The effectiveness of a cross-flow heat exchanger of the type used in forced-air systems is calculated in Chapter 3 and shown in Fig. 6.21.

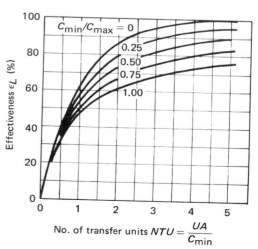

**FIGURE 6.21** Heat-exchanger effectiveness for cross-flow, unmixed-fluid load devices of the type used in forced-air solar-heating systems. From "Compact Heat Exchangers" by W. M. Kays and A. L. London. Copyright 1964 by McGraw-Hill Book Co. Used with permission of McGraw-Hill Book Co.

# Performance Prediction for Liquid-Based Solar-Heating Systems

Prediction of the energy delivery of a solar-heating system is most accurately carried out by a mathematical modeling approach as described in Chapter 5. The use of hourly solar or weather data for several years is necessary for an accurate estimate of long-term system performance. The f-chart approach of Klein (8a, 19) described above for water-heating systems can also be used for liquid, space-heating systems. This method has the essential advantage of an accuracy nearly as good as that of the hourly method but it permits the use of monthly solar and climatic data. *The space-heating f-chart applies only to the system type shown in Fig. 6.14.*

## The f-Chart for Space Heating

The f-chart for space-heating systems is shown in Fig. 6.11. It is identical to that for water-heating systems, except for the definition of the loss parameter $P_L$. The dimensionless solar $(P_s)$ and loss $(P_L)$ parameters for liquid-based heating systems are defined as

$$P_s = \frac{F_{hx}F_R A_c \overline{I_c}(\overline{\tau\alpha})}{L} \tag{6.20}$$

**TABLE 6.16** Nominal Values of Physical Parameters and Modifying Groups for f-Chart Use[a]

| Parameter | Nominal value | Modified parameter[b] |
|---|---|---|
| Flow rate[c] $\dfrac{(\dot{m}c_p)_c}{A_c}$ | 0.0128 liters $H_2O$ equivalent/ sec $\cdot$ $m_c^2$ | $P_L = P_{L,\,nom}\dfrac{F_R}{(F_{hx}F_R)_{nom}}\dfrac{F_{hx}F_R}{}$  $P_s = P_{s,\,nom}\dfrac{F_{hx}F_R}{(F_{hx}F_R)_{nom}}$ |
| Storage volume (water) $V_s = \left(\dfrac{M}{\rho A_c}\right)_s$ | 75 liters $H_2O/m_c^2$ | $P_L = P_{L,\,nom}\left(\dfrac{V_s}{75}\right)^{-0.25}$ |
| Load heat exchanger $\dfrac{\epsilon_L(\dot{m}c_p)_{air}}{Q_L}$ | 2.0 | $P_s = P_{s,\,nom}$ $\left\{0.393 + 0.651\exp\left[-0.139\dfrac{Q_L}{\epsilon_L(\dot{m}c_p)_{air}}\right]\right\}$ |

[a]Table prepared from data and equations presented in (8a, 19).
[b]Multiply basic definition of $P_s$ and $P_L$ in Eqs. (6.20) and (6.21) by factor for nonnominal group values; $(F_{hx}F_R)_{nom}$ refers to values of $F_{hx}F_R$ at collector rating or test conditions.
[c]In liquid systems the correction for flow rate is small and can usually be ignored if variation is no more than 50 percent below the nominal value.

and

$$P_L = \frac{F_{hx} F_R A_c U_c \, \Delta t}{L}(T_r - \bar{T}_a) \tag{6.21}$$

All parameters in these equations are defined in Table 6.14. $T_r$ is a constant reference temperature with a value of 100°C (212°F).

The $f$-chart in Fig. 6.11 is based on nominal values of collector flow rate, thermal storage mass, and load heat-exchanger effectiveness. It is possible to use the $f$-chart for other values of these important parameters if the loss parameter and a solar parameter are appropriately modified. Table 6.16 contains the nominal values used in generating the $f$-chart and the dimensionless group that is used to modify the loss or the solar parameter for other system values.

### EXAMPLE 6.6

Calculate the annual heating energy delivery of a solar space-heating system using a double-glazed flat-plate collector in Bismarck, North Dakota. The building and solar system specifications are given below. Use English units.

#### Building Specifications

Location: 47°N latitude (see Appendix 2, Table A2.2)
Space-heating load: 15,000 Btu/°F · day

#### Solar System Specifications

Collector loss coefficient: $F_R U_c = 0.80$ Btu/hr · ft$^2$ · °F
Collector optical efficiency (average): $F_R(\overline{\tau\alpha}) = 0.70$
Collector tilt: $\beta = L + 15° = 62°$
Collector area: $A_c = 650$ ft$^2$
Collector fluid flow rate: $\dot{m}_c/A_c = 11.4$ lb H$_2$O/hr · ft$_c$$^2$
Collector fluid heat capacity: $c_{p_c} = 0.9$ Btu/lb · °F (antifreeze)
Storage capacity: 17 lb H$_2$O/ft$_c$$^2$
Storage fluid flow rate: $\dot{m}_s/A_c = 20$ lb H$_2$O/hr ·ft$_c$$^2$
Storage fluid heat capacity: $c_{p_s} = 1$ Btu/lb · °F (water)
Heat-exchanger effectiveness: 0.75
Load heat exchanger: $\epsilon_L(\dot{m}c_p)_{\text{air}}/Q_L = 2.0$

#### Climatic Data

Climatic data from the NWS are tabulated in Table 6.17.

### SOLUTION

The $f$-chart method is amenable to a step-by-step application. The following order is suggested.

**TABLE 6.17**  Climatic and Solar Data for Bismarck, North Dakota

| Month | Avg. ambient temperature (°F) | Heating degree F-days | Horizontal solar radiation (langleys/day) |
|-------|------|------|------|
| Jan. | 8.2 | 1761 | 157 |
| Feb. | 13.5 | 1442 | 250 |
| Mar. | 25.1 | 1237 | 356 |
| Apr. | 43.0 | 660 | 447 |
| May | 54.4 | 339 | 550 |
| June | 63.8 | 122 | 590 |
| July | 70.8 | 18 | 617 |
| Aug. | 69.2 | 35 | 516 |
| Sep. | 57.5 | 252 | 390 |
| Oct. | 46.8 | 564 | 272 |
| Nov. | 28.9 | 1083 | 161 |
| Dec. | 15.6 | 1531 | 124 |

1. Calculate monthly collector-plane insolation for each month.
2. Calculate solar and loss parameters $P_s$ and $P_L$ [Eqs. (6.20) and (6.21)] for each month and heat-exchanger penalty factor $F_{hx}$ (Eq. 5.30).
3. Evaluate $f_s$ from the $f$-chart for each month.
4. Calculate total annual energy delivery from monthly totals.

Each of these steps is shown in Table 6.18. The heat-exchanger penalty factor $F_{hx} = 0.974$ [from Eq. (5.30)].

**TABLE 6.18**  The $f$-Chart Summary for Example 6.6

| Month | Collector-plane radiation (Btu/day · ft²) | Monthly energy demand (MMBtu) | $P_L$ [a] | $P_s$ [a] | $f_s$ |
|-------|------|------|------|------|------|
| Jan. | 1506 | 26.41 | 2.68 | 0.72 | 0.46 |
| Feb. | 1784 | 21.63 | 2.88 | 0.95 | 0.60 |
| Mar. | 1795 | 18.55 | 3.50 | 1.22 | 0.73 |
| Apr. | 1616 | 9.90 | 5.75 | 2.00 | 0.94 |
| May | 1606 | 5.08 | 10.78 | >3.00 | 1.00 |
| June | 1571 | 1.83 | >20.00 | >3.00 | 1.00 |
| July | 1710 | 0.27 | >20.00 | >3.00 | 1.00 |
| Aug. | 1712 | 0.52 | >20.00 | >3.00 | 1.00 |
| Sep. | 1721 | 3.78 | 13.76 | >3.00 | 1.00 |
| Oct. | 1722 | 8.46 | 6.79 | 2.58 | 1.00 |
| Nov. | 1379 | 16.24 | 3.79 | 1.04 | 0.61 |
| Dec. | 1270 | 22.96 | 2.98 | 0.70 | 0.43 |
| Annual | — | 135.66 | — | — | 0.65 |

[a] $P_s > 3.0$ or $P_L > 18.0$ implies $f_s = 1.0$; no correction for storage size and flow rates is required.

The results of the computer simulations on which the $f$-chart is based can be represented by an empirically derived function. One such polynomial expression (19) is

$$f_s = 1.029\, P_s - 0.065\, P_L - 0.245\, P_s{}^2 + 0.0018\, P_L{}^2 + 0.0215\, P_s{}^3 \qquad (6.19)$$

valid for the range

$$0 \leqslant P_s \leqslant 3.0$$

$$0 \leqslant P_L \leqslant 18.0$$

$$0 \leqslant f_s \leqslant 1.0$$

It is further required that $P_s > P_L/12$ to ensure that the monthly insolation is above the useful threshold and that thermal losses are below an upper bound at which energy absorbed is equal to energy lost from the absorber plate.

### The SLR *Method of Delivery Estimation*

Balcomb and Hedstrom (4) have developed a method for predicting monthly delivery of a solar-heating system based on hourly calculations for 25 U.S. locations. The method is slightly simpler to use than the $f$-chart method but it lacks the generality of the latter since it is not based on a nondimensionalization of the governing equations. Effects of storage size, heat-exchanger properties, collector parameters $[F_R U_c$ and $F_R(\overline{\tau\alpha})]$, fluid flow rates, and properties of the load device cannot be incorporated into the method in analytical form, whereas these important parameters can be changed analytically in the $f$-chart approach.

The *SLR* (solar load ratio) method is based on the empirical correlation of hourly computer model simulations of energy delivery of a "nominal" solar system. The nominal system is based on liquid design guidelines described in previous sections of this chapter. These specifications are not detailed herein, since it is not possible to adjust them for other situations. Monthly energy delivery is correlated in terms of the *SLR* defined as

$$SLR \equiv \frac{A_c \overline{I}_c}{L} \qquad (6.22)$$

where $A_c$ is the net collector area, $\overline{I}_c$ is the monthly collector-plane solar radiation, and $L$ is the monthly energy demand. The *SLR* is similar to the solar parameter $P_s$ of the $f$-chart but it does not include the important optical properties of the collector.

The monthly fraction of load $f_s$ delivered by the solar system can be expressed in terms of the *SLR* (for $SLR < 5.66$) by the empirical equation

$$f_s = 1.06 - 1.366e^{-0.55SLR} + 0.306e^{-1.05SLR} \qquad (6.23)$$

For $SLR > 5.66$, $f_s = 1.00$.

## EXAMPLE 6.7

Repeat the preceding example using the $SLR$ method. Compare the results with those of that example, in which the $f$-chart method was used.

## SOLUTION

Calculate the $SLR$ using $\bar{I}_c$ and $L$ values from the previous example (see Table 6.18) for $A_c = 650$ ft$^2$. The solution is presented in tabular form in Table 6.19.

The table shows that the $SLR$ method consistently underpredicts the $f$-chart during months of significant heating demand. Although the nominal $SLR$ system configuration differs from that in the example, it is not possible to make identical comparisons because of the restrictive nature of the $SLR$ assumptions. Neither the $f$-chart method nor the $SLR$ method has been compared with a sufficient number of actual buildings to know the accuracy of either in predicting the performance of buildings in the field. Until such comparisons are made, the $f$-chart is recommended as the method of choice, since it includes the effects of many more system characteristics than does the $SLR$ method.

**TABLE 6.19**  Comparison of $f$-Chart and $SLR$ Values for Example 6.7

| Month | Monthly energy demand $L$ [a] (MMBtu) | Avg. monthly collector-plate solar radiation $\bar{I}_c$ [a] (Btu/day · ft²) | SLR | $f_s$ (SLR) | $f_s$ (f-chart) |
|---|---|---|---|---|---|
| Jan. | 26.41 | 1506 | 1.15 | 0.43 | 0.46 |
| Feb. | 21.63 | 1784 | 1.50 | 0.52 | 0.60 |
| Mar. | 18.55 | 1795 | 1.95 | 0.63 | 0.73 |
| Apr. | 9.90 | 1616 | 3.18 | 0.83 | 0.94 |
| May | 5.08 | 1606 | >5.66 | 1.00 | 1.00 |
| June | 1.83 | 1571 | >5.66 | 1.00 | 1.00 |
| July | 0.27 | 1710 | >5.66 | 1.00 | 1.00 |
| Aug. | 0.52 | 1712 | >5.66 | 1.00 | 1.00 |
| Sep. | 3.78 | 1721 | >5.66 | 1.00 | 1.00 |
| Oct. | 8.46 | 1722 | 4.10 | 0.92 | 1.00 |
| Nov. | 16.24 | 1379 | 1.66 | 0.57 | 0.61 |
| Dec. | 22.96 | 1270 | 1.11 | 0.41 | 0.43 |
| Annual | 135.66 | — | — | 0.59 | 0.65 |

[a]See Table 6.18.

# SOLAR AIR-HEATING SYSTEMS

Air has been used as the working fluid in solar-heating systems since World War II. Although demonstrated in fewer buildings than liquid systems, air systems have several advantages that can lead to their use in smaller installations in single- and multifamily residences. In addition, air systems are well suited to crop drying and air preheating in certain processes.

Table 6.20 lists some of the advantages and disadvantages of air and liquid systems. Several of the disadvantages of an air system follow from the poor heat-transfer properties of air. Large space requirements for ducts preclude the use of air systems in large buildings because of space limitations. This is usually not a disadvantage in air systems used in residences. However, the thermal performance and costs for both systems are nearly identical if the air system does not have leaks.

In all air-based solar-heating systems a particle-bed storage device is used. Although water offers a higher storage energy density than most solids, the difficulty of economically exchanging heat from an air stream to a liquid storage fluid stream precludes its use. Storage controls and operating parameters are described in detail in a subsequent section.

Flat-plate collectors are usually required for air-heating systems to provide sufficient area to effectively transfer heat from the absorber plate to the air stream. Concentrators, with their higher absorber surface solar fluxes, present significant problems in effective heat removal from the absorber when air is used. A flat-plate collector provides a good thermodynamic match for temperature demands in space-heating applications. Air-cooled flat-plate collectors are described in Chapter 4.

## Heating System Physical Configuration

In many ways liquid-based and air-based solar systems are similar in operation. Figure 6.22 is a schematic diagram of a typical air-heating system. Similar flow regimes and collector orientations are used in both types of systems. The air system is simpler, however, since a collector heat exchanger and associated pumps, pipes, and expansion tanks are not present. The use of pebble-bed storage is advantageous because of the stratification that results. Stratification ensures a low inlet temperature to the collector; the collector inlet temperature is approximately the temperature of the cool zone of storage, which in turn is the building return air temperature ($\sim 15$–$20°C$). Cool collector inlet temperatures are essential to efficient operation of an air system. Since stratification cannot be achieved with isothermal phase-change storage media, they are not suitable for use with air-based systems. Frequently, more dampers are required than shown in Fig. 6.22 to prevent leakage and backflows through a cold collector via a leaky damper. Careful design of dampers and actuators is essential.

The collector-to-storage loop consists of insulated ducts, collector manifolds, and a hot-water preheat exchanger. These tube-and-fin exchangers provide some preheat to hot water if placed at the collector outlet to ensure their exposure to the

**TABLE 6.20** Advantages and Disadvantages of Air and Liquid Space-Heating Systems

| Air Systems | |
|---|---|
| Advantages | Disadvantages |
| No freezing problem | Space heating only |
| No internal corrosion problem with dry air | Large space requirements for ducts |
| Leaks of smaller consequence | Larger storage volume required for rocks |
| No heat exchanger between collector-storage and storage-building loops | Cannot store heat and heat building at the same time—major problem in low load seasons |
| No boiling or pressure problems | Low ($\rho c_p$) product for air |
| Easy for do-it-yourselfers | |
| Simple and reliable | |

| Liquid Systems | |
|---|---|
| Advantages | Disadvantages |
| Higher transport energy density | Freezing problem |
| Better heat-transfer properties | Leakage problem |
| Water storage has higher energy density | Corrosion problem—water chemistry needs monitoring |
| Suitable for space heating and cooling | Heat exchanger required for collector-storage and storage-building loops |
| Small fluid conduits | Boiling and fluid expansion provisions required |

hottest air available. Collector flow balancing is achieved by controlling the pressure drop through each collector and using equal duct lengths for each as shown in Figure 5.10.

Figure 6.23 shows a hot-air system. During the storage-charging mode (*b*), heated air flows through the rock bed at a low flow rate determined by the desired temperature rise in the collector, usually 2–3 ft$^3$/min · ft$_c$$^2$. As progressively more heat is stored, the interface between the hot and cold regions of storage moves downward in the storage bin. The air returning to the collector is at the temperature of the cool region of storage.

During daytime heating on a sunny day (*a*), air from the collector is diverted to the building instead of to storage. During sunny periods when no heating demand exists, storage is charged by warm air from the collector. During the nighttime or cloudy daytime heating modes, heat is removed from storage by a counterflow of air through the rock bed, as shown in (*c*). The outlet temperature from storage is close to the daytime collection (inlet) temperature, since the air being heated passes through the hottest zone of storage last. As progressively more heat is removed from storage, the interface between the hot and cold regions of storage moves upward.

The storage medium for air systems has typically been 1- to 2-in-diameter granite or river-bed rocks, which cost approximately $5.00–$10.00/ton. An air filter is required between the heated space and storage to eliminate dust buildup

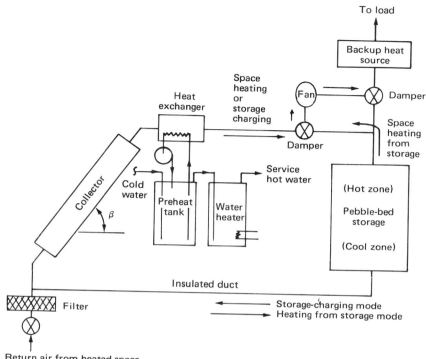

**FIGURE 6.22**  Typical air-based solar space-heating system showing directions of fluid flow and important components.

in the gravel bed (dust would reduce the heat-transfer coefficient to the rock pieces and increase the bed pressure drop). The recommended amount of storage to be used is roughly the same as for a liquid system—about 60 lb of rock (0.75 $ft^3/ft_c^2$, 0.25 $m^3/m_c^2$) per square foot of collector. Design information on the expected pressure drop through pebble-bed storage is contained in Chapter 7.

## Collector Designs

Several air-cooled collector designs are shown in Fig. 6.24. However, relatively few performance data for such collectors have been published to provide designers with the information necessary to predict system performance and determine optimal system configurations reliably.

One of the early computer-aided design studies that does provide guidelines for the designer of air-collector systems was conducted by Balcomb and his co-workers (6, 7). Figure 6.25 shows some of their practical results; Fig. 6.25a depicts the variation of solar delivery with storage volume. It is clear that storage volumes in excess of 60 lb/ft² improve the annual solar delivery by a negligible amount. Likewise, Fig. 6.25b indicates that collector air flow rates greater than 3 ft³/min · ft²

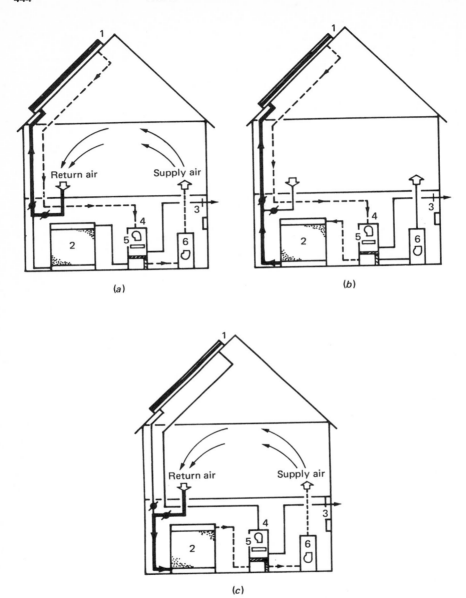

**FIGURE 6.23** Schematic diagram of operating modes of solar air system: (*a*) space heating from collector; (*b*) storing solar heat; (*c*) space heating from storage. Components are (1) collector, (2) storage, (3) control, (4) fan, (5) hot-water preheater (optional), (6) backup heater. Courtesy of the Solaron Corp.

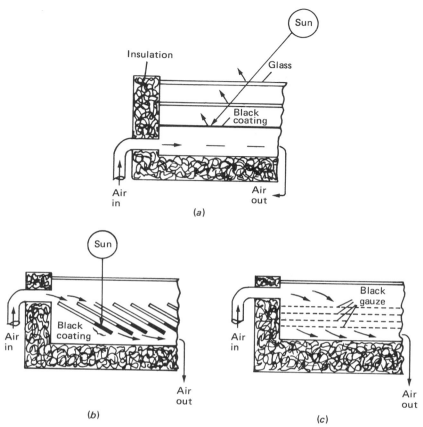

**FIGURE 6.24** Air-cooled solar collectors: (*a*) double-glazed absorber plate; (*b*) single-glazed overlapping glass plate; (*c*) single-glazed porous absorber. Adapted from *Proc. World Symp. on Appl. Solar Energy, Phoenix, Ariz., 1955*, published by Stanford Research Inst., Menlo Park, Calif., 1956.

increase delivery very little. Figure 6.25*c* and *d* shows temperature profiles for a particular rock bed modeled for various times of day for both the storage-charging and -discharging modes. Any properly sized storage will have similar temperature distributions; the key property is that the outlet temperature from storage should be low to permit the inherently inefficient air collector to operate at as low a temperature as possible. The optimal geometric configuration of a storage bed is roughly cubical in order to minimize the air pressure drop.

## Fluid Flow Rates

The flow rate through an air collector and storage loop is a compromise between pressure drop and energy transfer to the air stream. As shown in Figs. 6.25 and 6.26, the effect of air capacitance rate $(\dot{m}c_p)_c$ can be significant. Apart from the

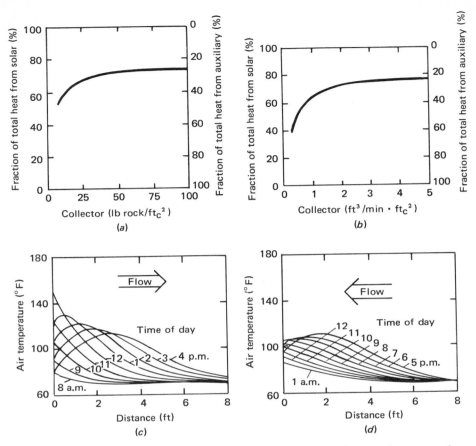

**FIGURE 6.25** Hot-air solar system operating parameters: (*a*) effect of storage size on annual energy delivery; (*b*) effect of collector air flow rate on annual energy delivery; (*c*) typical temperature distribution in rock bed during storage heating by solar energy; (*d*) typical temperature distribution in rock bed during heat removal from storage. Adapted from (6).

beneficial effect of increased flow rate on the heat-removal factor $F_R$, which is usually quite low for air collectors ($\sim 0.6$–$0.7$), increased flows tend to cause an offsetting decrease in storage bed stratification. Decreased stratification penalizes collector performance, since the collector inlet temperature is higher under these circumstances.

Nominal flow rates used in air collectors are about 10 liters/$m_c^2 \cdot$ sec (2 ft$^3$/min $\cdot$ ft$_c^2$). This capacitance rate is well below that for liquid systems but is a compromise between the offsetting effects of stratification reduction and improved heat transfer from the absorber plate which would result from higher flow rates. This effect is not present in liquid systems, since storage stratification is generally of little consequence in liquid systems. Figure 6.27 shows an air handler incorporating fan, motor, and all dampers needed for an air system.

## Thermal Storage

Solid-phase storage is discussed in detail in Chapter 5. In this section storage sizing for air systems is considered. About 300 kg/m$_c^2$ ($\sim$ 60 lb/ft$_c^2$) of ordinary, uniformly sized river gravel, 3–7 cm in diameter, has been used successfully for space-heating storage. This amount represents approximately 24 hr of carry-through in the absence of sunshine. In the subsequent section discussing the *f*-chart for air systems, the effect of storage size on air systems is described. To minimize the pressure drop through storage a short fluid path length is required. However, too short a bed precludes the stratification essential for successful operation of the air system.

One difficulty with the particle-bed storage required for air systems is that heat cannot be simultaneously removed from storage for space heating and added to storage by the collector loop. This problem occurs because air flows through storage in opposite directions in the two modes. The standard designs for liquid systems permits this simultaneous operation, however.

During periods of moderate heat load—spring and fall—this lack of flexibility in air systems can cause operational problems. If, for example, on a sunny day in fall, space heating is required, it will be done from the collector directly as shown in Fig. 6.23*a*. In fall the collector exit temperature can be 60°C (140°F) or above, a temperature that will satisfy the fairly small space-heating demand very quickly. This overly rapid response generally causes an overshoot in room temperature with consequent discomfort. In buildings equipped with air conditioning, this system may come on to lower room temperature to its set point. The spring and fall overshoot problem can be ameliorated by using a blending damper to mix solar-heated air with

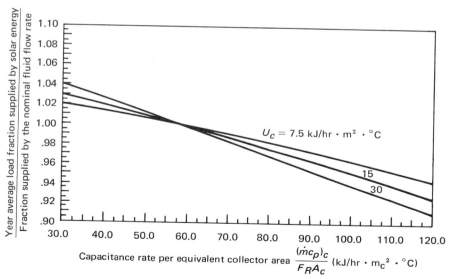

**FIGURE 6.26** Effect of collector fluid flow rate on annual energy delivery in an air-based solar-heating system. The effect of the collector loss coefficient $U_c$ is also shown. The nominal fluid flow rate is 58 kJ/m$_c^2$ · °C · hr based on the equivalent area $F_R A_c$. Adapted from (19).

**FIGURE 6.27** Air handler, valving, and control assembly for air-based solar-heating systems for small buildings. Courtesy of the Solar Control Corp.

lower temperature air to provide a more gradual room temperature rise. This feature adds mechanical and electronic complexity and cost to the system shown in Fig. 6.22, however.

Excess heat rejection in air systems is readily accomplished by venting the collectors to the environment. In dusty or industrial locations this may not be practical and a low flow of filtered air can be dumped to the atmosphere as a heat sink. This method has an advantage over a complete shutdown in that water preheating can continue through the low load season.

### Other Mechanical Components

Air systems require fewer auxiliary components than liquid systems. Controls in air systems are similar in function to those in liquid systems and are not described in detail here. The controller operates one blower, one pump (preheat), and two dampers, which are the analogs of valves in air systems. The dampers must seal tightly. For example, if the damper (leftmost damper in Fig. 6.22) controlling the source—collector or storage—of air flow to the heated space should leak, the fan could draw cold air from the nighttime collector through the damper to mix with warm air from storage. Such a leak could reduce the heating effect substantially.

Sizing of the blower and ducts in air systems can be done by conventional methods. Pressure-drop data for air ducts are presented in Appendix 5, in a convenient form for such calculations in Fig. A5.4. In some cases a two-speed blower is required in air systems if the system pressure drop in the collector-to-space heating mode is substantially different from the pressure drop through storage. Careful design can frequently avoid this problem, however.

## Performance Prediction for Air Solar-Heating Systems

The annual energy delivered by nominally sized air and liquid systems is nearly the same. Although working fluid properties, collector configurations, and flow loops differ significantly between the two systems, the net effect of the differences is small over the long term.

Klein (19) has prepared an $f$-chart for air systems using the simulation methodology described above. The air $f$-chart is shown in Fig. 6.28. Loss and solar parameters—$P_L$ and $P_s$—are identical to those for liquid systems and are defined in Eqs. (6.20) and (6.21). Since air systems do not use a collector-to-storage heat exchanger, the heat-exchanger factor is given by

$$F_{hx} \equiv 1.0 \qquad\qquad (6.24)$$

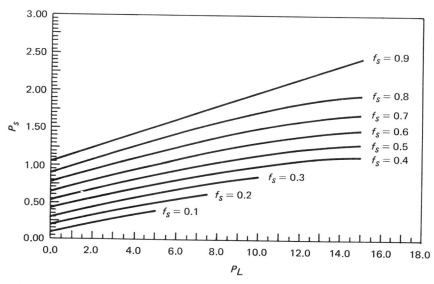

**FIGURE 6.28** The $f$-chart for air-based solar space-heating systems. From (19).

The data from which the $f$-chart was created can be expressed by an empirical equation. One such equation (8a) is

$$f_s = 1.040\, P_s - 0.065\, P_L - 0.159\, P_s^2 + 0.00187\, P_L^2 - 0.0095\, P_s^3 \qquad (6.25)$$

valid for the range

$$0 \leqslant P_s \leqslant 3.0 \qquad 0 \leqslant P_L \leqslant 18.0 \qquad 0 \leqslant f_s \leqslant 1.0$$

It is further required that $P_s > 0.07\, P_L$ to ensure that the monthly insolation is above the useful threshold level and that thermal losses are below an upper bound at which energy absorbed is equal to energy lost from the absorber plate.

The $f$-chart for air systems is based on nominal values of air flow rate and storage capacity. Table 6.21 shows the nominal value of these two parameters and the dimensionless group to be used to multiply the loss parameter to correct for flow rate or storage changes from the nominal value used to construct Fig. 6.28.

The *SLR* method described in the preceding section for liquid systems can be used without change for air systems. The $f$-chart includes the effects of more system variables than does the *SLR* method; therefore, the use of the former is recommended until experimental data on actual buildings are available.

## Durability of Air Systems

Although liquid-based solar-heating systems have existed from an earlier date than air systems, none have operated for an extended period of time. Therefore long-term durability data on components of liquid systems are not available. However, an air-based solar-heating system located in Denver, Colorado, has been in continuous operation since 1957 with no maintenance. This situation provides a unique insight into the durability of air systems.

**TABLE 6.21** Factors for Storage Capacity and Air Flow Rate for $f$-Chart[a]

| Parameter | Nominal value[b] | Loss parameter multiplier |
|---|---|---|
| Storage capacity $V_S$ | 0.25 m³/m$_c$² | $\left(\dfrac{0.25}{V_S}\right)^{0.3}$ |
| Fluid volumetric flow rate $Q_c$ | 10.1 liters/sec · m$_c$² | $\left(\dfrac{Q_c}{10.1}\right)^{0.28}$ |

[a]Adapted from (8a, 19).
[b]Based on net collector area; fluid volume at standard atmosphere conditions.

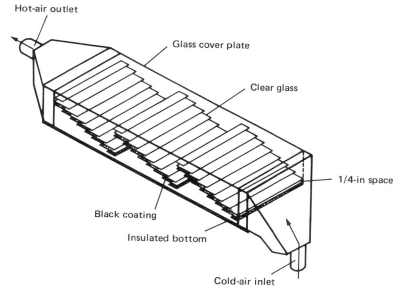

**FIGURE 6.29** Löf solar collector. Adapted from *Proc. World Symp. Appl. Solar Energy, Phoenix, Ariz., 1955*, published by Stanford Research Inst., Menlo Park, Calif., 1956.

The system is a residential solar-heating system designed and owned by G. O. G. Löf. It is described in Appendix 6 and a table of performance data measured during the 1959-1960 heating season is included. The collector design used in this building is shown in Fig. 6.29. The performance of this system was reevaluated during the 1974-1975 heating season and is summarized in Table 6.22 (38).

Figure 6.30 is a plot of useful energy collection versus a demand/solar radiation ratio expressed as degree C-days divided by GJ solar radiation per month on the collector plane. The annual data for 1959-1960 showed that 28.7 percent of the intercepted solar radiation was delivered as useful energy. In 1974-1975 20.6 percent was delivered. This indicates a drop of 28 percent over 15 yr or an annual rate of falloff of 1.9 percent. This small performance drop, which occurred with *no maintenance of any kind*, could be expected to be the *upper bound of the long-term deterioration rate* of air systems. Similar data for liquid systems are presently not available.

## PASSIVE SOLAR-HEATING SYSTEMS

### Introduction

The solar-heating systems described previously in this chapter are generally classified as *active*, mechanical systems. These systems are characterized by energy transport

**TABLE 6.22**  Comparison of Löf Air-System Performance during Two Heating Seasons[a]

| Performance criteria | 1959–1960[b] | 1974–1975[b] |
|---|---|---|
| Total solar radiation on 45° collector area during month (49.1 m²) | 118.02 | 122.56 |
| When collection cycles operated | 82.07 | 102.04 |
| Useful collected solar heat | 33.87 | 25.28 |
| Solar heat absorbed by | | |
| Rock storage tubes | 15.85 | 19.65 |
| Solar water preheater | 2.14 | 3.05 |
| Heat delivered by natural gas for | | |
| House heating | 107.29 | 98.30 |
| Water heating | 13.04 | 1.69 |
| Total heat load | | |
| House heating | | |
| Solar plus natural gas | 139.02 | 120.53 |
| Calculated from monthly degree-days | 139.03 | 120.53 |
| Water heating (solar plus natural gas) | 15.18 | 4.74 |
| House heating + water heating (solar + natural gas) | 154.20 | 125.27 |
| Electrical energy used by air blower electric motor | | 9.36 |
| Solar collector efficiency (%) | 41.3 | 24.8 |
| Water heating furnished by solar energy (%) | 14.1 | 64.3 |
| House heating furnished by solar energy (%) | 22.8 | 18.4 |
| Portion of total heat load from solar energy (%) | 22.0 | 20.2 |

[a]Adapted from (38). Energy units in the table are GJ.
[b]Dec. through Apr. total.

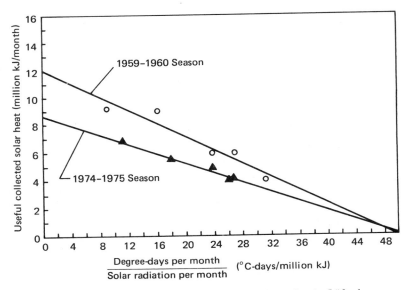

**FIGURE 6.30**  Comparison of the monthly energy delivery by the Löf solar-heating system for the years 1959–1960 and 1974–1975 expressed as a function of the energy demand/insolation ratio. From (38).

452

in fluid loops, active control systems that divert flows of fluids to various subsystems, and solar collectors in which forced convection is the mode of heat transfer.

Another class of solar-heating systems for buildings can be identified in which heat transfer is by the free convection and radiation modes—no pumps or blowers are required—and in which solar collection and storage subsystems are combined into one component. These systems are called *passive* systems and can range from all-glass greenhouses to massive adobe structures of the type used for centuries in the U.S. Southwest. Figure 6.31 shows the two generic types of space conditioning systems using both liquid and air.

Although the performance of passive solar-heating systems can be analyzed using principles of heat transfer, optics, and spherical trigonometry, little substantive analysis has been done. In addition, few experimental data exist for these systems, most of which have been built in the U.S. Southwest, the Mediterranean, the Middle East, and parts of Asia.

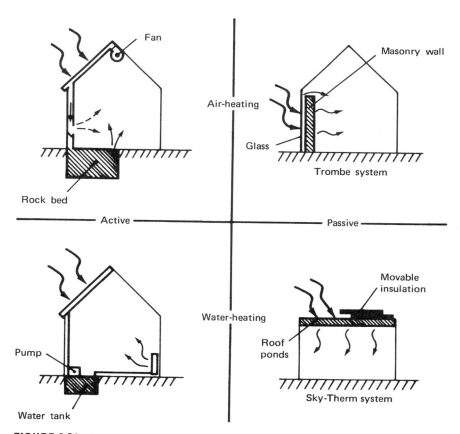

**FIGURE 6.31** Classes of passive and active solar-heating systems using liquid or air as working fluids.

Passive systems contain the five basic components of all solar systems described in Chapter 5. Typical passive realizations of these components are

1. Collector—windows, water ponds, dark walls.
2. Storage—walls, large interior thermal masses irradiated by sunlight in winter, water ponds. Frequently these are integrated with the collector.
3. Distribution system—radiation, free convection, simple circulation fans.
4. Controls—moving insulation panels to control building or collector heat loss, vents, and windows. These are frequently manual.
5. Backup system—any nonsolar heating system.

The design of passive systems requires the strategic placement of windows, storage masses, and occupied spaces. The fundamental principles of solar radiation geometry and tilt factors described in Chapter 2 are used to size and locate windows or other radiation intercepting devices. Storage masses are generally larger than those used in active systems and are frequently a part of the collection and distribution system. In the next section a lumped-parameter analysis of a typical passive system is described. The principles used there can be applied to any passive heating system.

### EXAMPLE 6.8

Calculate the effectiveness of a 10-ft$^2$, single-glazed window used as a passive solar collector in Tucson, Arizona, (latitude $= 32°13'$N) for a clear cold January 21. Use hourly clear-sky radiation data as tabulated in Appendix 2. Assume that the building interior is at an effective uniform temperature $T_i = 68°$F, and that the ambient temperature $T_a$ is as shown in Table 6.23.

### SOLUTION

Acquire hourly clear-sky solar radiation data $I_W$ on a vertical surface from Appendix 2 for 32°N latitude. These data will be used as an approximation for Tucson. Calculate the hourly energy gain $q_g$, loss $q_L$, and net heat flow $q_{net}$ through the window, assuming a window heat-loss coefficient $U_W$ of 1.1 Btu/hr $\cdot$ ft$^2$ $\cdot$ °F, that is,

$$q_L = U_W A_W (T_i - T_a) \tag{6.26}$$

$$q_{net} = q_g - q_L \tag{6.27}$$

where the rate of heat gain is given by

$$q_g = \tau_W I_W A_W \tag{6.28}$$

Use an average window transmittance of 0.85. The calculations are summarized in Table 6.23. The net gain for the window is 1122 Btu/day. Therefore, on a clear day a window is a net heat gain mechanism. Note, however, that if the solar radiation were only 10 percent less, the window would be a net heat loss

**TABLE 6.23** Summary of Calculations for Net Heat Gain of a Window

| Hour no. | Insolation (Btu/hr · ft²) | Rate of heat gain $q_g$ (Btu/hr) | Ambient temperature (°F) | Rate of heat loss $q_L$ (Btu/hr) |
|---|---|---|---|---|
| 1 | – | – | 10 | 638 |
| 2 | – | – | 10 | 638 |
| 3 | – | – | 12 | 616 |
| 4 | – | – | 13 | 605 |
| 5 | – | – | 14 | 594 |
| 6 | – | – | 15 | 583 |
| 7 | 1 | 8 | 15 | 583 |
| 8 | 115 | 978 | 17 | 561 |
| 9 | 181 | 1539 | 19 | 539 |
| 10 | 221 | 1879 | 20 | 528 |
| 11 | 245 | 2083 | 20 | 528 |
| 12 | 253 | 2151 | 21 | 517 |
| 13 | 245 | 2083 | 22 | 506 |
| 14 | 221 | 1879 | 20 | 528 |
| 15 | 181 | 1539 | 20 | 528 |
| 16 | 115 | 978 | 18 | 550 |
| 17 | 1 | 8 | 17 | 561 |
| 18 | – | – | 15 | 583 |
| 19 | – | – | 13 | 605 |
| 20 | – | – | 12 | 616 |
| 21 | – | – | 11 | 627 |
| 22 | – | – | 10 | 638 |
| 23 | – | – | 8 | 660 |
| 24 | – | – | 7 | 671 |
| Totals (Btu) | – | 15,125 | – | 14,003 |

device. If the window could be insulated at night, the energy gain could be increased, albeit at increased complexity and maintenance cost.

## Example Thermal Analysis of Passive System

Figure 6.32 is a sketch of a simple type of a small, passively heated structure. First developed by E. S. Morse[*] in the nineteenth century and recently revived by Trombe et al. (35), several examples, including an apartment building, have been built in the Pyrénéés-Orientales district of France and in the U.S. Southwest. A large concrete mass 40 cm or more in thickness is exposed to sunlight through large, south-facing windows. Sunlight absorbed on the surface of the thermal mass is either transferred to the interior of the storage mass by conduction or convection from the surface as

---

[*]For an interesting survey of historical applications of solar heat, see "Solar Heat–Its Practical Application," by C. H. Pope, published by the author, 1903; see p. 72 for the Morse idea.

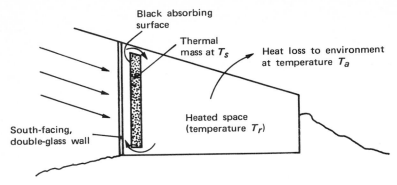

**FIGURE 6.32** Schematic diagram of a Trombe-type passive solar-heating system showing the passive collection device and storage mass. The south-facing transparent wall can be insulated during periods of no sun to reduce heat loss.

shown. In addition some reradiated thermal energy is transferred to the environment through the glass wall. The air heated by convection rises and passes into the heated space.

During periods of no sunlight the south-facing wall may be insulated to reduce potentially large heat losses through the glass. Heat stored in the thermal-mass wall is radiated and convected into the space to be heated. During the summer, vents at the top of the south-facing wall may be opened and the warm concrete wall can create a chimney effect to enhance ventilation.

Figure 6.33 is an equivalent thermal circuit for the building in Fig. 6.32. Three temperature nodes can be identified—room temperature, storage temperature, and ambient temperature. The circuit responds to climatic variables represented by a current injection $I_s$ (solar radiation) and by the ambient temperature $T_a$. The storage temperature $T_s$ and room temperature $T_r$ are determined by current flows in the equivalent circuit.

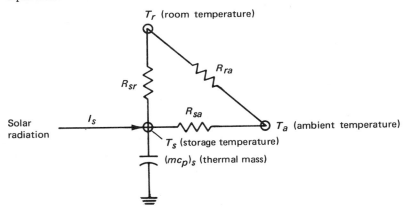

**FIGURE 6.33** Equivalent thermal circuit for passively heated solar building in Fig. 6.32.

Balcomb and Hedstrom (5) have evaluated the performance of a passive system represented by the circuit in Fig. 6.33. Hourly weather data for Los Alamos, New Mexico, a typical site for such a building, were used to predict performance. Several interesting conclusions were drawn from this preliminary work and are listed below.

1. For a given aperture area an optimal passive system using a vertical glass wall can perform nearly as well as an active system using a tilted collector.
2. Storage mass twice that for an active system should be used; storage should be exposed directly to sunshine during the heating season.
3. An optimal coupling of storage to the space $(R_{sr})$ exists; for too small a value the room overheats, while for too large a value, the storage mass becomes too warm and reradiation from the storage through the glass is too great.

In addition, storage should have a small Biot modulus so that absorbed energy can be conducted into the core of storage expeditiously during sunshine periods. A system of this type is called a high-conductance, Trombe-type passive system. A typical storage material for a high-conductance system is water, which has the required Biot modulus magnitude.

Figure 6.34 shows the effect of storage capacity on annual energy delivery for a high-conductance, Trombe-type passive system. It is to be noted that a single-glazed south-wall system with no night insulation (the "solar house" of the 1950s) is a net energy *loss* mechanism for negligibly small storage. Even for very large storage it is an ineffectual system at best. For the other configurations shown in the figure the energy delivery is at a useful level for storage above 30 Btu/ft$_g^2 \cdot °$F ($\sim 600$ kJ/m$_g^2 \cdot °$C).

A comparison of several simulated passive systems and a nominal active system is presented in Fig. 6.35. It is seen that a well-designed passive system using a vertical glass wall performs nearly as well as an active system tilted near the optimum angle. Although these results are preliminary and for only one location, the performance ability of a properly designed passive system is quite significant. Note that the passive system modeled had a less restrictive room temperature control criterion in that the permissible temperature range was $70 \pm 5°$F. This control design tends to favor the passive system.

Test data on several passively heated structures have been measured by Trombe et al. (35) for the buildings described in Table 6.24. They discovered that about one-third of the solar delivery to the building is via the thermocirculation of air, the balance being by conduction through the wall. Long-term collection efficiency for the 1967 prototype was 35 percent during the heating season and less in the off-season. In addition, they learned that the thermal storage wall, if made from masonry, should be 40–45 cm thick. If the thickness is too great ($\sim 60$ cm), heat passes through the wall too slowly; if too small ($\sim 35$ cm), heat transport is too fast. It was also learned that residual moisture in the incompletely cured masonry storage masses in the 1974 buildings caused a significant performance penalty because of the unavoidable partial use of solar heat to evaporate the moisture. A curing time of up to 2 yr was expected.

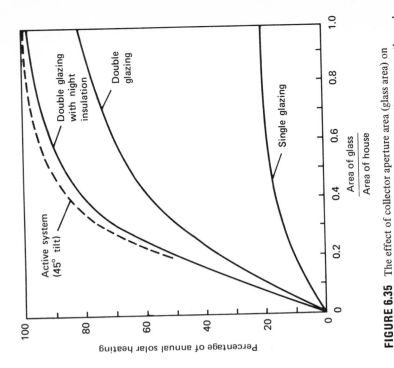

**FIGURE 6.35** The effect of collector aperture area (glass area) on annual energy delivery of active and passive solar-heating systems; thermal storage 30 Btu/ft$_g^2$ · °F ($\sim$ 600 kJ/m$_g^2$ · °C) for both systems. Curves shown are based on Los Alamos, New Mexico, data for 1972–1973. From (5).

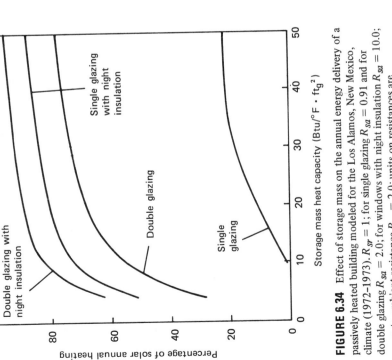

**FIGURE 6.34** Effect of storage mass on the annual energy delivery of a passively heated building modeled for the Los Alamos, New Mexico, climate (1972–1973). $R_{sr} = 1$; for single glazing $R_{sa} = 0.91$ and for double glazing $R_{sa} = 2.0$; for windows with night insulation $R_{sa} = 10.0$; room-to-ambient resistance $R_{ra} = 2.0$; units on resistances are hr · ft$_g^2$ · °F/Btu. From (5).

458

**TABLE 6.24** Important Characteristics of the CNRS Solar Buildings Employing the Trombe Wall Heating System[a]

| | 1967 Prototypes | | 1974 Solar houses | |
|---|---|---|---|---|
| Stories | 1 | 2 | 2 | 3 |
| No. of rooms | 4 | 6 | 6 | 7 |
| Floor area (m²) | 76 | 210 | 180 | 250 |
| Volume (m³) | 300 | 525 | 450 | 650 |
| Collector area (m²) | 48 | 55 | 45 | 65 |
| Collector area per m³ | 0.16 | 0.1 | 0.1 | 0.1 |
| Insulation[b] | | | | |
| $U$ (W/m² · °C) | 1.34 | 0.40 | id | id |
| $UA/V$ (W/m³ · °C) | 1.63 | 1.00 | id | id |
| Collector storage wall[c] thickness (m) | 0.60 | 0.37 | id | id |
| Auxiliary heating | Electricity | Electricity | Electricity | Electricity |

[a]Adapted from (35).

[b]$U$ is the equivalent heat-loss coefficient for the buildings based upon heat-loss area $A$; $V$ is the building volume.

[c]Heavy gravels, concrete.

In the passive systems described above the instantaneous efficiency is an inadequate measure of system performance and has not been used. Since systems with large thermal masses have long time constants, they rarely reach a steady state for which an efficiency can be defined. The proper index of performance is the integrated energy delivery on a daily, monthly, or annual basis.

## Movable-Insulation Heating and Cooling System

A movable-insulation heating and cooling system has been designed by Hay and Yellott (14a, 39). The concept of this passive system is based on the assumption that a water bag on top of a roof shielded by a movable curtain can act as collector, storage, and heat exchanger. The method of operation of this type of system is illustrated in Fig. 6.36. The roof is covered with approximately 20 cm (8 in) of water sealed in clear polyvinyl chloride (PVC) bags which are connected to inflatable air cells above the water for increased collection efficiency. These bags can be covered intermittently by sliding polyurethane insulation panels 5 cm (2 in) thick. For heat collection the water bags are exposed to the sun by removing the panels and storing them elsewhere. If no heat is wanted or no sun is available, the panels can be closed to reduce heat losses from the bags. To cool the water in the bags during the summer, the panels can be opened at night for radiative and convective cooling to the night sky. The cooling effect is enhanced by deflating the air cells during part of the cooling season. The panel control system intended to achieve these thermal objectives is shown in Fig. 6.37.

A solar test room using this type of passive heating and cooling was built in 1967 by Hay and Yellott and a three-bedroom house based on this principle was designed and built by Hay in Atascadero, California, in 1972. The house had a

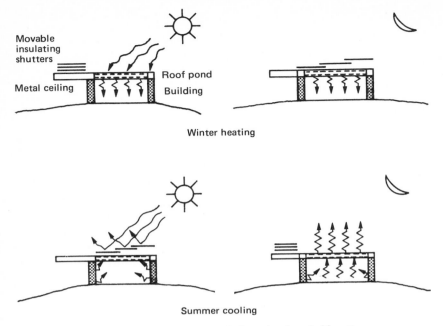

**FIGURE 6.36** Operating modes of a naturally heated and cooled house.

split-level slab floor and room heights of 2.4 and 3 m (8 and 10 ft); 35 percent of the external walls were lightweight concrete block. Half of these walls were 30 cm (12 in) thick with no insulation and half were 20 cm (8 in) thick with vermiculite-filled cavities. An equivalent amount of block with sand-filled cavities was used in making indoor partition walls. Forty percent of the wall area was frame constructed with fiberglass insulation of 8 cm (3 in). Double-glazed windows and doors constituted 25 percent of the wall area and vertical insulation (5 × 61 cm) was placed on

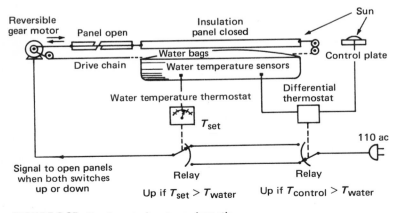

**FIGURE 6.37** Panel control system schematic.

the slab perimeter. The external walls, partition walls, and slab fllors were intended to act as thermal capacitors, which would help damp variations in room temperature resulting from variations of the water temperature in the bags.

The movable-insulation heating and cooling system was tested in 1975 during 9 months with normal family occupancy (28a). The performance test showed that if the system had operated during the entire winter it would have saved the equivalent of about 9 bbl of oil and could have delivered 21,000 MJ. The largest cooling load handled by the system was during July when the average daily heat removed from the house was 177 MJ. The total amount of heat removed during the cooling season was estimated to be 1900 MJ, and based on an air-conditioning coefficient of performance of 4.7 and an electrical generation of efficiency of 30 percent, this cooling would be equivalent to the savings of 2 bbl of oil during the entire cooling season. Based on these test results it is very questionable how cost-effective such a system would really be since the maintenance costs of the large number of movable parts required could be substantial unless the panels were moved by the occupants instead of a mechanical system. Moreover, this approach can be used only for cooling in dry climates that have cool and clear nights and for heating in nonfreezing climates.

## SOLAR–HEATING ECONOMICS

The principles of economic analysis to be used in selecting the cost-optimal solar system configuration are detailed in Chapter 5. The method of economic optimization described there includes effects of inflation, interest rate, taxes, insurance, maintenance, and depreciation. Since these principles apply to all solar systems, they are not rederived here. $F$-charts or other models are used to calculate the energy delivery production function required for the economic analysis.

Figure 6.38 contains several total annual cost curves for liquid-based solar space- and water-heating systems. These curves, based on the methodology of Chapter 5, illustrate the need for a careful analysis on a project-by-project basis. Rules of thumb regarding costs and collector areas are too simplistic to be useful and can lead to system configurations far from the cost-optimal design. The energy delivery calculation methods described in this chapter along with the economic methods in Chapter 5 provide a rational, step-by-step design procedure for all solar-heating systems.

## SOLAR–ASSISTED HEAT PUMPS

The heat pump concept was conceived by Lord Kelvin over 120 years ago. Since that time the use of such devices has often been advocated because heat pumps can reduce power consumption compared with conventional heating systems. Until recently, however, the heat pump was often viewed as a cooling system that could be used for heating, but it was not usually designed for that purpose (13). The design of heat pumps consequently favored cooling performance and applications for heating were not successful. Moreover, during the late 1950s and early 1960s, a combination of poor design, insufficient testing, and inadequate service triggered

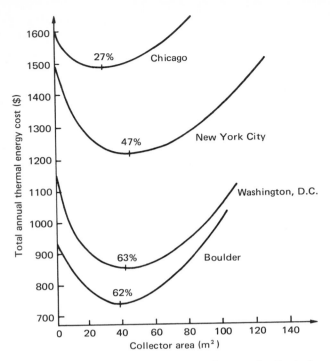

**FIGURE 6.38** Example total cost curves for space heating in four
U.S. cities. Utility rates are from the first quarter of 1976; a 5
percent differential inflation rate in fuels is assumed; the system
cost of $160/m² is amortized at 10 percent for 20 yr. Performance
is for a typical solar-heating system in a single-family dwelling of
conventional construction using an electrical backup.

reliability problems for homeowners and utilities, which gave the heat pump a bad
reputation. However, recent attempts to reduce the energy requirements for heating
homes have rekindled a nationwide interest in the use of heat pumps. Furthermore,
the performance of a heat pump can be augmented by using relatively inexpensive solar collectors to reduce both the annual heating cost and the consumption
of scarce fuels to the point where it has become competitive with conventional
heating systems as well as solar-thermal collector systems in many parts of the
world.

## Carnot Coefficient of Performance

A heat pump is a device that causes heat to flow from a region of lower
temperature to a region of higher temperature. Since this direction of heat flow is
opposite to the direction required by the second law of thermodynamics for
spontaneous transfer of heat, external energy must be added to "pump" heat
"uphill." The efficiency of a heat pump can be defined in accordance with the

desired effect that the system is designed to achieve. If the desired effect is thermal energy delivery at a higher temperature level to heat a house, the appropriate efficiency would be the ratio of the energy delivered to the house to the amount of energy required to operate the system. According to the first law of thermodynamics, the energy delivered at the higher temperature will be equal to the sum of that extracted from the cold region, or sink, plus the energy required by the heat pump to achieve the desired temperature level. The total energy delivered at the higher temperature level in a building will always be greater than the amount of energy required to run the system. This means that the heating efficiency of a heat pump, called the coefficient of performance (COP), will always be greater than 1 for an ideal device with no losses.

As will be shown in Chapter 7 in more detail, the same equipment used to operate a heat pump as a heating system can be used to remove heat from a building, that is, to air condition the building, by reversing the direction of operation. Thus, when the desired effect of the heat pump is refrigeration, the appropriate definition of its COP will be the ratio of the amount of heat removed from the building to the amount of energy required to achieve this desired effect.

The larger the temperature difference over which the heat must be pumped, the greater will be the amount of energy required to operate the system. According to the definitions

$$COP_{heating} = \frac{\text{energy delivered}}{\text{compressor work}} = \frac{Q_{delivered}}{W} \qquad (6.29)$$

and

$$COP_{cooling} = \frac{\text{energy removed}}{\text{compressor work}} = \frac{Q_{removed}}{W} \qquad (6.30)$$

For a reversible heat pump, just as for a reversible heat engine, the upper limit of efficiency is equal to that obtainable in a Carnot cycle between $T_h$ and $T_c$. The upper Carnot limit of the COP for heating is given by the expression

$$COP_{max,heating} = \frac{T_h}{T_h - T_l} \qquad (6.31)$$

whereas for cooling

$$COP_{max,cooling} = \frac{T_l}{T_h - T_l} = COP_{max,heating} - 1 \qquad (6.32)$$

where $T_h$ = high temperature level in K
$T_l$ = low temperature level of the cycle in K

### EXAMPLE 6.9

Determine the Carnot COP for a heat pump operating to maintain a house at 318 K (70°F) in a 294 K (30°F) environment.

### SOLUTION

The Carnot COP is equal to $318/24 = 13.25$. Note that if a solar collector could boost the sink temeprature only 12 K, the ideal efficiency would be doubled.

## Efficiency of Real Heat Pumps

The ideal value of 13.25 for the COP is somewhat misleading, because under real conditions a heat pump must operate across a wider temperature gap than the Carnot cycle indicates. In most heating systems heat is supplied to a central location and is distributed as sensible heat in a working fluid. For example, in a hot-air system house air is first heated by the working fluid and then forced through ducts in order to distribute the heat through the house. Two temperature differences are necessary for these processes. Moreover, the total rate of heat distributed is proportional to the product of the mass flow rate of the air and its temperature change. Thus, for a given rate of heat flow, the flow rate of air required increases as the temperature rise decreases. In a system using a heat pump efficiencies increase by minimizing the temperature difference between the high and low temperatures. For example, if hot air is delivered at 342 K (100°F) instead of 372 K (160°F), as in a combustion system, the heat pump COP improves. However, there are practical limits to lowering the delivery temperature, similar to those in a solar-thermal system described previously. Another, although smaller, temperature difference is necessary to extract heat from the air blown across the outside heat exchanger. These temperature differences are necessary for any heat pump no matter how well it is designed, and they may decrease the maximum obtainable COP from 13.25 to the order of 7.5. In order to transfer heat between the refrigerant fluid and the surrounding air, additional temperature differences are necessary. For a reasonably sized heat exchanger this temperature is on the order of 12 K (20°F). The total temperature difference required to operate a real heat pump will therefore be closer to 66 K (110°F) than the 24 K (40°F) indicated by the Carnot analysis above. As shown in Fig. 6.39, this yields a maximum COP on the order of 5.2 (13).

Real heat pumps have additional losses that cause them to operate at only approximately half this value. A COP of 5 could be obtained only by a system operating on the Carnot cycle. In a real cycle additional losses degrade the efficiency as shown in Fig. 6.40. The top line represents the previously calculated Carnot efficiency. The effect of using a conventional vapor-compression cycle is shown on the second line. Using this same ideal refrigeration cycle, but including losses resulting from motor and compressor inefficiencies, drops the COP even lower. The final estimate of the true COP expected must include all electrical power

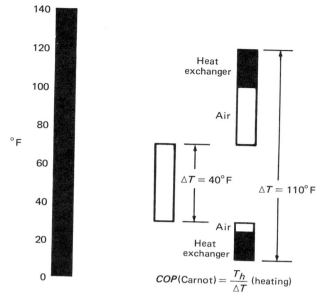

$$COP(\text{Carnot}) = \frac{T_h}{\Delta T} \text{ (heating)}$$

**FIGURE 6.39** Maximum heat pump efficiency calculations must be based on refrigerant temperatures rather than source and sink temperatures.

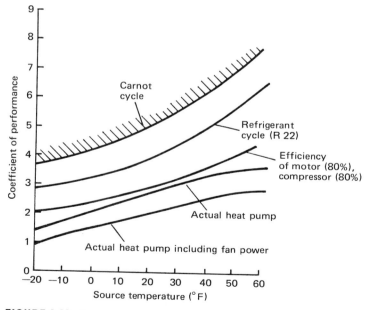

**FIGURE 6.40** Heat pump heating efficiency as it depends on source temperature for systems of different degrees of ideality. Data based on a residential General Electric heat pump.

requirements as well as losses in the lines resulting from fluid friction and valves. Depending on the source temperature, the heat pump heating efficiency will be closer to 2 or 3 than the ideal value of 13 calculated from the idealized Carnot cycle.

## Thermodynamic Analysis of Heat Pumps

The different types of heat pumps are classified according to the heat source and the heat sink that they use. The most common types that can be augmented by solar collectors are air-to-air, water-to-air, and water-to-water heat pumps.

By far the most widely used arrangement is the air-to-air heat pump. Its operation is illustrated in the simplified diagram shown in Fig. 6.41. The major components are a refrigerant working fluid, a compressor, a condenser, an evaporator, and a throttling device (often called a thermal expansion valve, TXV). In addition, a heat pump to be used for cooling as well as heating requires a reversing valve capable of reversing the flow of refrigerant in the system so that the indoor coils can act as an evaporator in the cooling cycle (Fig. 6.41a) and as a condenser on the heating cycle (Fig. 6.41b).

A thermodynamic analysis of a heat pump with a given working fluid can be made by tracing the steps in the cycle in sequence. The refrigerant is first pumped to a high temperature and pressure by the compressor and then directed by the reversing valve to the outdoor coil. In the *cooling mode* a fan forces outdoor air, which is at a lower temperature than the refrigerant, over the coil where heat is removed and the refrigerant is condensed. The refrigerant then flows through the TXV, where its pressure is reduced, and into the low-pressure indoor coil. The indoor coil is essentially a heat exchanger in which the refrigerant is evaporated by absorbing heat from the surroundings, which are cooled in the process. In absorbing this heat the liquid refrigerant boils, becomes a vapor, and flows toward the compressor where it is recompressed, and the cycle repeats.

The same equipment can be used to heat a building by reversing the direction of the flow of the refrigerant so that the functions of the indoor and outdoor coils are reversed. Since the outdoor coil acts as an evaporator in the *heating mode*, the air leaving it is colder than the entering air. When the outdoor temperature is low and the humidity is high, frost will therefore often form on the outdoor coil. This imposes a thermal resistance, which reduces the efficiency of the cycle and can impair the operation of the heat pump. To remove the frost, the heating cycle must be reversed back to the cooling cycle for a short time until the hot gas melts the frost and removes the additional thermal resistance on the heat exchanger.

Two other common types of heat pump use water as a heat source. In one mode of operation, called a water-to-air pump, water from a river, a lake, or a tank heated by a solar collector serves as the heat source while the interior heat exchanger inside the building is a coil in air. In this arrangement the indoor coil serves as a refrigerant condenser during the heating cycle and a refrigerant evaporator during the cooling cycle as shown in Fig. 6.41. The only difference between an air-to-air and a water-to-air heat pump is that in the latter the heat

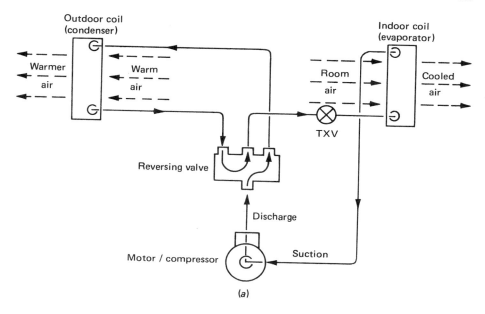

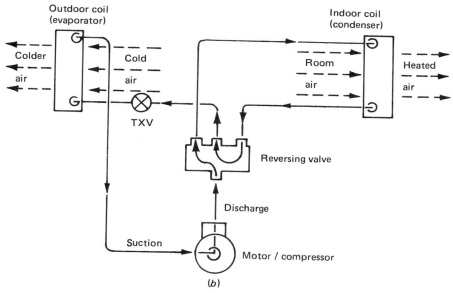

**FIGURE 6.41** Basic air-to-air heat pump system: (*a*) cooling and defrost cycle; (*b*) heating cycle.

source is a heat exchanger operating between water and the refrigerant. Just as in the air-to-air heat pump, the refrigerant flow can be reversed by a four-way valve to obtain a heating or cooling cycle.

The third type of heat pump of interest is the water-to-water arrangement. In this type both condenser and evaporator are water-to-refrigerant heat exchangers. Whereas in the previous two types the refrigerant flow must be switched, the water-to-water heat pump switches the water flow. This has certain advantages, because when space cooling is required, chilled water leaving the evaporator is directed to the space-conditioning equipment. On the other hand, when space heating is required, warm water leaving the condenser is directed to the space-conditioning equipment. As a result, the working fluid circuit is always the same and the condenser and evaporator design can be optimized for peak efficiency because they serve only one function. Water-to-water heat pumps used in conjunction with solar collectors in order to reduce the amount of energy required to heat the refrigerant are especially attractive for large buildings where they can be designed to operate efficiently.

For a quantitative analysis the basic refrigeration-heat pump cycle is conveniently displayed on a pressure enthalpy diagram, as shown in Fig. 6.42. The vertical scale is the absolute pressure and the horizontal scale is the enthalpy of the refrigerant. Property tables and charts such as shown in Fig. 6.42 are available from different sources, for example, the ASHRAE "Handbook of Fundamentals" (2). To the left of the saturated liquid line the refrigerant is completely liquid, whereas to the right of the saturated vapor line the refrigerant is completely in the vapor phase. Between the two lines the refrigerant is partly liquid and partly vapor, being mostly liquid to the left and mostly vapor to the right.

The cycle shown in Fig. 6.42 is simplified for illustrative purposes. Beginning at point A the compressor takes in saturated vapor and compresses it to point B. In this process work is added to the working fluid as its enthalpy, temperature, and pressure are increased. From point B to point C heat is removed from the working fluid, i.e., heat is delivered to the building during condensation. As the heat is removed from the working fluid the vapor is condensed and at point C the working fluid is completely in liquid form. The process from C to D is a throttling process at constant enthalpy during which the pressure and temperature of the working fluid are reduced. After the throttling the working fluid enters the evaporator as a mixture of about 75 percent liquid and 25 percent vapor. In the process from D to A the working fluid is evaporated as heat is transferred to it from the source. The change in enthalpy of the refrigerant is the refrigerating effect, indicated as $RE$. As the refrigerant boils and absorbs heat, it vaporizes and then enters the compressor as a vapor at point A, from which point the cycle repeats.

A solar-augmented heat pump cycle is shown schematically in Fig. 6.43. The thermodynamics of a solar-augmented heat pump cycle are shown by the dotted line in Fig. 6.42. A solar collector is used to increase the temperature level of the heat source so that the external work required to elevate the working fluid to the required temperature level is decreased from A–B to A′–B′. This decreases the energy required to operate the compressor and increases the cycle $COP$.

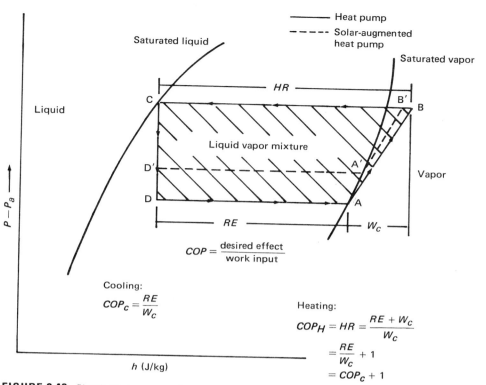

**FIGURE 6.42** Simple heat pump cycle.

$$COP = \frac{\text{desired effect}}{\text{work input}}$$

Cooling:

$$COP_c = \frac{RE}{W_c}$$

Heating:

$$COP_H = HR = \frac{RE + W_c}{W_c}$$

$$= \frac{RE}{W_c} + 1$$

$$= COP_c + 1$$

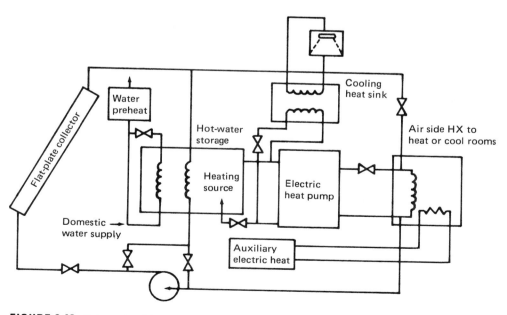

**FIGURE 6.43** Solar-assisted heat pump.

## Heat Pump Capacity

One undesirable feature of heat pumps is illustrated in Fig. 6.44, in which the delivery rate of potential energy is plotted as a function of the ambient temperature. Inspection of this figure shows that the ability of a heat pump to deliver heat to a house decreases as the outside temperature decreases. This trend is opposite to the heat demand of the building, which increases at colder temperatures. If the heat demand is plotted as a function of ambient temperature, it will intersect the delivery capacity of the heat pump at some point known as the *balance point*. At this temperature the heat pump can supply all the heat required by a house, but below this temperature some additional energy must be supplied to maintain the house temperature. In most cases this supplementary heat is obtained by electrical resistance heaters built directly into the indoor handler of the heat pump. Advanced heat pumps using two-speed compressors for periods of low temperature can have a flatter heat pump output curve than that shown in Fig. 6.44.

Since at ambient temperatures below the balance point supplementary electric heat is added to a house with a *COP* of unity, the total *COP* of the entire heating system is reduced. Thus, even if a heat pump operates at high *COP* above the balance point, and if the system requires a large amount of makeup heat, the

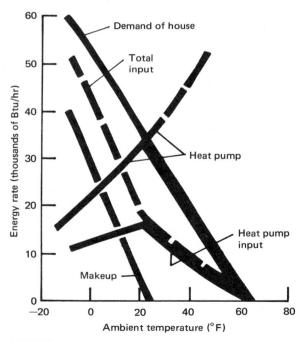

**FIGURE 6.44** Heat pump heating capacity, house demands, and supplementary heat requirements as functions of outdoor temperature. Data based on a residential General Electric heat pump.

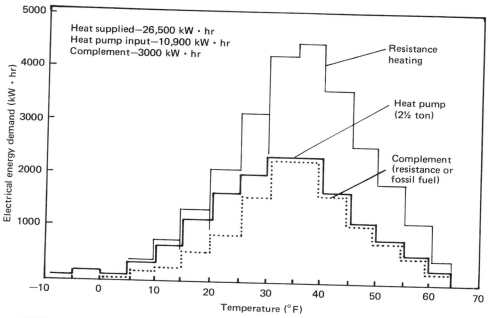

**FIGURE 6.45** Relationship between seasonal house heating demand and energy inputs for heating systems using a heat pump.

system efficiency throughout the year will be lowered. This shows that the performance of a heating system is a function of not only the heat pump *COP* but also the size of the heat pump used and the seasonal temperature distribution to which the building is exposed. Figure 6.45 shows the total heating demand of an 1800-ft² house in the Boston, Massachusetts, area with a 2.5-ton heat pump, which results in a theoretical balance point of 30°F. The area under the uppermost curve represents the total heat demand of the building and also indicates the amount of energy input that would be needed if the house were heated with an electrical resistance heater. The energy required by the heat pump system is shown as the area under the next lower curve. It includes the resistance makeup heat. The dotted curve is the energy input to the heat pump alone. Thus the difference between the heavy line and the dotted line indicates the supplementary heat required to fill the demand of the house. The totals listed in this figure indicate that the heat pump system required 13,900 kW · hr to supply 26,500 kW · hr of heat. The ratio of the total seasonal demand to the energy required is known as the seasonal performance factor (*SPF*). The *SPF* is the best measure of a heat pump system performance and for this example it equals 1.91.

In order to determine what combination of solar augmentation should be used with a given heat pump, it is necessary to conduct an optimization study similar to that described for solar-thermal heating systems. Since the cost of collectors constitutes the largest portion of the cost of a solar-augmented heat pump system, it is necessary to optimize the size that can be used economically.

**TABLE 6.25**  Calculation of Seasonal Performance Factor for Example 6.10

| | Sep. | Oct. | Nov. | Dec. | Jan. | Feb. | Mar. | Apr. | May | Year |
|---|---|---|---|---|---|---|---|---|---|---|
| Average temperature $T_a$ (°F) | 64.3 | 53.7 | 41.4 | 35.8 | 32.7 | 34.7 | 39.3 | 48.7 | 57.4 | — |
| Heating degree F-days $DD$ | 117 | 428 | 819 | 1035 | 1132 | 938 | 887 | 558 | 288 | — |
| House demand $D$ (MBtu/month) $= DD \times 650 \times 24/10^6$ | 1.8 | 6.7 | 12.8 | 16.1 | 17.7 | 14.6 | 13.8 | 8.7 | 4.5 | 96.7 |
| Horizontal insolation $\bar{I}_h$ (kBtu/month · ft²) | 46.3 | 34.9 | 25.0 | 20.5 | 22.6 | 30.1 | 45.0 | 51.7 | 51.7 | — |
| Effective tilt factor for total radiation $\bar{R}$ | 1.13 | 1.50 | 1.83 | 2.12 | 1.96 | 1.68 | 1.25 | 0.98 | 0.82 | — |
| Collectable solar energy $\bar{S}$ (MBtu/month) $= \bar{I}_h \times \bar{R} \times A_c \times \eta_c$ $= \bar{I}_h \times \bar{R} \times 750 \times 0.3$ | 11.8 | 11.8 | 10.3 | 9.8 | 10.0 | 11.4 | 12.7 | 11.4 | 9.5 | 98.7 |
| Solar energy utilized $\bar{Q}_u$ (MBtu/month) | 1.8 | 6.7 | 10.3 | 9.8 | 10.0 | 11.4 | 12.7 | 8.7 | 4.5 | 75.9 |
| Electrical energy required $UE$ (MBtu/month) $= 0.1 \times \bar{Q}_u$ | 0.18 | 0.67 | 1.03 | 0.98 | 1.00 | 1.14 | 1.27 | 0.87 | 0.45 | 7.59 |
| Energy deficiency $DEF$ (MBtu) $= D - (\bar{Q}_u + UE)$ | — | — | 1.5 | 5.3 | 6.7 | 2.1 | — | — | — | 15.6 |
| Energy used to operate heat pump (MBtu/month) ($HP$) $= DEF/1.6^a$ | — | — | 0.9 | 3.3 | 4.2 | 1.3 | — | — | — | 9.7 |

$$SPF = \frac{D}{HP + UE} = \frac{96.7}{9.7 + 7.59} = 5.6$$

[a]Long-term average $COP$ for a heat pump of a given capacity in Boulder, Colorado. The numerical constant (1.6) varies with the capacity rating of the heat pump. (1 MBtu = $10^6$ Btu).

## EXAMPLE 6.10

Calculate the *SPF* for a solar-heating system using a heat pump backup. This is not a solar-assisted heat pump system. The building is located in Boulder, Colorado, with a heat load of 650 Btu/hr · °F. The solar collector has an area of 750 ft$^2$ and is tilted at 55°. Calculate an approximate solar delivery assuming that the delivery is given by the monthly total collector-plane insolation multiplied by an average efficiency of 0.30. The *f*-chart could be used if more accuracy is needed.

## SOLUTION

The solution is carried out on a monthly basis in Table 6.25.

From a thermodynamic viewpoint it may appear desirable to use solar heat to increase the *SPF* of a heat pump. However, computer simulation studies have shown no clear-cut advantage to this method on an energy delivery per unit collector area basis (12, 17, 27). In some cases, the solar-assisted heat pump may be the less effective system because heat can be delivered from solar storage with an electrical *COP* of about 20 (heat delivered divided by pump energy). It is therefore inefficient to pass this thermal energy through a heat pump with a *COP* of 3 or 4.

For system simplicity it is easier to use a heat pump only as the auxiliary energy source for a solar building. In this role it acts as any other backup device.

## PROBLEMS

6.1. The no-load temperature of a building with internal heat sources is given by Eq. (6.11). How would this equation be modified to account for heat losses through the surface of an unheated slab, the heat losses being independent of ambient temperature?

6.2. An unheated garage is placed on the north wall of a building to act as a thermal buffer zone. If the garage has roof area $A_r$, window area $A_{wi}$, door area $A_d$, and wall area $A_{wa}$, what is the effective $U$ value for the north wall of the building if its area is $A_n$? The garage floor is well insulated and has negligible heat loss. Express the effective $U$ value in terms of the $U$ values and areas of the several garage surfaces.

6.3. What is the annual energy demand for a building in Denver, Colorado, if the peak heat load is 150,000 Btu/hr based on a design temperature difference of 75°F? Internal heat sources are estimated to be 20,000 Btu/hr and the design building interior temperature is 70°F.

6.4. What is the January solar load fraction for a water-heating system in Washington, D.C., using 100 m$^2$ of solar collector if the water demand is 4 m$^3$/day at 65°C with a source temperature of 12°C? No heat exchanger is used and the solar collector efficiency curve is given in Fig. 6.12; the solar collector is tilted at an angle equal to the latitude.

6.5. Repeat Problem 6.4 for Albuquerque, New Mexico, in July if the water source temperature is $17°C$.

6.6. Explain how the $f$-chart (Fig. 6.11) can be used *graphically* to determine the solar load fraction for a range of collector sizes once the solar and loss parameters have been evaluated for only one system size. *Hint*: consider a straight line passing through the origin and the point $(P_s, P_L)$.

6.7. The $f$-chart was generated using data for flat-plate collectors. What modifications would be necessary to use it for a compound parabolic concentrator collector? Describe the effect on each $f$-chart parameter in Table 6.14.

6.8. In an attempt to reduce cost a solar designer has proposed replacing the shell-and-tube heat exchanger in Fig. 6.14 with a tube coil immersed in the storage tank. The shell-and-tube heat exchanger originally specified had a surface area of 10 $m^2$ and a $U$ value of 2000 $W/m^2 \cdot K$ to be used with a 100-$m^2$ solar collector. Using Eq. (3.71) to estimate the $U$ value of the submerged coil, how many feet of 0.5-inch (1.27-cm) diameter copper pipe would be needed to achieve the same value of $UA$ product as the shell-and-tube heat exchanger? What percentage of the storage tank volume would be consumed by this coil if 50 kg of water is used per square meter of collector? Use a storage water temperature of $60°C$ and a collector water outlet temperature of $70°C$ for the calculations.

6.9. If a solar system delivers 2500 $MJ/m^2 \cdot yr$ with a water flow rate of 30 $kg/m_c^2 \cdot hr$ and a plate efficiency factor $F' = 0.93$, how much energy will it deliver if the flow rate is doubled? Neglect the effect of flow rate on $F'$; the collector has a heat-loss conductance of 4 $W/m^2 \cdot °C$.

6.10. How large (MJ/hr) should a heat-rejection system be if it must dump the entire heat production of a 1000-$m^2$ solar collector array operating in Denver, Colorado, on August 21 if the collector is at $100°C$ and the ambient temperature is $35°C$? Use solar collector data in Fig. 6.12 and hourly solar radiation data in Appendix 2.

6.11. Use the $f$-chart to determine the amount of solar energy that can be delivered in Little Rock, Arkansas, in January for the following solar and building conditions:

Building:
    Load: 40 million Btu/month
    Latitude: $35°N$
Solar System:
    Collector tilt: $55°$
    Area: 1000 $ft^2$
    Ambient temperature: $40.6°F$
    Collector efficiency curve: see Fig. 6.12
    No heat exchanger used
    Nominal storage, flow rate, and load heat-exchanger values used

6.12. Repeat Problem 6.11 if storage size is doubled and halved.

6.13. Repeat Problem 6.11 for an air-heating system for which $F_R U_c = 0.64$ Btu/hr $\cdot ft^2 \cdot °F$ and $F_R(\tau\alpha) = 0.50$ (typical commercial values). Do air or liquid collectors deliver more energy per square meter in this case?

6.14. Repeat Problem 6.13 if collector fluid flow rate is doubled.

6.15. Using the data in Problem 6.11, calculate system performance for a horizontal and for a vertical collector. Assume Eq. (6.18) applies to both cases.

6.16. Write the differential equation for the model of a passive system in Fig. 6.33. If the thermal capacitance is 125 kJ/m$^2 \cdot$ °C, and values of thermal resistances shown in Fig. 6.34 are used, what is the rise from the initial 35°C temperature in the storage in 1 hr if the insolation is 700 W/m$^2$ and the ambient temperature is 0°C. Use a typical room temperature for a residence and assume it is constant during the hour for simplicity.

6.17. Describe a method for using the f-chart to generate a solar production function (see Chapter 5) with the load and collector area as independent variables. Write a computer program to (1) use results of several f-chart runs for a given building in a given site using a given solar system and (2) correlate these results into logarithmic and power law production functions.

6.18. The f-chart is based on $F_R(\tau\alpha)$ and $F_R U_c$ values, which can be deduced from a plot of collector efficiency versus $(T_{f,in} - T_a)/I$. If such a plot is not available but (a) a plot of efficiency versus $(\overline{T}_f - T_a)/I$ is available or (b) a plot of efficiency versus $(T_{f,out} - T_a)/I$ is available, how can $F_R(\overline{\tau\alpha})$ and $F_R U_c$ be calculated from the slope and intercept of these two curves? Express your results in terms of the slopes, intercepts, and fluid capacitance rate $\dot{m}c_p/A_c$.

6.19. The schematic diagrams below illustrate the operation of a solar-assisted heat-pump system and a solar system augmented with a heat pump. Discuss the advantages and disadvantages of each system with respect to different climatic conditions.

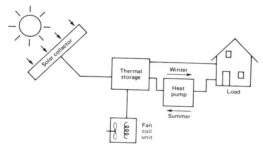

Solar-assisted heat-pump system

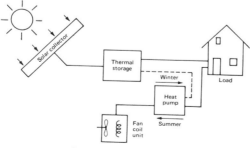

Solar system augmented with heat pump

# REFERENCES

1. Andreassy, S., "Proceedings of the UN Conference on New Sources of Energy," vol. 5, p. 20, New York, 1964; see also Root, D. E., A Simplified Engineering Approach to Swimming Pool Heating, *Sol. Energy*, vol. 3, p. 60, 1959.
2. ASHRAE, "Handbook of Fundamentals," American Society of Heating, Refrigerating, and Air Conditioning Engineers, New York, 1978.
3. ASHRAE, "Energy Conservation in New Building Design," Standard 90-75, American Society of Heating, Refrigerating, and Air Conditioning Engineers, New York, 1975.
4. Balcomb, J. D., and J. C. Hedstrom, A Simplified Method for Calculating Required Solar Array Size for Space Heating, in "Sharing the Sun," vol. 4, pp. 281–294, ISES, Winnipeg, 1976.
5. Balcomb, J. D., and J. D. Hedstrom, Simulation Analysis of Passive Solar Heated Buildings—Preliminary Results, *Los Alamos Rept.* LA-UR-76-89, 1976; see also, under the same title, paper in *Sol. Energy*, vol. 19, pp. 277–282, 1977.
6. Balcomb, J. D., et al., "Solar Handbook for Los Alamos," LASL, New Mexico, 1975; Balcomb, J. D., et al., "Design Considerations of Air-cooled Collector/ Rock-Bin Storage Solar Heating Systems," paper presented at the ISES Annual Meeting, Los Angeles, California, 1975.
7. Baumeister, T., ed. "Marks' Standard Handbook for Mechanical Engineers," 7th ed., McGraw-Hill Book Co., New York, 1967.
8. Beckman, W. A., J. A. Duffie, and S. A. Klein, Simulation of Solar Heating Systems, in "Applications of Solar Energy for Heating and Cooling Buildings," chap. 9, American Society of Heating, Refrigerating, and Air Conditioning Engineers, New York, 1977.
8a. Beckman, W. A., S. A. Klein, and J. A. Duffie, "Solar Heating Design by the F-Chart Method," John Wiley & Sons, New York, 1977.
9. Brace Research Institute, Numerous publications on efficient greenhouse design for mid-latitude climates.
10. Close, D. J., The Performance of Solar Water Heaters with Natural Circulation, *Sol. Energy*, vol. 6, p. 33, 1962.
11. deWinter, F., "How to Design & Build a Solar Swimming Pool Heater," Copper Development Association, Inc., New York, 1975.
12. Freeman, T. L., et al., "Computer Modeling of Heat Pumps and the Simulator of Solar-Heat Pump Systems," ASME Paper 75-WI/Sol-3, 1975.
13. Gilman, S. F., Solar Energy Heat Pump Systems for Heating and Cooling Buildings, *ERDA Rept.* C00-2560-1, 1975.
14. Gutierrez, G., et al., Simulation of Forced Circulation Water Heaters: Effects of Auxiliary Energy Supply Load Type and Storage Capacity, *Sol. Energy*, vol. 15, p. 287, 1974.
14a. Hay, H. R., and J. I. Yellott, Natural Air Conditioning with Roof-Ponds and Movable Insulation, *Trans. ASHRAE*, vol. 75, p. 165, 1969.
15. Hittman Associates, "Verification of the Time-Response Method for Heat Load Calculation," Rept. 2300-00259, U.S. Government Printing Office, Washington, D.C., 1973.
16. Jacobs, M., and S. R. Peterson, "Making the Most of Your Energy Dollars," National Bureau of Standards, Washington, D.C., 1974 (U.S. Government Printing Office Rept. 003-003-01446-0).

17. Karman, V. D., et al., Simulation Study of Solar Heat Pump Systems, in "Sharing the Sun," vol. 3, pp. 324–340, ISES, Winnipeg, 1976.
18. Kays, W. M., and A. L. London, "Compact Heat Exchangers," McGraw-Hill Book Co., New York, 1964.
19. Klein, S. A., "A Design Procedure for Solar Heating Systems," Ph.D. dissertation, University of Wisconsin, Madison, 1976; for an approach similar to the $f$-chart for other solar-thermal systems operating above a minimum temperature above that for space-heating ($\sim 20°C$), see Klein, S. A., and W. A. Beckman, A General Design Method for Closed Loop Solar Energy Systems, *Proc. 1977 ISES Meeting*, 1977.
20. Klein, S. A., W. A. Beckman, and J. A. Duffie, A Design Procedure for Solar Heating Systems, *Sol. Energy*, vol. 18, p. 113, 1976.
21. Klein, S. A., W. A. Beckman, and J. A. Duffie, "A Design Procedure for Solar Air Heating Systems," paper presented at the 1976 ISES Conference, American Section, Winnipeg, Manitoba, August 15–20, 1976.
22. Klein, S. A., et al. A Method for Simulation of Solar Processes and its Application, *Sol. Energy*, vol. 17, pp. 29–37, 1975.
23. Kreider, J. F., and F. Kreith, "Solar Heating and Cooling," Hemisphere Publ. Corp., Washington, D.C., revised 1st ed., 1977.
24. Kusuda, T., *NBSLD*, National Bureau of Standards, Washington, D.C., 1973.
25. Löf, G. O. G., and R. A. Tybout, Cost of House Heating with Solar Energy, vol. 14, p. 253, 1973.
26. Löf, G. O. G., and D. J. Close, Solar Water Heaters, in "Low Temperature Engineering Applications of Solar Energy," American Society of Heating, Refrigerating, and Air Conditioning Engineers, New York, 1967.
27. Marvin, W. C., and S. A. Mummer, Optimum Consideration of Solar Energy and the Heat Pump for Residential Heating, in "Sharing the Sun," vol. 3, pp. 321–323, ISES, Winnipeg, 1976.
28. Mutch, J. J., "Residential Water Heating," Maxwell Scientific International, Elmsford, N.Y., 1974.
28a. Nyles, P. W. B., Thermal Evaluation of a House Using Movable-Insulation Heating and Cooling System, *Sol. Energy*, vol. 18, pp. 412–419, 1976.
29. Phillips, W. F., and R. D. Cook, "Natural Circulation from a Flat Plate Collector to a Hot Liquid Storage Tank," ASME Paper 75-HT-53, 1975.
30. Sfeir, A., et al., A Numerical Model for a Solar Water Heater, in "Heliotechnique and Development" (COMPLES 1975) vol. 2, pp. 38–52, Development Analysis Associates, Cambridge, Massachusetts, 1976.
31. Thom, H. C. S., Normal Degree-Days Above Any Base by the Universal Truncation Coefficient, *Mon. Weather Rev.*, vol. 94, pp. 461–465, 1966.
32. Thom, H. C. S., Normal Degree-Days Below Any Base, *Mon. Weather Rev.*, vol. 82, pp. 111–115, 1954.
33. Thom, H. C. S., The Rational Relationship between Heating Degree-Days and Temperature, *Mon. Weather Rev.*, vol. 82, pp. 1–6, 1954.
34. Thom, H. C. S., Seasonal Degree-Day Statistics for the U.S., *Mon. Weather Rev.*, vol. 80, pp. 143–147, 1952.
35. Trombe, F., et al., "Some Characteristics of the CNRS Solar House Collectors," CNRS Solar Laboratory, Font Romeu/Odeillo, France, 1976.
36. TRW, Inc., Solar Heating and Cooling of Buildings (Phase O) Executive Summary, *NSF Rept.* NSF-R4-N-74-022A, 1974.

37. U.S. Department of Commerce, "Climatic Atlas of the United States," U.S. Government Printing Office, Washington, D.C., 1968.
38. Ward, J. C., and G. O. G. Löf, Long-term (18 years) Performance of a Residential Solar Heating System, *Sol. Energy*, vol. 18, pp. 301–308, 1976; see also by the same authors, Maintenance Costs of Solar Air Heating Systems, *Proc. 1977 ISES Meeting*, 1977.
39. Yellott, J. I., and H. R. Hay, Thermal Analysis of a Building with Natural Air Conditioning, *Trans. ASHRAE*, vol. 75, p. 78, 1969.

# SOLAR COOLING AND DEHUMIDIFICATION

*The real cycle you're working on is a cycle called yourself.*

Robert Pirsig

## SOLAR SPACE COOLING AND REFRIGERATION*

In some ways solar energy is better suited to space cooling and refrigeration than to space heating, but until recently this application of solar energy has received little attention. However, energy for air conditioning has been the fastest-growing segment of the energy consumption market in the United States and is currently using about 3 percent of the total energy output. Also, at the 1975 UNESCO Conference on Solar Energy several developing countries stated that small solar-driven refrigeration units are urgently needed for preservation of food in rural areas where little electric power is available. Consequently, accelerated progress in the application of solar energy to space cooling and refrigeration can be expected.

The seasonal variation of solar energy is extremely well suited to the space-cooling requirements of buildings. The principal factors affecting the temperature in a building are the average quantity of radiation received and the environmental air temperature. Since the warmest seasons of the year correspond to periods of high insolation, solar energy is most available when comfort cooling is most needed. Moreover, as we have seen in Chapter 4, the efficiency of solar collectors increases with increasing insolation and increasing environmental temperature. Consequently, in the summer the amount of energy delivered per unit surface area of collector can be larger than that in winter.

There are several approaches that can be taken to solar space cooling and refrigeration. Because of the very limited operating experience with solar-cooling systems, their design must be based on basic principles and experience with conventional cooling systems. The material presented in this chapter will therefore stress the fundamental principles of operation of refrigeration cycles and combine them with special features of the components in a solar system.

---

*In view of the fact that the space comfort industry in the United States uses English units exclusively, some of the examples in this chapter will be worked in these units to facilitate communication with practicing engineers.

The two principal methods of lowering air temperature for comfort cooling are refrigeration with actual removal of energy from the air or evaporative cooling of the air with adiabatic vaporization of moisture into it. Refrigeration systems can be used under any entering air condition, whereas evaporative cooling can be used only when the entering air has a comparatively low relative humidity. Insofar as the application of solar energy is concerned, evaporative coolers have solar combination possibilities, as, for example, in connection with a nocturnal cooling system which uses the storage rock bed from an air-based solar-heating system and various desiccant systems.

There are two main types of refrigeration systems for space cooling. The most widely used method employs a vapor-compression refrigeration cycle. The other uses an absorption refrigeration cycle similar to that of the gas refrigerator. Electric energy is used in the first type to operate a compressor that circulates a refrigerant vapor to an air-cooled or water-cooled condenser, from which the resulting liquid passes through a throttle valve to the evaporator or cooling unit, as shown in Fig. 7.1. Evaporation of the refrigerant results in heat removal from the water in the "chiller." The water is then used to cool the air in the building. The cycle is completed by vapor returning to the compressor. Heat is thus removed from a high-temperature source and discharged to the surrounding atmosphere or to cooling water by means of mechanical work input to the compressor, which raises the temperature of the refrigerant above the temperature of the heat sink. In a solar-driven vapor-compression cycle the shaft work to the compressor is supplied directly by a heat engine powered by solar energy.

In an absorption system the refrigerant is evaporated or distilled from a less

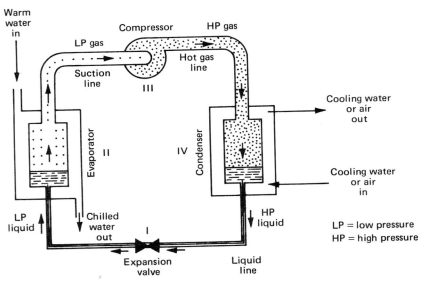

**FIGURE 7.1** Schematic diagram illustrating the basic refrigeration vapor-compression cycle.

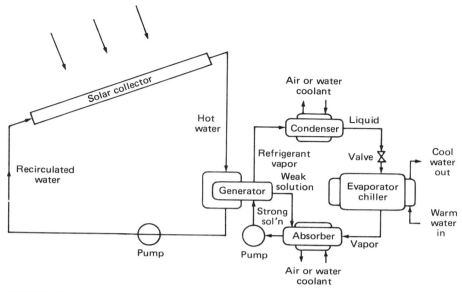

**FIGURE 7.2**  Schematic diagram of a solar-powered absorption refrigeration system.

volatile liquid absorbent, the vapor is condensed in a water- or air-cooled condenser, and the resulting liquid is passed through a reducing valve to the cooling section of the unit. There it chills the water as it evaporates, and the resulting vapor flows into a vessel, where it is reabsorbed in the stripped absorbing liquid and pumped back to the heated generator. The heat required to evaporate the refrigerant in the generator can be supplied directly from solar energy as shown in Fig. 7.2. Since, without concentration, solar energy cannot heat a working fluid much above 363 K (195°F), the efficiency of converting this energy into work is low compared with its direct utilization as heat to evaporate the refrigerant. Consequently, space cooling with flat-plate solar collectors seems to favor the use of an absorption process. However, if some concentration is used, as for example in a compound parabolic concentrator collector, the vapor compression cycle may turn out to be more satisfactory and cheaper in the long run.

In summary, solar cooling of buildings represents a potentially significant application of solar energy for building air conditioning in most sunny regions of the United States. Solar-cooling technology is presently not as advanced as solar-heating technology, but research work is expected to close the gap between the two by 1985 or so. Several viable, solar air-conditioning schemes are described in this chapter and methods for tentative system design are presented in detail.

## Cooling Requirements for Buildings

The cooling load of a building is the rate at which heat must be removed to maintain the air in a building at a given temperature and humidity. It is usually calculated on the basis of the peak load expected during the cooling season. For a given building the cooling load depends primarily on

1. Inside and outside dry-bulb temperatures
2. Inside and outside relative humidities
3. Solar-radiation heat load
4. Wind speed
5. Life-style of occupants, e.g., infiltration

A method of calculating the cooling load is presented in detail in Ref. (1).
The steps in calculating the cooling loads of a building are

1. Specify the building characteristics: wall area, type of construction, and surface characteristics; roof area, type of construction, and surface characteristics; window area, setback, and glass type; and building location and orientation.
2. Specify the outside and inside wet- and dry-bulb temperatures.
3. Specify the solar heat load and wind speed.
4. Calculate building cooling loads resulting from the following: heat transfer through windows; heat transfer through walls; heat transfer through roof; sensible heat gains resulting from infiltration and exfiltration; latent heat gains (water vapor); and internal heat sources, such as people, lights, etc.

Equations (7.1)–(7.7) may be used to calculate the various cooling loads for a building; cooling loads resulting from lights, building occupants, etc. may be estimated from Ref. (1). For unshaded or partially shaded windows, the load is

$$Q_{wi} = A_{wi}\left[F_{sh}\bar{\tau}_{b,wi}I_{h,b}\frac{\cos i}{\sin \alpha} + \bar{\tau}_{d,wi}I_{h,d} + \bar{\tau}_{r,wi}I_r + U_{wi}(T_{out} - T_{in})\right] \quad (7.1)$$

For shaded windows the load (neglecting diffuse sky radiation) is

$$Q_{wi,sh} = A_{wi,sh}U_{wi}(T_{out} - T_{in}) \quad (7.2)$$

For unshaded walls the load is

$$Q_{wa} = A_{wa}\left[\bar{\alpha}_{s,wa}\left(I_r + I_{h,d} + I_{h,b}\frac{\cos i}{\sin \alpha}\right) + U_{wa}(T_{out} - T_{in})\right] \quad (7.3)$$

For shaded walls the load (neglecting diffuse sky radiation) is

$$Q_{wa,sh} = A_{wa,sh}[U_{wa}(T_{out} - T_{in})] \quad (7.4)$$

For the roof the load is

$$Q_{rf} = A_{rf}\left[\bar{\alpha}_{s,rf}\left(I_{h,d} + I_{h,b}\frac{\cos i}{\sin \alpha}\right) + U_{rf}(T_{out} - T_{in})\right] \quad (7.5)$$

For sensible-heat infiltration and exfiltration the load is

$$Q_i = \dot{m}_a(h_{\text{out}} - h_{\text{in}}) \qquad (7.6)$$

For moisture infiltration and exfiltration the load is

$$Q_w = \dot{m}_a(W_{\text{out}} - W_{\text{in}})\lambda_w \qquad (7.7)$$

where

$Q_{\text{wi}}$ = heat flow through unshaded windows of area $A_{\text{wi}}$, Btu/hr

$Q_{\text{wi,sh}}$ = heat flow through shaded windows of area $A_{\text{wi,sh}}$, Btu/hr

$Q_{\text{wa}}$ = heat flow through unshaded walls of area $A_{\text{wa}}$, Btu/hr

$Q_{\text{wa,sh}}$ = heat flow through shaded walls of area $A_{\text{wa,sh}}$, Btu/hr

$Q_{rf}$ = heat flow through roof of area $A_{rf}$, Btu/hr

$Q_i$ = heat load resulting from infiltration and exfiltration, Btu/hr

$Q_w$ = latent heat load, Btu/hr

$I_{h,b}$ = beam component of insolation on horizontal surface, Btu/hr · ft$^2$

$I_{h,d}$ = diffuse component of insolation on horizontal surface, Btu/hr · ft$^2$

$I_r$ = ground-reflected component of insolation, Btu/hr · ft$^2$

$W_{\text{out}}, W_{\text{in}}$ = outside and inside humidity ratios, lb$_m$ H$_2$O/lb$_m$ dry air

$U_{\text{wi}}, U_{\text{wa}}, U_{rf}$ = overall heat-transfer coefficients for windows, walls, and roof, including radiation, Btu/hr · ft$^2$ · °F

$\dot{m}_a$ = net infiltration and exfiltration of dry air, lb$_m$/hr

$T_{\text{out}}$ = outside dry-bulb temperature, °F

$T_{\text{in}}$ = indoor dry-bulb temperature, °F

$F_{\text{sh}}$ = shading factor (1.0 = unshaded, 0.0 = fully shaded)

$\bar{\alpha}_{s,\text{wa}}$ = wall solar absorptance

$\bar{\alpha}_{s,rf}$ = roof solar absorptance

$i$ = solar-incidence angle on walls, windows, and roof, deg

$h_{\text{out}}, h_{\text{in}}$ = outside and inside air enthalpy, Btu/lb$_m$

$\alpha$ = solar-altitude angle, deg

$\lambda_w$ = latent heat of water vapor, Btu/lb$_m$

$\bar{\tau}_{b,\text{wi}}$ = window transmittance for beam (direct) insolation

$\bar{\tau}_{d,\text{wi}}$ = window transmittance for diffuse insolation

$\bar{\tau}_{r,\text{wi}}$ = window transmittance for ground-reflected insolation

## EXAMPLE 7.1

Determine the cooling load for a building in Phoenix, Arizona, with the specifications tabulated below.

## SOLUTION

To determine the cooling load for the building just described, calculate the following factors in the order listed.

| Factor | Description or specification |
|---|---|
| Building characteristics: | |
| Roof: | |
| Type of roof | Flat, shaded |
| Area $A_{rf,\text{sh}}$, ft² | 1700 |
| Walls (painted white): | |
| Size, north and south, ft | 8 × 60 (two) |
| Size, east and west, ft | 8 × 40 (two) |
| Area $A_{\text{wa}}$, north and south walls, ft² | $480 - A_{\text{wi}} = 480 - 40 = 440$ (two) |
| Area $A_{\text{wa}}$, east and west walls, ft² | $320 - A_{\text{wi}} = 320 - 40 = 280$ (two) |
| Absorptance $\bar{\alpha}_{s,\text{wa}}$ of white paint | 0.12 |
| Windows: | |
| Size, north and south, ft | 4 × 5 (two) |
| Size, east and west, ft | 4 × 5 (two) |
| Shading factor $F_{\text{sh}}$ | 0.20 |
| Insolation transmittance | $\bar{\tau}_{b,\text{wi}} = 0.60; \bar{\tau}_{d,\text{wi}} = 0.81; \bar{\tau}_{r,\text{wi}} = 0.60$ |
| Location and latitude | Phoenix, Ariz.; 33°N |
| Date | August 1 |
| Time and local-solar-hour angle $H_s$ | Noon; $H_s = 0$ |
| Solar declination $\delta_s$, deg | 18°14′ |
| Wall surface tilt from horizontal $\beta$ | 90° |
| Temperature, outside and inside, °F | $T_{\text{out}} = 100; T_{\text{in}} = 75$ |
| Insolation $I$, Btu/hr · ft² | $I_{h,b} = 185; I_{h,d} = 80; I_r = 70$ |
| $U$ factor for walls, windows, and roof | $U_{\text{wa}} = 0.19; U_{\text{wi}} = 1.09; U_{rf} = 0.061$ |
| Infiltration, lb$_m$ dry air/hr | Neglect |
| Exfiltration, lb$_m$ dry air/hr | Neglect |
| Internal loads | Neglect |
| Latent heat load $Q_w$, percent | 30 percent of wall sensible heat load[a] |

[a]Approximate rule of thumb for Phoenix.

1. Incidence angle for the south wall $i$

$$\cos i = \cos \delta_s \cos (L - \beta) + \sin \delta_s \sin (L - \beta) = 0.257$$

2. Solar altitude $\alpha$

$$\sin \alpha = \sin \delta_s \sin L + \cos \delta_s \cos L \cos H_s = \cos (L - \delta_s) = \cos 15° = 0.96$$

3. South-facing window load [from Eq. (7.1)]

$$Q_{\text{wi}} = 40 \left\{ (0.2 \times 0.6) \left[ 185 \frac{0.257}{0.96} + (0.81 \times 80) \right] + (0.60 \times 70) \right. $$
$$\left. + [1.09(100 - 75)] \right\} = 5600 \text{ Btu/hr}$$

4. Shaded-window load [from Eq. (7.2)]

$$Q_{\text{wi,sh}} = (3 \times 40)[1.09(100 - 75)] = 3270 \text{ Btu/hr}$$

5. South-facing wall load [from Eq. (7.3)]

$$Q_{wa} = (480 - 40) \left\{ 0.12 \left[ 70 + 80 + \left( 185 \frac{0.257}{0.96} \right) \right] + 0.19(100 - 75) \right\}$$

$$= 12{,}610 \text{ Btu/hr}$$

6. Shaded-wall load [from Eq. (7.4)]

$$Q_{wa,sh} = [(480 + 320 + 320) - (3 \times 40)][0.19(100 - 75)] = 4750 \text{ Btu/hr}$$

7. Roof load [from Eq. (7.5)]

$$Q_{rf} = 1700[\bar{a}_{s,rf} \times 0 + 0.061(100 - 75)] = 2600 \text{ Btu/hr}$$

8. Latent-heat load (30 percent of sensible wall load)

$$Q_{wa} = 0.3[(480 + 480 + 320 + 320) - (4 \times 40)][0.19(100 - 75)]$$

$$= 2050 \text{ Btu/hr}$$

9. Infiltration-exfiltration load

$$Q_i = 0$$

10. Total cooling load for the building described in the example

$$Q_{tot} = Q_{wi} + Q_{wi,sh} + Q_{wa} + Q_{wa,sh} + Q_{rf} + Q_w + Q_i$$

$$Q_{tot} = 30{,}880 \text{ Btu/hr}$$

## Vapor-Compression Cycle

The principle of operation of a vapor-compression refrigeration cycle can be illustrated conveniently with the aid of a pressure-enthalpy diagram as shown in Fig. 7.3. The ordinate is the pressure of the refrigerant in $N/m^2$ absolute and the abscissa its enthalpy in kJ/kg. The roman numerals in Fig. 7.3 correspond to the physical locations in the schematic diagram of Fig. 7.1.

Process I is a throttling process in which hot liquid refrigerant at the condensing pressure $p_c$ passes through the expansion valve, where its pressure is reduced to the evaporator pressure $p_e$. In this process some vapor is produced and the state of the mixture of liquid refrigerant and vapor entering the evaporator is shown by point A. Since the expansion process is isenthalpic (that is, constant enthalpy), the following relation holds:

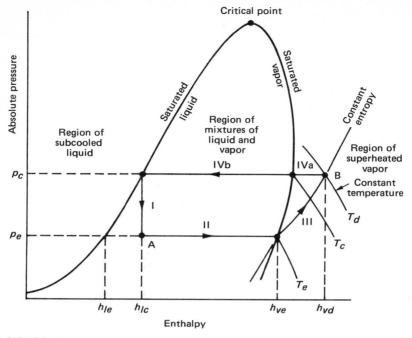

**FIGURE 7.3** Simple refrigeration cycle on pressure-enthalpy diagram.

$$
\begin{array}{c}
\text{Enthalpy of} \\
\text{mixture entering} \\
\text{evaporation}
\end{array}
=
\begin{array}{c}
\text{enthalpy} \\
\text{of vapor} \\
\text{at } p_e
\end{array}
=
\left(
\begin{array}{c}
\text{enthalpy} \\
\text{of liquid} \\
\text{at } p_e
\end{array}
\right)
(1-f)
=
\begin{array}{c}
\text{enthalpy of} \\
\text{liquid at condenser} \\
\text{pressure}
\end{array}
\quad (7.8)
$$

where $f$ is the percentage mass of liquid vaporized. The relation in equation form is

$$
h_{ve}f + h_{le}(1-f) = h_{lc} \tag{7.9}
$$

and

$$
f = \frac{h_{ec} - h_{le}}{h_{ve} - h_{le}} \tag{7.10}
$$

where subscripts $l$ and $v$ denote liquid and vapor, respectively.

Process II represents the vaporization of the remaining liquid. This is the process during which heat is removed from the chiller. Thus, the specific refrigeration effect per kilogram of refrigerant $q_r$ is

$$
q_r = h_{ve} - h_{lc} \quad \text{in kJ/kg (Btu/lb)} \tag{7.11}
$$

In the United States it is still common practice to measure refrigeration in terms of *tons*. One ton is the amount of cooling produced if 1 ton of ice is melted over a period of 24 hr. Since 1 ton = 907.2 kg and the latent heat of fusion of water is 334.9 kJ/kg

$$1 \text{ ton} = \frac{(907.2 \text{ kg}) \times (334.9 \text{ kJ/kg})}{(24 \text{ hr}) \times (3600 \text{ sec/hr})} = 3.516 \text{ kW} = 12{,}000 \text{ Btu/hr} \quad (7.12)$$

If the desired rate of refrigeration requires a heat-transfer rate of $q_{\text{refrig}}$ (kW), the rate of mass flow of refrigerant necessary $\dot{m}_r$ in kg/sec is

$$\dot{m}_r = \frac{q_{\text{refrig}}}{(h_{ve} - h_{lc})} \quad \text{in kW/kJ} \cdot \text{kg} \quad (7.13)$$

Process III in Fig. 7.3 represents the compression of refrigerant from pressure $p_e$ to $p_c$. The process requires work input from an external source and it is this work that is obtained from a solar-driven expander-turbine in a solar vapor refrigeration cycle. In general, if the heated vapor leaving the compressor is at the condition represented by point B in Fig. 7.3, the work of compression $W_c$ is

$$W_c = \dot{m}_r(h_{vd} - h_{ve}) \quad (7.14)$$

In idealized cycle analysis the compression process is usually assumed to be isentropic.

Process IV represents the condensation of the refrigerant. Actually, sensible heat is first removed in the subprocess IVa as the vapor is cooled at constant pressure from $T_d$ to $T_c$ and latent heat is removed at the condensation temperature $T_c$, corresponding to the saturation pressure $p_c$ in the condenser. The heat-transfer rate in the condenser $q_c$ in kW is

$$q_c = \dot{m}_r(h_{vd} - h_{lc}) \quad (7.15)$$

This heat must be rejected into the environment—either to cooling water or to the atmosphere if no water is available.

The overall performance of a refrigeration machine is usually expressed as the ratio of the heat transferred in the evaporator $q_r$ to the shaft work supplied to the compressor per kilogram of refrigerant. This ratio is called the *coefficient of performance (COP)*, defined by

$$COP = \frac{h_{ve} - h_{lc}}{h_{vd} - h_{ve}} = \frac{q_r}{W_c} \quad (7.16)$$

The highest coefficient of performance for any given evaporator and condenser temperatures would be obtained if the system were operating on a reversible Carnot cycle. Under these conditions (1)

$$COP \text{ (Carnot)} = \frac{T_e}{T_d - T_e} \tag{7.17}$$

However, frictional effects and irreversible heat losses reduce the *COP* of real cycles much below this optimum.

### EXAMPLE 7.2

Calculate the amount of shaft work to be supplied to a 1-ton (3.52-kW) refrigeration plant operation at evaporator and condenser temperatures of 273 K and 308 K, respectively, using Refrigerant 12 (R-12) as the working fluid. The properties of Refrigerant 12 are tabulated in Table 7.1. (More complete data are given in Appendix 7, Table A7.2.) Also calculate the *COP* and the mass flow rate of the refrigerant.

### SOLUTION

From the property table the enthalpies for process I are

saturated vapor at 273 K     $h_{ve} = 187.5 \text{ kJ/kg}$

saturated liquid at 308 K    $h_{ec} = \phantom{0}69.6 \text{ kJ/kg}$

saturated liquid at 273 K    $h_{le} = \phantom{0}36.1 \text{ kJ/kg}$

**TABLE 7.1**  Properties of Refrigerant 12 for Example 7.2

| Temperature (K) | Absolute pressure (kN/m²) | Vapor specific volume (m³/kg) | Liquid enthalpy (kJ/kg) | Vapor enthalpy (kJ/kg) | Vapor entropy (kJ/kg · K) |
|---|---|---|---|---|---|
| | | Saturated | | | |
| 273 | 308.6 | 0.0554 | 36.05 | 187.5 | 0.6966 |
| 278 | 362.6 | 0.0475 | 40.69 | 189.3 | 0.6943 |
| 308 | 847.7 | 0.0206 | 69.55 | 201.5 | 0.6839 |
| 313 | 960.7 | 0.0182 | 74.59 | 203.2 | 0.6825 |
| | | Superheated | | | |
| 278 | 308.6 | 0.0564 | – | 190.8 | 0.7081 |
| 313 | 847.7 | 0.0212 | – | 205.2 | 0.6959 |
| 318 | 847.7 | 0.0218 | – | 208.9 | 0.7078 |
| 323 | 847.7 | 0.0224 | – | 212.7 | 0.7196 |

Therefore, from Eq. 7.10

$$f = \frac{69.6 - 36.1}{187.5 - 36.1} = 0.22$$

The mass rate of flow of refrigerant $\dot{m}_r$ is obtained from Eq. (7.10) and the enthalpies above, or

$$\dot{m}_r = \frac{3.52 \text{ kW}}{(187.5 - 69.6) \text{ kJ/kg}} = 0.0298 \text{ kg/sec}$$

The specific shaft-work input required is

$$W_c = h_{vd} - h_{vc} = \frac{W_c}{\dot{m}_r}$$

The entropy $s_e$ of the saturated vapor entering the compressor at 273 K and 308.6 K $\cdot$ N/m$^2$ is 0.697 kJ/kg $\cdot$ K. From the property table, superheated vapor at a pressure of 847.7 kN/m$^2$ has an entropy of 0.697 at a temperature between 313 and 318 K. By linear interpolation between these two temperatures and their corresponding entropy values, the outlet temperature at the compressor is found to be 314.6 K with an enthalpy of 206.4 kJ/kg. Thus, the energy input to the working fluid by the compressor is

$$W_c = 0.0298(206.4 - 187.5) = 0.56 \text{ kW}$$

Finally, the heat-transfer rate from the refrigerant to the sink, or cooling water in the condenser, is from Eq. (7.15)

$$q_c = \dot{m}_r(h_{vd} - h_{lc}) = 0.298(206.4 - 69.6) = 4.08 \text{ kW}$$

The *COP* of the thermodynamic cycle is

$$COP = \frac{187.5 - 69.6}{206.4 - 187.5} = 6.3$$

whereas the Carnot *COP* is 273/35 or 7.8.

The above cycle has been idealized. In practice, the liquid entering the expansion valve is several degrees below the condensing temperature, while the vapor entering the compressor is several degrees above the evaporation temperature. In addition, pressure drops occur in the suction, discharge, and liquid pipelines and the compression is not truly isentropic. Finally, the work required to drive the compressor is somewhat larger than $W_c$ above, because of frictional losses. All of these factors must be taken into account in a realistic engineering design.

The vapor-compression cooling cycle described above can be combined with a solar-driven Rankine power cycle. One scheme suitable for building air conditioning is shown in Fig. 7.4. This system is split into two loops, but uses a single working fluid and a shared condenser. The preheater is an option that will enhance the performance at high boiler temperatures. In a single-fluid system, no shaft seals are required and the turb-compressor can be operated on fluid bearings. These features should give the system a long life with minimum maintenance compared with dual-fluid systems, which give slightly better performance.

The performance of a single-fluid solar cycle has been calculated by Burriss (4). Figure 7.5 shows the cycle *COP* as a function of the temperature in the solar boiler, where the working fluid is vaporized before passing through the turbine that drives the compressor as shown in Fig. 7.4. The improvement in performance by using higher boiler temperatures and a preheater is apparent. Burriss also estimated the solar collector areas required and his results are shown in Table 7.2.

A two-fluid, solar-driven organic Rankine cycle capable of providing 3 tons of air conditioning or 1 kW electric power has been built and tested by Barber (2). A schematic diagram of the system is shown in Fig. 7.6. Water from a flat-plate solar collector provides the heat input to a low-temperature organic Rankine cycle with R-113 as the working fluid. This working fluid passes through a high-speed (50,000 rpm) turbine, which drives a refrigeration compressor in a vapor cycle with R-12 as the working fluid.

At the design point the solar collector temperature is 375 K with a flow rate

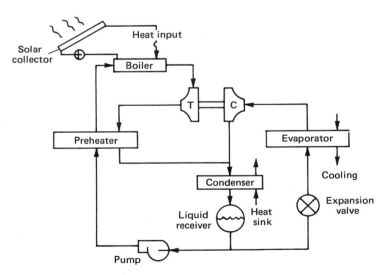

**FIGURE 7.4** Schematic diagram of a single-fluid, solar-powered vapor-compression air conditioner.

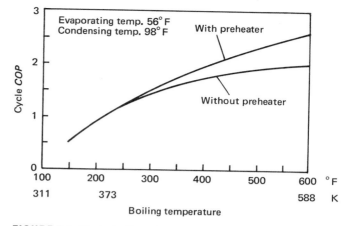

**FIGURE 7.5** Typical effect of cycle boil temperature on *COP*. From (4).

of 45 liters/min. Solar radiation evaporates R-113 in the boiler to provide a turbine inlet temperature of 355 K. A water-cooled condenser with 100 liters/min of 303 K cooling water is required to achieve an R-113 condensing temperature of 278 K. With a turbine efficiency of 80 percent, the Rankine cycle efficiency (turbine shaft power divided by solar-heat input to the working fluid) is 11.5 percent. With an 85 percent efficient R-12 compressor for the cooling machinery this cycle gives a *COP* of 0.71. The turbine provides power through a speed reducer to a conventional piston air-conditioning compressor and/or to a motor-generator. The interface makes it possible to provide supplemental energy at 220 V, 60-cycle ac, single phase during periods of solar outage and to generate up to 1 kW of power, depending on solar input, when solar radiation is available but no air conditioning is required. Operation with the motor alone can provide 3.17 tons of air conditioning while 2.0 hp is supplied to the compressor. Other solar-powered refrigeration cycles have been built with capacities up to 100 tons. They are all experimental in nature and were built to provide performance data on this new solar-powered air-conditioning concept.

**TABLE 7.2** Performance of Solar-Powered Refrigeration System[a]

|  | Solar collector type | |
| --- | --- | --- |
|  | Flat-plate | Concentrator |
| Collector efficiency, percent | 40 | 60 |
| Cycle boiling temperature, K | 344 | 588 |
| Cycle *COP* | 0.6 | 2.6 |
| Collector area, m²/ton | 12.4 | 1.67 |

[a]From (4).

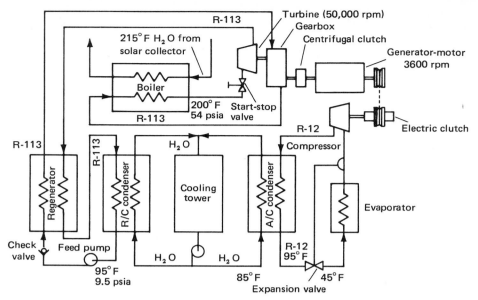

**FIGURE 7.6** Schematic diagram of a dual-fluid, solar-powered vapor-compression air conditioner. From (4).

## Absorption Air Conditioning

Absorption air conditioning is the only air-conditioning system compatible with the upper collection-temperature limits imposed by currently available flat-plate collectors (13). Home-sized absorption air-conditioning units are more expensive than vapor-compression air-conditioning units, but to date only absorption air-conditioning units compatible with solar equipment are commercially available.

Presently, two types of absorption air-conditioning systems are widely marketed in the United States: the lithium bromide-water (LiBr-$H_2O$) system and the ammonia-water (NH$_3$-$H_2O$) system. An absorption air-conditioning system is shown in Fig. 7.7. Absorption air conditioning differs from vapor-compression air conditioning only in the positive pressure gradient stage (right of the dashed line in Fig. 7.7). In absorption air-conditioning systems, the pressurization is accomplished by first dissolving the refrigerant in a liquid (the absorbent) in the *absorber section*, then pumping the solution to a high pressure with an ordinary liquid pump. The low-boiling refrigerant is then driven from solution by the addition of heat in the *generator*. By this means the refrigerant vapor is compressed without the large input of high-grade shaft work that the vapor-compression air conditioning demands. The remainder of the system consists of a condenser, expansion valve, and evaporator, identical in function to those used in a vapor-compression air-conditioning system.

Of the two common absorption air-conditioning systems, the LiBr-$H_2O$ is simpler, since a rectifying column is not needed. In the NH$_3$-$H_2O$ system a rectifying column assures that no water vapor, mixed with $NH_3$, enters the evaporator where it could freeze. In the LiBr-$H_2O$ system water vapor is the refrigerant. In addition, the $NH_3$-$H_2O$ system requires higher generator temperatures (250-300°F) than a flat-plate solar collector can provide. On the other hand the LiBr-$H_2O$ system operates satisfactorily at a generator temperature of 190-200°F, which is achievable by a flat-plate collector. Also, the LiBr-$H_2O$ system has a larger *COP* than the $NH_3$-$H_2O$ system. The disadvantage of LiBr-$H_2O$ systems is that evaporators cannot operate at temperatures much below 40°F since the refrigerant is water vapor.

The effective performance of an absorption cycle depends on the two materials that comprise the refrigerant-absorbent pair. Desirable characteristics for the refrigerant-absorbent pair are

1. The absence of a solid-phase absorbent
2. A refrigerant more volatile than the absorbent so that separation from the absorbent occurs easily in the generator
3. An absorbent that has a small affinity for the refrigerant
4. A high degree of stability for long-term operations

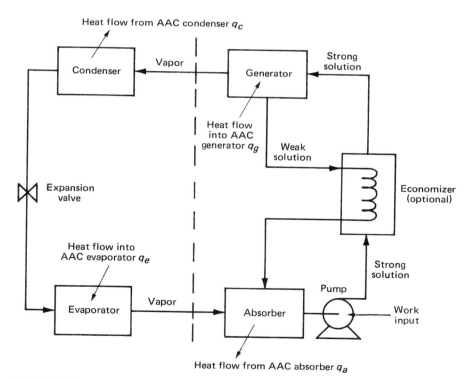

**FIGURE 7.7**  Diagram of heat and fluid flow of absorption air conditioner, with economizer. From (10b).

5. A refrigerant that has a large latent heat so that the circulation rate can be kept at the minimum
6. A low corrosion rate and nontoxicity for safety reasons

A problem with the LiBr-H$_2$O pair is possible crystallization in the generator.

Large absorption air conditioners are manufactured by many of the air conditioning manufacturers in the United States but only one makes a residential-sized (3–5 ton) LiBr-H$_2$O unit for solar installations.

If the pump work is neglected, the *COP* of an absorption air conditioner can be calculated from Fig. 7.7:

$$COP = \frac{\text{cooling effect}}{\text{heat input}} = \frac{q_e}{q_g} \tag{7.18}$$

The *COP* values for absorption air conditioning range from 0.5 for a small, single-stage unit to 0.85 for a double-stage, steam-fired unit. These values are about 15 percent of the *COP* values that can be achieved by a vapor-compression air conditioner. It is difficult to compare the *COP* of an absorption air conditioner with that of a vapor-compression air conditioner directly, because the efficiency of electric power generation or transmission is not included in the *COP* of the vapor-compression air conditioning. However, the limited operating experience with absorption cooling systems using solar flat-plate collectors to supply heat to the generator has not demonstrated to date the economic viability of this concept, especially for single-family residences. The year-round *COP* of absorption systems often suffers from transient effects as well as from the thermodynamic limitations imposed by the upper temperature limits of the collectors. The most serious problem, however, is that the pumping power required to achieve the same cooling effect, although relatively small compared to the power required for a vapor compressor, must be multiplied by the ratio of the vapor-compression *COP* to the absorption-cycle *COP* in order to account for the larger amount of refrigerant flow. On that basis the electric power input to an absorption system can be of the same order of magnitude as for a vapor-compression system. Moreover, since normally, as is shown later, for economic reasons only a fraction of the total heat load is supplied by the solar collectors, the total energy input, i.e., electrical power to pumps and controls plus the auxiliary energy requirements of the generator, can exceed the electrical power input required to operate a Rankine vapor-compression system to provide the same amount of cooling.

### EXAMPLE 7.3

A water-lithium bromide, absorption refrigeration system such as that shown in Fig. 7.8 is to be analyzed for the following requirements:

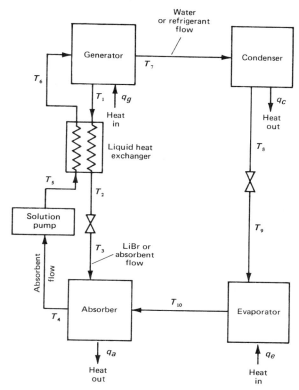

**FIGURE 7.8** Lithium bromide-water, absorption refrigeration cycle (see Table 7.3). From (10b).

1. The machine is to provide 100 tons of refrigeration with an evaporator temperature of 40°F, an absorber outlet temperature of 90°F, and a condenser temperature of 110°F.
2. The approach at the low-temperature end of the liquid heat exchanger is to be 10°F.
3. The generator is heated by a flat-plate solar collector capable of providing a temperature level of 192°F for the evaporation of the refrigerant.

Determine the *COP*, absorbent and refrigerant flow rates, and heat input required for a 100-ton unit.

**SOLUTION**

Analytical evaluation of the LiBr-H$_2$O cycle requires that several simplifying factors be assumed, for example,

1. At those points in the cycle for which temperatures are specified, the refrigerant and absorbent phases are in equilibrium.
2. With the exception of pressure reductions across the expansion device between points 8 and 9 in Fig. 7.8, pressure reductions in the lines and heat exchangers are neglected.
3. The temperature difference at the inlet to the liquid heat exchanger is 10°F.
4. Pressures at the evaporator and condenser are equal to the vapor pressure of the refrigerant, i.e., water, as found in steam tables.
5. Enthalpies for LiBr-$H_2O$ mixtures are given in Fig. 7.9.

As the first step in solving the problem, set up a table similar to Table 7.3; enter values of pressure in the appropriate table columns, enthalpy, and weight

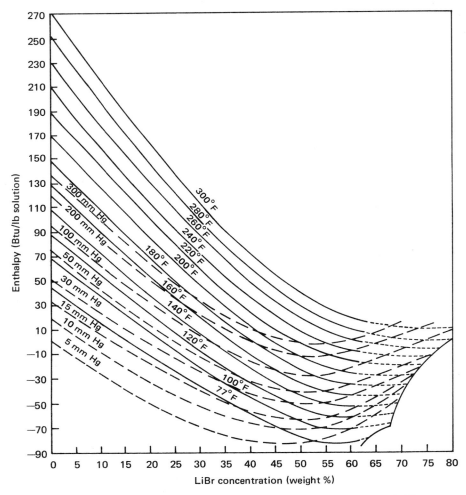

**FIGURE 7.9** Enthalpy-concentration diagram for LiBr-$H_2O$ combination. From (10b).

**TABLE 7.3** Thermodynamic Properties of Refrigerant and Absorbent for Fig. 7.8[a]

| Condition no. in fig. 7.8 | Temperature (°F) | Pressure (mm Hg) | LiBr weight fraction (%) | Flow (lb/lb H$_2$O) | Enthalpy (Btu/lb) |
|---|---|---|---|---|---|
| $T_1$ | 192 | 66.0 | 0.61 | 11.2 | −30 |
| $T_2$ | 100 | 66.0 | 0.61 | 11.2 | −70 |
| $T_3$ | 100 | 6.3 | 0.61 | 11.2 | −70 |
| $T_4$ | 90 | 6.3 | 0.56 | 12.2 | −75 |
| $T_5$ | 90 | 66.0 | 0.56 | 12.2 | −75 |
| $T_6$ | 163 | 66.0 | 0.56 | 12.2 | −38.8 |
| $T_7$ | 192 | 66.0 | 0 | 1.0 | 1147 |
| $T_8$ | 110 | 66.0[b] | 0 | 1.0 | 78 |
| $T_9$ | 40 | 6.3[b] | 0 | 1.0 | 78 |
| $T_{10}$ | 40 | 6.3 | 0 | 1.0 | 1079 |

[a]From (10b).
[b]These values were derived from (10a).

fraction for which sufficient information is available. For example, at point 8 the temperature is 110°F, and the vapor pressure of steam corresponding to this pressure in the condenser is 1.28 psia (66 mm Hg).

**Mass Balance Equations**

Relative flow rates for the absorbent (LiBr) and the refrigerant (H$_2$O) are obtained from material balances. A total material balance on the generator gives

$$\dot{m}_6 = \dot{m}_1 + \dot{m}_7$$

while a LiBr balance gives

$$\dot{m}_6 X_r = \dot{m}_1 X_{ab}$$

where $X_{ab}$ = concentration of LiBr in absorbent, lb/lb of solution
$X_r$ = concentration of LiBr in refrigerant, lb/lb of solution
Substituting $(\dot{m}_1 + \dot{m}_7)$ for $\dot{m}_6$ gives

$$\dot{m}_1 X_r + \dot{m}_7 X_r = \dot{m}_1 X_{ab}$$

Since the fluid entering the condenser is pure refrigerant, that is, water, $\dot{m}_7$ is the same as the flow rate of the refrigerant $\dot{m}_r$:

$$\frac{\dot{m}_1}{\dot{m}_7} = \frac{X_r}{X_{ab} - X_r} = \frac{\dot{m}_{ab}}{\dot{m}_r}$$

where $\dot{m}_{ab}$ = flow rate of absorbent, lb/hr
$\dot{m}_r$ = flow rate of refrigerant, lb/hr

Substituting for $X_r$ and $X_{ab}$ from the table gives the ratio of absorbent-to-refrigerant flow rate:

$$\frac{\dot{m}_{ab}}{\dot{m}_r} = \frac{0.56}{0.61 - 0.56} = 11.2$$

The ratio of the refrigerant-absorbent solution flow rate $\dot{m}_s$ to the refrigerant-solution flow rate $\dot{m}_r$ is

$$\frac{\dot{m}_s}{\dot{m}_r} = \frac{\dot{m}_{ab} + \dot{m}_r}{\dot{m}_r} = 11.2 + 1 = 12.2$$

**Energy Balance Equations**

The enthalpy of the refrigerant-absorbent solution leaving the liquid heat exchanger at point 6 is obtained from an overall energy balance on the unit, or

$$\dot{m}_s h_5 + \dot{m}_{ab} h_1 = \dot{m}_{ab} h_2 + \dot{m}_s h_6$$

Hence

$$h_6 = h_5 + \left[ \frac{\dot{m}_{ab}}{\dot{m}_s} (h_1 - h_2) \right] = -75 + \frac{11.2}{12.2} [-30 - (-70)]$$

$$= -38.2 \text{ Btu/lb of solution}$$

The temperature corresponding to this value of enthalpy and a pressure of 66 mm Hg is found from Fig. 7.9 to be 163°F.

The flow rate of refrigerant required to produce the desired 100 tons of refrigeration (equivalent to 1,200,000 Btu/hr) is obtained from an energy balance about the evaporator,

$$q_e = \dot{m}_r (h_9 - h_{10})$$

where $q_e$ is the cooling effect produced by the refrigeration unit, and

$$\dot{m}_r = \frac{1,200,000}{1079 - 78} = 1200 \text{ lb/hr}$$

The flow rate of the absorbent is

$$\dot{m}_{ab} = \frac{\dot{m}_{ab}}{\dot{m}_r} \dot{m}_r = 11.2 \times 1200 = 13,400 \text{ lb/hr}$$

while the flow rate of the solution is

$$\dot{m}_s = \dot{m}_{ab} + \dot{m}_r = 13{,}400 + 1200 = 14{,}600 \text{ lb/hr}$$

The rate at which heat must be supplied to the generator $q_g$ is obtained from the heat balance

$$q_g = \dot{m}_r h_7 + \dot{m}_{ab} h_1 - \dot{m}_{ab} h_6$$

$$= [(1200 \times 1147) + (13{,}400 \times -30)] - (14{,}600 \times -38.2)$$

$$= 1{,}540{,}000 \text{ Btu/hr}$$

This requirement, which determines the size of the solar collector, probably represents the maximum heat load that the refrigeration unit must supply during the hottest part of the day.

The coefficient of performance $COP$ is

$$COP = \frac{q_e}{q_g} = \frac{1{,}200{,}000}{1{,}540{,}000} = 0.78$$

The rate of heat transfer in the other three heat-exchanger units—the liquid heat exchanger, the water condenser, and the generator—is obtained from heat balances. For the liquid heat exchanger this gives

$$q_{1-2} = \dot{m}_{ab}(h_1 - h_2) = 13{,}400[(-30) - (-70)] = 540{,}000 \text{ Btu/hr}$$

where $q_{1-2}$ is heat transferred from the absorbent stream to the refrigerant-absorbent stream. For the water condenser the rate of heat transfer $q_{7-8}$ rejected to the environment is

$$q_{7-8} = \dot{m}_r(h_7 - h_8) = 1200(1147 - 78) = 1{,}280{,}000 \text{ Btu/hr}$$

The rate of heat removal from the absorber can be calculated from an overall heat balance on this system:

$$q_a = q_{7-8} - q_{sup} - q_{refrig} = 1{,}280{,}000 - 1{,}540{,}000 - 1{,}200{,}000$$

$$= -1{,}460{,}000 \text{ Btu/hr}$$

Explicit procedures for the mechanical and thermal design as well as the sizing of the heat exchangers are presented in standard heat-transfer texts. In large commercial units it may be possible to use higher concentrations of LiBr, operate at a higher absorber temperature, and thus save on heat-exchanger cost. In a solar-driven unit this approach would require a concentrator-type absorber because flat-plate solar collectors cannot achieve a sufficiently high temperature to raise the temperature level in the absorber of an absorption air conditioner much above 190°F.

## Comparison of Mechanical and Absorption Refrigeration Systems

Absorption refrigeration systems operate on cycles in which the primary fluid, a gaseous refrigerant, which has been vaporized in an evaporator, is absorbed by a secondary fluid, called the absorbent. Absorption refrigeration cycles can be viewed thermodynamically as a combination of a heat-engine cycle and a vapor-compression refrigeration cycle, which are also the two components of a mechanical refrigeration system. Simplified diagrams for the two methods of providing refrigeration are shown in Figs. 7.10 and 7.11, respectively. A comparison of these two schematic diagrams indicates similarities between the main components in the absorption cycle and the components in a heat-engine cycle driving a mechanical refrigeration cycle.

In both cycles, heat from a high-temperature source is transferred in a heat exchanger to obtain a relatively high-pressure vapor. In the absorption cycle, the heat input occurs in a generator, from which streams of refrigerant and absorbent emanate. In the heat-engine cycle, the heat input occurs at the boiler, from which a vapor is produced that drives a turbine. The condenser in the absorption cycle is equivalent to the refrigerant condenser in the mechanical refrigeration cycle. In both heat exchanges, heat is transferred from the refrigerant at relatively high pressures.

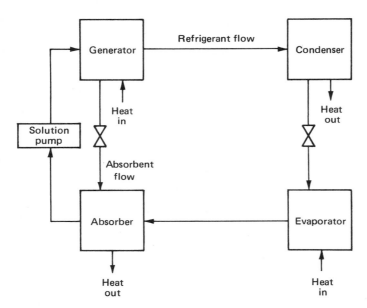

**FIGURE 7.10** Basic absorption refrigeration cycle without economizer. From (10b).

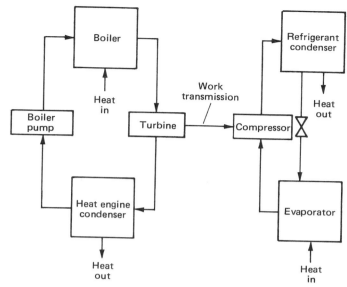

**FIGURE 7.11**  Combination of heat-engine cycle and mechanical refrigeration cycle. From (10b).

In both methods of refrigeration, the high-pressure refrigerant (from which heat has been removed in the condenser) is passed through an expansion valve that reduces the temperature and pressure of the refrigerant before it enters the evaporator. In both methods, heat is transferred to the refrigerant in the evaporator, where as a result of this heat transfer, the refrigerant is vaporized at relatively low pressures. It is the evaporator that absorbs the heat and provides the refrigeration effect in both methods.

The absorber in the absorption refrigeration cycle corresponds to the heat-engine condenser in the mechanical refrigeration cycle. Heat is transferred out of the absorber in the absorption cycle, and out of the heat-engine condenser in the combination method, to an intermediate temperature sink to facilitate the conversion of relatively low-pressure vapor to the liquid state. In the absorption method, the absorbent is mixed with the refrigerant in the absorber. The final similarity between the two systems is the solution pump and the boiler pump. In both systems a small amount of work is necessary to increase the pressure of the liquid before entering the boiler or generator of the cycle.

The turbine, which extracts heat energy from the high-temperature vapor from the boiler in the combination cycle thereby transforming heat into work to drive a compressor, does not have a counterpart in the absorption method. In the absorption method, the energy input occurs in the form of heat into the generator, and hence the generator can operate at a temperature of less than 200°F. In some absorption cycles, the heat supply can be provided by flat-plate solar collectors. On the other hand, in the combination heat-engine and mechanical refrigeration cycle, the turbine drive requires a relatively high-temperature vapor for efficient operation and it is difficult to obtain good performance with a flat-plate solar collector. A concentrator-type collector is therefore required.

The relationship between work and heat for an ideal heat-engine operating on a Carnot cycle is

$$W = q_g \frac{T_h - T_{hs}}{T_h} \qquad (7.19)$$

where $W$ = work output rate
$\quad q_g$ = heat input rate
$\quad T_h$ = temperature of heat source
$\quad T_{hs}$ = temperature of heat sink

The relationship between the work required and the refrigeration load for an ideal mechanical refrigeration machine operating on the reverse Carnot cycle is

$$-W = q_e \frac{T_{hs} - T_1}{T_1} \qquad (7.20)$$

where $-W$ = work input rate
$\quad q_e$ = rate of refrigeration
$\quad T_1$ = temperature of the refrigeration load
$\quad T_{hs}$ = temperature of the heat sink

The coefficient of performance for the combination of this engine cycle and the mechanical refrigeration machine is given by

$$COP = \frac{q_e}{q_g} = \frac{T_1(T_h - T_{hs})}{T_h(T_{hs} - T_1)} \qquad (7.21)$$

Equation (7.21) applies to the ideal absorption refrigeration process as shown earlier as well as to the combination heat-engine and mechanical refrigeration cycle. The coefficient of performance is plotted as a function of sink temperature for various temperatures of the heat source for a Carnot cycle with a refrigeration load at 40°F in Fig. 7.12.

## Nonmechanical Systems

### Australian Rock System

A nocturnal cooling-storage system first tried in 1955 in a desert in the southwest United States has been more recently tested and developed in Australia. The system consists of a large bed of rocks cooled by drawing cool night air across them and exposing them to night sky radiation. During the day, warm inside air may be cooled by circulating it through the rock bed. Augmented cooling can be achieved by drawing the night air through a porous surface having a high emittance at low

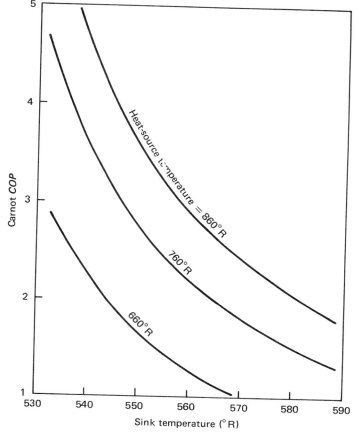

**FIGURE 7.12**  Ideal coefficient of performance for absorption refrigeration cycles with evaporator (load) temperature $T_l$ of 500°R. From (10b).

temperature and facing the night sky. Such a system operates best in a desert climate, where the night skies are clear, humidity is low, and the diurnal temperature variation may be 45°F. Although this system is not an active solar-cooling system, it uses the same pebble-bed storage that an air-cooled solar-heating system uses for heat storage.

The use of a pebble bed to provide energy storage for both heating and cooling cycles has been described by Close (5) and Dunkle (7). Figure 5.1 shows how a solar air heater can be combined with pebble-bed storage to provide a heat source for a building. The cycles in Fig. 7.13 illustrate the operation of the same pebble bed when it is used for cooling. During the night, cool air from the outside is brought through an evaporative cooler into the pebble bed and is cooled at approximately wet-bulb temperature to a condition approaching saturation. The entire bed is eventually brought to this temperature by passing the cooled air

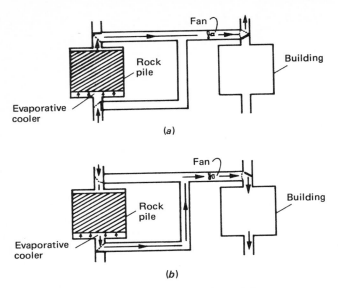

**FIGURE 7.13** Operation of pebble-bed thermal storage as a
source of air conditioning. From (10b).

through it for several hours, as shown in Fig. 7.13a. On the next day, when cooling
is required in the building, outside air is drawn vertically downward through the bed
and cooled to the bed temperature. As the air leaves the bed it may be cooled
further by evaporation and is then passed into the building as shown in Fig. 7.13b.

The mode of operation of the cycle just described can be illustrated quantitatively by means of an example. The psychometric chart in Fig. 7.14 indicates the

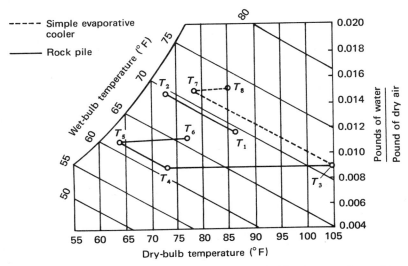

**FIGURE 7.14** Cooling cycles for pebble bed with and without evaporative cooler.
From (10b).

state points of the air passing through the cycle during a 24-hr period with environmental conditions corresponding to a night with a wet-bulb temperature of 70°F and a dry-bulb temperature of 86°F, and a day with a dry-bulb temperature of 105°F and wet-bulb temperature of 71°F. In Fig. 7.14, the line $(T_1 - T_2)$ corresponds to evaporative cooling of the nighttime air to 80 percent of saturation, which will cool the rocks in the bed to 73°F if steady state can be achieved at the lowest temperature.

When the building requires cooling the next day, air is introduced into the pebble bed under conditions corresponding to point $T_3$ in Fig. 7.14, cooled in the bed to 73°F, corresponding to point $T_4$, and then cooled evaporatively at constant wet-bulb temperature to a dry-bulb temperature of 63°F, corresponding again to 80 percent saturation. This air is then passed into the building to maintain conditions of 77°F and about 57 percent relative humidity. The increase in internal energy of the air, corresponding to the amount of heat transferred from the hot interior of the building to the coolant, is 3.5 Btu/lb. Thus, for a cooling load of 30,000 Btu/hr (2.5 tons), the required air circulation rate is about

$$\frac{(30,000 \text{ Btu/hr})(14 \text{ ft}^3/\text{lb of air})}{(3.5 \text{ Btu/lb of air})(60 \text{ min/hr})} = 2000 \text{ ft}^3/\text{min}$$

For comparison, the dotted line in Fig. 7.14 corresponds to a simple evaporative cooler. To maintain a temperature of 85°F in the building, the required air flow rate is about 30 percent larger than that for a rock bed that maintains the building at 77°F.

To determine the size of the pebble bed necessary for a cooling period of 12 hr, a heat balance must be made on the rocks, or

$$V_r \rho_r c_r (T_{a,\text{in}} - T_{a,\text{out}}) = \dot{m}_a c_a (T_{a,\text{in}} - T_{a,\text{out}})\theta \tag{7.22}$$

where  $V_r$ = volume of the rock in the bed, ft$^3$
$\rho_r$ = density of the rock, lb/ft$^3$
$c_r$ = specific heat of the rock, Btu/lb · °F
$T_{a,\text{in}}$ = temperature of air entering bed, °F
$T_{a,\text{out}}$ = temperature of air leaving bed, °F
$\dot{m}_a$ = air flow rate, lb/hr
$c_a$ = specific heat of air, Btu/lb · °F
$\theta$ = cooling period, hr

Equation (7.22) assumes that all the thermal energy stored can be extracted at the maximum temperature potential. For less efficient operation the size of the rock pile must be increased.

The size of the pile is approximately

$$V_r = \frac{(2000 \times 0.24)(T_{a,\text{in}} - T_{a,\text{out}})(12 \times 60)}{(85 \times 0.21)(T_{a,\text{in}} - T_{a,\text{out}})14} = 1400 \text{ ft}^3 \tag{7.23}$$

However, since the pile has empty spaces, the actual volume will be larger by the inverse of the empty-space fraction, defined as

$$\frac{\text{Volume of rock pile} - \text{volume of voids}}{\text{Volume of rock pile}}$$

In order to design a rock-pile storage system completely the friction factor for packed beds and the heat-transfer coefficient for air passing through the pile must be known. In Fig. 7.15 the friction factor for packed beds $f$ (defined as 2 times the pressure drop $\Delta p$ divided by $\rho_f v_f{}^2$ times the ratio $d_r/l$, where $l$ is the packed bed length) is plotted as a function of the particle Reynolds number $\text{Re}_p$ (defined as $v_f \rho_f d_r / \mu_f$). In Fig. 7.15 $v_f$ is the superficial air velocity (air flow rate per pile

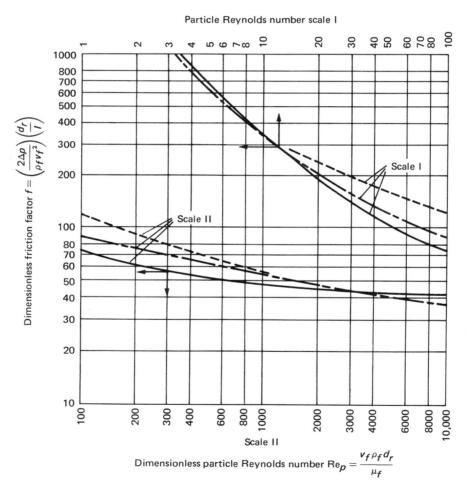

**FIGURE 7.15** Friction factors for packed beds as functions of particle Reynolds number. Adapted from (7).

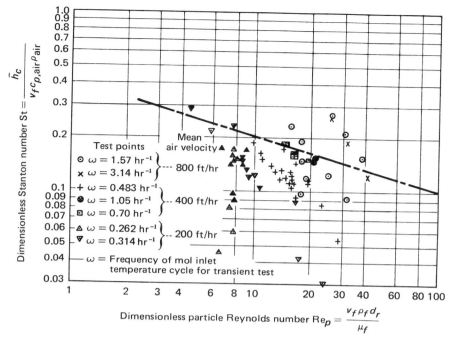

**FIGURE 7.16** Comparison of measured heat-transfer coefficients for packed beds as a function of particle Reynolds number. Adapted from (5).

cross-sectional area), $d_r$ is the equivalent spherical diameter of the rocks $(6\,V_p/\pi)^{1/3}$, $V_p$ is the rock-particle volume, $\mu_f$ is the viscosity of the air in the pile, $\rho_f$ is the density of the air, and $\Delta p$ is the pressure drop across the pile. These parameters must be in a consistent set of units to make the abscissa and the ordinate dimensionless. This correlation, proposed by Dunkle (7), agrees well with other correlations but is simpler to use. Experimentally measured heat-transfer coefficients from Close (5) are shown in Fig. 7.16 as a function of the particle Reynolds number. In Fig. 7.16 the Stanton number St, defined as a heat-transfer coefficient divided by the product of the air density, the velocity, and the specific heat, is plotted again as a function of the particle Reynolds number. Figure 7.17 shows the pressure decrease in inches of water per foot of the depth of the pile and the heat-transfer coefficient as a function of the superficial air velocity in feet per minute for various rock sizes. The following example illustrates the calculations for a rock-pile storage system. For a more detailed analysis the variation in temperature during a 24-hr period must be taken into account.

## EXAMPLE 7.4

A rock pile is required to store 1 million Btu. Charging and extraction are at constant rates, and each lasts approximately 10 hr. The temperature difference is to be 60° (70°F minimum and 130°F maximum). The maximum allowable pressure decrease is 0.1 in of water. Approximately 80 percent of the energy stored in the pile can be extracted. Determine the fluid flow rate and ratio of energy storage to energy required to move the air through the bed.

**SOLUTION**

Assume the rock has a density of 85 lb/ft³ and a specific heat of 0.21 Btu/lb · °F. The total volume of rock required for the storage would then be 934 ft³. If this storage is to be charged in 10 hr, the mass flow rate of air $\dot{m}_a$ may be obtained from a heat balance. If it can be assumed that all the heat is extracted at the stored temperature

$$\dot{m}_a = \frac{10^6 \text{ Btu}}{(10 \text{ hr})(0.24 \text{ Btu/lb} \cdot {}^\circ\text{F})(60^\circ\text{F})} = 7000 \text{ lb/hr}$$

The volumetric flow rate for air at 70°F is thus about 1600 ft³/min. If the maximum allowable pressure decrease is 0.1 in of water, the ratio of heat stored to energy required to operate the pile is approximately 400:1.

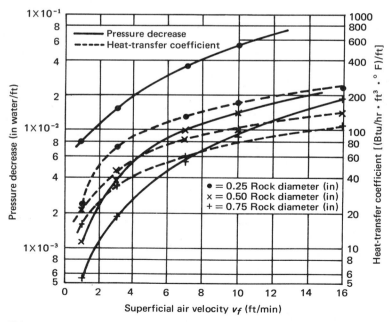

**FIGURE 7.17** Heat-transfer coefficients and pressure drops for various air flows and rock sizes in a pebble bed. From (10b).

**TABLE 7.4** Dimensions and Air Velocity for Various Rock Beds[a]

| Bed dimensions and air velocity | Rock diameter (in) | | | Design basis |
|---|---|---|---|---|
| | 0.25 | 0.5 | 0.75 | |
| Bed height, ft | 3.4 | 5.9 | 7.4 | Maximum allowable pressure drop (maximum height) |
| Bed area, ft² | 275.5 | 159 | 126 | |
| Air velocity, ft/min | 6 | 10.5 | 13.1 | |
| Bed height, ft | 2.75 | 4.0 | 5.0 | Intermediate height |
| Bed area, ft² | 340 | 233.5 | 186.8 | |
| Air velocity, ft/min | 4.85 | 7.07 | 8.8 | |
| Bed height, ft | 2 | 2.0 | 2.0 | Minimum height |
| Bed area, ft² | 467 | 467 | 467 | |
| Air velocity, ft/min | 3.53 | 3.53 | 3.53 | |

[a]From (10b).

Table 7.4 summarizes the dimensions and air velocities for various pebble beds as compiled by Close (5) for engineering design. It should be noted that best performance is obtained when the rocks are as small as possible and the bed is sized for the maximum allowable pressure decrease. The smaller the rocks, however, the more the pile becomes subject to blockage by dust, and the installation of a screen for dust removal may be necessary.

## Radiation Cooling of Terrestrial Surfaces

Infrared radiation exchanges occur continuously between the lower layers of the atmosphere and the earth's surface. Since the effective sky temperature for radiation is below that of terrestrial surfaces, a net heat loss occurs for these surfaces. Although this effect is small when compared with solar radiation exchanges, it can provide a means of rejecting heat to the atmosphere (cooling) directly by radiation in some climatic zones.

A brief analysis of terrestrial infrared radiation transport is presented in this section with supporting empirical data. The net radiation lost for a terrestrial surface in steady state $q_{ir}$ is the algebraic sum of outgoing radiation and incoming radiation from the atmosphere (mostly water vapor radiation emission). As shown in Ref. (6), $q_{ir}$ can be calculated empirically from the equation

$$q_{ir} = \epsilon_s \sigma T_a^4 (0.39 - 0.0096e)(1 - a_3 CC) + 4\epsilon_s \sigma T_a^3 (T_s - T_a) \qquad (7.24)$$

where $\epsilon_s$ is the surface infrared emittance, $T_a$ is the ambient temperature, $T_s$ is the surface temperature, $e$ is the vapor pressure of atmospheric water vapor in mm Hg, and $CC$ is the cloud cover expressed in tenths. The value of the single empirical constant $a_3$ is a function of latitude as shown in Table 7.5 ($\sigma = 118 \times 10^{-9}$ cal/cm² · day · K⁴).

**TABLE 7.5** Values of $a_3$ for the Infrared Flux
Equation (7.24)

| Latitude | Coefficient $a_3$ |
|----------|-------------------|
| 5        | 0.50              |
| 10       | 0.52              |
| 15       | 0.55              |
| 20       | 0.59              |
| 30       | 0.63              |
| 40       | 0.68              |
| 50       | 0.72              |
| 60       | 0.76              |
| 70       | 0.80              |
| 80       | 0.84              |
| 85       | 0.86              |

## EXAMPLE 7.5

An owner of a swimming pool in Los Angeles, California, ($L = 34°$N) wishes to cool an overheated pool in summer by using an unglazed solar swimming pool heater for 10 hr after sundown. If the emittance of the collector is 0.90, the convection coefficient from the bare collector $h_c$ is 15 W/m$^2$ · K, and the area $A_c$ is 50 m$^2$, what cooling effect can be experienced for a night during which the cloud cover is 25 percent? The average ambient temperature is 26°C (79°F), the atmospheric water-vapor pressure is 3.0 mm Hg, and the pool temperature $T_p$ is 34°C (93°F).

## SOLUTION

As a first-order estimate, the collector surface temperature is assumed to be the same as the temperature of the circulating pool water. The heat loss $Q_c$ from the bare collector during 10 hr is

$$Q_c = A_c[h_c(T_p - T_a) + q_{ir}] \times 10 \text{ hr} \tag{7.25}$$

Using Eq. (7.24), $q_{ir}$ on an *hourly* basis is

$$q_{ir} = \frac{(0.9)(118 \times 10^{-9})}{24}(299)^4(0.39 - 0.0096 \times 3)(1.0 - 0.66 \times 0.25)$$

$$+ \frac{(4 \times 0.9)(118 \times 10^{-9})}{24}(299)^3(34 - 26) = 14.4 \text{ cal/cm}^2 \cdot \text{hr}$$

$$= 168 \text{ W/m}^2 \ (53 \text{ Btu/hr} \cdot \text{ft}^2)$$

Substituting $q_{ir}$ in Eq. (7.25) gives

$$Q_c = 50[15(34 - 26) + 168] \times 10 = 144 \text{ kW} \cdot \text{hr}$$

The nocturnal radiation heat loss is more than half of the convection cooling effect in Los Angeles.

Repeating the example for a site in the Arizona desert for which $CC = 0.0$, $e = 0.1$ mm Hg, and the temperature drops to $18°C$ ($64°F$) at night gives

$$q_{ir} = \frac{(0.9)(118 \times 10^{-9})}{24}(291)^4(0.39 - 0.0096 \times 0.1)(1.0 - 0.0)$$

$$+ \frac{(4 \times 0.9)(118 \times 10^{-9})}{24}(291)^3(34 - 18) = 19.3 \text{ cal/cm}^2 \cdot \text{hr}$$

$$= 0.238 \text{ W/m}^2 \ (0.073 \text{ Btu/hr} \cdot \text{ft}^2)$$

The nocturnal radiation effect can be quite large in the favorable desert climate.

## Sky-Therm System*

H. R. Hay built a home in Atascadero, California, which uses his patented, passive, solar-heating and solar-cooling arrangement, as described in detail in Chapter 6. For the summer function such a system relies on nocturnal radiation and convection losses to cool the water on the roof. Such losses are generally on the order of 100–200 Btu/ft$^2$ during one night. Thus, to provide the equivalent of 1 ton of cooling approximately 1000 ft$^2$ of roof area would be required. It appears therefore that in the Southwest, where the system has been proposed for application, additional daytime cooling might be required unless a substantial temperature swing inside the house is acceptable to the occupants.

## The Economics of Solar Air Conditioning

A detailed economic analysis of solar air conditioning today suffers from the lack of knowledge of costs and performance of the LiBr-H$_2$O system when the system is modified for solar use. Several investigators have studied the system in detail, and a performance picture is beginning to emerge. Löf and Tybout (12) have modeled a combined solar-heated and solar-cooled residence by computer using current estimates on costs and performance of a small (3- to 5-ton) LiBr-H$_2$O absorption air conditioner. Their analysis is much like the approach described in Chapter 6 for optimizing solar-heating systems. Figure 7.18 shows a detailed schematic diagram, including all components of a typical solar-cooling system.

*Registered trademark, Sky-Therm Processes and Engineering, Los Angeles, Calif.

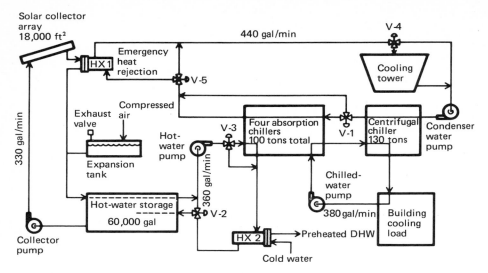

Solar cooling and domestic hot-water system, solar elementary school at Dade County, Florida.
Solar system designer: Mueller Associates Inc., Baltimore, Maryland.

**FIGURE 7.18** Schematic diagram of the Dade County, Florida, solar project showing all major components of a reliable cooling system. Courtesy of Andrew Parker, P. E., Mueller Associates, Inc.

Table 7.6 shows the additional costs in the solar system used by the authors. The model optimization runs showed that

1. In climates where both air conditioning and heating are required, a *combined solar-heating–air-conditioning system is more economical* than either a solar-heating or solar-cooling system by itself. Figure 7.19 shows one of the plots presented in the Löf and Tybout study with more realistic costs substituted for their values.

**TABLE 7.6** Costs Assumed in Calculations of Fig. 7.19[a]

| Item | Cost ($) | Comments |
|---|---|---|
| Storage costs | 0.05/lb water | Per pound of water stored per square foot of collector area (including tank, etc.) |
| Controls | 150 | Fixed cost |
| Pipes, fittings | $100 + 0.10A_{coll}$ | Fixed plus variable cost |
| Motors, pumps | $50 + 0.20A_{coll}$ | Fixed plus variable cost |
| Heat exchangers | $75 + 0.15A_{coll}$ | Fixed plus variable cost |
| Collector | $12.00A_{coll}$ | Projection based on economics of scale |
| LiBr-$H_2O$ air conditioning | 1000 | Cost above mechanical air conditioning |

[a]Equipment useful life is assumed to be 20 yr; discount rate is assumed to be 8% ($CRF = 0.102$). For more realistic costs see Fig. 7.19. Adapted from (12).

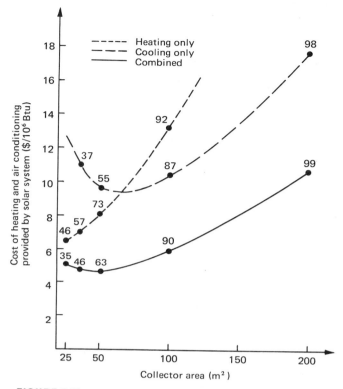

**FIGURE 7.19** Annual cost of solar heating, solar cooling, and combined solar heating and solar cooling for a model home in Albuquerque, New Mexico. Numbers on the curves are percentages of load carried by solar under the following conditions: collector tilt $\beta$ = latitude $L$ = 35.05°N, heating demand = 25,000 Btu/°F · day, system cost amortized for 20 yr at 8 percent, collector cost with two glass covers = $12/ft, absorption air conditioner $1,000 more costly than standard mechanical system. Adapted from (12).

2. *Optimal storage was found to be about 10–15 $lb_m/ft^2$,* the same value determined for the optimal heating-only system described in Chapter 6.
3. *Between 30 and 70 percent of the combined loads could be met by the solar-optimal system,* modeled in the eight climatic zones studied, at a cost of $1.75–$3.00 per million delivered Btu. (More realistic cost is $4.00–$6.00/million Btu.)
4. Solar energy costs less in nearly all locations than electrical energy used for heating and cooling. The costs were computed in 1970 dollars.

In this study the optimal mix of solar and auxiliary energy was not determined. The optimal mix may be obtained by determining the mix at the point where marginal costs of solar and auxiliary energy are equal.

## Summary

This section has dealt with various methods by which solar energy can be used to provide building cooling. The first part presented the procedure for calculating the cooling load on a building during the cooling season. A design study of the cooling load for a typical building illustrated the various parameters that affect the cooling requirements.

Several thermodynamic cooling cycles have been considered in detail in the remainder of the section. Performance criteria for the Rankine cycle-powered vapor-compression cycle were described. The absorption cycle was also discussed and the method of analyzing the cycle thermodynamically was shown in detail for a lithium bromide system. A nonmechanical cooling system that uses pebble-bed storage as a means of summer cooling in climates where cool nights are frequent was described. Design correlations for pressure drop and heat-transfer coefficient for this system were given. Economic considerations that determine the viability of solar-cooling systems were also summarized.

## SOLAR DEHUMIDIFICATION

In regions of the world experiencing significant building cooling demands (6), solar energy may be used to provide the fraction of the cooling load resulting from dehumidification of humid outdoor air. In such locations from one-third to one-quarter of the cooling system output is typically consumed in reducing the wet-bulb temperature to comfort levels. A few solar dehumidification systems have been built using either liquid or solid desiccants, which are regenerated with solar energy. This section summarizes the proposed methods.

### Solid-Phase Desiccants

Materials that have a high affinity for water vapor can be used to dehumidify moist air (8). The two conversion materials that have been used in solar systems are silica gel and the molecular sieve, a selective absorber. Figure 7.20 shows the equilibrium adsorption capacity of several substances. Note that the molecular sieve has the highest capacity up to 30 percent humidity, and silica gel is optimal between 30 and 75 percent—the typical humidity range for buildings.

Figure 7.21 is a schematic diagram of an adsorption device, which achieves both dehumidification and cooling, under test by the Institute for Gas Technology. It consists of two unmixed streams of air that interact through two counterrotating wheels—a heat-exchange wheel (regenerator) and a drying wheel consisting of appropriately mounted molecular sieve material in which the inlet stream is heated by the heat of adsorption released to the adsorbent. In the mode shown, outside air enters the unit and is dried adiabatically in the hot adsorbent wheel to the conditions shown. This hot dry air is cooled first in the regenerator and then in an evaporative cooler to its delivery condition.

Air removed from the conditioned space is used to cool the regenerator after passing through an evaporative cooler. This air stream, which is used ultimately to

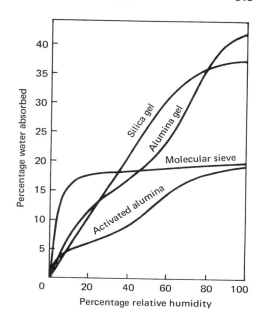

**FIGURE 7.20** Equilibrium capacities of common water adsorbents.

regenerate (dry) the adsorbent, is heated by solar energy in a heat exchanger. Although solar heat is usually adequate to regenerate silica gel, some topping—provided by fuel—is usually needed for part (one-third or so) of the stream used for molecular sieve regeneration depending on the cooling load. The hot air drives water vapor from the adsorbent wheel and is then exhausted to the environment. It is to be noted that only 25°F of the 150°F temperature rise required for sieve regeneration is provided by solar energy. Hence, solar energy has less application for a molecular sieve design than for a silica gel design, in which the adsorbent can be completely regenerated at air temperatures achievable by common solar systems.

The unit shown in Fig. 7.21 can also be operated with no outside air (recirculation mode) by causing the inlet to the left-hand stream to come from the conditioned space rather than from out-of-doors. Likewise, the inlet air for the left-hand stream is taken from outdoors rather than indoors in the recirculation mode.

The system in Fig. 7.21 can also be modified to achieve only dehumidification. This is done by removing the evaporative coolers that are frequently troublesome in regions where hard water must be used. The cooling effect can then be achieved by a solar-powered absorption machine, for example. In addition, several smaller wheels in series may be used to reduce the physical size of the unit. This is usually necessary for silica gel, since this adsorbent is very sensitive to operating temperature and is ideally operated isothermally. Isothermal operation is best achieved by several adsorption steps with intercooling (regeneration) using relatively thin adsorbent wheels.

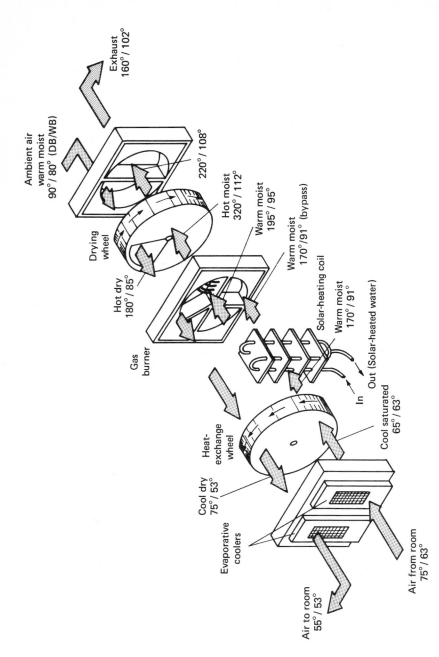

**FIGURE 7.21** Schematic diagram of a solid-phase solar-fired adsorption-dehumidification system that uses a molecular sieve (Munter's Environmental Control System). An evaporative cooling stage is also shown.

Exhaust 160° / 102°

Ambient air warm moist 90° / 80° (DB/WB)

220° / 108°

Drying wheel

Hot moist 320° / 112°

Warm moist 195° / 95°

Warm moist 170° / 91° (bypass)

Hot dry 180° / 85°

Solar-heating coil

Warm moist 170° / 91°

Gas burner

Out (Solar-heated water)

In

Cool saturated 65° / 63°

Heat-exchange wheel

Cool dry 75° / 53°

Evaporative coolers

Air to room 55° / 53°

Air from room 75° / 63°

## Liquid-Phase Desiccants

A number of hydrophilic liquids can be used to dehumidify air for space conditioning. Figure 7.22 shows schematically the fluid streams for one liquid-based concept using triethylene glycol (TEG). In order to cause rapid absorption of water vapor, the TEG is atomized to increase its surface area in the absorption chamber. The dehumidified air can be cooled by a standard air conditioner or by an evaporative cooler downstream of the absorption chamber.

The TEG absorbent is regenerated in a stripping chamber by the use of hot air from a solar collector. The hot air drives moisture from the atomized TEG in the stripper and the strong TEG collects in the stripper, is cooled, and returns to the absorption chamber spray nozzles. Heat exchangers (regenerators) can be inserted at several points between weak and strong streams to improve the thermodynamic efficiency of the process.

One difficulty with the TEG system described is the possibility of entrainment of droplets in the air streams. Such entrainment can be controlled by careful design of eliminator sections at the exit of each chamber. Liquid glycol storage is usually

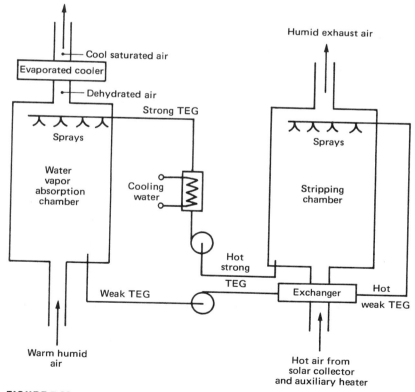

**FIGURE 7.22** Schematic diagram of solar-fired dehumidification system using triethylene glycol (TEG).

done in the absorption chamber when the system is not operating. Although fuel-fired liquid desiccant machines are commercially available, only one solar-fired experimental unit (11) has been built prior to 1975.

## SUMMARY

The techniques useful for solar cooling, refrigeration, and dehumidification have been described in this chapter. The absorption method is the most suitable thermodynamic match of flat-plate collector to cooling machine because of the usability of low temperatures by the LiBr absorption method. Concentrating collection systems can be coupled with heat engines to power vapor-compression refrigeration machines at somewhat higher *COP* levels than are achievable by the absorption–flat-plate systems (3).

Solar regeneration of desiccants is also a method whereby the full cooling load can be met by solar energy. Both solid and liquid desiccants have been regenerated. A small cooling effect can be achieved by exposing warm surfaces to the cool night sky. Some radiation loss from such surfaces can be used for lowering building temperature in some arid climates with large day-night temperature excursions.

Because of the infancy of solar-cooling technology, few design procedures analogous to those used for sizing of heating systems exist (13). The only design tool presently acceptable for accurate component sizing is the detailed computer model making hour-by-hour calculations of energy demands and system delivery. The economic optimization methods of Chapter 5 can then be used for optimization of the complete system.

## PROBLEMS

7.1. Consider the absorption refrigeration cycle, shown in the line diagram below, that uses lithium bromide as carrier and water as refrigerant. By using steam tables and the chart giving the properties of lithium bromide and water, calculate first

1. Heat removed from the absorber
2. Heat removed from the condenser
3. Heat added to the evaporator
4. *COP* of the cycle

Then calculate, for a flat-plate collector with an $\alpha\tau$ intercept of 0.81 and a $U$ factor of 3 W/m$^2$ · K, the area required for operation in Arizona during August for a 3-ton unit. Assume that the enthalpy of the water vapor leaving the condenser can be approximated by the equation

$$h_{vc} = 2463 + 1.9\ T_c \text{ kJ/kg}$$

and that the enthalpy of the liquid water is

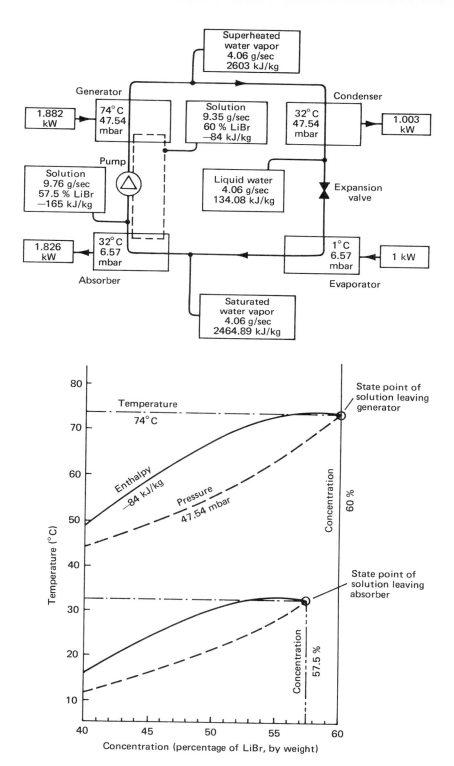

$$h_{lc} = 4.2 \, T_c \text{ kJ/kg}$$

where $T_c$ is the temperature of evaporation in °C. In the analysis assume that evaporation occurs at 1°C and condensation at 32°C. Answers for numbers 1, 2, 3, and 4 of the list above: 1.83 kW, 1.0 kW, 1.00 kW, 0.53, respectively.

7.2. Make a preliminary design for a solar driven Rankine refrigeration machine to provide the temperature environment required below the surface of a 20 × 40 m ice skating rink that is to operate all year in the vicinity of Denver, Colorado. State all your assumptions.

7.3. An inventor has proposed a desiccant heat pump to augment the heat delivered by a solar collector, and a sketch of the system is shown below. The collector bottom is covered with a desiccant, such as silica gel or zeolite, that contains water, and during the day the solar radiation heats up the desiccant bed and vaporizes a significant fraction of the water. The water vapor thus driven off passes through a heat exchanger in the building, and the vapor condenses with the release of heat to the building. At night, the liquid water stored in a tank passes to a heat exchanger outside the building where it absorbs heat and evaporates. The vapor produced then condenses in the desiccant at a temperature of 150°F, and air from the building is circulated between the building and collector to supply heat to the building at night. Comment on the feasibility of the system proposed and estimate the "effective *COP*."

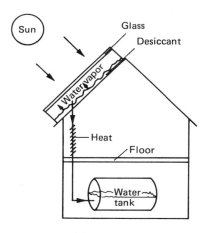

(a) Daytime operation

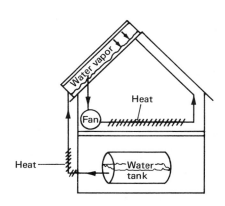

(b) Nighttime operation

# REFERENCES

1. ASHRAE, "Handbook of Fundamentals," American Society of Heating, Refrigerating, and Air Conditioning Engineers, New York, 1972.
2. Barber, R., Solar Organic Rankine Cycle Powered Three-Ton Cooling System, in Francis deWinter (ed.), "Solar Cooling for Buildings," pp. 206–214, *USGPO* 3800-00189, U.S. Government Printing Office, Washington, D.C., 1975.
3. Beckman, W. A., and J. A. Duffie, "Modeling of Solar Heating and Air Conditioning," National Technical Information Service, Springfield, Va., 1973.
4. Burriss, W., Solar Powered Rankine Cycle Cooling System, in Francis deWinter (ed.), "Solar Cooling for Buildings," pp. 186–189, *USGPO* 3800-0189, U.S. Government Printing Office, Washington, D.C., 1975.
5. Close, D. J., Rock Pile Thermal Storage for Comfort Air Conditioning, *Mech. Chem. Eng. Trans. Inst. Eng. (Australia)*, vol. MC-1, p. 11, 1965.
6. deJong, B., "Net Radiation Received by a Horizontal Surface on Earth," Delft University Press, Delft, 1973.
7. Dunkle, R. V., A Method of Solar Air Conditioning, *Mech. Chem. Eng. Trans. Inst. Eng. (Australia)*, vol. MC-1, p. 74, 1965.
8. Felske, D., et al., Solar Dehumidification of Citicorp Center. The Use of Solar Energy for the Cooling of Buildings, *ERDA Rept.* SAN/1122-76/2, 1976 (available from NTIS).
9. Hay, H. R., and J. I. Yellott, A Naturally Air Conditioned Building, *Mech. Eng.*, vol. 92, p. 19, 1970.
10. Hay, H. R., Energy Technology and Solarchitecture, *Mech. Eng.*, vol. 94, p. 18, 1973.
10a. Keenan, J. H., and F. G. Keyes, "Thermodynamic Properties of Steam," John Wiley & Sons, New York, 1936.
10b. Kreider, J. F., and F. Kreith, "Solar Heating and Cooling," revised 1st ed., Hemisphere Publ. Corp., 1977.
11. Löf, G. O. G., Cooling with Solar Energy, *Proc. World Symp. Applied Solar Energy (1955)*, Johnson Reprint Corp., New York, 1956.
12. Löf, G. O. G., and R. A. Tybout, Design and Cost of Optimal Systems for Residential Heating and Cooling by Solar Energy, *Sol. Energy*, vol. 16, p. 9, 1974.
13. Löf, G. O. G., "Design and Construction of a Residential Solar Heating and Cooling System," National Technical Information Service, Springfield, Va., 1974.

# SOLAR ELECTRIC POWER AND PROCESS HEAT

*The illusion of unlimited powers, nourished by astonishing scientific and technological achievements, has produced the concurrent illusion of having solved the problem of production.*

E. F. Schumacher

This chapter presents an overview of the state of the art of solar energy utilization for electric power generation and industrial process heat. In contrast to solar heating and cooling, which is largely a matter of economics, solar power generation on a large scale is still a developing technology. There are two basic approaches to solar electric power generation. One is by photovoltaic processes, often called direct solar energy conversion. Photovoltaic solar cells have been used successfully for many years in the space program, where efficiency and cost are not major factors. At the present time, however, photovoltaic power generation is not practical on a large scale because available systems have relatively low efficiency and are very expensive. Single crystal silicon cells can achieve a net efficiency of 15 percent, compared with a theoretical upper limit of about 23 percent. Cadmium sulfide and polycrystalline silicon, which are potentially much cheaper per square foot of surface, can achieve efficiencies of only about 5–8 and 1.5–2 percent, respectively. But at the present time the initial investment per kilowatt output is still 10–20 times larger than that for conventional electric power plants, and until dramatic reductions in the manufacturing cost of direct solar power systems can be achieved, this means of solar energy utilization will be restricted to applications where cost is not a deterrent.

For the near term solar thermal power is economically more promising. One drawback at this time is that no large-scale system has ever been built and all projections of cost and technology are based on small-scale experiments or analytical models. There are two basic systems for solar thermal power. One type is the distributed collector system in which a series of individual collectors heat a fluid. The collectors are connected by ducts that feed the heated working fluid to a central station for power generation. The main drawback of this approach is the cost of piping and the thermal losses in transporting the hot working fluid. The second type consists of the central receiver with many reflectors directing solar energy to a single collector located on top of a tower. In this approach heat losses are minimized, but each reflector, or heliostat, must be individually steered to track the sun. This requirement adds substantially to the cost (15, 26, 35). A third method, intermediate between these two systems, includes the spherical, stationary reflector, tracking absorber, which may offer advantages for units in the 5–100 MW range. The last type of system is the solar pond in which solar radiation is used to heat the upper layers of a shallow pond with a salt concentration gradient (21, 33).

Solar energy can also be used to supply industrial heat, as in a solar still, or it can provide low-pressure steam, which is used in many industries. Solar collectors could also be used as preheaters to reduce consumption of fossil fuel in industrial applications (12).

As noted previously, in order to achieve high temperatures it is necessary to use concentrating collectors. Good efficiency can be attained with concentrating collectors only in areas where the ratio of beam to diffuse solar radiation is high. This requirement imposes severe limitations on the siting of solar power plants. In addition, storage at high temperature is difficult and expensive. All of these factors pose technical as well as economic problems for the future of solar electric power. However, the technical obstacles to solar electric power do not appear insurmountable if sufficient energy and capital are made available.

## HISTORICAL PERSPECTIVE

Attempts to harness the sun's energy for power production date back to at least 1774 (35), when the French chemist Lavoisier and the English scientist Joseph Priestley discovered oxygen and developed the theory of combustion by concentrating the rays of the sun on mercuric oxide in a test tube, collecting the gas produced with the aid of solar energy, and burning a candle in the gas. Also during the same year an impressive picture of Lavoisier was published in which he stands on a platform near the focus of a large glass lens and is carrying out other experiments with focused sunlight (see schematic, Fig. 8.1).

A century later, in 1878, a small solar power plant was exhibited at the World's Fair in Paris (Fig. 8.2). To drive this solar steam engine, sunlight was focused from a parabolic reflector onto a steam boiler located at its focus; this produced the steam that operated a small reciprocating steam engine that ran a printing press. In 1901 a 10-hp solar steam engine was operated by A. G. Eneas in Pasadena, California (9). It used a 700-ft$^2$ focusing collector the shape of a truncated cone as shown in Fig. 8.3. Between 1907 and 1913 the American engineer F. Shuman developed solar-driven hydraulic pumps; in 1913 he built, jointly with C. V. Boys, a 50-hp solar engine for pumping irrigation water from the Nile near Cairo in Egypt (Fig. 8.4). This device used long parabolic troughs that focused solar radiation onto a central pipe with a concentration ratio of 4.5:1.

With the increasing availability of low-cost oil and natural gas, interest in solar energy for power production waned. Except for C. G. Abbott, who exhibited in 1936 a $\frac{1}{2}$-hp solar-powered engine at an International Power Conference in Washington, D.C., and in 1938 in Florida, an improved, somewhat smaller version

**FIGURE 8.1**  Solar furnace used by Lavoisier in 1774. Photograph courtesy of Bibliothèque Nationale de Paris. Lavoisier, *Oeuvres*, vol. 3.

**FIGURE 8.2**  Parabolic collector powered a printing press at the 1878 Paris Exposition.

**FIGURE 8.3** Irrigation pumps were run by a solar-powered steam engine in Arizona in the early 1900s. The system consisted of an inverted cone that focused rays of the sun on a boiler.

with a flash boiler, there was very little activity in the field of solar power between 1915 and 1950. Interest in solar power revived in 1949 when at the centennial meeting of the American Association for the Advancement of Science in Washington, D.C., one session was devoted to future energy sources. At that time, the potentials as well as the economic problems of solar energy utilization were clearly presented by Daniels (8). Some important conferences that considered solar power generation were held by UNESCO in 1954, the Association for Applied Solar Energy in 1955, the U.S. National Academy of Sciences in 1961, and the United Nations in 1961. In addition, a research and development program supported by the National Aeronautics and Space Administration to build a solar electric power system capable of supplying electricity for the U.S. space program was undertaken in the 1960s. However, widespread interest developed only after research funds

**FIGURE 8.4** Solar irrigation pump (50-hp) operating in 1913 in Egypt.

became available for the development of earth-bound solar electric power and process heat increased after the oil embargo in 1973.

## SOLAR-THERMAL POWER PRINCIPLES

Practically all electrical energy is generated in thermal power plants that operate on a Rankine cycle. A schematic diagram of the components required for such a plant is given in Fig. 8.5. To operate a power plant, heat must be supplied to the boiler where the working fluid, usually water, is vaporized and then superheated. The superheated steam enters the turbine, where energy is extracted by passing it through the nozzles of a turbine wheel whose shaft drives an electric generator. The working fluid leaving the turbine is still a vapor, but at low pressure and relatively low temperature. It is returned to the liquid state in the condenser, where its latent heat is extracted. The heat removed from the working fluid must be rejected to a heat sink, which may be a body of water, if this is available, or the atmosphere. The heat transfer in the condenser, an essential step in any thermal power cycle, is the main cause of thermal pollution.

After passing through the condenser, the liquid is pumped at high pressure into the boiler and the cycle repeats. The contribution that solar energy can make in such a scheme is to provide all or some of the heat required by the boiler. Since solar energy is not available continuously, one must either provide thermal storage sufficient to supply the heat required by the boiler at night and during bad weather, or use a standby conventional energy source to provide the heat when solar energy is unavailable. The latter type of system is called a hybrid system. Since an "all-solar system" requires a very large and expensive capacity for thermal storage,

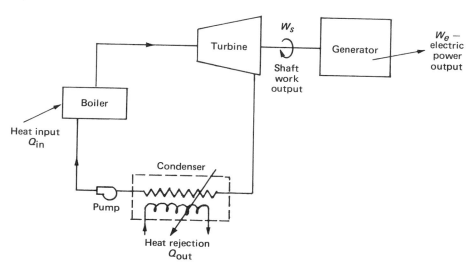

**FIGURE 8.5** Schematic diagram of power plant using the Rankine cycle.

while a hybrid system requires complex and fast-acting controls, a combination of the two, that is, a hybrid system with a small capacity for thermal storage to damp out thermal cycles, has been proposed as a compromise. Figure 8.6 shows schematically the manner in which the solar energy contribution may vary with changing environmental conditions during a 24-hr period in a hybrid system.

Theoretically the most efficient cycle for thermal power generation is the Carnot cycle. It is not used in practice because heat and friction losses in real machines would reduce the efficiency of a Carnot engine below the level attainable in a Rankine cycle. However, to gain a feel for some of the variables that determine the efficiency of a solar-thermal power plant, it will be convenient to examine an ideal solar-driven Carnot system. First, consider the efficiency of the power plant alone. If frictional losses and pumping energy are neglected, the efficiency of the power cycle is (19)

$$\eta = \frac{W_s}{Q_{in}} = \frac{Q_{in} - Q_{out}}{Q_{in}} \tag{8.1}$$

where $W_s$ = shaft work output from the cycle
$\quad Q_{in}$ = heat input to the cycle
$\quad Q_{out}$ = heat rejected by the cycle
In the theoretically limiting Carnot cycle the heat-addition and heat-rejection processes occur at constant temperatures, $T_1$ and $T_2$, respectively, while compression and expansion of the working fluid take place at constant entropy. Since the isothermal addition of heat is $T_1 \Delta s$ and the isothermal rejection of heat is equal to $T_2 \Delta s$, where $\Delta s$ is the change in entropy, the Carnot engine efficiency, $\eta_E$ is given by

$$\eta_E = \frac{T_1 - T_2}{T_1} \tag{8.2}$$

Examination of the above equation shows that to attain a high cycle efficiency the temperature of the working fluid should be as high as possible and the temperature of the sink into which the heat is rejected should be as low as possible. The sink temperature is essentially determined by the environment, whose temperature level is usually fixed by conditions beyond control. The only choice is whether water or air is to be used as the sink. Conventional power plants that use fossil fuel, which is easily transportable, are generally located along rivers or on the shores of a large water body, because heat rejection to a water sink is less expensive and can be accomplished at a lower temperature than heat rejection to atmospheric air. For solar-driven power plants, however, other considerations enter the site selection. Solar power can be provided economically only where land is cheap and the insolation high. Most of these regions are located in areas where water is scarce, for example, the Sahara desert. Consequently, heat rejection to an air-cooled condenser may be necessary. The main problem with air-cooled condensers is that because of

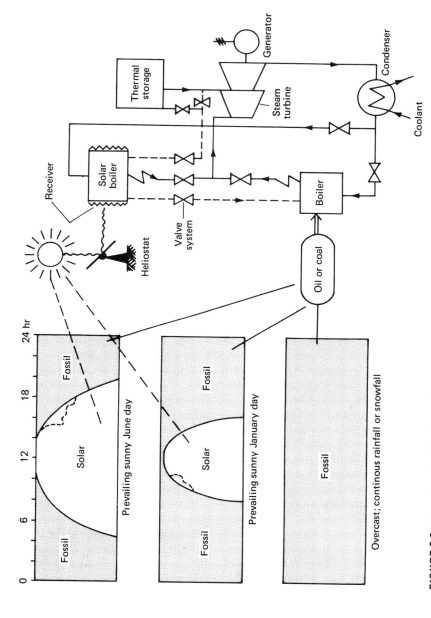

**FIGURE 8.6** Operation of a hybrid central receiver-type solar-thermal power plant.

529

the smaller convection heat-transfer coefficient attainable with air than with water, either a larger temperature difference between the working fluid and the sink or a larger surface area is required in order to obtain the same rate of heat transfer as in a water-cooled condenser. The first option would decrease the thermal cycle efficiency, while the latter would increase the initial cost of the system. Thus either of these approaches will increase the cost of energy. At the present time several studies are under way to ascertain the most favorable sites for solar power installations. The factors mentioned above are of course not the only ones considered in a site selection. However, they illustrate some of the new elements to be considered when contemplating a shift from conventional fuel to solar energy for production of thermal electric power.

The second requirement for attaining a high cycle efficiency is to heat the working fluid to a high temperature. The main problems with this approach are practical difficulties in achieving a high temperature, the heat-transfer coefficient obtainable with high-temperature working fluids in a receiver, the temperature limitations of the working fluids and the materials of construction, and the cost of tracking reflectors.

The performance of any solar power plant has another important limitation: the thermal efficiency characteristics of solar collectors. It was shown in Chapter 4 that the performance of solar collectors decreases with increasing temperature. To achieve high temperatures in a solar collector it is important to use selective surfaces for the receiver and to concentrate the incoming solar radiation to achieve a high flux. But irrespective of the kind of improvements in the solar collector, its efficiency will decrease with increasing temperature. Thus, for a combination scheme of power generation using a solar collector and a thermal cycle, the overall efficiency will be the product of the collector efficiency $\eta_c$ and the cycle efficiency $\eta_E$ or

$$\eta_{\text{overall}} = \eta_c \eta_E \qquad (8.3)$$

Since $\eta_c$ decreases with increasing temperature and $\eta_E$ increases with increasing temperature, there will be an optimum temperature for a solar-driven power plant. This is illustrated more quantitatively in the analysis below, where the efficiency of a concentrator-collector is approximated by a simple model (Fig. 8.7).

## Analysis for Low Concentration Ratios

For low concentration ratios the heat loss from a collector with receiver area $A_r$ and aperture area $A_a$ may be linearized, as shown for a flat-plate collector in Chapter 4. If $U_r$ is the heat-loss conductance for the receiver and $q_{s,r}$ is the solar radiation incident on $A_r$, the efficiency is given by the relation

$$\eta_c = \frac{\tau_r \alpha_r A_r q_{s,r} - U_r A_r (T_r - T_a)}{A_a I_s} \qquad (8.4)$$

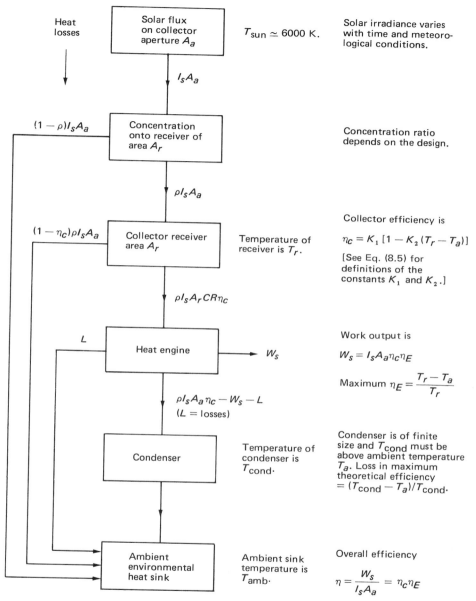

**FIGURE 8.7** First and second law considerations for solar heat engine performance.

or

$$\eta_c = \tau_r \alpha_r \left[ \frac{q_{s,r}}{I_s CR} - \frac{U_r A_r}{\tau_r \alpha_r I_s A_a} (T_r - T_a) \right] \tag{8.5}$$

For simplicity, the temperature of the fluid in Eq. (8.5) is approximated by $T_r$, although in the collector the working fluid may undergo not only a phase change at constant temperature, but also an increase in temperature level while being superheated. The environmental temperature level for the solar collector has been set equal to the sink temperature for the engine, $T_2$, a reasonable assumption.

As mentioned previously, a solar collector for electric power must be combined with a concentrator of some sort in order to increase the energy flux incident on the receiver surface. In the most general view, the key parameter for this step is the concentration ratio $CR$, defined as the ratio of the aperture area that receives solar radiation directly $A_a$ to the receiver area on which the solar radiation is concentrated $A_r$, or $CR = A_a/A_r$.

Since no surface is perfectly reflective, there will be some reflection losses even for a geometrically perfect system. These reflective losses are directly proportional to the reflectance of the reflector or surface $\rho$. A detailed analysis of reflection losses can be made only after a specific reflector-receiver combination has been selected. For this initial view we shall set the solar flux incident on the receiver equal to

$$q_{s,r} = I_s \rho CR \tag{8.6}$$

Combining this relation with Carnot efficiency gives the overall efficiency relation

$$\eta_{\text{overall}} = \left( \frac{T_r - T_a}{T_r} \right) \left[ 1 - \frac{U_r}{\tau_r \alpha_r \rho I_s CR} (T_r - T_a) \right] \tau_r \alpha_r \rho = \frac{\tau_r \alpha_r \rho}{T_r} \left( \Delta T - \frac{U_r \Delta T^2}{\tau_r \alpha_r \rho I_s CR} \right) \tag{8.7}$$

For a given environmental temperature the maximum value of $\eta_{\text{overall}}$ can be obtained by differentiating the right-hand side of Eq. (8.7) and solving for $T_r$. This gives for the receiver temperature at maximum efficiency

$$T_r(\eta_{\text{max}}) = \sqrt{\left( I_s CR \frac{\rho \tau_r \alpha_r}{U_r} + T_a \right) T_a} \tag{8.8}$$

It is apparent that in order to attain a high efficiency, large values of concentration and small overall heat-transfer coefficients for the receiver are necessary. Under favorable conditions the average annual incident solar flux is on the order of 0.8 kW/m$^2$ at 32°N latitude on the earth's surface. With an overall efficiency of 20 percent, a land area on the order of 1 mi$^2$ would be required for a solar power plant with an electric output of 100 MWe.

## EXAMPLE 8.1

In a desert at 32°N latitude the average annual incident solar flux is estimated to be 0.8 kW/m$^2$ and the environmental temperature 305 K. Calculate the optimum collector temperature for a good flat-plate collector with $U_r = 5$ W/m$^2 \cdot$ K and $(\alpha\tau) = 0.8$, and for a compound parabolic concentrator (*CPC*) with a concentration ratio $= 3.0$, $U_r = 2$ W/m$^2 \cdot$ K, and $\rho = 0.9$. Then calculate the Carnot efficiency for these conditions and the maximum overall efficiency for both systems.

## SOLUTION

From Eq. (8.8) the optimum collector temperature for the flat-plate collector is

$$T_{r,\text{opt}} = \sqrt{\left(\frac{800 \times 0.8}{5} + 305\right) 305} \text{ K} = 363.5 \text{ K}$$

whereas for the *CPC* with $CR = 3.0$

$$T_{r,\text{opt}} = \sqrt{\left(\frac{800 \times 3 \times 0.8 \times 0.9}{2} + 305\right) 305} \text{ K} = 597 \text{ K}$$

For the flat-plate collector the Carnot efficiency is

$$\eta_{c,FP} = \frac{363.5 - 305}{363.5} = 16.1 \text{ percent}$$

and for the *CPC* it is

$$\eta_{c,CPC} = \frac{597 - 305}{597} = 49 \text{ percent}$$

Then, the maximum overall efficiency for the flat-plate–Carnot system is

$$\eta_{FP,\text{max}} = 16.1 \times \left(1 - \frac{5 \times 58.5}{0.8 \times 800}\right) \times 0.8 = 7 \text{ percent}$$

and for the *CPC* it is

$$\eta_{CPC, \max} = 49 \times \left(1 - \frac{2 \times 317}{0.8 \times 0.9 \times 800 \times 3}\right) \times 0.72 = 22 \text{ percent}$$

It is thus apparent that flat-plate collectors are not suitable for power generation.

## Analysis for High Concentration Ratios

As shown above, the maximum temperature attainable with flat-plate collectors is too low to achieve a reasonable efficiency. If higher temperatures are desired, it is necessary to increase the radiation flux by optical focusing. This can be achieved by numerous devices that cover the range of concentration ratios from about 2 up to the order of 1000, as shown in Chapter 4. However, the acceptance angle of focusing collectors decreases with increasing concentration ratio. Most focusing collectors must therefore track the sun with a degree of precision that increases with concentration ratio. Also, most concentrating devices can utilize effectively only the direct solar radiation that restricts the geographic location at which they can operate over any reasonable fraction of time during the year.

Figure 8.8 shows the temperature and pressure regimes of conventional turbogenerators (26). It is apparent that for operating modern steam turbines temperatures on the order of 700 K (800°F) are necessary. To evaluate the efficiency and the concentration ratio necessary to achieve such temperatures a simplified thermal analysis is presented below.

Assuming that the main heat-transfer mechanism by which the receiver loses heat is radiation, the useful energy delivered by a concentrating collector is given by the relation

$$q_u = A_a I_s \rho(\tau\alpha)_r - \epsilon_r \sigma A_r (T_r^4 - T_a^4) \tag{8.9}$$

The efficiency of the collector, defined previously as the ratio between the useful energy delivered and the incident solar radiation, is then given by

$$\eta_c = \rho(\tau\alpha)_r - \frac{\epsilon_r \sigma}{I_s CR} (T_r^4 - T_a^4) \tag{8.10}$$

It is apparent that this efficiency is very dependent on the collector temperature and the concentration ratio. To minimize collector radiation losses it is necessary to have as high a concentration ratio as possible and a low infrared emittance of the receiver surface $\epsilon_r$. Figure 8.9 shows efficiencies of collectors for various values of the parameter $A_a/\epsilon_r A_r$, calculated from Eq. (8.10). Another interpretation of Eq. (8.10) is presented in Fig. 8.10, which shows the relationship between the collector-receiver temperature $T_r$ and the ratio $A_a/\epsilon_r A_r$. The zero-efficiency curve corresponds to the equilibrium temperature at which the incident radiation is exactly equal to the radiation losses from the receiver surface. The shaded area in

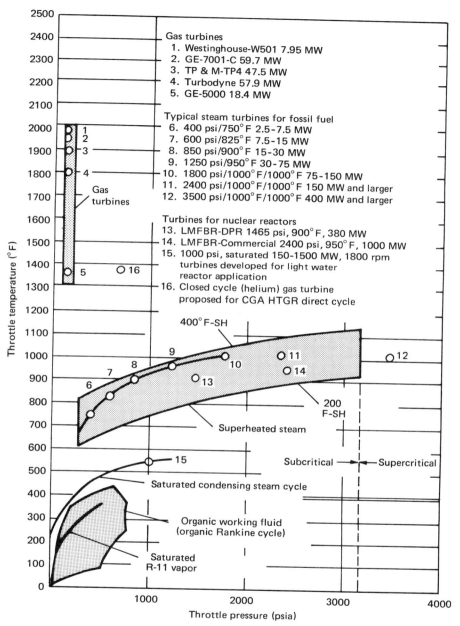

**FIGURE 8.8**  Temperature and pressure regimes of conventional turbomachinery.

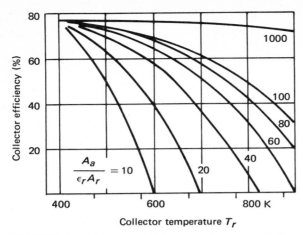

**FIGURE 8.9** Efficiencies of high-temperature concentrating collector versus collector-receiver temperature for various values of $A_a/\epsilon_r A_r$. Calculated from Eq. (8.10) with $\rho = 0.95$; $\tau = 0.9$; $I_s = 800$ W/m$^2$; $T_a = 400$ K; $(A_a/A_r \geqslant 10)$.

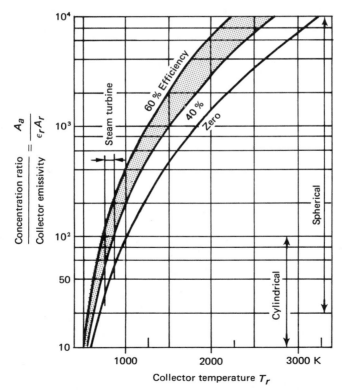

**FIGURE 8.10** Relationship between emittance-modified concentration ratio $A_a/\epsilon_r A_r$ and collector-receiver temperature $T_r$ for various collector efficiencies.

this figure corresponds to an operating range between 40 and 60 percent efficiency for the collector. The concentration ratio of cylindrical focusing devices is limited to roughly 100, as shown in Chapter 4. This means that operating temperatures for cylindrical collectors are expected to be below 800 K. However, under real conditions concentration ratios will not exceed 30. This means that the emittance of the surface would have to be less than 0.1 to achieve temperatures between 800 and 900 K (18).

The overall efficiency of a power system is obtained by multiplying the collector efficiency, obtained from Eq. (8.10), by the Carnot efficiency $(T_r - T_a)/T_r$, multiplied by a factor to account for irreversibility. Figure 8.11 shows the upper limit of efficiency obtainable by a Carnot-type power plant with a solar collector having an efficiency given by Eq. (8.10). It is obvious that to obtain high efficiency, it is necessary to attain a reasonably high collector temperature, although for any given value of $A_a/\epsilon_r A_r$, there is an upper limit beyond which the radiation losses from the receiver outweigh the improvement in efficiency that can be obtained from the power plant.

Two methods of obtaining reasonable concentration in cylindrical systems would be to use a *CPC* collector or a two-dimensional Fresnel lens as solar concentrator with a vacuum-insulated heat pipe covered by a selective coating at the focal line of the collector. The second method is illustrated in Fig. 8.12. However, the degree of concentration achievable with this system is limited and therefore spherical focusing, which does not require the use of large spherical or parabolic mirrors, seems to be preferable for high temperatures. This approach has been utilized in the solar furnace of the Centre National de la Recherche Scientifique,

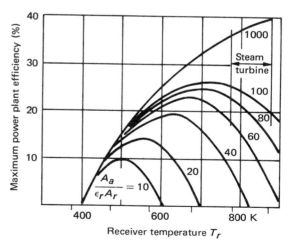

**FIGURE 8.11** Theoretical maximum efficiency of a solar power plant. Calculated from $\eta = \eta_c[(T_r - T_a)/T_r]$, where $\eta_c$ is the collector efficiency [Eq. (8.10)] and the sink temperature $T_a$ is assumed to be 400 K.

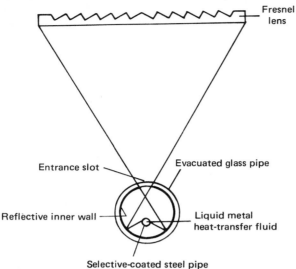

**FIGURE 8.12** Cross section of a linear Fresnel focusing collector with an evacuated-tube type of receiver.

where 63 spherical-plane mirrors are used to reflect solar radiation onto a fixed paraboloid with a surface of 2000 m² (36). From there the radiation is concentrated on a receiver where heat fluxes on the order of 1600 W/cm², corresponding to a black-body radiation temperature of 4100 K, can be achieved. An extension of this approach is the solar power tower shown in Fig. 8.13, which was discussed in Chapter 4. Several concepts for large-scale solar power stations using the power tower approach have been proposed. One of these (15) describes a 100-MW power station consisting of many thousands of heliostats, each 4 × 6 m in size, distributed over an area of approximately 4 km² with a coverage of 40 percent. Radiation is collected by a receiver at the top of a 300-m tower where operating temperatures

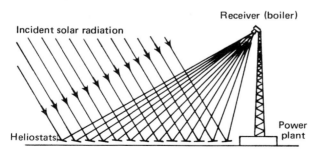

**FIGURE 8.13** Schematic diagram of the solar tower concept. Individually steerable mirrors (heliostats) reflect the solar radiation onto a receiver where superheated steam is generated.

up to 1200 K are expected. Another configuration for a 100-MW solar power plant, proposed by Blake et al. (7), uses a cavity receiver to produce superheated steam at about 770 K and 85 bar pressure to power a 100-MW turbine. The power tower approach has the advantage of comparatively small transmission losses.

## SOLAR INDUSTRIAL PROCESS HEAT

Industrial process heat consumed over 15 quadrillion ($10^{15}$) Btu's of energy in 1972 and this amount is expected to double before the end of the century. A study conducted for the Energy Research and Development Administration by the InterTechnology Corporation (13) indicated that solar energy has the potential of providing about 20 percent of this energy. The economic outlook for industrial solar heat appears to be extremely favorable because process heat solar collectors could be used throughout the year and each system can be designed to fit the temperature level required for its specific applications, which is particularly important in the use of process heat. About 5 percent of the industrial heat is used in processes requiring a temperature of about 373 K (212°F) and about 24 percent, temperatures below 455 K (350°F). The requirements of the former sector could be satisfied by simple flat-plate collectors, whereas the latter would need some sort of concentrating collectors, of which the nontracking *CPC* and the tracking parabolic trough varieties appear to be promising.

Solar systems for process heat could also be combined with other types of energy sources in such a way that the solar system supplies the heat needed in order to reach a certain temperature below the minimum required for the process, and fossil or nuclear energy supplies the rest of the energy needed to obtain the desired final temperature (14). With this preheat scheme solar energy might supply up to 27 percent of the industrial process heat in the former sector and up to 40 percent in the latter. Industries that appear to have significant potential for using solar-thermal energy include mining (drying of various minerals), food processing (large quantities of hot water and low-temperature steam), tobacco manufacturing (drying and rehydrating tobacco with low-grade steam), textiles (hot water and hot air), furniture manufacturing (warm air), certain chemical industries (synthetic rubber, soap, processing plastic, toilet goods, and organic dyes), petroleum refining (preheat of crude oil prior to distillation), asphalt paving mixtures (hot air for drying), leather tanning, ready-mix concrete, curing concrete block, foundries, and metals fabricating. In the chemical industries there are many chemical reactions that are carried out at moderate temperatures for which the reactants must be heated initially. Solar heating could provide a substantial amount of the heat for these processes too.

With an assumed baseline configuration for an industrial process heat system delivering an assumed constant load of hot working fluid as shown in Fig. 8.14, the influence of climatic region on system performance has been investigated in 90 locations throughout the United States (13, 14). Quantities calculated included the collector area required and the useful collector output in Btu per square foot per year for various regions. Although annual average daily insolation was the primary

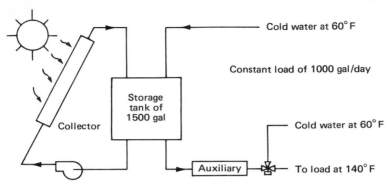

**FIGURE 8.14** Schematic of system assumed for analysis of solar industrial process heat in Ref. (14).

climatic variable influencing solar system performance, it was not the only one. Constant performance regions must be defined on the basis of performance calculation rather than simply on the basis of constant average insolation lines. Table 8.1 presents a partial summary of the relative costs for solar process heat in seven regions in the United States, showing the collector area required for the baseline system; the collector output; and relative cost as a function of region, temperature, and percentage annual load provided by the solar system. The low-temperature calculations were made for a flat-plate collector with a single-glass cover and the high-temperature calculations, for a simple concentrating device.

A simplified analysis of the optimum concentration ratio for a *CPC* solar collector system suitable for delivering process heat was made by Arthur D. Little, Inc., for Argonne National Laboratory (3). Although the optimum concentration ratio depends on many factors such as the specific application of the system, the cost of the components, and properties of the collector material, the approximate analysis shown below is useful for indicating how these factors affect the optimum concentration ratio for a *CPC* collector suitable for industrial heat production.

## Optimum Economic Concentration Ratio for Conventional *CPC*s

The useful power per unit area of collector aperture area is given approximately by the relation

$$\frac{q_u}{A_a} = (\alpha\tau)_r I_s - \frac{U_r(T_r - T_a)}{CR} \tag{8.11}$$

where $q_u/A_a$ = useful power per unit aperture area
$\alpha_r$ = absorptance of the receiver-absorber for solar radiation
$\tau_r$ = transmittance of solar radiation by the cover pane(s) and reflectors system
$I_s$ = insolation per unit aperture area

$U_r$ = overall transmittance between collector surface and the ambient air environment

$T_r$ = receiver temperature

$T_a$ = ambient air temperature

$CR$ = concentration ratio

The total cost of the collector per unit aperture area is given by

**TABLE 8.1**  Relative Costs for Solar Process Heat by Region and Percentage Annual Load[a,b]

| Region | Performance for 50 percent of annual load by solar | | | Performance for 75 percent of annual load by solar | | |
|---|---|---|---|---|---|---|
| | Average collector area (ft²) | Average useful collector output (Btu/ft² · yr) | Relative cost of heat | Average collector area (ft²) | Average useful collector output (Btu/ft² · yr) | Relative cost of heat |
| Low-temperature heat—140°F | | | | | | |
| I | 698 | 171,000 | 2.7 | 1892 | 99,500 | 4.6 |
| II | 533 | 224,000 | 2.1 | 1132 | 158,000 | 2.9 |
| III | 436 | 274,000 | 1.7 | 840 | 214,000 | 2.2 |
| IV | 382 | 313,000 | 1.5 | 706 | 254,000 | 1.8 |
| V | 326 | 367,000 | 1.3 | 578 | 310,000 | 1.5 |
| VI | 281 | 425,000 | 1.1 | 488 | 367,000 | 1.3 |
| VII | 259 | 461,000 | 1.0 | 444 | 404,000 | 1.1 |
| Intermediate-temperature heat—200°F | | | | | | |
| I | 1448 | 146,000 | 3.2 | 5393 | 66,000 | 7.9 |
| II | 1071 | 197,000 | 2.4 | 2675 | 120,000 | 3.9 |
| III | 849 | 247,000 | 1.9 | 1842 | 172,000 | 2.7 |
| IV | 739 | 284,000 | 1.6 | 1504 | 210,000 | 2.2 |
| V | 638 | 328,000 | 1.4 | 1235 | 255,000 | 1.8 |
| VI | 541 | 388,000 | 1.2 | 1009 | 313,000 | 1.5 |
| VII | 495 | 425,000 | 1.1 | 902 | 350,000 | 1.3 |

[a]From (14).

| [b]Region | Annual average insolation (langley/day) | Representative city |
|---|---|---|
| I | 300 ± 20 | Schenectady, NY |
| II | 340 ± 20 | Madison, WI |
| III | 380 ± 20 | Lincoln, NE |
| IV | 420 ± 20 | Stillwater, OK |
| V | 460 ± 20 | Fort Worth, TX |
| VI | >480 | El Paso, TX |

$$\frac{C}{A_a} = ZR + L + M + \frac{S}{CR} \tag{8.12}$$

where $C$ = total cost of collector

    $Z$ = ratio of reflector to aperture area ($Z \cong 0.4 + 0.6\ CR$)

    $R$ = cost of reflector per unit area

    $L$ = cost of transparent cover per unit aperture area

    $M$ = mounting cost per unit aperture area

    $S$ = absorber cost per unit area

Combining Eqs. (8.11) and (8.12) gives

$$\frac{q_u}{A_a} = \frac{ZR + L + M + S/CR}{[(\alpha\tau)_r I_s - U_r(T_r - T_a)]/CR} \tag{8.13}$$

The cost per unit of power delivered is large at low concentration ratios when the application requires a high collector temperature and when the concentration ratio is so large that the material costs outweigh the improved performance because of the increase in concentration ratio. The optimum concentration ratio for a given application and specified operating conditions can be obtained by taking the derivative with respect to $CR$ and solving for $CR_{opt}$. This yields

$CR_{opt}$

$$= \frac{U_r(T_r - T_a)}{(\alpha\tau)_r I_s} \left\{ 1 + \sqrt{1 + \frac{(\alpha\tau)_r I_s^{\,2}}{U_r(T_r - T_a)} \left[ \frac{U_r(T_r - T_a)}{(\alpha\tau)_r I_s}(L + M + 0.4R) + S \right] \frac{1.67}{ZR}} \right\} \tag{8.14}$$

Using the above relation for $CR_{opt}$, the A. D. Little Company analyzed the optimum $CPC$ design and performance with a good selective surface absorber. The nominal parameters used in this analysis and the calculated values of $CR_{opt}$ for the nominal values of $(T_r - T_a)$ are given in Table 8.2.

## SOLAR DISTILLATION OF SALINE WATER

Solar distillation for the production of potable water from saline water has been practiced for many years. A solar distillation plant, covering 51,000 ft$^2$ of land, was built in Las Salinas, Chile, in 1872, to provide fresh water from salt water for use at a nitrate mine (34). Single-glass-covered flat-plate collectors with salt water flowing downward over slanting roofs were used to vaporize some of the water, which was then condensed on the air-cooled underside of the roof. This plant ran effectively for over 40 yr and produced up to 6000 gal of fresh water per day until the nitrate mine was exhausted.

The stills used to date are called roof-type stills and Fig. 8.15 illustrates their

**TABLE 8.2** Analysis for Optimum Concentration Ratio ($CR_{opt}$) Based on Minimum Installed Cost Per Peak Watt for Collector Components and Nominal Parameters[a]

| Variable | Nominal value | Values of independent variable | | | | | | | | | |
|---|---|---|---|---|---|---|---|---|---|---|---|
| $(T_r - T_a)$ | 200°F | 100 | 400 | 600 | | | | | | | |
| $(L + M)$ | $4.00 | | | | 2.50 | 6.50 | | | | | |
| $R$ | $1.00 | | | | | | 0.50 | 1.50 | | | |
| $S$ | $5.00 | | | | | | | | 1.00 | 2.50 | 10.00 |
| $CR_{opt}$ | 3.55 | 3.22 | 4.21 | 4.86 | 3.44 | 3.74 | 4.87 | 2.94 | 1.97 | 2.83 | 4.64 |

[a]

| Parameter | Nominal value | | |
|---|---|---|---|
| $(\alpha\tau)_r$ | 0.80 | | |
| $I_s$ (Btu/hr · ft²) | 300 | | |
| $U_r$ (Btu/hr · ft² · °F) | 0.36 | | |
| $(T_r - T_a)(°F)$ | 100 | 200 | 400, 600 |
| $(L + M)$ [Component costs ($/ft²) | 2.50 | 4.00 | 6.00 |
| $R$    including manufacturing | 0.50 | 1.00 | 1.50 |
| $S$    markup] | 1.00, 2.50 | 5.00 | 10.00 |

method of operation. The still is irradiated by direct and diffuse solar radiation $I_s$ as well as some infrared radiation $I_i$ from the surroundings. The long-wavelength radiation is absorbed by the glass, but $\tau_{s,g}I_s$ of the solar radiation reaches the saline water; $\tau_{s,w}\tau_{s,g}I_s$ reaches the bottom of the trough where $\alpha_{s,t}\tau_{s,g}\tau_{s,w}I_s$ is absorbed per unit still area. A value of 0.8 has been suggested as a reasonable approximation for $\alpha_{s,t}\tau_{s,w}\tau_{s,g}$ if good glass, a blackened tray surface, and a 0.2-m thick layer of water are used (10). Typical roof stills are 1.5 m wide, 2–25 m long, and approximately 0.2 m deep; the glass roof should have a slope of 10–15°.

Of the energy absorbed at the bottom, one part $q_{k,s}$ is lost through the insulation by conduction while the other part will be transferred to the saline water in the still tray above. Of the latter portion, some will heat the water, if the water

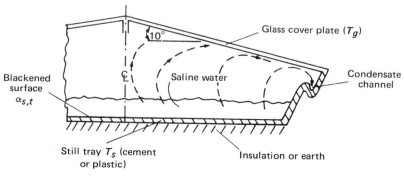

**FIGURE 8.15** Sketch of roof-type solar still.

temperature is less than the tray temperature, while the rest is transferred from the surface of the water by free convection, radiation, and evaporation to the underside of the glass cover. Some of this heat passes through the glass by conduction and is transferred from the outer surface of the glass cover by convection and radiation to the surrounding atmosphere. The thermal circuit for this system is shown in Fig. 8.16.

To obtain the greatest yield, the rate of evaporation of water, which is proportional to $q_e$, should be as large as possible. An inspection of the thermal circuit shows that to achieve large values of $q_e$, $\tau_{s,g}\tau_{s,w}\alpha_{s,t}I_s$ should be as large as possible, while $q_{c,s}$, $q_{r,s}$, and $q_{k,s}$ should be as small as possible. To achieve these results the transmittance of the glass cover and the absorptance of the tray surface must be large, and the bottom of the tray should be well insulated. Obviously, the amount of solar irradiation should be as large as possible.

In order to analyze the thermal performance, heat balances must be written for the glass cover and the tray and the resulting equations must be solved simultaneously to obtain the tray and glass cover temperatures. The objective is to obtain

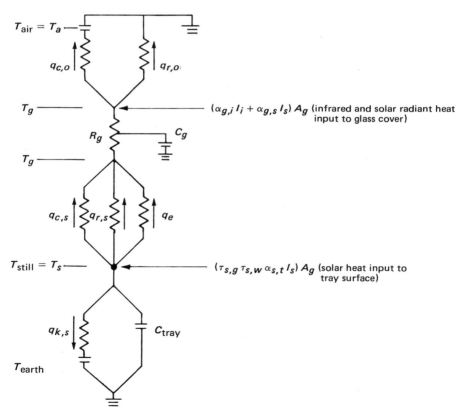

**FIGURE 8.16** Thermal circuit for solar still.

the thermal efficiency, defined as the ratio of the energy used in evaporating saline water to the incident solar radiation.

If steady-state conditions prevail, the glass resistance $R_g$ is neglected, and the water temperature is assumed equal to the tray temperature, a heat balance on the glass cover gives

$$q_e + q_{r,s} + q_{c,s} + \alpha_{g,i} I_i A_g + \alpha_{g,s} I_s A_g = q_{c,o} + q_{r,o} \tag{8.15}$$

where $q_{r,s}$ = rate of heat transfer by radiation from the water in the tray to the inner surface of the glass = $A\bar{h}_{r,s}(T_s - T_g)$

$\bar{h}_{r,s} = (1/\epsilon_w + 1/\epsilon_g - 1)/\sigma(T_s^2 + T_g^2)(T_s + T_g)$

$q_{c,s}$ = rate of heat transfer by convection from water to inner surface of the glass = $A\bar{h}_{c,s}(T_s - T_g)$

$q_{c,o}$ = rate of convection heat loss from the glass to the ambient air = $A\bar{h}_{c,o}(T_g - T_a)$

$q_{r,o}$ = rate of radiation heat loss from the glass = $A\bar{h}_{r,o}(T_g - T_a)$

$\bar{h}_{r,o} \cong 1/\epsilon_g \alpha(T_g^2 + T_a^2)(T_g + T_a)$

Similarly, a heat balance on the saline water gives

$$\alpha_{s,t} \tau_{s,g} \tau_{s,w} I_s = q_k + q_{c,s} + q_{r,s} + q_e \tag{8.16}$$

In the enclosed space between the water surface and the glass cover, heat and mass transfer occur simultaneously as brine is evaporated, condenses on the lower surface of the cooler glass, and finally flows into the condensate channel. Since the glass cover is nearly horizontal, the correlations developed in Chapter 3 for free convection between parallel plates can be used to calculate the Nusselt number and heat-transfer coefficient. However, because heat and mass transfer occur simultaneously in the still, the buoyancy term in the Grashof number must be modified to take account of the density gradient resulting from composition as well as temperature. As shown in Ref. (22), in horizontal enclosed air spaces in the range $3 \times 10^5 < \mathrm{Gr}_L < 10^7$, the relationship

$$\mathrm{Nu}' = \frac{\bar{h}_c' L}{k} = 0.075 \left( \frac{L^3 \rho^2 g \beta_T \Delta T'}{\mu^2} \right)^{1/3} \tag{8.17}$$

can be used to calculate the heat-transfer coefficient if $\Delta T'$ is considered an equivalent temperature difference between the water and the glass cover and includes the molecular weight difference resulting from the change in vapor concentration in evaluating the buoyancy. For an air-water system $\Delta T'$ is given by (10)

$$\Delta T' = (T_s - T_g) + \left( \frac{p_{w,s} - p_{w,g}}{39 - p_w} \right) T_s \tag{8.18}$$

where $p_{w,s}$ = partial pressure of water (in psia) at the temperature of the water surface $T_s$

$p_{w,g}$ = vapor pressure of water in psia at the temperature of the glass cover $T_g$ Since the size of the still does not affect the heat-transfer coefficient in the Grashof number range of interest, $\bar{h}_c$ in Eq. (8.17) can be approximated by evaluating all physical constants at an average still temperature, or

$$\bar{h}_c' = 0.13 \left[ (T_s - T_g) + \left( \frac{p_{w,s} - p_{w,g}}{39 - p_{w,s}} \right) T_s \right]^{1/3} \tag{8.19}$$

The rate of heat transfer by convection between the water and the glass is then

$$q_c = \bar{h}_c' A_s (T_s - T_g) \tag{8.20}$$

and using the analogy between heat and mass transfer, the rate of mass transfer $q_m$ (in lb/hr) is

$$q_m = 0.2 \bar{h}_c' A_s (p_{w,s} - p_{w,g}) \tag{8.21}$$

The heat-transfer rate resulting from evaporation $q_e$ (in Btu/hr) equals the rate of mass transfer times $h_{f,g}$, the latent heat of evaporation at $T_g$, or

$$q_e' = 0.2 \bar{h}_c' A_s (p_{w,s} - p_{w,g}) h_{f,g} \tag{8.22}$$

If the tray of the still is resting on the ground, the conduction loss is difficult to estimate because it depends on the conductivity of the earth, which varies considerably with moisture content. If the bottom of the tray is insulated and raised above the ground, the preferable arrangement unless too expensive, the bottom heat loss is given by

$$q_k = U_b A_s (T_s - T_a) \tag{8.23}$$

where the bottom conductance $U_b$ is composed of the tray insulation and the free convection elements in series and can be calculated by the methods given in Chapter 3.

Since $p_{w,s}$ and $p_{w,g}$ in Eqs. (8.21) and (8.22) are functions of $T_s$ and $T_g$, respectively, Eqs. (8.15) and (8.16) can be solved simultaneously for $T_s$ and $T_g$ provided $I_s$ and all the physical properties are known. However, this approach is more complex than assuming a still temperature and solving for $T_g$ and $I_s$.

## EXAMPLE 8.2

For a solar still operating at a brine temperature of 155°F (410 K) in ambient air at 76°F (298 K) with a bottom conductance $U_b = 0.3$ Btu/hr · ft² · °F, and $(1/\epsilon_g + 1/\epsilon_w - 1) = 0.9$, calculate the rates of heat transfer by convection $q_c$, radiation $q_r$, and conduction $q_k$, and the rate of evaporation per unit still area $q_e$. Then calculate the still efficiency.

## SOLUTION

First the temperature of the glass cover is calculated from Eq. (8.15) [the partial pressure of water vapor was obtained from Ref. (18a)]. This temperature is found to be 135°F (390.5 K) and the various heat fluxes are shown below:

$q_{c,s}/A_s$, convection flux from brine   = 10 Btu/hr $\cdot$ ft$^2$ (31.5 W/m$^2$)
$q_e/A_s$, rate of evaporation per unit area = 158 Btu/hr $\cdot$ ft$^2$ (497 W/m$^2$)
$q_{r,s}/A_s$, radiation flux from brine   = 28 Btu/hr $\cdot$ ft$^2$ (88 W/m$^2$)
$q_{k,s}/A_s$, conduction back loss flux   = 23 Btu/hr $\cdot$ ft$^2$ (72 W/m$^2$)

$\alpha_{s,t}\, \tau_{s,g}\, \tau_{s,w}\, I_s$       = 219 Btu/hr $\cdot$ ft$^2$ (689 W/m$^2$)
$I_s$                = 219/0.8 = 274 Btu/hr $\cdot$ ft$^2$
                    (274 W/m$^2$)

The thermal efficiency is then

$$\eta = \frac{q_e}{I_s} = \frac{158}{274} = 58 \text{ percent}$$

A factor omitted in the preceding analysis that can be important is the effect of air leakage. It increases with increasing temperature and can also be accelerated by high wind. Leakage should be minimized for good performance.

A still similar to the one analyzed above was built and tested at the University of California Field Station located at a latitude of 38°. The production of fresh water during 1958 is shown in Fig. 8.17 as a function of solar altitude for clear and cloudy days and in Fig. 8.18 as a function of insolation for sunny days. Maximum efficiencies were on the order of 35–40 percent. One main reason for the efficiency being less than that predicted was that the tray insulation was inadequate and 15 percent of the total energy reaching the tray was lost through the bottom.

The maximum performance of a still is limited by the heat of vaporization of water and the solar insolation. Under favorable conditions with $I_s = 250$ Btu/hr $\cdot$ ft$^2$, $\eta_s = 40$ percent, and $h_{f,g} = 1000$ Btu/lb, one square foot of still can provide 0.1 lb of water/hour or about 1 lb/day. Thus, approximately 10 ft$^2$ is required for a production rate of 1 gal/day and 27 ft$^2$ of still would be required to produce 1000 gal/yr.

## Large Basin-Type Stills

Figure 8.19 is a plot of the productivity of several relatively large basin-type solar stills as a function of solar radiation. These curves represent several different still

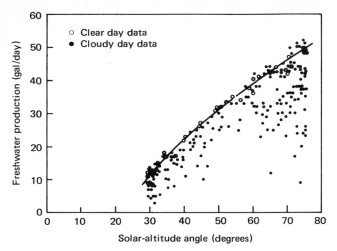

**FIGURE 8.17** Production rates for 1958 for University of California solar still no. 16.

designs and sizes at a number of locations. They are not all directly comparable, however, because of many differences in the conditions under which the data were obtained. Such factors as experimental adjustments during operation, design differences in portions of the same installations, degree of attention to maintenance, etc., render these results only roughly comparable. Superiority of any particular design cannot be inferred from these curves. Indeed, the data show much similarity in performance for considerably different designs and highly variable conditions.

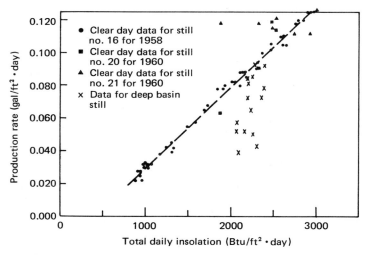

**FIGURE 8.18** Production versus insolation for various University of California solar stills.

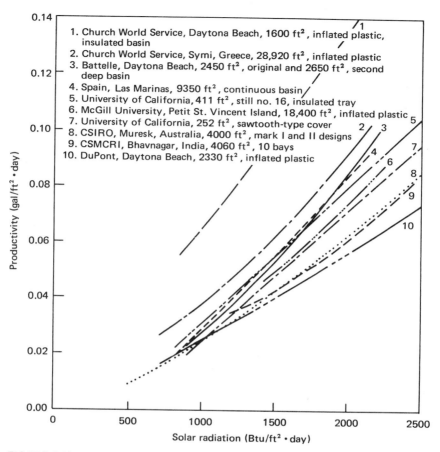

**FIGURE 8.19** Productivity of large basin-type solar stills. From (11).

The average productivity of the nine lower curves is represented quite well by the following equation:

$$P = (1.1 \times 10^{-3}) \left(\frac{I_s}{100}\right)^{1.4} \text{ gal/ft}^2 \cdot \text{day} \qquad (8.24)$$

where $I_s$ = total insolation in Btu/ft$^2$ · day. The variation in productivities from this average is approximately ±25 percent at an insolation level of 2000 Btu/ft$^2$ · day.

## Economics of Solar Distillation

A relation for estimating the cost of potable water from saline water by solar distillation has been proposed by a United Nations Panel (30). The total annual cost of providing solar-distilled water for use equals the sum of the total fixed costs as an annual percentage of capital investment, the total cost of supplying salt water to the still, and the total operating labor and supervision expense. This amount divided by the total number of gallons of distilled water produced (plus annual collection and storage of rainwater) equals the cost of each gallon of potable water produced. If all of this water can be used, this figure is also the cost of each gallon of water used. The following equation, in which $\overline{IA}$, $\overline{MR}$, and $\overline{TI}$ represent the average value over the estimated life of the installation, gives the cost of potable water:

$$C = \frac{10\,I\,(\overline{IA} + \overline{MR} + \overline{TI}) + 1000\,(Oc + S)}{A_s(Y_D + Y_R)} \qquad (8.25)$$

where $C$ = cost of water (\$/1000 gal)

$\quad I$ = total capital investment (\$)

$\quad \overline{IA}$ = annual interest and amortization payments (percentage of investment; see Chapter 5)

$\quad \overline{MR}$ = annual maintenance and repair labor and materials (percentage of investment)

$\quad \overline{TI}$ = annual taxes and insurance charges (percentage of investment)

$\quad O$ = annual operating labor (work-hr)

$\quad c$ = operating labor wage (\$/work-hr)

$\quad Y_D$ = annual unit yield of distilled water (gal/ft$^2$)

$\quad Y_R$ = annual unit yield of collected rainwater (gal/ft$^2$)

$\quad A_s$ = area of still on which yields are based (ft$^2$)

$\quad S$ = total cost (fixed and operating) of saltwater supply

Since some of the items in this equation depend on gross distiller size and some do not, an actual cost figure must be based on a specific total size. The major part of the total cost, however, is in the fixed costs related to capital investment, and since capital investment is nearly directly proportional to distiller size (after the relatively constant costs of pumps, piping, and accessories have been greatly exceeded by those costs proportional to area), most of the water cost in sizable plants (that is, with average annual capacities of 10,000 gal/day or more) can be determined without reference to a specific distiller size. Thus, if the total annual fixed charges are 10 percent and the still costs \$0.60–\$2.00/ft$^2$ to construct (including storage and auxiliaries), and if the annual productivity is 25 gal/ft$^2$, the resulting fixed cost of \$3–\$6/1000 gal represents nearly the entire cost of the solar-distilled water produced. These cost estimates could be reduced by rainwater collection.

Table 8.3 shows a breakdown of costs in the United States for the latest variation of the OSW deep-basin Daytona Beach design. The table shows the construction labor in terms of work-hours, rather than dollars, in order that they can be used in regions where labor costs are quite different from those in the United States. The approximate plant size on which these figures are based would be in the range of 50,000 ft$^2$ and above. It is seen that if a labor rate of \$4/hr is assumed, the total cost is approximately \$1.25/ft$^2$. If, however, the labor wage rate

**TABLE 8.3**  Typical Basin-type Still Costs within the United States[a]

| Item | Material ($/1000 ft² of basin area) | Labor (work-hr/1000 ft² of basin area) |
|------|-----------------------|-----------------------|
| Layout, grading, compacting, and soil sterilization | 10 | 20 |
| Asphalt-mat liner | 80 | 40 |
| Concrete blocks | 60 | 10 |
| Precast concrete beams | 190 | 25 |
| Glass and asphaltic cement | 210 | 25 |
| Distillate-trough materials | 50 | 5 |
| Miscellaneous piping and pumps | 70 | 5 |
| Storage tank | 50 | 5 |
| Total | 720 | 135 |

[a]In terms of 1970 dollars. From (14).

comparable to that in some of the less-developed areas were used, say $0.50/work-hr, the cost would be only about $0.80/ft².

The economic potential of solar distillation was summarized by an ad hoc panel of the National Academy of Sciences in 1976 (30). Solar distillation requires relatively large capital investment per unit of capacity and, in properly designed and constructed systems, a minimum of operating and maintenance costs. Product-water costs, therefore, depend primarily on still productivity, service life, capital cost of the installation, and amortization and interest rates.

Since productivity of a solar still is dependent on the intensity and duration of the insolation it receives, it is subject to both diurnal and seasonal variations, which must be taken into account in any analysis of needs and costs. Productivity of a solar still is conveniently referred to, in round numbers, as being typically 0.1 gal/ft² · day (40 liters/m² · day) for a clear summer day, with seasonal fluctuations expected to range down to half that on clear days in mid-latitudes (15–45°). Thus, experience shows that a still will yield about 25 gal/ft² (10,000 liters/m²) annually, with some variations dependent on climate and design. A typical lifetime for a still constructed of concrete, glass, and other durable materials is 20 yr or more with little or no maintenance. Some stills, however, use less durable materials that must be replaced periodically.

The lack of more general use of solar stills is almost entirely related to the high capital investment required and the resulting high cost of water produced. In recent years, durable solar stills have been built for a unit cost of $1.50–$3/ft² ($16–$32/m²). If a durable solar still can be constructed for $1/ft² ($10.75/m²), including materials and labor, typical yields would produce water at a cost of about $4/1000 gal ($1/1000 liters). These figures may be compared with typical fresh-water costs of $0.05–$0.25/1000 gal ($0.01–$0.05/1000 liters), and costs of de-salted water from large fuel-fired evaporation plants in the vicinity of $1–$1.50/1000 gal ($0.25/1000 liters). There seems to be little prospect, therefore, that large solar stills can be competitive with large nuclear or coal-fired desalting plants unless fuel prices escalate greatly. However, a significant advantage of the solar still is its flexibility in choice of size. In situations where a community or an industry requires small quantities of water—for example, less than 50,000 gal/day (200,000 liters/day)—the solar still may be more economical than the conventional desalting plant.

Table 8.4 shows the estimated capital costs of a large basin still constructed in a developing country having an hourly wage rate of $0.50. The total investment of about $0.90/ft$^2$ and an annual productivity of about 25 gal/ft$^2$ of distilled water and 8 gal/ft$^2$ of rainwater would be equivalent to $10 per average daily gallon production capacity. In a plant of this size, this level of cost should be readily attainable. Developments that have occurred since the above estimate was reported may actually reduce these costs.

## NONCONVECTING SOLAR PONDS

The nonconvecting solar pond is a horizontal-surfaced solar collector using the absorption of solar radiation at the bottom of a 1- or 2-m-deep body of water to generate low-temperature heat. Since heat storage is an integral part of ponds, they have promise in some parts of the world for continuous energy delivery to process or space-conditioning systems. Modern solar ponds were first studied scientifically in Israel by Bloch, Tabor, et al. (32, 33). This section describes ponds in which a temperature substantially above ambient—by 50 K or more—may be achieved. The ocean is also a type of solar pond, but it is treated in Chapter 9 under natural solar systems.

### Introduction

When solar energy enters a pond, the infrared component is absorbed within a few centimeters, near the surface, since water is opaque to long-wave radiation. The visible and ultraviolet components of sunlight can penetrate clear water to a depth

**TABLE 8.4**  Basin-type Solar Still Capital Costs[a]

| Item | Cost ($) |
|---|---|
| Basic still materials, $0.72/ft$^2$ | 396,000 |
| Erection and assembly, $0.50/worker-hr | 39,600 |
| Instruments | 5,000 |
| Feedwater supply | 500 |
| Contingencies | 20,000 |
| Engineering | 4,000 |
| Construction interest | 25,600 |
| Site | 1,500 |
| Total plant investment | 492,200 |

[a]In terms of 1970 dollars. Capacity = 50,000 gal/day; 550,000 ft$^2$; distillate plus rainwater = 33 gal/ft$^2$ · yr. Abstracted from Ref. (30).

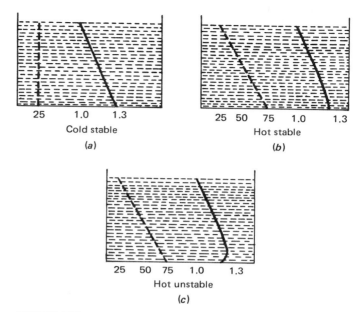

**FIGURE 8.20**  Density gradients in a solar pond. If the gradient is
positive when cold (*a*), it becomes less positive when heated (*b*); if the
lowest layers become too hot, the density profile may reverse (*c*) and
the onset of convection will follow. Dashed lines represent
temperature profiles (units are °C and g/cm³).

of several meters. These radiation components can be absorbed at the bottom of the
pond by a dark-colored surface. The lowest layer of water is then the hottest and
would tend to rise because of its relatively lower density if measures were not taken
to prevent it.

In nonconvecting solar ponds, the water at the bottom is made heavier than
that at the top by dissolving salt in the water as shown in Fig. 8.20. The
concentration of salt is decreased from bottom to top so that the natural tendency
of ponds to mix by the creation of convection currents is effectively eliminated if
the density gradient is adequate.

Since stationary water is quite an effective insulator, it is possible for the
lowest layers of a well-designed solar pond to boil. Boiling, of course, must be
avoided, for it destroys the stable density gradient. Therefore, the design of a solar
pond for heat generation must involve a mechanism for useful heat removal in
sufficient quantity to avoid boiling (Fig. 8.21).

Since solar ponds are usually envisioned to be on the order of hectares in size, a
heat-exchanger pipe network using a separate working fluid is impractical. However,
hydrodynamic principles predict that a layer of fluid could be removed slowly from
the bottom of the pond without disturbing the main body of water. This is
evidenced on a large scale by the ability of ocean currents—Gulf Stream, Benguela
Current, etc.—to retain their identity over thousands of miles. The fluid in the

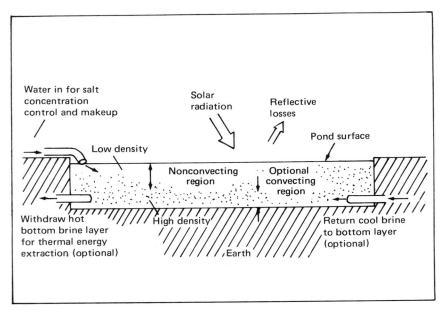

**FIGURE 8.21** Schematic diagram of a nonconvecting solar pond showing conduits for heat withdrawal; surface washing; and an optional convecting zone near the bottom (see text).

removed layer of a pond is then passed through a heat exchanger for useful heat removal and returned to the bottom of the pond. Since the returned fluid is cooler than the extracted fluid, the required density gradient is maintained. In practice, the horizontal flows can disturb the hydrostatic equilibrium somewhat, depending on the Reynolds and Froude numbers, and a mixed layer can exist at the pond's bottom. In this layer, convection can occur, although it can be confined to a zone on the order of centimeters. Nielsen and Rabl (25) have described a pond in which a convective zone is an intentional feature of the design proposed for space heating.

Several practical difficulties arise during long-term use of solar ponds in the field. Since a salinity gradient exists, diffusion of salt from regions of high concentration to low concentration occurs naturally. Hence, the density gradient required for hydrostatic stability tends to destroy itself. This diffusion can amount to 60 tons/km$^2$ · day (32). To maintain the gradient, it is necessary to supply salt to the bottom of the pond and to wash the surface with a weak brine solution. The supply of salt to a pond is a major maintenance cost for this type of solar collector. Tabor (32) has suggested an idea whereby the bottom layer of a pond is decanted, partially evaporated, and returned to the pond bottom. Simultaneously, water is added to the top of the pond to replace that evaporated from the bottom layer. As a result, the net flow of water is downward, and the *relative* flow (diffusion) of salt is upward. By proper matching of these rates, the salt can remain stationary in a fixed frame of reference.

The replacement of water that evaporates naturally from the surface of a pond is a second practical consideration. If water is not replaced at the top, a local reversal of the stable salinity gradient will occur and a convective layer will form at the surface. This is to be avoided, since it destroys part of the insulation effect of the stagnant pond. Nielsen and Rabl (25) have proposed the use of a plastic cover over the pond to help maintain the proper gradient. Since reflection of solar radiation from plastic is greater than that for water (see Chapter 3), they also suggest maintaining a thin water layer over the plastic cover.

Other practical matters that must be considered are the effects of wind, waves, rain, and leaks on the pond gradient (halocline). Waves can destroy the halocline and therefore must be controlled. An economical and effective method for this task has yet to be developed. Very severe rainstorms and wind can destroy the surface gradient to a depth of 35-50 cm (24). A plastic cover could presumably avoid the problem if it could stay intact during severe storms. The problem of dirt in a pond depends on the location of the particles. Either on the surface or suspended within the pond, they reduce solar transmittance and must be removed by withdrawing, filtering, and replacing a layer of the pond. Dirt on the bottom can likewise be removed but may not cause a major problem if left in place. Leaves in a pond should be avoided, since they may float, may lose their color to the water, and may plug fluid circulation machinery and conduits. Algae growth in ponds should be readily controllable by chemical means.

A final feature of a solar pond determines its geographic limits. Since the ponds are horizontal and the sun is low in the sky north or south of the mid-latitudes in winter, ponds must be used near the equator if winter yields are not to be curtailed sharply. Other climatic effects in the tropics—monsoons or storms—must also be considered in finding the best sites for large-scale pond usage.

## Solar Pond Stability Criteria

One of the principal costs of a solar pond is the dissolved salt required to establish the stable density gradient. For economical application it is essential that no more salt be used than necessary to assure hydrostatic stability. Figure 8.22 shows the solubility of several common salts in water as a function of temperature. Salts like $NH_4NO_3$ or $KNO_3$, whose solutions are as transparent as water, are best for solar ponds, since solubility increases significantly with temperature thereby ensuring a significant density gradient. Salts such as $Na_2SO_4$ with the opposite solubility characteristic would not be suitable for ponds.

Weinberger (37) has established the criteria for the hydrostatic and hydrodynamic stability of large solar ponds. To avoid convection, the fluid density must decrease upward from the bottom, that is,

$$\frac{d\rho}{dz} = \frac{\partial \rho}{\partial s}\frac{ds}{dz} + \frac{\partial \rho}{\partial T}\frac{dT}{dz} \geqslant 0 \qquad (8.26)$$

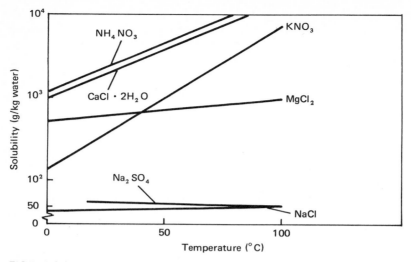

**FIGURE 8.22**  Solubility of some common inorganic salts usable in solar ponds.

where $s$ is the salt concentration, $\rho$ is the density, $T$ is the temperature, and $z$ is the vertical coordinate, increasing downward.

However, to avoid the growth of oscillatory motion, a stronger condition is required (25):

$$\left(\frac{\nu + K_s}{\rho}\right)\frac{\partial \rho}{\partial s}\frac{ds}{dz} + \left(\frac{\nu + K_T}{\rho}\right)\frac{\partial \rho}{\partial T}\frac{dT}{dz} \geqslant 0 \qquad (8.27)$$

where $\nu$, $K_s$, and $K_T$ are momentum, salt, and thermal diffusivities, respectively. Based on this condition, a convective layer of thickness $h$ will grow if (25)

$$q_{net} > 0.89\, h(\rho c) \sqrt{\frac{K_T}{t}} \left(-\frac{dT}{dz} - \frac{\nu + K_s}{\nu + K_T}\frac{\partial \rho/\partial s}{\partial \rho/\partial T}\frac{ds}{dz}\right) \qquad (8.28)$$

where $t$ is the time during which heat is absorbed at rate $q_{net}$ in a region of heat capacity $(\rho c)$ per unit volume.

If the net heat rate is above that required for convective layer growth, the layer grows in thickness as (20)

$$h^2 = Cq_{net} + h^2\,(t = 0) \qquad (8.29)$$

where

$$C = \frac{2}{(\rho c)[ds/dz\,(\partial \rho/\partial s)/(\partial \rho/\partial T) + dT/dz]} \qquad (8.30)$$

**TABLE 8.5**  The Spectral Absorption of Sunlight in Water[a]

| Wavelength (μm) | Layer depth | | | | |
|---|---|---|---|---|---|
| | 0 | 1 cm | 10 cm | 1 m | 10 m |
| 0.2–0.6 | 23.7 | 23.7 | 23.6 | 22.9 | 17.2 |
| 0.6–0.9 | 36 | 35.3 | 36.0 | 12.9 | 0.9 |
| 0.9–1.2 | 17.9 | 12.3 | 0.8 | 0.0 | 0.0 |
| 1.2 and over | 22.4 | 1.7 | 0.0 | 0.0 | 0.0 |
| Total | 100.0 | 73.0 | 54.9 | 35.8 | 18.1 |

[a]Numbers in the table give the percentage of sunlight in the wavelength band passing through water of the indicated thickness.

If the heat rate $q_{net}$ is less than the critical value from Eq. (8.28), the convective layer may not decrease as is the case in many other hydrodynamic phenomena. This is a result of the much greater numerical value of thermal diffusivity compared with salt diffusivity. Nielsen and Rabl (25) have shown that the salinity gradient to stabilize a convective zone is about five times that required to maintain a stagnant zone. A density difference of 25 percent has successfully stabilized ponds in practice.

## Thermal Performance of Solar Ponds

In a solar pond, solar radiation is partially reflected at the surface, partially absorbed in the water, and partially absorbed at the bottom. The absorption of solar energy by water does not follow a simple Bouger's law since the absorption phenomena differ widely with wavelength. Table 8.5 is a summary of absorption of sunlight in water as a function of wavelength. Absorption in solutions of inorganic salts used in solar ponds is expected to be nearly the same.

The data from Table 8.5 are plotted in Fig. 8.23. An adequate curve fit of these data is represented by the sum of several exponential terms. If $\tau(x)$ is the transmittance of water of depth $x$, it can be related to $x$ by (25)

$$\tau(x) = \sum_{i=1}^{4} a_i e^{-b_i x} \qquad (8.31)$$

where the regression coefficients $a_i$ and $b_i$ are

| Wavelength (μm) | $a_i$ | $b_i$ ($\times 10^{-3}$ cm$^{-1}$) |
|---|---|---|
| 0.2 –0.6 | 0.237 | 0.32 |
| 0.6 –0.75 | 0.193 | 4.5 |
| 0.75–0.90 | 0.167 | 30.0 |
| 0.90–1.20 | 0.179 | 350.0 |

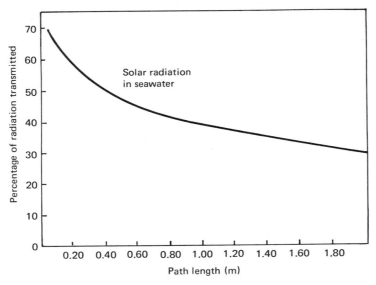

**FIGURE 8.23** Transmittance of water to solar radiation as a function of the thickness of the water layer; see Table 8.5 for numerical values.

Equation (8.31) does not include the infrared spectrum ($\lambda > 1.2$ $\mu$m), since it is of no interest in solar pond analysis. As shown in Refs. (16, 27, 37), a detailed analysis of heat transfer in a solar pond is very complex and must include effects of volumetric absorption, variation of density, and conductivity with salinity.

Figure 8.23 shows that about 30 percent of the incident radiation reaches the absorbing bottom surface of a 2-m solar pond. This represents the upper limit of collection efficiency of a pond. But since the bottom is an imperfect absorber and heat losses occur from the top and bottom surfaces, the thermal efficiency will rarely exceed 25 percent. However, as described in Chapter 5, thermal efficiency is an inadequate index of solar system viability. Economics must be considered. Since solar ponds are presently not in commercial use, their economic analyses are not reliable. However, Rabl and Nielsen (27) have predicted that solar ponds for building heating may be more economical than fuel oil heating in some areas. Styris et al. (31) have analyzed solar ponds for process heat as well.

An additional consequence of the data shown in Fig. 8.23 should be noted. The insulating effect of the nonconvecting layer in a solar pond increases linearly with depth. In most practical cases (16, 27, 37) this insulating effect increases more rapidly than does the attenuation of sunlight because of increased depth. For an example, a 1-m-deep pond that can achieve a no-load temperature of 100°C, could achieve a temperature of about 180°C if it were 2 m deep. Of course, brine costs increase with pond depth and pond depth should therefore be selected to match the task for which the heat is required.

In summary, solar ponds afford a simple method for solar-thermal conversion with collector and storage integrated into a single component. Temperatures on the order of 100°C have been achieved. Unresolved problems related to wave and weather disturbances, salt diffusion, and heat extraction must be solved before widespread use of ponds for process heat or space conditioning can be expected.

## PHOTOVOLTAIC CONVERSION

The direct conversion of sunlight to electricity by means of the photovoltaic effect does not require an intervening heat engine and generator and therefore offers a direct and potentially economic method for solar power production. Moreover, the theoretical efficiencies attainable are comparable to those achievable by a solar collector–engine-generator system. Solar cells have been used on a fairly large scale by the U.S. space program in well over 1000 satellites. In addition, solar system exploratory vehicles have reliably used solar cells as one source of their electrical energy.

Processes for economic mass production of solar cells have yet to be developed. In an attempt to improve the cost position of solar cells Ravi, Mlavsky, and co-workers (29) have demonstrated a continuous laboratory process for producing silicon ribbons for solar cells. The expected refinement of this process and other continuous processes could result in economical production of electricity from the sun. It is ultimately the large-scale demand for these devices that will determine the rate at which prices of solar cells decrease to or below the level of other energy sources in the future.

### Introduction

Becquerel (6) first demonstrated in 1839 that a voltage could be developed when light was absorbed by an electrode immersed in an electrolyte. Adams and Day (1) demonstrated an analogous effect in solids using the element selenium in 1877. It was not until nearly a century after these discoveries that this laboratory phenomenon was considered a possible source of electrical energy. A group of RCA scientists demonstrated the photovoltaic effect in 1954 in a semiconductor junction of the type developed for use in diodes and transistors.

The most common substance used currently in photovoltaic devices is silicon. Each atom of silicon has four outer-orbit or valence electrons, which form covalent bonds with adjacent silicon atoms in a crystalline lattice. If an impurity such as phosphorus with five valence electrons is introduced into a silicon lattice (usually called doping), the resultant material contains excess electrons. These electrons are free to migrate about the lattice and to act as current carriers. A semiconductor with excess electrons is called an $n$-type. Similarly, if silicon were doped with an impurity such as boron with three valence electrons, there would be a deficiency of one electron per boron atom. A substance with electron vacancies or "holes" is called a $p$-type. Holes are also free to act as positive-charge current carriers by analogy with free electrons in a phosphorus-doped silicon lattice.

Semiconductor $p$–$n$ junctions are formed in practice by joining $p$- and $n$-type materials into a single crystal. Junctions are usually formed using diffusion or ion im-plantation techniques to insert impurities of the $n$-type into $p$-type lattices or vice versa.

The photovoltaic effect is the generation of an electrical potential when absorbed radiation ionizes atoms in the vicinity of a potential barrier (for example, a $p$–$n$ junction). If the radiation energy level is sufficient, separated electron-hole pairs are created, in turn creating an electromotive force capable of causing a current flow through an electrical load. The incident photon energy must equal or exceed the valence-to-conduction band gap in order to produce hole-pairs.

Figure 8.24 shows the photovoltaic process schematically. An incident photon of the proper energy dislodges an electron in the silicon crystal. This electron migrates to the ohmic contact or electrode in the $n$ layer; simultaneously, the hole created by the photon-electron energy transfer migrates toward the $p$ layer. A current flow from $n$ to $p$ electrodes is thereby induced through the external circuit. Photons with energy below the band gap are partially absorbed as heat, whereas photons with energy above the band gap both create an electron-hole pair and generate heat. Although the conversion efficiency of a solar cell is not limited by the Carnot efficiency, as are thermal solar-to-electrical conversion methods, other factors limit the theoretical efficiency of photocells to less than 25 percent.

Table 8.6 exhibits the band-gap energy of common photocell materials along with their theoretical and measured efficiencies. It will be demonstrated below that the measured efficiencies shown may be increased by the concentration of radiation onto the cell surface. If $I_c$ denotes the insolation on the cell collection area $A_c$ the efficiency of a solar cell $\eta_p$ is defined as

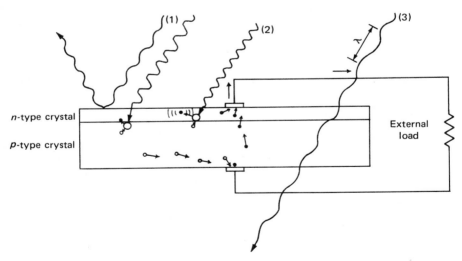

**FIGURE 8.24** Schematic diagram of the photovoltaic process in silicon. A photon of the proper wavelength (1) creates a hole-electron pair (small circle and dot); these carriers in turn migrate to the respective electrodes. Photons of shorter wavelength (2) create a pair as well as heat while longer-wavelength photons (3) pass through the photocell.

**TABLE 8.6**  Characteristics of Common Photocells

| Material | Band gap $V_g$ (eV) | Theoretical efficiency | Measured efficiency |
|---|---|---|---|
| Silicon (SI) | 1.11 | 24 | 18 |
| Indium phosphide (InP) | 1.25 | 23 | 3 |
| Gallium arsenide (GaAs) | 1.35 | 24 | 11 |
| Cadmium telluride (CdTe) | 1.45 | 21 | 7 |
| Gallium phosphide (GaP) | 2.25 | 17 | 1 |
| Cadmium sulfide (CdS) | 2.40 | 16 | 7 |

$$\eta_p \equiv \frac{(IV)_m}{I_c A_c} \qquad (8.32)$$

where the subscript $m$ denotes the point of maximum power. The maximum-power point is defined as that point located on the current-voltage characteristic (Fig. 8.25) representing the maximum current-voltage product. Geometrically it is the area of the largest rectangle bounded by the $I$-$V$ curve. The open circuit voltage $V_{oc}$ and closed circuit current $I_s$ are also shown. These parameters can be used to characterize the full $I$-$V$ curve as shown in the next section. The denominator of the efficiency definition represents the total incident radiation on the aperture of the photovoltaic converter.

As described in Chapter 2, solar radiation has a spectral distribution encompassing the band-gap energies shown in Table 8.6. Therefore, some incident radiation is not sufficiently energetic to liberate an electron. The energy contained in a photon $E_p$ is given by

$$E_p = h\nu \qquad (8.33)$$

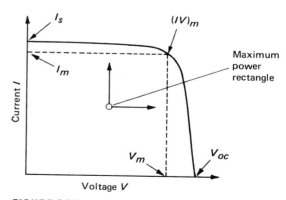

**FIGURE 8.25**  Voltage and current output from an illuminated solar cell.

where the frequency $\nu$ is related to the wavelength $\lambda$ and the speed of light $c$ by

$$\nu = \frac{c}{\lambda} \tag{8.34}$$

## EXAMPLE 8.3

Calculate the wavelength of light capable of forming an electron-hole pair in silicon.

## SOLUTION

The band-gap energy of silicon is 1.11 eV. By combining Eqs. (8.33) and (8.34) it is easy to show that

$$\lambda = \frac{hc}{E} \tag{8.35}$$

For $c = 3 \times 10^8$ m/sec, $h = 6.625 \times 10^{-34}$ J · sec, and 1 eV $= 1.6 \times 10^{-19}$ J, Eq. (8.35) gives the required wavelength

$$\lambda = \frac{(6.625 \times 10^{-34} \text{ J} \cdot \text{sec})(3 \times 10^8 \text{ m/sec})}{(1.11)(1.6 \times 10^{-19} \text{ J})} = 1.12 \ \mu m$$

Therefore, any radiation with wavelength less than 1.12 $\mu m$ can theoretically generate a photovoltaic current in silicon. Figure 8.26 shows the region of available energy for silicon as a function of wavelength.

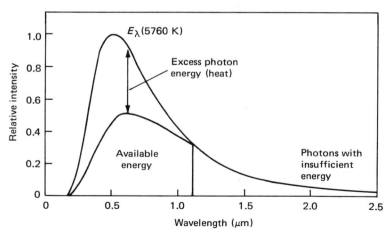

**FIGURE 8.26** Energy utilization for a silicon photocell as a function of wavelength. Adapted from (38).

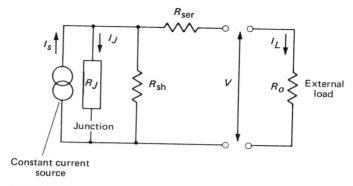

**FIGURE 8.27**  Equivalent circuit of an illuminated *p–n* photocell with internal series and shunt resistances and nonlinear junction impedance $R_J$.

## Thermal and Electronic Analysis of Photocells

In this section the principal parameters governing the rate of energy conversion of a photocell are described. For more detailed analyses the reader is referred to Refs. (2, 17, 23, 28).

The photocell can be modeled as a macroscopic equivalent circuit as shown in Fig. 8.27. It consists of a current source $I_s$ and a group of resistors. A simple network analysis neglecting the series and shunt resistances $R_{ser}$ and $R_{sh}$, respectively, shows that the load current $I_L$ is

$$I_L = I_s - I_J \tag{8.36}$$

where $I_J$ is the junction current leakage of holes and electrons and $I_s$ is the short-circuit current.

The power $P_L$ generated in the load $R_L$ is given by

$$P_L = I_L V \tag{8.37}$$

where $V$ is the load voltage drop (cell voltage gain). As shown by Krenz (19), the semiconductor junction current leakage $I_J$ is

$$I_J = I_0(e^{e_o V/kT} - 1) \tag{8.38}$$

where $k$ is Boltzmann's constant ($1.38 \times 10^{-23}$ J/K) and $T$ is the junction temperature (K). $I_o$ is the saturation current proportional to $(T^4 e^{-V_g/kT})$ where $V_g$ is the band gap, always greater than the load voltage $V$. Hence, Eq. (8.38) indicates that junction leakage $I_J$ increases rapidly with temperature, causing the load current $I_L$ to decrease since $I_s$ is nearly temperature independent. Numerical examples of the strong temperature effect will be shown below. The power can then be written from

Eqs. (8.37) and (8.38) as

$$P_L = [I_s - I_o(e^{e_o V/kT} - 1)]\, V \tag{8.39}$$

If the expression for $P_L$ is differentiated with respect to $V$, the independent variable, and equated to zero, the following criterion for the maximum-power voltage $V_m$ gives:

$$(e^{e_o V_m/kT}) \left(1 + \frac{e_o V_m}{kT}\right) = 1 + \frac{I_s}{I_o} \tag{8.40}$$

from which

$$I_{L,m} = \frac{e_o V_m/kT}{1 + e_o V_m/kT}\,(I_o + I_s) \tag{8.41}$$

In all practical photovoltaic devices $I_o \ll I_s$; therefore the peak efficiency $\eta_{p,m}$ can be written from Eq. (8.41) as

$$\eta_{p,m} \approx \frac{(e_o V_m/kT)V_m I_s}{(1 + e_o V_m/kT)I_c A_c} \tag{8.42}$$

where $V_m$ is determined by trial and error from Eq. (8.40) for a given $I_o$. The number of photons with energy above the band-gap energy decreases with increasing band-gap voltage (decreasing wavelength) as shown in Chapter 2. However, since the output voltage increases with band gap, these two offsetting effects cause a peak efficiency to exist at a specific band-gap value. Figure 8.28 shows the theoretical efficiency–band-gap curve. It is noted that the most common photocell material, silicon, is not the ideal, peak-efficiency material.

As described earlier, the operating temperature must be kept low for efficient operation. The data in Fig. 8.28 show this strong temperature dependence. This effect is particularly important for cells illuminated by concentrated sunlight. It is unlikely that the efficiency of energy conversion by photovoltaic means will ever exceed 25 percent at ambient temperatures.

The efficiency of photovoltaic conversion depends linearly on the short-circuit current $I_s$ [Eq. (8.42)] produced in the junction. The current in turn depends directly on the number of photons $n_p$ converted to electrons per unit time as shown in Eq. (8.38).

$$I_s = \eta_o\,[(1 - \rho_c)\alpha]\,e_o n_p \tag{8.43}$$

where $\rho_c$ is the cell-cover reflectance (see Chapter 2) and $\alpha$ is the absorptance calculated from Bouger's law:

$$\alpha = 1 - e^{-K_B L} \tag{8.44}$$

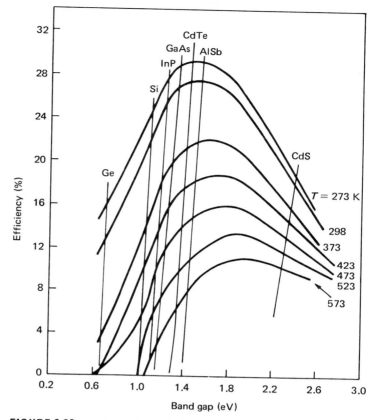

**FIGURE 8.28** Maximum theoretical efficiency of photovoltaic converters as a function of band-gap energy. Reprinted from "Sunlight to Electricity: Prospects for Solar Energy Conversion by Photovoltaics" by J. A. Merrigan (23) by permission of The M.I.T. Press, Cambridge, Massachusetts.

where $K_B$ is the extinction coefficient and $L$ is the layer depth. The collector carrier efficiency $\eta_o$ is the ratio of carriers in motion to the number created from sufficiently energetic photons $n_p$. Its value is in the range 0.50–0.65 for well-designed cells.

The only quantity not readily calculable in the short-circuit current equation is the carrier efficiency. Although it is beyond the scope of this book to present the details of its derivation, it depends on three parameters to the lowest order:

1. Bouger's law absorption constant $K_B$
2. Minority carrier lifetime—for example, lifetime of a hole in the $n$ layer
3. Surface recombination rate of minority carriers

The optimal combination of these three properties requires very thin surface layers (small $L \approx K_B^{-1}$), low surface recombination rates achievable by etching, and minority carrier lifetimes less than $10^{-6}$ sec. Silicon solar cells are about $250 \,\mu m$ thick with a $p$-$n$ junction $0.5$-$2.0 \,\mu m$ below the illuminated surface.

The preceding analysis has assumed that the shunt and series resistances $R_{sh}$ and $R_{ser}$, respectively, of Fig. 8.27 are small. In practice $R_{sh}$ can be made small to minimize parasitic losses. However, the series resistance cannot be made negligibly small. The need to minimize $R_{ser}$ would indicate high doping levels and deep junctions, which reduce the short-circuit current $I_s$. Most cells compromise on the conflicting requirements using very carefully designed ohmic contacts between electrodes and semiconductors and by sizing the contact "fingers" to be very small and closely spaced. The series resistance is a critical parameter in the performance of concentrator-photovoltaic devices where large fluxes (large $n_p$) cause large short-circuit currents and large $I_s^2 R_{ser}$ power losses with considerable cell heating. Figure 8.29 shows the effect of series resistance on the $I$-$V$ characteristic of a silicon cell.

Figure 8.30 summarizes the losses described above experienced in photovoltaic conversion of sunlight to electricity in silicon cells. Major reduction in losses will require new methods not now known.

## Solar Cells with Concentrators

One of the most promising methods for reducing the cost of solar cell arrays is to replace expensive semiconductor area with relatively inexpensive reflecting or refracting concentrator area. Silicon cells have shown the ability to increase current output with increasing flux density and to withstand increased internal heat generation. In this section the important principles of configurations of concentrator-solar cells are described. In addition, test data are presented. Complete descriptions of several concentrators are given in Chapter 4.

Three criteria must be met for a concentrating system to be viable:

1. *Optical efficiency and concentration effect per unit cost* must be high.
2. Cells must have a *low internal series resistance* $R_{ser}$.
3. The array must have an *efficient and economical cooling system*.

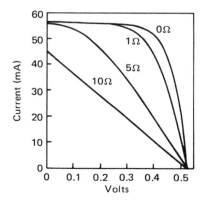

**FIGURE 8.29** The effects of series resistance on measured silicon solar cell curves. Tungsten light, 100 mW/cm² ; cell area = 2 cm². From (17).

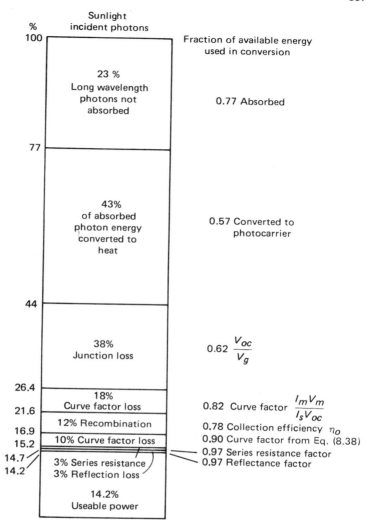

**FIGURE 8.30** Photovoltaic conversion factors in silicon cells. Adapted from (23).

Figure 8.31 shows the effect of intensity of radiation on the $I$–$V$ characteristic of CdS photocells. These curves represent data measured on a cell at $25°C$. They show that open circuit voltage $V_{oc}$ increases only slightly but short-circuit current increases nearly linearly with intensity for a given radiation spectrum as expected from Eq. (8.40).

The increase of current with intensity may result in modest efficiency gains at low levels of irradiation but the parasitic $I^2 R_{ser}$ losses are important for higher current levels. It is, therefore, essential that cells used with concentrators be

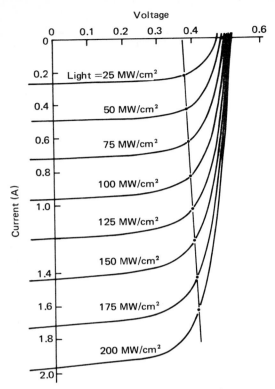

**FIGURE 8.31** Effect of illumination intensity on
CdS photocell current-voltage characteristic curve.
From Shirland, F. A., *Adv. Energy Convers.*, vol. 6,
pp. 201–221, October 1966; see also (5).

designed with excellent semiconductor-to-electrode contacts and very low series
resistance $R_{ser}$. In addition, a cooling system must be provided in order to avoid the
potentially severe degradation of efficiency with high temperature.

Spectrolab and Arizona State University (ASU) have conducted experiments on
several configurations of concentrators and silicon solar cells. The cells were
fabricated using special methods to reduce series resistance by means of many very
narrow contact fingers, thereby minimizing the opaque ohmic area and maximizing
contact perimeter. Figure 8.32 shows a typical efficiency curve produced in the
study. Efficiency is seen to peak at nearly 12 percent for $CR \approx 13$. Above this
concentration, the $I^2 R_{ser}$ loss nullifies any current generation increases. The ASU
study showed that the efficiency gain achievable is not easily predicted and that it
varies with concentrator type, cell ohmic contact configuration, and operating
temperature.

The optimum configuration of concentrator-solar cells must be evaluated by
minimizing the cost of the combined system. Procedures for economic analyses are
described in detail in Chapter 5. Merrigan (23) has carried out such a study for

silicon cells cooled by simple radiation, by air in free convection, or by forced convection with water. Figure 8.33 shows the significant effects of cooling mode on cost. The most effective cooling method is by means of water; radiation levels up to 25,000 W/m² (concentration of 25-35X) can be used effectively as shown. The economic analysis shows that simply selecting a configuration on the basis of maximum efficiency or maximum power will result in an uneconomical system.

## Simple Photocell Radiometer

The linear relationship between short-circuit current $I_s$ in a photocell and incident radiation [Eq. (8.43)] can be used to construct a simple solar radiometer. If a small resistance is placed across a photocell as shown in Fig. 8.34 it will operate in the short-circuit mode. Part of the current can be shunted through a milliammeter, which serves as the insolation readout.

For precision components, the accuracy of this simple radiometer is about ±5 percent. It can be calibrated against a standard pyranometer at several levels of illumination to convert mA readings to W/m² or Btu/hr · ft². Since silicon has optical properties that differ from those of glass, the accuracy of the simple radiometer diminishes at incidence angles departing from normal incidence.

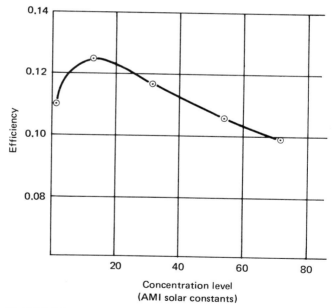

**FIGURE 8.32** Effect of concentration of solar radiation on efficiency of a silicon photocell 2 × 2 cm with 80 ohmic contact fingers. Adapted from (4).

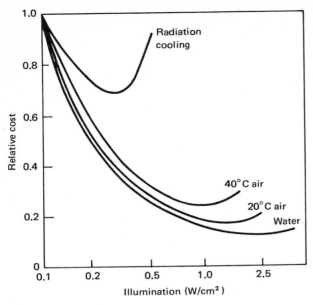

**FIGURE 8.33**  Operating temperatures and relative costs of
silicon photovoltaic converters provided with sunlight
concentrators and cooling. Reprinted from "Sunlight to
Electricity: Prospects for Solar Energy Conversion by
Photovoltaics" by J. A. Merrigan (23) by permission of The
M.I.T. Press, Cambridge, Massachusetts.

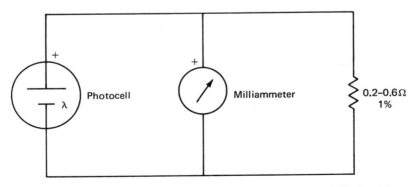

**FIGURE 8.34**  Simple radiometer circuit. Precision, temperature stabilized resistor,
and meter are required for accuracy; the milliammeter range should be matched to
the aperture area of the solar cell.

## Practical Use of Solar Cells

The principles governing the behavior of solar cells are for the most part well understood. However, the practical realization of solar cell power has yet to be achieved. Three distinct problems require solving before widespread, economical use will exist:

1. *Increase efficiency* to near the limit of 25 percent achievable theoretically. This increase is necessary for the solar cells to be competitive with other conversion schemes.
2. *Reduce costs* of production by a factor of about 50–100 by mass production and other means.
3. Ensure a *useful lifetime* of 15–20 yr.

The solutions to these problems will require significant research and development in both the academic and industrial sectors. This development in turn depends on the existence of a significant demand for direct solar-produced electricity on a worldwide basis.

## PROBLEMS

8.1. A solar electric engine operating between 40 and 200°F has an efficiency equal to one-half the Carnot efficiency. This engine is to drive a 5-hp pump. If the collector has an efficiency of 50 percent, calculate the area needed for operation in Egypt where the insolation averages about 250 Btu/ft² · hr during the day. State any additional assumptions.

8.2. The graph below shows the Rankine cycle efficiency as a function of maximum cycle temperature for different working fluids. Prepare a thermal

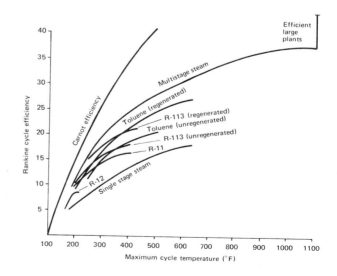

analysis matching a line focusing collector, a paraboloid disk collector, and a *CPC* collector with a concentration ratio of 3:1 to a suitable working fluid in Washington, D.C., and Albuquerque, New Mexico. Comment on storage needs in both locations. Low-temperature organic and steam efficiencies computed by Barber (5a) with the following assumptions: expander efficiency, 80 percent; pump efficiency, 50 percent; mechanical efficiency, 95 percent; condensing temperature, 95°F; regeneration efficiency, 80 percent; high side pressure loss, 5 percent; low side pressure loss, 8 percent. High-temperature steam efficiencies from (23a).

8.3. The schematic diagram shown below for a solar-driven irrigation pump was developed by Battelle Memorial Institute and uses tracking parabolic trough collectors. Solar energy is to heat the water in the collectors to 423 K, which then vaporizes a Freon working fluid that powers the pump-turbine. Calculate the surface area needed to power a 50-hp pump capable of delivering up to 38,000 liters/min of water at noon in Albuquerque, New Mexico. (Answer: $510 \text{ m}^2$.)

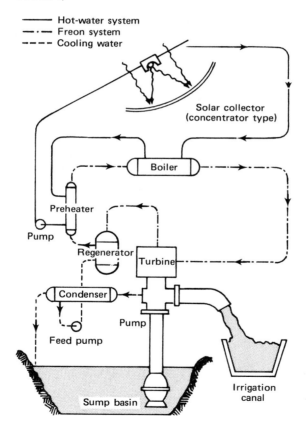

8.4. Design a simple radiometer of the type shown in Fig. 8.34 to measure the radiation incident on the absorber plane of a 3:1 *CPC* solar collector. Specify the meter range if the silicon photocell has an area of 3 cm².

8.5. Calculate the peak efficiency of a 1-cm$^2$ silicon solar cell for the following conditions:

| | |
|---|---|
| Insolation | $I_c = 1030$ W/m$^2$ |
| Saturation current | $I_o = 5 \times 10^{-13}$ A |
| Conversion efficiency | $\eta_o = 61$ percent |
| Surface refractive index | $n = 3.54$ |
| Active layer thickness | $L = 10$ $\mu$m |
| Extinction coefficient of active layer | $K = 10^3$/cm |
| Operating temperature | $T = 300$ K |
| Number of photons | $n_p = 3 \times 10^{17}$ photons/sec with band gap $> 1.11$ eV |
| Electron charge | $e_o = 1.6 \times 10^{-19}$ eV |

8.6. What is the average energy of photons in near-earth space?

8.7. Derive the maximum-power voltage equation (8.40).

8.8. Calculate the efficiency of a photocell with properties given in Problem 8.5 with one change: use a low reflectance coating with an index of refraction of 1.2.

8.9. Write a closed-form expression for the work output of a solar-powered heat engine if the energy delivery of the solar collector at high temperature is given approximately by the expression

$$q_u = (\tau\alpha)_{\text{eff}}I_c - \frac{\sigma\epsilon\overline{T}_f{}^4}{CR}$$

(convection and conduction losses are neglected) where $\overline{T}_f$ is the average fluid temperature and $CR$ is the concentration ratio. Two heat engines are to be evaluated: (1) Carnot cycle:

$$\text{Cycle efficiency } \eta_c = 1 - \frac{T_\infty}{\overline{T}_f}$$

and (2) Brayton cycle:

$$\text{Cycle efficiency } \eta_B = 1 - \frac{C_B T_\infty}{\overline{T}_f} \qquad C_B \geqslant 1$$

where $C_B = (r_p)^{(k-1)/k}$
$\quad r_p =$ the compressor pressure ratio
$\quad k =$ the specific heat ratio of the working fluid

Write an equation with $\overline{T}_f$ as the independent variable that, when solved, will specify the value of $\overline{T}_f$ to be used for maximum work output as a function of concentration ratio, surface emittance and $(\tau\alpha)_{\text{eff}}$ product, insolation level, and $C_B$. Optional: solve the equation derived above for $CR = 100$, $\epsilon = 0.5$, and $(\tau\alpha)_{\text{eff}} = 0.70$, at an insolation level of 1 kW/m$^2$. What is the efficiency of a solar-powered Carnot cycle and a Brayton cycle for which $C_B = 2$?

# REFERENCES

1. Adams, W. H., and Day, The Action of Light on Selenium, *Proc. R. Soc. London Ser. A*, vol. 25, p. 113, 1877.
2. Angrist, S. W., "Direct Energy Conversion," 3rd ed. Allyn and Bacon, Boston, 1976.
3. Anon., Goals Study for Technical Development and Economic Evaluation of the Compound Parabolic Concentrator Concept for Solar Energy Collector Applications, prepared for Argonne National Laboratory by Arthur D. Little, Inc., Cambridge, Massachusetts, under PO 31-109-38-3190, July 1975.
4. Arizona State University and Spectrolab, Terrestrial Photovoltaic Power Systems with Sunlight Concentrators, *ERDA Rept.* COO-2590-1, 1976.
5. Backus, C. E., ed., "Solar Cells," Institute of Electrical and Electronics Engineers, New York, 1976.
5a. Barber, R. E., "Solar Powered Rankine Cycle Engine Cycles—Characteristics and Costs," June 1976.
6. Becquerel, E., On Electrical Effects under the Influence of Solar Radiation, *C. R. Acad. Sci.*, vol. 9, p. 561, 1839.
7. Blake, F. A., T. R. Tracey, J. D. Walton, and S. Bouner, One MWth Bench Model Cavity Receiver Steam Generator, *Sol. Energy*, vol. 18, pp. 513–523, 1976.
8. Daniels, F., Solar Energy, *Science*, vol. 109, pp. 51–57, 1949.
9. Daniels, F., "Direct Use of the Sun's Energy," Yale University Press, New Haven, 1964.
10. Dunkle, R. V., Solar Water Distillation: The Roof Type Still and a Multiple Effect Diffusion Still, *Proc. 1961 International Heat Transfer Conference*, American Society of Mechanical Engineers, New York, 1961.
11. Eibling, J. A., S. G. Talbert, and G. O. G. Löf, "Digest of Technology of Basin-Type Solar Stills," Paper 5/69, presented at the International Solar Energy Conference, Melbourne, Australia, 1970; by the same authors, Solar Stills for Community Use, *Sol. Energy*, vol. 13, pp. 263–276, 1971.
12. "Energy for Rural Development," National Academy of Sciences, Washington, D.C., 1976.
13. Fraser, M. D., InterTechnology Corporation Assesses Industrial Heat Potential, Solar Engineering, vol. 1, pp. 11–12, September 1976.
14. Fraser, M. D., "Survey of the Applications of Solar Thermal Energy to Industrial Process Heat," paper presented at the Conference of the International Solar Energy Society, Winnipeg, Manitoba, 1976.
15. Hildebrandt, A. F., and L. L. Vant-Hull, Solar Thermal Power System Based on Optical Transmission, *Sol. Energy*, vol. 18, pp. 31–39, 1976.
16. Hipser, M. S., and R. F. Roehm, "Heat Transfer Considerations of a Non-Convecting Solar Pond Heat Exchanger," ASME Paper 76-WA/Sol-4, 1976.
17. Hovel, H. J., "Solar Cells," Academic Press, New York, 1975.
18. Kauer, E., R. Kersten, and F. Mahdjuri, Photothermal Conversion, *Acta Electron.* vol. 18, pp. 295–304, 1975.
18a. Keenan, J. H., and F. G. Keys, "Thermodynamic Properties of Steam," John Wiley & Sons, New York, 1956.
19. Krenz, J. H., "Energy Conversion and Utilization," Allyn and Bacon, Boston, 1976.

20. Leshuk, J. P., R. J. Zaworski, D. L. Styris, and O. K. Harling, Solar Pond Stability Experiments, in "Sharing the Sun," vol. 5, pp. 188–202, ISES, Winnipeg, 1976.
21. Loferski, J. J., Recent Research on Photovoltaic Solar Energy Converters, *Proc. IEEE*, vol. 51, pp. 667–674, 1963.
22. McAdams, W. H., "Heat Transmission," 3rd ed., McGraw-Hill Book Co., New York, 1954.
23. Merrigan, J. A., "Sunlight to Electricity: Prospects for Solar Energy Conversion by Photovoltaics, The M.I.T. Press, Cambridge, 1975.
23a. National Power Survey Advisory Panel, Report on Energy Conservation, p. 102, December 1974.
24. Nielsen, C. E., Experience with a Prototype Solar Pond for Space Heating, in "Sharing the Sun," vol. 5, pp. 169–182, ISES, Winnipeg, 1976.
25. Nielsen, C. E., and A. Rabl, Salt Requirement and Stability of Solar Ponds, in "Sharing the Sun," vol. 5, pp. 183–187, ISES, Winnipeg, 1976.
26. Powell, C. J., E. Fourakis, J. M. Hammer, G. A. Smith, J. C. Grosskreutz, and E. McBride, Dynamic Conversion of Solar Generated Heat into Electricity, *NASA Rept.* CR-134724, 1974.
27. Rabl, A., and C. E. Nielsen, Solar Ponds for Space Heating, *Sol. Energy*, vol. 17, pp. 1–12, 1975.
28. Rappaport, P., The Photovoltaic Effect and Its Utilization, *RCA Rev.*, vol. 20, p. 373, 1959.
29. Ravi, K. V., and A. I. Mlavsky, EFG Growth of Silicon Ribbon for Solar Cells, in "Sharing the Sun," vol. 6, pp. 23–33, ISES, Winnipeg, 1976.
30. "Solar Distillation as a Means of Meeting Small Scale Water Demands," United Nations, New York, 1970.
31. Styris, D. L., O. K. Harling, R. J. Zaworski, and J. Leshuk, The Nonconvecting Solar Pond Applied to Building and Process Heating, *Sol. Energy*, vol. 18, pp. 245–251, 1976.
32. Tabor, H., Solar Ponds, *Sol. Energy*, vol. 7, pp. 189–194, 1963.
33. Tabor, H., Solar Ponds, *Sci. J.*, pp. 66–71, 1966.
34. Talbert, S. G., J. A. Eibling, and G. O. G. Löf, Manual on Solar Distillation of Saline Water, *Office of Saline Water, Research and Development Progress Rept.* 546, U.S. Department of Interior, Washington, D.C., 1970.
35. Trombe, F., High Temperature Furnaces, *Proc. World Symposium on Applied Solar Energy, Phoenix, Arizona, 1955*, Stanford Research Institute, Menlo Park, California, 1956.
36. Trombe, F., and L. Phat Vinh, Thousand KW Solar Furnace Built by the National Center of Scientific Research in Odeillo (France), *Sol. Energy*, vol. 15, pp. 57–61, 1973.
37. Weinberger, H., The Physics of the Solar Pond, *Sol. Energy*, vol. 8, pp. 45–56, 1964.
38. Wolf, M., Limitations and Possibilities for Improvement of Photovoltaic Solar Energy Converters, *Proc. IRE*, vol. 48, pp. 1246–1263, 1960.

# NATURAL SOLAR CONVERSION SYSTEMS

*I dreamed my genesis in sweat of death, fallen twice in the feading sea, grown stale of Adam's brine until, vision of new man strength, I seek the sun.*

Dylan Thomas

The field of solar energy utilization is conveniently divided into mechanical and natural processes. In the preceding chapters the technical and economic aspects of the mechanical options have been presented. This chapter is devoted to the natural processes, all of which use one or more parts of the biosphere as a means of storage and/or transfer of energy. The first part of the chapter deals with wind energy conversion, the second part with ocean energy conversion, and the third part with photosynthetic and biomass conversion.

## WIND ENERGY CONVERSION SYSTEMS

Of the total incident solar flux absorbed by the earth, approximately 20 percent or $2 \times 10^{16}$ W is absorbed by the earth's atmosphere. A small portion of the absorbed energy is transformed into kinetic energy of the atmosphere. It has been estimated that the atmospheric and ocean convective currents combined account for a power of $3.7 \times 10^{14}$ W (22), while according to Putnam (41), the winds account for a power of $2 \times 10^{13}$ W. Although this is only a small portion of the incident solar energy, it is three times the 1972 energy-consumption rate for the world, and if 1 percent of the power in the wind could be utilized, it could supply $2 \times 10^{11}$ W or approximately 3 percent of the current world energy-consumption rate. If wind power were used for the generation of electricity, its thermal equivalent could be as high as 8–9 percent of the total. Technical and economic problems related to the location and variability of wind intensity tend to limit the attractiveness of wind energy conversion systems, but the potential extractable wind power is not insignificant.

### Types of Wind Machines

Windmill technology developed 10 centuries ago. Horizontal mills were used in Persia as early as the tenth century and more conventional vertical sail-type mills were used to grind corn and pump water as early as the twelfth century in Europe.

**577**

Windmills with outputs as great as 50 hp were used for pumping of water in the drainage system employed for the reclamation of portions of the Netherlands. Smaller windmills, typically 5 m in diameter with outputs of less than 1 hp, were used in rural America for irrigation and to drive small electric generators used to charge the batteries that provided current for electric lighting. An excellent description of European windmills is given by Reynolds (44) and a bibliography of technical work published prior to 1945 is presented in Refs. (16) and (36).

Since 1920 numerous attempts have been made to design technically feasible and economically viable windmills for large-scale power generation. Generally, two approaches were taken in the design of wind-power machines: horizontal-axis turbines whose axis of rotation is parallel to the direction of the wind and vertical-axis machines whose axis of rotation is perpendicular to the direction of the wind.

The Savonius rotor is an example of a vertical-axis wind turbine (VAWT). In its most simplified form it appears as a vertical cylinder sliced in half from top to bottom, with the two halves being displaced as shown in Fig. 9.1. It works on the same principle as a cup anemometer with the addition that wind can also pass between the bent sheets. Torque is produced by the pressure difference between the concave and convex surfaces of the half facing the wind and also by recirculation effects on the convex surface. The Savonius design can be fairly efficient, reaching a maximum of around 31 percent. However, it was found to be very heavy per unit power output. A Savonius rotor requires approximately 30 times more surface for the same power as a conventional rotor blade wind turbine. It is therefore not expected that Savonius rotors will be of interest for large-scale power production.

The Darrieus rotor shown in Fig. 9.2 is another example of a vertical-axis machine. The performance of a Darrieus rotor is not nearly as good as that of a propeller-type rotor and the machine requires an external power input for starting. Nevertheless, the simplicity of its design and the associated potential for low-cost production make it a promising candidate for economical power production.

A vertical-axis Darrieus wind turbine, 17 m in diameter and 19 m high (55 × 61 ft), was built in 1977 and is undergoing performance tests at the Sandia Laboratories. The turbine produces 60 kW of electricity in a wind of 45 km/hr (28

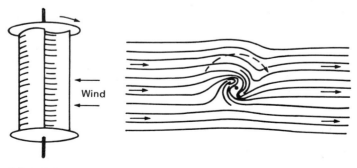

**FIGURE 9.1**   Savonius rotor.

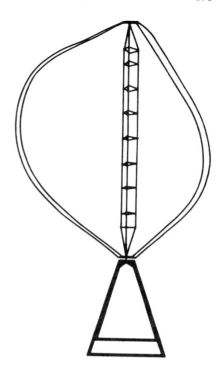

**FIGURE 9.2**  Darrieus rotor.

mi/hr) and 30 kW in a wind of 35 km/hr (22 mi/hr). The VAWT has a total height of 34 m (110 ft), including a 10-m (34-ft) anemometer tower at the top and a 5-m (15-ft) high base, which houses the transmission and electric generator. The electrical system produces a 60-cycle alternating current that can be fed into an electric utility line.

The key element of the VAWT is its blades, which have a cross section similar to that of an airplane wing. The aerodynamic lift produced by the wind on the blades causes the turbine to rotate regardless of the wind direction. This characteristic obviates the need for a mechanism to turn the turbine into the wind. Another potential cost-saving feature is that no pitch control is needed to limit the rotational speed under high-wind conditions because the VAWT automatically stalls at high wind speeds when under load. This means that by selecting an appropriately sized generator, the VAWT will not produce more power than the generator rating under any wind conditions. Moreover, the equipment used for power generating can be attached to the drive shaft at ground level, which reduces the structural cost of the tower.

The most famous horizontal-axis machine is the Smith-Putnam windmill, installed at Grandpa's Knob in Vermont in 1940. This wind turbine was designed to generate 1 MW. It had a 170-ft-diameter rotor consisting of two stainless steel blades that used NACA 4418 airfoil sections. The rotor and generator weighed about 250 tons and were supported by a 100-ft tower. The pitch control was automatic and kept the blades at a constant speed of 28.7 rpm at wind velocities of 18 mi/hr and above. As the wind velocity increased, the blades would begin to feather by turning edgewise. The blades were designed with an ability to cone up to 20° to guard against sudden gusts and still maintain a reasonably constant speed. The coning itself was damped by oil-filled cylinders so that the power plant could withstand winds up to 120 mi/hr, or those up to 100 mi/hr with 6 in of ice on the leading edge. The turbine, shown in Fig. 9.3, operated as a test unit until February 1943, when the 24-in main bearing failed. Because of the war a replacement could not be secured for 2 yr. In 1945 one of the blades came off and ended experimentation with this windmill. In spite of the structural failure of the blade, the Smith-Putnam design illustrated the potential of electric power generation by large-scale wind turbines. Recently a two-bladed propeller unit with 40-m-diameter blades has been constructed for the Energy Research and Development Administration and is undergoing tests in Ohio.

In addition to the conventional Smith-Putnam windmill several other experimental wind-axis machines have been built. One of the more interesting is the Enfield-Andreau ducted rotor shown in Fig. 9.4, which was built in England in 1954. Unlike conventional machines, it has no coupling between the propeller and generator and its blades are hollow. As the blades are turned by the winds, centrifugal forces pull air from the hollow tower through the blade tips. At the

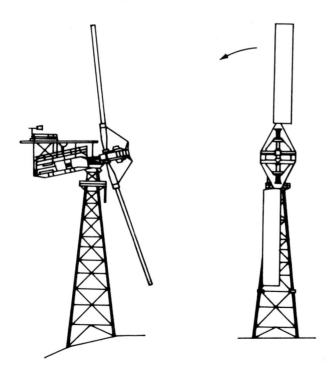

**FIGURE 9.3**  Smith-Putnam wind turbine.

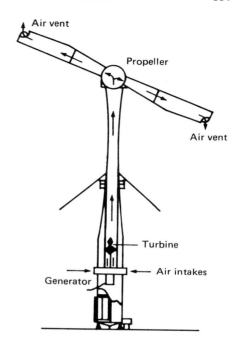

**FIGURE 9.4**  Enfield-Andreau
ducted rotor.

same time, the pressure difference between the tip of the rotor and the blade pedestal also draws air up through the semivacuum created in the tower. As air flows through the tower it passes through a turbine that drives the generator. The blade in the experimental unit was 80 ft in diameter and capable of producing 100 kW in a wind of 30 mi/hr at 95 rpm. In order to maintain constant rotor speed, hydraulic motors were used to vary the blade pitch. This design was effective at wind speeds up to 60 mi/hr; the blades were constructed so that they could flap under wind pressure of heavy gusts. The motion of the rotor to face into the wind was aided and controlled by a power-operated system. The main advantage of the ducted rotor wind machine is that the power-generating equipment is not supported aloft. A manual for the aerodynamic design of wind-power machines has recently been prepared by Wilson and Lissaman (55). Figure 9.5 shows their performance comparison of the various types of rotors that have been constructed to date.

## Performance of Wind Generators

Of the varied windmill designs conceived, the vast majority use screw propellers that are similar to airplane propellers but operate in reverse, that is, with a negative torque and a negative thrust. In fact, a windmill propeller is exactly equivalent to an airplane propeller advancing through a stationary fluid with the speed of the wind but in an opposite direction. However, whereas an airplane propeller is designed to produce thrust, the thrust of a windmill is not important, except to calculate the loads on bearings and the stresses in the supporting structure. The

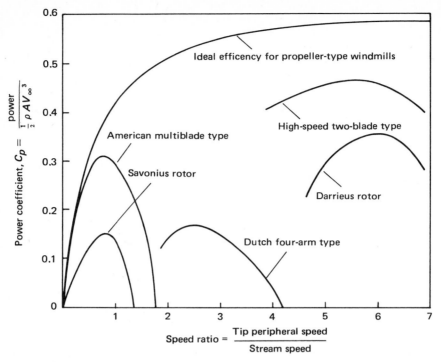

**FIGURE 9.5** Typical performance of wind-power machines.

efficiency of an ordinary propeller is equal to the thrust it develops times the speed with which it moves (kg · m/sec or ft · lb$_f$/sec) divided by the power input necessary to drive it. However, the thrust of a windmill does no work because the propeller remains stationary. It is therefore necessary to frame another definition of efficiency. The efficiency of a windmill $\eta_w$ is defined as the work done per unit time $P$ divided by the kinetic energy per unit time of the fluid passing through the area swept by the rotor, or

$$\eta_w = \frac{P}{\frac{1}{2}V_\infty^2 \dot{m}} = \frac{nT}{\frac{1}{2}\rho V_\infty^2 V_\infty A} = \frac{nT}{\frac{1}{2}\rho V_\infty^3 (\pi D^2/4)} = \frac{8nT}{\pi\rho V_\infty^3 D^2} \tag{9.1}$$

where $V_\infty$ = the wind velocity in m/sec (ft/sec)
$D$ = the propeller diameter in m (ft)
$n$ = the rotational speed in rad/sec (rev/sec)
$\rho$ = the density of the air in kg/m$^3$ (lb$_m$/ft$^3$)
$T$ = the torque in kg · m (ft · lb$_f$)
$\dot{m}$ = mass flow rate in kg/sec (lb$_m$/sec)

In terms of the parameters used in conventional propeller design, that is, the nondimensional propeller torque coefficient $Q_c = T/\rho V^2 D^3$ and the so-called advance ratio $J = V/nD$, the efficiency is

$$\eta_w = \frac{8}{\pi} \frac{Q_c}{J} \tag{9.2}$$

Experimental data for propellers can be used to obtain a relation between $Q_c$ and $J$ for a given design and twist of the blades, and much of the experience gained with screw propellers can be transferred to the design of windmills. Here, however, we shall not discuss the aerodynamic design of wind generators, but rather concentrate on evaluating the performance under various environmental and meteorological conditions.

To illustrate the important parameters in the performance analysis of wind machines, we will first derive an expression for the maximum efficiency attainable by a windmill. This value is an upper bound for an ideal wind machine operating without any irrotational fluid mechanical losses and is related to a real aerogenerator in the same way as the Carnot efficiency of an ideal heat engine is related to the actual efficiency of a heat engine. The efficiency of a real aerogenerator will then be evaluated empirically and related to the ideal value.

*Free Propeller Analysis*

To develop this simplified theory replace the real propeller by a large number of hypothetical blades having the same diameter and solidity, so that the hypothetical blades will occupy the same percentage of the swept area. In the field of fluid mechanics this is called an actuator. The wind approaches this actuator with a free-stream velocity $V_\infty$ and free-stream static pressure $p_\infty$. Upstream of the actuator the cross-sectional area of the stream tubes passing through the swept area is somewhat smaller than the area of the actuator disk where the axial velocity is $U$. The pressure just upstream of the disk is $p^+$, but since a force is applied to the actuator in order to rotate the blades, the pressure downstream is reduced to $p^-$. Far downstream from the actuator the velocity in the stream tube is $V_2$ and since the streamlines there are straight again, no pressure gradient normal to them can exist; thus, the pressure is again $p_\infty$.

The function of a wind turbine is to extract energy from an air stream and to produce mechanical energy that may be transformed into other forms of energy. In a real system, in addition to the energy extracted, energy losses also occur because of the rotational motion of the fluid that is imparted by the blades and frictional drag. To determine the maximum possible output of a wind turbine, we shall assume that

1. Blades operate without frictional drag.
2. A well-defined slipstream separates the flow passing through the actuator from that outside the actuator disk.
3. The static pressures inside and outside the slipstream tube far ahead of and behind the rotor are equal to the undisturbed free-stream static pressure $(p_2 = p_\infty)$.
4. The thrust loading is uniform over the actuator disk.
5. No rotation is imparted to the flow by the actuator.

Conservation of momentum for the control volume shown in Fig. 9.6 gives for the thrust force $F$

$$F = \dot{m}(V_\infty - V_2) = \rho A U(V_\infty - V_2) \tag{9.3a}$$

Using the pressures in front of and behind the actuator, the thrust can also be expressed as

$$F = A(p^+ - p^-) \tag{9.3b}$$

Applying Bernoulli's equation to flow upstream of the wind turbine, we obtain

$$\tfrac{1}{2}\rho V_\infty^2 + p_\infty = \tfrac{1}{2}\rho U^2 + p^+ \tag{9.4}$$

and, downstream of the wind turbine, we get

$$\tfrac{1}{2}\rho V_2^2 + p_\infty = \tfrac{1}{2}\rho U^2 + p^- \tag{9.5}$$

Subtracting Eq. (9.5) from Eq. (9.4) gives

$$p^+ - p^- = \tfrac{1}{2}\rho(V_\infty^2 - V_2^2)$$

Substituting this into Eq. (9.3b) we obtain the thrust as

$$F = \tfrac{1}{2}\rho A(V_\infty^2 - V_2^2) \tag{9.6}$$

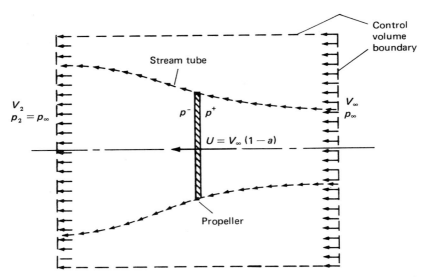

**FIGURE 9.6** Schematic diagram showing control volume, stream tube, velocities, and pressures for idealized aerogenerator analysis.

Equating the thrust from Eq. (9.3$a$) with Eq. (9.6) gives

$$\tfrac{1}{2}\rho A(V_\infty{}^2 - V_2{}^2) = \rho A U(V_\infty - V_2)$$

which shows that $U = (V_\infty + V_2)/2$, that is, that the velocity through the turbine is the average of the wind velocity ahead of the turbine and wake velocity behind the turbine.

An axial induction factor $a$ is defined by

$$U \equiv V_\infty(1 - a) \qquad\qquad (9.7)$$

Using the definition of $a$ to eliminate $U$ in Eq. (9.7) gives

$$V_\infty(1 - a) = \frac{V_\infty + V_2}{2}$$

The wake velocity can then be expressed as

$$V_2 = V_\infty(1 - 2a)$$

Thus, the wake-induced velocity is twice that of the induced velocity in the plane of the rotor, and therefore

$$a = 1 - \frac{V_\infty + V_2}{2V_\infty} \qquad\qquad (9.8)$$

This implies that, when the rotor would absorb all the wind energy, that is, when $V_2 = 0$, then $a$ will have a maximum value of $\tfrac{1}{2}$. Since power $P_w$ is equal to mass flow rate times the change in kinetic energy,

$$P_w = \rho A U \left( \frac{V_\infty{}^2}{2} - \frac{V_2{}^2}{2} \right) = \frac{1}{2}\rho A V_\infty{}^3 4a(1 - a)^2$$

or

$$P_w = 2\rho A V_\infty{}^3 a(1 - a)^2$$

To determine the value of $a$ at which maximum power occurs, we differentiate $P_w$ with respect to $a$ and equate the result to zero:

$$\frac{dP_w}{da} = 2\rho A V_\infty{}^3(1 - 4A + 3a^2) = 0$$

It is apparent that maximum power occurs when $a = \tfrac{1}{3}$ and thus

$$P_{max} = \tfrac{16}{27}(\tfrac{1}{2}\rho A V_\infty{}^3)$$          (9.9)

The power coefficient $C_p$ is defined as

$$C_p = \frac{\text{power}}{\tfrac{1}{2}\rho A V_\infty{}^3}$$          (9.10)

and its maximum value is 0.593 ($\tfrac{16}{27}$), or approximately 0.6. The instantaneous power output of a real windmill $P_w$ can be expressed empirically by the relation

$$P_w = 0.6C_w \left(\frac{V_\infty{}^2}{2}\right)\dot{m} = C_p\left(\frac{\rho A V_\infty{}^3}{2}\right)$$          (9.11)

where $C_w$ is a dimensionless efficiency factor and accounts for frictional and rotational losses.

### Shrouded Actuator

Shrouds or ducts can be used to increase the static thrust of aerogenerator propellers, because a properly designed shroud can reduce the slipstream contraction of a thrusting propeller, regain some of the kinetic energy, and thus increase the thrust-to-power ratio of a stationary propeller. Since the wind velocity of the blade is higher in a shrouded system than on a free propeller, smaller blades can be used and tip losses can be reduced. To calculate the performance of a shrouded propeller it is usual to assume that the flow leaves the shroud exit at free-stream static pressure, so that there is no further change in slipstream velocity and the shroud exit area is the ultimate wake cross section. For a static-free propeller the ultimate slipstream is one-half the propeller area. Thus, any shroud that reduces the final slipstream contraction to less than this value will increase the thrust to power ratio of the system. A shrouded windmill is expected to have superior performance because of the improvement in thrusting propeller performance due to the shroud. A comprehensive analysis of shrouded windmills is given by Lilley and Rainbird (30) and Kogan and Nissim (25).

For a free windmill the optimal wake cross section should be about twice that of the windmill disc. By increasing the optimal wake cross section while keeping the wake axial induced flow at the optimal level of two-thirds the free-stream velocity, the power coefficient will exceed the free rotor limit of 0.593. From a physical viewpoint, a shroud increases the wake expansion, causes more flow to be drawn through the rotor, and thereby increases its power extraction capacity. Unlike the case of the free rotor, a momentum-type analysis cannot be made on this device without assumptions that must be verified experimentally.

Figure 9.7 shows a typical shrouded windmill system. Assuming the mass flow through the system is $\dot{m}$, the power extracted is

$$P = \dot{m}V^2 2a(1 - a)$$          (9.12)

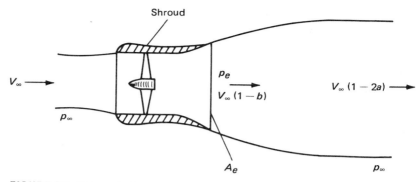

**FIGURE 9.7** Schematic diagram of shrouded aerogenerator.

The force on the entire rotor and shroud system may be written as

$$F = \dot{m} V_\infty 2a \tag{9.13}$$

We note that these equations are not closed because no expression for $\dot{m}$ is available. In the free actuator theory the remaining equation was obtained by stating that the force on the system is the force on the actuator ($F = A\Delta p$), which gives the result $\dot{m} = \rho A V(1 - a)$. For a shrouded system, it is still true that the propeller force is given by $A\Delta p$, but the shroud force cannot be determined by simple momentum theory without knowing the pressure on the outside of the shroud.

The velocity at the shroud exit is shown as $V_\infty(1 - b)$. Assuming, as is done in powered shrouded propeller theory, that the pressure at the shroud exit is free-stream static, $b = 2a$ and the mass flow can be determined as $\dot{m} = A_e V_\infty(1 - 2a)$. Then, basing the power coefficient on the shroud exit area $A_e$, we obtain

$$C_p = \frac{P}{\rho A V_\infty^3 / 2} = 4a(1 - a)(1 - 2a) \tag{9.14}$$

Maximizing this expression gives $C_{p\,\text{max}} = 0.385$ with $a = 0.211$. Writing the power coefficient in terms of rotor area, we get

$$C_p = \frac{4a(1 - a)(1 - 2a)A_e}{A} \tag{9.15}$$

If the shroud-to-rotor-area ratio exceeds 1.54, the power coefficient of the shrouded system, based on rotor area, will exceed that of the free rotor.

At this point all performance characteristics are determined by the assumption made for the flow condition at the duct exit. This can be expressed by writing the power coefficient (based on the exit area) and the shroud exit pressure coefficient

$C_p^*$, which gives

$$C_p = 4a(1 - a)(1 - b) \tag{9.16}$$

and

$$C_p^* = (2a - b)(2 - 2a - b) \tag{9.17}$$

If the exit pressure is lower than free-stream static, which must be the case, the wake expands downstream of the shroud with a higher mass flow and higher power coefficient. Available experimental data (24, 25) suggest that shrouded aerogenerators can achieve better performance than free propellers. However, no large systems have so far been built and tested.

### Yearly Performance

To determine the energy output per unit area swept by the rotor over a year $E_{yr}/A$, the instantaneous output must be integrated over the year. For the ideal case we get

$$\frac{E_{yr}}{A} = 0.3\rho \int_{t=0}^{t=1 \text{ yr}} V_\infty(t)^3 \, dt \tag{9.18}$$

The dependence of the useful energy delivered on the cube of the velocity is noteworthy, because for a real aerogenerator the frictional losses reduce the available energy and a minimum wind speed is required to obtain any power output. Past experience has shown that if a mean wind velocity, averaged over the entire year, were used in Eq. (9.18) and the energy delivered by the ideal machine multiplied by the efficiency factor $C_w$, the predictions for the useful energy delivered can sometimes be larger than the energy actually delivered. The reason for this discrepancy becomes apparent when the variation in wind speed at a given location is plotted as a frequency spectrum, as shown in Fig. 9.8 for a location in Australia, and the power output is plotted as a function of wind speed, as shown in Fig. 9.9 for an aerogenerator with $C_p = 0.5$ and a rotor diameter of 50 m. An appreciable percentage of the kinetic energy in the wind over the year comes at velocities below 4 m/sec, but little useful energy can be delivered at this low wind speed because of frictional losses. At wind speeds above 8 m/sec (18 mi/hr), the unit could deliver 1 MW of power, but a 50-m rotor in such a wind experiences very large stresses that can cause mechanical failure. Past experience in Europe has shown that utilization of wind energy is economically viable only when the mean annual wind velocity is above 4 m/sec, corresponding to an available energy specific density of at least 500 kW · hr/m² · yr. At the same time, however, regions in which very high winds occur occasionally are not suitable for wind power, because all the structural supports must then be designed to withstand the worst wind condition, while the available power output corresponds to much lower wind velocities.

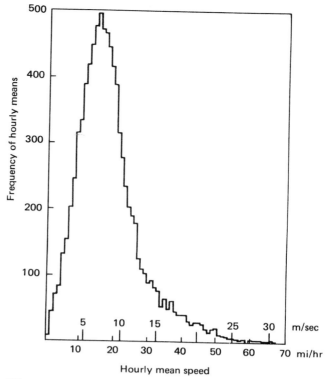

**FIGURE 9.8** Typical annual velocity-frequency diagram of wind [data for South Australia, cited in Ref. (6)]. From Mullett, L. F., *J. Inst. Eng. Aust.*, p. 69, March 1957.

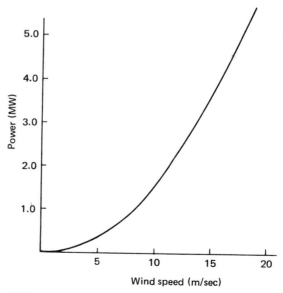

**FIGURE 9.9** Power output versus average wind speed for a 100-m diameter rotor with an efficiency of 50 percent of that of an ideal machine.

## EXAMPLE 9.1

Calculate for two locations the useful power delivered by an aerogenerator having a rotor diameter of 50 m and an efficiency factor of 0.5. In location A the wind energy is approximately constant at 6 m/sec during the year, while in location B the wind energy is 3 m/sec over 6 months and 9 m/sec over the rest of the year.

## SOLUTION

From Eq. (9.11) the power output (with an air density $\rho = 1.2$ kg/m$^3$) is

$$P_w = (0.15)(1.2 \text{ kg/m}^3)\left(\frac{\pi}{4}2500 \text{ m}^2\right)[V_\infty^3 \text{ (m/sec)}^3]\ (\text{sec}^2/\text{m}) = 0.354\ V_\infty^3\ \text{W}$$

For the three wind speeds we get

|  | $V_\infty$ (m/sec) | | |
|---|---|---|---|
|  | 3 | 6 | 9 |
| $[(1/2g_c)\rho V_\infty^2]$ (kg/m$^2$) | 10.8 | 43.2 | 97 |
| $V_\infty^3$ [(m/sec)$^3$] | 27 | 216 | 729 |
| $P_w$ (kW) | 9.7 | 76.5 | 258 |

The total energy delivered is obtained from Eq. (9.18). Using 8760 hr/yr, in location A

$$E_{yr} = \int_0^{1 \text{ yr}} P_w(t)\, dt = 76.5 \text{ kW} \times 8760 = 6.7 \times 10^5 \text{ kW} \cdot \text{hr/yr}$$

while in location B

$$E_{yr} = (9.7)(4380) + (258)(4380) = 11.7 \times 10^5 \text{ kW} \cdot \text{hr/yr}$$

This example illustrates the need for accurate wind data in predicting yearly output.

## Economics of Wind Power

The economics of wind power has characteristics similar to the economics of solar-thermal power. Although the energy source is free, a substantial initial investment is required and the yearly amount of principal and interest, as well as maintenance costs, have to be compared with the dollar value of the useful energy delivered. Hütter (23) has shown that the cost of constructing machines having the same propeller diameter but different installed output capacities has an optimum, so that designs with too small as well as too large a possible output are uneconomical. In Fig. 9.10 the ratio of the cost of manufacture of a wind electric plant to the output of useful energy (in kilowatt-hours) over a 10-yr period is plotted against the energy production density (in watts per square meter of swept area) for two installations, one in a valley with low average wind speeds and one on a mountain top. It can be seen that the economic optimum of the plant size depends on the local wind conditions over the year in a manner similar to the dependence of the optimum collector size on the solar energy available during a year.

Using the similarity of wind frequency distribution, we can plot the useful annual energy output per unit of propeller-swept area as a function of the specific power output (that is, inherent power capacity per unit area) for given values of the annual average wind speed. The results are shown in Fig. 9.11. It is apparent that no advantage is gained by increasing the inherent power capacity per unit of propeller-swept area beyond a certain point, the value of which varies with wind speed. Under normal circumstances in areas with wind speeds between 4 and 6 m/sec it is hardly worthwhile to employ units with a specific output larger than 100 $W/m^2$, because the useful energy output increases at a much smaller rate than the manufacturing cost. Only in selected areas with high average wind speeds in excess of 7 or 8 m/sec are machines with larger inherent power capacity economical. The economic efficiency of a wind power plant will, therefore, be primarily determined by the rotor size, rather than by the built-in power capacity. There are, however, many regions of the world where the wind near ground level averages more than 7 m/sec (15 mi/hr) during the year and wind energy will become an economically viable power source in the near future.

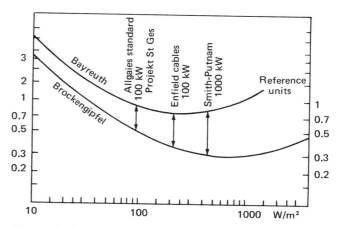

**FIGURE 9.10** Relative cost factor (cost of manufacture over 10 yr of kilowatt output) versus specific power capability for a valley location and a mountain top in Europe. From (23).

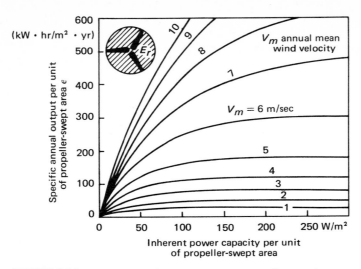

**FIGURE 9.11**  Specific annual power output per propeller-swept area versus specific inherent power capacity for various average wind speeds. From (23).

**TABLE 9.1**  Calculation of Mean Annual Wind Velocity and Mean Annual Energy Wind Velocity[a]

| $n$ | Average velocity | | Measured frequency distribution $f_n$ | Col. 2 × col. 4 (m/sec) | $V_\infty{}^3$ [(knots)³] | 0.1365 × col. 4 × col. 6 [(m/sec)³] |
|---|---|---|---|---|---|---|
| | knots | m/sec | | | | |
| | 2 | 3 | 4 | 5 | 6 | 7 |
| 1 | 0 | 0 | 0.009 | 0 | 0 | — |
| 2 | 1 | 0.515 | 0.023 | 0.118 | 1 | — |
| 3 | 3 | 1.545 | 0.056 | 0.865 | 27 | 0.21 |
| 4 | 5 | 2.575 | 0.091 | 0.234 | 125 | 1.56 |
| 5 | 7 | 3.605 | 0.141 | 0.510 | 343 | 6.61 |
| 6 | 9 | 4.635 | 0.133 | 0.616 | 729 | 13.2 |
| 7 | 11 | 5.665 | 0.145 | 0.821 | 1331 | 26.3 |
| 8 | 13 | 6.695 | 0.116 | 0.778 | 2197 | 34.9 |
| 9 | 15 | 7.725 | 0.080 | 0.618 | 3375 | 36.8 |
| 10 | 17 | 8.755 | 0.062 | 0.545 | 4913 | 41.5 |
| 11 | 19 | 9.785 | 0.050 | 0.490 | 6859 | 46.8 |
| 12 | 21 | 10.815 | 0.037 | 0.400 | 9261 | 46.6 |
| 13 | 23 | 11.845 | 0.025 | 0.296 | 12167 | 41.2 |
| 14 | 25 | 12.875 | 0.014 | 0.180 | 15625 | 29.9 |
| 15 | 28 | 14.420 | 0.018 | 0.260 | 21952 | 53.8 |
| $\sum_{n=1}^{n=15}$ | | | 1.000 | 5.84 | | 381.0 |

[a]Velocities at 146-ft level of WKY–TV tower.

## EXAMPLE 9.2

The wind frequency distribution measured on top of a 146-ft-tall television tower in Oklahoma is given in Table 9.1. Determine:

1. The mean annual wind velocity.
2. The annual mean energy wind velocity.
3. The average power per square meter for the maximum power capacity and for a power capacity of 0.4 kW/m$^2$.
4. The energy output as a function of propeller diameter assuming that only winds above 3 m/sec are usable and that the combined efficiency for the aerodynamic, mechanical, and electrical components is 75 percent.
5. The maximum permissible cost of the installation to compete with electricity at \$0.5/kW $\cdot$ hr, assuming a life of 20 yr and an interest rate of 8 percent. Also assume that no storage is required, that the windmill has a diameter of 18 m, and that operating expenses such as maintenance are 50 percent of the capital investment expense.

## SOLUTION

1. To determine the mean annual wind velocity multiply the wind speed, given in column 2 of Table 9.1, by the percentage of time during the year that this wind speed was observed, given in column 4. This result is shown in column 5. Adding the velocities gives the mean annual wind speed

$$\bar{V} = \sum_{n=1}^{n=15} V_n f_n = 5.84 \text{ m/sec}$$

2. The annual mean energy wind velocity is obtained from the relation

$$\bar{V}_a = \left( \sum_{n=1}^{n=15} f_n V_n{}^3 \right)^{1/3} = \sqrt[3]{381} = 7.25 \text{ m/sec}$$

3. To obtain the average power per square meter of swept area use Eq. (9.11). With $\rho = 1.22$ kg/m$^3$ and $C_w = 0.75$,

$$\frac{P_w}{A} = (0.6) \left( 0.75 \frac{\bar{V}_a{}^3}{2g_c} \right) = 0.105 \text{ kW/m}^2$$

The above is true only if the generating system is capable of absorbing all the wind energy at the maximum wind speeds. This would require the capacity of handling winds up to 14.4 m/sec, or a power density of $(0.0037)(14.4)^3 = 1.1$ kW/m$^2$. Such a large power density would be excessive from a cost-effectiveness standpoint, because the stresses on the blades and the load on the supporting structure would be very large, and there is not enough power in wind at this velocity during the year to justify the investment. Using a more realistic installed capacity of 0.4 kW/m² for the "ideal" machine, only winds below 10.25 m/sec will be usable. From the tabulation, the annual mean energy is then reduced to 6.8 m/sec and the average power per square meter is reduced to 0.117 kW. Assuming that only winds above 3 m/sec are used will merely reduce the available power from 0.116 to 0.115 kW because very little power could be extracted at these low velocities even by an ideal machine.

From the above, the yearly power density output (8760 hr/yr) is $(0.115)(8760)(0.82)(0.75) = 625$ kW · hr/m² · yr. The factor 0.82 derives from the fact that only 82 percent of the time wind velocities fall between 3 and 10.25 m/sec during the year.

4. On the basis of the above calculations of power density, the power output for windmills with various diameters can be obtained as shown in Table 9.2.

5. From the above tabulation, the output of an 18-m diameter windmill would be 159,000 kW · hr/yr. At 8 percent interest over 10 yr, the yearly cost is 10 percent of the initial capital invested; to this cost must be added the maintenance cost. Thus, the acceptable initial investment to compete over 20 yr with electrical cost at $0.03/kW · hr that would be economically justifiable is

$$\text{Initial investment} = \frac{(159,000 \text{ kW} \cdot \text{hr/yr})(0.03 \text{ \$/kW} \cdot \text{hr})}{0.15 \text{ percent/yr}} = \$31,000$$

This cost seems to be within the realm of possibility once aerogenerators are in mass production.

**TABLE 9.2** Power Output for Windmills of Various Diameters

| Diameter | | Swept area (m²) | Installed capacity (kW) | Energy output (kW · hr/yr) |
|---|---|---|---|---|
| m | ft | | | |
| 4 | 13.1 | 12.5 | 5.0 | 7,900 |
| 6 | 19.7 | 28.2 | 11.3 | 17,600 |
| 8 | 26.2 | 50.2 | 20.1 | 31,400 |
| 10 | 32.8 | 78.5 | 31.0 | 49,000 |
| 12 | 39.4 | 113.0 | 45.2 | 70,500 |
| 14 | 46.0 | 153.9 | 61.5 | 96,000 |
| 16 | 52.5 | 201.0 | 80.4 | 126,000 |
| 18 | 59.0 | 254.3 | 102.0 | 159,000 |
| 20 | 65.6 | 314.0 | 125.8 | 196,000 |
| 22 | 72.2 | 380.0 | 152.2 | 237,000 |
| 24 | 78.7 | 452.0 | 181.1 | 283,000 |

As with solar-thermal conversion, the intermittency of the wind poses special problems in storage and distribution. When aerogenerators are used for pumping water, such as for the dikes of Holland, the intermittency does not create a problem because the pumping can be done whenever the wind blows. The situation is quite different, however, when wind energy is used to generate electric power. Until now, when aerogenerators were used to generate electricity, the energy was stored in lead-acid batteries. But this is not a viable solution for the large-scale use of wind energy, because the world supply of lead is limited and the cost is prohibitive.

It has been suggested that wind generators be combined in a large electric power grid that could even out local fluctuation in wind. This scheme, used in combination with nuclear or solar power, appears feasible and is economically effective, because the distribution grid would act as storage. There are at present several studies under way to evaluate this scheme in more detail.

A promising storage scheme suggested by Bergey (5) utilizes a combination of a windmill and a water turbine. Water could be pumped to a high reservoir during periods of high wind and the potential energy could then be used during periods of low wind. Assuming a constant output from the wind-water generating system as the basis for sizing the water generator and the storage system, the surplus and deficit can be calculated as shown below.

Assuming an overall efficiency of 50 percent for the conservative case in which no supplemental water is available, it is possible to approximate the level output for the 18-m mill at 0.068 kW/m² of blade area or approximately 600 kW · hr/yr. The 50 percent efficiency is intended to include losses from evaporation as well as pumping and generating.

Table 9.3 shows the wind generator output throughout the velocity range and the weighted deficit or surplus at each wind velocity.

This wind power deficit or surplus is in agreement with the assumed conversion efficiency of approximately 50 percent. Thus, the water generator would be operating at wind speeds below 6.2 m/sec to make up the power deficit. At higher wind speeds, the excess wind generator capacity would be used to pump water to the reservoir. Hence, the total annual output of the combined system would be: $(0.068)(8760)(254.3) = 152,000$ kW · hr, which represents an overall loss in efficiency of only 4.5 percent when compared with the 159,000 kW · hr available if all wind power were fed into a power grid.

However, the water generator capacity would have to be at least $(0.068)(254.3) = 17.3$ kW compared with the installed capacity of the wind generator of $(0.400)(254.3) = 102.0$ kW. Provisions for water storage and the duplication in generators, control, and switching circuits would add substantially to the cost of the basic wind generator installation.

Another frequently proposed application of wind power is provision of electricity for individual homes. Krenz (26) estimates that the average per capita residential consumption rate for electrical energy is approximately 300 W. A typical family of four thus uses an average of 1.2 kW (28.8 kW · hr/day or 864 kW · hr/month). Assuming a wind speed of 10 m/sec, a 75 percent efficiency, and a

**TABLE 9.3**  Weighted Wind Power Deficit or Surplus at Various Wind Velocities

| Wind velocity | | | | Deficit or | Weighted |
|---|---|---|---|---|---|
| k | m/sec | Frequency | Output (kW/m$^2$) | surplus$^a$ (kW/m$^2$) | deficit or surplus (kW/m$^2$) |
| 0 | – | 0.009 | 0 | −0.0680 | −0.00061 |
| 1 | 0.515 | 0.023 | 0 | −0.0680 | −0.00156 |
| 3 | 1.545 | 0.056 | 0 | −0.0680 | −0.00381 |
| 5 | 2.575 | 0.091 | 0 | −0.0680 | −0.00619 |
| 7 | 3.605 | 0.141 | 0.0130 | −0.0550 | −0.00775 |
| 9 | 4.635 | 0.133 | 0.0275 | −0.0405 | −0.00539 |
| 11 | 5.665 | 0.145 | 0.0505 | −0.0175 | −0.00254 |
| 13 | 6.695 | 0.116 | 0.0831 | +0.0151 | +0.00175 |
| 15 | 7.725 | 0.080 | 0.1280 | +0.0600 | +0.00480 |
| 17 | 8.755 | 0.062 | 0.1865 | +0.1185 | +0.00735 |
| 19 | 9.785 | 0.050 | 0.2600 | +0.1920 | +0.00960 |
| 21 | 10.815 | 0.037 | 0.3500 | +0.2820 | +0.01040 |
| 23 | 11.845 | 0.025 | 0.4000 | +0.3320 | +0.00830 |
| 25 | 12.875 | 0.014 | 0.4000 | +0.3320 | +0.00465 |
| 28 | 14.420 | 0.018 | 0.4000 | +0.3320 | +0.00600 |

$\Sigma = -0.02785$ (for k = 0 through 11)

$\Sigma = +0.05285$ (for k = 13 through 28)

$^a$Annual deficit = (0.02785)(8760)(254.3) = 62,000 kW · hr; annual surplus = (0.05285)(8760)(254.3) = 118,000 kW · hr.

wind factor of one-fourth, a windmill of modest size (5 m in diameter) would be required. If energy is to be stored in lead-acid batteries, and 4 days' storage capacity is deemed necessary, approximately 120 kW · hr of storage battery capacity is required. A lead-acid battery can be charged and discharged about 1000 complete cycles during a typical battery life. Electricity stored by a 1-kW · hr battery costing $50 would thus cost $0.05/kW · hr, based on this typical battery life. The cost of the windmill must be added to the investment in batteries. For the idealized case considered previously, a unit capable of delivering 10,000 kW · hr/yr would be competitive with electric energy at $0.06/kW · hr at an initial cost of

$$C_0 = \frac{(10,000 \text{ kW} \cdot \text{hr})(\$0.06/\text{kW} \cdot \text{hr})}{0.15} = \$4,000$$

This is not unreasonable, but the total cost of power, including storage, would then be $0.11/kW · hr. Thus, small-scale windmill-battery storage systems appear to be economically viable only in areas where other forms of energy are very expensive or not obtainable.

## Wind Power Climatology of the United States

In order to estimate the economic viability of wind power, it is necessary to know the available wind power in the region where wind machines are to be installed. As shown previously, the power available theoretically in a stream tube of unit area is equal to $\frac{1}{2}\rho V_\infty^3$, where $V_\infty^3$ is the wind speed and $\rho$ the air density. To evaluate sites for possible wind energy conversion it is necessary to know the geographic distribution of wind power. Such a wind atlas for the United States was prepared at Sandia Laboratories by Reed (43) using the data available in the archives of the National Climatic Center. All data were obtained by means of a "standard anemometer," 10 m (33 ft) above ground level and far from any obstructions. Results of this study are presented in Fig. 9.12, which gives the annual average of the available wind power. Areas of high wind power usually have their highest winds in the winter and much less wind power in the summer and fall seasons (43).

In addition to geographic location, the height above ground affects the wind power density. In the atmospheric boundary layer above the ground, wind speed increases with height in a complex way. The wind profile depends on the thermodynamic stability of the air mass and on the roughness of the terrain. For long-time averages, however, it is often assumed that wind speed is proportional to the one-seventh power of height above ground. Figure 9.13 shows a comparison of the experimental data for two meteorological research towers at Albuquerque, New Mexico, and Hanford, Washington, with this approximate relationship. In this figure the tower data wind power is plotted against height. The relation of wind power with height above atmosphere follows the three-sevenths slope on a log-log plot. It is therefore reasonable to estimate wind power as a function of height with the so-called one-seventh law for wind speed. In very rough terrain, however, this approximate relationship may not be reliable.

Radice (42) extended Reed's siting study for wind machines by investigating not only the areas with high wind power but also sites which were in close proximity to user needs. To construct a site-selection map, Radice computed power factors. The yearly average wind power map was divided into a grid network, each grid measuring 5° latitude by 5° longitude. A power factor for each grid was obtained by multiplying the average wind power for the grid by the area fraction of the standard metropolitan statistical areas (SMSAs) lying within the grid. The average grid wind power was obtained by weighting the yearly wind power by area. Thus, the power factor $PF$ was computed by the following equation:

$$PF = \frac{\Sigma(PY_i A_i)}{A_t^2} A_s \qquad (9.19)$$

where $PF$ = power factor in $W/m^2$
  $PY_i$ = yearly average wind power of power level $i$ in $W/m^2$
  $A_i$ = area included within the $i$th power level contour
  $A_t$ = total area of grid
  $A_s$ = sum of the SMSAs contained within the grid
Grids with power factors larger than 20 $W/m^2$ were considered as optimal sites for placement of wind-driven power-conversion devices.

The results of Radice's analysis are summarized in Fig. 9.14. On the basis of this study, it appears that the most favorable locations for the use of wind power

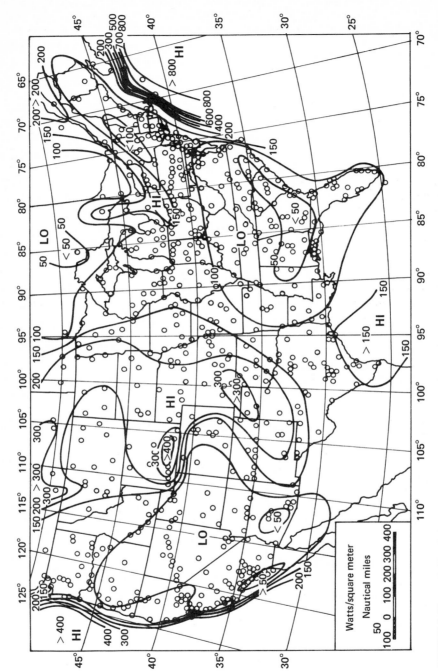

**FIGURE 9.12** Available wind power, annual average. From (43).

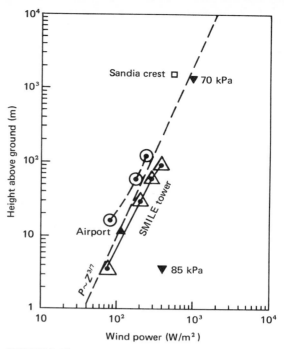

**FIGURE 9.13** Annual average wind power versus height.
△ Albuquerque, New Mexico; ⊙ Hanford, Washington.

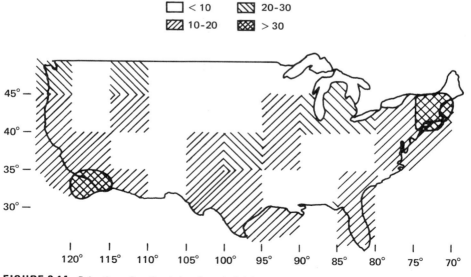

**FIGURE 9.14** Selection of optimal sites for wind-driven apparatus. From (42).

are in the northeastern and southwestern parts of the United States. It should be noted that offshore sites, which tend to have the highest wind power but not always the optimal power factors, have the advantage of not occupying valuable land area. However, a careful feasibility and economic evaluation of offshore stations must be made before such sites can be considered for development of wind power.

## OCEAN THERMAL GRADIENT AND OCEAN WAVE POWER

In this section two natural solar conversion schemes that utilize the ocean as receiver, storage, and transmitter of solar energy are discussed. The first scheme relies on absorption of solar radiation to heat the upper layers of the ocean and thereby create a temperature differential across which a conventional heat engine can operate. The second scheme relies on the transmission of wind energy to create waves whose energy can be utilized to produce power.

### Ocean Thermal Energy Conversion System

Tropical oceans have a temperature difference between the surface and deep water on the order of 20 K. Figure 9.15 shows a typical ocean temperature profile. The surface layer of the ocean, which is about 200 m deep, is heated by the sun and stays at a temperature of about 298 K. At lower depths the water comes from the arctic regions and its temperature can be as low as 278 K. The possibility of using this temperature difference to drive a thermal engine has seemed attractive for many years because the magnitude of this natural solar resource is extremely large. Moreover, there is no need to build a collector system since the oceans themselves act not only as natural collectors for solar energy but also as thermal storage vaults. Unfortunately, although the amount of energy available from ocean gradients is large, there are considerable practical problems in the conversion of this resource into useful work. The available temperature difference is small and consequently the thermal conversion efficiency of the power cycle is low. This means that very large areas are necessary in order to transfer the heat to the working fluid of the cycle; huge heat exchangers are therefore required. Other problems are related to the corrosive action of seawater and the difficulty in achieving stable mooring of a plant in a storm-prone environment. Also, except for the Gulf of Mexico and the Gulf Stream near Florida, ocean sites with large thermal gradients are far removed from the electric energy demand centers in the United States.

A 40-kW ocean thermal energy conversion system (OTEC) was built by Claude on the coast of Cuba in the late 1920s (11). He succeeded in erecting a power plant capable of delivering useful work, but he experienced considerable practical difficulties primarily because he used an open thermal cycle in which the seawater was the working fluid and the dissolved air in the water thus could not be easily removed. Another problem was that the power plant was placed on land. This location necessitated the use of a long cold water inlet pipe, which was subsequently wrecked in a storm.

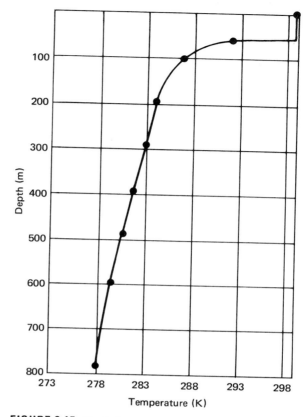

**FIGURE 9.15**   Typical ocean temperature profile.

A method to overcome some of the difficulties encountered by Claude was suggested by Anderson and Anderson in 1965 (2). They proposed using a closed Rankine cycle with an organic fluid as the working fluid, as shown in Fig. 9.16, with a vertical pipe to bring the cooling water to the condenser. Almost all subsequent work on OTECs has been based on that scheme. Figure 9.17 is a sketch of one scheme proposed for an OTEC power plant mooring system.

The operation of the solar sea power plant shown in Fig. 9.16 is quite simple in concept. Warm seawater is pumped through the boiler (or evaporator) where the working fluid for the closed Rankine cycle is vaporized. The working fluid could be ammonia, propane, or an organic fluid such as refrigerant R-12 or R-31. The high-pressure vapor passes through the turbine where shaft work is extracted. The condenser is maintained at the lowest possible temperature by pumping deep cold water at about 278 K through it. After condensation in the condenser the liquid working fluid is pumped back into the evaporator and the cycle is closed.

A solar sea power plant is inefficient because of the small temperature

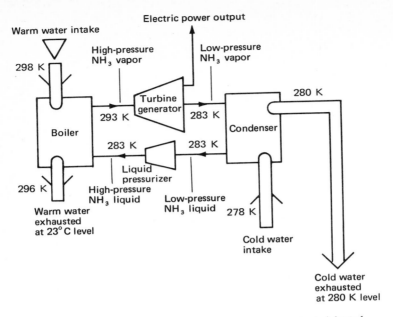

**FIGURE 9.16** The basic components of a solar sea power plant. Adapted from (29).

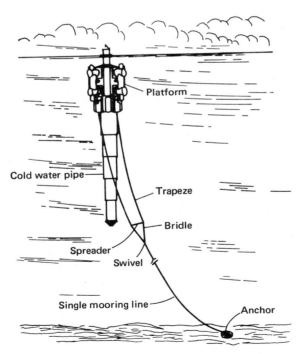

**FIGURE 9.17** Schematic of Lockheed OTEC mooring system.

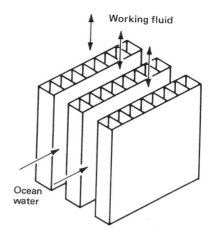

**FIGURE 9.18** Schematic of plate-fin heat exchanger. From (42).

difference available. Its efficiency is on the order of 3 percent compared with 35 percent for a modern coal-fired steam power plant. Solar sea power boilers must therefore deliver 10 times as much heat as the boiler in a conventional plant, and this requires 10 times more surface area in the boiler tube core as well as in the condenser. There is, fortunately, a compensating factor. The vapor pressure of liquids rises exponentially with temperature, and since sea power plants must operate at lower temperatures, they also operate under lower pressures. This permits construction of the boiler and condenser heat exchangers with thinner tubes and less materials (see Fig. 9.18). The design and the cost of such huge heat exchangers is probably the key problem for the successful implementation of an OTEC power plant.

Except for a small land-based demonstration, no working model of an OTEC power plant has been built to date, but several detailed design studies have been completed. Table 9.4 shows the design site locations assumed by five major investigation teams. The TRW and the Lockheed studies are based on essentially current engineering know-how and state-of-the-art practice whereas the University of

**TABLE 9.4** Design Site Location and Characteristics of Various Investigators

| Investigator | Location | Temperature difference (°C) | Depth of cold water inlet (or length of cold water pipe) (m) |
|---|---|---|---|
| SSPI | Off coast of Florida or Carribean | 20 | 610 |
| UM | Gulf Stream off Miami | 17.8 | 340 (320) |
| CMU | Tropical (within 20° of equator) | 22.2 | 610 |
| TRW | Pacific Ocean | 21.8 | 1220 |
| Lockheed | Open ocean (south of Hawaii) | 18.5 | 490 (305) |

Massachusetts (UM), Carnegie-Mellon University (CMU), and Sea Solar Power, Inc. (SSPI) investigations are based on novel and as yet commercially unavailable heat-exchanger systems. One of the proposed heat exchangers, a compact cross flow, is shown in Fig. 9.18. Table 9.5 lists the baseline design data for the TRW and Lockheed studies while Table 9.6 lists the key items for the OTEC system proposed by the UM team, which is by far the most optimistic (20).

The two major subcomponents in all the proposed systems are the heat exchangers (evaporator and condenser). Because of the small temperature difference and the large volumes of seawater that must pass through these exchangers, they are large. In their design particular attention must be paid not only to thermal phenomena but also to all of the internal pressure drops. This point is illustrated in Fig. 9.19 from the UM design study, which shows how the heat exchanger volume

**TABLE 9.5** Comparison of TRW and Lockheed OTEC Baseline Designs

| System item | TRW | Lockheed |
|---|---|---|
| | **Designs** | |
| | **Characteristics** | |
| Power plant and module size | 100 MWe net (125 MWe gross) 4 modules of 25 MWe net | 160 MWe net (240 MWe gross) 4 modules of 40 MWe net |
| Working fluid | Ammonia | Ammonia |
| Evaporators | 4 shell-and-tube units (15.2-m diameter) 65,400 titanium tubes (13.1 m long, 0.035-m OD) | 4 shell-and-tube units (22.2-m diameter) 120,000 titanium tubes (16 m long, 0.051-m OD) |
| Condensers | 4 shell-and-tube units (15.2-m diameter) 75,900 titanium tubes (13.1 m long, 0.038-m OD) | 4 shell-and-tube units (22.2-m diameter) 120,000 titanium tubes (16 m long, 0.051-m OD) |
| Turbine-generator | 4 units—6.4-m-diameter vertical shaft, diffuser mounted on condenser | 8 units—3.5-m-diameter vertical shaft, diffusers (in pairs) mounted on condensers |
| Hull | Cylindrical surface vessel, 104-m diameter, 52 m high, 32-m draft, reinforced and prestressed concrete | Semisubmerged, spar-type platform, 75-m maximum diameter, 180 m high, main hull 30.5 m below surface (4 separate power modules), reinforced and prestressed concrete |
| Cold water pipe | Fiberglass-reinforced plastic, 15.2-m ID, 220 m long | Prestressed concrete, 305 m long, 5 telescoping sections (each 61 m long), 10 ranging from 31.1 to 38.4 m |
| Mooring-positioning | Dynamic positioning system, warm water effluent manifolded in two steerable nozzles, cold water effluent manifolded in one fixed nozzle | Single-point mooring (up to 6100-m depth) consists of trapeze, spreader, bridle, swivel, mooring line, and anchor |

**TABLE 9.6**  Key Items of the OTEC System Proposed by the
University of Massachusetts

| Item | Characteristics |
|---|---|
| Power plant and module size | 400 MWe net (480 MWe gross) 16 modules of 25 MWe net |
| Working fluid | Propane |
| Evaporators | 16 operational (2 spares) plate-fin units (9.2 m high, 27.5 m wide, 8.8 m deep) 90-10 Cu/Ni, 2.54-mm walls, 1.27-mm fins |
| Condensers | 16 plate-fin units (8.7 m high, 17.3 m wide, 8.8 m deep) 90-10 Cu/Ni, 2.54-mm walls, 1.27-mm fins |
| Turbine-generator | 16 units, 13.5-m diameter, horizontal shaft |
| Hull | Submersible–twin cylindrical hull, reinforced concrete with copper nickel sheathing, evaporators in separate modules above hull |
| Cold water pipe | Aluminum–neutrally buoyant, elliptical–26.4-m hydraulic diameter, hinged between hulls, 340 m long |
| Mooring | Single point mooring–neutrally buoyant tether (using 18–9.5-m diameter HT monel clad steel sphere links with buoyancy chambers) |

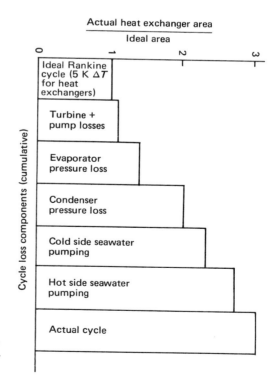

**FIGURE 9.19**  Effect of cycle
irreversibilities on relative heat-
exchanger size.

must be increased to compensate for the thermal irreversibilities in the real cycle as compared to an ideal Rankine cycle.

## EXAMPLE 9.3

A 100-MW ocean thermal gradient power plant is to be designed for a location where the ocean surface temperature is 300 K (80°F) and water at a lower depth is available at 278 K (40°F). If the heat exchangers are sized to operate a power plant using R-22 as the working fluid between 294 and 284 K (70 and 50°F), calculate the flow rates of R-22 and water required, assuming that the condenser heat exchangers have an overall conductance of 1000 W/m² · K (176 Btu/hr · ft² · °F) and an effectiveness of 100 percent. Also calculate the plant efficiency and the surface area of the condenser. The saturation pressure of R-22 at 294 K is $9.38 \times 10^5$ N/m² (136.1 psia), the enthalpy is 256.8 kJ/kg (110.4 Btu/lb$_m$), and the entropy is 0.900 kJ/kg · K (0.215 Btu/lb$_m$ · °R). The enthalpy of the saturated liquid at 284 K is 56.45 kJ/kg (24.27 Btu/lb) and the heat of vaporization is 197.0 kJ/kg (84.68 Btu/lb$_m$). The saturation pressure at 284 K is $6.81 \times 10^5$ N/m² (98.73 psia), while the entropy at 284 K of the saturated liquid is 0.217 kJ/kg · K (0.0519 Btu/lb$_m$ · °R) and the change in entropy during change of phase from liquid to vapor is 0.695 kJ/kg · K (0.166 Btu/lb$_m$ · °R). Assume efficiencies for the pump and turbine are 80 and 90 percent, respectively. The specific volume of the saturated R-22 liquid at 284 K is 0.000799 m³/kg (0.0128 ft³/lb). A schematic diagram of the system is shown below.

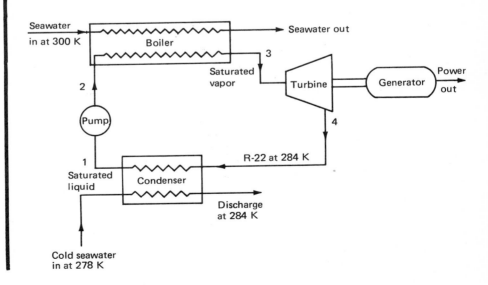

## SOLUTION

For an isentropic expansion through the turbine the entropy at points 3 and 4 is the same. Denoting the quality of the working fluid by $x$ we can set the entropies equal:

$$s_3 = 0.900 = s_{f4} + x_4 h_{fg} = 0.217 + 0.695x$$

Thus,

$$x = \frac{0.683}{0.695} = 0.98$$

and the enthalpy of the fluid leaving the turbine is

$$h_4 = 56.45 + x(197.0) = 56.45 + (0.98)(197.0) = 249.5 \text{ kJ/kg}$$

The turbine output $W_T$ is $\eta_t(h_3 - h_4)$ or

$$W_T = 0.9(256.8 - 249.5) = 6.57 \text{ kJ/kg (2.83 Btu/lb}_m)$$

The pumping power $W_p$ required between points 1 and 2 is

$$W_p = \frac{v \Delta p}{\eta_p} = \frac{(0.0799 \times 10^{-2})[(9.38 - 6.81) \times 10^5]}{0.80}$$

$$= 0.257 \text{ kJ/kg (0.11 Btu/lb}_m)$$

The net power output of the plant is

$$W_{net} = W_T - W_p = 6.57 - 0.26 = 6.31 \text{ kJ/kg (2.71 Btu/lb}_m)$$

For a 100-MW plant the mass flow rate of the working fluid is

$$\dot{m} = \frac{(100 \text{ MW})(10^5 \text{ kJ/sec})(3600 \text{ sec/hr})}{6.31 \text{ kJ/kg}}$$

$$= 5.7 \times 10^7 \text{ kg/hr (or } 1.2 \times 10^8 \text{ lb}_m/\text{hr)}$$

The total heat input in the boiler equals

$$Q_{in} = h_3 - h_2 = 256.8 - (56.45 + 0.26) = 200 \text{ kJ/kg}$$

so that the efficiency is

$$\eta = \frac{W_{net}}{Q_{in}} = \frac{6.31}{200} = 3.16 \text{ percent}$$

To determine the size of the heat exchanger, calculate the rate of heat transfer in the condenser $q_c$

$$q_c = \dot{m}(h_4 - h_1) = (5.7 \times 10^7 \text{ kg/hr})[(249.5 - 56.45) \text{ kJ/kg}]$$

$$= 1.1 \times 10^{10} \text{ kJ/hr} (1.04 \times 10^{10} \text{ Btu/hr})$$

However,

$$q_c = UA \, \Delta T_{mean} = A(1.0 \text{ kJ/sec} \cdot \text{m}^2 \cdot \text{K})(3 \text{ K})$$

Thus

$$A = \frac{1.1 \times 10^{10} \text{ kJ/hr}}{(3 \text{ kJ/sec} \cdot \text{m}^2)(3600 \text{ sec/hr})} = 1,020,000 \text{ m}^2 \ (11,000,000 \text{ ft}^2)$$

In addition to this large heat exchanger, provisions must be made to pump cold seawater at a rate of approximately $4.4 \times 10^8$ kg/hr ($10^9$ lb$_m$/hr) from a considerable depth to the surface and then through the condenser. Although the costs of these tasks are high and create serious questions regarding the economic viability of OTEC systems, one must remember that the 1977 projections of capital cost for nuclear fission power plants to be operational in 1985 are about $1,250/kWe delivered. It has been proposed that a full-sized OTEC system be designed and built in no more than 6 yr (28); once such a plant is actually operating, performance and cost estimates will be subject to considerably less uncertainty.

Heat-exchanger configurations proposed to date fall into three basic categories:

1. Conventional shell-and-tube designs (TRW and Lockheed)
2. Shell-and-tube design with augmenting heat-transfer surfaces or fins (CMU)
3. Plate-fin of sandwich construction (compact cross-flow-type) as shown in Fig. 9.18 (UM and SSPI)

In addition to these basic designs, various heat-exchanger materials have been recommended (titanium, aluminum, copper-nickel, or plastic). Also, ocean water fouling controls and different locations relative to the ocean surface and other cycle components have been considered.

Since no practical system of any size has ever been built, the cost and performance estimates of OTEC systems depend on the assumptions made for the analysis. Table 9.7 gives a comparison of the thermal cycle parameters of the key

**TABLE 9.7**  Comparison of Thermal Cycle Parameters of Various Investigators

| Parameter | Investigator | | | | |
|---|---|---|---|---|---|
| | SSPI | UM | CMU | TRW | Lockheed |
| Net power output (MW) | 100 | 400 | 100 | 100 | 160 |
| Gross power output (MW) | 125 | 480 | 140 | 125 | 240 |
| No. of modules | 4 | 16 | 4 | 4 | 4 |
| Working fluid | R-12/41 | Propane | Ammonia | Ammonia | Ammonia |
| Specific working fluid flow rate (kg/sec · MW net) | 234 | 116 | 34.5 | 32.6 | 38.4 |
| Site $\Delta T$ (°C) | 20.0 | 17.8 | 22.2 | 21.8 | 18.5 |
| Working fluid $\Delta T$ (°C) | 10.0 | 9.4 | 11.1 | 11.3 | 12.0 |
| Turbine efficiency (percent) | 90 | 90 | 90 | 90 | 90 |
| Net cycle efficiency (percent) | 2.3 | 2.4 | 2.3 | 2.4 | 2.1 |
| Cold water flow (m³/sec) | 330 | 1470 | 430 | 370 | 1800 |
| Cold water $\Delta T$ (°C) | 3.3 | 2.6 | 2.0 | 2.7 | 1.0 |
| Hot water flow (m³/sec) | 620 | 4520 | 515 | 440 | 1370 |
| Hot water $\Delta T$ (°C) | 1.7 | 0.9 | 2.0 | 2.3 | 1.35 |

investigations and Fig. 9.20 shows estimates of capital costs. The cost estimates vary over a wide range, but the major variable is the cost of the heat exchanger. OTEC plants in the tropics can operate with power-generating parameters more favorable than those in the Gulf Stream. In the tropics the heat-exchanger area required is only one-half to one-fourth the size per kilowatt delivered than in the Gulf Stream, but OTEC plants in those areas would require much more expensive energy delivery systems (31).

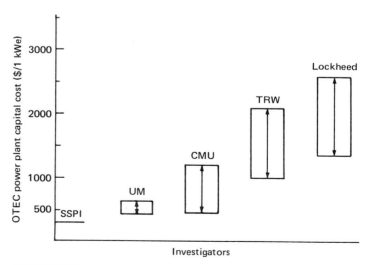

**FIGURE 9.20**  Estimates of capital cost.

## Ocean Wave Power

Oceans, in addition to storing solar radiation as thermal energy in their upper layers, store energy from the sun and the winds in their waves. The idea of converting this kind of natural solar energy is quite old, but no serious research effort was devoted to the development of wave power conversion systems until 1974 when a study of energy options for the United Kingdom elevated wave energy to the third priority, after nuclear power and coal. A one-tenth-scale unit has been designed and became operational in 1977. Since no actual operating experience is yet available, fundamental principles and preliminary design will be treated here.

The amount of energy in a wave train can be estimated by calculating the change in potential energy as water in a wave above the level of the sea falls into the trough in front of the wave. If the density of seawater is $\rho$, then the mass of water in the half-sinusoid above sea level per unit width of a wave with wavelength $\lambda$ is

$$\text{Mass} = \rho \, \frac{\lambda}{2} \frac{H_{tc}}{2\sqrt{2}} \tag{9.20}$$

using the notation shown in Fig. 9.21. The center of gravity is a distance $H_{tc}/2\sqrt{2}$ above sea level and falls to an equal distance below at its lowest point. The change in potential energy as a result of this motion is therefore

$$\Delta PE = g\rho \, \frac{\lambda}{2} \left( \frac{H_{tc}}{2\sqrt{2}} \right)^2 = \frac{\rho g H_{tc}^2}{16} \tag{9.21}$$

It can be shown [see Ref. (27)] that the frequency of gravity waves in deep water is $(1/\lambda)(g\lambda/2\pi)^{1/2}$; therefore, the rate of transfer of potential energy, which in the ideal case is equal to the power $P$, is given by the relation

$$P = \frac{\rho g H_{tc}^2 (g\lambda)^{1/2}}{16\sqrt{2\pi}} \tag{9.22}$$

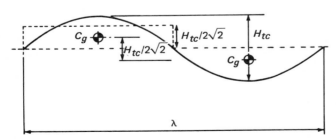

**FIGURE 9.21** Part of a progressive train of sinusoidal gravity waves in deep water. $\lambda$, wavelength; $H_{tc}$, height from trough to crest; $C_g$, center of gravity.

As waves progress along the sea they transport energy, and if this rate of energy transport is measured as it passes along a stationary line, one can specify the power density per meter of frontage in kilowatts per meter.

In a real case the amount of power available depends on the average height of the waves. Oceanographic measurements usually report the period of the waves and a significant height $H_s$, defined as the average height of the highest one-third of all the waves. Salter (45) has shown that $H_{tc}$ is related to $H_s$ by

$$H_{tc}^2 = \frac{H_s^2}{2} \tag{9.23}$$

and that $\lambda$ can be related to $T$ by

$$\lambda = \frac{T^2 g}{2\pi} \tag{9.24}$$

Thus the power from a continuous wave train of width $W$ is also given by

$$P = \frac{W\rho g^2 T H_s^2}{64\pi} \tag{9.25}$$

Salter (45) analyzed all the waves observed during 13 yr at a station near the Hebrides (Station India, 59°N, 19°W) and his results are summarized in Fig. 9.22. This figure shows the number of occurrences, in parts per thousand, of waves of given height and period. Applying Eq. (9.21) to each cell in Fig. 9.22 and summing all the power outputs, Salter found that the average power over a whole year is 77 kW/m of frontage. Model tests in a water tunnel showed that the fraction of energy that can be extracted depends on the depth of the installation. Above a depth $d$, this fraction is $[1 - \exp(-8\pi^2 d/T^2 g)]$. The available power can be calculated for various depths, with each cell in the array containing the contribution to that depth. Figure 9.22 shows that some of the power comes from very powerful waves. To avoid unduly large stresses in a practical installation the system needs to be underdriven, that is, submerged, during the most severe weather. Table 9.8 shows the annual power density for combined depth and power limitations; Fig. 9.23 shows graphically the annual power density as a function zero-crossing period $T_z(s)$.

The key problem in extracting power from ocean waves is to find a method of converting dispersed, random, and alternating forces into a concentrated direct force with a mechanism that is efficient at low power levels and that can withstand large waves. This makes a rigid connection to the seabed essentially impossible for a large installation. The installation should be floating, much of the equipment must be submerged below the surface to avoid concentration in stresses, and power should be smoothly extractable. These requirements suggest that the use of an object with an up-and-down motion would be less suitable than one with a to-and-fro motion. One shape that meets these requirements is shown schematically in Fig. 9.24. It

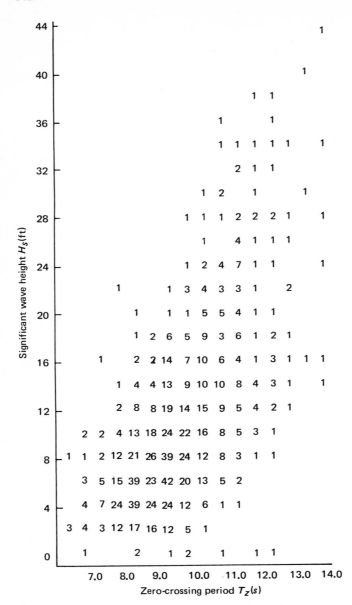

**FIGURE 9.22** Scatter diagram for whole-year observations at station India. From (45).

**TABLE 9.8** Annual Power Density for Different Combinations of Depth and Power Limit[a]

| Depth (m) | Power limit (kW/m) | | |
|:---:|:---:|:---:|:---:|
| | 100 | 200 | 300 |
| 5 | 22.7 | 24.0 | 24.1 |
| 7.5 | 29.0 | 32.3 | 33.0 |
| 10 | 33.5 | 38.6 | 40.0 |
| 15 | 39.1 | 47.0 | 49.8 |
| 20 | 42.4 | 51.9 | 56.0 |
| 25 | 44.1 | 55.0 | 59.8 |

[a]Station India (59°N, 19°W); total power = 77 kW/m. From (45).

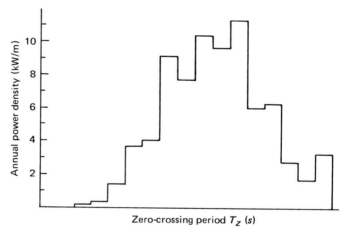

**FIGURE 9.23** Annual power density as a function of zero-crossing period for Station India. From (45).

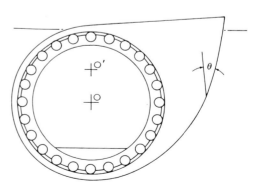

**FIGURE 9.24** Cross section of a duck.

rotates about its center O and absorbs power from waves coming from the right. Its rear is a half-cylinder centered at O, but from the bottom point it enlarges into another cylinder centered about O'. This shape continues until it reaches an angle $\theta$, at which point it develops into a tangent that is continued above the surface. When this vane moves there is no change in displacement of water behind it, but the changes in front extend from zero at the bottom to amounts close to those in the approaching wave. Thus, the vane shape appears to meet all requirements. Salter (45) claims that a good vane of this type should be able to extract nearly all the power contained in the layer of water above its depth of immersion. Figure 9.24 shows a cross section of a single vane, called a "Salter duck," while Fig. 9.25 is an artist's impression of a string of ducks for full-scale operation at sea.

A means of developing continuous power in full-scale equipment is described in Refs. (45) and (46). The random rotations of each duck in a string produce unidirectional pulses of water that flow through a special pump that recirculates the water through the system and also acts as a check valve. Details of the design of an individual duck are presented in Refs. (45) and (46).

In addition to the design of the individual duck, the reference against which the waves can act presents a problem. Since the equipment must be located at sea, if a connection to land were required, this would restrict the places where the equipment can operate. The land connection is, however, not necessary if many ducks are attached to each another and form a floating structure 0.5–1 km long. As the length is increased the installation utilizes a wider range of wave phases and thus becomes more stable. Each duck can be filled with concrete and the combined action of the inertia of the ballast weight, the drag of the deeper portions in quieter water, and the coupling of the machine to waves of opposing phase gives stability to the system.

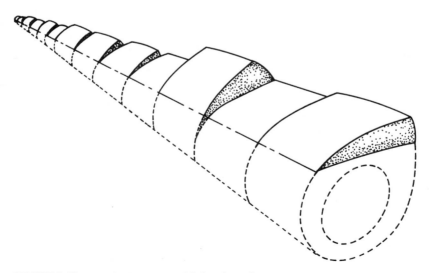

**FIGURE 9.25**  Artist's impression of full-scale equipment at sea.

The power calculated by means of Eq. (9.25) assumes that all waves can be characterized by a single average size and will therefore not reveal annual variations in power output. To calculate the seasonal power distribution a more detailed spectral analysis is required. If $H_{rms}$ is the root mean square displacement of the water surface, that is, the standard deviation, $T$ is the wave period, and $k = \rho g^2 / 4\pi \cong 7.82$ kW/m$^3$ · sec, the power for a mixed wave train of spectrum of $dS$ ($H_{rms}^2 = \int dS$) is given from linear theory (40) by

$$P = k\int TDS = kT_e H_{rms}^2 \qquad (9.26)$$

where $T_e$ is the so-called energy period. The experimentally measured seasonal distribution of $T_e$ and $H_{rms}$ for Station India is shown in Fig. 9.26 with contours of $P$ calculated from Eq. (9.26). With this information and performance data for the ducks, the appropriate duck size can be selected by an optimization procedure. Figure 9.27 shows the distribution of power density plotted against the wave frequency for the entire year, the summer, and the winter (32). Also shown are experimental curves of efficiency versus frequency for ducks of 6, 10, and 16 m diameter. It is apparent that although all duck efficiency curves decrease at lower frequencies, larger ducks are more efficient at lower frequencies than smaller ducks, because the portion of the energy traveling in the upper water layer increases at lower frequencies.

The average power transferred from the sea to a duck can be obtained by multiplying the appropriate curves in Fig. 9.27 and integrating, provided no power limitations are imposed. For example, to obtain the power transfer for a 10-m duck during the winter one combines the top curve of power density with the 10-m efficiency curve. The portion of the time during which power outputs might exceed a given level was calculated by using the spectra in Fig. 9.27. The results are shown in Fig. 9.28, where for comparison the total available power is also shown. The mean available power with simple limits on power output imposed is equal to the area under the relevant curves of Fig. 9.28 to the left of the desired power limit. The results for winter and summer are shown in Fig. 9.29 for no output limits and for limits of 50, 75, 100, 200 kW/m. Since waves are larger in winter than in summer, the power limit is more severe in winter. However, even the lowest limit (50 kW/m) gives a substantially higher load factor ($\sim$80 percent for $d \geqslant 12$ m), suggesting that economic optimization would favor a relatively low-rated transmission line rather than aiming the design for maximum average output.

Ocean waves cannot approach solar radiation in total amounts of energy, but they can provide greater power density than aerogenerators. One might view an ocean wave installation as the second stage of a windmill of which the open sea is the first stage. Average power density in the North Atlantic is between 80 and 90 kW/m. A 500-km front could supply all of the electric power for England. Favorable sites for ocean wave power (OWP) in the United States are located in areas where solar radiation is sparse, for example, Maine and Oregon. The annual distribution of ocean waves is complementary to the availability of direct solar radiation. The dynamic storage mechanism of the ocean produces maximum power

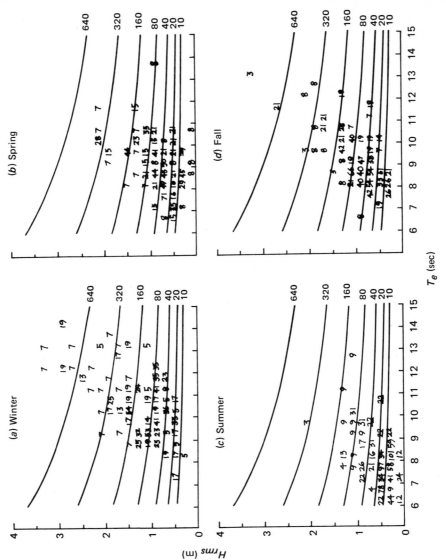

**FIGURE 9.26** Frequency of occurrence at O.W.S. India. From (32).

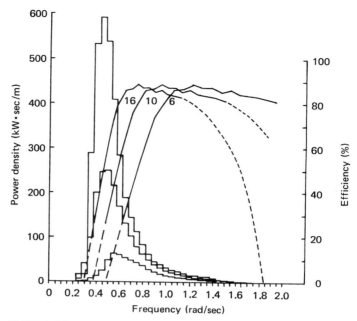

**FIGURE 9.27** Distribution of power frequency for the whole year. From (32).

in winter, minimum in summer, and intermediate in spring and autumn. The energy payback period for OWP is favorable, requiring about 8000 hr of operation to deliver the energy consumed in its construction, according to Salter et al. (46). Thus, the solar energy storage vault of the ocean may be able to deliver useful energy at a reasonable cost in locations favorable for OWP installations.

## PHOTOSYNTHETIC, BIOMASS, AND PHOTOCHEMICAL SOLAR ENERGY UTILIZATION

*The force that through the green fire drives the flower.*

**Dylan Thomas**

In preceding sections the transformation of solar radiation into heat or electricity by means of thermal, fluid mechanical, or direct photovoltaic conversion processes was discussed. There is also the photochemical method, the basic conversion process of solar energy into usable chemical form. This process has been occurring in nature as photosynthesis since the beginning of life on earth, and has in the course of millions of years produced all fossil fuels. It has been studied extensively, but so far all attempts to exploit photochemical reactions technologically, that is, without a

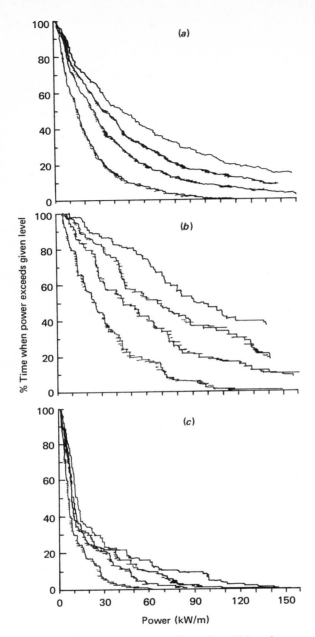

**FIGURE 9.28** The proportion of time for which each power level is exceeded for: (*a*) whole year, (*b*) winter, and (*c*) summer. In each case the top curve refers to the total power available; the other solid curves, in descending order, to predicted power output from ducks of 16, 10, and 6 m diameter. From (32).

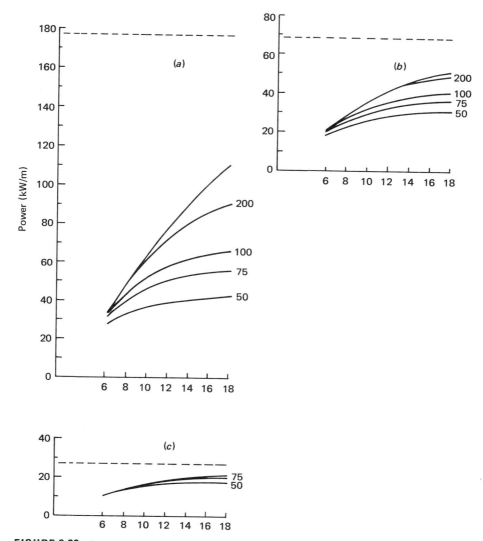

**FIGURE 9.29** Average power output as a function of duck diameters for: (*a*) winter, (*b*) spring, and (*c*) summer, without output limits (top curve) and with output limits of 50, 75, 100, and 200 kW/m. From (32).

living plant, have been commercially unsuccessful. In this section we will consider photosynthetic processes as well as photochemical principles of solar energy conversion.

The main drawback of photosynthetic and photochemical conversion is their low efficiency. Whereas thermal conversion may be on the order of 60 percent, photosynthetic and photochemical conversion efficiencies are an order of magnitude less (see Table 9.9). However, photosynthetic processes are less capital intensive

**TABLE 9.9** Efficiency of the Transformation of Solar Radiation by Means of Various Devices or Processes into Heat or Electricity[a]

| Method of conversion or device for transformation | Efficiency of device or process (%) | Efficiency of transformation into heat (%) | Efficiency of transformation into electricity (%) |
|---|---|---|---|
| Transformation of radiation into heat in a solar boiler with a reflector system | 65–75 | 65–75 | 5–20 |
| Transformation of the heat of steam into mechanical energy in steam engine or turbine | 10–30 | | |
| Transformation of mechanical energy into electricity | 90–95 | | |
| Steam of low pressure (2–5 atm) in a solar boiler with a parabolocylindrical reflector | 45–55 | 45–55 | 2–6 |
| Transformation of the energy of the steam into mechanical energy | 5–10 | | |
| Transformation of mechanical energy into electricity | 90–95 | | |
| Solar heaters of the flat-plate type for heating water up to 350 K (transformation of this heat into mechanical energy not worthwhile) | 30–45 | 30–45 | – |
| Transformation of solar radiation into heat for thermoelectrogenerators in installations with paraboloidal reflectors | 60–70 | – | 2.5–3.5 |
| Transformation of heat in thermoelectro-generators into electricity | 4–6 | | |
| Direct transformation of solar radiation into electricity in photocells | 5–15 | – | 5–15 |
| Growing of forests in natural conditions for firewood | 0.5–1.5 | 0.35–1.2 | 0.03–0.34 |
| Burning of wood in furnace of boiler | 70–80 | | |
| Mechanical energy obtained from steam | 10–30 | | |
| Transformation into electricity | 90–95 | | |
| Natural processes of the transformation of radiant energy into hydroenergy obtained at hydropower stations | 0.002 | – | 0.002 |
| Transformation of solar radiation by plants into potential chemical energy—natural photosynthesis | 0.251 | | 0.0075–0.05 |
| Transformation of the chemical energy of green fodder into muscular energy by animals (oxen, horses) | 35 | | |

[a]Abstracted from (4).

than thermal conversion processes and there are no theoretical grounds why their efficiency cannot be improved. Photochemical conversion is still in the laboratory stage, but has some promise for the long-term future.

## Photosynthesis and Biomass Conversion

All living beings, plants and animals alike, need organic food for their subsistence, growth, and multiplication. Green plants, however, differ from most other organisms because they are able to reduce atmospheric carbon dioxide to the level of sugar by means of radiant energy from the sun. They are thus independent of any external supply of oxidizable fuel. The process by which green plants harness the energy of the sun and build organic compounds from $CO_2$ and water is called photosynthesis. Of foremost importance in these plants' capability of utilizing solar energy is the chloroplast, which is the seat of photosynthetic activity of the green cells and the site of all the chlorophyll and accessory pigments associated with photosynthetic activity. In cells of the higher plant, the chloroplast is a disc-shaped body, roughly 4–8 $\mu$m in diameter by about 1 $\mu$m in width. From 1 to 100 such bodies are present in each cell. When light of appropriate wavelength is absorbed by the chloroplast, $CO_2$ is reduced to the level of sugar, and a volume of gaseous oxygen equal to the $CO_2$ reduced is liberated. The direction of these chemical reactions is exactly the reverse of those occurring during the oxidation of foodstuffs in respiration. Consequently, plants are important in the balance of nature, because they produce food for other organisms and restore to the air the oxygen needed by most organisms for respiration.

Six units of $CH_2O$ would yield $C_6H_{12}O_6$, which is glucose. Using the formula $CH_2O$ to designate the basic unit of a carbohydrate, the photosynthetic equation can be written as follows:

$$CO_2 + H_2O \xrightarrow[\text{energy}]{\text{light}} CH_2O + O_2 \tag{9.27}$$

The equation, although apparently properly balanced, gives an erroneously over-simplified impression of the mechanism by which the photosynthetic reaction occurs (51). It has been demonstrated that the oxygen released in photosynthesis comes not from $CO_2$ but from water. In fact, the photolysis of water is a key to the understanding of photosynthesis, because this is one of the processes in which light energy actually does chemical work. Since there are two atoms of oxygen produced in the above reaction equation and each water molecule contains only one atom of oxygen, at least two water molecules must enter into the reaction. Therefore, a balanced equation that more correctly represents the mechanism of the overall reaction would appear as follows:

$$CO_2 + 2H_2O \xrightarrow[\text{energy}]{\text{light}} CH_2O + O_2 + H_2O \tag{9.28}$$

The oxygen evolved in the process is derived from water entering the reaction and the water molecule that is formed differs from either of the two split in the photolytic process. Figure 9.30 shows a scheme that may help in visualizing the basic aspects of the reaction. According to this scheme, light breaks the water molecule and releases oxygen. The hydrogen that is also produced is used to reduce $CO_2$ to $CH_2O$ and to produce a new water molecule. The schematic diagram in Fig. 9.30 is still somewhat simplified, and biochemists have shown that there are many additional substeps in the reaction that play a part in the total process.

The essential features of this scheme, however, have been verified by studying photosynthetic microorganisms. Some of these are algae, whose biochemical pathways are similar to green plants, but they can be made to use hydrogen gas instead of water to reduce $CO_2$ to the $CH_2O$ level of carbohydrate:

$$CO_2 + 2H_2 \rightarrow CH_2O + H_2O \tag{9.29}$$

This scheme also indicates that light energy is used to split (photolyze) the hydrogen donor, and the reducing power thereby released can reduce $CO_2$ to $CH_2O$.

For the photosynthetic utilization of solar energy, the potential upper limit of the efficiency of the photosynthetic process is important. In land plants approximately 0.5–2.0 percent of the light falling on any ecosystem such as a forest or a field is transformed into chemical energy. This efficiency is important for agriculture and biomass conversion, but it does not indicate the theoretical upper limit of photosynthesis. Most of the incident radiation from the sun is lost by reflection, transmission, and ineffective absorption by pigments other than chlorophyll. Therefore, only a fraction of the incoming solar radiation is actually absorbed by the active pigment. The efficiency with which this latter fraction is utilized gives an indication of the upper limit of the theoretical efficiency of the photochemical process because it measures the minimum number of light quanta necessary to bring one molecule of $CO_2$ to the energy level of a carbohydrate. Detailed studies have shown that 8 quanta of red light, containing 320 kg · cal of energy, is sufficient to

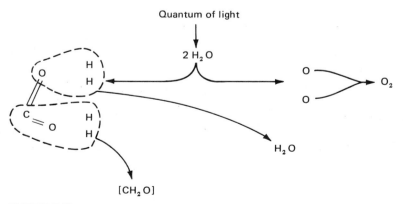

**FIGURE 9.30**  A simplified view of the photosynthetic process.

store the equivalent of 112 kg · cal in the form of organic matter. These figures indicate that the net thermodynamic efficiency of photosynthesis is approximately 35 percent. However, this is not the efficiency that properly relates total incoming radiant energy to useful energy production on the same basis as that used in previous chapters to evaluate the efficiency of thermal processes and calculate the cost-effectiveness of solar energy systems (13). The difference in the definition of efficiency has sometimes caused difficulties in assessing the upper limit of the economic potential of biomass conversion. The outlook for bioconversion consistent with the proper definition of efficiency is presented below.

Under present conditions, the average acre of land under cultivation yields approximately 1 ton of total dry organic matter per year because much of the light falling on the ecosystem is not trapped by chlorophyll and is therefore lost for photosynthesis. The scope for increasing the utilization of photosynthetic products and for using photosynthesis in natural and artificial processes would be enormous if we could learn to increase the energy output-to-input ratios by improving productivity, changing crop types, and harvesting efficiently. To gain an idea of the potential, it should be noted that of the total solar irradiation on the earth's surface, which is approximately $3 \times 10^{24}$ J/yr, each year about $3 \times 10^{21}$ J of energy is fixed as carbon by photosynthesis. This represents stored solar energy and is equivalent to 10 times the world's 1970 energy consumption despite the fact that the overall solar energy conversion efficiency of photosynthesis is only about 0.1 of 1 percent. Thus, the potential for improvement seems great, but there are many problems that must be overcome (48).

The world's population of $4 \times 10^9$ people consumes approximately $1.5 \times 10^{19}$ J/yr in food, which represents about 0.5 percent of the photosynthetically fixed carbon (19). Table 9.10 shows average to good annual yields of dry matter produced under various climatic conditions (54), and Fig. 9.31 shows the expected annual plant yield as a function of the solar radiation for various photosynthetic efficiencies (17, 49). Inspection of Table 9.10 shows that at the present time photosynthetic efficiencies range between 0.6 to 1.6 percent. It will be instructive to estimate what the maximum obtainable efficiencies of photosynthesis might be. Table 9.11 presents the results of such an analysis. About 50 percent of the incoming radiation is unavailable for photosynthetic activity because it falls outside the range of wavelengths that plants can utilize. It seems impossible to reduce reflection, absorption, and transmission of radiation through leaves and ecostands to less than 20 percent. A 77 percent loss represents the quantum efficiency requirements for $CO_2$ fixation in 680 $\mu$m of light assuming that 10 quanta per $CO_2$ molecule is necessary and that the energy content of red light is the radiation peak of visible light. Adding to that a 40 percent loss due to respiration, we obtain a maximum possible photosynthetic efficiency of about 5.5 percent. At the present time, photosynthesis under favorable conditions in the field yields an efficiency on the order of 1 percent on an annual basis, although certain crops, such as sugar cane grown in the southwest of Texas, can obtain an efficiency as high as 3 percent. Table 9.12 shows some high short-term dry-weight yields of certain crops that might be obtainable under conditions in the field on an annual basis if productivity could

**TABLE 9.10**  Average to Good Annual Yields of Dry Matter[a]

| Dry matter | Yield | | Photosynthetic efficiency (% of total radiation) |
|---|---|---|---|
| | tons/hectare · yr | g/m² · day | |
| Tropical | | | |
| Napier grass | 88 | 24 | 1.6 |
| Sugar cane | 66 | 18 | 1.2 |
| Reed swamp | 59 | 16 | 1.1 |
| Annual crops | 30 | — | — |
| Perrenial crops | 75–80 | — | — |
| Rain forest | 35–50 | | |
| Temperate (Europe) | | | |
| Perennial crops | 29 | 8 | 1.0 |
| Annual crops | 22 | 6 | 0.8 |
| Grassland | 22 | 6 | 0.8 |
| Evergreen forest | 22 | 6 | 0.8 |
| Deciduous forest | 15 | 4 | 0.6 |
| Savanna | 11 | 3 | — |
| Desert | 1 | 0.3 | 0.02 |

[a]From (17).

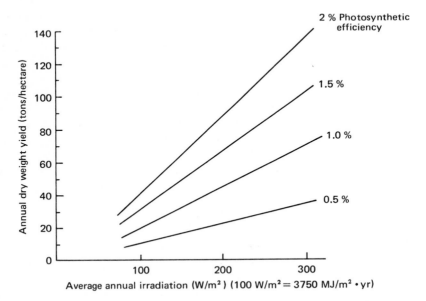

**FIGURE 9.31**  Expected annual plant yields as a function of solar irradiation for various photosynthetic efficiencies.

**TABLE 9.11** Photosynthetic Efficiency and Energy Losses[a]

| Energy losses | Available light energy (%) |
|---|---|
| At sea level | 100 |
| 50 percent, as a result of 400–700-nm light being the photosynthetically usable wavelengths | 50 |
| 20 percent, because of reflection, absorption and transmission by leaves | 40 |
| 77 percent, representing quantum efficiency requirements for $CO_2$ fixation in 680-nm light (assuming 10 quanta/$CO_2$,[b] and that the energy content of 575-nm red light is the radiation peak of visible light) | 9.2 |
| 40 percent, resulting from respiration | 5.5 |

[a]From (17).

[b]If the minimum quantum requirement is 8 quanta/$CO_2$, then this loss factor becomes 72 percent (instead of 77 percent) giving a final photosynthetic efficiency of 6.7 percent (instead of 5.5 percent).

be maintained by growing the plants under optimal conditions (48). If the annual photosynthetic efficiency of plants could be increased, yields would increase, and the prospects for photosynthetic product utilization would improve substantially, as shown in Fig. 9.31.

Under controlled environments (greenhouse-type conditions) it may be possible to grow plants with high productivity. Maximum energy output must be obtained with a minimum energy input so that designs of greenhouse structures must be radically different from those used at present. In both temperate and hot climates optimization of the building structures will be crucial. In the past optimization has not usually been obtained and therefore excess energy has been required to produce a crop. Under greenhouse conditions it might be possible to manipulate the $CO_2$ concentration of the atmosphere. This has been shown to greatly increase the yield of crops, because in greenhouses sufficient $CO_2$ is available for the full growth potential of the plants. Additionally, under high $CO_2$ growth conditions the inherent nitrogen fixation capabilities of legumes are increased manyfold. It has recently been shown that the limitation in nitrogen fixation is not the bacteria but the lack of photosynthetic efficiency and fixed carbon products being translocated to the bacteria in the roots. This knowledge thus suggests a way to increase yield and decrease nitrogen fertilizer input (49).

Once photosynthetic efficiencies have been sufficiently increased, the concept of energy plantation, that is, of growing crops solely for their fuel content, does not seem to be unreasonable. The control of plant growth and the utilization of the most appropriate end product to obtain the highest energy output-to-input ratio would be the objective of such a plan. The end product could be converted to liquid gases, methane, and solid fuels by pyrolysis or fermentation. Cellulose could also be degraded for the production of sugar or other similar processes (49). There are presently chemicals available that can increase the size of the stomata in the

**TABLE 9.12**  Short-term Dry-Weight Yields of Crops with Short-term
Photosynthetic Efficiencies[a]

| Crop | Country | Yield (g/m² · day) | Photosynthetic efficiency (% of total radiation) |
|---|---|---|---|
| Temperate | | | |
| Tall fescue | U.K. | 43 | 3.5 |
| Rye grass | U.K. | 28 | 2.5 |
| Cocksfoot | U.K. | 40 | 3.3 |
| Sugar beet | U.K. | 31 | 4.3 |
| Kale | U.K. | 21 | 2.2 |
| Barley | U.K. | 23 | 1.8 |
| Maize | U.K. | 24 | 3.4 |
| Wheat | Netherlands | 18 | 1.7 |
| Peas | Netherlands | 20 | 1.9 |
| Red clover | New Zealand | 23 | 1.9 |
| Maize | New Zealand | 29 | 2.7 |
| Maize | U.S., Kentucky | 40 | 3.4 |
| Subtropical | | | |
| Alfalfa | U.S., California | 23 | 1.4 |
| Potato | U.S., California | 37 | 2.3 |
| Pine | Australia | 41 | 2.7 |
| Cotton | U.S., Georgia | 27 | 2.1 |
| Rice | S. Australia | 23 | 1.4 |
| Sugar cane | U.S., Texas | 31 | 2.8 |
| Sudan grass | U.S., California | 51 | 3.0 |
| Maize | U.S., California | 52 | 2.9 |
| Algae | U.S., California | 24 | 1.5 |
| Tropical | | | |
| Cassava | Malaysia | 18 | 2.0 |
| Rice | Tanzania | 17 | 1.7 |
| Rice | Philippines | 27 | 2.9 |
| Palm oil | Malaysia (whole year) | 11 | 1.4 |
| Napier grass | El Salvador | 39 | 4.2 |
| Bullrush millet | Australia, N.T. | 54 | 4.3 |
| Sugar cane | Hawaii | 37 | 3.8 |
| Maize | Thailand | 31 | 2.7 |

[a]From (12).

leaves of plants; by this chemical manipulation it would be possible to increase the yield per acre provided sufficient water were available. One method of increasing the efficiency of photosynthesis that has been put into practice is the artificial cultivation of tiny aquatic algae in thin layers of fertilized water. This process will be discussed in detail later.

## Solar Energy Plantation

An energy plantation is an ecosystem grown specifically for its fuel value. Feasibility studies of such plantations, including the solar energy plantation (SEP), have been carried out for the United States and Australia (8, 15), and some of the advantages claimed for this type of conversion are listed below:

1. SEPs are capable of storing energy for use at will.
2. SEPs are renewable.
3. SEP technology is available, is not capital intensive, and can be developed on a large scale within a decade or two.
4. SEPs could be developed with present human and material resources.
5. SEPs are reasonably priced.
6. SEPs are ecologically free of hazards except for fire risks.


Plants used for SEP conversion should grow fast, give maximum ground coverage throughout the whole year, and be tolerant of the lower temperatures and light intensities that occur in the winter. Factors such as cultivation systems, cropping systems, post-harvest deterioration, and transport systems should be considered. Crops that could fix their own nitrogen and require little added fertilizer would be a great advantage.

Energy plantations could consist of deciduous or coniferous forests, which give photosynthetic efficiencies of between 0.4 and 0.8 percent on an annual basis. However, shrubs, weeds, crops, grasses, and algae should also be considered. The Australian study identified five species, namely, *Eucalyptus*, cassava (*Manihot*), *Hibiscus*, Napier grass (*Pennisetum purpureum*), and sugar cane (*Saccharinum*), as being potentially desirable high-yield crops that can be harvested over the whole year. It has been estimated (1) that Brazil could obtain all its current energy needs by growing cassava on about 1 percent of its total land area. Cassava grows well on poor soils, and yields greater than 50 tons/hectare have been obtained. Starch is the principal product and this can be easily fermented to alcohol, producing a fuel that could compete economically with oil. A similar study for the United States (52) suggested that fast-growing hybrid poplars would be most economical. These are deciduous trees that resprout from stumps when cut. They show a yearly solar energy conversion efficiency of about 0.6 percent. With a rainfall of 0.38 m or more annually, about 12,200 hectares would be sufficient to fuel a 400-MW power plant. The cost estimates of Ref. (54) indicate that energy from SEPs could compete with oil, but not with coal at current prices. Tree farms are economically more attractive than single season crops because more biomass can be stored per harvest.

Different types of fuel could be produced by energy plantations. Solid fuel may be the most desirable, but since harvested plants often contain large amounts of water and thus require predrying (which consumes energy unless done by a solar process), it may be better to ferment or pyrolyze the plants. In pyrolysis the end products could be gaseous, liquid, or solid. Pyrolysis systems already exist on a fairly large scale for waste disposal and the production of charcoal. Fermentation of plant material yields mainly alcohols, acetic acid, or methane. Another possibility is chemical reduction at high temperatures and pressures. Pilot plants for such processes already exist, and they have given yields of high calorific value.

## Cellulose Utilization and Waste Disposal

Cellulose is a major component of plants and occurs in large quantities in crops and in waste from agriculture and food processing. About $10^{11}$ tons are produced annually, equivalent to about 70 kg per person in the world per year. Cellulose is the most abundant single organic compound on earth and could be exploited as a source of energy, food, and industrial chemicals (9, 21).

Figure 9.32 shows schematically a well-known technology for converting cellulose to glucose. This method uses an enzymatic hydrolysis treatment to break down the cellulose. The cellulose digested to sugars can subsequently be used as a source of food, feedstock, or liquid fuels (for example, alcohols), after suitable fermentation. Good sources of cellulose would be grasses, cereal straw, *Hibiscus*, *Eucalyptus*, and poplars.

It has been proposed (3) that in southwest U.S. deserts, crops yielding high cellulose could be produced in enclosed environments that conserve water and maximize photosynthesis by enhanced (0.2 percent) $CO_2$ concentrations; yields corresponding to a 5 percent photosynthetic conversion efficiency, and even higher, are envisaged. Table 9.13 shows the two- to threefold increases in photosynthetic yields obtainable with increased $CO_2$ levels.

An average household produces about 15 kg of refuse per week. About 60 percent of this mass consists of paper and vegetable material, which could be the source of cellulose to be converted to alcohol, acetone, and chemical feedstocks. One ton of waste paper will produce about one-half ton of glucose, which could yield 250 liters of alcohol (48). Thus, agricultural and urban wastes could make a sizeable contribution to meet energy requirements if economical systems were developed for their local collection and treatment. It has been estimated that about

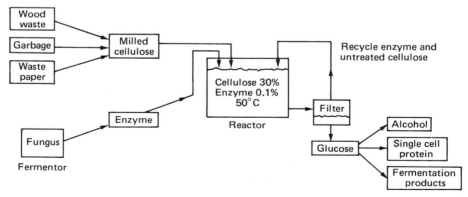

**FIGURE 9.32**  Fermentation of cellulose to produce glucose. Adapted from (30a).

**TABLE 9.13**  Photosynthetic Yields at Normal and Enhanced $CO_2$ Gas Levels[a]

|  | At normal $CO_2$ levels | At enhanced $CO_2$ levels |
|---|---|---|
| Maize, sorghum, sugar cane | 60–75 | 100 |
| Rice | 40–75 | 135 |
| Sunflower | 50–65 | 130 |
| Cotton | 40–50 | 100 |
| Soybean, sugar beet | 30–40 | 56 |
| Oats, wheat, barley | 30–35 | 66 |
| Tobacco | 20–25 | 67 |
| Tomato, cucumber, lettuce | 20–25 | 50 |
| Tree species, grapes, ornamentals, citrus | 10–20 | 40 |

[a]Milligrams of $CO_2$ fixed per square centimeter per hour. From (56).

6 percent of the current U.S. electrical requirements could be supplied from readily available agricultural and urban waste (10).

## Solar Energy Conversion by Algae

The possibility of using certain types of algae as direct solar energy converters is of interest because this photosynthetic process could become a supplemental source of protein or fuel, and many liquid and semisolid wastes from dwellings, industries, and farms are suitable for the growth of photosynthetic algae (18). The harvested algae may be fed directly to animals, burned in order to generate heat, or fermented in order to produce methane. At the same time, however, the process can serve to dispose of waste products, and it can even purify water. Figure 9.33 shows a

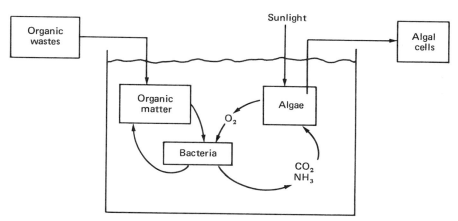

**FIGURE 9.33**  Scheme for simultaneous waste disposal and algae culture.

scheme proposed for simultaneous waste disposal and algae culture, which according to its designer could reduce the cost of waste disposal in California by 25-50 percent (37). In California average daily yields of dry algae in excess of 100 kg/hectare have been obtained by Oswald (34, 38) in an algae pond of $10^6$ liters at a 3 percent yearly overall photosynthetic efficiency. Large feeding systems for cattle and chickens have been provided with algal ponds into which the animal waste is fed directly. About 40 percent of the nitrogen is recovered in the algae, which is subsequently refed to the animals.

One way to view the algae energy conversion is to separate it into two sequential processes: photosynthesis and secondary synthesis. The end result of the second step is an increase in cell mass, leading to division into larger numbers of cells whose growth rate can be expressed mathematically as

$$N(t) = N_0 e^{kt} \tag{9.30}$$

where $N$ = the number of cells or mass at time $t$

$N_0$ = the initial mass

$k$ = a rate constant

There are many potential candidates for algal conversions, but so far most experimental work has been done with strains of *Chlorella*, because it is readily available and cultivated. The alga is grown in dilute salt solution, which may be industrial wastes, and by regulating the salt concentration, particularly the amount of fixed nitrogen, much faster rates of growth than under natural conditions can be obtained. In addition to salt, $CO_2$ must be supplied, because 1 kg of dry *Chlorella* requires 1.8 kg of $CO_2$, which is much more than can be obtained by natural processes from the ambient air.

Algal suspensions exposed to direct sunlight must transfer considerable amounts

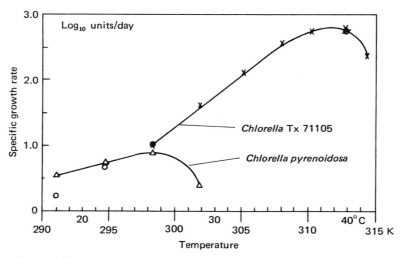

**FIGURE 9.34** Growth rate of *Chlorella* as a function of temperature. From (47).

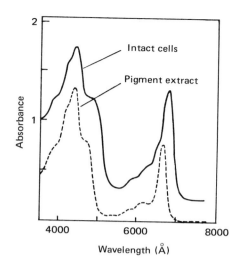

**FIGURE 9.35** Spectral absorption curves for *Chlorella*. The curve for intact cells is partly corrected for scattering effects. Redrawn from (39).

of heat to maintain a suitable equilibrium temperature. Figure 9.34 shows the growth rates of two species of *Chlorella* as a function of temperature. The standard *Chlorella*, as well as most strains commonly used in experiments, cannot survive prolonged exposure to temperatures above 303 K (85°F), although some high-temperature strains have tolerated 312 K (33, 34, 50). In areas where solar insolation is high, air and soil temperatures are also high. Thus special cooling systems may be necessary to avoid overheating.

The spectral response of *Chlorella*, as shown in Fig. 9.35, extends over the entire visible spectrum between 0.4 and 0.7 μm, and about 50 percent of the solar irradiation at the earth's surface is available for photosynthetic algal growth. However, the rate of growth increases linearly with irradiation flux only up to a certain point and then levels off to a constant rate, the so-called light saturation level (35). This behavior is illustrated qualitatively in Fig. 9.36. The light saturation level occurs for *Chlorella* at an irradiation flux of only about 5 W/m² compared with a maximum available solar flux of up to 100 W/m². Since irradiation above the saturation level will not increase the yield, a geometric arrangement that will maximize the utilization of the incident light must be sought. As shown in Fig. 9.36 the radiation in the culture will decrease exponentially as in any transparent medium. Thus the depth of the culture and the cell concentration are reciprocal functions. It has been found (35) that if the thickness of the culture is reduced to about 0.01 m, the cell concentration may be raised to about 50 g/liter. In deep ponds a maximum concentration of no more than 1 g/liter can be expected. The cell concentration attainable is of course important, because it determines the economics of the harvesting procedure.

Another consequence of the exponential decay in radiation flux is a reduction in the efficiency with which the incident radiation can be actually used. Laboratory experiments have shown that *Chlorella* can convert approximately 20 percent of the visible radiation into energy of biomass (34). This means that the production of

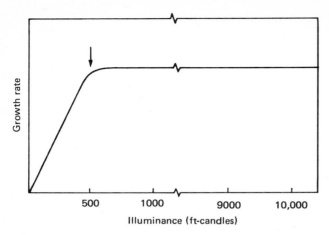

**FIGURE 9.36** Growth rate as a function of illuminance for a thin suspension of *Chlorella pyrenoidosa*. Redrawn diagrammatically from (35).

5.6 J requires 28 J of absorbed radiation in the visible spectrum. However, since only 50 percent of the total solar radiation impinging on earth is in the visible spectrum, the conversion efficiency (calculated on the same basis as that used in thermal processes, that is, total available flux) is only about 4 percent. Unfortunately, even this figure cannot be achieved in any practical installation because of the light saturation level mentioned previously.

The cells at the surface are exposed to an irradiation flux of 100 W/m$^2$ (10,000 ft-candles), but can grow only at a rate equivalent to 5 W/m$^2$. Full use of the available irradiation will thus not occur until the absorption has been reduced to the level of light saturation. This is illustrated in Fig. 9.37, where the shaded area under the curve represents the amount of the total solar irradiation that can be used at maximum efficiency, whereas the total area under the curve represents all the solar energy available in the algal culture. This physiological limitation reduced the yield by a factor of 5–10 times below that which would be theoretically possible if no light saturation occurred. In practice, an overall efficiency of no more than 2 or 3 percent can therefore be achieved in algal energy conversion at the present time.

Efforts to circumvent the reduction in efficiency caused by the light saturation phenomenon have followed two directions:

1. The use of appropriate turbulent mixing of algal cultures in order to produce suitable time-averaged irradiation levels for all the cells.
2. The search for algae with good conversion efficiencies but higher saturation levels.

A bench-scale open-circulation system for algal production was built in Japan some years ago. It consisted of a round pond, 5 m in diameter and 20 cm in depth, with a small island at the center. The culture (2000 liters in volume) was circulated by a $\frac{1}{2}$-hp pump from one corner of the pond through a pipe placed underground to the two rotatable arms extending from the center to the periphery of the pond. The arms had many holes, spaced 3 cm from each other, oriented diagonally downward. These arms were submerged in the culture solution, the clearance between the arms and the bottom of the pond being 4 cm. Because of water flux from the holes, the arms rotated slowly (at approximately 180°/min) and circulated the solution evenly in the pond; at the same time they could stir and sweep intermittently the algal cells in the solution. Near the outlet from the pump, a small tube for feeding the air enriched with $CO_2$ was inserted into the circulating pipe, in order to supply the culture evenly with $CO_2$.

The advantages claimed for this culture system are (53):

1. Only a small amount of energy is needed for the circulation of the culture and the supply of $CO_2$.
2. The precipitation of algal cells on the bottom of the culture pond can be prevented.
3. No cooling equipment is necessary (as for a closed system).
4. Mixing of the whole culture, which is necessary on such occasions as seeding of fresh algal cells, and the regulation of composition, or acidity of the culture solution, can readily be achieved.

Care must be taken to regulate the water quantity after a heavy rain and to remove occasionally any accumulated dust.

Figure 9.38 shows the growth of biomass as a function of time during a rainy 2-month period, Fig. 9.39 shows the average daily yield for an entire year (as well

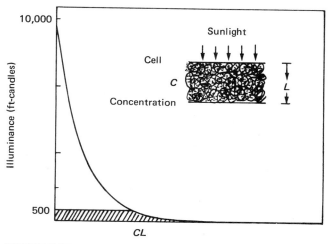

**FIGURE 9.37**  Decrease in illuminance in a dense culture exposed to full sunlight.

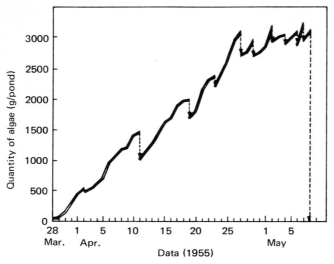

**FIGURE 9.38**  An example of a culture run by the open circulation system. The alga was a strain of *Chlorella.* Black parts of the curve indicate cloudy or rainy days; the occasional drop of the curve (vertical dashed lines) results from the harvest of algal cells. From (53).

as the environmental conditions), and Fig. 9.40 shows the average monthly energy conversion efficiency based on the total solar radiation received. The higher efficiencies during periods of low solar irradiation are consistent with the analysis presented above. The results of this experimental program indicate that production at an average of 12.4 g dried algal cells per square meter per day throughout the year (or 18 metric tons per acre per year) can be achieved in practice under conditions in Japan.

Figure 9.41 compares the cost of protein derived from *Chlorella* with the cost of protein in other foods used in Japan. It should be noted, however, that attempts to use algal cells for human food encountered some problems because of an unpleasant aftertaste and discoloration of conventional foods. These should not be a serious problem if the protein were used to feed animals.

A more speculative scheme using algae for energy conversion was outlined by Bockris (6). He proposed to produce hydrogen by irradiating water in the presence of a special alga, *Anabaena cylindrica.* These algae contain cells, called heterocysts, that protect the enzymes of the plant from oxygen and that can produce hydrogen from water without noticeable decrease over a day. However, the hydrogen production mechanism is not understood and no production system has as yet been devised.

The possibility of using solar bioconversion processes to provide energy or heat, and to produce fuels such as methanol, is presently under intense study all over the world. A survey of currently operating bioconversion plants has recently been

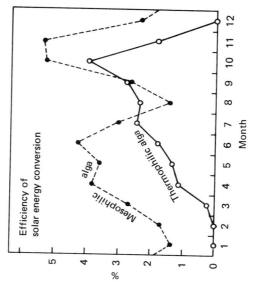

**FIGURE 9.40** Monthly variation of the efficiency of solar energy conversion effected by mesophilic and thermophilic *Chlorella* strains. Efficiency was calculated on a total radiation basis and the energy content of algal cells was 5.5 kcal/g. From (53).

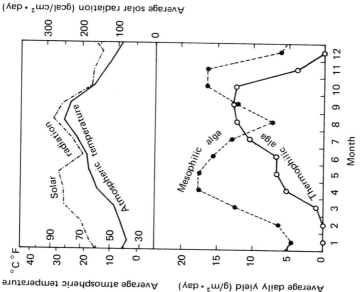

**FIGURE 9.39** Monthly variation of average solar energy (including infrared radiation), atmospheric temperature, and the average daily yield obtained with mesophilic and thermophilic *Chlorella* strains. The experiment ran from February 1954 to January 1955. From (53).

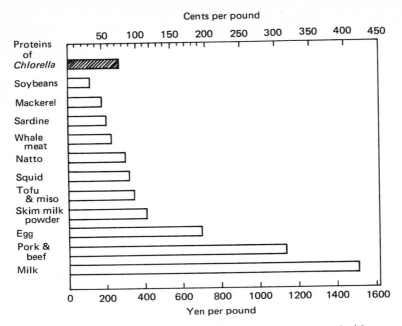

**FIGURE 9.41**   Estimated market price of *Chlorella* protein compared with prices of proteins of other foods (in Japan, 1954). From (53).

conducted at the Solar Energy Research Institute and the results of this survey are summarized in Table 9.14, where 10 processes are described. An examination of these processes shows the variety of bioconversion processes and also indicates the wide range in cost at which current bioconversion processes can deliver useful energy.

## Photochemical Principles

The photochemical process requires the absorption of photons, that is, a combination of suitable molecules with photons of the appropriate energy contents, to produce a chemical change. Solar radiation is composed of a large number of photons, each with a definite amount of energy known as a quantum. Although sunlight is received abundantly, a photon has a chance of being absorbed by a molecule and changing it chemically only if the receptor molecule can be rearranged to absorb an amount of energy nearly equal to that of the impinging photon.

As shown in Chapter 3, the energy associated with a photon $E$ is given by Eq. (3.75):

$$E = h\nu$$

where $h$ = Planck's constant ($6.6256 \times 10^{-34}$ J $\cdot$ sec)
$\nu$ = frequency of light in $sec^{-1}$

**TABLE 9.14** Thermal and Biological Conversion of Biomass to Fuel and Energy

| Process | Plant cost ($) | Plant size (tons/day) | Input form | Typical fuel cost ($/MMBtu) | Daily energy output (MMBtu/day) | Selling price ($/MMBtu) | Present U.S. capacity (MMBtu/yr) | Operating units | Capital energy cost ($/MMBtu · yr) |
|---|---|---|---|---|---|---|---|---|---|
| Compaction: for collection and improved storage, shipment, and use (e.g., wood) | 750K | 250 | Waste biomass | 0.41 | 4500 (solid fuel pellets) | 1.22 | $4.9 \times 10^6$ | 3 | 0.45 |
| Combustion of solid municipal wastes | 35M | 1000 | Solid municipal wastes | −1.55 | 5400 (high power steam) | <2.00 | $2.0 \times 10^6$ | 1 (U.S.) 20 (Europe) | 17.75 |
| Pyrolysis oil (e.g., Tech-Air, Atlanta, GA, Div. AmCanCo) | 1M | 85 | Waste biomass | 0.62 | 816 | <2.40 | $0.30 \times 10^6$ | 1 | 3.35 |
| Gasifiers: Medium Btu gas (municipal solid waste to 350 Btu/SCF) (e.g., Purox) | 4M | 300 | Solid municipal wastes | −1.11 | 1890 | 1.80 | $0.68 \times 10^6$ | 1 | 5.79 |
| Gasifiers: Low Btu gas (e.g., Forest Fuels, Keene, N.H.) | 70K | 14.4 | Mill waste | 0.62 | 264 | | 4300 | 6 | 1.37 |
| Home stoves and furnaces | 300 | | Round wood | 3.12 | 1.2 | 5.67 | $350 \times 10^6$ | 7.8MM | 0.68 |
| Power generation: from sugar cane wastes; supplies 10% of Hawaiian electricity | 37M | 4570 | Bagasse | <0 | 3840 | 1.50 | $10 \times 10^6$ (Hawaii only) | 12 | 3.27 |
| Power and heat from waste liquor | | | | | | <2.00 | $500 \times 10^6$ | 200 | |
| Sewage to gas and electricity (Los Angeles Hyperion plant) | | 350 | | 0 | 2250 | 0.36 | $>0.82 \times 10^6$ | Not determined | |
| Anaerobic digestion: small manure digestor (high Btu gas) | 25K | 1.25 | Wet cow manure | 0 | 2.5 ($925$ Btu/ft$^3$) | Used on site | 912 | 1 (5 on order) | 27.39 |

In the initial photochemical act one photon is absorbed by each activated molecule. The absorption of a sufficiently energetic quantum by a molecule raises this molecule into an electronically excited state from which it may either decompose spontaneously or interact with some other molecule. The photochemical reaction process can be represented schematically by the following equations:

$$R_1 + h\nu = R_1^*$$ (9.31)

$$R_1^* \rightarrow P_1$$ (9.32)

$$R_1^* + R_2 \rightarrow P_2 + P_3$$ (9.33)

where  $R_1$ = the reactant molecule that absorbs the photon
$R_1^*$ = the reactant molecule in the excited state
$P_1$ = the product of a unimolecular reaction
$P_2$ and $P_3$ = the products of a bimolecular reaction

If the product molecules possess more energy than the reactant, the process is called *endergonic* (energy storing). If the product molecule is less energetic, the process is called *exergonic* (energy releasing). The two types of reactions are illustrated schematically in Fig. 9.42 for a simple generalized photochemical reaction $R \xrightarrow{h\nu} P$.

The difference in energy level between $R$ and $P$ is a result of the different amount of energy stored in the chemical bonds of $R$ and $P$. Only endergonic photochemical reactions are of interest for solar energy conversion, because they result in the storage of some of the energy of the absorbed photons in the chemical bond of the product $P$ and can therefore be used to convert solar energy into

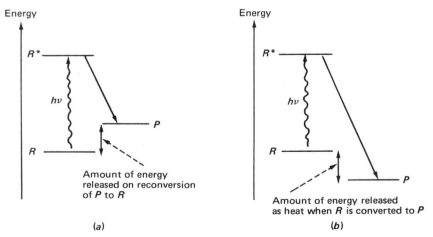

**FIGURE 9.42**  Energy levels in the reaction $R \xrightarrow{h\nu} P$ for (*a*) an endergonic photochemical reaction and (*b*) an exergonic photochemical reaction.

chemical energy. Exergonic reactions simply dissipate the energy of the absorbed photons into heat so that the product $P$ has actually less energy than the reactant $R$.

The majority of solar photochemical processes proposed for possible utilization are cyclic systems. They start with an endergonic reaction in which chemical energy is stored in the product $P$. At a later time this is followed by the reverse process $P \rightarrow R$ in which energy is released. The reactant $R$ thus acts essentially as a material in which energy can be stored by means of a photochemical transformation, provided the product $P$ is kinetically inert at normal temperatures. $P$ can thus be viewed as a storable fuel. In practice it is more convenient to deal with the energy associated with one Avogadro's number ($6.02 \times 10^{23}$) of photons, called an Einstein of radiation, than with a single photon. The energy associated with one Einstein of radiation is $Nh\nu$, or

$$Nh\nu = \frac{1.2 \times 10^{16}}{(m)} \text{ kJ} = \frac{11.35}{(A)} \text{ Btu}$$

After absorbing a photon, the molecule may undergo a transition in vibrational, rotational, and electronic energy states, depending on the energy of the absorbed photon. Absorption of energy from photons in the solar spectrum (ultraviolet or visible) is usually accomplished by a displacement of an electron from its normal position in the molecule. In the infrared, the absorption of energy occurs by a displacement of atoms within the molecule or by rotation of the molecule. But only electronic transitions are sufficiently energetic to produce chemical reactions. Such transitions in the ultraviolet and visible range correspond to energy levels of about 40 kcal per Einstein at $7 \times 10^{-7}$ $\mu$m. After activation of a molecule by a photon in this energy range one of the following phenomena may occur:

1. The activated molecule may transmit its energy to other molecules by collision, increase their translational energy, and raise their temperature. In this manner radiant energy can be converted directly into heat without any chemical change. This is the usual effect produced by solar energy.
2. The activation may break a chemical bond in the molecule and decompose it into atoms or groups of atoms.
3. The electronically activated molecule may transfer its energy to a molecule of some other chemical compound by collision and the other molecule may undergo a chemical change. A loose combination between the photoactivated molecule and the chemically reacting molecules for a certain minimum period of time is necessary for this process to occur.
4. Fluorescence emission of the absorbed radiation may occur, but the emitted light will be of a different wavelength than the exciting radiation.
5. The activation may be sufficiently energetic to drive an electron from the molecule and cause ionization.

These different mechanisms of disposing of the absorbed energy are called the photochemical primary processes and are in each case limited to a maximum quantum yield $\Phi$ defined as

$$\Phi = \frac{\text{number of molecules reacting}}{\text{number of photons absorbed}}$$

Although there can be no photochemical reaction if there is no photon absorption, and although one photon only activates one molecule, the quantum yield varies greatly in different reactions because of secondary reactions that often follow the primary process. For example, the photoactivation of a mixture of $H_2$ and $Cl_2$ results in the primary event

$$Cl \xrightarrow{h\nu} 2Cl \qquad (9.34)$$

but each chlorine atom can react with a hydrogen molecule to start a chain reaction of the kind

$$Cl + H_2 \rightarrow HCl + H \qquad (9.35)$$

$$H + Cl_2 \rightarrow HCl + Cl \qquad (9.36)$$

with the final quantum yield of HCl being very large. Conversely, processes such as deactivation, fluorescence, recombination of the primary products in a reverse reaction, or side reactions forming new products will result in a low quantum yield.

## EXAMPLE 9.4

Calculate the number of kilograms of photochemical product per square kilometer per 24-hr day, assuming a molecular weight of 100, a quantum yield of 1.0, and a photochemical absorption of 1.0 percent of the incident sunlight at an average wavelength of $3 \times 10^{-7}$ $\mu$m. Assume that during a 24-hr day there are 500 min of sunlight with an average intensity of 4.186 $J/m^2 \cdot min$.

## SOLUTION

The total amount of solar energy received during one 24-hr day is $(4.186$ $J/cm^2 \cdot min)(500 \times 10^{10}$ $cm^2/km^2) = 2.09 \times 10^{13}$ $J/km^2$. One hundredth of this is absorbed and can produce chemical changes.

The energy of 1 photon (or 1 quantum) is given by $E = h\nu$, where $h = 6.6256 \times 10^{-34}$ $J \cdot sec$ and $\nu = (3 \times 10^8$ $m/sec)$ per wavelength in meters, or $E = (6.6256 \times 10^{-34}$ $J \cdot sec)(3 \times 10^8/5 \times 10^{-7})$ and $\nu = 3.97 \times 10^{-19}$ J. Then, the number of quanta per day is

$$\frac{2.09 \times 10^{11} \text{ J/km}^2}{3.97 \times 10^{-19} \text{ J}} = 5.28 \times 10^{29} \text{ km}^{-2}$$

With a quantum yield of unity, this is equivalent to $5.28 \times 10^{29}$ molecules/$\text{km}^2 \cdot$ day of photochemical products. Since there are $6.02 \times 10^{23}$ molecules in 1 mole, this is equivalent to

$$\frac{5.28 \times 10^{29} \text{ molecules/km}^2 \cdot \text{day}}{6.02 \times 10^{23} \text{ molecules/mole}} = 8.77 \times 10^5 \text{ moles/km}^2 \cdot \text{day}$$

For a molecular weight of 100, the yield would be

$$(8.77 \times 10^5 \text{ moles/day})(100 \text{ g/mole}) \left(\frac{1}{1000} \text{ kg/g}\right) = 8.77 \times 10^4 \text{ kg/km}^2 \cdot \text{day}$$

In the field of agricultural engineering the yield is usually expressed in tons per acre per day. In these units, the yield of the hypothetical photochemical reaction would be

$$(8.77 \times 10^4 \text{ g/km}^2 \cdot \text{day})(4.05 \times 10^{-3} \text{ km}^2/\text{acre})\left(\frac{1}{9.08 \times 10^8} \text{ ton/kg}\right)$$

$$= 0.3 \text{ tons/acre} \cdot \text{day}$$

Unfortunately, no known photochemical reaction can approach this productivity. Photosynthesis in agriculture usually has a productivity on the order of 1 ton of dry plant material per acre *per year*!

The mere absorption of a photon does not necessarily produce chemical reaction. For a primary reaction to occur, the energy of the excited molecule must be greater than a quantity called the activation energy $E^*$. If the excited molecule has an energy larger than $E^*$, it can undergo decomposition. Quantum statistical considerations show that the rate of reaction $k$ is

$$k = \frac{RT}{Nh\nu} e^{-\Delta F^*/RT} \tag{9.37}$$

where $\Delta F^*$ is the standard free energy of activation. In terms of standard enthalpy and entropy of activation,

$$\Delta F^* = \Delta H^* - T\Delta S^* \tag{9.38}$$

and it follows that

**TABLE 9.15**  Endothermic Photochemical Reactions Initiated by Absorption of Solar Radiation[a]

| Overall chemical reaction | Effective wavelengths (Å) | Quantum yield Φ (product) | $\Delta F^\circ_{298}$ (kcal/mole) | Percent efficiency $Q$ |
|---|---|---|---|---|
| $NO_2 \rightarrow NO + \frac{1}{2}O_2$ | <4350 | 0.046(NO) | 9.0 | 0.6 |
| | 4050 | 0.36 | | 4.6 |
| | 3660 | 0.77 | | 8.8 |
| $NOCl \rightarrow NO + \frac{1}{2}Cl_2$ | <6370 | 2.0(NO) | 4.9 | 21.8 |
| | 3650 | 2.0 | | 12.5 |
| $NO_3^- \rightarrow NO_2^- + \frac{1}{2}O_2$ (water solution) | <3350 | 0.014($NO_2^-$) | 18.0 | 0.20 |
| | 2804 | 0.024 | | 0.42 |
| | 2700 | 0.07 | | 1.2 |
| | 2540 | 0.30 | | 4.8 |
| $2Fe^{++} + I_3^- \rightarrow 2Fe^{3+} + 3I^-$ (water solution) | <5460 | 2.0($Fe^{3+}$) | 9.4 | 9.0 |
| | 3660 | | | |
| $I_2 + NO_2^- + H_2O \rightarrow NO_3^- + 2HI$ | <5790 | 0.1($NO_3^-$) | 13.8 | 6.0 |
| $Fe^{++} + \frac{1}{2}$ thionine $\rightarrow Fe^{3+} + \frac{1}{2}$ leucothionine (acid solution, pH 2) | 4360 visible | | 9.5 | 2.1 |
| | ~6000 | ~0.5($Fe^{3+}$) | | 9.9 |
| $H_2O \rightarrow H_2 + \frac{1}{2}O_2$ ($Ce^{3+}$, $Ce^{4+}$, $HClO_4$ solution) | <3500 | 0.0007($H_2$) | 56.7 | 0.035 |
| | 2537 | | | |
| 2 Anthracene → Dianthracene (organic solution) | <3800 | | 15.6 ($\Delta H^\circ_{298}$) | |
| | 3660 | 0.25(Dian) | | 5.0 |
| | 3130 | 0.23 | | 3.9 |
| $AgCl(s) \rightarrow Ag(s) + \frac{1}{2}Cl_2$ | <4050 | 1.0(Ag) | 26.2 | 37.1 |
| $AgBr(s) \rightarrow Ag(s) + \frac{1}{2}Br_2$ | <4600 | 1.0(Ag) | 22.9 | 37.0 |
| $AgI(s) \rightarrow Ag(s) + \frac{1}{2}I_2$ | <4400 | 1.0(Ag) | 15.8 | 24.3 |
| $AgCl(s) + \frac{1}{2}H_2O \rightarrow Ag(s) + HCl + \frac{1}{4}O_2$ (water suspension) | <4050 | (1.0?) | 23.2 | (32.9) |
| | <4000 | | | |
| $H_2O + \frac{1}{2}O_2 \rightarrow H_2O_2$ (water suspension of ZnO) | 3660 | 0.09($H_2O_2$) | 25 | 2.9 |
| | 3130 | 0.13 | | 3.6 |
| $H_2O + \frac{1}{2}O_2 \rightarrow H_2O_2$ (water suspension of CdS) | <5460 | (0.02) | 25 | (1.0) |
| | 4360 | (0.09) | | (3.4) |
| | 3650 | (0.11) | | (3.5) |
| $H_2O + \frac{1}{2}O_2 \rightarrow H_2O_2$ (water suspension of CdSe) | <5780 | (0.007) | 25 | (0.4) |
| | 5460 | (0.024) | | (1.1) |
| | 3650 | (0.024) | | (0.8) |

[a]Solar radiation in the visible or near-ultraviolet part of the spectrum. From "Introduction to the Utilization of Solar Energy," by A. M. Zarem and D. D. Erway, eds. Copyright 1963 by McGraw-Hill Book Co. Used with permission of McGraw-Hill Book Co.

$$k = \frac{RT}{Nh\nu} e^{\Delta S^*/R} e^{-\Delta H^*/RT} \tag{9.39}$$

The net energy change, and hence the recoverable energy, is measured in terms of the standard free energy of the reaction $\Delta F^\circ$, where the reactants and products are taken at ground-state levels. If $\Delta F^\circ$ is positive, the products possess greater chemical energy than the reactants and there has been a net uptake of energy. Such endothermic reactions are of prime interest in solar conversion. Calvert (7) has compiled the following list of characteristics for an ideal photochemical reaction to convert solar radiation into chemical energy:

1. The reactant molecules should absorb all the radiation in the visible and ultraviolet parts of the spectrum.
2. A given reaction follows the absorption of light with perfect efficiency, that is, the primary quantum yield is unity.
3. The overall chemical change in the reaction system involves a large increase in free energy.
4. The products formed absorb no light in the visible and ultraviolet regions.
5. The exothermic reaction of the products is slow under ordinary conditions, but can proceed readily under specially controlled conditions, for example, at elevated temperature or in a battery, to liberate energy.
6. The reactants are cheap and do not require expensive equipment for the conversion.

Table 9.15 lists some endothermic photochemical reactions that can be initiated by the absorption of solar radiation in the visible or near-ultraviolet part of the spectrum. It is apparent that none of the reactions in this table comes close to the ideal reaction characteristics. The advantages and limitation of each reaction as a converter of solar energy into chemical energy are discussed in Ref. (7). None of the known reactions is really a promising candidate for commercialization at this time, but continued research may produce a viable photochemical solar conversion scheme that does not involve nature's intervention.

## PROBLEMS

9.1. It has recently been proposed that ocean thermal energy conversion could be used to produce ammonia, a chemical currently manufactured from natural gas at a total energy consumption of about 2.5 percent of the U.S. natural gas consumption during 1975. Since the manufacture of ammonia could be accomplished anywhere, the plant could be located in a favorable location, for example off the coast of Brazil, where the difference in temperature between the upper layer of the ocean and the layer at the 600-m depth is approximately 20 K. Using an OTEC power plant of 500 MWe capacity, estimate the number of tons of ammonia that could be produced per year. For a price of $165/ton of ammonia, estimate the time required to regain the capital investment in the OTEC production plant and compare your estimate with the claim that the capital could be repaid in 2.5 yr.

9.2. Aluminum sells today at approximately $800/ton. The production of aluminum is one of the most energy-intensive processes in the metallurgical industry, and it has recently been proposed that an OTEC plant could be used for the production of aluminum. Metallurgical estimates indicate that approximately 18 kW · hr of energy is required in the refining process for each kilogram of aluminum produced. Assuming a plant load factor of 85 percent, calculate the size of an OTEC plant required to produce 250,000 tons/yr of aluminum, if there is a temperature gradient similar to that in the previous problem. Then calculate the time required to repay an investment cost in the plant and compare your results with the estimate of 3 yr obtained in the literature.

9.3. Estimate the size of a forest that would be required to supply under steady-state conditions the fuel to a 100-MW power plant that burns wood. State your assumption of the photosynthetic conversion efficiency, the heating value of wood, the energy required for cutting the trees and transporting the wood to the boiler, and make a realistic assessment of the efficiency of the power plant fed by photosynthetic solar radiation. Also indicate the location at which you propose to grow the trees and estimate the cost of the electric power, using a realistic value for the land occupied by the forest used to supply the fuel.

9.4. If the sea power plant described in Example 9.3 is to deliver power at $8/10^6$ Btu, estimate the maximum permissible cost of the condenser and evaporator heat-exchanger surface in dollars per square foot, assuming a 20-yr life, 10 percent discount rate, and 7 percent yearly fuel inflation.

9.5. The tabulation below gives the annual power density of different combinations of depth and power limit as measured at stations A and B. Prepare a graph with the annual power density plotted against the zero-crossing period and calculate the total power (Answers: 36.8 kW/m for A and 25.8 kW/m for B).

| Station | Depth (m) | Power limit (kW/m) | | |
|---------|-----------|------|------|------|
|         |           | 100  | 200  | 300  |
| A | 5 | 16.4 | 16.5 | 16.5 |
|   | 7.5 | 20.8 | 21.5 | 21.5 |
|   | 10 | 23.8 | 25.1 | 25.1 |
|   | 15 | 27.1 | 29.4 | 29.8 |
|   | 20 | 28.7 | 31.8 | 32.4 |
|   | 25 | 29.5 | 33.0 | 33.9 |
| B | 5 | 10.9 | 11.0 | 11.0 |
|   | 7.5 | 14.2 | 14.5 | 14.5 |
|   | 10 | 16.6 | 17.1 | 17.2 |
|   | 15 | 19.3 | 20.7 | 20.8 |
|   | 20 | 20.9 | 22.7 | 22.9 |
|   | 25 | 21.8 | 23.9 | 24.2 |

# REFERENCES

1. Abelson, P. M., Energy Alternatives for Brazil, *Science*, vol. 189, p. 417, 1975.
2. Anderson, J. H., and J. H. Anderson, Jr., "Large Scale Sea Thermal Power," ASME Paper 65-WA/SOL-6, December 1965.
3. Bassham, J. A., "The Energetics of Cellulose Production by Green Plants," paper presented at AAAS Meeting, New York, January 1975.
4. Baum, V. A., Prospects for the Application of Solar Energy, and Some Research Results in the USSR, *Proc. World Symp. Applied Solar Energy, Phoenix, Arizona, 1955*, Stanford Research Institute, Menlo Park, California, 1965.
5. Bergey, K. H., "Feasibility of Wind Power Generation for Central Oklahoma," University of Oklahoma, Norman, Oklahoma, 1971.
6. Bockris, J. O'M., "Energy: The Solar-Hydrogen Alternative," The Architectural Press, Ltd., London, 1976.
7. Calvert, J. G., Photochemical Processes for Utilization of Solar Energy, in A. M. Zarem and D. D. Erway, eds., "Introduction to the Utilization of Solar Energy," pp. 190–211, McGraw-Hill Book Co., New York, 1963.
8. Calvin, M., Solar Energy by Photosynthesis, *Science*, vol. 184, pp. 375–381, 1973.
9. Chedd, G., Cellulose from Sunlight, *New Sci.*, pp. 572–576, March 6, 1975.
10. Cherry, W. R., and F. H. Morse, Conclusions and Recommendations of the U.S. Solar Energy Panel, Abstr. E39, "Sun in the Service of Mankind," UNESCO, Paris, 1973.
11. Claude, G., Power from the Tropical Sea, *Mech. Eng.*, vol. 52, no. 12, p. 1039, December 1930.
12. Cooper, J. P., "Photosynthesis and Productivity in Different Environments," pp. 593–621, Cambridge University Press, Cambridge, England, 1975.
13. Daniels, F., "Direct Use of the Sun's Energy," Yale University Press, New Haven, 1964.
14. Draper, L., and E. M. Squire, Waves at Ocean Weather Ship Station India (59N., 190W.), *Trans. Inst. Nav. Archit.*, vol. 109, pp. 85–93, 1967.
15. Gifford, R. M., and R. J. Millington, Energetics of Agriculture and Food Production with Special Emphasis on the Australian Situation, in "Energy and How We Live," UNESCO Mass and Biosphere Symposium, Adelaide, Australia, 1976.
16. Golding, E. W., "Generation of Electricity by Wind," Philosophical Library, New York, 1956.
17. Hall, D. O., Control and Manipulation of Photosynthesis, *Proc. Conf. Solar Energy Agriculture*, U.K. Section, International Solar Energy Society, The Royal Institution, London, 1976.
18. Hall, D. O., Photosynthesis—A Practical Energy Source, *Proc. CENTO Solar Energy Conf., Teheran, Iran, October 1976*.
19. Hall, D. O., and K. K. Rao, "Photosynthesis," Edward Arnold, London, 1972.
20. Heronemus, W. E., et al., Summary of University of Massachusetts Research on Gulf Stream Based Ocean Thermal Power Plants, *Proc. Third Workshop Ocean Thermal Energy Conversion, Houston, Texas, May 1975*, Johns Hopkins University, APL/JHU, SR 75-2, August 1975.

21. Heslop-Harrison, J., Crops, Commodities and Energy Capture, *J. Inst. Biol.*, vol. 22, pp. 60–67, 1975.

22. Hubbert, M. K., The Energy Resources of the Earth, *Sci. Am.*, vol. 224, pp. 60–70, 1971.

23. Hütter, U., Planning and Balancing of Energy of Small-Output Wind Power Plants, *Wind and Solar Energy, Proc. New Delhi Symp.*, UNESCO, pp. 76–89, October 1954.

24. Igra, O., Design and Performance Characteristics of a Shrouded Aerogenerator Turbine, *University of the Negev, Department of Mechanical Engineering Rept.*, Israel, 1972.

25. Kogan, A., and E. Nissim, Shrouded Aerogenerator Design Study, parts I and II, *Bull. Res. Counc. Isr., Sect. C*, vol. 11, 1962 and 1963.

26. Krenz, J. H., "Energy, Conversion and Utilization," Allyn and Bacon, Boston, 1976.

27. Lamb, H., "Hydrodynamics," Dover, New York, 1945.

28. Lavi, A., Solar Sea Power Project, *Carnegie Mellon University Final Rept.* NSF/RANN/SE/GI-39114/PR/74/6, January 1975.

29. Lavi, A., and C. Zener, Plumbing the Ocean Depths: A New Source of Power, *IEEE Spectrum*, vol. 10, p. 22, 1973.

30. Lilley, G. M., and W. J. Rainbird, A Preliminary Report on the Design and Performance of Ducted Windmills, *Cranfield, CoA. Rept.* 102, 1956.

30a. Mandels, M., L. Hontz, and J. Bystrom, Enzymatic Hydrolysis of Waste Cellulose, *Biotechnol. Bioeng.*, vol. 16, pp. 1471–1493, 1974.

31. McGowan, J. G., Ocean Thermal Energy Conversion—A Significant Solar Resource, *Sol. Energy*, vol. 18, pp. 81–92, 1976.

32. Mollison, D., O. P. Buneman, and S. H. Salter, Wave Power Availability in the North-East Atlantic, *Nature*, vol. 263, pp. 223–226, 1976.

33. Myers, J., Physiology of the Algae, *Annu. Rev. Microbiol.*, vol. 5, pp. 157–180, 1951.

34. Myers, J., Algae as an Energy Converter, *Proc. U.N. Conf. New Sources of Energy*, pp. 227–230, United Nations, New York, 1955.

35. Myers, J., J. N. Phillips, and J. R. Graham, On the Mass Culture of Algae, *Plant Physiol.*, vol. 26, pp. 539–548, 1951.

36. New York University, Final Report on the Wind Turbine, Office of Production, Research and Development, War Production Board, PB25370, Washington, D.C., 1946.

37. Oswald, W. J., Productivity of Algae in Sewage Disposal, *Sol. Energy*, vol. 15, pp. 107–117, 1974.

38. Oswald, W. J., Determinants of Feasibility in Bioconversion of Solar Energy, *Proc. 8th Int. Congr. Photobiology*, Plenum Press, London, 1977.

39. Phillips, J. N., and J. Myers, Growth Rate of *Chlorella* in Flashlight, *Plant Physiol.*, vol. 29, pp. 152–161, 1954.

40. Pierson, W. J., Jr., and L. Moskowitz, Proposed Spectral Form for Fully Developed Wind Seas Based on Similarity Theory of S. A. Kitaigorodskii, *J. Geophys. Res.*, vol. 69, pp. 5181–5190, 1964.

41. Putnam, P. C., "Power from the Wind," Van Nostrand Reinhold Co., New York, 1948.

42. Radice, F. C., Jr., Siting of Wind Driven Apparatus, *Proc. 11th IECEC Conf.*, pp. 1736–1740, 1976.

43. Reed, J. W., "Wind Power Climatology of the United States," *Sandia Laboratories* SAND 74-0348, Albuquerque, New Mexico, 1975.
44. Reynolds, J., "Windmills and Watermills," Praeger, New York, 1970.
45. Salter, S. H., Wave Power, *Nature*, vol. 249, pp. 720–724, 1974.
46. Salter, S. H., D. C. Jeffrey, and J. R. M. Taylor, "Wave Power—Nodding Duck Wave Energy Extraction," paper presented at Energy from Ocean Conference, North Carolina State University, Raleigh, N.C., January 1976.
47. Shibata, K., A. A. Benson, and M. Calvin, The Absorption Spectra of Suspensions of Living Micro-organisms, *Biochim. Biophys. Acta*, vol. 15, pp. 461–470, 1954.
48. "Solar Energy—A UK Assessment," U.K.–ISES, The Royal Institution, London, 1976.
49. "Solar Energy in Agriculture," Joint Conf. Proceedings, U.K.–ISES, The Royal Institution, London, 1976.
50. Sorokin, C., and J. Myers, A High Temperature Strain of *Chlorella*, *Science*, vol. 117, pp. 330–331, 1953.
51. Steward, F. C., "Plants at Work," Addison-Wesley Publ. Co., Reading, Mass., 1964.
52. Szego, G. C., and C. C. Kemp, The Energy Plantation, *Proc. 79th Meeting Am. Inst. Chem. Eng.*, American Society of Chemical Engineers, New York, 1975.
53. Tamiya, H., Growing *Chlorella* for Food and Feed, *Proc. World Symp. Applied Solar Energy (1955)*, Johnson Reprint Corp., New York, 1964.
54. Westlake, D. F., Comparisons of Plant Productivity, *Biol. Rev.*, vol. 38, pp. 385–425, 1963.
55. Wilson, R. E., and Peter B. S. Lissaman, "Applied Aerodynamics of Wind Power Machines," NSF/RANN Grant No. GI 41840, Oregon State University, Corvallis, Oregon, 1974.
56. Wittwer, S. H., Maximum Production Capacity for Food Crops, *BioScience*, vol. 24, pp. 216–224, 1974.
57. Zarem, A. M., and Duane D. Erway, eds., "Introduction to the Utilization of Solar Energy," McGraw-Hill Book Co., New York, 1963.

# THE INTERNATIONAL SYSTEM OF UNITS, FUNDAMENTAL CONSTANTS, AND CONVERSION FACTORS

Appendix 1

The International System of Units (SI) is based on seven base units. Other derived units can be related to these base units through governing equations. The base units with the recommended symbols are listed in Table A1.1. Derived units of interest in solar engineering are given in Table A1.2.

Standard prefixes can be used in the SI system to designate multiples of the basic units and thereby conserve space. The standard prefixes are listed in Table A1.3.

Table A1.4 lists some physical constants that are frequently used in solar engineering, together with their values in the SI system of units.

Conversion factors between the SI and English systems for commonly used quantities are given in Table A1.5.

**TABLE A1.1**  The Seven SI Base Units

| Quantity | Name of unit | Symbol |
|---|---|---|
| Length | Meter | m |
| Mass | Kilogram | kg |
| Time | Second | sec |
| Electric current | Ampere | A |
| Thermodynamic temperature | Kelvin | K |
| Luminous intensity | Candela | cd |
| Amount of a substance | Mole | mol |

**TABLE A1.2**  SI Derived Units

| Quantity | Name of unit | Symbol |
|---|---|---|
| Acceleration | Meters per second squared | $m/sec^2$ |
| Area | Square meters | $m^2$ |
| Density | Kilogram per cubic meter | $kg/m^3$ |
| Dynamic viscosity | Newton-second per square meter | $N \cdot sec/m^2$ |
| Force | Newton ($= 1 \ kg \cdot m/sec^2$) | N |
| Frequency | Hertz | Hz |
| Kinematic viscosity | Square meter per second | $m^2/sec$ |
| Plane angle | Radian | rad |
| Potential difference | Volt | V |
| Power | Watt ($= 1 \ J/s$) | W |
| Pressure | Pascal ($= 1 \ N/m^2$) | Pa |
| Radiant intensity | Watts per steradian | W/sr |
| Solid angle | Steradian | sr |
| Specific heat | Joules per kilogram-Kelvin | $J/kg \cdot K$ |
| Thermal conductivity | Watts per meter-Kelvin | $W/m \cdot K$ |
| Velocity | Meters per second | m/sec |
| Volume | Cubic meter | $m^3$ |
| Work, energy, heat | Joule ($= 1 \ N \cdot m$) | J |

**TABLE A1.3**  SI Prefixes

| Multiplier | Symbol | Prefix |
|---|---|---|
| $10^{12}$ | T | Tera |
| $10^9$ | G | Giga |
| $10^6$ | m | Mega |
| $10^3$ | k | Kilo |
| $10^2$ | h | Hecto |
| $10^1$ | da | Deka |
| $10^{-1}$ | d | Deci |
| $10^{-2}$ | c | Centi |
| $10^{-3}$ | m | Milli |
| $10^{-6}$ | $\mu$ | Micro |
| $10^{-9}$ | n | Nano |
| $10^{-12}$ | p | Pico |
| $10^{-15}$ | f | Femto |
| $10^{-18}$ | a | Atto |

**TABLE A1.4**  Physical Constants in SI Units

| Quantity | Symbol | Value |
|---|---|---|
| Avogadro constant | $N$ | $6.022169 \times 10^{26} \ kmol^{-1}$ |
| Boltzmann constant | $k$ | $1.380622 \times 10^{-23} \ J/K^1$ |
| First radiation constant | $C_1 = 2\pi hc^2$ | $3.741844 \times 10^{-16} \ W \cdot m^2$ |
| Gas constant | $R$ | $8.31434 \ \times 10^3 \ J/kmol \cdot K$ |
| Planck constant | $h$ | $6.626196 \times 10^{-34} \ J \cdot sec$ |
| Second radiation constant | $C_2 = hc/k$ | $1.438833 \times 10^{-2} \ m \cdot K$ |
| Speed of light in a vacuum | $c$ | $2.997925 \times 10^8 \ m/sec^1$ |
| Stefan-Boltzmann constant | $\sigma$ | $5.66961 \ \times 10^{-8} \ W/m^2 \cdot K^4$ |

**TABLE A1.5** Conversion Factors

| Physical quantity | Symbol | Conversion factor |
|---|---|---|
| Area | $A$ | 1 ft$^2$ = 0.0929 m$^2$ |
| | | 1 in$^2$ = 6.452 × 10$^{-4}$ m$^2$ |
| Density | $\rho$ | 1 lb$_m$/ft$^3$ = 16.018 kg/m$^3$ |
| | | 1 slug/ft$^3$ = 515.379 kg/m$^3$ |
| Heat, energy, or work | $Q$ or $W$ | 1 Btu = 1055.1 J |
| | | 1 cal = 4.186 J |
| | | 1 ft · lb$_f$ = 1.3558 J |
| | | 1 hp · hr = 2.685 × 10$^6$ J |
| Force | $F$ | 1 lb$_f$ = 4.448 N |
| Heat flow rate | $q$ | 1 Btu/hr = 0.2931 W |
| | | 1 Btu/sec = 1055.1 W |
| Heat flux | $q/A$ | 1 Btu/hr · ft$^2$ = 3.1525 W/m$^2$ |
| Heat-transfer coefficient | $h$ | 1 Btu/hr · ft$^2$ · F = 5.678 W/m$^2$ · K |
| Length | $L$ | 1 ft = 0.3048 m |
| | | 1 in = 2.54 cm |
| | | 1 mi = 1.6093 km |
| Mass | $m$ | 1 lb$_m$ = 0.4536 kg |
| | | 1 slug = 14.594 kg |
| Mass flow rate | $\dot{m}$ | 1 lb$_m$/hr = 0.000126 kg/sec |
| | | 1 lb$_m$/sec = 0.4536 kg/sec |
| Power | $\dot{W}$ | 1 hp = 745.7 W |
| | | 1 ft · lb$_f$/sec = 1.3558 W |
| | | 1 Btu/sec = 1055.1 W |
| | | 1 Btu/hr = 0.293 W |
| Pressure | $p$ | 1 lb$_f$/in$^2$ = 6894.8 Pa (N/m$^2$) |
| | | 1 lb$_f$/ft$^2$ = 47.88 Pa (N/m$^2$) |
| | | 1 atm = 101,325 Pa (N/m$^2$) |
| Radiation | $I$ | 1 langley = 41,860 J/m$^2$ |
| Specific heat capacity | $c$ | 1 Btu/lb$_m$ · °F = 4187 J/kg · K |
| Internal energy or enthalpy | $e$ or $h$ | 1 Btu/lb$_m$ = 2326.0 J/kg |
| | | 1 cal/g = 4184 J/kg |
| Temperature | $T$ | $T(°R) = (9/5)T(K)$ |
| | | $T(°F) = [T(°C)](9/5) + 32$ |
| | | $T(°F) = [T(K) - 273.15](9/5) + 32$ |
| Thermal conductivity | $k$ | 1 Btu/hr · ft · °F = 1.731 W/m · K |
| Thermal resistance | $R_{th}$ | 1 hr · °F/Btu = 1.8958 K/W |
| Velocity | $V$ | 1 ft/sec = 0.3048 m/sec |
| | | 1 mi/hr = 0.44703 m/sec |
| Viscosity, dynamic | $\mu$ | 1 lb$_m$/ft · sec = 1.488 N · sec/m$^2$ |
| | | 1 cP = 0.00100 N · sec/m$^2$ |
| Viscosity, kinematic | $\nu$ | 1 ft$^2$/sec = 0.09029 m$^2$/sec |
| | | 1 ft$^2$/hr = 2.581 × 10$^{-5}$ m$^2$/sec |
| Volume | $V$ | 1 ft$^3$ = 0.02832 m$^3$ |
| | | 1 in$^3$ = 1.6387 × 10$^{-5}$ m$^3$ |
| | | 1 gal (U.S. liq.) = 0.003785 m$^3$ |
| Volumetric flow rate | $\dot{Q}$ | 1 ft$^3$/min = 0.000472 m$^3$/sec |

**TABLE A2.1a**  Solar Irradiance for Different Air Masses[a]

| Wavelength | Air mass; $\alpha = 0.66$; $\beta = 0.085$[b] | | | | |
|---|---|---|---|---|---|
| | 0 | 1 | 4 | 7 | 10 |
| 0.290 | 482.0 | 0.0 | 0.0 | 0.0 | 0.0 |
| 0.295 | 584.0 | 0.0 | 0.0 | 0.0 | 0.0 |
| 0.300 | 514.0 | 4.1 | 0.0 | 0.0 | 0.0 |
| 0.305 | 603.0 | 11.4 | 0.0 | 0.0 | 0.0 |
| 0.310 | 689.0 | 30.5 | 0.0 | 0.0 | 0.0 |
| 0.315 | 764.0 | 79.4 | 0.1 | 0.0 | 0.0 |
| 0.320 | 830.0 | 202.6 | 2.9 | 0.0 | 0.0 |
| 0.325 | 975.0 | 269.5 | 5.7 | 0.1 | 0.0 |
| 0.330 | 1059.0 | 331.6 | 10.2 | 0.3 | 0.0 |
| 0.335 | 1081.0 | 383.4 | 17.1 | 0.8 | 0.0 |
| 0.340 | 1074.0 | 431.3 | 24.9 | 1.8 | 0.1 |
| 0.345 | 1069.0 | 449.2 | 33.3 | 2.5 | 0.2 |
| 0.350 | 1093.0 | 480.5 | 40.8 | 3.5 | 0.3 |
| 0.355 | 1083.0 | 498.0 | 48.4 | 4.7 | 0.5 |
| 0.360 | 1068.0 | 513.7 | 57.2 | 6.4 | 0.7 |
| 0.365 | 1132.0 | 561.3 | 68.4 | 8.3 | 1.0 |
| 0.370 | 1181.0 | 603.5 | 80.5 | 10.7 | 1.4 |
| 0.375 | 1157.0 | 609.4 | 89.0 | 13.0 | 1.9 |
| 0.380 | 1120.0 | 608.0 | 97.2 | 15.6 | 2.5 |
| 0.385 | 1098.0 | 609.8 | 104.5 | 17.9 | 3.1 |
| 0.390 | 1098.0 | 623.9 | 114.5 | 21.0 | 3.9 |
| 0.395 | 1189.0 | 691.2 | 135.8 | 26.7 | 5.2 |
| 0.400 | 1429.0 | 849.9 | 178.8 | 37.6 | 7.9 |
| 0.405 | 1644.0 | 992.8 | 218.7 | 48.2 | 10.6 |
| 0.410 | 1751.0 | 1073.7 | 247.5 | 57.1 | 13.2 |

Note: See page 657 for footnotes.

**TABLE A2.1***a*  Solar Irradiance for Different Air
Masses*a* (*Continued*)

| Wavelength | Air mass; $\alpha = 0.66$; $\beta = 0.085^b$ | | | | |
|---|---|---|---|---|---|
| | 0 | 1 | 4 | 7 | 10 |
| 0.415 | 1774.0 | 1104.5 | 266.5 | 64.3 | 15.5 |
| 0.420 | 1747.0 | 1104.3 | 278.9 | 70.4 | 17.8 |
| 0.425 | 1693.0 | 1086.5 | 287.2 | 78.9 | 20.1 |
| 0.430 | 1639.0 | 1067.9 | 295.4 | 81.7 | 22.6 |
| 0.435 | 1663.0 | 1100.1 | 318.4 | 92.2 | 26.7 |
| 0.440 | 1810.0 | 1215.5 | 368.2 | 111.5 | 33.8 |
| 0.445 | 1922.0 | 1310.4 | 415.3 | 131.6 | 41.7 |
| 0.450 | 2006.0 | 1388.4 | 460.3 | 152.6 | 50.6 |
| 0.455 | 2057.0 | 1434.8 | 486.9 | 165.2 | 56.1 |
| 0.460 | 2066.0 | 1452.2 | 504.4 | 175.2 | 60.8 |
| 0.465 | 2048.0 | 1450.7 | 515.7 | 183.3 | 65.1 |
| 0.470 | 2033.0 | 1451.2 | 527.9 | 192.0 | 69.8 |
| 0.475 | 2044.0 | 1470.3 | 547.3 | 203.7 | 75.8 |
| 0.480 | 2074.0 | 1503.4 | 572.6 | 218.1 | 83.1 |
| 0.485 | 1976.0 | 1443.3 | 562.4 | 219.2 | 85.4 |
| 0.490 | 1950.0 | 1435.2 | 572.2 | 228.2 | 91.0 |
| 0.495 | 1960.0 | 1453.6 | 592.9 | 241.9 | 98.7 |
| 0.500 | 1942.0 | 1451.2 | 605.6 | 252.7 | 105.5 |
| 0.505 | 1920.0 | 1440.1 | 607.6 | 256.4 | 108.2 |
| 0.510 | 1882.0 | 1416.8 | 604.4 | 257.8 | 110.0 |
| 0.515 | 1833.0 | 1384.9 | 597.3 | 257.6 | 111.1 |
| 0.520 | 1833.0 | 1390.0 | 606.1 | 264.3 | 115.2 |
| 0.525 | 1852.0 | 1409.5 | 621.3 | 273.9 | 120.7 |
| 0.530 | 1842.0 | 1406.9 | 626.9 | 279.4 | 124.5 |
| 0.535 | 1818.0 | 1393.6 | 627.7 | 282.8 | 127.4 |
| 0.540 | 1783.0 | 1371.7 | 624.5 | 284.4 | 129.5 |
| 0.545 | 1754.0 | 1354.2 | 623.2 | 286.8 | 132.0 |
| 0.550 | 1725.0 | 1336.6 | 621.7 | 289.2 | 134.5 |
| 0.555 | 1720.0 | 1335.7 | 625.5 | 293.0 | 137.3 |
| 0.560 | 1695.0 | 1319.2 | 622.0 | 293.3 | 138.3 |
| 0.565 | 1705.0 | 1330.0 | 631.3 | 299.6 | 142.2 |
| 0.570 | 1712.0 | 1338.4 | 639.5 | 305.6 | 146.0 |
| 0.575 | 1719.0 | 1346.9 | 647.8 | 311.6 | 149.6 |
| 0,580 | 1715.0 | 1346.7 | 652.0 | 315.7 | 152.8 |
| 0.585 | 1712.0 | 1347.3 | 656.6 | 320.0 | 156.0 |
| 0.590 | 1700.0 | 1340.7 | 657.7 | 322.6 | 158.3 |
| 0.595 | 1682.0 | 1329.4 | 656.4 | 324.1 | 160.0 |
| 0.600 | 1660.0 | 1319.6 | 655.8 | 325.9 | 162.0 |
| 0.605 | 1647.0 | 1311.0 | 661.3 | 333.6 | 168.2 |
| 0.610 | 1635.0 | 1307.9 | 669.6 | 342.8 | 175.5 |
| 0.620 | 1602.0 | 1294.2 | 682.4 | 359.9 | 189.7 |
| 0.630 | 1570.0 | 1280.9 | 695.6 | 377.8 | 205.2 |
| 0.640 | 1544.0 | 1272.1 | 711.4 | 397.9 | 222.5 |
| 0.650 | 1511.0 | 1257.1 | 723.9 | 416.9 | 240.1 |
| 0.660 | 1486.0 | 1244.2 | 730.2 | 428.6 | 251.6 |

| Wavelength | Air mass; $\alpha = 0.66; \beta = 0.085$[b] | | | | |
|---|---|---|---|---|---|
| | 0 | 1 | 4 | 7 | 10 |
| 0.670 | 1456.0 | 1226.8 | 733.8 | 438.9 | 262.5 |
| 0.680 | 1427.0 | 1209.9 | 737.4 | 449.5 | 273.9 |
| 0.690 | 1402.0 | 1196.2 | 742.9 | 461.3 | 286.5 |
| 0.698 | 1374.6 | 1010.3 | 546.1 | 311.8 | 181.6 |
| 0.700 | 1369.0 | 1175.3 | 743.7 | 470.6 | 297.7 |
| 0.710 | 1344.0 | 1157.4 | 739.2 | 472.1 | 301.5 |
| 0.720 | 1314.0 | 1135.1 | 731.7 | 471.6 | 304.0 |
| 0.728 | 1295.5 | 1003.1 | 582.3 | 351.7 | 212.5 |
| 0.730 | 1290.0 | 1117.8 | 727.1 | 479.0 | 307.7 |
| 0.740 | 1260.0 | 1095.1 | 718.9 | 471.9 | 309.8 |
| 0.750 | 1235.0 | 1076.6 | 713.2 | 472.4 | 313.0 |
| 0.762 | 1205.5 | 794.0 | 357.1 | 163.6 | 69.1 |
| 0.770 | 1185.0 | 1039.2 | 700.8 | 472.7 | 318.8 |
| 0.780 | 1159.0 | 1019.4 | 693.6 | 472.0 | 321.1 |
| 0.790 | 1134.0 | 1000.3 | 686.7 | 471.4 | 323.6 |
| 0.800 | 1109.0 | 981.2 | 679.4 | 470.5 | 325.8 |
| 0.806 | 1095.1 | 874.4 | 547.7 | 355.9 | 234.4 |
| 0.825 | 1048.0 | 931.6 | 654.3 | 459.6 | 322.8 |
| 0.830 | 1036.0 | 921.8 | 649.3 | 457.3 | 322.1 |
| 0.835 | 1024.5 | 912.4 | 644.4 | 455.2 | 321.5 |
| 0.846 | 998.1 | 476.2 | 181.0 | 85.9 | 44.2 |
| 0.860 | 968.0 | 506.4 | 212.0 | 107.4 | 58.3 |
| 0.870 | 947.0 | 453.8 | 174.7 | 84.0 | 43.8 |
| 0.875 | 436.5 | 449.2 | 173.4 | 83.6 | 43.7 |
| 0.887 | 912.5 | 448.6 | 178.3 | 87.7 | 46.7 |
| 0.900 | 891.0 | 448.9 | 183.7 | 92.3 | 50.0 |
| 0.907 | 882.8 | 455.2 | 190.9 | 97.6 | 53.7 |
| 0.915 | 874.5 | 461.5 | 198.5 | 103.2 | 57.5 |
| 0.925 | 863.5 | 279.0 | 73.6 | 28.0 | 12.1 |
| 0.930 | 858.0 | 221.8 | 46.9 | 15.4 | 6.0 |
| 0.940 | 847.0 | 313.4 | 95.0 | 39.6 | 18.5 |
| 0.950 | 837.0 | 296.5 | 86.3 | 35.0 | 16.0 |
| 0.955 | 828.5 | 321.1 | 102.3 | 44.1 | 21.2 |
| 0.965 | 811.5 | 344.4 | 120.4 | 55.1 | 27.8 |
| 0.975 | 794.0 | 576.9 | 346.0 | 224.6 | 150.1 |
| 0.985 | 776.0 | 544.6 | 316.1 | 201.2 | 132.4 |
| 1.018 | 719.2 | 617.5 | 391.0 | 247.5 | 156.7 |
| 1.082 | 620.0 | 512.9 | 290.4 | 164.4 | 93.1 |
| 1.094 | 602.0 | 464.1 | 303.1 | 210.8 | 149.9 |
| 1.098 | 596.0 | 503.7 | 304.1 | 183.6 | 110.9 |
| 1.101 | 591.8 | 504.8 | 362.7 | 267.3 | 198.8 |
| 1.128 | 560.5 | 135.1 | 27.7 | 9.1 | 3.6 |
| 1.131 | 557.0 | 152.2 | 35.3 | 12.6 | 5.3 |
| 1.137 | 550.1 | 143.1 | 31.7 | 11.0 | 4.5 |
| 1.144 | 542.0 | 191.2 | 57.4 | 24.2 | 11.6 |

**TABLE A2.1a** Solar Irradiance for Different Air Masses[a] (*Continued*)

| Wavelength | Air mass; $\alpha = 0.66$; $\beta = 0.085$[b] | | | | |
|---|---|---|---|---|---|
| | 0 | 1 | 4 | 7 | 10 |
| 1.147 | 538.5 | 174.5 | 48.2 | 19.3 | 8.8 |
| 1.178 | 507.0 | 399.3 | 195.1 | 95.4 | 46.6 |
| 1.189 | 496.0 | 402.2 | 214.5 | 114.4 | 61.0 |
| 1.193 | 492.0 | 424.0 | 310.8 | 233.3 | 176.6 |
| 1.222 | 464.3 | 391.8 | 235.3 | 141.3 | 84.9 |
| 1.236 | 451.2 | 390.8 | 254.1 | 165.2 | 107.4 |
| 1.264 | 426.5 | 329.2 | 209.7 | 140.0 | 94.3 |
| 1.276 | 416.7 | 342.6 | 238.6 | 172.6 | 126.3 |
| 1.288 | 406.8 | 347.3 | 216.1 | 134.4 | 83.7 |
| 1.314 | 386.1 | 298.3 | 137.6 | 63.5 | 29.3 |
| 1.335 | 369.7 | 190.6 | 85.0 | 46.7 | 27.7 |
| 1.384 | 343.7 | 5.7 | 0.1 | 0.0 | 0.0 |
| 1.432 | 321.0 | 44.6 | 5.4 | 1.3 | 0.4 |
| 1.457 | 308.6 | 85.4 | 20.6 | 7.7 | 3.3 |
| 1.472 | 301.4 | 77.4 | 17.4 | 6.2 | 2.6 |
| 1.542 | 270.4 | 239.3 | 165.9 | 115.0 | 79.7 |
| 1.572 | 257.3 | 222.6 | 168.1 | 130.4 | 102.1 |
| 1.599 | 245.4 | 216.0 | 166.7 | 131.5 | 104.5 |
| 1.608 | 241.5 | 208.5 | 157.4 | 122.1 | 95.7 |
| 1.626 | 233.6 | 206.7 | 160.7 | 127.5 | 101.9 |
| 1.644 | 225.6 | 197.9 | 152.4 | 120.1 | 95.5 |
| 1.650 | 223.0 | 195.7 | 150.9 | 119.1 | 94.7 |
| 1.676 | 212.1 | 181.9 | 114.8 | 72.4 | 45.7 |
| 1.732 | 187.9 | 161.5 | 102.5 | 65.1 | 41.3 |
| 1.782 | 166.6 | 136.7 | 75.6 | 41.8 | 23.1 |
| 1.862 | 138.2 | 4.0 | 0.1 | 0.0 | 0.0 |
| 1.955 | 112.9 | 42.7 | 14.5 | 6.8 | 3.6 |
| 2.008 | 102.0 | 69.4 | 35.8 | 17.7 | 6.4 |
| 2.014 | 101.2 | 74.7 | 45.5 | 28.8 | 17.8 |
| 2.057 | 95.6 | 69.5 | 41.3 | 25.3 | 14.8 |
| 2.124 | 87.4 | 70.0 | 35.9 | 18.4 | 9.5 |
| 2.156 | 83.8 | 66.0 | 32.3 | 15.8 | 7.7 |
| 2.201 | 78.9 | 66.1 | 49.1 | 38.0 | 29.7 |
| 2.266 | 72.4 | 61.6 | 46.8 | 36.8 | 29.3 |
| 2.320 | 67.6 | 57.2 | 43.2 | 33.8 | 26.8 |
| 2.338 | 66.3 | 54.7 | 39.9 | 30.4 | 23.4 |
| 2.356 | 65.1 | 52.0 | 36.3 | 26.5 | 19.6 |
| 2.388 | 62.8 | 36.0 | 18.7 | 11.7 | 7.8 |
| 2.415 | 61.0 | 32.5 | 15.8 | 9.4 | 6.0 |
| 2.453 | 58.3 | 29.6 | 13.7 | 7.9 | 5.0 |
| 2.494 | 55.4 | 20.3 | 6.8 | 3.2 | 1.7 |
| 2.537 | 52.4 | 4.6 | 0.4 | 0.1 | 0.0 |
| 2.900 | 35.0 | 2.9 | 0.2 | 0.0 | 0.0 |
| 2.941 | 33.4 | 6.0 | 1.0 | 0.3 | 0.1 |
| 2.954 | 32.8 | 5.7 | 0.9 | 0.3 | 0.1 |

**TABLE A2.1a** Solar Irradiance for Different Air Masses[a] (*Continued*)

| Wavelength | Air mass; $\alpha = 0.66$; $\beta = 0.085$[b] | | | | |
|---|---|---|---|---|---|
| | 0 | 1 | 4 | 7 | 10 |
| 2.973 | 32.1 | 8.7 | 2.2 | 0.9 | 0.4 |
| 3.005 | 30.8 | 7.8 | 1.8 | 0.7 | 0.3 |
| 3.045 | 28.8 | 4.7 | 0.7 | 0.2 | 0.1 |
| 3.056 | 28.2 | 4.9 | 0.8 | 0.2 | 0.1 |
| 3.097 | 26.2 | 3.2 | 0.4 | 0.1 | 0.0 |
| 3.132 | 24.9 | 6.8 | 1.7 | 0.7 | 0.3 |
| 3.156 | 24.1 | 18.7 | 12.6 | 8.9 | 6.3 |
| 3.204 | 22.5 | 2.1 | 0.2 | 0.0 | 0.0 |
| 3.214 | 22.1 | 3.4 | 0.5 | 0.1 | 0.0 |
| 3.245 | 21.1 | 3.9 | 0.7 | 0.2 | 0.1 |
| 3.260 | 20.6 | 3.7 | 0.6 | 0.2 | 0.1 |
| 3.285 | 19.7 | 14.2 | 8.5 | 5.1 | 2.8 |
| 3.317 | 18.8 | 12.9 | 6.9 | 3.5 | 1.3 |
| 3.344 | 18.1 | 4.2 | 0.9 | 0.3 | 0.1 |
| 3.403 | 16.5 | 12.3 | 7.8 | 5.1 | 3.2 |
| 3.450 | 15.6 | 12.5 | 8.9 | 6.7 | 5.0 |
| 3.507 | 14.5 | 12.5 | 9.9 | 8.1 | 6.7 |
| 3.538 | 14.2 | 11.8 | 8.8 | 6.9 | 5.5 |
| 3.573 | 13.8 | 10.9 | 5.4 | 2.6 | 1.3 |
| 3.633 | 13.1 | 10.8 | 8.3 | 6.7 | 5.5 |
| 3.673 | 12.6 | 9.1 | 6.1 | 4.6 | 3.5 |
| 3.696 | 12.3 | 10.4 | 8.2 | 6.7 | 5.6 |
| 3.712 | 12.2 | 10.9 | 9.0 | 7.6 | 6.5 |
| 3.765 | 11.5 | 9.5 | 7.2 | 5.9 | 4.8 |
| 3.812 | 11.0 | 8.9 | 6.7 | 5.4 | 4.4 |
| 3.888 | 10.4 | 8.1 | 5.6 | 4.0 | 2.9 |
| 3.923 | 10.1 | 8.0 | 5.6 | 4.2 | 3.1 |
| 3.948 | 9.9 | 7.8 | 5.5 | 4.0 | 3.0 |
| 4.045 | 9.1 | 6.7 | 4.1 | 2.6 | 1.5 |
| Total $W \cdot m^2$ | 1353 | 889.2 | 448.7 | 255.2 | 153.8 |

[a]$W/m^2 \cdot \mu m$; $H_2O$ 20 min; $O_3$ 3.4 mm. From Thekaekara, M. P., Data on Incident Solar Energy, "The Energy Crisis and Energy from the Sun," Institute for Environmental Sciences, 1974.

[b]The parameters $\alpha$ and $\beta$ are measures of turbidity of the atmosphere. They are used in the atmospheric transmittance equation $\bar{\tau}_{atm} = e^{-(C_1 + C_2)m}$; $C_1$ includes Rayleigh and ozone attenuation; $C_2 \equiv \beta/\lambda^\alpha$.

**TABLE A2.1***b*  Direct Solar Irradiance as a Function of Air Mass for Varying Atmospheric Parameters[a]

| | | | | Solar irradiance | | | | |
|---|---|---|---|---|---|---|---|---|
| | | | | For air mass | | | | |
| | | | | 1 | 1.5 | 2 | 3 | 5 |
| Atmospheric parameters | | | | For solar zenith angles | | | | |
| $H_2O_{cm}$ | $O_{3\,cm}$ | $\beta$ | $\alpha$ | 0 | 48.2° | 60° | 70.5° | 78.5° |
| 2.0 | 0.20 | 0.02 | 1.3 | 961 | 876 | 805 | 693 | 534 |
| 2.0 | 0.20 | 0.17 | 0.66 | 804 | 672 | 568 | 414 | 232 |
| 2.0 | 0.34 | 0.02 | 1.3 | 956 | 870 | 798 | 684 | 523 |
| 2.0 | 0.34 | 0.17 | 0.66 | 800 | 668 | 563 | 409 | 228 |
| 2.0 | 0.55 | 0.02 | 1.3 | 949 | 861 | 788 | 672 | 508 |
| 2.0 | 0.55 | 0.17 | 0.66 | 795 | 661 | 556 | 402 | 222 |
| 0.5 | 0.34 | 0.02 | 1.3 | 1024 | 943 | 875 | 767 | 612 |
| 0.5 | 0.34 | 0.17 | 0.66 | 859 | 727 | 622 | 464 | 273 |
| 1.0 | 0.34 | 0.02 | 1.3 | 992 | 909 | 840 | 729 | 571 |
| 1.0 | 0.34 | 0.17 | 0.66 | 832 | 700 | 595 | 439 | 252 |
| 5.0 | 0.34 | 0.02 | 1.3 | 898 | 807 | 732 | 615 | 454 |
| 5.0 | 0.34 | 0.17 | 0.66 | 750 | 616 | 513 | 363 | 193 |

[a]In W/m$^2$ ; see p. 657 for definition of $\alpha$ and $\beta$.

**TABLE A2.2***a*  Monthly Averaged, Daily Extraterrestrial Insolation on a Horizontal Surface[a]

| Latitude (deg) | Jan. | Feb. | Mar. | Apr. | May | June | July | Aug. | Sep. | Oct. | Nov. | Dec. |
|---|---|---|---|---|---|---|---|---|---|---|---|---|
| 20 | 2346 | 2656 | 3021 | 3297 | 3417 | 3438 | 3414 | 3321 | 3097 | 2748 | 2404 | 2238 |
| 25 | 2105 | 2458 | 2896 | 3262 | 3460 | 3517 | 3476 | 3316 | 3003 | 2571 | 2173 | 1988 |
| 30 | 1854 | 2242 | 2748 | 3204 | 3480 | 3576 | 3516 | 3288 | 2887 | 2377 | 1931 | 1728 |
| 35 | 1594 | 2012 | 2579 | 3122 | 3478 | 3613 | 3534 | 3237 | 2748 | 2165 | 1678 | 1462 |
| 40 | 1329 | 1769 | 2391 | 3018 | 3455 | 3631 | 3532 | 3164 | 2589 | 1939 | 1418 | 1193 |
| 45 | 1061 | 1515 | 2185 | 2893 | 3412 | 3631 | 3511 | 3070 | 2410 | 1700 | 1154 | 925 |
| 50 | 797 | 1255 | 1963 | 2748 | 3351 | 3616 | 3474 | 2957 | 2214 | 1450 | 890 | 664 |
| 55 | 541 | 991 | 1727 | 2585 | 3277 | 3591 | 3424 | 2826 | 2001 | 1192 | 632 | 417 |
| 60 | 305 | 727 | 1478 | 2407 | 3194 | 3567 | 3371 | 2683 | 1773 | 931 | 388 | 197 |
| 65 | 106 | 472 | 1219 | 2217 | 3116 | 3568 | 3331 | 2531 | 1533 | 670 | 172 | 31 |

[a]In Btu/day · ft$^2$. From Kreider, J. F., and F. Kreith, "Solar Heating and Cooling," revised 1st ed., Hemisphere Publ. Corp., 1977.

**TABLE A2.2b** Monthly Averaged, Daily Extraterrestrial Insolation on a Horizontal Surface[a]

| Latitude (deg) | Jan. | Feb. | Mar. | Apr. | May | June | July | Aug. | Sep. | Oct. | Nov. | Dec. |
|---|---|---|---|---|---|---|---|---|---|---|---|---|
| 20 | 7415 | 8397 | 9552 | 10422 | 10801 | 10868 | 10794 | 10499 | 9791 | 8686 | 7598 | 7076 |
| 25 | 6656 | 7769 | 9153 | 10312 | 10936 | 11119 | 10988 | 10484 | 9494 | 8129 | 6871 | 6284 |
| 30 | 5861 | 7087 | 8686 | 10127 | 11001 | 11303 | 11114 | 10395 | 9125 | 7513 | 6103 | 5463 |
| 35 | 5039 | 6359 | 8153 | 9869 | 10995 | 11422 | 11172 | 10233 | 8687 | 6845 | 5304 | 4621 |
| 40 | 4200 | 5591 | 7559 | 9540 | 10922 | 11478 | 11165 | 10002 | 8184 | 6129 | 4483 | 3771 |
| 45 | 3355 | 4791 | 6909 | 9145 | 10786 | 11477 | 11099 | 9705 | 7620 | 5373 | 3648 | 2925 |
| 50 | 2519 | 3967 | 6207 | 8686 | 10594 | 11430 | 10981 | 9347 | 6998 | 4583 | 2815 | 2100 |
| 55 | 1711 | 3132 | 5460 | 8171 | 10358 | 11352 | 10825 | 8935 | 6325 | 3770 | 1999 | 1320 |
| 60 | 963 | 2299 | 4673 | 7608 | 10097 | 11276 | 10657 | 8480 | 5605 | 2942 | 1227 | 623 |
| 65 | 334 | 1491 | 3855 | 7008 | 9852 | 11279 | 10531 | 8001 | 4846 | 2116 | 544 | 97 |

[a]In W · hr/m².

**TABLE A2.3** Pyranometer and Pyrheliometer Stations of the U.S. National Weather Service

| | | Pyranometer stations | | | |
|---|---|---|---|---|---|
| State | Station | WBAN no. | Type of record | Chart | Period of record |
| AK | Annette | 25308 | Daily | Cir. | 1949–Pres. |
| | Barrow<br>(Hourly data worked but not processed.) | 27502 | Daily | Roll | 1951–Pres. |
| | Bethel<br>(Hourly data worked but not processed.) | 26615 | Daily | Roll | 1949–Pres. |
| | Fairbanks<br>(Hourly data worked but not processed.) | 26411 | Daily | Roll | 1942–Pres. |
| | Matanuska<br>(Hourly data worked but not processed.) | X5733 | Daily | Roll | 1954–Pres. |
| AZ | Phoenix<br>(Hourly radiation discontinued per RAO 1967.) | 23182 | Daily | Roll | 1949–Pres. |
| | Page<br>(Hourly data worked WRPC–NCC 1959–1970.) | X6180 | Daily | Roll | 1959–Pres. |
| AR | Little Rock | 13963 | Daily | Cir. | 1949–Pres. |
| CA | Fresno<br>(Hourly data discontinued per RAO 1967.) | 93193 | Daily | Roll | 1953–Pres. |
| | Los Angeles WBAS<br>(Hourly data 1962–1967?) | 23174 | Daily | Cir.<br>Cir.<br>Roll<br>Cir. | 1951–Pres.<br>1951–1961<br>1962–1968<br>1968–1973 |
| | Los Angeles Civic Center | 93134 | Daily | Cir.[a]<br><br>Roll[b] | 1949–Pres.<br>1949–1968<br>1968–Pres. |
| | Santa Maria<br>(Hourly data discontinued per RAO 1967.) | 23273 | Daily | Roll | 1949–Pres. |
| CO | Grand Junction | 23066 | Daily | Cir. | 1949–Pres. |
| DC | Washington-Sterling | 93734 | Daily/hourly | Roll | 1950–Pres. |
| | WBO Washington | 93725 | Daily/hourly | Roll<br>Cir. | 1950–1952<br>1953–1958 |
| | Silver Hill | 93722 | Daily/hourly | Roll | 1953–1960 |
| | Sterling | 93734 | Daily/hourly | Roll | 1960–Pres. |
| FL | Apalachicola | 12832 | Daily/hourly | Roll | 1949–Pres. |

Note: See page 663 for footnotes.

**TABLE A2.3** Pyranometer and Pyrheliometer Stations of the U.S. National Weather Service (*Continued*)

| | Pyranometer stations | | | | |
|---|---|---|---|---|---|
| State | Station | WBAN no. | Type of record | Chart | Period of record |
| FL | Lakeland | 12883 | Daily | Cir. | 1963–Pres. |
| | Miami | 12839 | Daily/hourly | Roll | 1949–Pres. |
| | Tallahassee | 93805 | Daily | Roll | 1968–Pres. |
| | (Hourly data worked and can be punched.) | | | | |
| | Tampa | 12842 | Daily | Cir. | 1949–Pres. |
| GA | Atlanta | 13874 | Daily | Cir. | 1949–Pres. |
| HI | Mauna Loa | X6198 | Daily | None | ?– |
| | (Hourly data worked and can be punched.) | | | | |
| ID | Boise | 24131 | Daily | Cir. | 1949–Pres. |
| IN | Indianapolis | 93819 | Daily | Cir. | 1949–Pres. |
| IA | Ames (closed) | X0201 | Daily | Roll | 1959–1972 |
| | (Hourly data worked and can be punched.) (Number of charts in house unknown.) | | | | |
| LA | Lake Charles | 03937 | Daily/hourly | Roll | 1949–Pres. |
| ME | Caribou | 14607 | Daily/hourly | Roll | 1949–Pres. |
| | Portland | 14764 | Daily | Cir. | 1949–Pres. |
| MA | Blue Hill | 14753 | Daily/hourly | Roll | 1954–Pres. |
| | (Charts missing 1969–1971.) | | | | |
| | Boston (closed) | 94701 | ? | Roll | 1945–1968 |
| MI | Sault Ste. Marie | 14847 | Daily/? | Cir. | 1950–Pres. |
| | (Roll 1950–1958; cir. 1958–1973) | | | | |
| MN | St. Cloud | 14926 | Daily | Cir. | 1949–Pres. |
| MO | Columbia | 03945 | Daily/hourly | Roll | 1950–Pres. |
| MT | Glasgow | 94008 | Daily | Cir. | 1950–Pres. |
| | Great Falls | 24143 | Daily | Roll | 1949–Pres. |
| | (Hourly data discontinued per RAO 1967.) | | | | |
| NE | Lincoln (closed) | 14971 | Daily/hourly | ? | 1942–1959 |
| | North Omaha | 94918 | Daily/hourly | Roll | 1956–Pres. |
| NV | Ely | 23154 | Daily | Roll | 1949–Pres. |
| | (Hourly data discontinued per RAO 1967.) | | | | |

| Pyranometer stations | | | | |
|---|---|---|---|---|
| State | Station | WBAN no. | Type of record | Chart | Period of record |

| State | Station | WBAN no. | Type of record | Chart | Period of record |
|---|---|---|---|---|---|
| NV | Las Vegas | 23169 | Daily | Cir. | 1949–Pres. |
| | Reno | 23185 | Daily | None | ? |
| NM | Albuquerque | 23050 | Daily/hourly | Roll | 1949–Pres. |
| NY | New York Central Park | 94728 94706 | Daily/hourly | Roll | 1942–Pres. |
| NC | Cape Hatteras | 93729 | Daily/hourly | Roll | 1947–Pres. |
| | Greensboro | 13723 | Daily | Cir. | 1949–Pres. |
| ND | Bismarck | 24011 | Daily/hourly | Roll | 1950–Pres. |
| OH | Cleveland (Roll 1949–1953; cir. 1955–1973.) | 14820 | Daily | Cir. | 1949–Pres. |
| OK | Oklahoma City | 13967 | Daily | Cir. | 1949–Pres. |
| OR | Astoria | 94224 | Daily | Cir. | 1949–Pres. |
| | Medford (Hourly data discontinued per RAO 1967–cir. 1949–1954; roll 1955–Present.) | 24225 | Daily | Roll | 1949–Pres. |
| SC | Charleston | 13880 | Daily/hourly | Roll | 1949–Pres. |
| SD | Rapid City | 24090 | Daily | Cir. | 1949–Pres. |
| TN | Nashville | 13897 | Daily/hourly | Roll | 1942–Pres. |
| | Oak Ridge (Hourly data worked and can be punched. Corrected data available from station.) | 03841 | Daily | None | 1949–Pres. |
| TX | Brownsville | 12919 | Daily/hourly | Roll | 1949–Pres. |
| | El Paso | 23044 | Daily/hourly | Roll | 1949–Pres. |
| | Ft. Worth | 03927 | Daily/hourly | Roll | 1949–Pres. |
| | Midland | 23023 | Daily | Cir. | 1953–Pres. |
| | San Antonio | 12921 | Daily | Cir. | 1949–Pres. |
| UT | Flaming Gorge (Hourly data worked WRPC and NCC, 1959–1970. Charts for this period only.) | X2864 | Daily | Roll | 1959–Pres. |
| | Salt Lake City (Hourly data not worked.) | 24124 | Daily | Roll | 1966–Pres. |
| VT | Burlington | 14742 | Daily | Cir. | 1962–Pres. |

**TABLE A2.3** Pyranometer and Pyrheliometer Stations of the U.S. National Weather Service (*Continued*)

| Pyranometer stations | | | | | |
|---|---|---|---|---|---|
| State | Station | WBAN no. | Type of record | Chart | Period of record |
| WA | Seattle-Tacoma (Hourly data discontinued per RAO 1967.) | 24233 | Daily | Roll | 1950–Pres. |
| | Spokane | 24157 | Daily | Cir. | 1949–Pres. |
| WI | Madison | 14837 | Daily/hourly | Roll | 1951–Pres. |
| WY | Lander | 24021 | Daily | Cir. | 1949–Pres. |
| PAC | Canton Island (closed) | 60703 | Daily/hourly | Roll | 1950–1967 |
| | Guam (closed) (charts only) | 41415 | — | Roll | 1957–1969 |
| | Wake Island (Hourly data worked and can be punched.) | 41606 | Daily | Roll | 1950–Pres. |
| W IND | Swan Island (Hourly data worked and can be punched.) | 11807 | Daily | Roll | 1949–Pres. |

| Pyrheliometer stations (normal incidence, N.I.) | | | | |
|---|---|---|---|---|
| State | Station | WBAN no. | Period of record | Chart information |
| AZ | Tucson | X8815 | 1961–Pres. | No charts |
| DC | Washington | 93725 | 1954–1956 | |
| HI | Mauna Loa | X6198 | 1961–Pres. | No charts |
| MA | Blue Hill | 14753 | 1937–1959 | 2 yr with charts[c] |
| NE | Lincoln | 14971 | 1942, 1946–1955 | |
| | North Omaha | 94918 | 1957–1972[d] | Not all charts on hand |
| NM | Albuquerque | 23060 | 1961–Pres. | Recorder runs every day |
| WI | Madison | 14837 | 1961–1972[d] | |

[a]Daily record poor for this period.
[b]Hourly data may be worked for period.
[c]Same N.I. may be recorded on hemispheric recordings.
[d]Station equipment was temporarily out of service; there was a large break in the record.

**TABLE A2.4** Mean Daily Solar Radiation, Langleys,[a] and Years of Record Used[b]

| State | Location | Jan | Yr | Feb | Yr | Mar | Yr | Apr | Yr | May | Yr | Jun | Yr | Jul | Yr | Aug | Yr | Sep | Yr | Oct | Yr | Nov | Yr | Dec | Yr | Ann |
|---|---|---|---|---|---|---|---|---|---|---|---|---|---|---|---|---|---|---|---|---|---|---|---|---|---|---|
| AK | Annette | 63 | 6 | 115 | 6 | 236 | 7 | 364 | 7 | 437 | 7 | 438 | 6 | 438 | 6 | 341 | 6 | 258 | 6 | 122 | 7 | 59 | 7 | 41 | 7 | 243 |
| | Barrow | d | | 38 | 8 | 180 | 8 | 380 | 8 | 513 | 8 | 528 | 8 | 429 | 9 | 255 | 9 | 115 | 10 | 41 | 10 | d | 9 | d | 9 | 206 |
| | Bethel | 38 | 9 | 108 | 10 | 282 | 10 | 444 | 9 | 457 | 10 | 454 | 10 | 376 | 10 | 252 | 10 | 202 | 10 | 115 | 10 | 44 | 9 | 22 | 9 | 233 |
| | Fairbanks | 16 | 25 | 71 | 27 | 213 | 25 | 376 | 28 | 461 | 28 | 504 | 29 | 434 | 28 | 317 | 28 | 180 | 29 | 82 | 30 | 26 | 26 | 6 | 26 | 224 |
| | Matanuska | 32 | 6 | 92 | 6 | 242 | 6 | 356 | 4 | 436 | 7 | 462 | 6 | 409 | 6 | 314 | 6 | 198 | 3 | 100 | 3 | 38 | 6 | 15 | 7 | 224 |
| AZ | Page | 300 | 2 | 382 | 3 | 526 | 3 | 618 | 3 | 695 | 3 | 707 | 11 | 680 | 3 | 596 | 3 | 516 | 3 | 402 | 3 | 310 | 3 | 243 | 3 | 498 |
| | Phoenix | 301 | 11 | 409 | 5 | 526 | 5 | 638 | 5 | 724 | 5 | 739 | 5 | 658 | 6 | 613 | 6 | 566 | 11 | 449 | 11 | 334 | 11 | 281 | 11 | 520 |
| | Tucson | 315 | 5 | 391 | 5 | 540 | 4 | 655 | 5 | 729 | 5 | 699 | 5 | 626 | 6 | 588 | 6 | 570 | 6 | 442 | 6 | 356 | 6 | 305 | 6 | 518 |
| AR | Little Rock | 188 | 9 | 260 | 9 | 353 | 10 | 446 | 5 | 523 | 5 | 559 | 5 | 556 | 6 | 518 | 8 | 439 | 7 | 343 | 8 | 244 | 10 | 187 | 10 | 385 |
| CA | Davis | 174 | 18 | 257 | 17 | 390 | 17 | 528 | 18 | 625 | 18 | 694 | 18 | 682 | 18 | 612 | 18 | 493 | 18 | 347 | 19 | 222 | 19 | 148 | 19 | 431 |
| | Fresno | 184 | 31 | 289 | 31 | 427 | 31 | 552 | 31 | 647 | 31 | 702 | 32 | 682 | 32 | 621 | 32 | 510 | 31 | 376 | 32 | 250 | 31 | 161 | 32 | 450 |
| | Inyokern (China Lake) | 306 | 11 | 412 | 11 | 562 | 11 | 683 | 11 | 772 | 11 | 819 | 11 | 772 | 11 | 729 | 11 | 635 | 10 | 467 | 9 | 363 | 11 | 300 | 12 | 568 |
| | LaJolla | 244 | 19 | 302 | 18 | 397 | 19 | 457 | 20 | 506 | 19 | 487 | 21 | 497 | 22 | 464 | 22 | 389 | 22 | 320 | 21 | 277 | 20 | 221 | 20 | 380 |
| | Los Angeles WBAS | 248 | 10 | 331 | 10 | 470 | 10 | 515 | 10 | 572 | 9 | 596 | 9 | 641 | 10 | 581 | 10 | 503 | 10 | 373 | 10 | 289 | 10 | 241 | 10 | 463 |
| | Los Angeles WBO | 243 | 9 | 327 | 9 | 436 | 9 | 483 | 9 | 555 | 9 | 584 | 9 | 651 | 9 | 581 | 10 | 500 | 10 | 362 | 10 | 281 | 10 | 234 | 10 | 436 |
| | Riverside[e] | 275 | 8 | 367 | 11 | 478 | 8 | 541 | 9 | 623 | 9 | 680 | 11 | 673 | 11 | 618 | 11 | 535 | 11 | 407 | 9 | 319 | 9 | 270 | 9 | 483 |
| | Santa Maria | 263 | 8 | 346 | 11 | 482 | 11 | 552 | 11 | 635 | 11 | 694 | 11 | 680 | 11 | 613 | 11 | 524 | 11 | 419 | 11 | 313 | 11 | 252 | 11 | 481 |
| | Soda Springs | 223 | 4 | 316 | 4 | 374 | 4 | 551 | 4 | 615 | 3 | 691 | 4 | 760 | 5 | 681 | 3 | 510 | 3 | 357 | 4 | 248 | 4 | 182 | 3 | 459 |
| CO | Boulder | 201 | 9 | 268 | 9 | 401 | 9 | 460 | 10 | 460 | 9 | 525 | 9 | 520 | 5 | 439 | 5 | 412 | 5 | 310 | 10 | 222 | 10 | 182 | 4 | 367 |
| | Grand Junction[c] | 227 | 9 | 324 | 9 | 434 | 9 | 546 | 8 | 615 | 9 | 708 | 8 | 676 | 8 | 595 | 8 | 514 | 8 | 373 | 10 | 260 | 6 | 212 | 10 | 456 |
| | Grand Lake (Granby) | 212 | 6 | 313 | 7 | 423 | 7 | 512 | 8 | 552 | 8 | 633 | 8 | 600 | 8 | 505 | 8 | 476 | 6 | 361 | 7 | 234 | 7 | 184 | 7 | 417 |
| | Washington (C.O.) | 174 | 3 | 266 | 3 | 344 | 3 | 411 | 2 | 551 | 3 | 494 | 2 | 536 | 2 | 446 | 2 | 375 | 3 | 299 | 3 | 211 | 3 | 166 | 3 | 356 |
| | American University | 158 | 39 | 231 | 39 | 322 | 39 | 398 | 39 | 467 | 39 | 510 | 39 | 496 | 39 | 440 | 38 | 364 | 38 | 278 | 38 | 192 | 39 | 141 | 39 | 333 |
| | Silver Hill | 177 | 7 | 247 | 7 | 342 | 7 | 438 | 7 | 513 | 7 | 555 | 7 | 511 | 7 | 457 | 7 | 391 | 8 | 293 | 8 | 202 | 7 | 156 | 6 | 357 |
| FL | Apalachicola | 298 | 11 | 367 | 11 | 441 | 10 | 535 | 10 | 603 | 9 | 578 | 9 | 529 | 9 | 511 | 9 | 456 | 9 | 413 | 9 | 332 | 11 | 262 | 10 | 444 |
| | Belle Isle | 297 | 10 | 330 | 10 | 412 | 10 | 463 | 10 | 483 | 10 | 464 | 10 | 488 | 11 | 461 | 10 | 400 | 10 | 366 | 11 | 313 | 11 | 291 | 10 | 397 |
| | Gainesville | 267 | 11 | 343 | 11 | 427 | 12 | 517 | 12 | 579 | 12 | 521 | 12 | 488 | 10 | 483 | 10 | 418 | 9 | 347 | 8 | 300 | 10 | 233 | 10 | 410 |
| | Miami Airport | 349 | 10 | 415 | 9 | 489 | 9 | 540 | 10 | 553 | 10 | 532 | 10 | 532 | 10 | 505 | 10 | 440 | 10 | 384 | 10 | 353 | 10 | 316 | 10 | 451 |
| | Tallahassee | 274 | 2 | 311 | 2 | 423 | 3 | 499 | 3 | 547 | 3 | 521 | 3 | 508 | 3 | 542 | 2 | c | 2 | c | | 292 | 2 | 230 | 2 | g |
| | Tampa | 327 | 8 | 391 | 8 | 474 | 22 | 539 | 22 | 596 | 11 | 574 | 11 | 534 | 10 | 494 | 10 | 452 | 10 | 400 | 9 | 356 | 9 | 300 | 9 | 453 |
| GA | Atlanta | 218 | 11 | 290 | 11 | 380 | 10 | 488 | 11 | 533 | 11 | 562 | 11 | 532 | 11 | 508 | 10 | 416 | 9 | 344 | 11 | 268 | 11 | 211 | 11 | 396 |
| | Griffin | 234 | 9 | 295 | 9 | 385 | 10 | 522 | 11 | 570 | 11 | 577 | 11 | 556 | 11 | 522 | 11 | 535 | 11 | 368 | 11 | 283 | 11 | 201 | 11 | 413 |
| HI | Honolulu | 363 | 9 | 422 | 4 | 516 | 4 | 559 | 5 | 617 | 5 | 615 | 5 | 615 | 5 | 612 | 8 | 573 | 5 | 507 | 5 | 426 | 5 | 371 | 5 | 516 |
| | Mauna Loa Obs. | 522 | 2 | 576 | 2 | 680 | 2 | 689 | 3 | 727 | 3 | c | | 703 | 3 | 642 | 2 | 602 | 2 | 560 | 2 | 504 | 3 | 481 | 3 | g |
| | Pearl Harbor | 359 | 5 | 400 | 4 | 487 | 4 | 529 | 4 | 573 | 5 | 566 | 5 | 598 | 5 | 567 | 5 | 539 | 5 | 466 | 5 | 386 | 5 | 343 | 5 | 484 |
| ID | Boise | 138 | 20 | 236 | 20 | 342 | 20 | 485 | 20 | 585 | 20 | 636 | 18 | 670 | 20 | 576 | 20 | 460 | 10 | 301 | 11 | 182 | 11 | 124 | 11 | 395 |
| | Twin Falls | 163 | 20 | 240 | 20 | 355 | 19 | 462 | 21 | 552 | 20 | 592 | 18 | 602 | 20 | 540 | 20 | 432 | 20 | 286 | 20 | 176 | 20 | 131 | 11 | 378 |
| IL | Chicago | 96 | 19 | 147 | 19 | 227 | 19 | 331 | 19 | 424 | 19 | 458 | 18 | 473 | 19 | 403 | 18 | 403 | 19 | 207 | 20 | 120 | 20 | 76 | 20 | 273 |
| | Lemont | 170 | 6 | 242 | 6 | 340 | 10 | 402 | 6 | 506 | 6 | 553 | 6 | 540 | 6 | 498 | 6 | 398 | 6 | 275 | 5 | 165 | 5 | 138 | 6 | 352 |
| IN | Indianapolis | 144 | 10 | 213 | 10 | 316 | 10 | 396 | 10 | 488 | 9 | 543 | 11 | 541 | 10 | 490 | 11 | 405 | 11 | 293 | 11 | 177 | 11 | 132 | 11 | 345 |

Note: See page 666 for footnotes.

**664**

Solar radiation data table (monthly means with years of record as superscript; footnote markers: c, g). Values read left to right as 13 period columns.

| State | Station | C1 | C2 | C3 | C4 | C5 | C6 | C7 | C8 | C9 | C10 | C11 | C12 | C13 |
|---|---|---|---|---|---|---|---|---|---|---|---|---|---|---|
| IA | Ames | 174⁵ | 253⁵ | 326⁵ | 403⁵ | 480⁵ | 541⁵ | 436⁶ | 460⁶ | 367⁶ | 274⁷ | 187⁷ | 143⁷ | 345 |
| KS | Dodge City | 255⁷ | 316⁷ | 418⁷ | 528⁷ | 568⁷ | 650⁷ | 642⁸ | 592⁹ | 493⁹ | 380⁹ | 285¹⁰ | 234¹⁰ | 447 |
| | Manhattan | 192⁹ | 264⁹ | 345⁹ | 433³ | 527³ | 551⁴ | 531⁴ | 526⁴ | 410⁴ | 292⁴ | 227⁴ | 156⁴ | 371 |
| KY | Lexington | 172⁹ | 263⁹ | 357⁹ | 480¹¹ | 581¹¹ | 628¹¹ | 617¹¹ | 563¹¹ | 494¹¹ | 357¹¹ | 245⁹ | 174¹¹ | 411 |
| LA | Lake Charles | 245¹¹ | 306¹¹ | 397¹¹ | 481¹¹ | 555¹¹ | 591¹¹ | 526¹¹ | 511¹¹ | 449¹¹ | 402¹¹ | 300²⁰ | 250²⁰ | 418 |
| | New Orleans | 214¹⁴ | 259¹⁴ | 335¹⁴ | 412¹⁶ | 449¹⁶ | 443¹⁴ | 417¹³ | 416¹⁵ | 383¹⁵ | 357¹³ | 278¹³ | 198¹³ | 347 |
| | Shreveport | 232³ | 292³ | 384³ | 446³ | 558⁴ | 557⁵ | 578⁴ | 528⁴ | 414³ | 354⁴ | 254⁴ | 205⁴ | 400 |
| ME | Caribou | 133⁸ | 231⁸ | 364⁸ | 400⁸ | 476¹⁰ | 470¹⁰ | 508¹¹ | 448¹¹ | 336¹¹ | 212¹¹ | 111¹¹ | 107⁹ | 316 |
| | Portland | 152⁷ | 235⁷ | 352⁷ | 409⁷ | 514⁹ | 539⁹ | 561⁹ | 488⁸ | 383⁷ | 278⁷ | 157⁸ | 137⁹ | 350 |
| MA | Amherst | 116² | [c] | 300² | [c] | 431² | 514² | 502² | 449² | 354² | 266² | 152² | 124² | [g] |
| | Blue Hill | 153²⁷ | 228²⁷ | 319²⁶ | 389²⁶ | 469²⁷ | 510²⁷ | 502²⁶ | 449²⁷ | 354²⁸ | 266²⁸ | 162²⁸ | 135²⁸ | 328 |
| | Boston | 129¹⁶ | 194¹⁷ | 290¹⁷ | 350¹⁷ | 445¹⁶ | 483¹⁶ | 486¹⁶ | 411¹⁶ | 334¹⁷ | 235¹⁷ | 136¹⁶ | 115¹⁵ | 301 |
| | Cambridge | 153⁴ | 235³ | 323³ | 400³ | 420³ | 476³ | 482⁴ | 464⁴ | 367⁴ | 253⁴ | 164⁴ | 124⁴ | 322 |
| | East Wareham | 140¹³ | 218¹³ | 305¹² | 385¹⁴ | 452¹⁴ | 508¹⁴ | 495¹⁴ | 436¹⁴ | 365¹³ | 258¹⁴ | 163¹⁴ | 140¹³ | 322 |
| | Lynn | 118² | 209² | 300² | 394² | 454² | 549² | 528⁴ | 432³ | 341³ | 241³ | 135³ | 107³ | 317 |
| MI | East Lansing | 121¹⁰ | 210¹¹ | 309¹¹ | 359¹¹ | 483¹¹ | 547¹¹ | 540¹¹ | 466³ | 373¹¹ | 255¹¹ | 136¹¹ | 108¹¹ | 311 |
| | Sault Ste. Marie | 130⁹ | 225⁹ | 356⁹ | 416⁸ | 523¹⁰ | 557¹¹ | 573¹¹ | 472¹⁰ | 322¹⁰ | 216⁹ | 105⁹ | 96⁹ | 333 |
| MN | St. Cloud | 168⁸ | 260⁸ | 356⁸ | 426⁸ | 496⁸ | 535⁸ | 557⁹ | 486⁸ | 366⁸ | 237⁷ | 146⁸ | 124⁸ | 348 |
| MO | Columbia (C.O.) | 173¹⁰ | 251¹⁰ | 340¹⁰ | 434¹¹ | 530¹¹ | 574¹¹ | 574¹⁰ | 522⁹ | 453⁹ | 322⁷ | 225⁸ | 158⁸ | 380 |
| | University of Missouri | 166⁵ | 248⁶ | 324⁶ | 429⁶ | 501⁶ | 560⁶ | 583⁶ | 509⁶ | 417⁶ | 324⁵ | 177⁵ | 146⁵ | 365 |
| MT | Glasgow | 154⁶ | 258⁸ | 385⁷ | 466⁷ | 568⁸ | 605⁸ | 645⁹ | 531¹⁰ | 410¹⁰ | 267⁸ | 154⁸ | 116⁷ | 388 |
| | Great Falls | 140⁸ | 232⁹ | 366⁹ | 434⁷ | 528⁸ | 583⁸ | 639⁹ | 532⁹ | 407¹⁰ | 264¹⁰ | 154¹⁰ | 112¹⁰ | 366 |
| | Summit | 122² | 162² | 268² | 414³ | 462³ | 493³ | 560³ | 510² | 354² | 216² | 102² | 76² | 312 |
| NE | Lincoln | 188³⁹ | 259³⁹ | 365³⁹ | 416³⁹ | 494⁴⁰ | 544³⁸ | 568³⁸ | 484³⁸ | 396³⁸ | 296³⁶ | 199⁴⁰ | 159⁴⁰ | 363 |
| | North Omaha | 193³ | 299³ | 365³ | 463³ | 516³ | 546⁴ | 568⁴ | 519⁴ | 410⁴ | 298⁴ | 204⁴ | 170⁴ | 379 |
| NV | Ely | 236⁷ | 339⁷ | 468⁹ | 563⁹ | 625⁹ | 712¹⁰ | 647¹⁰ | 618¹¹ | 518¹¹ | 394¹¹ | 289¹⁰ | 218¹⁰ | 469 |
| | Las Vegas | 277¹¹ | 384¹¹ | 519¹¹ | 621¹¹ | 702¹¹ | 748¹¹ | 675¹¹ | 627¹¹ | 551¹¹ | 429¹¹ | 318¹¹ | 258¹¹ | 509 |
| NJ | Seabrook | 157⁸ | 227⁸ | 318⁸ | 403⁸ | 482⁸ | 527⁸ | 509⁸ | 455⁹ | 385⁹ | 278⁷ | 192⁸ | 140⁸ | 339 |
| NH | Mt. Washington | 117² | 218² | 238² | [c] | 686² | 726² | 683² | 626² | 554² | 438² | 334² | 96² | [g] |
| NM | Albuquerque | 303¹³ | 386¹³ | 511¹³ | 618¹³ | 686¹³ | 726¹² | 683¹² | 626¹³ | 554¹⁴ | 438¹⁴ | 334¹⁵ | 276¹⁴ | 512 |
| NY | Ithaca | 116²² | 194²¹ | 272²³ | 334²³ | 440²⁴ | 501²³ | 515²³ | 453²¹ | 346²¹ | 231²³ | 120²³ | 96¹⁴ | 302 |
| | Central Park | 130³⁴ | 199³⁴ | 290³³ | 369³⁵ | 432³⁵ | 470³⁴ | 459³⁵ | 389³⁶ | 331³⁶ | 242³⁶ | 147³⁶ | 115²³ | 298 |
| | Sayville | 160¹¹ | 249¹¹ | 335¹⁰ | 415¹⁰ | 494¹⁰ | 565¹⁰ | 543¹⁰ | 462⁸ | 385⁸ | 289⁸ | 186¹⁰ | 142¹¹ | 352 |
| | Schenectady | 130⁹ | 200⁸ | 273⁸ | 338⁹ | 413⁹ | 448⁸ | 441⁸ | 397⁸ | 299⁸ | 218⁸ | 128⁸ | 104⁸ | 282 |
| | Upton | 155⁸ | 232⁹ | 339⁸ | 428⁹ | 502⁹ | 573⁸ | 543⁷ | 475⁷ | 391¹⁰ | 293⁶ | 182⁸ | 146⁷ | 355 |
| NC | Greensboro | 200⁷ | 276⁷ | 354⁸ | 469⁸ | 531⁸ | 564⁸ | 544⁷ | 485⁸ | 406¹⁰ | 322¹⁰ | 243¹⁰ | 197⁸ | 383 |
| | Hatteras | 238⁷ | 317⁹ | 426⁹ | 569⁹ | 635⁹ | 652⁹ | 625¹⁰ | 562¹⁰ | 471¹¹ | 358¹¹ | 282¹¹ | 214¹¹ | 443 |
| | Raleigh | 235³ | 302³ | [c] | 466³ | 494³ | 564⁹ | 535⁹ | 476¹¹ | 379¹¹ | 307¹¹ | 235¹¹ | 199¹⁰ | [g] |
| ND | Bismarck | 157⁷ | 250⁸ | 356⁸ | 447⁷ | 550⁸ | 590⁹ | 617⁹ | 516¹¹ | 390¹¹ | 272¹¹ | 161¹¹ | 124¹⁰ | 369 |
| OH | Cleveland | 125⁸ | 183⁶ | 303⁷ | 286⁷ | 502⁸ | 562⁸ | 562⁸ | 494⁴ | 278⁴ | 289⁸ | 141¹¹ | 115¹⁰ | 335 |
| | Columbus | 128⁷ | 200⁷ | 297⁷ | 391⁷ | 471⁷ | 562⁸ | 542⁹ | 477⁴ | 422⁴ | 286⁸ | 176⁹ | 129⁹ | 340 |
| | Put-in-Bay | 126¹⁰ | 204⁹ | 302⁹ | 386¹⁰ | 468¹⁰ | 544¹¹ | 561¹⁰ | 487¹⁰ | 382¹¹ | 275⁹ | 144¹¹ | 109¹¹ | 332 |

**TABLE A2.4** Mean Daily Solar Radiation, Langleys,[a] and Years of Record Used[b] (Continued)

| Location | Jan | Yr | Feb | Yr | Mar | Yr | Apr | Yr | May | Yr | Jun | Yr | Jul | Yr | Aug | Yr | Sep | Yr | Oct | Yr | Nov | Yr | Dec | Yr | Ann |
|---|---|---|---|---|---|---|---|---|---|---|---|---|---|---|---|---|---|---|---|---|---|---|---|---|---|
| OK Oklahoma City | 251 | 10 | 319 | 10 | 409 | 9 | 494 | 10 | 536 | 10 | 615 | 7 | 610 | 8 | 593 | 8 | 487 | 9 | 377 | 10 | 291 | 9 | 240 | 9 | 436 |
| Stillwater | 205 | 8 | 289 | 8 | 390 | 9 | 454 | 9 | 504 | 9 | 600 | 10 | 596 | 10 | 545 | 10 | 455 | 11 | 354 | 10 | 269 | 9 | 209 | 8 | 405 |
| OR Astoria | 90 | 7 | 162 | 8 | 270 | 8 | 375 | 8 | 492 | 8 | 469 | 8 | 539 | 8 | 461 | 8 | 354 | 7 | 209 | 8 | 111 | 8 | 79 | 8 | 301 |
| Corvallis | 89 | 2 | c | | 287 | 3 | 406 | 3 | 517 | 3 | 570 | 3 | 676 | 3 | 558 | 4 | 397 | 4 | 235 | 4 | 144 | 4 | 80 | 4 | g |
| Medford | 116 | 11 | 215 | 11 | 336 | 11 | 482 | 11 | 592 | 11 | 652 | 11 | 698 | 10 | 605 | 11 | 447 | 11 | 279 | 11 | 149 | 11 | 93 | 11 | 389 |
| PA Pittsburgh | 94 | 6 | 169 | 5 | 216 | 6 | 317 | 6 | 429 | 6 | 491 | 6 | 497 | 7 | 409 | 7 | 339 | 6 | 207 | 5 | 118 | 6 | 77 | 5 | 280 |
| State College | 133 | 19 | 201 | 19 | 295 | 20 | 380 | 20 | 456 | 20 | 518 | 20 | 511 | 20 | 444 | 20 | 358 | 20 | 256 | 20 | 149 | 20 | 118 | 20 | 318 |
| RI Newport | 155 | 23 | 232 | 22 | 334 | 23 | 405 | 23 | 477 | 23 | 527 | 24 | 513 | 24 | 455 | 24 | 377 | 24 | 271 | 24 | 176 | 24 | 139 | 24 | 338 |
| SC Charleston | 252 | 11 | 314 | 11 | 388 | 11 | 512 | 11 | 551 | 11 | 564 | 11 | 520 | 11 | 501 | 11 | 404 | 11 | 338 | 11 | 286 | 11 | 225 | 11 | 404 |
| SD Rapid City | 183 | 11 | 277 | 11 | 400 | 11 | 482 | 11 | 532 | 11 | 585 | 11 | 590 | 11 | 541 | 11 | 435 | 11 | 315 | 10 | 204 | 10 | 158 | 10 | 392 |
| TN Nashville | 149 | 18 | 228 | 19 | 322 | 19 | 432 | 19 | 503 | 18 | 551 | 18 | 530 | 17 | 473 | 17 | 403 | 17 | 308 | 19 | 208 | 18 | 150 | 19 | 355 |
| Oak Ridge | 161 | 10 | 239 | 11 | 331 | 11 | 450 | 11 | 518 | 11 | 551 | 11 | 526 | 11 | 478 | 11 | 416 | 11 | 318 | 11 | 213 | 10 | 163 | 11 | 364 |
| TX Brownsville | 297 | 10 | 341 | 10 | 402 | 10 | 456 | 10 | 564 | 10 | 610 | 9 | 627 | 8 | 568 | 8 | 475 | 11 | 411 | 11 | 296 | 11 | 263 | 10 | 442 |
| El Paso | 333 | 11 | 430 | 11 | 547 | 10 | 654 | 11 | 714 | 11 | 729 | 11 | 666 | 11 | 640 | 10 | 576 | 11 | 460 | 11 | 372 | 11 | 313 | 11 | 536 |
| Ft. Worth | 250 | 11 | 320 | 11 | 427 | 11 | 488 | 11 | 562 | 11 | 651 | 11 | 613 | 11 | 593 | 11 | 503 | 11 | 403 | 11 | 306 | 11 | 245 | 9 | 445 |
| Midland | 283 | 7 | 358 | 8 | 476 | 9 | 550 | 8 | 611 | 8 | 617 | 8 | 608 | 7 | 574 | 8 | 522 | 9 | 396 | 9 | 325 | 8 | 275 | 8 | 466 |
| San Antonio | 279 | 9 | 347 | 9 | 417 | 9 | 445 | 9 | 541 | 9 | 612 | 9 | 639 | 9 | 585 | 9 | 493 | 10 | 398 | 10 | 295 | 10 | 256 | 10 | 442 |
| UT Flaming Gorge | 238 | 2 | 298 | 2 | 443 | 2 | 522 | 2 | 565 | 2 | 650 | 2 | 599 | 2 | 538 | 3 | 425 | 3 | 352 | 3 | 262 | 3 | 215 | 3 | 426 |
| Salt Lake City | 163 | 8 | 256 | 8 | 354 | 8 | 479 | 8 | 570 | 7 | 621 | 7 | 620 | 6 | 551 | 7 | 446 | 8 | 316 | 8 | 204 | 8 | 146 | 9 | 394 |
| VA Mt. Weather | 172 | 2 | 274 | 2 | 338 | 2 | 414 | 2 | 508 | 2 | 525 | 3 | 510 | 3 | 430 | 3 | 375 | 3 | 281 | 2 | 202 | 2 | 168 | 2 | 350 |
| WA North Head | c | | 167 | 7 | 257 | 7 | 432 | 2 | 509 | 2 | 487 | 3 | 486 | 3 | 436 | 3 | 321 | 3 | 205 | 2 | 122 | 2 | 77 | 3 | g |
| Friday Harbor | 87 | 8 | 157 | 7 | 274 | 8 | 418 | 8 | 514 | 8 | 578 | 10 | 586 | 10 | 507 | 11 | 351 | 8 | 194 | 10 | 102 | 10 | 75 | 8 | 320 |
| Prosser | 117 | 4 | 222 | 4 | 351 | 4 | 521 | 5 | 616 | 4 | 680 | 4 | 707 | 4 | 604 | 4 | 458 | 4 | 274 | 4 | 136 | 4 | 100 | 4 | 399 |
| Pullman | 121 | 4 | 205 | 4 | 304 | 2 | 462 | 2 | 558 | 4 | 653 | 5 | 699 | 5 | 562 | 5 | 410 | 4 | 245 | 5 | 146 | 5 | 96 | 5 | 372 |
| University of Washington | 67 | 9 | 126 | 9 | 245 | 10 | 364 | 9 | 445 | 10 | 461 | 10 | 496 | 11 | 435 | 10 | 299 | 8 | 170 | 9 | 93 | 9 | 59 | 9 | 272 |
| Seattle-Tacoma | 75 | 9 | 139 | 9 | 265 | 9 | 403 | 9 | 503 | 9 | 511 | 9 | 566 | 9 | 452 | 10 | 324 | 10 | 188 | 10 | 104 | 9 | 64 | 10 | 300 |
| Spokane | 119 | 8 | 204 | 8 | 321 | 8 | 474 | 9 | 563 | 9 | 596 | 9 | 665 | 9 | 556 | 9 | 404 | 10 | 225 | 9 | 131 | 9 | 75 | 7 | 361 |
| WI Madison[f] | 148 | 46 | 220 | 46 | 313 | 45 | 394 | 47 | 466 | 47 | 514 | 47 | 531 | 47 | 452 | 47 | 348 | 47 | 241 | 47 | 145 | 44 | 115 | 46 | 324 |
| WY Lander | 226 | 8 | 324 | 8 | 452 | 9 | 548 | 9 | 587 | 11 | 678 | 11 | 651 | 11 | 586 | 10 | 472 | 8 | 354 | 9 | 239 | 9 | 196 | 9 | 443 |
| Laramie | 216 | 3 | 295 | 3 | 424 | 3 | 508 | 3 | 554 | 3 | 643 | 3 | 606 | 3 | 536 | 3 | 438 | 3 | 324 | 3 | 229 | 3 | 186 | 4 | 408 |
| **Island Stations** | | | | | | | | | | | | | | | | | | | | | | | | | |
| Canton Island | 588 | 9 | 626 | 7 | 634 | 7 | 604 | 7 | 561 | 9 | 549 | 8 | 550 | 9 | 597 | 9 | 640 | 9 | 651 | 9 | 600 | 8 | 572 | 8 | 597 |
| San Juan, P.R. | 404 | 5 | 481 | 4 | 580 | 4 | 622 | 4 | 519 | 5 | 536 | 6 | 639 | 6 | 549 | 6 | 531 | 8 | 460 | 6 | 411 | 6 | 411 | 6 | 512 |
| Swan Island | 442 | 6 | 496 | 7 | 615 | 6 | 646 | 6 | 625 | 6 | 544 | 8 | 588 | 8 | 591 | 7 | 535 | 8 | 457 | 7 | 394 | 8 | 382 | 8 | 526 |
| Wake Island | 438 | 7 | 518 | 7 | 577 | 7 | 627 | 7 | 642 | 8 | 656 | 6 | 629 | 7 | 623 | 7 | 587 | 6 | 525 | 7 | 482 | 7 | 421 | 7 | 560 |

[a] Langley is the unit used to denote 1 gram calorie per square centimeter.

[b] From "Climatic Atlas of the United States," U.S. Government Printing Office, 1968.

[c] Only one year of data for the month; no means computed.

[d] Barrow is in darkness during the winter months.

[e] Riverside data prior to March 1952 not used because of instrumental difficulties.

[f] Madison data after 1957 not used due to exposure influences

[g] Indicates no data for the month (or incomplete data for the year).

**TABLE A2.5** Solar Collector Tilt Factor[a]

| Month | $L = 20°$ $\beta = 20°$ | $\beta = 40°$ | $L = 30°$ $\beta = 30°$ | $\beta = 50°$ | $L = 40°$ $\beta = 40°$ | $\beta = 60°$ | $L = 50°$ $\beta = 50°$ | $\beta = 70°$ |
|---|---|---|---|---|---|---|---|---|
| Jan | 1.36 | 1.52 | 1.68 | 1.88 | 2.28 | 2.56 | 3.56 | 3.94 |
| Feb | 1.22 | 1.28 | 1.44 | 1.52 | 1.80 | 1.90 | 2.49 | 2.62 |
| Mar | 1.08 | 1.02 | 1.20 | 1.15 | 1.36 | 1.32 | 1.65 | 1.62 |
| Apr | 1.00 | 0.83 | 1.00 | 0.84 | 1.05 | 0.90 | 1.16 | 1.00 |
| May | 0.92 | 0.70 | 0.87 | 0.66 | 0.88 | 0.66 | 0.90 | 0.64 |
| Jun | 0.87 | 0.63 | 0.81 | 0.58 | 0.79 | 0.60 | 0.80 | 0.56 |
| Jul | 0.89 | 0.66 | 0.83 | 0.62 | 0.82 | 0.64 | 0.84 | 0.62 |
| Aug | 0.95 | 0.78 | 0.93 | 0.76 | 0.96 | 0.78 | 1.02 | 0.83 |
| Sep | 1.04 | 0.95 | 1.11 | 1.00 | 1.24 | 1.12 | 1.44 | 1.32 |
| Oct | 1.17 | 1.20 | 1.36 | 1.36 | 1.62 | 1.64 | 2.10 | 2.14 |
| Nov | 1.30 | 1.44 | 1.60 | 1.76 | 2.08 | 2.24 | 3.16 | 3.32 |
| Dec | 1.39 | 1.60 | 1.76 | 1.99 | 2.48 | 2.80 | 4.04 | 4.52 |

[a]The solar collector tilt factor is the ratio of monthly beam insolation on a tilted surface to monthly beam insolation on a horizontal surface. Here $\beta$ = collector tilt angle and $L$ = collector latitude. From Kreider, J. F., and F. Kreith, "Solar Heating and Cooling," revised 1st ed., Hemisphere Publ. Corp., 1977.

**TABLE A2.6a**  Solar Position and Insolation Values for 24 Degrees North Latitude[a]

| Date | Solar time AM | Solar time PM | Solar position Alt | Solar position Azm | Normal | Horiz. | 14 | 24 | 34 | 44 | 90 |
|------|------|------|------|------|------|------|------|------|------|------|------|
| | | | | | | | **South facing surface angle with horiz.** | | | | |
| Jan 21 | 7 | 5 | 4.8 | 65.6 | 71 | 10 | 17 | 21 | 25 | 28 | 31 |
| | 8 | 4 | 16.9 | 58.3 | 239 | 83 | 110 | 126 | 137 | 145 | 127 |
| | 9 | 3 | 27.9 | 48.8 | 288 | 151 | 188 | 207 | 221 | 228 | 176 |
| | 10 | 2 | 37.2 | 36.1 | 308 | 204 | 246 | 268 | 282 | 287 | 207 |
| | 11 | 1 | 43.6 | 19.6 | 317 | 237 | 283 | 306 | 319 | 324 | 226 |
| | 12 | | 46.0 | 0.0 | 320 | 249 | 296 | 319 | 332 | 336 | 232 |
| | Surface daily totals | | | | 2766 | 1622 | 1984 | 2174 | 2300 | 2360 | 1766 |
| Feb 21 | 7 | 5 | 9.3 | 74.6 | 158 | 35 | 44 | 49 | 53 | 56 | 46 |
| | 8 | 4 | 22.3 | 67.2 | 263 | 116 | 135 | 145 | 150 | 151 | 102 |
| | 9 | 3 | 34.4 | 57.6 | 298 | 187 | 213 | 225 | 230 | 228 | 141 |
| | 10 | 2 | 45.1 | 44.2 | 314 | 241 | 273 | 286 | 291 | 287 | 168 |
| | 11 | 1 | 53.0 | 25.0 | 321 | 276 | 310 | 324 | 328 | 323 | 185 |
| | 12 | | 56.0 | 0.0 | 324 | 288 | 323 | 337 | 341 | 335 | 191 |
| | Surface daily totals | | | | 3036 | 1998 | 2276 | 2396 | 2436 | 2424 | 1476 |
| Mar 21 | 7 | 5 | 13.7 | 83.3 | 194 | 60 | 63 | 64 | 62 | 59 | 27 |
| | 8 | 4 | 27.2 | 76.8 | 267 | 141 | 150 | 152 | 149 | 142 | 64 |
| | 9 | 3 | 40.2 | 67.9 | 295 | 212 | 226 | 229 | 225 | 214 | 95 |
| | 10 | 2 | 52.3 | 54.8 | 309 | 266 | 285 | 288 | 283 | 270 | 120 |
| | 11 | 1 | 61.9 | 33.4 | 315 | 300 | 322 | 326 | 320 | 305 | 135 |
| | 12 | | 66.0 | 0.0 | 317 | 312 | 334 | 339 | 333 | 317 | 140 |
| | Surface daily totals | | | | 3078 | 2270 | 2428 | 2456 | 2412 | 2298 | 1022 |
| Apr 21 | 6 | 6 | 4.7 | 100.6 | 40 | 7 | 5 | 4 | 4 | 3 | 2 |
| | 7 | 5 | 18.3 | 94.9 | 203 | 83 | 77 | 70 | 62 | 51 | 10 |
| | 8 | 4 | 32.0 | 89.0 | 256 | 160 | 157 | 149 | 137 | 122 | 16 |
| | 9 | 3 | 45.6 | 81.9 | 280 | 227 | 227 | 220 | 206 | 186 | 46 |
| | 10 | 2 | 59.0 | 71.8 | 292 | 278 | 282 | 275 | 259 | 237 | 61 |
| | 11 | 1 | 71.1 | 51.6 | 298 | 310 | 316 | 309 | 293 | 269 | 74 |
| | 12 | | 77.6 | 0.0 | 299 | 321 | 328 | 321 | 305 | 280 | 79 |
| | Surface daily totals | | | | 3036 | 2454 | 2458 | 2374 | 2228 | 2016 | 488 |
| May 21 | 6 | 6 | 8.0 | 108.4 | 86 | 22 | 15 | 10 | 9 | 9 | 5 |
| | 7 | 5 | 21.2 | 103.2 | 203 | 98 | 85 | 73 | 59 | 44 | 12 |
| | 8 | 4 | 34.6 | 98.5 | 248 | 171 | 159 | 145 | 127 | 106 | 15 |
| | 9 | 3 | 48.3 | 93.6 | 269 | 233 | 224 | 210 | 190 | 165 | 16 |
| | 10 | 2 | 62.0 | 87.7 | 280 | 281 | 275 | 261 | 239 | 211 | 22 |
| | 11 | 1 | 75.5 | 76.9 | 286 | 311 | 307 | 293 | 270 | 240 | 34 |
| | 12 | | 86.0 | 0.0 | 288 | 322 | 317 | 304 | 281 | 250 | 37 |
| | Surface daily totals | | | | 3032 | 2556 | 2447 | 2286 | 2072 | 1800 | 246 |
| Jun 21 | 6 | 6 | 9.3 | 111.6 | 97 | 29 | 20 | 12 | 12 | 11 | 7 |
| | 7 | 5 | 22.3 | 106.8 | 201 | 103 | 87 | 73 | 58 | 41 | 13 |
| | 8 | 4 | 35.5 | 102.6 | 242 | 173 | 158 | 142 | 122 | 99 | 16 |
| | 9 | 3 | 49.0 | 98.7 | 263 | 234 | 221 | 204 | 182 | 155 | 18 |
| | 10 | 2 | 62.6 | 95.0 | 274 | 280 | 269 | 253 | 229 | 199 | 18 |
| | 11 | 1 | 76.3 | 90.8 | 279 | 309 | 300 | 283 | 259 | 227 | 19 |
| | 12 | | 89.4 | 0.0 | 281 | 319 | 310 | 294 | 269 | 236 | 22 |
| | Surface daily totals | | | | 2994 | 2574 | 2422 | 2230 | 1992 | 1700 | 204 |

[a]From Kreider, J. F., and F. Kreith, "Solar Heating and Cooling," revised 1st ed., Hemisphere Publ. Corp., 1977.

**TABLE A2.6a** Solar Position and Insolation Values for 24 Degrees North Latitude[a] (*Continued*)

| Date | Solar time AM | Solar time PM | Solar position Alt | Solar position Azm | Normal | Horiz. | South facing surface angle with horiz. 14 | 24 | 34 | 44 | 90 |
|------|----|----|------|-------|--------|--------|------|------|------|------|------|
| Jul 21 | 6 | 6 | 8.2 | 109.0 | 81 | 23 | 16 | 11 | 10 | 9 | 6 |
| | 7 | 5 | 21.4 | 103.8 | 195 | 98 | 85 | 73 | 59 | 44 | 13 |
| | 8 | 4 | 34.8 | 99.2 | 239 | 169 | 157 | 143 | 125 | 104 | 16 |
| | 9 | 3 | 48.4 | 94.5 | 261 | 231 | 221 | 207 | 187 | 161 | 18 |
| | 10 | 2 | 62.1 | 89.0 | 272 | 278 | 270 | 256 | 235 | 206 | 21 |
| | 11 | 1 | 75.7 | 79.2 | 278 | 307 | 302 | 287 | 265 | 235 | 32 |
| | 12 | | 86.6 | 0.0 | 280 | 317 | 312 | 298 | 275 | 245 | 36 |
| | Surface daily totals | | | | 2932 | 2526 | 2412 | 2250 | 2036 | 1766 | 246 |
| Aug 21 | 6 | 6 | 5.0 | 101.3 | 35 | 7 | 5 | 4 | 4 | 4 | 2 |
| | 7 | 5 | 18.5 | 95.6 | 186 | 82 | 76 | 69 | 60 | 50 | 11 |
| | 8 | 4 | 32.2 | 89.7 | 241 | 158 | 154 | 146 | 134 | 118 | 16 |
| | 9 | 3 | 45.9 | 82.9 | 265 | 223 | 222 | 214 | 200 | 181 | 39 |
| | 10 | 2 | 59.3 | 73.0 | 278 | 273 | 275 | 268 | 252 | 230 | 58 |
| | 11 | 1 | 71.6 | 53.2 | 284 | 304 | 309 | 301 | 285 | 261 | 71 |
| | 12 | | 78.3 | 0.0 | 286 | 315 | 320 | 313 | 296 | 272 | 75 |
| | Surface daily totals | | | | 2864 | 2408 | 2402 | 2316 | 2168 | 1958 | 470 |
| Sep 21 | 7 | 5 | 13.7 | 83.8 | 173 | 57 | 60 | 60 | 59 | 56 | 26 |
| | 8 | 4 | 27.2 | 76.8 | 248 | 136 | 144 | 146 | 143 | 136 | 62 |
| | 9 | 3 | 40.2 | 67.9 | 278 | 205 | 218 | 221 | 217 | 206 | 93 |
| | 10 | 2 | 52.3 | 54.8 | 292 | 258 | 275 | 278 | 273 | 261 | 116 |
| | 11 | 1 | 61.9 | 33.4 | 299 | 291 | 311 | 315 | 309 | 295 | 131 |
| | 12 | | 66.0 | 0.0 | 301 | 302 | 323 | 327 | 321 | 306 | 136 |
| | Surface daily totals | | | | 2878 | 2194 | 2342 | 2366 | 2322 | 2212 | 992 |
| Oct 21 | 7 | 5 | 9.1 | 74.1 | 138 | 32 | 40 | 45 | 48 | 50 | 42 |
| | 8 | 4 | 22.0 | 66.7 | 247 | 111 | 129 | 139 | 144 | 145 | 99 |
| | 9 | 3 | 34.1 | 57.1 | 284 | 180 | 206 | 217 | 223 | 221 | 138 |
| | 10 | 2 | 44.7 | 43.8 | 301 | 234 | 265 | 277 | 282 | 279 | 165 |
| | 11 | 1 | 52.5 | 24.7 | 309 | 268 | 301 | 315 | 319 | 314 | 182 |
| | 12 | | 55.5 | 0.0 | 311 | 279 | 314 | 328 | 332 | 327 | 188 |
| | Surface daily totals | | | | 2868 | 1928 | 2198 | 2314 | 2364 | 2346 | 1442 |
| Nov 21 | 7 | 5 | 4.9 | 65.8 | 67 | 10 | 16 | 20 | 24 | 27 | 29 |
| | 8 | 4 | 17.0 | 58.4 | 232 | 82 | 108 | 123 | 135 | 142 | 124 |
| | 9 | 3 | 28.0 | 48.9 | 282 | 150 | 186 | 205 | 217 | 224 | 172 |
| | 10 | 2 | 37.3 | 36.3 | 303 | 203 | 244 | 265 | 278 | 283 | 204 |
| | 11 | 1 | 43.8 | 19.7 | 312 | 236 | 280 | 302 | 316 | 320 | 222 |
| | 12 | | 46.2 | 0.0 | 315 | 247 | 293 | 315 | 328 | 332 | 228 |
| | Surface daily totals | | | | 2706 | 1610 | 1962 | 2146 | 2268 | 2324 | 1730 |
| Dec 21 | 7 | 5 | 3.2 | 62.6 | 30 | 3 | 7 | 9 | 11 | 12 | 14 |
| | 8 | 4 | 14.9 | 55.3 | 225 | 71 | 99 | 116 | 129 | 139 | 130 |
| | 9 | 3 | 25.5 | 46.0 | 281 | 137 | 176 | 198 | 214 | 223 | 184 |
| | 10 | 2 | 34.3 | 33.7 | 304 | 189 | 234 | 258 | 275 | 283 | 217 |
| | 11 | 1 | 40.4 | 18.2 | 314 | 221 | 270 | 295 | 312 | 320 | 236 |
| | 12 | | 42.6 | 0.0 | 317 | 232 | 282 | 308 | 325 | 332 | 243 |
| | Surface daily totals | | | | 2624 | 1474 | 1852 | 2058 | 2204 | 2286 | 1808 |

[b]1 Btu/hr · ft² = 3.152 W/m². Ground reflection not included on normal or horizontal surfaces.

**TABLE A2.6b**  Solar Position and Insolation Values for 32 Degrees North Latitude[a]

| Date | Solar time AM | PM | Solar position Alt | Azm | BTUH/sq. ft. total insolation on surfaces[b] Normal | Horiz. | South facing surface angle with horiz. 22 | 32 | 42 | 52 | 90 |
|------|----|----|------|------|--------|--------|------|------|------|------|------|
| Jan 21 | 7 | 5 | 1.4 | 65.2 | 1 | 0 | 0 | 0 | 0 | 1 | 1 |
|  | 8 | 4 | 12.5 | 56.5 | 203 | 56 | 93 | 106 | 116 | 123 | 115 |
|  | 9 | 3 | 22.5 | 46.0 | 269 | 118 | 175 | 193 | 206 | 212 | 181 |
|  | 10 | 2 | 30.6 | 33.1 | 295 | 167 | 235 | 256 | 269 | 274 | 221 |
|  | 11 | 1 | 36.1 | 17.5 | 306 | 198 | 273 | 295 | 308 | 312 | 245 |
|  | 12 |  | 38.0 | 0.0 | 310 | 209 | 285 | 308 | 321 | 324 | 253 |
|  | Surface daily totals |  |  |  | 2458 | 1288 | 1839 | 2008 | 2118 | 2166 | 1779 |
| Feb 21 | 7 | 5 | 7.1 | 73.5 | 121 | 22 | 34 | 37 | 40 | 42 | 38 |
|  | 8 | 4 | 19.0 | 64.4 | 247 | 95 | 127 | 136 | 140 | 141 | 108 |
|  | 9 | 3 | 29.9 | 53.4 | 288 | 161 | 206 | 217 | 222 | 220 | 158 |
|  | 10 | 2 | 39.1 | 39.4 | 306 | 212 | 266 | 278 | 283 | 279 | 193 |
|  | 11 | 1 | 45.6 | 21.4 | 315 | 244 | 304 | 317 | 321 | 315 | 214 |
|  | 12 |  | 48.0 | 0.0 | 317 | 255 | 316 | 330 | 334 | 328 | 222 |
|  | Surface daily totals |  |  |  | 2872 | 1724 | 2188 | 2300 | 2345 | 2322 | 1644 |
| Mar 21 | 7 | 5 | 12.7 | 81.9 | 185 | 54 | 60 | 60 | 59 | 56 | 32 |
|  | 8 | 4 | 25.1 | 73.0 | 260 | 129 | 146 | 147 | 144 | 137 | 78 |
|  | 9 | 3 | 36.8 | 62.1 | 290 | 194 | 222 | 224 | 220 | 209 | 119 |
|  | 10 | 2 | 47.3 | 47.5 | 304 | 245 | 280 | 283 | 278 | 265 | 150 |
|  | 11 | 1 | 55.0 | 26.8 | 311 | 277 | 317 | 321 | 315 | 300 | 170 |
|  | 12 |  | 58.0 | 0.0 | 313 | 287 | 329 | 333 | 327 | 312 | 177 |
|  | Surface daily totals |  |  |  | 3012 | 2084 | 2378 | 2403 | 2358 | 2246 | 1276 |
| Apr 21 | 6 | 6 | 6.1 | 99.9 | 66 | 14 | 9 | 6 | 6 | 5 | 3 |
|  | 7 | 5 | 18.8 | 92.2 | 206 | 86 | 78 | 71 | 62 | 51 | 10 |
|  | 8 | 4 | 31.5 | 84.0 | 255 | 158 | 156 | 148 | 136 | 120 | 35 |
|  | 9 | 3 | 43.9 | 74.2 | 278 | 220 | 225 | 217 | 203 | 183 | 68 |
|  | 10 | 2 | 55.7 | 60.3 | 290 | 267 | 279 | 272 | 256 | 234 | 95 |
|  | 11 | 1 | 65.4 | 37.5 | 295 | 297 | 313 | 306 | 290 | 265 | 112 |
|  | 12 |  | 69.6 | 0.0 | 297 | 307 | 325 | 318 | 301 | 276 | 118 |
|  | Surface daily totals |  |  |  | 3076 | 2390 | 2444 | 2356 | 2206 | 1994 | 764 |
| May 21 | 6 | 6 | 10.4 | 107.2 | 119 | 36 | 21 | 13 | 13 | 12 | 7 |
|  | 7 | 5 | 22.8 | 100.1 | 211 | 107 | 88 | 75 | 60 | 44 | 13 |
|  | 8 | 4 | 35.4 | 92.9 | 250 | 175 | 159 | 145 | 127 | 105 | 15 |
|  | 9 | 3 | 48.1 | 84.7 | 269 | 233 | 223 | 209 | 188 | 163 | 33 |
|  | 10 | 2 | 60.6 | 73.3 | 280 | 277 | 273 | 259 | 237 | 208 | 56 |
|  | 11 | 1 | 72.0 | 51.9 | 285 | 305 | 305 | 290 | 268 | 237 | 72 |
|  | 12 |  | 78.0 | 0.0 | 286 | 315 | 315 | 301 | 278 | 247 | 77 |
|  | Surface daily totals |  |  |  | 3112 | 2582 | 2454 | 2284 | 2064 | 1788 | 469 |
| Jun 21 | 6 | 6 | 12.2 | 110.2 | 131 | 45 | 26 | 16 | 15 | 14 | 9 |
|  | 7 | 5 | 24.3 | 103.4 | 210 | 115 | 91 | 76 | 59 | 41 | 14 |
|  | 8 | 4 | 36.9 | 96.8 | 245 | 180 | 159 | 143 | 122 | 99 | 16 |
|  | 9 | 3 | 49.6 | 89.4 | 264 | 236 | 221 | 204 | 181 | 153 | 19 |
|  | 10 | 2 | 62.2 | 79.7 | 274 | 279 | 268 | 251 | 227 | 197 | 41 |
|  | 11 | 1 | 74.2 | 60.9 | 279 | 306 | 299 | 282 | 257 | 224 | 56 |
|  | 12 |  | 81.5 | 0.0 | 280 | 315 | 309 | 292 | 267 | 234 | 60 |
|  | Surface daily totals |  |  |  | 3084 | 2634 | 2436 | 2234 | 1990 | 1690 | 370 |

[a]From Kreider, J. F., and F. Kreith, "Solar Heating and Cooling," revised 1st ed., Hemisphere Publ. Corp., 1977.

**TABLE A2.6b** Solar Position and Insolation Values for 32 Degrees North Latitude[a] (*Continued*)

| Date | Solar time | | Solar position | | BTUH/sq. ft. total insolation on surfaces[b] | | | | | | |
|------|-----|-----|------|------|--------|--------|------|------|------|------|------|
| | AM | PM | Alt | Azm | | | South facing surface angle with horiz. | | | | |
| | | | | | Normal | Horiz. | 22 | 32 | 42 | 52 | 90 |
| Jul 21 | 6 | 6 | 10.7 | 107.7 | 113 | 37 | 22 | 14 | 13 | 12 | 8 |
| | 7 | 5 | 23.1 | 100.6 | 203 | 107 | 87 | 75 | 60 | 44 | 14 |
| | 8 | 4 | 35.7 | 93.6 | 241 | 174 | 158 | 143 | 125 | 104 | 16 |
| | 9 | 3 | 48.4 | 85.5 | 261 | 231 | 220 | 205 | 185 | 159 | 31 |
| | 10 | 2 | 60.9 | 74.3 | 271 | 274 | 269 | 254 | 232 | 204 | 54 |
| | 11 | 1 | 72.4 | 53.3 | 277 | 302 | 300 | 285 | 262 | 232 | 69 |
| | 12 | | 78.6 | 0.0 | 279 | 311 | 310 | 296 | 273 | 242 | 74 |
| | Surface daily totals | | | | 3012 | 2558 | 2422 | 2250 | 2030 | 1754 | 458 |
| Aug 21 | 6 | 6 | 6.5 | 100.5 | 59 | 14 | 9 | 7 | 6 | 6 | 4 |
| | 7 | 5 | 19.1 | 92.8 | 190 | 85 | 77 | 69 | 60 | 50 | 12 |
| | 8 | 4 | 31.8 | 84.7 | 240 | 156 | 152 | 144 | 132 | 116 | 33 |
| | 9 | 3 | 44.3 | 75.0 | 263 | 216 | 220 | 212 | 197 | 178 | 65 |
| | 10 | 2 | 56.1 | 61.3 | 276 | 262 | 272 | 264 | 249 | 226 | 91 |
| | 11 | 1 | 66.0 | 38.4 | 282 | 292 | 305 | 298 | 281 | 257 | 107 |
| | 12 | | 70.3 | 0.0 | 284 | 302 | 317 | 309 | 292 | 268 | 113 |
| | Surface daily totals | | | | 2902 | 2352 | 2388 | 2296 | 2144 | 1934 | 736 |
| Sep 21 | 7 | 5 | 12.7 | 81.9 | 163 | 51 | 56 | 56 | 55 | 52 | 30 |
| | 8 | 4 | 25.1 | 73.0 | 240 | 124 | 140 | 141 | 138 | 131 | 75 |
| | 9 | 3 | 36.8 | 62.1 | 272 | 188 | 213 | 215 | 211 | 201 | 114 |
| | 10 | 2 | 47.3 | 47.5 | 287 | 237 | 270 | 273 | 268 | 255 | 145 |
| | 11 | 1 | 55.0 | 26.8 | 294 | 268 | 306 | 309 | 303 | 289 | 164 |
| | 12 | | 58.0 | 0.0 | 296 | 278 | 318 | 321 | 315 | 300 | 171 |
| | Surface daily totals | | | | 2808 | 2014 | 2288 | 2308 | 2264 | 2154 | 1226 |
| Oct 21 | 7 | 5 | 6.8 | 73.1 | 99 | 19 | 29 | 32 | 34 | 36 | 32 |
| | 8 | 4 | 18.7 | 64.0 | 229 | 90 | 120 | 128 | 133 | 134 | 104 |
| | 9 | 3 | 29.5 | 53.0 | 273 | 155 | 198 | 208 | 213 | 212 | 153 |
| | 10 | 2 | 38.7 | 39.1 | 293 | 204 | 257 | 269 | 273 | 270 | 188 |
| | 11 | 1 | 45.1 | 21.1 | 302 | 236 | 294 | 307 | 311 | 306 | 209 |
| | 12 | | 47.5 | 0.0 | 304 | 247 | 306 | 320 | 324 | 318 | 217 |
| | Surface daily totals | | | | 2696 | 1654 | 2100 | 2208 | 2252 | 2232 | 1588 |
| Nov 21 | 7 | 5 | 1.5 | 65.4 | 2 | 0 | 0 | 0 | 1 | 1 | 1 |
| | 8 | 4 | 12.7 | 56.6 | 196 | 55 | 91 | 104 | 113 | 119 | 111 |
| | 9 | 3 | 22.6 | 46.1 | 263 | 118 | 173 | 190 | 202 | 208 | 176 |
| | 10 | 2 | 30.8 | 33.2 | 289 | 166 | 233 | 252 | 265 | 270 | 217 |
| | 11 | 1 | 36.2 | 17.6 | 301 | 197 | 270 | 291 | 303 | 307 | 241 |
| | 12 | | 38.2 | 0.0 | 304 | 207 | 282 | 304 | 316 | 320 | 249 |
| | Surface daily totals | | | | 2406 | 1280 | 1816 | 1980 | 2084 | 2130 | 1742 |
| Dec 21 | 8 | 4 | 10.3 | 53.8 | 176 | 41 | 77 | 90 | 101 | 108 | 107 |
| | 9 | 3 | 19.8 | 43.6 | 257 | 102 | 161 | 180 | 195 | 204 | 183 |
| | 10 | 2 | 27.6 | 31.2 | 288 | 150 | 221 | 244 | 259 | 267 | 226 |
| | 11 | 1 | 32.7 | 16.4 | 301 | 180 | 258 | 282 | 298 | 305 | 251 |
| | 12 | | 34.6 | 0.0 | 304 | 190 | 271 | 295 | 311 | 318 | 259 |
| | Surface daily totals | | | | 2348 | 1136 | 1704 | 1888 | 2016 | 2086 | 1794 |

[b]1 Btu/hr · ft² = 3.152 W/m². Ground reflection not included on normal or horizontal surfaces.

**TABLE A2.6c** Solar Position and Insolation Values for 40 Degrees North Latitude[a]

| Date | Solar time AM | Solar time PM | Solar position Alt | Solar position Azm | Normal | Horiz. | 30 | 40 | 50 | 60 | 90 |
|------|----|----|------|------|--------|--------|------|------|------|------|------|
| | | | | | | | South facing surface angle with horiz. | | | | |
| Jan 21 | 8 | 4 | 8.1 | 55.3 | 142 | 28 | 65 | 74 | 81 | 85 | 84 |
| | 9 | 3 | 16.8 | 44.0 | 239 | 83 | 155 | 171 | 182 | 187 | 171 |
| | 10 | 2 | 23.8 | 30.9 | 274 | 127 | 218 | 237 | 249 | 254 | 223 |
| | 11 | 1 | 28.4 | 16.0 | 289 | 154 | 257 | 277 | 290 | 293 | 253 |
| | 12 | | 30.0 | 0.0 | 294 | 164 | 270 | 291 | 303 | 306 | 263 |
| | Surface daily totals | | | | 2182 | 948 | 1660 | 1810 | 1906 | 1944 | 1726 |
| Feb 21 | 7 | 5 | 4.8 | 72.7 | 69 | 10 | 19 | 21 | 23 | 24 | 22 |
| | 8 | 4 | 15.4 | 62.2 | 224 | 73 | 114 | 122 | 126 | 127 | 107 |
| | 9 | 3 | 25.0 | 50.2 | 274 | 132 | 195 | 205 | 209 | 208 | 167 |
| | 10 | 2 | 32.8 | 35.9 | 295 | 178 | 256 | 267 | 271 | 267 | 210 |
| | 11 | 1 | 38.1 | 18.9 | 305 | 206 | 293 | 306 | 310 | 304 | 236 |
| | 12 | | 40.0 | 0.0 | 308 | 216 | 306 | 319 | 323 | 317 | 245 |
| | Surface daily totals | | | | 2640 | 1414 | 2060 | 2162 | 2202 | 2176 | 1730 |
| Mar 21 | 7 | 5 | 11.4 | 80.2 | 171 | 46 | 55 | 55 | 54 | 51 | 35 |
| | 8 | 4 | 22.5 | 69.6 | 250 | 114 | 140 | 141 | 138 | 131 | 89 |
| | 9 | 3 | 32.8 | 57.3 | 282 | 173 | 215 | 217 | 213 | 202 | 138 |
| | 10 | 2 | 41.6 | 41.9 | 297 | 218 | 273 | 276 | 271 | 258 | 176 |
| | 11 | 1 | 47.7 | 22.6 | 305 | 247 | 310 | 313 | 307 | 293 | 200 |
| | 12 | | 50.0 | 0.0 | 307 | 257 | 322 | 326 | 320 | 305 | 208 |
| | Surface daily totals | | | | 2916 | 1852 | 2308 | 2330 | 2284 | 2174 | 1484 |
| Apr 21 | 6 | 6 | 7.4 | 98.9 | 89 | 20 | 11 | 8 | 7 | 7 | 4 |
| | 7 | 5 | 18.9 | 89.5 | 206 | 87 | 77 | 70 | 61 | 50 | 12 |
| | 8 | 4 | 30.3 | 79.3 | 252 | 152 | 153 | 145 | 133 | 117 | 53 |
| | 9 | 3 | 41.3 | 67.2 | 274 | 207 | 221 | 213 | 199 | 179 | 93 |
| | 10 | 2 | 51.2 | 51.4 | 286 | 250 | 275 | 267 | 252 | 229 | 126 |
| | 11 | 1 | 58.7 | 29.2 | 292 | 277 | 308 | 301 | 285 | 260 | 147 |
| | 12 | | 61.6 | 0.0 | 293 | 287 | 320 | 313 | 296 | 271 | 154 |
| | Surface daily totals | | | | 3092 | 2274 | 2412 | 2320 | 2168 | 1956 | 1022 |
| May 21 | 5 | 7 | 1.9 | 114.7 | 1 | 0 | 0 | 0 | 0 | 0 | 0 |
| | 6 | 6 | 12.7 | 105.6 | 144 | 49 | 25 | 15 | 14 | 13 | 9 |
| | 7 | 5 | 24.0 | 96.6 | 216 | 214 | 89 | 76 | 60 | 44 | 13 |
| | 8 | 4 | 35.4 | 87.2 | 250 | 175 | 158 | 144 | 125 | 104 | 25 |
| | 9 | 3 | 46.8 | 76.0 | 267 | 227 | 221 | 206 | 186 | 160 | 60 |
| | 10 | 2 | 57.5 | 60.9 | 277 | 267 | 270 | 255 | 233 | 205 | 89 |
| | 11 | 1 | 66.2 | 37.1 | 283 | 293 | 301 | 287 | 264 | 234 | 108 |
| | 12 | | 70.0 | 0.0 | 284 | 301 | 312 | 297 | 274 | 243 | 114 |
| | Surface daily totals | | | | 3160 | 2552 | 2442 | 2264 | 2040 | 1760 | 724 |
| Jun 21 | 5 | 7 | 4.2 | 117.3 | 22 | 4 | 3 | 3 | 2 | 2 | 1 |
| | 6 | 6 | 14.8 | 108.4 | 155 | 60 | 30 | 18 | 17 | 16 | 10 |
| | 7 | 5 | 26.0 | 99.7 | 216 | 123 | 92 | 77 | 59 | 41 | 14 |
| | 8 | 4 | 37.4 | 90.7 | 246 | 182 | 159 | 142 | 121 | 97 | 16 |
| | 9 | 3 | 48.8 | 80.2 | 263 | 233 | 219 | 202 | 179 | 151 | 47 |
| | 10 | 2 | 59.8 | 65.8 | 272 | 272 | 266 | 248 | 224 | 194 | 74 |
| | 11 | 1 | 69.2 | 41.9 | 277 | 296 | 296 | 278 | 253 | 221 | 92 |
| | 12 | | 73.5 | 0.0 | 279 | 304 | 306 | 289 | 263 | 230 | 98 |
| | Surface daily totals | | | | 3180 | 2648 | 2434 | 2224 | 1974 | 1670 | 610 |

[a]From Kreider, J. F., and F. Kreith, "Solar Heating and Cooling," revised 1st ed., Hemisphere Publ. Corp., 1977.

**TABLE A2.6c** Solar Position and Insolation Values for 40 Degrees North Latitude[a] (*Continued*)

| Date | Solar time AM | Solar time PM | Solar position Alt | Solar position Azm | Normal | Horiz. | South facing surface angle with horiz. 30 | 40 | 50 | 60 | 90 |
|------|-----|-----|------|-------|--------|--------|------|------|------|------|------|
| Jul 21 | 5 | 7 | 2.3 | 115.2 | 2 | 0 | 0 | 0 | 0 | 0 | 0 |
|  | 6 | 6 | 13.1 | 106.1 | 138 | 50 | 26 | 17 | 15 | 14 | 9 |
|  | 7 | 5 | 24.3 | 97.2 | 208 | 114 | 89 | 75 | 60 | 44 | 14 |
|  | 8 | 4 | 35.8 | 87.8 | 241 | 174 | 157 | 142 | 124 | 102 | 24 |
|  | 9 | 3 | 47.2 | 76.7 | 259 | 225 | 218 | 203 | 182 | 157 | 58 |
|  | 10 | 2 | 57.9 | 61.7 | 269 | 265 | 266 | 251 | 229 | 200 | 86 |
|  | 11 | 1 | 66.7 | 37.9 | 275 | 290 | 296 | 281 | 258 | 228 | 104 |
|  | 12 | | 70.6 | 0.0 | 276 | 298 | 307 | 292 | 269 | 238 | 111 |
|  | Surface daily totals | | | | 3062 | 2534 | 2409 | 2230 | 2006 | 1728 | 702 |
| Aug 21 | 6 | 6 | 7.9 | 99.5 | 81 | 21 | 12 | 9 | 8 | 7 | 5 |
|  | 7 | 5 | 19.3 | 90.9 | 191 | 87 | 76 | 69 | 60 | 49 | 12 |
|  | 8 | 4 | 30.7 | 79.9 | 237 | 150 | 150 | 141 | 129 | 113 | 50 |
|  | 9 | 3 | 41.8 | 67.9 | 260 | 205 | 216 | 207 | 193 | 173 | 89 |
|  | 10 | 2 | 51.7 | 52.1 | 272 | 246 | 267 | 259 | 244 | 221 | 120 |
|  | 11 | 1 | 59.3 | 29.7 | 278 | 273 | 300 | 292 | 276 | 252 | 140 |
|  | 12 | | 62.3 | 0.0 | 280 | 282 | 311 | 303 | 287 | 262 | 147 |
|  | Surface daily totals | | | | 2916 | 2244 | 2354 | 2258 | 2104 | 1894 | 978 |
| Sep 21 | 7 | 5 | 11.4 | 80.2 | 149 | 43 | 51 | 51 | 49 | 47 | 32 |
|  | 8 | 4 | 22.5 | 69.6 | 230 | 109 | 133 | 134 | 131 | 124 | 84 |
|  | 9 | 3 | 32.8 | 57.3 | 263 | 167 | 206 | 208 | 203 | 193 | 132 |
|  | 10 | 2 | 41.6 | 41.9 | 280 | 211 | 262 | 265 | 260 | 247 | 168 |
|  | 11 | 1 | 47.7 | 22.6 | 287 | 239 | 298 | 301 | 295 | 281 | 192 |
|  | 12 | | 50.0 | 0.0 | 290 | 249 | 310 | 313 | 307 | 292 | 200 |
|  | Surface daily totals | | | | 2708 | 1788 | 2210 | 2228 | 2182 | 2074 | 1416 |
| Oct 21 | 7 | 5 | 4.5 | 72.3 | 48 | 7 | 14 | 15 | 17 | 17 | 16 |
|  | 8 | 4 | 15.0 | 61.9 | 204 | 68 | 106 | 113 | 117 | 118 | 100 |
|  | 9 | 3 | 24.5 | 49.8 | 257 | 126 | 185 | 195 | 200 | 198 | 160 |
|  | 10 | 2 | 32.4 | 35.6 | 280 | 170 | 245 | 257 | 261 | 257 | 203 |
|  | 11 | 1 | 37.6 | 18.7 | 291 | 199 | 283 | 295 | 299 | 294 | 229 |
|  | 12 | | 39.5 | 0.0 | 294 | 208 | 295 | 308 | 312 | 306 | 238 |
|  | Surface daily totals | | | | 2454 | 1348 | 1962 | 2060 | 2098 | 2074 | 1654 |
| Nov 21 | 8 | 4 | 8.2 | 55.4 | 136 | 28 | 63 | 72 | 78 | 82 | 81 |
|  | 9 | 3 | 17.0 | 44.1 | 232 | 82 | 152 | 167 | 178 | 183 | 167 |
|  | 10 | 2 | 24.0 | 31.0 | 268 | 126 | 215 | 233 | 245 | 249 | 219 |
|  | 11 | 1 | 28.6 | 16.1 | 283 | 153 | 254 | 273 | 285 | 288 | 248 |
|  | 12 | | 30.2 | 0.0 | 288 | 163 | 267 | 287 | 298 | 301 | 258 |
|  | Surface daily totals | | | | 2128 | 942 | 1636 | 1778 | 1870 | 1908 | 1686 |
| Dec 21 | 8 | 4 | 5.5 | 53.0 | 89 | 14 | 39 | 45 | 50 | 54 | 56 |
|  | 9 | 3 | 14.0 | 41.9 | 217 | 65 | 135 | 152 | 164 | 171 | 163 |
|  | 10 | 2 | 20., | 29.4 | 261 | 107 | 200 | 221 | 235 | 242 | 221 |
|  | 11 | 1 | 25.0 | 15.2 | 280 | 134 | 239 | 262 | 276 | 283 | 252 |
|  | 12 | | 26.6 | 0.0 | 285 | 143 | 253 | 275 | 290 | 296 | 263 |
|  | Surface daily totals | | | | 1978 | 782 | 1480 | 1634 | 1740 | 1796 | 1646 |

[b]1 Btu/hr · ft² = 3.152 W/m². Ground reflection not included on normal or horizontal surfaces.

**TABLE A2.6d** Solar Position and Insolation Values for 48 Degrees North Latitude[a]

| Date | Solar time AM | Solar time PM | Solar position Alt | Solar position Azm | BTUH/sq. ft. total insolation on surfaces[b] Normal | Horiz. | South facing surface angle with horiz. 38 | 48 | 58 | 68 | 90 |
|------|----|----|------|------|--------|-------|------|------|------|------|------|
| Jan 21 | 8 | 4 | 3.5 | 54.6 | 37 | 4 | 17 | 19 | 21 | 22 | 22 |
|  | 9 | 3 | 11.0 | 42.6 | 185 | 46 | 120 | 132 | 140 | 145 | 139 |
|  | 10 | 2 | 16.9 | 29.4 | 239 | 83 | 190 | 206 | 216 | 220 | 206 |
|  | 11 | 1 | 20.7 | 15.1 | 261 | 107 | 231 | 249 | 260 | 263 | 243 |
|  | 12 | | 22.0 | 0.0 | 267 | 115 | 245 | 264 | 275 | 278 | 255 |
|  | Surface daily totals | | | | 1710 | 596 | 1360 | 1478 | 1550 | 1578 | 1478 |
| Feb 21 | 7 | 5 | 2.4 | 72.2 | 12 | 1 | 3 | 4 | 4 | 4 | 4 |
|  | 8 | 4 | 11.6 | 60.5 | 188 | 49 | 95 | 102 | 105 | 106 | 96 |
|  | 9 | 3 | 19.7 | 47.7 | 251 | 100 | 178 | 187 | 191 | 190 | 167 |
|  | 10 | 2 | 26.2 | 33.3 | 278 | 139 | 240 | 251 | 255 | 251 | 217 |
|  | 11 | 1 | 30.5 | 17.2 | 290 | 165 | 278 | 290 | 294 | 288 | 247 |
|  | 12 | | 32.0 | 0.0 | 293 | 173 | 291 | 304 | 307 | 301 | 258 |
|  | Surface daily totals | | | | 2330 | 1080 | 1880 | 1972 | 2024 | 1978 | 1720 |
| Mar 21 | 7 | 5 | 10.0 | 78.7 | 153 | 37 | 49 | 49 | 47 | 45 | 35 |
|  | 8 | 4 | 19.5 | 66.8 | 236 | 96 | 131 | 132 | 129 | 122 | 96 |
|  | 9 | 3 | 28.2 | 53.4 | 270 | 147 | 205 | 207 | 203 | 193 | 152 |
|  | 10 | 2 | 35.4 | 37.8 | 287 | 187 | 263 | 266 | 261 | 248 | 195 |
|  | 11 | 1 | 40.3 | 19.8 | 295 | 212 | 300 | 303 | 297 | 283 | 223 |
|  | 12 | | 42.0 | 0.0 | 298 | 220 | 312 | 315 | 309 | 294 | 232 |
|  | Surface daily totals | | | | 2780 | 1578 | 2208 | 2228 | 2182 | 2074 | 1632 |
| Apr 21 | 6 | 6 | 8.6 | 97.8 | 108 | 27 | 13 | 9 | 8 | 7 | 5 |
|  | 7 | 5 | 18.6 | 86.7 | 205 | 85 | 76 | 69 | 59 | 48 | 21 |
|  | 8 | 4 | 28.5 | 74.9 | 247 | 142 | 149 | 141 | 129 | 113 | 69 |
|  | 9 | 3 | 37.8 | 61.2 | 268 | 191 | 216 | 208 | 194 | 174 | 115 |
|  | 10 | 2 | 45.8 | 44.6 | 280 | 228 | 268 | 260 | 245 | 223 | 152 |
|  | 11 | 1 | 51.5 | 24.0 | 286 | 252 | 301 | 294 | 278 | 254 | 177 |
|  | 12 | | 53.6 | 0.0 | 288 | 260 | 313 | 305 | 289 | 264 | 185 |
|  | Surface daily totals | | | | 3076 | 2106 | 2358 | 2266 | 2114 | 1902 | 1262 |
| May 21 | 5 | 7 | 5.2 | 114.3 | 41 | 9 | 4 | 4 | 4 | 3 | 2 |
|  | 6 | 6 | 14.7 | 103.7 | 162 | 61 | 27 | 16 | 15 | 13 | 10 |
|  | 7 | 5 | 24.6 | 93.0 | 219 | 118 | 89 | 75 | 60 | 43 | 13 |
|  | 8 | 4 | 34.7 | 81.6 | 248 | 171 | 156 | 142 | 123 | 101 | 45 |
|  | 9 | 3 | 44.3 | 68.3 | 264 | 217 | 217 | 202 | 182 | 156 | 86 |
|  | 10 | 2 | 53.0 | 51.3 | 274 | 252 | 265 | 251 | 229 | 200 | 120 |
|  | 11 | 1 | 59.5 | 28.6 | 279 | 274 | 296 | 281 | 258 | 228 | 141 |
|  | 12 | | 62.0 | 0.0 | 280 | 281 | 306 | 292 | 269 | 238 | 149 |
|  | Surface daily totals | | | | 3254 | 2482 | 2418 | 2234 | 2010 | 1728 | 982 |
| Jun 21 | 5 | 7 | 7.9 | 116.5 | 77 | 21 | 9 | 9 | 8 | 7 | 5 |
|  | 6 | 6 | 17.2 | 106.2 | 172 | 74 | 33 | 19 | 18 | 16 | 12 |
|  | 7 | 5 | 27.0 | 95.8 | 220 | 129 | 93 | 77 | 59 | 39 | 15 |
|  | 8 | 4 | 37.1 | 84.6 | 246 | 181 | 157 | 140 | 119 | 95 | 35 |
|  | 9 | 3 | 46.9 | 71.6 | 261 | 225 | 216 | 198 | 175 | 147 | 74 |
|  | 10 | 2 | 55.8 | 54.8 | 269 | 259 | 262 | 244 | 220 | 189 | 105 |
|  | 11 | 1 | 62.7 | 31.2 | 274 | 280 | 291 | 273 | 248 | 216 | 126 |
|  | 12 | | 65.5 | 0.0 | 275 | 287 | 301 | 283 | 258 | 225 | 133 |
|  | Surface daily totals | | | | 3312 | 2626 | 2420 | 2204 | 1950 | 1644 | 874 |

[a]From Kreider, J. F., and F. Kreith, "Solar Heating and Cooling," revised 1st ed., Hemisphere Publ. Corp., 1977.

**TABLE A2.6d**  Solar Position and Insolation Values for 48 Degrees North Latitude[a] (*Continued*)

| Date | Solar time AM | PM | Solar position Alt | Azm | Normal | Horiz. | South facing surface angle with horiz. 38 | 48 | 58 | 68 | 90 |
|------|---------------|-----|--------------------|-----|--------|--------|---------|------|------|------|------|
| Jul 21 | 5 | 7 | 5.7 | 114.7 | 43 | 10 | 5 | 5 | 4 | 4 | 3 |
|  | 6 | 6 | 15.2 | 104.1 | 156 | 62 | 28 | 18 | 16 | 15 | 11 |
|  | 7 | 5 | 25.1 | 93.5 | 211 | 118 | 89 | 75 | 59 | 42 | 14 |
|  | 8 | 4 | 35.1 | 82.1 | 240 | 171 | 154 | 140 | 121 | 99 | 43 |
|  | 9 | 3 | 44.8 | 68.8 | 256 | 215 | 214 | 199 | 178 | 153 | 83 |
|  | 10 | 2 | 53.5 | 51.9 | 266 | 250 | 261 | 246 | 224 | 195 | 116 |
|  | 11 | 1 | 60.1 | 29.0 | 271 | 272 | 291 | 276 | 253 | 223 | 137 |
|  | 12 |  | 62.6 | 0.0 | 272 | 279 | 301 | 286 | 263 | 232 | 144 |
|  | Surface daily totals |  |  |  | 3158 | 2474 | 2386 | 2200 | 1974 | 1694 | 956 |
| Aug 21 | 6 | 6 | 9.1 | 98.3 | 99 | 28 | 14 | 10 | 9 | 8 | 6 |
|  | 7 | 5 | 19.1 | 87.2 | 190 | 85 | 75 | 67 | 58 | 47 | 20 |
|  | 8 | 4 | 29.0 | 75.4 | 232 | 141 | 145 | 137 | 125 | 109 | 65 |
|  | 9 | 3 | 38.4 | 61.8 | 254 | 189 | 210 | 201 | 187 | 168 | 110 |
|  | 10 | 2 | 46.4 | 45.1 | 266 | 225 | 260 | 252 | 237 | 214 | 146 |
|  | 11 | 1 | 52.2 | 24.3 | 272 | 248 | 293 | 285 | 268 | 244 | 169 |
|  | 12 |  | 54.3 | 0.0 | 274 | 256 | 304 | 296 | 279 | 255 | 177 |
|  | Surface daily totals |  |  |  | 2898 | 2086 | 2300 | 2200 | 2046 | 1836 | 1208 |
| Sep 21 | 7 | 5 | 10.0 | 78.7 | 131 | 35 | 44 | 44 | 43 | 40 | 31 |
|  | 8 | 4 | 19.5 | 66.8 | 215 | 92 | 124 | 124 | 121 | 115 | 90 |
|  | 9 | 3 | 28.2 | 53.4 | 251 | 142 | 196 | 197 | 193 | 183 | 143 |
|  | 10 | 2 | 35.4 | 37.8 | 269 | 181 | 251 | 254 | 248 | 236 | 185 |
|  | 11 | 1 | 40.3 | 19.8 | 278 | 205 | 287 | 289 | 284 | 269 | 212 |
|  | 12 |  | 42.0 | 0.0 | 280 | 213 | 299 | 302 | 296 | 281 | 221 |
|  | Surface daily totals |  |  |  | 2568 | 1522 | 2102 | 2118 | 2070 | 1966 | 1546 |
| Oct 21 | 7 | 5 | 2.0 | 71.9 | 4 | 0 | 1 | 1 | 1 | 1 | 1 |
|  | 8 | 4 | 11.2 | 60.2 | 165 | 44 | 86 | 91 | 95 | 95 | 87 |
|  | 9 | 3 | 19.3 | 47.4 | 233 | 94 | 167 | 176 | 180 | 178 | 157 |
|  | 10 | 2 | 25.7 | 33.1 | 262 | 133 | 228 | 239 | 242 | 239 | 207 |
|  | 11 | 1 | 30.0 | 17.1 | 274 | 157 | 266 | 277 | 281 | 276 | 237 |
|  | 12 |  | 31.5 | 0.0 | 278 | 166 | 279 | 291 | 294 | 288 | 247 |
|  | Surface daily totals |  |  |  | 2154 | 1022 | 1774 | 1860 | 1890 | 1866 | 1626 |
| Nov 21 | 8 | 4 | 3.6 | 54.7 | 36 | 5 | 17 | 19 | 21 | 22 | 22 |
|  | 9 | 3 | 11.2 | 42.7 | 179 | 46 | 117 | 129 | 137 | 141 | 135 |
|  | 10 | 2 | 17.1 | 29.5 | 233 | 83 | 186 | 202 | 212 | 215 | 201 |
|  | 11 | 1 | 20.9 | 15.1 | 255 | 107 | 227 | 245 | 255 | 258 | 238 |
|  | 12 |  | 22.2 | 0.0 | 261 | 115 | 241 | 259 | 270 | 272 | 250 |
|  | Surface daily totals |  |  |  | 1668 | 596 | 1336 | 1448 | 1518 | 1544 | 1442 |
| Dec 21 | 9 | 3 | 8.0 | 40.9 | 140 | 27 | 87 | 98 | 105 | 110 | 109 |
|  | 10 | 2 | 13.6 | 28.2 | 214 | 63 | 164 | 180 | 192 | 197 | 190 |
|  | 11 | 1 | 17.3 | 14.4 | 242 | 86 | 207 | 226 | 239 | 244 | 231 |
|  | 12 |  | 18.6 | 0.0 | 250 | 94 | 222 | 241 | 254 | 260 | 244 |
|  | Surface daily totals |  |  |  | 1444 | 446 | 1136 | 1250 | 1326 | 1364 | 1304 |

[b]1 Btu/hr · ft² = 3.152 W/m². Ground reflection not included on normal or horizontal surfaces.

**TABLE A2.6e** Solar Position and Insolation Values for 56 Degrees North Latitude[a]

| Date | Solar time AM | Solar time PM | Solar position Alt | Solar position Azm | BTUH/sq. ft. total insolation on surfaces[b] Normal | Horiz. | South facing surface angle with horiz. 46 | 56 | 66 | 76 | 90 |
|------|----|----|------|-------|--------|--------|-----|-----|-----|-----|-----|
| Jan 21 | 9 | 3 | 5.0 | 41.8 | 78 | 11 | 50 | 55 | 59 | 60 | 60 |
|  | 10 | 2 | 9.9 | 28.5 | 170 | 39 | 135 | 146 | 154 | 156 | 153 |
|  | 11 | 1 | 12.9 | 14.5 | 207 | 58 | 183 | 197 | 206 | 208 | 201 |
|  | 12 | | 14.0 | 0.0 | 217 | 65 | 198 | 214 | 222 | 225 | 217 |
|  | Surface daily totals | | | | 1126 | 282 | 934 | 1010 | 1058 | 1074 | 1044 |
| Feb 21 | 8 | 4 | 7.6 | 59.4 | 129 | 25 | 65 | 69 | 72 | 72 | 69 |
|  | 9 | 3 | 14.2 | 45.9 | 214 | 65 | 151 | 159 | 162 | 161 | 151 |
|  | 10 | 2 | 19.4 | 31.5 | 250 | 98 | 215 | 225 | 228 | 224 | 208 |
|  | 11 | 1 | 22.8 | 16.1 | 266 | 119 | 254 | 265 | 268 | 263 | 243 |
|  | 12 | | 24.0 | 0.0 | 270 | 126 | 268 | 279 | 282 | 276 | 255 |
|  | Surface daily totals | | | | 1986 | 740 | 1640 | 1716 | 1742 | 1716 | 1598 |
| Mar 21 | 7 | 5 | 8.3 | 77.5 | 128 | 28 | 40 | 40 | 39 | 37 | 32 |
|  | 8 | 4 | 16.2 | 64.4 | 215 | 75 | 119 | 120 | 117 | 111 | 97 |
|  | 9 | 3 | 23.3 | 50.3 | 253 | 118 | 192 | 193 | 189 | 180 | 154 |
|  | 10 | 2 | 29.0 | 34.9 | 272 | 151 | 249 | 251 | 246 | 234 | 205 |
|  | 11 | 1 | 32.7 | 17.9 | 282 | 172 | 285 | 288 | 282 | 268 | 236 |
|  | 12 | | 34.0 | 0.0 | 284 | 179 | 297 | 300 | 294 | 280 | 246 |
|  | Surface daily totals | | | | 2586 | 1268 | 2066 | 2084 | 2040 | 1938 | 1700 |
| Apr 21 | 5 | 7 | 1.4 | 108.8 | 0 | 0 | 0 | 0 | 0 | 0 | 0 |
|  | 6 | 6 | 9.6 | 96.5 | 122 | 32 | 14 | 9 | 8 | 7 | 6 |
|  | 7 | 5 | 18.0 | 84.1 | 201 | 81 | 74 | 66 | 57 | 46 | 29 |
|  | 8 | 4 | 26.1 | 70.9 | 239 | 129 | 143 | 135 | 123 | 108 | 82 |
|  | 9 | 3 | 33.6 | 56.3 | 260 | 169 | 208 | 200 | 186 | 167 | 133 |
|  | 10 | 2 | 39.9 | 39.7 | 272 | 201 | 259 | 251 | 236 | 214 | 174 |
|  | 11 | 1 | 44.1 | 20.7 | 278 | 220 | 292 | 284 | 268 | 245 | 200 |
|  | 12 | | 45.6 | 0.0 | 280 | 227 | 303 | 295 | 279 | 255 | 209 |
|  | Surface daily totals | | | | 3024 | 1892 | 2282 | 2186 | 2038 | 1830 | 1458 |
| May 21 | 4 | 8 | 1.2 | 125.5 | 0 | 0 | 0 | 0 | 0 | 0 | 0 |
|  | 5 | 7 | 8.5 | 113.4 | 93 | 25 | 10 | 9 | 8 | 7 | 6 |
|  | 6 | 6 | 16.5 | 101.5 | 175 | 71 | 28 | 17 | 15 | 13 | 11 |
|  | 7 | 5 | 24.8 | 89.3 | 219 | 119 | 88 | 74 | 58 | 41 | 16 |
|  | 8 | 4 | 33.1 | 76.3 | 244 | 163 | 153 | 138 | 119 | 98 | 63 |
|  | 9 | 3 | 40.9 | 61.6 | 259 | 201 | 212 | 197 | 176 | 151 | 109 |
|  | 10 | 2 | 47.6 | 44.2 | 268 | 231 | 259 | 244 | 222 | 194 | 146 |
|  | 11 | 1 | 52.3 | 23.4 | 273 | 249 | 288 | 274 | 251 | 222 | 170 |
|  | 12 | | 54.0 | 0.0 | 275 | 255 | 299 | 284 | 261 | 231 | 178 |
|  | Surface daily totals | | | | 3340 | 2374 | 2374 | 2188 | 1962 | 1682 | 1218 |
| Jun 21 | 4 | 8 | 4.2 | 127.2 | 21 | 4 | 2 | 2 | 2 | 2 | 1 |
|  | 5 | 7 | 11.4 | 115.3 | 122 | 40 | 14 | 13 | 11 | 10 | 8 |
|  | 6 | 6 | 19.3 | 103.6 | 185 | 86 | 34 | 19 | 17 | 15 | 12 |
|  | 7 | 5 | 27.6 | 91.7 | 222 | 132 | 92 | 76 | 57 | 38 | 15 |
|  | 8 | 4 | 35.9 | 78.8 | 243 | 175 | 154 | 137 | 116 | 92 | 55 |
|  | 9 | 3 | 43.8 | 64.1 | 257 | 212 | 211 | 193 | 170 | 143 | 98 |
|  | 10 | 2 | 50.7 | 46.4 | 265 | 240 | 255 | 238 | 214 | 184 | 133 |
|  | 11 | 1 | 55.6 | 24.9 | 269 | 258 | 284 | 267 | 242 | 210 | 156 |
|  | 12 | | 57.5 | 0.0 | 271 | 264 | 294 | 276 | 251 | 219 | 164 |
|  | Surface daily totals | | | | 3438 | 2526 | 2388 | 2166 | 1910 | 1606 | 1120 |

[a]From Kreider, J. F., and F. Kreith, "Solar Heating and Cooling," revised 1st ed., Hemisphere Publ. Corp., 1977.

**676**

**TABLE A2.6e**  Solar Position and Insolation Values for 56 Degrees North Latitude[a] (*Continued*)

| Date | Solar time AM | Solar time PM | Solar position Alt | Solar position Azm | BTUH/sq. ft. total insolation on surfaces[b] Normal | Horiz. | South facing surface angle with horiz. 46 | 56 | 66 | 76 | 90 |
|------|----|----|------|-------|--------|-------|-----|-----|-----|-----|-----|
| Jul 21 | 4 | 8 | 1.7 | 125.8 | 0 | 0 | 0 | 0 | 0 | 0 | 0 |
| | 5 | 7 | 9.0 | 113.7 | 91 | 27 | 11 | 10 | 9 | 8 | 6 |
| | 6 | 6 | 17.0 | 101.9 | 169 | 72 | 30 | 18 | 16 | 14 | 12 |
| | 7 | 5 | 25.3 | 89.7 | 212 | 119 | 88 | 74 | 58 | 41 | 15 |
| | 8 | 4 | 33.6 | 76.7 | 237 | 163 | 151 | 136 | 117 | 96 | 61 |
| | 9 | 3 | 41.4 | 62.0 | 252 | 201 | 208 | 193 | 173 | 147 | 106 |
| | 10 | 2 | 48.2 | 44.6 | 261 | 230 | 254 | 239 | 217 | 189 | 142 |
| | 11 | 1 | 52.9 | 23.7 | 265 | 248 | 283 | 268 | 245 | 216 | 165 |
| | 12 | | 54.6 | 0.0 | 267 | 254 | 293 | 278 | 255 | 225 | 173 |
| | Surface daily totals | | | | 3240 | 2372 | 2342 | 2152 | 1926 | 1646 | 1186 |
| Aug 21 | 5 | 7 | 2.0 | 109.2 | 1 | 0 | 0 | 0 | 0 | 0 | 0 |
| | 6 | 6 | 10.2 | 97.0 | 112 | 34 | 16 | 11 | 10 | 9 | 7 |
| | 7 | 5 | 18.5 | 84.5 | 187 | 82 | 73 | 65 | 56 | 45 | 28 |
| | 8 | 4 | 26.7 | 71.3 | 225 | 128 | 140 | 131 | 119 | 104 | 78 |
| | 9 | 3 | 34.3 | 56.7 | 246 | 168 | 202 | 193 | 179 | 160 | 126 |
| | 10 | 2 | 40.5 | 40.0 | 258 | 199 | 251 | 242 | 227 | 206 | 166 |
| | 11 | 1 | 44.8 | 20.9 | 264 | 218 | 282 | 274 | 258 | 235 | 191 |
| | 12 | | 46.3 | 0.0 | 266 | 225 | 293 | 285 | 269 | 245 | 200 |
| | Surface daily totals | | | | 2850 | 1884 | 2218 | 2118 | 1966 | 1760 | 1392 |
| Sep 21 | 7 | 5 | 8.3 | 77.5 | 107 | 25 | 36 | 36 | 34 | 32 | 28 |
| | 8 | 4 | 16.2 | 64.4 | 194 | 72 | 111 | 111 | 108 | 102 | 89 |
| | 9 | 3 | 23.3 | 50.3 | 233 | 114 | 181 | 182 | 178 | 168 | 147 |
| | 10 | 2 | 29.0 | 34.9 | 253 | 146 | 236 | 237 | 232 | 221 | 193 |
| | 11 | 1 | 32.7 | 17.9 | 263 | 166 | 271 | 273 | 267 | 254 | 223 |
| | 12 | | 34.0 | 0.0 | 266 | 173 | 283 | 285 | 279 | 265 | 233 |
| | Surface daily totals | | | | 2368 | 1220 | 1950 | 1962 | 1918 | 1820 | 1594 |
| Oct 21 | 8 | 4 | 7.1 | 59.1 | 104 | 20 | 53 | 57 | 59 | 59 | 57 |
| | 9 | 3 | 13.8 | 45.7 | 193 | 60 | 138 | 145 | 148 | 147 | 138 |
| | 10 | 2 | 19.0 | 31.3 | 231 | 92 | 201 | 210 | 213 | 210 | 195 |
| | 11 | 1 | 22.3 | 16.0 | 248 | 112 | 240 | 250 | 253 | 248 | 230 |
| | 12 | | 23.5 | 0.0 | 253 | 119 | 253 | 263 | 266 | 261 | 241 |
| | Surface daily totals | | | | 1804 | 688 | 1516 | 1586 | 1612 | 1588 | 1480 |
| Nov 21 | 9 | 3 | 5.2 | 41.9 | 76 | 12 | 49 | 54 | 57 | 59 | 58 |
| | 10 | 2 | 10.0 | 28.5 | 165 | 39 | 132 | 143 | 149 | 152 | 148 |
| | 11 | 1 | 13.1 | 14.5 | 201 | 58 | 179 | 193 | 201 | 203 | 196 |
| | 12 | | 14.2 | 0.0 | 211 | 65 | 194 | 209 | 217 | 219 | 211 |
| | Surface daily totals | | | | 1094 | 284 | 914 | 986 | 1032 | 1046 | 1016 |
| Dec 21 | 9 | 3 | 1.9 | 40.5 | 5 | 0 | 3 | 4 | 4 | 4 | 4 |
| | 10 | 2 | 6.6 | 27.5 | 113 | 19 | 86 | 95 | 101 | 104 | 103 |
| | 11 | 1 | 9.5 | 13.9 | 166 | 37 | 141 | 154 | 163 | 167 | 164 |
| | 12 | | 10.6 | 0.0 | 180 | 43 | 159 | 173 | 182 | 186 | 182 |
| | Surface daily totals | | | | 748 | 156 | 620 | 678 | 716 | 734 | 722 |

[b]1 Btu/hr · ft² = 3.152 W/m². Ground reflection not included on normal or horizontal surfaces.

677

**TABLE A2.6f** Solar Position and Insolation Values for 64 Degrees North Latitude[a]

| Date | Solar time AM | PM | Solar position Alt | Azm | BTUH/sq. ft. total insolation on surfaces[b] Normal | Horiz. | South facing surface angle with horiz. 54 | 64 | 74 | 84 | 90 |
|------|----|----|------|------|--------|--------|-----|-----|-----|-----|-----|
| Jan 21 | 10 | 2 | 2.8 | 28.1 | 22 | 2 | 17 | 19 | 20 | 20 | 20 |
|  | 11 | 1 | 5.2 | 14.1 | 81 | 12 | 72 | 77 | 80 | 81 | 81 |
|  | 12 | | 6.0 | 0.0 | 100 | 16 | 91 | 98 | 102 | 103 | 103 |
|  | Surface daily totals | | | | 306 | 45 | 268 | 290 | 302 | 306 | 304 |
| Feb 21 | 8 | 4 | 3.4 | 58.7 | 35 | 4 | 17 | 19 | 19 | 19 | 19 |
|  | 9 | 3 | 8.6 | 44.8 | 147 | 31 | 103 | 108 | 111 | 110 | 107 |
|  | 10 | 2 | 12.6 | 30.3 | 199 | 55 | 170 | 178 | 181 | 178 | 173 |
|  | 11 | 1 | 15.1 | 15.3 | 222 | 71 | 212 | 220 | 223 | 219 | 213 |
|  | 12 | | 16.0 | 0.0 | 228 | 77 | 225 | 235 | 237 | 232 | 226 |
|  | Surface daily totals | | | | 1432 | 400 | 1230 | 1286 | 1302 | 1282 | 1252 |
| Mar 21 | 7 | 5 | 6.5 | 76.5 | 95 | 18 | 30 | 29 | 29 | 27 | 25 |
|  | 8 | 4 | 20.7 | 62.6 | 185 | 54 | 101 | 102 | 99 | 94 | 89 |
|  | 9 | 3 | 18.1 | 48.1 | 227 | 87 | 171 | 172 | 169 | 160 | 153 |
|  | 10 | 2 | 22.3 | 32.7 | 249 | 112 | 227 | 229 | 224 | 213 | 203 |
|  | 11 | 1 | 25.1 | 16.6 | 260 | 129 | 262 | 265 | 259 | 246 | 235 |
|  | 12 | | 26.0 | 0.0 | 263 | 134 | 274 | 277 | 271 | 258 | 246 |
|  | Surface daily totals | | | | 2296 | 932 | 1856 | 1870 | 1830 | 1736 | 1656 |
| Apr 21 | 5 | 7 | 4.0 | 108.5 | 27 | 5 | 2 | 2 | 2 | 1 | 1 |
|  | 6 | 6 | 10.4 | 95.1 | 133 | 37 | 15 | 9 | 8 | 7 | 6 |
|  | 7 | 5 | 17.0 | 81.6 | 194 | 76 | 70 | 63 | 54 | 43 | 37 |
|  | 8 | 4 | 23.3 | 67.5 | 228 | 112 | 136 | 128 | 116 | 102 | 91 |
|  | 9 | 3 | 29.0 | 52.3 | 248 | 144 | 197 | 189 | 176 | 158 | 145 |
|  | 10 | 2 | 33.5 | 36.0 | 260 | 169 | 246 | 239 | 224 | 203 | 188 |
|  | 11 | 1 | 36.5 | 18.4 | 266 | 184 | 278 | 270 | 255 | 233 | 216 |
|  | 12 | | 97.6 | 0.0 | 268 | 190 | 289 | 281 | 266 | 243 | 225 |
|  | Surface daily totals | | | | 2982 | 1644 | 2176 | 2082 | 1936 | 1736 | 1594 |
| May 21 | 4 | 8 | 5.8 | 125.1 | 51 | 11 | 5 | 4 | 4 | 3 | 3 |
|  | 5 | 7 | 11.6 | 112.1 | 132 | 42 | 13 | 11 | 10 | 9 | 8 |
|  | 6 | 6 | 17.9 | 99.1 | 185 | 79 | 29 | 16 | 14 | 12 | 11 |
|  | 7 | 5 | 24.5 | 85.7 | 218 | 117 | 86 | 72 | 56 | 39 | 28 |
|  | 8 | 4 | 30.9 | 71.5 | 239 | 152 | 148 | 133 | 115 | 94 | 80 |
|  | 9 | 3 | 36.8 | 56.1 | 252 | 182 | 204 | 190 | 170 | 145 | 128 |
|  | 10 | 2 | 41.6 | 38.9 | 261 | 205 | 249 | 235 | 213 | 186 | 167 |
|  | 11 | 1 | 44.9 | 20.1 | 265 | 219 | 278 | 264 | 242 | 213 | 193 |
|  | 12 | | 46.0 | 0.0 | 267 | 224 | 288 | 274 | 251 | 222 | 201 |
|  | Surface daily totals | | | | 3470 | 2236 | 2312 | 2124 | 1898 | 1624 | 1436 |
| Jun 21 | 3 | 9 | 4.2 | 139.4 | 21 | 4 | 2 | 2 | 2 | 2 | 1 |
|  | 4 | 8 | 9.0 | 126.4 | 93 | 27 | 10 | 9 | 8 | 7 | 6 |
|  | 5 | 7 | 14.7 | 113.6 | 154 | 60 | 16 | 15 | 13 | 11 | 10 |
|  | 6 | 6 | 21.0 | 100.8 | 194 | 96 | 34 | 19 | 17 | 14 | 13 |
|  | 7 | 5 | 27.5 | 87.5 | 221 | 132 | 91 | 74 | 55 | 36 | 23 |
|  | 8 | 4 | 34.0 | 73.3 | 239 | 166 | 150 | 133 | 112 | 88 | 73 |
|  | 9 | 3 | 39.9 | 57.8 | 251 | 195 | 204 | 187 | 164 | 137 | 119 |
|  | 10 | 2 | 44.9 | 40.4 | 258 | 217 | 247 | 230 | 206 | 177 | 157 |
|  | 11 | 1 | 48.3 | 20.9 | 262 | 231 | 275 | 258 | 233 | 202 | 181 |
|  | 12 | | 49.5 | 0.0 | 263 | 235 | 284 | 267 | 242 | 211 | 189 |
|  | Surface daily totals | | | | 3650 | 2488 | 2342 | 2118 | 1862 | 1558 | 1356 |

[a]From Kreider, J. F., and F. Kreith, "Solar Heating and Cooling," revised 1st ed., Hemisphere Publ. Corp., 1977.

**TABLE A2.6f** Solar Position and Insolation Values for 64 Degrees North Latitude[a] (Continued)

| Date | Solar time AM | Solar time PM | Solar position Alt | Solar position Azm | BTUH/sq. ft. total insolation on surfaces[b] Normal | Horiz. | South facing surface angle with horiz. 54 | 64 | 74 | 84 | 90 |
|------|----|----|------|-------|--------|-------|-----|-----|-----|-----|-----|
| Jul 21 | 4 | 8 | 6.4 | 125.3 | 53 | 13 | 6 | 5 | 5 | 4 | 4 |
| | 5 | 7 | 12.1 | 112.4 | 128 | 44 | 14 | 13 | 11 | 10 | 9 |
| | 6 | 6 | 18.4 | 99.4 | 179 | 81 | 30 | 17 | 16 | 13 | 12 |
| | 7 | 5 | 25.0 | 86.0 | 211 | 118 | 86 | 72 | 56 | 38 | 28 |
| | 8 | 4 | 31.4 | 71.8 | 231 | 152 | 146 | 131 | 113 | 91 | 77 |
| | 9 | 3 | 37.3 | 56.3 | 245 | 182 | 201 | 186 | 166 | 141 | 124 |
| | 10 | 2 | 42.2 | 39.2 | 253 | 204 | 245 | 230 | 208 | 181 | 162 |
| | 11 | 1 | 45.4 | 20.2 | 257 | 218 | 273 | 258 | 236 | 207 | 187 |
| | 12 | | 46.6 | 0.0 | 259 | 223 | 282 | 267 | 245 | 216 | 195 |
| | Surface daily totals | | | | 3372 | 2248 | 2280 | 2090 | 1864 | 1588 | 1400 |
| Aug 21 | 5 | 7 | 4.6 | 108.8 | 29 | 6 | 3 | 3 | 2 | 2 | 2 |
| | 6 | 6 | 11.0 | 95.5 | 123 | 39 | 16 | 11 | 10 | 8 | 7 |
| | 7 | 5 | 17.6 | 81.9 | 181 | 77 | 69 | 61 | 52 | 42 | 35 |
| | 8 | 4 | 23.9 | 67.8 | 214 | 113 | 132 | 123 | 112 | 97 | 87 |
| | 9 | 3 | 29.6 | 52.6 | 234 | 144 | 190 | 182 | 169 | 150 | 138 |
| | 10 | 2 | 34.2 | 36.2 | 246 | 168 | 237 | 229 | 215 | 194 | 179 |
| | 11 | 1 | 37.2 | 18.5 | 252 | 183 | 268 | 260 | 244 | 222 | 205 |
| | 12 | | 38.3 | 0.0 | 254 | 188 | 278 | 270 | 255 | 232 | 215 |
| | Surface daily totals | | | | 2808 | 1646 | 2108 | 1008 | 1860 | 1662 | 1522 |
| Sep 21 | 7 | 5 | 6.5 | 76.5 | 77 | 16 | 25 | 25 | 24 | 23 | 21 |
| | 8 | 4 | 12.7 | 72.6 | 163 | 51 | 92 | 92 | 90 | 85 | 81 |
| | 9 | 3 | 18.1 | 48.1 | 206 | 83 | 159 | 159 | 156 | 147 | 141 |
| | 10 | 2 | 22.3 | 32.7 | 229 | 108 | 212 | 213 | 209 | 198 | 189 |
| | 11 | 1 | 25.1 | 16.6 | 240 | 124 | 246 | 248 | 243 | 230 | 220 |
| | 12 | | 26.0 | 0.0 | 244 | 129 | 258 | 260 | 254 | 241 | 230 |
| | Surface daily totals | | | | 2074 | 892 | 1726 | 1736 | 1696 | 1608 | 1532 |
| Oct 21 | 8 | 4 | 3.0 | 58.5 | 17 | 2 | 9 | 9 | 10 | 10 | 10 |
| | 9 | 3 | 8.1 | 44.6 | 122 | 26 | 86 | 91 | 93 | 92 | 90 |
| | 10 | 2 | 12.1 | 30.2 | 176 | 50 | 152 | 159 | 161 | 159 | 155 |
| | 11 | 1 | 14.6 | 15.2 | 201 | 65 | 193 | 201 | 203 | 200 | 195 |
| | 12 | | 15.5 | 0.0 | 208 | 71 | 207 | 215 | 217 | 213 | 208 |
| | Surface daily totals | | | | 1238 | 358 | 1088 | 1136 | 1152 | 1134 | 1106 |
| Nov 21 | 10 | 2 | 3.0 | 28.1 | 23 | 3 | 18 | 20 | 21 | 21 | 21 |
| | 11 | 1 | 5.4 | 14.2 | 79 | 12 | 70 | 76 | 79 | 80 | 79 |
| | 12 | | 6.2 | 0.0 | 97 | 17 | 89 | 96 | 100 | 101 | 100 |
| | Surface daily totals | | | | 302 | 46 | 266 | 286 | 298 | 302 | 300 |
| Dec 21 | 11 | 1 | 1.8 | 13.7 | 4 | 0 | 3 | 4 | 4 | 4 | 4 |
| | 12 | | 2.6 | 0.0 | 16 | 2 | 14 | 15 | 16 | 17 | 17 |
| | Surface daily totals | | | | 24 | 2 | 20 | 22 | 24 | 24 | 24 |

[b]1 Btu/hr · ft² = 3.152 W/m². Ground reflection not included on normal or horizontal surfaces.

**TABLE A2.7**  Reflectivity Values for Fifteen Characteristic Surfaces
(Integrated over Solar Spectrum and Angle of Incidence)[a]

| Surface | Average reflectivity |
|---|---|
| Snow (freshly fallen or with ice film) | 0.75 |
| Water surfaces (relatively large incidence angles) | 0.07 |
| Soils (clay, loam, etc.) | 0.14 |
| Earth roads | 0.04 |
| Coniferous forest (winter) | 0.07 |
| Forests in autumn, ripe field crops, plants | 0.26 |
| Weathered blacktop | 0.10 |
| Weathered concrete | 0.22 |
| Dead leaves | 0.30 |
| Dry grass | 0.20 |
| Green grass | 0.26 |
| Bituminous and gravel roof | 0.13 |
| Crushed rock surface | 0.20 |
| Building surfaces, dark (red brick, dark paints, etc.) | 0.27 |
| Building surfaces, light (light brick, light paints, etc.) | 0.60 |

[a]From Hunn, B. D., and D. O. Calafell, Determination of Average Ground Reflectivity for Solar Collectors, *Sol. Energy*, vol. 19, p. 87, 1977; see also R. J. List, "Smithsonian Meteorological Tables," 6th ed., Smithsonian Institution Press, pp. 442–443, 1949.

The altitude and azimuth of the sun are given by

$$\sin a = \sin \phi \sin \delta + \cos \phi \cos \delta \cos h \qquad (1)$$

and

$$\sin \alpha = -\cos \delta \sin h / \cos a \qquad (2)$$

where $a$ = altitude of the sun (angular elevation above the horizon)

$\phi$ = latitude of the observer

$\delta$ = declination of the sun

$h$ = hour angle of sun (angular distance from the meridian of the observer)

$\alpha$ = azimuth of the sun (measured eastward from north)

From Eqs. (1) and (2) it can be seen that the altitude and azimuth of the sun are functions of the latitude of the observer, the time of day (hour angle), and the date (declination).

Figure A2.1(*b–g*) provides a series of charts, one for each 5° of latitude (except 5°, 15°, 75°, and 85°) giving the altitude and azimuth of the sun as a function of the true solar time and the declination of the sun in a form originally suggested by Hand. Linear interpolation for intermediate latitudes will give results within the accuracy to which the charts can be read.

On these charts, a point corresponding to the projected position of the sun is determined from the heavy lines corresponding to declination and solar time.

**To find the solar altitude and azimuth:**

1. Select the chart or charts appropriate to the latitude.
2. Find the solar declination $\delta$ corresponding to the date.
3. Determine the *true solar time* as follows:
   (a) To the *local standard time* (zone time) add 4' for each degree of longitude the station is east of the standard meridian or subtract 4' for each degree west of the standard meridian to get the *local mean solar time.*
   (b) To the *local mean solar time* add algebraically the equation of time; the sum is the required *true solar time.*
4. Read the required altitude and azimuth at the point determined by the declination and the true solar time. Interpolate linearly between two charts for intermediate latitudes.

It should be emphasized that the solar altitude determined from these charts is the true geometric position of the center of the sun. At low solar elevations terrestrial refraction may considerably alter the apparent position of sun. Under average atmospheric refraction the sun will appear on the horizon when it actually is about 34' below the horizon; the effect of refraction decreases rapidly with increasing solar elevation. Since sunset or sunrise is defined as the time when the upper limb of the sun appears on the horizon, and the semidiameter of the sun is 16', sunset or sunrise occurs under average atmospheric refraction when the sun is 50' below the horizon. In polar regions especially, unusual atmospheric refraction can make considerable variation in the time of sunset or sunrise.

The 90°N chart is included for interpolation purposes; the azimuths lose their directional significance at the pole.

**Altitude and azimuth in southern latitudes.** To compute solar altitude and azimuth for southern latitudes, change the sign of the solar declination and proceed as above. The resulting azimuths will indicate angular distance from *south* (measured eastward) rather than from north.

(a)

**FIGURE A2.1** Description of method for calculating true solar time, together with accompanying meteorological charts, for computing solar-altitude and azimuth angles. (*a*) Description of method; (*b*) chart, 25°N latitude; (*c*) chart, 30°N latitude; (*d*) chart, 35°N latitude; (*e*) chart, 40°N latitude; (*f*) chart, 45°N latitude; (*g*) chart, 50°N latitude. Description and charts reproduced from the "Smithsonian Meteorological Tables" with permission from the Smithsonian Institute, Washington, D.C.

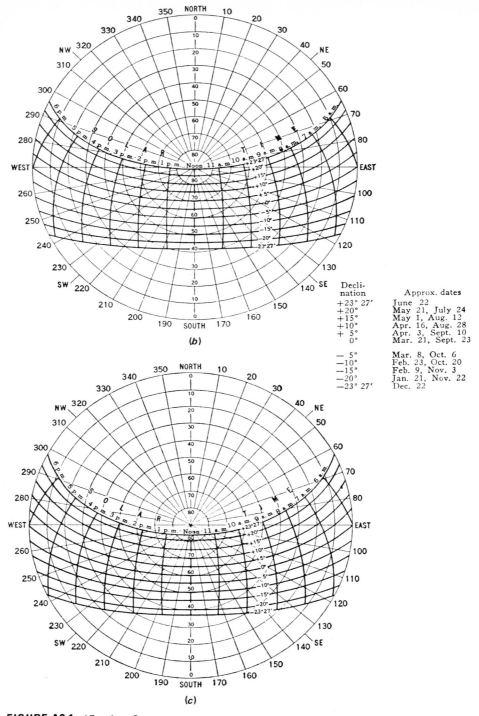

| Decli-nation | Approx. dates |
|---|---|
| +23° 27' | June 22 |
| +20° | May 21, July 24 |
| +15° | May 1, Aug. 12 |
| +10° | Apr. 16, Aug. 28 |
| + 5° | Apr. 3, Sept. 10 |
| 0° | Mar. 21, Sept. 23 |
| — 5° | Mar. 8, Oct. 6 |
| —10° | Feb. 23, Oct. 20 |
| —15° | Feb. 9, Nov. 3 |
| —20° | Jan. 21, Nov. 22 |
| —23° 27' | Dec. 22 |

(b)

(c)

**FIGURE A2.1** (*Continued*)

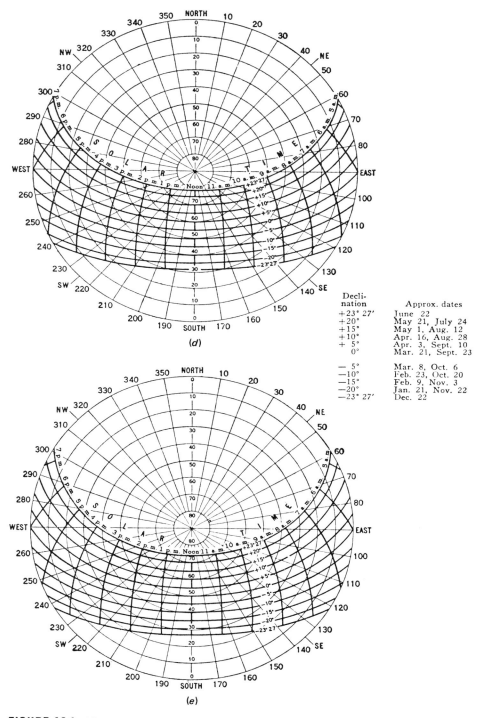

| Declination | Approx. dates |
|---|---|
| +23° 27′ | June 22 |
| +20° | May 21, July 24 |
| +15° | May 1, Aug. 12 |
| +10° | Apr. 16, Aug. 28 |
| + 5° | Apr. 3, Sept. 10 |
| 0° | Mar. 21, Sept. 23 |
| — 5° | Mar. 8, Oct. 6 |
| —10° | Feb. 23, Oct. 20 |
| —15° | Feb. 9, Nov. 3 |
| —20° | Jan. 21, Nov. 22 |
| —23° 27′ | Dec. 22 |

(d)

(e)

**FIGURE A2.1** (*Continued*)

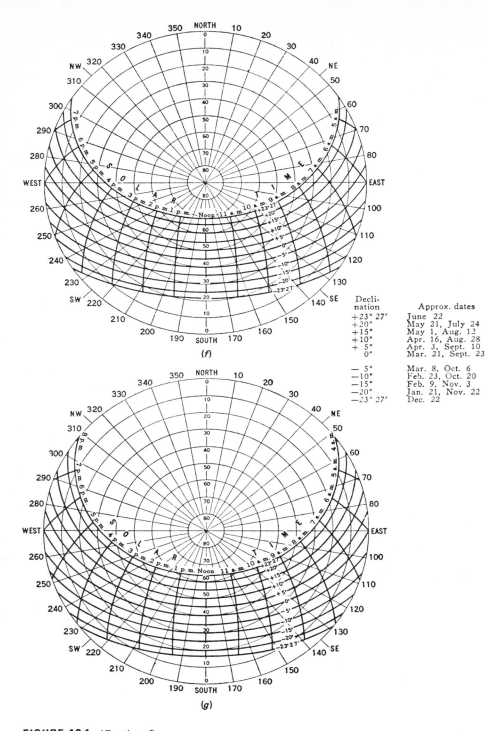

| Decli-nation | Approx. dates |
|---|---|
| +23° 27' | June 22 |
| +20° | May 21, July 24 |
| +15° | May 1, Aug. 12 |
| +10° | Apr. 16, Aug. 28 |
| + 5° | Apr. 3, Sept. 10 |
| 0° | Mar. 21, Sept. 23 |
| — 5° | Mar. 8, Oct. 6 |
| —10° | Feb. 23, Oct. 20 |
| —15° | Feb. 9, Nov. 3 |
| —20° | Jan. 21, Nov. 22 |
| —23° 27' | Dec. 22 |

(f)

(g)

**FIGURE A2.1** (*Continued*)

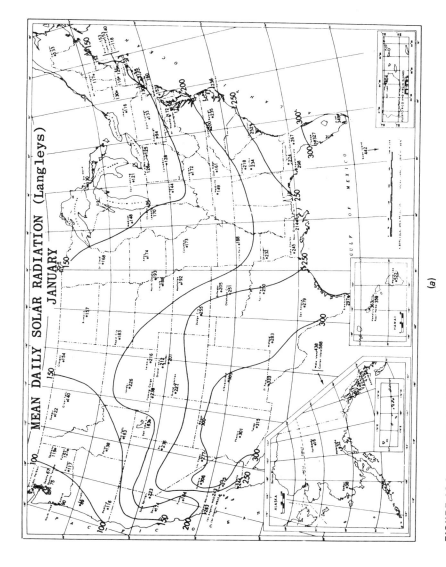

**FIGURE A2.2** Mean daily insolation data for the United States. (*a*) January; (*b*) February; (*c*) March; (*d*) April; (*e*) May; (*f*) June; (*g*) July; (*h*) August; (*i*) September; (*j*) October; (*k*) November; (*l*) December. Reproduced with permission from the "Climatic Atlas of the United States," U.S. Government Printing Office, 1968.

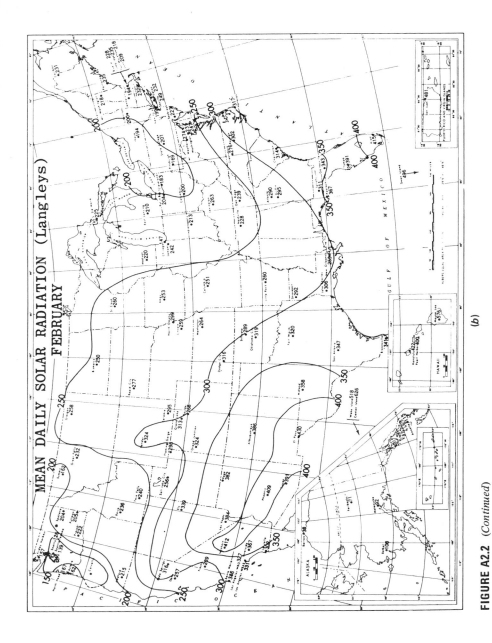

**FIGURE A2.2** *(Continued)*

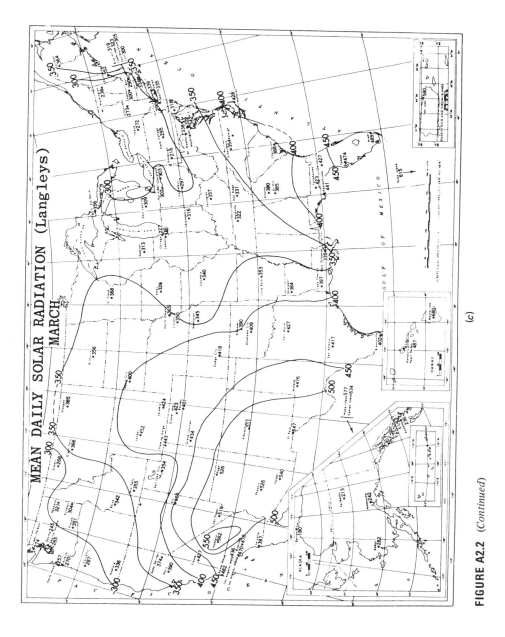

**FIGURE A2.2** *(Continued)*

687

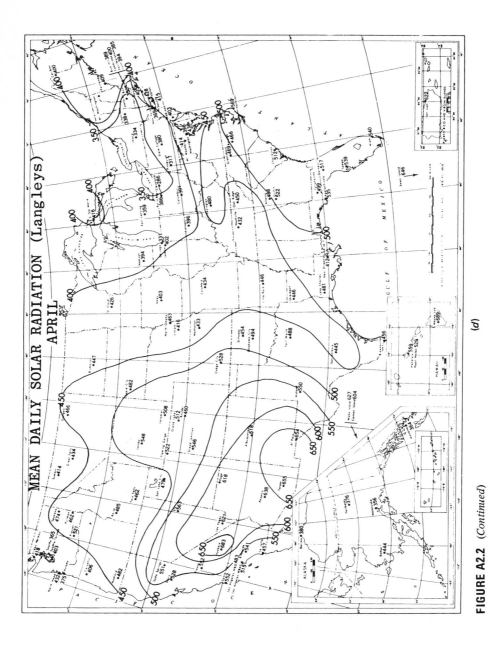

FIGURE A2.2 (Continued)

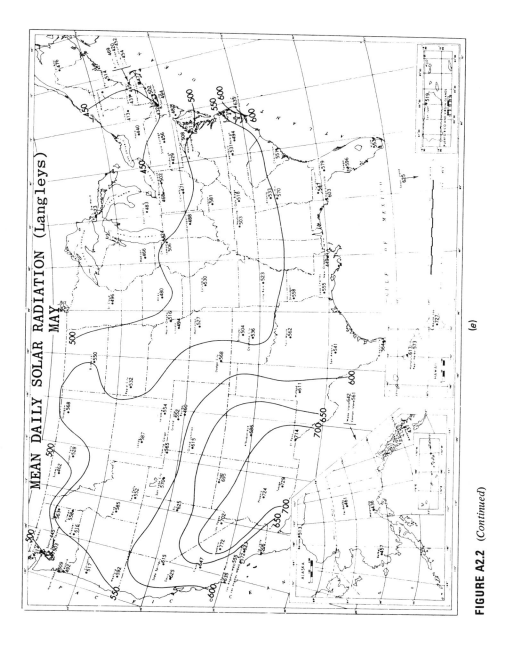

**FIGURE A2.2** (*Continued*)

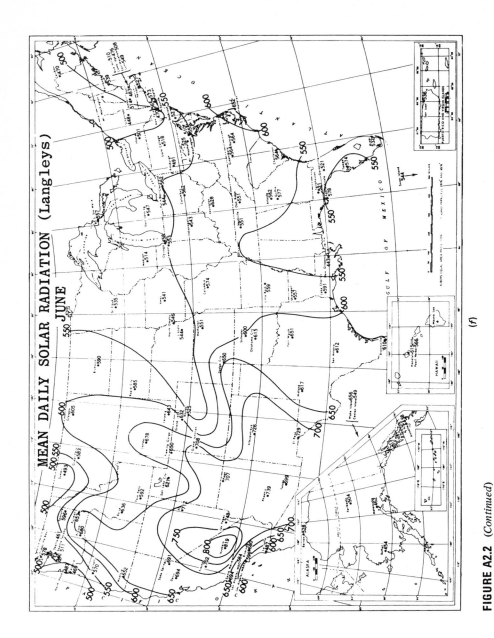

**FIGURE A2.2** (Continued)

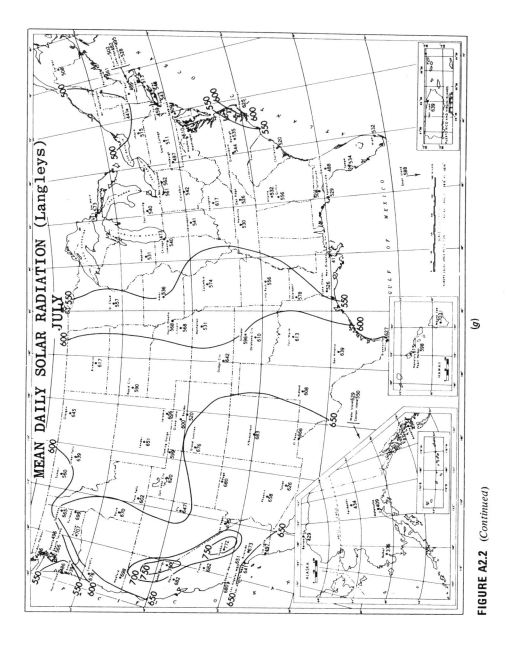

**FIGURE A2.2** (Continued)

691

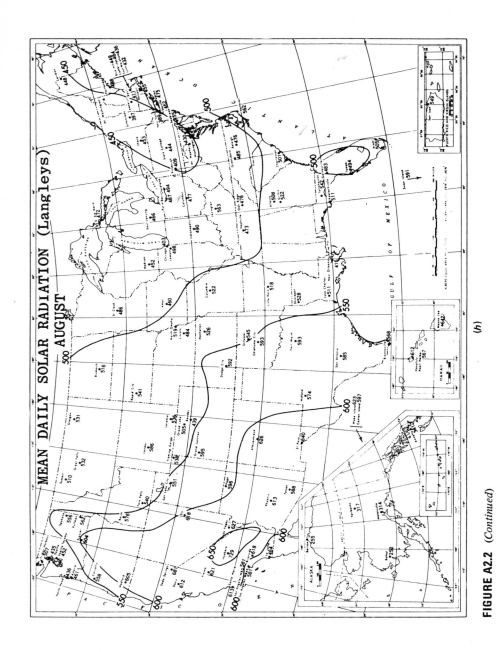

**FIGURE A2.2** *(Continued)*

692

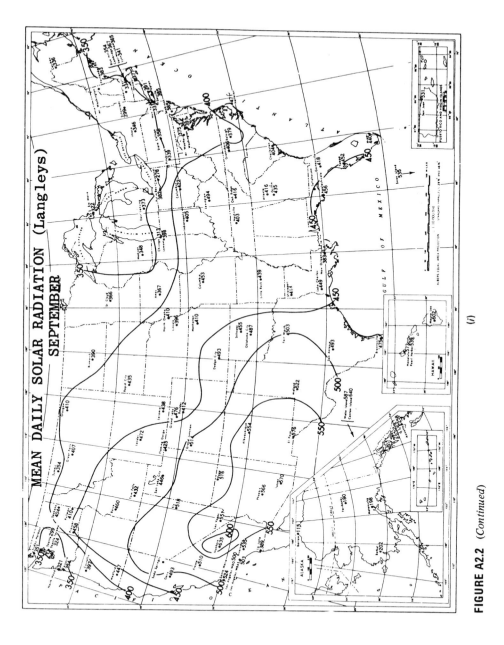

**FIGURE A2.2** *(Continued)*

FIGURE A2.2 (Continued)

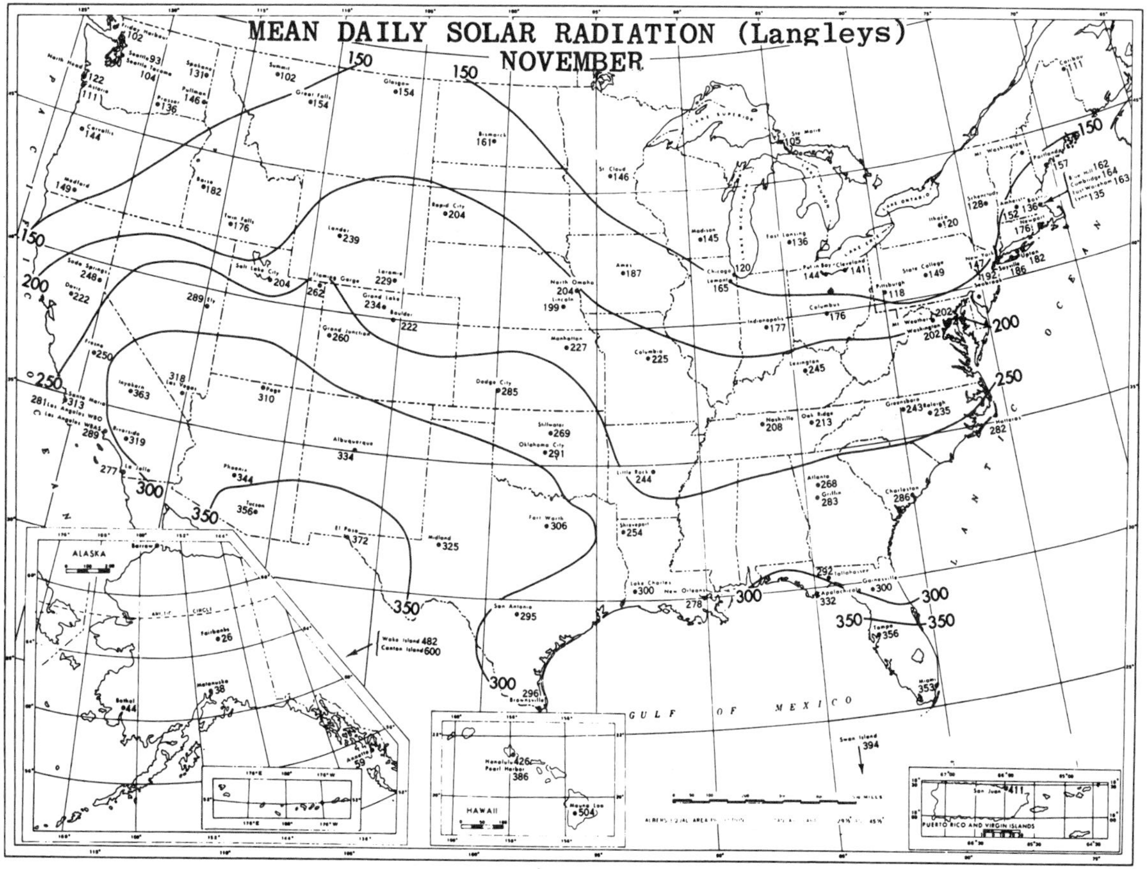

MEAN DAILY SOLAR RADIATION (Langleys)
NOVEMBER

FIGURE A2.2 (Continued)

695

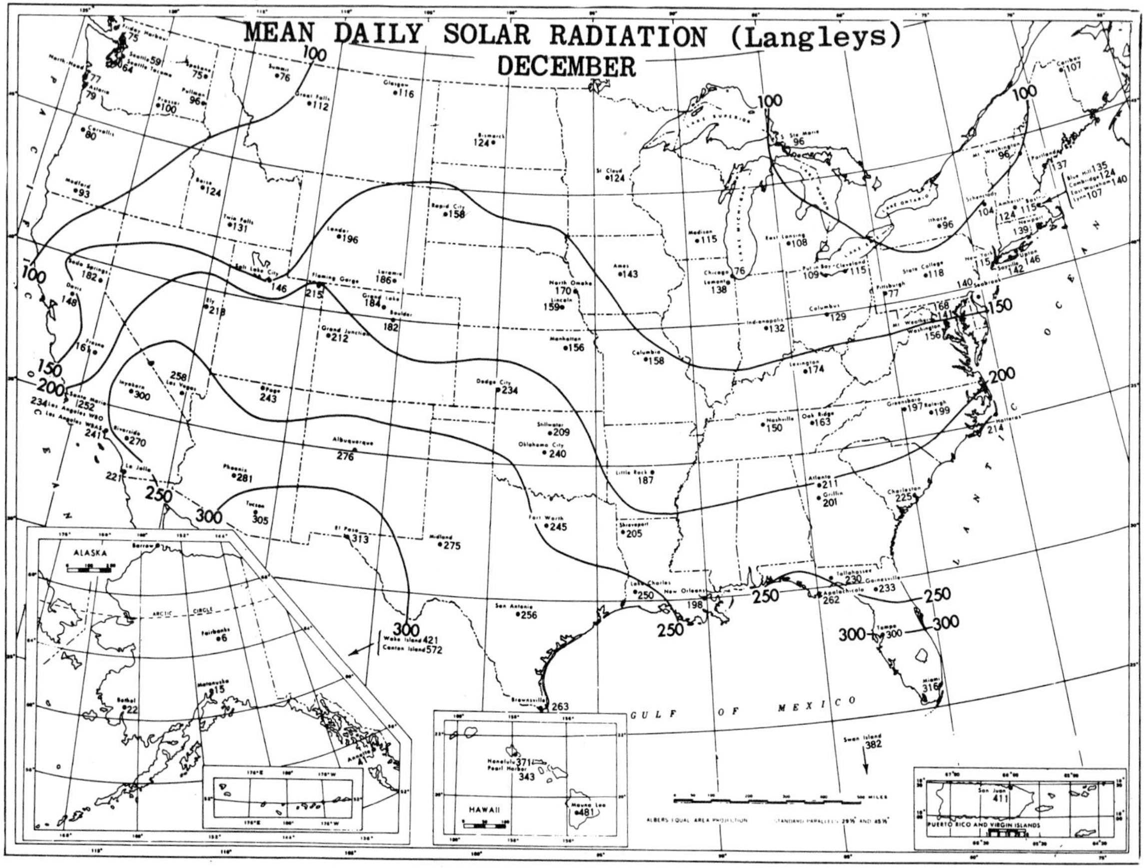

**FIGURE A2.2** *(Continued)*

# SUPPLEMENTARY MATERIAL
## FOR CHAPTER 3

**3** Appendix

**TABLE A3.1** Properties of Dry Air at Atmospheric Pressures between 250 and 1000 K[a]

| $T^b$ (K) | $\rho$ (kg/m³) | $c_p$ (kJ/kg · K) | $\mu$ (kg/m · sec × 10⁵) | $\nu$ (m²/sec × 10⁶) | $k$ (W/m · K) | $\alpha$ (m²/sec × 10⁴) | Pr |
|---|---|---|---|---|---|---|---|
| 250 | 1.4128 | 1.0053 | 1.488 | 9.49 | 0.02227 | 0.13161 | 0.722 |
| 300 | 1.1774 | 1.0057 | 1.983 | 15.68 | 0.02624 | 0.22160 | 0.708 |
| 350 | 0.9980 | 1.0090 | 2.075 | 20.76 | 0.03003 | 0.2983 | 0.697 |
| 400 | 0.8826 | 1.0140 | 2.286 | 25.90 | 0.03365 | 0.3760 | 0.689 |
| 450 | 0.7833 | 1.0207 | 2.484 | 28.86 | 0.03707 | 0.4222 | 0.683 |
| 500 | 0.7048 | 1.0295 | 2.671 | 37.90 | 0.04038 | 0.5564 | 0.680 |
| 550 | 0.6423 | 1.0392 | 2.848 | 44.34 | 0.04360 | 0.6532 | 0.680 |
| 600 | 0.5879 | 1.0551 | 3.018 | 51.34 | 0.04659 | 0.7512 | 0.680 |
| 650 | 0.5430 | 1.0635 | 3.177 | 58.51 | 0.04953 | 0.8578 | 0.682 |
| 700 | 0.5030 | 1.0752 | 3.332 | 66.25 | 0.05230 | 0.9672 | 0.684 |
| 750 | 0.4709 | 1.0856 | 3.481 | 73.91 | 0.05509 | 1.0774 | 0.686 |
| 800 | 0.4405 | 1.0978 | 3.625 | 82.29 | 0.05779 | 1.1951 | 0.689 |
| 850 | 0.4149 | 1.1095 | 3.765 | 90.75 | 0.06028 | 1.3097 | 0.692 |
| 900 | 0.3925 | 1.1212 | 3.899 | 99.3 | 0.06279 | 1.4271 | 0.696 |
| 950 | 0.3716 | 1.1321 | 4.023 | 108.2 | 0.06525 | 1.5510 | 0.699 |
| 1000 | 0.3524 | 1.1417 | 4.152 | 117.8 | 0.06752 | 1.6779 | 0.702 |

[a]From *Natl. Bureau Standards (U.S.) Circ. 564*, 1955.

[b]Symbols: K = absolute temperature, degrees Kelvin; $\nu = \mu/\rho$; $\rho$ = density; $c_p$ = specific heat capacity; $\alpha = c_p\rho/k$; $\mu$ = viscosity; $k$ = thermal conductivity; Pr = Prandtl number, dimensionless. The values of $\mu$, $k$, $c_p$, and Pr are not strongly pressure-dependent and may be used over a fairly wide range of pressures.

**TABLE A3.2**  Properties of Water (Saturated Liquid) between 273 and 533 K[a]

| K | °F | °C | $c_p$ (kJ/kg · °C) | $\rho$ (kg/m³) | $\mu$ (kg/m · sec) | $k$ (W/m · °C) | Pr | $\dfrac{g\beta\rho^2 c_p}{\mu k}$ (m⁻³ · °C⁻¹) |
|---|---|---|---|---|---|---|---|---|
| 273 | 32 | 0 | 4.225 | 999.8 | $1.79 \times 10^{-3}$ | 0.566 | 13.25 | |
| 277.4 | 40 | 4.44 | 4.208 | 999.8 | 1.55 | 0.575 | 11.35 | $1.91 \times 10^9$ |
| 283 | 50 | 10 | 4.195 | 999.2 | 1.31 | 0.585 | 9.40 | $6.34 \times 10^9$ |
| 288.6 | 60 | 15.56 | 4.186 | 998.6 | 1.12 | 0.595 | 7.88 | $1.08 \times 10^{10}$ |
| 294.1 | 70 | 21.11 | 4.179 | 997.4 | $9.8 \times 10^{-4}$ | 0.604 | 6.78 | $1.46 \times 10^{10}$ |
| 299.7 | 80 | 26.67 | 4.179 | 995.8 | 8.6 | 0.614 | 5.85 | $1.91 \times 10^{10}$ |
| 302.2 | 90 | 32.22 | 4.174 | 994.9 | 7.65 | 0.623 | 5.12 | $2.48 \times 10^{10}$ |
| 310.8 | 100 | 37.78 | 4.174 | 993.0 | 6.82 | 0.630 | 4.53 | $3.3 \times 10^{10}$ |
| 316.3 | 110 | 43.33 | 4.174 | 990.6 | 6.16 | 0.637 | 4.04 | $4.19 \times 10^{10}$ |
| 322.9 | 120 | 48.89 | 4.174 | 988.8 | 5.62 | 0.644 | 3.64 | $4.89 \times 10^{10}$ |
| 327.4 | 130 | 54.44 | 4.179 | 985.7 | 5.13 | 0.649 | 3.30 | $5.66 \times 10^{10}$ |
| 333.0 | 140 | 60 | 4.179 | 983.3 | 4.71 | 0.654 | 3.01 | $6.48 \times 10^{10}$ |
| 338.6 | 150 | 65.55 | 4.183 | 980.3 | 4.3 | 0.659 | 2.73 | $7.62 \times 10^{10}$ |
| 342.1 | 160 | 71.11 | 4.186 | 977.3 | 4.01 | 0.665 | 2.53 | $8.84 \times 10^{10}$ |
| 349.7 | 170 | 76.67 | 4.191 | 973.7 | 3.72 | 0.668 | 2.33 | $9.85 \times 10^{10}$ |
| 355.2 | 180 | 82.22 | 4.195 | 970.2 | 3.47 | 0.673 | 2.16 | $1.09 \times 10^{11}$ |
| 360.8 | 190 | 87.78 | 4.199 | 966.7 | 3.27 | 0.675 | 2.03 | |
| 366.3 | 200 | 93.33 | 4.204 | 963.2 | 3.06 | 0.678 | 1.90 | |
| 377.4 | 220 | 104.4 | 4.216 | 955.1 | 2.67 | 0.684 | 1.66 | |
| 388.6 | 240 | 115.6 | 4.229 | 946.7 | 2.44 | 0.685 | 1.51 | |
| 399.7 | 260 | 126.7 | 4.250 | 937.2 | 2.19 | 0.685 | 1.36 | |
| 410.8 | 280 | 137.8 | 4.271 | 928.1 | 1.98 | 0.685 | 1.24 | |
| 421.9 | 300 | 148.9 | 4.296 | 918.0 | 1.86 | 0.684 | 1.17 | |
| 449.7 | 350 | 176.7 | 4.371 | 890.4 | 1.57 | 0.677 | 1.02 | |
| 477.4 | 400 | 204.4 | 4.467 | 859.4 | 1.36 | 0.665 | 1.00 | |
| 505.2 | 450 | 232.2 | 4.585 | 825.7 | 1.20 | 0.646 | 0.85 | |
| 533.0 | 500 | 260 | 4.731 | 785.2 | 1.07 | 0.616 | 0.83 | |

[a]Adapted from Brown, A. I., and S. M. Marco, "Introduction to Heat Transfer," 3d ed., McGraw-Hill Book Company, New York, 1958.

**TABLE A3.3**  Emittances and Absorptances of Materials[a]

| Substance | Short-wave absorptance | Long-wave emittance | $\dfrac{\alpha}{\epsilon}$ |
|---|---|---|---|
| Class I substances: Absorptance to emittance ratios less than 0.5 | | | |
| Magnesium carbonate, $MgCO_3$ | 0.025–0.04 | 0.79 | 0.03–0.05 |
| White plaster | 0.07 | 0.91 | 0.08 |
| Snow, fine particles, fresh | 0.13 | 0.82 | 0.16 |
| White paint, 0.017 in, on aluminum | 0.20 | 0.91 | 0.22 |
| Whitewash on galvanized iron | 0.22 | 0.90 | 0.24 |
| White paper | 0.25–0.28 | 0.95 | 0.26–0.29 |
| White enamel on iron | 0.25–0.45 | 0.9 | 0.28–0.5 |
| Ice, with sparse snow cover | 0.31 | 0.96–0.97 | 0.32 |
| Snow, ice granules | 0.33 | 0.89 | 0.37 |
| Aluminum oil base paint | 0.45 | 0.90 | 0.50 |
| White powdered sand | 0.45 | 0.84 | 0.54 |

Note: See page 700 for footnote.

**TABLE A3.3**  Emittances and Absorptances of Materials[a] (*Continued*)

| Substance | Short-wave absorptance | Long-wave emittance | $\dfrac{\alpha}{\epsilon}$ |
|---|---|---|---|
| Class II substances: Absorptance to emittance ratios between 0.5 and 0.9 | | | |
| Asbestos felt | 0.25 | 0.50 | 0.50 |
| Green oil base paint | 0.5 | 0.9 | 0.56 |
| Bricks, red | 0.55 | 0.92 | 0.60 |
| Asbestos cement board, white | 0.59 | 0.96 | 0.61 |
| Marble, polished | 0.5–0.6 | 0.9 | 0.61 |
| Wood, planed oak | — | 0.9 | — |
| Rough concrete | 0.60 | 0.97 | 0.62 |
| Concrete | 0.60 | 0.88 | 0.68 |
| Grass, green, after rain | 0.67 | 0.98 | 0.68 |
| Grass, high and dry | 0.67–0.69 | 0.9 | 0.76 |
| Vegetable fields and shrubs, wilted | 0.70 | 0.9 | 0.78 |
| Oak leaves | 0.71–0.78 | 0.91–0.95 | 0.78–0.82 |
| Frozen soil | — | 0.93–0.94 | — |
| Desert surface | 0.75 | 0.9 | 0.83 |
| Common vegetable fields and shrubs | 0.72–0.76 | 0.9 | 0.82 |
| Ground, dry plowed | 0.75–0.80 | 0.9 | 0.83–0.89 |
| Oak woodland | 0.82 | 0.9 | 0.91 |
| Pine forest | 0.86 | 0.9 | 0.96 |
| Earth surface as a whole (land and sea, no clouds) | 0.83 | — | — |
| Class III substances: Absorptance to emittance ratios between 0.8 and 1.0 | | | |
| Grey paint | 0.75 | 0.95 | 0.79 |
| Red oil base paint | 0.74 | 0.90 | 0.82 |
| Asbestos, slate | 0.81 | 0.96 | 0.84 |
| Asbestos, paper | | 0.93–0.96 | — |
| Linoleum, red-brown | 0.84 | 0.92 | 0.91 |
| Dry sand | 0.82 | 0.90 | 0.91 |
| Green roll roofing | 0.88 | 0.91–0.97 | 0.93 |
| Slate, dark grey | 0.89 | — | — |
| Old grey rubber | — | 0.86 | — |
| Hard black rubber | — | 0.90–0.95 | — |
| Asphalt pavement | 0.93 | — | — |
| Black cupric oxide on copper | 0.91 | 0.96 | 0.95 |
| Bare moist ground | 0.9 | 0.95 | 0.95 |
| Wet sand | 0.91 | 0.95 | 0.96 |
| Water | 0.94 | 0.95–0.96 | 0.98 |
| Black tar paper | 0.93 | 0.93 | 1.0 |
| Black gloss paint | 0.90 | 0.90 | 1.0 |
| Small hole in large box, furnace, or enclosure | 0.99 | 0.99 | 1.0 |
| "Hohlraum," theoretically perfect black body | 1.0 | 1.0 | 1.0 |

**TABLE A3.3** Emittances and Absorptances of Materials[a] (*Continued*)

| Substance | Short-wave absorptance | Long-wave emittance | $\dfrac{\alpha}{\epsilon}$ |
|---|---|---|---|
| Class IV substances: Absorptance to emittance ratios greater than 1.0 | | | |
| Black silk velvet | 0.99 | 0.97 | 1.02 |
| Alfalfa, dark green | 0.97 | 0.95 | 1.02 |
| Lampblack | 0.98 | 0.95 | 1.03 |
| Black paint, 0.017 in, on aluminum | 0.94–0.98 | 0.88 | 1.07–1.11 |
| Granite | 0.55 | 0.44 | 1.25 |
| Graphite | 0.78 | 0.41 | 1.90 |
| High ratios, but absorptances less than 0.80 | | | |
| Dull brass, copper, lead | 0.2–0.4 | 0.4–0.65 | 1.63–2.0 |
| Galvanized sheet iron, oxidized | 0.8 | 0.28 | 2.86 |
| Galvanized iron, clean, new | 0.65 | 0.13 | 5.0 |
| Aluminum foil | 0.15 | 0.05 | 3.00 |
| Magnesium | 0.3 | 0.07 | 4.3 |
| Chromium | 0.49 | 0.08 | 6.13 |
| Polished zinc | 0.46 | 0.02 | 23.0 |
| Deposited silver (optical reflector) untarnished | 0.07 | 0.01 | |
| Class V substances: Selective surfaces[b] | | | |
| Plated metals:[c] | | | |
| Black sulfide on metal | 0.92 | 0.10 | 9.2 |
| Black cupric oxide on sheet aluminum | 0.08–0.93 | 0.09–0.21 | |
| Copper ($5 \times 10^{-5}$ cm thick) on nickel or silver-plated metal | | | |
| Cobalt oxide on platinum | | | |
| Cobalt oxide on polished nickel | 0.93–0.94 | 0.24–0.40 | 3.9 |
| Black nickel oxide on aluminum | 0.85–0.93 | 0.06–0.1 | 14.5–15.5 |
| Black chrome | 0.87 | 0.09 | 9.8 |
| Particulate coatings: | | | |
| Lampblack on metal | | | |
| Black iron oxide, 47 $\mu$m grain size, on aluminum | | | |
| Geometrically enhanced surfaces:[d] | | | |
| Optimally corrugated greys | 0.89 | 0.77 | 1.2 |
| Optimally corrugated selectives | 0.95 | 0.16 | 5.9 |
| Stainless-steel wire mesh | 0.63–0.86 | 0.23–0.28 | 2.7–3.0 |
| Copper, treated with $NaClO_2$ and NaOH | 0.87 | 0.13 | 6.69 |

[a]From Anderson, B., "Solar Energy," McGraw-Hill Book Company, 1977, with permission.

[b]Selective surfaces absorb most of the solar radiation between 0.3 and 1.9 $\mu$m, and emit very little in the 5–15 $\mu$m range—the infrared.

[c]For a discussion of plated selective surfaces, see Daniels, "Direct Use of the Sun's Energy," especially chapter 12.

[d]For a discussion of how surface selectivity can be enhanced through surface geometry, see K. G. T. Hollands, Directional Selectivity Emittance and Absorptance Properties of Vee Corrugated Specular Surfaces, *J. Sol. Energy Sci. Eng.*, vol. 3, July 1963.

**TABLE A3.4** Thermal Properties of Metals and Alloys[a]

| Material | $k$, Btu/(hr)(ft)(°F) 32°F | $k$, Btu/(hr)(ft)(°F) 212°F | $k$, Btu/(hr)(ft)(°F) 572°F | $k$, Btu/(hr)(ft)(°F) 932°F | $c$, Btu/(lb$_m$)(°F) 32°F | $\rho$, lb$_m$/ft³ 32°F | $\alpha$, ft²/hr 32°F |
|---|---|---|---|---|---|---|---|
| Metals | | | | | | | |
| Aluminum | 117 | 119 | 133 | 155 | 0.208 | 169 | 3.33 |
| Bismuth | 4.9 | 3.9 | . . . | . . . | 0.029 | 612 | 0.28 |
| Copper, pure | 224 | 218 | 212 | 207 | 0.091 | 558 | 4.42 |
| Gold | 169 | 170 | . . . | . . . | 0.030 | 1203 | 4.68 |
| Iron, pure | 35.8 | 36.6 | . . . | . . . | 0.104 | 491 | 0.70 |
| Lead | 20.1 | 19 | 18 | . . . | 0.030 | 705 | 0.95 |
| Magnesium | 91 | 92 | . . . | . . . | 0.232 | 109 | 3.60 |
| Mercury | 4.8 | . . . | . . . | . . . | 0.033 | 849 | 0.17 |
| Nickel | 34.5 | 34 | 32 | . . . | 0.103 | 555 | 0.60 |
| Silver | 242 | 238 | . . . | . . . | 0.056 | 655 | 6.6 |
| Tin | 36 | 34 | . . . | . . . | 0.054 | 456 | 1.46 |
| Zinc | 65 | 64 | 59 | . . . | 0.091 | 446 | 1.60 |
| Alloys | | | | | | | |
| Admiralty metal | 65 | 64 | | | | | |
| Brass, 70% Cu, 30% Zn | 56 | 60 | 66 | . . . | 0.092 | 532 | 1.14 |
| Bronze, 75% Cu, 25% Sn | 15 | . . . | . . . | . . . | 0.082 | 540 | 0.34 |
| Cast iron | | | | | | | |
| Plain | 33 | 31.8 | 27.7 | 24.8 | 0.11 | 474 | 0.63 |
| Alloy | 30 | 28.3 | 27 | . . . | 0.10 | 455 | 0.66 |
| Constantan, 60% Cu, 40% Ni | 12.4 | 12.8 | . . . | . . . | 0.10 | 557 | 0.22 |
| 18-8 Stainless steel, | | | | | | | |
| Type 304 | 8.0 | 9.4 | 10.9 | 12.4 | 0.11 | 488 | 0.15 |
| Type 347 | 8.0 | 9.3 | 11.0 | 12.8 | 0.11 | 488 | 0.15 |
| Steel, mild, 1% C | 26.5 | 26 | 25 | 22 | 0.11 | 490 | 0.49 |

[a]From Kreith, F., "Principles of Heat Transfer," Intext Educational Publishers, New York, 1973.

**TABLE A3.5** Thermal Properties of Some Nonmetals[a]

| Material | Average temperature, °F | $k$, Btu/(hr)(ft)(°F) | $c$, Btu/$(lb_m)$(°F) | $\rho$, $lb_m$/ft³ | $\alpha$, ft²/hr |
|---|---|---|---|---|---|
| Insulating materials | | | | | |
| Asbestos | 32 | 0.087 | 0.25 | 36 | ~0.01 |
| | 392 | 0.12 | ... | 36 | ~0.01 |
| Cork | 86 | 0.025 | 0.04 | 10 | ~0.006 |
| Cotton, fabric | 200 | 0.046 | | | |
| Diatomaceous earth, | | | | | |
| powdered | 100 | 0.030 | 0.21 | 14 | ~0.01 |
| | 300 | 0.036 | ... | | |
| | 600 | 0.046 | ... | | |
| Molded pipe covering | 400 | 0.051 | ... | 26 | |
| | 1600 | 0.088 | ... | | |
| Glass wool | | | | | |
| Fine | 20 | 0.022 | ... | | |
| | 100 | 0.031 | ... | 1.5 | |
| | 200 | 0.043 | ... | | |
| Packed | 20 | 0.016 | ... | | |
| | 100 | 0.022 | ... | 6.0 | |
| | 200 | 0.029 | ... | | |
| Hair felt | 100 | 0.027 | ... | 8.2 | |
| Kaolin insulating | | | | | |
| brick | 932 | 0.15 | ... | 27 | |
| | 2102 | 0.26 | ... | | |
| Kaolin insulating | | | | | |
| firebrick | 392 | 0.05 | ... | 19 | |
| | 1400 | 0.11 | ... | | |
| 85% magnesia | 32 | 0.032 | ... | 17 | |
| | 200 | 0.037 | ... | 17 | |
| Rock wool | 20 | 0.017 | ... | 8 | |
| | 200 | 0.030 | ... | | |
| Rubber | 32 | 0.087 | 0.48 | 75 | 0.0024 |
| Building materials | | | | | |
| Brick | | | | | |
| Fire-clay | 392 | 0.58 | 0.20 | 144 | 0.02 |
| | 1832 | 0.95 | | | |
| Masonry | 70 | 0.38 | 0.20 | 106 | 0.018 |
| Zirconia | 392 | 0.84 | ... | 304 | |
| | 1832 | 1.13 | ... | | |
| Chrome brick | 392 | 0.82 | ... | 246 | |
| | 1832 | 0.96 | ... | | |
| Concrete | | | | | |
| Stone | ~70 | 0.54 | 0.20 | 144 | 0.019 |
| 10% Moisture | ~70 | 0.70 | ... | 140 | ~0.025 |
| Glass, window | ~70 | ~0.45 | 0.2 | 170 | 0.013 |
| Limestone, dry | 70 | 0.40 | 0.22 | 105 | 0.017 |
| Sand | | | | | |
| Dry | 68 | 0.20 | ... | 95 | |
| 10% $H_2O$ | 68 | 0.60 | ... | 100 | |
| Soil | | | | | |
| Dry | 70 | ~0.20 | 0.44 | ... | ~0.01 |
| Wet | 70 | ~1.5 | ... | ... | ~0.03 |
| Wood | | | | | |
| Oak ⊥ to grain | 70 | 0.12 | 0.57 | 51 | 0.0041 |
| ‖ to grain | 70 | 0.20 | 0.57 | 51 | 0.0069 |
| Pine ⊥ to grain | 70 | 0.06 | 0.67 | 31 | 0.0029 |
| ‖ to grain | 70 | 0.14 | 0.67 | 31 | 0.0067 |
| Ice | 32 | 1.28 | 0.46 | 57 | 0.048 |

[a] From Kreith, F., "Principles of Heat Transfer," Intext Educational Publishers, New York, 1973.

**TABLE A3.6** Thermal and Radiative Properties of Collector Cover Materials[h]

| Material name | Index of refraction (n) | τ (solar)[g] (%) | τ (solar)[a] (%) | τ (infrared)[b] (%) | Expansion coefficient (in/in · °F) | Temperature limits (°F) | Weatherability (comment) | Chemical resistance (comment) |
|---|---|---|---|---|---|---|---|---|
| Lexan (polycarbonate) | 1.586 (D 542)[c] | 125 mil 64.1 (± 0.8) | 125 mil 72.6 (± 0.1) | 125 mil 2.0 (est)[d] | $3.75 (10^{-5})$ (H 696) | 250–270 service temperature | Good: 2 yr exposure in Florida caused yellowing; 5 yr caused 5% loss in τ | Good: comparable to acrylic |
| Plexiglas (acrylic) | 1.49 (D 542) | 125 mil 89.6 (± 0.3) | 125 mil 79.6 (± 0.8) | 125 mil 2.0 (est)[e] | $3.9 (10^{-9})$ at 60°F; 4.6 $(10^{-6})$ at 100°F | 180–200 service temperature | Average to good: based on 20 yr testing in Arizona, Florida, and Pennsylvania | Good to excellent: resists most acids and alkalis |
| Teflon F.E.P. (fluorocarbon) | 1.343 (D 542) | 5 mil 92.3 (± 0.2) | 5 mil 89.8 (± 0.4) | 5 mil 25.6 (± 0.5) | $5.9 (10^{-5})$ at 160°F; 9.0 $(10^{-5})$ at 212°F | 400 continuous use; 475 short-term use | Good to excellent: based on 15 yr exposure in Florida environment | Excellent: chemically inert |
| Tedlar P.V.F. (fluorocarbon) | 1.46 (D 542) | 4 mil 92.2 (± 0.1) | 4 mil 88.3 (± 0.9) | 4 mil 20.7 (± 0.2) | $2.8 (10^{-5})$ (D 696) | 225 continuous use; 350 short-term use | Good to excellent: 10 yr exposure in Florida with slight yellowing | Excellent: chemically inert |
| Mylar (polyester) | 1.64–1.67 (D 542) | 5 mil 86.9 (± 0.3) | 5 mil 80.1 (± 0.1) | 5 mil 17.8 (± 0.5) | $0.94 (10^{-5})$ (D 696–44) | 300 continuous use; 400 short-term use | Poor: ultraviolet degradation great | Good to excellent: comparable to Tedlar |
| Sunlite[f] (fiberglass) | 1.54 (D 542) | 25 mil (P) 86.5 (± 0.2) 25 mil (R) 87.5 (± 0.2) | 25 mil (P) 75.4 (± 0.1) 25 mil (R) 77.1 (± 0.7) | 25 mil (P) 7.6 (± 0.1) 25 mil (R) 3.3 (± 0.3) | $1.4 (10^{-5})$ (D 696) | 200 continuous use causes 5% loss in τ | Fair to good: regular, 7 yr solar life; premium, 20 yr solar life | Good: inert to chemical atmospheres |
| Float glass (glass) | 1.518 (D 542) | 125 mil 84.3 (± 0.1) | 125 mil 78.6 (± 0.2) | 125 mil 2.0 (est)[d] | $4.8 (10^{-5})$ (D 696) | 1350 softening point; 100 thermal shock | Excellent: time proved | Good to excellent: time proved |

Note: See page 704 for footnotes.

703

**TABLE A3.6** Thermal and Radiative Properties of Collector Cover Materials[h] (*Continued*)

| Material name | Index of refraction (n) | τ (solar)[g] (%) | τ (solar)[a] (%) | τ (infrared)[b] (%) | Expansion coefficient (in/in · °F) | Temperature limits (°F) | Weatherability (comment) | Chemical resistance (comment) |
|---|---|---|---|---|---|---|---|---|
| Temper glass (glass) | 1.518 (D 542) | 125 mil 84.3 (± 0.1) | 125 mil 78.6 (± 0.2) | 125 mil 2.0 (est)[d] | 4.8 ($10^{-6}$) (D 696) | 450–500 continuous use; 500–550 short-term use | Excellent: time proved | Good to excellent: time proved |
| Clear lime sheet glass (low iron oxide glass) | 1.51 (D 542) | Insufficient data provided by ASG | 125 mil 87.5 (± 0.5) | 125 mil 2.0 (est) | 5.0 ($10^{-6}$) (D 696) | 400 for continuous operation | Excellent: time proved | Good to excellent: time proved |
| Clear lime temper glass (low iron oxide glass) | 1.51 (D 542) | Insufficient data provided by ASG | 125 mil 87.5 (± 0.5) | 125 mil 2.0 (est) | 5.0 ($10^{-6}$) (D 696) | 400 for continuous operation | Excellent: time proved | Good to excellent: time proved |
| Sunadex white crystal glass (0.01% iron oxide glass) | 1.50 (D 542) | Insufficient data provided by ASG | 125 mil 91.5 (± 0.2) | 125 mil 2.0 (est) | 4.7 ($10^{-6}$) (D 696) | 400 for continuous operation | Excellent: time proved | Good to excellent: time proved |

[a]Numerical integration ($\Sigma \tau_{avg} F_{\lambda_1 T - \lambda_2 T}$) for $\lambda = 0.2$–$4.0 \ \mu M$.

[b]Numerical integration ($\Sigma \tau_{avg} F_{\lambda_1 T - \lambda_2 T}$) for $\lambda = 3.0$–$50.0 \ \mu M$.

[c]All parenthesized numbers refer to ASTM test codes.

[d]Data not provided; estimate of 2% to be used for 125 mil samples.

[e]Degrees differential to rupture $2 \times 2 \times \frac{1}{4}$ in samples. Glass specimens heated and then quenched in water bath at 70° F.

[f]Sunlite premium data denoted by (P); Sunlite regular data denoted by (R).

[g]Compiled data based on ASTM Code E 424 Method B.

[h]Abstracted from Ratzel, A. C., and R. B. Bannerot, Optimal Material Selection for Flat-Plate Solar Energy Collectors Utilizing Commercially Available Materials, presented at ASME–AIChE Natl. Heat Transfer Conf., 1976.

**TABLE A3.7** Insulation Data for Mineral Wool, Industrial Felt, and Fiberglas[k]

| Material[a] | Nominal density (lb/ft³) | Temperature limitation (°F) | Mean temperature thermal conductivity | | Federal specification compliance | Producer | Standard sizes (variable) | Cost[c] ($/bd · ft) |
|---|---|---|---|---|---|---|---|---|
| | | | °F | Btu · in/ hr · ft² · °F[b] | | | | |
| | | | | Mineral wool | | | | |
| #10 Insulation | 10.0 | 1200 | 200 350 500 | 0.26 0.32 0.375 | HH-I-558 B form A, class 4 | Forty-Eight Insulations | 2' X 4' (Board) THK: 1–3" ($\frac{1}{2}$" inc) | 0.125–0.14 (dist)[d] Carload: 30,000 bd · ft |
| LTR Insulation | 8.0 | 1000 | 200 350 500 | 0.27 0.32 0.385 | HH-I-558 B form A, class 4 | Forty-Eight Insulations | 2' X 4' (Board) THK: 1–4" ($\frac{1}{2}$" inc) | 0.105–0.115 (dist) Carload: 30,000 bd · ft |
| I-T Insulation | 6.0 | 850 | 200 350 500 | 0.27 0.34 0.45 | HH-I-558 B form A, class 3 | Forty-Eight Insulations | 2' X 4' (Board) THK: 1–4" ($\frac{1}{2}$" inc) | 0.095–0.10 (dist) Carload: 30,000 bd · ft |
| MT-board (MT-10) | 10.0 | 1050 | 200 350 500 | 0.25 0.333 0.445 | HH-I-558 B form A, class 1, 2, 3 | Eagle-Picher | 2' X 4' (Board) THK: 1–3" ($\frac{1}{2}$" inc) | 0.13–0.14 (dist) Carload: 36,000 bd · ft |
| MT-board (MT-8) | 8.0 | 1050 | 200 350 500 | 0.255 0.350 0.470 | HH-I-558 B form A, class 1, 2, 3 | Eagle-Picher | 2' X 4' (Board) THK: 1–4" ($\frac{1}{2}$" inc) | 0.107–0.12 (dist) Carload: 36,000 bd · ft |
| MT-board (MT-6) | 6.0 | 1050 | 200 350 500 | 0.270 0.373 0.495 | HH-I-558 B form A, class 1, 2, 3 | Eagle-Picher | 2' X 4' (Board) THK: 1–4" ($\frac{1}{2}$" inc) | 0.085–0.10 (dist) Carload: 36,000 bd · ft |

Note: See page 708 for footnotes.

**TABLE A3.7** Insulation Data for Mineral Wool, Industrial Felt, and Fiberglas[k] (*Continued*)

| Material[a] | Nominal density (lb/ft³) | Temperature limitation (°F) | Mean temperature thermal conductivity | | Federal specification compliance | Producer | Standard sizes (variable) | Cost[c] ($/bd · ft) |
|---|---|---|---|---|---|---|---|---|
| | | | °F | Btu · in/ hr · ft² · °F[b] | | | | |
| | | | | | Industrial felt[e] | | | |
| Thermafiber (SF-234) | 8.0 | 1000 | 200<br>350<br>500 | 0.27<br>0.36<br>0.48 | HH-I-558 B form A, class 1, 2, 3 | United States Gypsum | THK: 1–2″ (NJ) Length: 60″ THK: 1–2½″ (IND) Length: 48″ | 0.131 (dist) 7,000–38,000 bd · ft |
| Thermafiber (SF-240) | 6.0 | 1000 | 200<br>350<br>500 | 0.27<br>0.37<br>0.50 | HH-I-558 B form A, class 1, 2, 3 | United States Gypsum | THK: 1–2½″ (TEX) Length: 90″ THK: 1–3⅓″ (IND) Length: 48″ | 0.095–0.113 (dist) 7,000–38,000 bd · ft |
| Thermafiber (SF-250) | 4.5 | 800 | 200<br>350<br>500 | 0.29<br>0.415<br>0.55 | HH-I-558 B form A, class 1, 2 | United States Gypsum | THK: 1–4″ (TEX) Length: 90″ THK: 1–5″ (IND) Length: 48″ | 0.081–0.10 (dist) 7,000–38,000 bd · ft |
| Thermafiber (SF-252) | 4.0 | 800 | 200<br>350<br>500 | 0.30<br>0.435<br>0.59 | HH-I-558 B form A, class 1, 2 | United States Gypsum | THK: 1–4″ (TEX) Length: 90″ THK: 1–5″ (IND) Length: 48″ | 0.07–0.087 (dist) 7,000–38,000 bd · ft |
| Thermafiber (SF-256) | 3.5 | 600 | 200<br>350<br>500 | 0.33<br>0.47<br>0.62 | HH-I-558 B form A, class 1, 2 | United States Gypsum | THK: 1–4″ (TEX) Length: 90″ THK: 1–6″ (IND) Length: 48″ | 0.066–0.084 (dist) 7,000–38,000 bd · ft |

| Product | Density | Max temp | Temp (°F) | Thermal conductivity | Specification | Manufacturer | Dimensions | Cost/packaging |
|---|---|---|---|---|---|---|---|---|
| Thermafiber (SF-260) | 3.0 | 500 | 200 / 350 / 500 | 0.35 / 0.50 / 0.65 | HH-I-558 B form A, class 1, 2 | United States Gypsum | THK: 1-4" (TEX) Length: 90" THK: 1-6" (IND) Length: 48" | 0.064–0.82 (dist) 7,000–38,000 bd · ft |
| Thermafiber (SF-270) | 2.5 | 400 | 200 / 350 / 500 | 0.39 / 0.56 / — | No data provided | United States Gypsum | THK: 1-4" (TEX) Length: 90" THK: 1-5" (IND) Length: 48" | 0.06–0.078 (dist) 7,000–38,000 bd · ft |

Foamglas and Fiberglas insulation[f]

| Product | Density | Max temp | Temp (°F) | Thermal conductivity | Specification | Manufacturer | Dimensions | Cost/packaging |
|---|---|---|---|---|---|---|---|---|
| Foamglas[g] | 8.5 | 600 | 200 / 350 / 500 | 0.46[j] / 0.58 / 0.74 | HH-I-551 D (Fed) ASTM C 552-73 | Pittsburgh Corning | 1' × 1.5' (Board) 1½' × 2' (Board) THX: 1½–4" (½" inc) | 0.22–0.24 (Corning)[h] Carload: 36,000 bd · ft |
| 701 Fiberglas[i] | 1.6 | 450 | 200 / 350 | 0.33 / 0.51 | HH-I-558 B form A, class 1 HH-I-558 B, type 1 form B, class 7 | Owens-Corning Fiberglas | 2' × 4' (Board) THK: 1½–4" (½" inc) | 0.07–0.08 (dist) Carload: 30,000–35,000 bd · ft |
| 703 Fiberglas[i] | 3.0 | 450 | 200 / 350 | 0.30 / 0.41 | HH-I-558 B form A, class 1, 2 | Owens-Corning Fiberglas | 2' × 4' (Board) THK: 1-2" (½" inc) | 0.14–0.15 (dist) Carload: 30,000–35,000 bd · ft |
| 705 Fiberglas[i] | 6.0 | 450 | 200 / 350 | 0.27 / 0.38 | HH-I-558 B form A, class 1, 2 | Owens-Corning Fiberglas | 2' × 4' (Board) THK: 1-2" (½" inc) | 0.25–0.27 (dist) Carload: 30,000–35,000 bd · ft |
| Thermal insulating wool type I[i] | 1.25 | 1000 | 200 / 350 / 500 | 0.41 / 0.65 / 0.85 | HH-I-558 B form B, type 1, class 6 | Owens-Corning Fiberglas | Rolls Width: 2' or 3' THK: 2", 3", 4" Length: 76', 52', 38' | 0.04–0.06 (Corning) ½ Carload: 35,000 bd · ft |

**TABLE A3.7**  Insulation Data for Mineral Wool, Industrial Felt, and Fiberglas[k] (Continued)

| Material[a] | Nominal density (lb/ft³) | Temperature limitation (°F) | Mean temperature thermal conductivity | | Producer | Federal specification compliance | Standard sizes (variable) | Cost[c] ($/bd·ft) |
|---|---|---|---|---|---|---|---|---|
| | | | °F | Btu·in/hr·ft²·°F[b] | | | | |
| Thermal insulating wool type II[f] | 2.4 | 1000 | 200 | 0.30[j] | Owens-Corning Fiberglas | HH-I-558 B form B, type 1, class 7, 8 | 2' × 8'<br>2' × 4' (Board)<br>THK: 1–3"<br>(½" inc) | 0.08–0.09 (Corning)<br>0.14–0.15 (dist)<br>Carload: 35,000 bd·ft |
| | | | 350 | 0.44 | | | | |
| | | | 500 | 0.60 | | | | |
| IS board[i] | 4.0 | 800 | 200 | 0.30 | Owens-Corning Fiberglas | HH-I-558 B form A, class 3 | 2' × 4', 3' × 4'<br>4' × 8' (Board)<br>THK: 1–6"<br>(½" inc) | 0.10–0.13 (Corning)<br>0.18–0.20 (dist)<br>Carload: 35,000 bd·ft |
| | | | 350 | 0.44 | | | | |
| | | | 500 | 0.61 | | | | |

[a]All insulations listed will not cause or aggravate corrosion and will absorb less than 1% moisture. All insulations listed appear as semirigid board composed of silica base refractory fibers bonded with special binders for service in indicated temperature ranges.

[b]Units are consistently employed within the insulation industry. Conductivity measurements consider a test specimen 1 in thick and 1 ft² normal area.

[c]Cost data current through October 30, 1975. Costs are based on carload purchases and include freight where necessary to move insulation to Houston.

[d]Cost from Houston distributors noted by (dist).

[e]Industrial felt is preformed mineral fiber felt, which will not cause or sustain corrosion. It absorbs less than 1% moisture by weight and is rated noncombustible. Insulation to be ordered in varying thicknesses and lengths; standard width of 24 in employed (see column 8).

[f]All codes in specification compliance column are federal specifications unless otherwise noted.

[g]Foamglas is an impermeable, incombustible, rigid insulation composed of completely sealed glass cells with no binder material. Its rigid form may allow for Foamglas being implemented as the collector box.

[h]Cost from Corning Houston warehouse noted by (Corning); cost from Houston distributor noted by (dist).

[i]Insulations are made of inorganic glass fibers preformed into semirigid to rigid rectangular boards (T1W I in blankets). Insulations will not accelerate nor cause corrosion and will absorb less than 1% moisture (by volume).

[j]The units of the mean temperature thermal conductivity for Foamglas and Fiberglas insulation are Btu·in/hr·ft·°F.

[k]Abstracted from Ratzel, R. C., and R. B. Bannerot, Optimal Material Selection for Flat-Plate Solar Energy Collectors Utilizing Commercially Available Materials, presented at ASME–AIChE Natl. Heat Transfer Conf., 1976.

**TABLE A3.8**  Superheated Steam–SI Units[a]

| Pressure (MN/m²) (saturation temperature)[b] | | 50°C 323.15 K | 100°C 373.15 K | 150°C 423.15 K | 200°C 473.15 K | 300°C 573.15 K | 400°C 673.15 K | 500°C 773.15 K | 700°C 973.15 K | 1000°C 1273.15 K | 1300°C 1573.15 K |
|---|---|---|---|---|---|---|---|---|---|---|---|
| 0.001 (6.98°C) (280.13 K) | $v$ | 149.093 | 172.187 | 195.272 | 218.352 | 264.508 | 310.661 | 356.814 | 449.117 | 587.571 | 726.025 |
| | $u$ | 2445.4 | 2516.4 | 2588.4 | 2661.6 | 2812.2 | 2969.0 | 3132.4 | 3479.6 | 4053.0 | 4683.7 |
| | $h$ | 2594.5 | 2688.6 | 2783.6 | 2880.0 | 3076.8 | 3279.7 | 3489.2 | 3928.7 | 4640.6 | 5409.7 |
| | $s$ | 9.2423 | 9.5129 | 9.7520 | 9.9671 | 10.3443 | 10.6705 | 10.9605 | 11.4655 | 12.1019 | 12.6438 |
| 0.002 (17.50°C) (290.65 K) | $v$ | 74.524 | 86.081 | 97.628 | 109.170 | 132.251 | 155.329 | 178.405 | 224.558 | 293.785 | 363.012 |
| | $u$ | 2445.2 | 2516.3 | 2588.3 | 2661.6 | 2812.2 | 2969.0 | 3132.4 | 3479.6 | 4053.0 | 4683.7 |
| | $h$ | 2594.3 | 2688.4 | 2783.6 | 2879.9 | 3076.7 | 3279.7 | 3489.2 | 3928.7 | 4640.6 | 5409.7 |
| | $s$ | 8.9219 | 9.1928 | 9.4320 | 9.6471 | 10.0243 | 10.3506 | 10.6406 | 11.1456 | 11.7820 | 12.3239 |
| 0.004 (28.96°C) (302.11 K) | $v$ | 37.240 | 43.028 | 48.806 | 54.580 | 66.122 | 77.662 | 89.201 | 112.278 | 146.892 | 181.506 |
| | $u$ | 2444.9 | 2516.1 | 2588.1 | 2661.5 | 2812.2 | 2969.0 | 3132.3 | 3479.6 | 4053.0 | 4683.7 |
| | $h$ | 2593.9 | 2688.2 | 2783.4 | 2879.8 | 3076.7 | 3279.6 | 3489.2 | 3928.7 | 4640.6 | 5409.7 |
| | $s$ | 8.6009 | 8.8724 | 9.1118 | 9.3271 | 9.7044 | 10.0307 | 10.3207 | 10.8257 | 11.4621 | 12.0040 |
| 0.006 (36.16°C) (309.31 K) | $v$ | 24.812 | 28.676 | 32.532 | 36.383 | 44.079 | 51.774 | 59.467 | 74.852 | 97.928 | 121.004 |
| | $u$ | 2444.6 | 2515.9 | 2588.1 | 2661.4 | 2812.2 | 2969.0 | 3132.3 | 3479.6 | 4053.0 | 4683.7 |
| | $h$ | 2593.4 | 2688.0 | 2783.3 | 2879.7 | 3076.6 | 3279.6 | 3489.1 | 3928.7 | 4640.6 | 5409.7 |
| | $s$ | 8.4128 | 8.6847 | 8.9244 | 9.1398 | 9.5172 | 9.8435 | 10.1336 | 10.6386 | 11.2750 | 11.8168 |
| 0.008 (41.51°C) (314.66 K) | $v$ | 18.598 | 21.501 | 24.395 | 27.284 | 33.058 | 38.829 | 44.599 | 56.138 | 73.446 | 90.753 |
| | $u$ | 2444.2 | 2515.7 | 2588.0 | 2661.4 | 2812.1 | 2969.0 | 3132.3 | 3479.6 | 4053.0 | 4683.7 |
| | $h$ | 2593.0 | 2687.7 | 2783.1 | 2879.6 | 3076.6 | 3279.6 | 3489.1 | 3928.7 | 4640.6 | 5409.7 |
| | $s$ | 8.2790 | 8.5514 | 8.7914 | 9.0069 | 9.3844 | 9.7107 | 10.0008 | 10.5058 | 11.1422 | 11.6841 |
| 0.010 (45.81°C) (318.96 K) | $v$ | 14.869 | 17.196 | 19.512 | 21.825 | 26.445 | 31.063 | 35.679 | 44.911 | 58.757 | 72.602 |
| | $u$ | 2443.9 | 2515.5 | 2587.9 | 2661.3 | 2812.1 | 2968.9 | 3132.3 | 3479.6 | 4053.0 | 4683.7 |
| | $h$ | 2592.6 | 2687.5 | 2783.0 | 2879.5 | 3076.5 | 3279.6 | 3489.1 | 3928.7 | 4640.6 | 5409.7 |
| | $s$ | 8.1749 | 8.4479 | 8.6882 | 8.9038 | 9.2813 | 9.6077 | 9.8978 | 10.4028 | 11.0393 | 11.5811 |

Note: See page 711 for footnotes.

**TABLE A3.8** Superheated Steam–SI Units$^a$ (*Continued*)

| Pressure (MN/m²) (saturation temperature)$^b$ | | 50°C 323.15 K | 100°C 373.15 K | 150°C 423.15 K | 200°C 473.15 K | 300°C 573.15 K | 400°C 673.15 K | 500°C 773.15 K | 700°C 973.15 K | 1000°C 1273.15 K | 1300°C 1573.15 K |
|---|---|---|---|---|---|---|---|---|---|---|---|
| | | | | | | Temperature | | | | | |
| 0.020 (60.06°C) (333.21 K) | v | 7.412 | 8.585 | 9.748 | 10.907 | 13.219 | 15.529 | 17.838 | 22.455 | 29.378 | 36.301 |
| | u | 2442.2 | 2514.6 | 2587.3 | 2660.9 | 2811.9 | 2968.8 | 3132.2 | 3479.5 | 4053.0 | 4683.7 |
| | h | 2590.4 | 2686.6 | 2782.3 | 2879.1 | 3076.3 | 3279.4 | 3489.0 | 3928.6 | 4640.6 | 5409.7 |
| | s | 7.8498 | 8.1255 | 8.3669 | 8.5831 | 8.9611 | 9.2876 | 9.5778 | 10.0829 | 10.7193 | 11.2612 |
| 0.040 (75.87°C) (349.02 K) | v | 3.683 | 4.279 | 4.866 | 5.448 | 6.606 | 7.763 | 8.918 | 11.227 | 14.689 | 18.151 |
| | u | 2438.8 | 2512.6 | 2586.2 | 2660.2 | 2811.5 | 2968.6 | 3132.1 | 3479.4 | 4052.9 | 4683.6 |
| | h | 2586.1 | 2683.8 | 2780.8 | 2878.1 | 3075.8 | 3279.1 | 3488.8 | 3928.5 | 4640.5 | 5409.6 |
| | s | 7.5192 | 7.8003 | 8.0444 | 8.2617 | 8.6406 | 8.9674 | 9.2577 | 9.7629 | 10.3994 | 10.9412 |
| 0.060 (85.94°C) (359.09 K) | v | 2.440 | 2.844 | 3.238 | 3.628 | 4.402 | 5.174 | 5.944 | 7.484 | 9.792 | 12.100 |
| | u | 2435.3 | 2510.6 | 2585.1 | 2659.5 | 2811.2 | 2968.4 | 3131.9 | 3479.4 | 4052.9 | 4683.6 |
| | h | 2581.7 | 2681.3 | 2779.4 | 2877.2 | 3075.3 | 3278.8 | 3488.6 | 3928.4 | 4640.4 | 5409.6 |
| | s | 7.3212 | 7.6079 | 7.8546 | 8.0731 | 8.4528 | 8.7799 | 9.0704 | 9.5757 | 10.2122 | 10.7541 |
| 0.080 (93.50°C) (366.65 K) | v | 1.8183 | 2.127 | 2.425 | 2.718 | 3.300 | 3.879 | 4.458 | 5.613 | 7.344 | 9.075 |
| | u | 2431.7 | 2508.7 | 2583.9 | 2658.8 | 2810.8 | 2968.1 | 3131.7 | 3479.3 | 4052.8 | 4683.5 |
| | h | 2577.2 | 2678.8 | 2777.9 | 2876.2 | 3074.8 | 3278.5 | 3488.3 | 3928.3 | 4640.4 | 5409.5 |
| | s | 7.1775 | 7.4698 | 7.7191 | 7.9388 | 8.3194 | 8.6468 | 8.9374 | 9.4428 | 10.0794 | 10.6213 |
| 0.100 (99.63°C) (372.78 K) | v | 1.4450 | 1.6958 | 1.9364 | 2.172 | 2.639 | 3.103 | 3.565 | 4.490 | 5.875 | 7.260 |
| | u | 2428.2 | 2506.7 | 2582.8 | 2658.1 | 2810.4 | 2967.9 | 3131.6 | 3479.2 | 4052.9 | 4683.5 |
| | h | 2572.7 | 2676.2 | 2776.4 | 2875.3 | 3074.3 | 3278.2 | 3488.1 | 3928.2 | 4640.3 | 5409.5 |
| | s | 7.0633 | 7.3614 | 7.6134 | 7.8343 | 8.2158 | 8.5435 | 8.8342 | 9.3398 | 9.9764 | 10.5183 |
| 0.200 (120.23°C) (393.38 K) | v | 0.6969 | 0.8340 | 0.9596 | 1.0803 | 1.3162 | 1.5493 | 1.7814 | 2.244 | 2.937 | 3.630 |
| | u | 2409.5 | 2496.3 | 2576.9 | 2654.4 | 2808.6 | 2966.7 | 3130.8 | 3478.8 | 4052.5 | 4683.2 |
| | h | 2548.9 | 2663.1 | 2768.8 | 2870.5 | 3071.8 | 3276.6 | 3487.1 | 3927.6 | 4640.0 | 5409.3 |
| | s | 6.6844 | 7.0135 | 7.2795 | 7.5066 | 7.8926 | 8.2218 | 8.5133 | 9.0194 | 9.6563 | 10.1982 |

| Pressure (MPa) | | | | | | | | | | | |
|---|---|---|---|---|---|---|---|---|---|---|---|
| 0.300 (133.55°C) (406.70 K) | $v$ | 0.4455 | 0.5461 | 0.6339 | 0.7163 | 0.8753 | 1.0315 | 1.1867 | 1.4957 | 1.9581 | 2.4201 |
| | $u$ | 2389.1 | 2485.4 | 2570.8 | 2650.7 | 2806.7 | 2965.6 | 3130.0 | 3478.4 | 4052.3 | 4683.0 |
| | $h$ | 2522.7 | 2649.2 | 2761.0 | 2865.6 | 3069.3 | 3275.0 | 3486.0 | 3927.1 | 4639.7 | 5409.0 |
| | $s$ | 6.4319 | 6.7965 | 7.0778 | 7.3115 | 7.7022 | 8.0330 | 8.3251 | 8.8319 | 9.4690 | 10.0110 |
| 0.400 (143.63°C) (416.78 K) | $v$ | 0.3177 | 0.4017 | 0.4708 | 0.5342 | 0.6548 | 0.7726 | 0.8893 | 1.1215 | 1.4685 | 1.8151 |
| | $u$ | 2366.3 | 2473.8 | 2564.5 | 2646.8 | 2804.8 | 2964.4 | 3129.2 | 3477.9 | 4052.0 | 4682.8 |
| | $h$ | 2493.4 | 2634.5 | 2752.8 | 2860.5 | 3066.8 | 3273.4 | 3484.9 | 3926.5 | 4639.4 | 5408.8 |
| | $s$ | 6.2248 | 6.6319 | 6.9299 | 7.1706 | 7.5662 | 7.8985 | 8.1913 | 8.6987 | 9.3360 | 9.8780 |
| 0.500 (151.86°C) (425.01 K) | $v$ | | 0.3146 | 0.3729 | 0.4249 | 0.5226 | 0.6173 | 0.7109 | 0.8969 | 1.1747 | 1.4521 |
| | $u$ | | 2461.5 | 2557.9 | 2642.9 | 2802.9 | 2963.2 | 3128.4 | 3477.5 | 4051.8 | 4682.5 |
| | $h$ | | 2618.7 | 2744.4 | 2855.4 | 3064.2 | 3271.9 | 3483.9 | 3925.9 | 4639.1 | 5408.6 |
| | $s$ | | 6.4945 | 6.8111 | 7.0592 | 7.4599 | 7.7938 | 8.0873 | 8.5952 | 9.2328 | 9.7749 |

[a]From Bolz, R. E., and G. L. Tuve, eds., "CRC Handbook of Tables for Applied Engineering Science," 2nd ed., Chemical Rubber Co., Cleveland, Ohio, 1973.
[b]Symbols: $v$ = specific volume, m³/kg; $u$ = specific internal energy, kJ/kg; $h$ = specific enthalpy, kJ/kg; $s$ = specific entropy, kJ/K · kg.

711

**TABLE A3.9** Saturated Steam and Water—SI Units[a]

| Temperature (K) | Pressure (MN/m²) | Specific volume (m³/kg) | | Specific internal energy (kJ/kg) | | Specific enthalpy (kJ/kg) | | | Specific entropy (kJ/kg · K) | |
|---|---|---|---|---|---|---|---|---|---|---|
| | | $v_f$ | $v_g$ | $u_f$ | $u_g$ | $h_f$ | $h_{fg}$ | $h_g$ | $s_f$ | $s_g$ |
| 273.15 | 0.0006109 | 0.0010002 | 206.278 | −0.03 | 2375.3 | −0.02 | 2501.4 | 2501.3 | −0.0001 | 9.1565 |
| 273.16 | 0.0006113 | 0.0010002 | 206.136 | 0 | 2375.3 | +0.01 | 2501.3 | 2501.4 | 0 | 9.1562 |
| 278.15 | 0.0008721 | 0.0010001 | 147.120 | +20.97 | 2382.3 | 20.98 | 2489.6 | 2510.6 | +0.0761 | 9.0257 |
| 280.13 | 0.0010000 | 0.0010002 | 129.208 | 29.30 | 2385.0 | 29.30 | 2484.9 | 2514.2 | 0.1059 | 8.975 |
| 283.15 | 0.0012276 | 0.0010004 | 106.379 | 42.00 | 2389.2 | 42.01 | 2477.7 | 2519.8 | 0.1510 | 8.9008 |
| 286.18 | 0.0015000 | 0.0010007 | 87.980 | 54.71 | 2393.3 | 54.71 | 2470.6 | 2525.3 | 0.1957 | 8.8279 |
| 288.15 | 0.0017051 | 0.0010009 | 77.926 | 62.99 | 2396.1 | 62.99 | 2465.9 | 2528.9 | 0.2245 | 8.7814 |
| 290.65 | 0.0020000 | 0.0010013 | 67.004 | 73.48 | 2399.5 | 73.48 | 2460.0 | 2533.5 | 0.2607 | 8.7237 |
| 293.15 | 0.002339 | 0.0010018 | 57.791 | 83.95 | 2402.9 | 83.96 | 2454.1 | 2538.1 | 0.2966 | 8.6672 |
| 297.23 | 0.0030000 | 0.0010027 | 45.665 | 101.04 | 2408.5 | 101.05 | 2444.5 | 2545.5 | 0.3545 | 8.5776 |
| 298.15 | 0.003169 | 0.0010029 | 43.360 | 104.88 | 2409.8 | 104.89 | 2442.3 | 2547.2 | 0.3674 | 8.5580 |
| 302.11 | 0.004000 | 0.0010040 | 34.800 | 121.45 | 2415.2 | 121.46 | 2432.9 | 2554.4 | 0.4226 | 8.4746 |
| 303.15 | 0.004246 | 0.0010043 | 32.894 | 125.78 | 2416.6 | 125.79 | 2430.5 | 2556.3 | 0.4369 | 8.4533 |
| 306.03 | 0.005000 | 0.0010053 | 28.192 | 137.81 | 2420.5 | 137.82 | 2423.7 | 2561.5 | 0.4764 | 8.3951 |
| 308.15 | 0.005628 | 0.0010060 | 25.216 | 146.67 | 2423.4 | 146.68 | 2418.6 | 2565.3 | 0.5053 | 8.3531 |
| 309.31 | 0.006000 | 0.0010064 | 23.739 | 151.53 | 2425.0 | 151.53 | 2415.9 | 2567.4 | 0.5210 | 8.3304 |
| 312.15 | 0.007000 | 0.0010074 | 20.530 | 163.39 | 2428.8 | 163.40 | 2409.1 | 2572.5 | 0.5592 | 8.2758 |
| 313.15 | 0.007384 | 0.0010078 | 19.523 | 167.56 | 2430.1 | 167.57 | 2406.7 | 2574.3 | 0.5725 | 8.2570 |
| 314.66 | 0.008000 | 0.0010084 | 18.103 | 173.87 | 2432.2 | 173.88 | 2403.1 | 2577.0 | 0.5926 | 8.2287 |
| 316.91 | 0.009000 | 0.0010094 | 16.203 | 183.27 | 2435.2 | 183.29 | 2397.7 | 2581.0 | 0.6224 | 8.1872 |
| 318.15 | 0.009593 | 0.0010099 | 15.258 | 188.44 | 2436.8 | 188.45 | 2394.8 | 2583.2 | 0.6387 | 8.1648 |
| 318.96 | 0.010000 | 0.0010102 | 14.674 | 191.82 | 2437.9 | 191.83 | 2392.8 | 2584.7 | 0.6493 | 8.1502 |
| 323.15 | 0.012349 | 0.0010121 | 12.032 | 209.32 | 2443.5 | 209.33 | 2382.7 | 2592.1 | 0.7038 | 8.0763 |
| 327.12 | 0.015000 | 0.0010141 | 10.022 | 225.92 | 2448.7 | 225.94 | 2373.1 | 2599.1 | 0.7549 | 8.0085 |
| 328.15 | 0.015758 | 0.0010146 | 9.568 | 230.21 | 2450.1 | 230.23 | 2370.7 | 2600.9 | 0.7679 | 7.9913 |

| | | | | | | | | | | |
|---|---|---|---|---|---|---|---|---|---|---|
| 333.15 | 0.019940 | 0.0010172 | 7.671 | 251.11 | 2456.6 | 251.13 | 2358.5 | 2609.6 | 0.8312 | 7.9096 |
| 333.21 | 0.020000 | 0.0010172 | 7.649 | 251.38 | 2456.7 | 251.40 | 2358.3 | 2609.7 | 0.8320 | 7.9085 |
| 338.15 | 0.025030 | 0.0010199 | 6.197 | 272.02 | 2463.1 | 272.06 | 2346.2 | 2618.3 | 0.8935 | 7.8310 |
| 342.25 | 0.030000 | 0.0010223 | 5.229 | 289.20 | 2468.4 | 289.23 | 2336.1 | 2625.3 | 0.9439 | 7.7686 |
| 343.15 | 0.031190 | 0.0010228 | 5.042 | 292.95 | 2469.6 | 292.98 | 2333.8 | 2626.8 | 0.9549 | 7.7553 |
| 348.15 | 0.038580 | 0.0010259 | 4.131 | 313.90 | 2475.9 | 313.93 | 2221.4 | 2635.3 | 1.0155 | 7.6824 |
| 349.02 | 0.040000 | 0.0010265 | 3.993 | 317.53 | 2477.0 | 317.58 | 2319.2 | 2636.8 | 1.0259 | 7.6700 |
| 353.15 | 0.047390 | 0.0010291 | 3.407 | 334.86 | 2482.2 | 334.91 | 2308.8 | 2643.7 | 1.0753 | 7.6122 |
| 354.48 | 0.050000 | 0.0010300 | 3.240 | 340.44 | 2483.9 | 340.49 | 2305.4 | 2645.9 | 1.0910 | 7.5939 |
| 358.15 | 0.057830 | 0.0010325 | 2.828 | 355.84 | 2488.4 | 355.90 | 2296.0 | 2651.9 | 1.1343 | 7.5445 |
| 359.09 | 0.060000 | 0.0010331 | 2.732 | 359.79 | 2489.6 | 359.86 | 2293.6 | 2653.5 | 1.1453 | 7.5320 |
| 363.10 | 0.070000 | 0.0010360 | 2.365 | 376.63 | 2494.5 | 376.70 | 2283.3 | 2660.0 | 1.1919 | 7.4797 |
| 363.15 | 0.070140 | 0.0010360 | 2.361 | 376.85 | 2494.5 | 376.92 | 2283.2 | 2660.1 | 1.1925 | 7.4791 |
| 366.65 | 0.080000 | 0.0010386 | 2.087 | 391.58 | 2498.8 | 391.66 | 2274.1 | 2665.8 | 1.2329 | 7.4346 |
| 368.15 | 0.084550 | 0.0010397 | 1.9819 | 397.88 | 2500.6 | 397.96 | 2270.2 | 2668.1 | 1.2500 | 7.4159 |

[a]Subscripts: $f$ refers to a property of liquid in equilibrium with vapor; $g$ refers to a property of vapor in equilibrium with liquid; $fg$ refers to a change by evaporation. Table from Bolz, R. E., and G. L. Tuve, eds., "CRC Handbook of Tables for Applied Engineering Science," 2nd ed., Chemical Rubber Co., Cleveland, Ohio, 1973.

**TABLE A4.1** Normal Distribution Function

$$F(z) = \frac{1}{\sqrt{2\pi}} \int_{-\infty}^{z} e^{-1/2\,t^2}\,dt$$

| z | 0.00 | 0.01 | 0.02 | 0.03 | 0.04 | 0.05 | 0.06 | 0.07 | 0.08 | 0.09 |
|---|------|------|------|------|------|------|------|------|------|------|
| 0.0 | 0.5000 | 0.5040 | 0.5080 | 0.5120 | 0.5160 | 0.5199 | 0.5239 | 0.5279 | 0.5319 | 0.5359 |
| 0.1 | 0.5398 | 0.5438 | 0.5478 | 0.5517 | 0.5557 | 0.5596 | 0.5636 | 0.5675 | 0.5714 | 0.5753 |
| 0.2 | 0.5793 | 0.5832 | 0.5871 | 0.5910 | 0.5948 | 0.5987 | 0.6026 | 0.6064 | 0.6103 | 0.6141 |
| 0.3 | 0.6179 | 0.6217 | 0.6255 | 0.6293 | 0.6331 | 0.6368 | 0.6406 | 0.6443 | 0.6480 | 0.6517 |
| 0.4 | 0.6554 | 0.6591 | 0.6628 | 0.6664 | 0.6700 | 0.6736 | 0.6772 | 0.6808 | 0.6844 | 0.6879 |
| 0.5 | 0.6915 | 0.6950 | 0.6985 | 0.7019 | 0.7054 | 0.7088 | 0.7123 | 0.7157 | 0.7190 | 0.7224 |
| 0.6 | 0.7257 | 0.7291 | 0.7324 | 0.7357 | 0.7389 | 0.7422 | 0.7454 | 0.7486 | 0.7517 | 0.7549 |
| 0.7 | 0.7580 | 0.7611 | 0.7642 | 0.7673 | 0.7704 | 0.7734 | 0.7764 | 0.7794 | 0.7823 | 0.7852 |
| 0.8 | 0.7881 | 0.7910 | 0.7939 | 0.7967 | 0.7995 | 0.8023 | 0.8051 | 0.8078 | 0.8106 | 0.8133 |
| 0.9 | 0.8159 | 0.8186 | 0.8212 | 0.8238 | 0.8264 | 0.8289 | 0.8315 | 0.8340 | 0.8365 | 0.8389 |
| 1.0 | 0.8413 | 0.8438 | 0.8461 | 0.8485 | 0.8508 | 0.8531 | 0.8554 | 0.8577 | 0.8599 | 0.8621 |
| 1.1 | 0.8643 | 0.8665 | 0.8686 | 0.8708 | 0.8729 | 0.8749 | 0.8770 | 0.8790 | 0.8810 | 0.8830 |
| 1.2 | 0.8849 | 0.8869 | 0.8888 | 0.8907 | 0.8925 | 0.8944 | 0.8962 | 0.8980 | 0.8997 | 0.9015 |
| 1.3 | 0.9032 | 0.9049 | 0.9066 | 0.9082 | 0.9099 | 0.9115 | 0.9131 | 0.9147 | 0.9162 | 0.9177 |
| 1.4 | 0.9192 | 0.9207 | 0.9222 | 0.9236 | 0.9251 | 0.9265 | 0.9279 | 0.9292 | 0.9306 | 0.9319 |
| 1.5 | 0.9332 | 0.9345 | 0.9357 | 0.9370 | 0.9382 | 0.9394 | 0.9406 | 0.9418 | 0.9429 | 0.9441 |
| 1.6 | 0.9452 | 0.9463 | 0.9474 | 0.9484 | 0.9495 | 0.9505 | 0.9515 | 0.9525 | 0.9535 | 0.9545 |
| 1.7 | 0.9554 | 0.9564 | 0.9573 | 0.9582 | 0.9591 | 0.9599 | 0.9608 | 0.9616 | 0.9625 | 0.9633 |
| 1.8 | 0.9641 | 0.9649 | 0.9656 | 0.9664 | 0.9671 | 0.9678 | 0.9686 | 0.9693 | 0.9699 | 0.9706 |
| 1.9 | 0.9713 | 0.9719 | 0.9726 | 0.9732 | 0.9738 | 0.9744 | 0.9750 | 0.9756 | 0.9761 | 0.9767 |
| 2.0 | 0.9772 | 0.9778 | 0.9783 | 0.9788 | 0.9793 | 0.9798 | 0.9803 | 0.9808 | 0.9812 | 0.9817 |
| 2.1 | 0.9821 | 0.9826 | 0.9830 | 0.9834 | 0.9838 | 0.9842 | 0.9846 | 0.9850 | 0.9854 | 0.9857 |
| 2.2 | 0.9861 | 0.9864 | 0.9868 | 0.9871 | 0.9875 | 0.9878 | 0.9881 | 0.9884 | 0.9887 | 0.9890 |
| 2.3 | 0.9893 | 0.9896 | 0.9898 | 0.9901 | 0.9904 | 0.9906 | 0.9909 | 0.9911 | 0.9913 | 0.9916 |
| 2.4 | 0.9918 | 0.9920 | 0.9922 | 0.9925 | 0.9927 | 0.9929 | 0.9931 | 0.9932 | 0.9934 | 0.9936 |
| 2.5 | 0.9938 | 0.9940 | 0.9941 | 0.9943 | 0.9945 | 0.9946 | 0.9948 | 0.9949 | 0.9951 | 0.9952 |
| 2.6 | 0.9953 | 0.9955 | 0.9956 | 0.9957 | 0.9959 | 0.9960 | 0.9961 | 0.9962 | 0.9963 | 0.9964 |
| 2.7 | 0.9965 | 0.9966 | 0.9967 | 0.9968 | 0.9969 | 0.9970 | 0.9971 | 0.9972 | 0.9973 | 0.9974 |
| 2.8 | 0.9974 | 0.9975 | 0.9976 | 0.9977 | 0.9977 | 0.9978 | 0.9979 | 0.9979 | 0.9980 | 0.9981 |
| 2.9 | 0.9981 | 0.9982 | 0.9982 | 0.9983 | 0.9984 | 0.9984 | 0.9985 | 0.9985 | 0.9986 | 0.9986 |
| 3.0 | 0.9987 | 0.9987 | 0.9987 | 0.9988 | 0.9988 | 0.9989 | 0.9989 | 0.9989 | 0.9990 | 0.9990 |
| 3.1 | 0.9990 | 0.9991 | 0.9991 | 0.9991 | 0.9992 | 0.9992 | 0.9992 | 0.9992 | 0.9993 | 0.9993 |
| 3.2 | 0.9993 | 0.9993 | 0.9994 | 0.9994 | 0.9994 | 0.9994 | 0.9994 | 0.9995 | 0.9995 | 0.9995 |
| 3.3 | 0.9995 | 0.9995 | 0.9995 | 0.9996 | 0.9996 | 0.9996 | 0.9996 | 0.9996 | 0.9996 | 0.9997 |
| 3.4 | 0.9997 | 0.9997 | 0.9997 | 0.9997 | 0.9997 | 0.9997 | 0.9997 | 0.9997 | 0.9997 | 0.9998 |

# ECONOMIC AND SYSTEMS' COMPONENT DATA

**TABLE A5.1** Capital-Recovery Factors[a]

| | Annual mortgage interest rate | | | | | | | | | | |
|---|---|---|---|---|---|---|---|---|---|---|---|
| $n$ | 7 | $7\frac{1}{2}$ | 8 | $8\frac{1}{2}$ | 9 | $9\frac{1}{2}$ | 10 | $10\frac{1}{2}$ | 11 | $11\frac{1}{2}$ | 12 |
| 1 | 1.070 | 1.075 | 1.080 | 1.085 | 1.090 | 1.095 | 1.100 | 1.105 | 1.110 | 1.115 | 1.120 |
| 2 | 0.553 | 0.557 | 0.561 | 0.565 | 0.568 | 0.572 | 0.576 | 0.580 | 0.584 | 0.588 | 0.592 |
| 3 | 0.381 | 0.385 | 0.388 | 0.392 | 0.395 | 0.399 | 0.402 | 0.406 | 0.409 | 0.413 | 0.416 |
| 4 | 0.295 | 0.299 | 0.302 | 0.305 | 0.309 | 0.312 | 0.315 | 0.319 | 0.322 | 0.326 | 0.329 |
| 5 | 0.244 | 0.247 | 0.250 | 0.254 | 0.257 | 0.260 | 0.264 | 0.267 | 0.271 | 0.274 | 0.277 |
| 6 | 0.210 | 0.213 | 0.216 | 0.220 | 0.223 | 0.226 | 0.230 | 0.233 | 0.236 | 0.240 | 0.243 |
| 7 | 0.186 | 0.189 | 0.192 | 0.195 | 0.199 | 0.202 | 0.205 | 0.209 | 0.212 | 0.216 | 0.219 |
| 8 | 0.167 | 0.171 | 0.174 | 0.177 | 0.181 | 0.184 | 0.187 | 0.191 | 0.194 | 0.198 | 0.201 |
| 9 | 0.153 | 0.157 | 0.160 | 0.163 | 0.167 | 0.170 | 0.174 | 0.177 | 0.181 | 0.184 | 0.188 |
| 10 | 0.142 | 0.146 | 0.149 | 0.152 | 0.156 | 0.159 | 0.163 | 0.166 | 0.170 | 0.173 | 0.177 |
| 11 | 0.133 | 0.137 | 0.140 | 0.143 | 0.147 | 0.150 | 0.154 | 0.158 | 0.161 | 0.165 | 0.168 |
| 12 | 0.126 | 0.129 | 0.133 | 0.136 | 0.140 | 0.143 | 0.147 | 0.150 | 0.154 | 0.158 | 0.161 |
| 13 | 0.120 | 0.123 | 0.127 | 0.130 | 0.134 | 0.137 | 0.141 | 0.144 | 0.148 | 0.152 | 0.156 |
| 14 | 0.114 | 0.118 | 0.121 | 0.125 | 0.128 | 0.132 | 0.136 | 0.139 | 0.143 | 0.147 | 0.151 |
| 15 | 0.110 | 0.113 | 0.117 | 0.120 | 0.124 | 0.128 | 0.131 | 0.135 | 0.139 | 0.143 | 0.147 |
| 16 | 0.106 | 0.109 | 0.113 | 0.117 | 0.120 | 0.124 | 0.128 | 0.132 | 0.136 | 0.139 | 0.143 |
| 17 | 0.102 | 0.106 | 0.110 | 0.113 | 0.117 | 0.121 | 0.125 | 0.129 | 0.132 | 0.136 | 0.140 |
| 18 | 0.099 | 0.103 | 0.107 | 0.110 | 0.114 | 0.118 | 0.122 | 0.126 | 0.130 | 0.134 | 0.138 |
| 19 | 0.097 | 0.100 | 0.104 | 0.108 | 0.112 | 0.116 | 0.120 | 0.124 | 0.128 | 0.132 | 0.136 |
| 20 | 0.094 | 0.098 | 0.102 | 0.106 | 0.110 | 0.113 | 0.117 | 0.121 | 0.126 | 0.130 | 0.134 |

[a] $n$ is the mortgage term in years.

**TABLE A5.2**  Interest Fraction of Mortgage Payment

| Yr left on mortgage | Annual mortgage interest rate | | | | | | | | | | |
|---|---|---|---|---|---|---|---|---|---|---|---|
| | 7 | $7\frac{1}{2}$ | 8 | $8\frac{1}{2}$ | 9 | $9\frac{1}{2}$ | 10 | $10\frac{1}{2}$ | 11 | $11\frac{1}{2}$ | 12 |
| 20 | 0.742 | 0.765 | 0.785 | 0.804 | 0.822 | 0.837 | 0.851 | 0.864 | 0.876 | 0.887 | 0.896 |
| 19 | 0.723 | 0.747 | 0.768 | 0.788 | 0.806 | 0.822 | 0.836 | 0.850 | 0.862 | 0.874 | 0.884 |
| 18 | 0.704 | 0.728 | 0.750 | 0.770 | 0.788 | 0.805 | 0.820 | 0.834 | 0.847 | 0.859 | 0.870 |
| 17 | 0.683 | 0.708 | 0.730 | 0.750 | 0.769 | 0.786 | 0.802 | 0.817 | 0.830 | 0.843 | 0.854 |
| 16 | 0.661 | 0.686 | 0.708 | 0.729 | 0.748 | 0.766 | 0.782 | 0.798 | 0.812 | 0.825 | 0.837 |
| 15 | 0.638 | 0.662 | 0.685 | 0.706 | 0.725 | 0.744 | 0.761 | 0.776 | 0.791 | 0.805 | 0.817 |
| 14 | 0.612 | 0.637 | 0.660 | 0.681 | 0.701 | 0.719 | 0.737 | 0.753 | 0.768 | 0.782 | 0.795 |
| 13 | 0.585 | 0.609 | 0.632 | 0.654 | 0.674 | 0.693 | 0.710 | 0.727 | 0.742 | 0.757 | 0.771 |
| 12 | 0.556 | 0.580 | 0.603 | 0.624 | 0.644 | 0.663 | 0.681 | 0.698 | 0.714 | 0.729 | 0.743 |
| 11 | 0.525 | 0.549 | 0.571 | 0.592 | 0.612 | 0.631 | 0.650 | 0.667 | 0.683 | 0.698 | 0.713 |
| 10 | 0.492 | 0.515 | 0.537 | 0.558 | 0.578 | 0.596 | 0.614 | 0.632 | 0.648 | 0.663 | 0.678 |
| 9 | 0.456 | 0.478 | 0.500 | 0.520 | 0.540 | 0.558 | 0.576 | 0.593 | 0.609 | 0.625 | 0.639 |
| 8 | 0.418 | 0.439 | 0.460 | 0.479 | 0.498 | 0.516 | 0.533 | 0.550 | 0.566 | 0.581 | 0.596 |
| 7 | 0.377 | 0.397 | 0.417 | 0.435 | 0.453 | 0.470 | 0.487 | 0.503 | 0.518 | 0.533 | 0.548 |
| 6 | 0.334 | 0.352 | 0.370 | 0.387 | 0.404 | 0.420 | 0.436 | 0.451 | 0.465 | 0.480 | 0.493 |
| 5 | 0.287 | 0.303 | 0.319 | 0.335 | 0.350 | 0.365 | 0.379 | 0.393 | 0.407 | 0.420 | 0.433 |
| 4 | 0.237 | 0.251 | 0.265 | 0.278 | 0.292 | 0.304 | 0.317 | 0.329 | 0.341 | 0.353 | 0.364 |
| 3 | 0.184 | 0.195 | 0.206 | 0.217 | 0.228 | 0.238 | 0.249 | 0.259 | 0.269 | 0.279 | 0.288 |
| 2 | 0.127 | 0.135 | 0.143 | 0.151 | 0.158 | 0.166 | 0.174 | 0.181 | 0.188 | 0.196 | 0.203 |
| 1 | 0.065 | 0.070 | 0.074 | 0.078 | 0.083 | 0.087 | 0.091 | 0.095 | 0.099 | 0.103 | 0.107 |

**TABLE A5.3** Present-Worth Factors

| Year | Discount rate (%) | | | | | | | | | | | | | |
|---|---|---|---|---|---|---|---|---|---|---|---|---|---|---|
|  | 0 | 1 | 2 | 3 | 4 | 5 | 6 | 7 | 8 | 9 | 10 | 11 | 12 | 13 |
| 1 | 1.000 | 0.990 | 0.980 | 0.971 | 0.962 | 0.952 | 0.943 | 0.935 | 0.926 | 0.917 | 0.909 | 0.901 | 0.893 | 0.885 |
| 2 | 1.000 | 0.980 | 0.961 | 0.943 | 0.925 | 0.907 | 0.890 | 0.873 | 0.857 | 0.842 | 0.826 | 0.812 | 0.797 | 0.783 |
| 3 | 1.000 | 0.971 | 0.942 | 0.915 | 0.889 | 0.864 | 0.840 | 0.816 | 0.794 | 0.772 | 0.751 | 0.731 | 0.712 | 0.693 |
| 4 | 1.000 | 0.961 | 0.924 | 0.888 | 0.855 | 0.823 | 0.792 | 0.763 | 0.735 | 0.708 | 0.683 | 0.659 | 0.636 | 0.613 |
| 5 | 1.000 | 0.951 | 0.906 | 0.863 | 0.822 | 0.784 | 0.747 | 0.713 | 0.681 | 0.650 | 0.621 | 0.593 | 0.567 | 0.543 |
| 6 | 1.000 | 0.942 | 0.888 | 0.837 | 0.790 | 0.746 | 0.705 | 0.666 | 0.630 | 0.596 | 0.564 | 0.535 | 0.507 | 0.480 |
| 7 | 1.000 | 0.933 | 0.871 | 0.813 | 0.760 | 0.711 | 0.665 | 0.623 | 0.583 | 0.547 | 0.513 | 0.482 | 0.452 | 0.425 |
| 8 | 1.000 | 0.923 | 0.853 | 0.789 | 0.731 | 0.677 | 0.627 | 0.582 | 0.540 | 0.502 | 0.467 | 0.434 | 0.404 | 0.376 |
| 9 | 1.000 | 0.914 | 0.837 | 0.766 | 0.703 | 0.645 | 0.592 | 0.544 | 0.500 | 0.460 | 0.424 | 0.391 | 0.361 | 0.333 |
| 10 | 1.000 | 0.905 | 0.820 | 0.744 | 0.676 | 0.614 | 0.558 | 0.508 | 0.463 | 0.422 | 0.386 | 0.352 | 0.322 | 0.295 |
| 11 | 1.000 | 0.896 | 0.804 | 0.722 | 0.650 | 0.585 | 0.527 | 0.475 | 0.429 | 0.388 | 0.350 | 0.317 | 0.287 | 0.261 |
| 12 | 1.000 | 0.887 | 0.788 | 0.701 | 0.625 | 0.557 | 0.497 | 0.444 | 0.397 | 0.356 | 0.319 | 0.286 | 0.257 | 0.231 |
| 13 | 1.000 | 0.879 | 0.773 | 0.681 | 0.601 | 0.530 | 0.469 | 0.415 | 0.368 | 0.326 | 0.290 | 0.258 | 0.229 | 0.204 |
| 14 | 1.000 | 0.870 | 0.758 | 0.661 | 0.577 | 0.505 | 0.442 | 0.388 | 0.340 | 0.299 | 0.263 | 0.232 | 0.205 | 0.181 |
| 15 | 1.000 | 0.861 | 0.743 | 0.642 | 0.555 | 0.481 | 0.417 | 0.362 | 0.315 | 0.275 | 0.239 | 0.209 | 0.183 | 0.160 |
| 16 | 1.000 | 0.853 | 0.728 | 0.623 | 0.534 | 0.458 | 0.394 | 0.339 | 0.292 | 0.252 | 0.218 | 0.188 | 0.163 | 0.141 |
| 17 | 1.000 | 0.844 | 0.714 | 0.605 | 0.513 | 0.436 | 0.371 | 0.317 | 0.270 | 0.231 | 0.198 | 0.170 | 0.146 | 0.125 |
| 18 | 1.000 | 0.836 | 0.700 | 0.587 | 0.494 | 0.416 | 0.350 | 0.296 | 0.250 | 0.212 | 0.180 | 0.153 | 0.130 | 0.111 |
| 19 | 1.000 | 0.828 | 0.686 | 0.570 | 0.475 | 0.396 | 0.331 | 0.277 | 0.232 | 0.194 | 0.164 | 0.138 | 0.116 | 0.098 |
| 20 | 1.000 | 0.820 | 0.673 | 0.554 | 0.456 | 0.377 | 0.312 | 0.258 | 0.215 | 0.178 | 0.149 | 0.124 | 0.104 | 0.087 |

**TABLE A5.4** Toxicological Properties of Common Glycols Used as Antifreezes in Solar Systems[a]

| | Single oral LD50 dose in rats (ml/kg) | Repeated oral feeding in rats acceptable level in diet and duration | Single skin penetration LD50 dose in rabbits (ml/kg) | Single inhalation concentrated vapor (8 hr) in rats | Primary skin irritation in rabbits | Eye injury in rabbits |
|---|---|---|---|---|---|---|
| Ethylene glycol | 7.40[b] | 0.18 g/kg/day (30 days) | >20 | Killed none of 6 | None | None |
| Diethylene glycol | 28.3 | 0.18 g/kg/day (30 days) | 11.9 | Killed none of 6 | None | None |
| Triethylene glycol | 28.2 | 0.83 g/kg/day (30 days) | >20 | Killed none of 6 | None | None |
| Tetraethylene glycol | 28.9 | 1.88 g/kg/day (2 yr) | >20 | Killed none of 6 | None | None |
| Propylene glycol | 34.6 | 2.0 g/kg/day (2 yr)[c] | >20 | Killed none of 6 | None | Trace |
| Dipropylene glycol | 14.8 | — | >20 | Killed none of 6 | None | Trace |
| Hexylene glycol | 4.06 | 0.31 g/kg/day (90 days) | 8.56 | Killed none of 6 | Trace | Minor |
| 2-Ethyl-1,3-hexanediol | 6.50 | 0.48 g/kg/day (90 days) | 15.2 | Killed none of 6 | Trace | Moderate |
| 1,5-Pentanediol | 5.89[d] | — | >20 | Killed none of 6 | None | Trace |

[a]The term LD50 refers to that quantity of chemical that kills 50 percent of dosed animals within 14 days. For uniformity, dosage is expressed in grams or milliliters per kilogram of body weight. Single skin penetration refers to a 24-hr covered skin contact with the liquid chemical. Single inhalation refers to the continuous breathing of a certain concentration of chemical for the stated period of time. Primary irritation refers to the skin response 24 hr after application of 0.01-ml amounts to uncovered skin. Eye injury refers to surface damage produced by the liquid chemical. Table from Union Carbide, Glycols, F-41515A 7/71-12M, p. 68, 1971.
[b]Single dose oral toxicity to humans is greater.
[c]Dogs.
[d]g/kg.

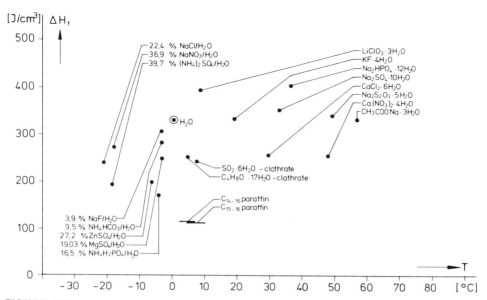

**FIGURE A5.1** Enthalpy of fusion and phase change temperatures of low-temperature phase-change storage media. From Schröder, J., Philips GmbH Forschungslaboratorium (Aachen) Report, 1977.

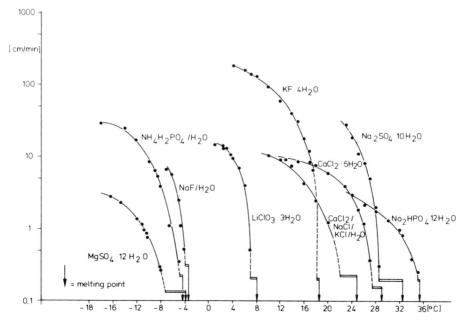

**FIGURE A5.2** Rate of crystallization of several low-temperature hydrates and eutectics. Note that $Na_2SO_4 \cdot 10H_2O$ and $CaCl_2/NaCl/KCl/H_2O$ have very inadequate rates without substantial subcooling. From Schröder, J., Philips GmbH Forschungslaboratorium (Aachen) Report, 1977.

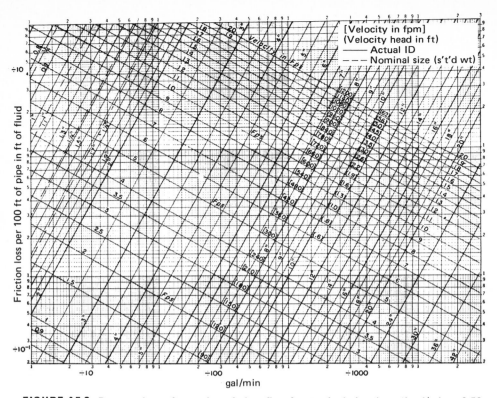

**FIGURE A5.3** Pressure drop of water in turbulent flow for standard pipe sizes. (1 gal/min = 3.79 × 10⁻³ m³/min; 1 ft = 0.305 m; 1 in = 2.54 cm.) From Potter, Philip J., "Power Plant Theory and Design," 2nd ed. Copyright © 1959 The Ronald Press Company, New York.

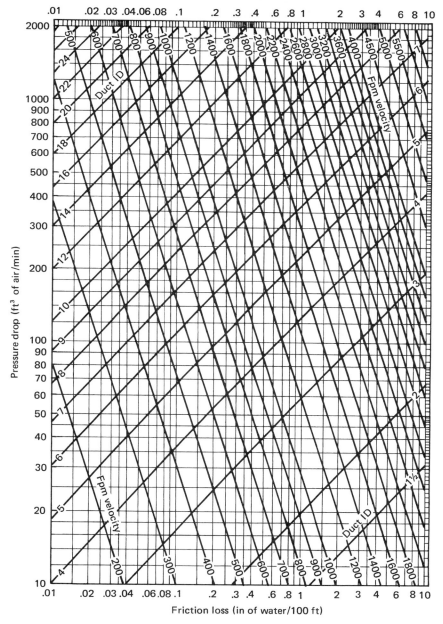

**FIGURE A5.4** Pressure drop of air flowing in ducts; velocities shown for circular ducts. (1 in $H_2O$ = 249 N/m²; 1 ft³ = 2.83 × 10⁻² m³; 1 in = 2.54 cm; 1 ft = 0.305 m.) From ASHRAE, "Handbook of Fundamentals," American Society of Heating, Refrigerating, and Air Conditioning Engineers, New York, 1972.

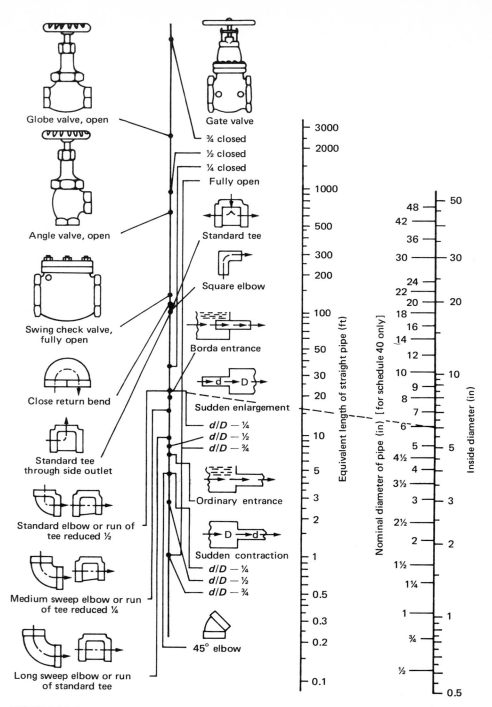

**FIGURE A5.5** Equivalent lengths of pipe, standard pipe fittings. (1 ft = 0.305 m; 1 in = 2.54 cm.) For sudden enlargements and sudden contractions the equivalent length is in feet of pipe of the smaller diameter $d$. Dashed line shows determination of equivalent length of a 6-in standard elbow. From the Crane Co. Technical Paper 410, "Flow of Fluids," 1957.

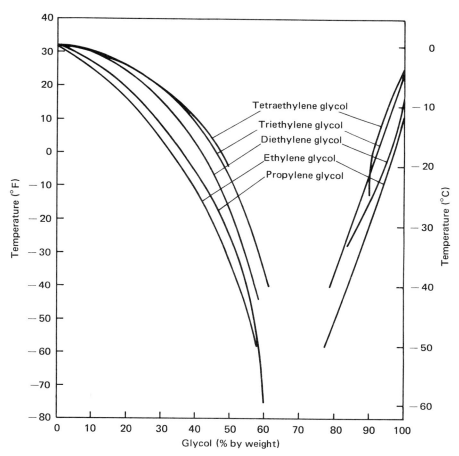

**FIGURE A5.6** Freezing-point depression of aqueous solutions of common glycol antifreezes. From Union Carbide, Glycols, F-41515A 7/71-12M, p. 17, 1971.

**TABLE A6.1** Weather Data for Selected U.S. Cities: Annual Summary[a]

| | Location | Sun total, % | Wind, avg, mi/hr | Rain total, in. | Relative humidity, % 7 A.M. | Relative humidity, % 1 P.M. | Precipitation, days | Snow per season, in. | 32°F or less, days | Particulate pollution, $\mu g/m^3$ |
|---|---|---|---|---|---|---|---|---|---|---|
| AL | Mobile | 61 | 9.5 | 68.13 | 85 | 57 | 123 | 0.4 | 21 | |
| AK | Juneau | 31 | 8.6 | 54.62 | 87 | 81 | 220 | 102.7 | 149 | |
| AZ | Phoenix | 86 | 5.8 | 7.20 | 55 | 33 | 35 | | 16 | 123 |
| AR | Little Rock | 62 | 8.2 | 48.66 | 84 | 56 | 102 | 5.5 | 68 | |
| CA | Los Angeles | 73 | 7.3 | 12.63 | 77 | 61 | 36 | | | 104 |
| | Sacramento | 78 | 8.5 | 16.29 | 83 | 65 | 57 | | 20 | |
| | San Francisco | 67 | 10.5 | 18.69 | 84 | 69 | 62 | | 4 | 59 |
| CO | Denver | 70 | 9.1 | 14.81 | 68 | 40 | 87 | 58.3 | 164 | 135 |
| CT | Hartford | 57 | 9.1 | 42.92 | 78 | 53 | 125 | 56.3 | 138 | 69 |
| DE | Wilmington | 58 | 9.0 | 44.56 | 79 | 55 | 116 | 22.6 | 104 | 127 |
| DC | Washington | 58 | 9.4 | 40.78 | 73 | 51 | 111 | 18.2 | 82 | 77 |
| FL | Jacksonville | 60 | 8.9 | 53.36 | 87 | 55 | 116 | 0.1 | 11 | |
| | Miami | 66 | 9.0 | 59.76 | 84 | 61 | 128 | | | 69 |
| GA | Atlanta | 61 | 9.2 | 47.14 | 83 | 57 | 115 | 1.8 | 62 | 85 |
| HI | Honolulu | 69 | 11.5 | 21.89 | 75 | 71 | 101 | | | 42 |
| ID | Boise | 67 | 9.0 | 11.43 | 70 | 53 | 90 | 21.7 | 127 | 105 |
| IL | Chicago | 57 | 10.3 | 33.18 | 75 | 58 | 120 | 38.4 | 119 | 145 |
| | Peoria | 58 | 10.4 | 34.84 | 83 | 62 | 109 | 22.4 | 133 | |
| IN | Indianapolis | 59 | 9.8 | 39.25 | 83 | 61 | 121 | 20.6 | 122 | 118 |
| IA | Des Moines | 60 | 11.3 | 30.37 | 82 | 63 | 105 | 31.9 | 140 | 103 |
| KS | Wichita | 65 | 12.8 | 28.41 | 80 | 55 | 83 | 14.3 | 112 | 67 |
| KY | Louisville | 58 | 8.3 | 41.32 | 82 | 59 | 122 | 18.0 | 98 | 127 |
| LA | New Orleans | 61 | 8.4 | 55.90 | 88 | 63 | 112 | 0.2 | 12 | 75 |
| ME | Portland | 59 | 8.7 | 42.85 | 80 | 60 | 126 | 73.7 | 162 | |
| MD | Baltimore | 58 | 9.8 | 43.05 | 77 | 53 | 111 | 25.1 | 103 | 118 |
| MA | Boston | 60 | 13.2 | 42.77 | 72 | 55 | 128 | 42.2 | 100 | |

Note: See page 728 for footnote.

**TABLE A6.1** Weather Data for Selected U.S. Cities: Annual Summary[a] (*Continued*)

| Location | | Sun total, % | Wind, avg, mi/hr | Rain total, in. | Relative humidity, % 7 A.M. | Relative humidity, % 1 P.M. | Precipitation, days | Snow per season, in. | 32°F or less, days | Particulate pollution, μg/m³ |
|---|---|---|---|---|---|---|---|---|---|---|
| MI | Detroit | 54 | 10.1 | 30.95 | 78 | 58 | 131 | 31.6 | 125 | 129 |
| | Sault Ste. Marie | 47 | 9.8 | 31.22 | 86 | 67 | 164 | 104.0 | 179 | |
| MN | Duluth | 55 | 11.8 | 28.97 | 79 | 62 | 134 | 76.8 | 186 | 76 |
| | Minneapolis–St. Paul | 58 | 10.7 | 24.78 | 81 | 61 | 112 | 44.4 | 160 | 74 |
| MS | Jackson | 59 | 7.9 | 50.82 | 90 | 58 | 106 | 1.8 | 54 | 80 |
| MO | Kansas City | 64 | 10.0 | 34.07 | 77 | 57 | 99 | 20.1 | 100 | 116 |
| | St. Louis | 59 | 9.5 | 35.31 | 81 | 57 | 105 | 17.4 | 107 | 213 |
| MT | Great Falls | 64 | 13.2 | 14.07 | 66 | 50 | 99 | 55.2 | 150 | |
| NE | Omaha | 62 | 11.0 | 27.56 | 79 | 59 | 98 | 32.0 | 137 | 115 |
| NV | Reno | 80 | 6.4 | 7.15 | 69 | 44 | 48 | 25.3 | 188 | |
| NH | Concord | 54 | 6.7 | 38.80 | 80 | 53 | 125 | 62.7 | 169 | 35 |
| NJ | Atlantic City | 56 | 10.9 | 42.36 | 80 | 54 | 112 | 17.8 | 121 | |
| NM | Albuquerque | 77 | 8.8 | 8.13 | 57 | 37 | 58 | 9.8 | 118 | |
| NY | Albany | 54 | 8.8 | 35.08 | 79 | 56 | 133 | 63.9 | 146 | |
| | Buffalo | 53 | 12.5 | 35.65 | 79 | 63 | 165 | 87.2 | 138 | |
| | New York | 59 | 9.5 | 42.37 | 72 | 56 | 121 | 29.9 | 82 | 131 |
| NC | Charlotte | 66 | 7.5 | 43.38 | 84 | 53 | 110 | 5.4 | 76 | 108 |
| | Raleigh | 61 | 7.9 | 43.58 | 84 | 52 | 113 | 7.4 | 88 | |
| ND | Bismarck | 62 | 10.8 | 15.15 | 77 | 55 | 96 | 38.1 | 188 | 82 |
| OH | Cincinnati | 58 | 7.1 | 39.51 | 80 | 57 | 132 | 19.0 | 99 | 117 |
| | Cleveland | 52 | 10.9 | 35.35 | 79 | 62 | 154 | 50.3 | 128 | 128 |
| | Columbus | 55 | 8.5 | 36.67 | 81 | 59 | 134 | 28.0 | 125 | 106 |
| OK | Oklahoma City | 67 | 13.2 | 30.82 | 82 | 56 | 82 | 9.5 | 85 | 70 |
| OR | Portland | 46 | 7.7 | 37.18 | 87 | 73 | 153 | 8.4 | 44 | 84 |
| PA | Philadelphia | 58 | 9.6 | 42.48 | 76 | 54 | 116 | 21.9 | 108 | 129 |
| | Pittsburgh | 53 | 9.4 | 36.14 | 79 | 57 | 149 | 46.0 | 128 | 165 |
| PR | San Juan | 64 | 8.5 | 64.21 | 82 | 66 | 205 | | | |
| RI | Providence | 57 | 11.0 | 42.13 | 73 | 54 | 124 | 40.9 | 126 | 80 |
| SC | Columbia | 64 | 7.0 | 46.82 | 86 | 50 | 109 | 1.2 | 55 | |
| SD | Sioux Falls | 63 | 11.1 | 25.16 | 81 | 59 | 93 | 41.6 | 173 | 69 |
| TN | Memphis | 65 | 9.2 | 49.73 | 82 | 57 | 103 | 6.1 | 60 | 76 |
| | Nashville | 58 | 7.8 | 45.15 | 85 | 57 | 118 | 11.8 | 75 | 98 |
| TX | Dallas | 65 | 10.9 | 34.55 | 79 | 55 | 80 | 2.1 | 37 | 84 |
| | El Paso | 83 | 9.9 | 7.89 | 52 | 35 | 44 | 4.4 | 63 | |
| | Houston | 62 | 10.8 | 45.95 | 87 | 60 | 103 | 0.4 | 13 | 88 |
| UT | Salt Lake City | 70 | 8.6 | 13.90 | 67 | 47 | 87 | 55.1 | 141 | 90 |
| VT | Burlington | 52 | 8.9 | 33.21 | 77 | 59 | 149 | 73.8 | 167 | 46 |
| VA | Norfolk | 63 | 10.6 | 44.94 | 79 | 58 | 115 | 7.5 | 56 | 94 |
| | Richmond | 61 | 7.7 | 44.21 | 82 | 53 | 113 | 15.0 | 86 | |
| WA | Seattle | 45 | 9.5 | 38.94 | 84 | 75 | 161 | 15.8 | 32 | 61 |
| | Spokane | 57 | 8.4 | 17.19 | 76 | 62 | 115 | 54.6 | 139 | |
| WV | Charleston | 48 | 6.6 | 44.43 | 82 | 55 | 147 | 28.4 | 101 | 213 |
| WI | Milwaukee | 56 | 11.8 | 29.51 | 81 | 64 | 120 | 43.4 | 149 | 124 |
| WY | Cheyenne | 64 | 12.7 | 15.06 | 63 | 40 | 99 | 51.6 | 171 | 32 |

[a]Federal government data, published 1971. For monthly data see "Statistical Abstract of the United States: 1971," 92nd ed., U.S. Bureau of the Census, 1971.

**TABLE A6.2**  Weather Conditions for Environmental Design[a]

| Location | Lati-tude, deg[b] | Eleva-tion, ft | Degree-days heating[c] | Winter[d] 99% | Winter[d] 97.5% | Summer[e] 1% DB | Summer[e] 2½%DB | Summer[e] 5% DB | Summer[e] 5% WB |
|---|---|---|---|---|---|---|---|---|---|
| **USA** | | | | | | | | | |
| Alabama, Birmingham | 33 | 610 | 2,551 | 19 | 22 | 97 | 94 | 93 | 77 |
| Alaska, Anchorage | 61 | 90 | 10,864 | −25 | −20 | 73 | 70 | 67 | 59 |
| Alaska, Juneau | 58 | 17 | 9,075 | −7 | −4 | 75 | 71 | 68 | 62 |
| Arizona, Phoenix | 34 | 1,117 | 1,765 | 31 | 34 | 108 | 106 | 104 | 75 |
| Arkansas, Little Rock | 35 | 257 | 3,219 | 19 | 23 | 99 | 96 | 94 | 78 |
| California, Los Angeles | 34 | 312 | 2,061 | 42 | 44 | 94 | 90 | 87 | 69 |
| California, Sacramento | 39 | 17 | 2,502 | 30 | 32 | 100 | 97 | 94 | 69 |
| California, San Francisco | 38 | 52 | 3,015 | 42 | 44 | 80 | 77 | 73 | 61 |
| Colorado, Denver | 40 | 5,283 | 6,283 | −2 | 3 | 92 | 90 | 89 | 63 |
| Connecticut, Hartford | 42 | 15 | 6,235 | 1 | 5 | 90 | 88 | 85 | 74 |
| Delaware, Wilmington | 40 | 78 | 4,930 | 12 | 15 | 93 | 90 | 87 | 76 |
| District of Columbia | 39 | 14 | 4,224 | 16 | 19 | 94 | 92 | 90 | 76 |
| Florida, Jacksonville | 31 | 24 | 1,239 | 29 | 32 | 96 | 94 | 92 | 79 |
| Florida, Miami | 26 | 7 | 214 | 44 | 47 | 92 | 90 | 89 | 79 |
| Florida, Tampa | 28 | 19 | 683 | 36 | 39 | 92 | 91 | 90 | 79 |
| Georgia, Atlanta | 34 | 1,005 | 2,961 | 18 | 23 | 95 | 92 | 90 | 76 |
| Hawaii, Honolulu | 21 | 7 | 0 | 60 | 62 | 87 | 85 | 84 | 73 |
| Idaho, Boise | 44 | 2,842 | 5,809 | 4 | 10 | 96 | 93 | 91 | 65 |
| Illinois, Chicago | 42 | 594 | 6,639 | −3 | 1 | 94 | 91 | 88 | 75 |
| Indiana, Indianapolis | 40 | 793 | 5,699 | 0 | 4 | 93 | 91 | 88 | 76 |
| Iowa, Des Moines | 42 | 948 | 6,588 | −7 | −3 | 95 | 92 | 89 | 76 |
| Kansas, Topeka | 39 | 877 | 5,182 | 3 | 6 | 99 | 96 | 94 | 77 |
| Kentucky, Louisville | 38 | 474 | 4,660 | 8 | 12 | 96 | 93 | 91 | 77 |
| Louisiana, New Orleans | 30 | 3 | 1,385 | 32 | 35 | 93 | 91 | 90 | 79 |
| Maine, Portland | 44 | 61 | 7,511 | −5 | 0 | 88 | 85 | 81 | 71 |
| Maryland, Baltimore | 39 | 146 | 4,654 | 12 | 15 | 94 | 91 | 89 | 77 |
| Massachusetts, Boston | 42 | 15 | 5,634 | 6 | 10 | 91 | 88 | 85 | 73 |
| Michigan, Detroit, (Met.) | 42 | 633 | 6,232 | 4 | 8 | 92 | 88 | 85 | 74 |
| Michigan, Sault Ste. Marie | 47 | 721 | 9,048 | −12 | −8 | 83 | 81 | 78 | 69 |
| Minnesota, Duluth | 47 | 1,426 | 10,000 | −19 | −15 | 85 | 82 | 79 | 69 |
| Minnesota, Minneapolis/ St. Paul | 45 | 822 | 8,382 | −14 | −10 | 92 | 89 | 86 | 74 |
| Mississippi, Jackson | 32 | 330 | 2,239 | 21 | 24 | 98 | 96 | 94 | 78 |
| Missouri, Kansas City | 39 | 742 | 4,711 | 4 | 8 | 100 | 97 | 94 | 76 |
| Missouri, St. Louis | 39 | 465 | 4,900 | 7 | 11 | 96 | 94 | 92 | 77 |
| Montana, Billings | 46 | 3,567 | 7,049 | −10 | −6 | 94 | 91 | 88 | 65 |
| Montana, Butte | 46 | 5,526 | − | −24 | −16 | 86 | 83 | 80 | 57 |
| Nebraska, Omaha | 41 | 978 | 6,612 | −5 | −1 | 97 | 94 | 91 | 76 |
| Nevada, Las Vegas | 36 | 2,162 | 2,709 | 23 | 26 | 108 | 106 | 104 | 70 |
| New Hampshire, Man-chester | 43 | 253 | − | −5 | 1 | 92 | 89 | 86 | 73 |
| New Jersey, Atlantic City | 40 | 11 | 4,812 | 14 | 18 | 91 | 88 | 85 | 76 |
| New Mexico, Albuquerque | 35 | 5,310 | 4,348 | 14 | 17 | 96 | 94 | 92 | 64 |
| New York, Albany | 43 | 277 | 6,875 | −5 | 0 | 91 | 88 | 85 | 73 |
| New York, Buffalo | 43 | 705 | 7,062 | 3 | 6 | 88 | 86 | 83 | 72 |
| New York, NYC-LaGuardia | 41 | 19 | 4,811 | 12 | 16 | 93 | 90 | 87 | 75 |
| North Carolina, Durham | 36 | 406 | − | 15 | 19 | 94 | 92 | 89 | 76 |
| North Dakota, Bismarck | 47 | 1,647 | 8,851 | −24 | −19 | 95 | 91 | 88 | 70 |
| Ohio, Cincinnati | 39 | 761 | 4,410 | 8 | 12 | 94 | 92 | 90 | 76 |
| Ohio, Cleveland | 41 | 777 | 6,351 | 2 | 7 | 91 | 89 | 86 | 74 |
| Oklahoma, Tulsa | 36 | 650 | 3,860 | 12 | 16 | 102 | 99 | 96 | 77 |
| Oregon, Portland | 46 | 57 | 4,635 | 26 | 29 | 91 | 88 | 84 | 67 |
| Pennsylvania, Philadelphia | 40 | 7 | 5,144 | 11 | 15 | 93 | 90 | 87 | 76 |
| Pennsylvania, Pittsburgh | 41 | 749 | 5,987 | 7 | 11 | 90 | 88 | 85 | 73 |

Note: See page 730 for footnotes.

**TABLE A6.2** Weather Conditions for Environmental Design[a] (*Continued*)

| Location | Lati-tude, deg | Eleva-tion, ft | Degree-days heating | Winter 99% | Winter 97.5% | Summer 1% DB | Summer 2½% DB | Summer 5% DB | Summer 5% WB |
|---|---|---|---|---|---|---|---|---|---|
| **USA** *(cont.)* | | | | | | | | | |
| Rhode Island, Providence | 42 | 55 | 5,954 | 6 | 10 | 89 | 86 | 83 | 74 |
| South Carolina, Columbia | 34 | 217 | 2,484 | 20 | 23 | 98 | 96 | 94 | 78 |
| South Dakota, Sioux Falls | 44 | 1,420 | 7,839 | −14 | −10 | 95 | 92 | 89 | 74 |
| Tennessee, Memphis | 35 | 263 | 3,232 | 17 | 21 | 98 | 96 | 94 | 78 |
| Tennessee, Nashville | 36 | 577 | 3,578 | 12 | 16 | 97 | 95 | 92 | 77 |
| Texas, Dallas | 33 | 481 | 2,363 | 19 | 24 | 101 | 99 | 97 | 78 |
| Texas, Houston | 30 | 158 | 1,396 | 29 | 33 | 96 | 94 | 92 | 79 |
| Texas, San Antonio | 30 | 792 | 1,546 | 25 | 30 | 99 | 97 | 96 | 76 |
| Utah, Salt Lake City | 41 | 4,220 | 6,052 | 5 | 9 | 97 | 94 | 92 | 65 |
| Vermont, Burlington | 45 | 331 | 8,269 | −12 | −7 | 88 | 85 | 83 | 71 |
| Virginia, Richmond | 38 | 162 | 3,865 | 14 | 18 | 96 | 93 | 91 | 77 |
| Washington, Seattle | 48 | 14 | 4,424 | 28 | 32 | 81 | 79 | 76 | 64 |
| Washington, Spokane | 48 | 2,357 | 6,655 | −2 | 4 | 93 | 90 | 87 | 63 |
| West Virginia, Charleston | 38 | 939 | 4,476 | 9 | 14 | 92 | 90 | 88 | 74 |
| Wisconsin, LaCrosse | 44 | 652 | 7,589 | −12 | −8 | 90 | 88 | 85 | 75 |
| Wisconsin, Milwaukee | 43 | 672 | 7,635 | −6 | −2 | 90 | 87 | 84 | 73 |
| Wyoming, Cheyenne | 41 | 6,126 | 7,381 | −6 | −2 | 89 | 86 | 83 | 61 |
| **CANADA** | | | | | | | | | |
| Alberta, Edmonton | 54 | 2,219 | 10,268 | −29 | −26 | 86 | 83 | 80 | 65 |
| Br. Columbia, Vancouver | 49 | 16 | 5,515 | 15 | 19 | 80 | 78 | 76 | 65 |
| Manitoba, Winnipeg | 50 | 786 | 10,679 | −28 | −25 | 90 | 87 | 84 | 72 |
| Ontario, Ottawa | 45 | 339 | 8,735 | −17 | −13 | 90 | 87 | 84 | 73 |
| Ontario, Toronto | 44 | 578 | 6,827 | −3 | 1 | 90 | 87 | 85 | 73 |
| Quebec, Montreal | 46 | 98 | 8,203 | −16 | −10 | 88 | 86 | 84 | 73 |
| Quebec, Quebec | 47 | 245 | 9,372 | −19 | −13 | 86 | 82 | 79 | 71 |
| **OTHER COUNTRIES** | | | | | | | | | |
| Argentina, Buenos Aires | 35 | 89 | — | 32 | 34 | 91 | 89 | 86 | 75 |
| Australia, Sydney | 34 | 138 | — | 40 | 42 | 89 | 84 | 80 | 72 |
| Brazil, Sao Paulo | 24 | 2,608 | — | 42 | 46 | 86 | 84 | 82 | 74 |
| France, Paris | 49 | 164 | — | 22 | 25 | 89 | 86 | 83 | 67 |
| Germany, Berlin | 52 | 187 | — | 7 | 12 | 84 | 81 | 78 | 66 |
| Hong Kong, Hong Kong | 22 | 109 | — | 48 | 50 | 92 | 91 | 90 | 80 |
| India, Calcutta | 23 | 21 | — | 52 | 54 | 98 | 97 | 96 | 82 |
| Italy, Rome | 42 | 377 | — | 30 | 33 | 94 | 92 | 89 | 72 |
| Japan, Tokyo | 36 | 19 | — | 26 | 28 | 91 | 89 | 87 | 79 |
| Mexico, Mexico City | 19 | 7,575 | — | 37 | 39 | 83 | 81 | 79 | 59 |
| Netherlands, Amsterdam | 52 | 5 | — | 20 | 23 | 79 | 76 | 73 | 63 |
| Soviet Union, Moscow | 56 | 505 | — | −11 | −6 | 84 | 81 | 78 | 65 |
| Spain, Madrid | 40 | 2,188 | — | 25 | 28 | 93 | 91 | 89 | 67 |
| Sweden, Stockholm | 59 | 146 | — | 5 | 8 | 78 | 74 | 72 | 60 |
| United Kingdom, London | 51 | 149 | — | 24 | 26 | 82 | 79 | 76 | 65 |

[a]Condensed from ASHRAE "Handbook of Fundamentals" and "Guide and Data Book," American Society of Heating, Refrigerating, and Air-Conditioning Engineers, 1972 and 1968, respectively. Extended data for over 1000 locations are given in these sources.

[b]Latitudes are given to the nearest degree.

[c]Degree-days are the yearly totals, 65°F base, i.e., for all days when the mean temperature was below 65°F. For monthly and yearly degree-days see Table A6.3.

[d]Winter temperatures are the temperatures equaled or exceeded 99 percent (and 97.5 percent) of the time during the coldest months. For average monthly temperatures see Table A6.4.

[e]Summer temperatures represent the highest 1 percent (or 2.5 or 5 percent) hourly dry-bulb (DB) or wet-bulb (WB) temperatures during the warmest months. In a normal season there would be about 30 hr above the 1 percent design temperature and 150 hr above the 5 percent dry-bulb design temperature. In most locations the 1 percent design wet-bulb temperature is about 2° above the 5 percent dry-bulb value.

# TABLE A6.3 Normal Total Heating Degree-Days, Base 65°F[a, b]

| | Location | Jul | Aug | Sep | Oct | Nov | Dec | Jan | Feb | Mar | Apr | May | Jun | Ann |
|---|---|---|---|---|---|---|---|---|---|---|---|---|---|---|
| AL | Birmingham | 0 | 0 | 6 | 93 | 363 | 555 | 592 | 462 | 363 | 108 | 9 | 0 | 2,551 |
| | Huntsville | 0 | 0 | 12 | 127 | 426 | 663 | 694 | 557 | 434 | 138 | 19 | 0 | 3,070 |
| | Mobile | 0 | 0 | 0 | 22 | 213 | 357 | 415 | 300 | 211 | 42 | 0 | 0 | 1,560 |
| | Montgomery | 0 | 0 | 0 | 68 | 330 | 527 | 543 | 417 | 316 | 90 | 0 | 0 | 2,291 |
| AK | Anchorage | 245 | 291 | 516 | 930 | 1,284 | 1,572 | 1,631 | 1,316 | 1,293 | 879 | 592 | 315 | 10,864 |
| | Annette | 242 | 208 | 327 | 567 | 738 | 899 | 949 | 837 | 843 | 648 | 490 | 321 | 7,069 |
| | Barrow | 803 | 840 | 1,035 | 1,500 | 1,971 | 2,362 | 2,517 | 2,332 | 2,468 | 1,944 | 1,445 | 957 | 20,174 |
| | Barter Is. | 735 | 775 | 987 | 1,482 | 1,944 | 2,337 | 2,536 | 2,369 | 2,477 | 1,923 | 1,373 | 924 | 19,862 |
| | Bethel | 319 | 394 | 612 | 1,042 | 1,434 | 1,866 | 1,903 | 1,590 | 1,655 | 1,173 | 806 | 402 | 13,196 |
| | Cold Bay | 474 | 425 | 525 | 772 | 918 | 1,122 | 1,153 | 1,036 | 1,122 | 951 | 791 | 591 | 9,880 |
| | Cordova | 366 | 391 | 522 | 781 | 1,017 | 1,221 | 1,299 | 1,086 | 1,113 | 864 | 660 | 444 | 9,764 |
| | Fairbanks | 171 | 332 | 642 | 1,203 | 1,833 | 2,254 | 2,359 | 1,901 | 1,739 | 1,068 | 555 | 222 | 14,279 |
| | Juneau | 301 | 338 | 483 | 725 | 921 | 1,135 | 1,237 | 1,070 | 1,073 | 810 | 601 | 381 | 9,075 |
| | King Salmon | 313 | 322 | 513 | 908 | 1,290 | 1,606 | 1,600 | 1,333 | 1,411 | 966 | 673 | 408 | 11,343 |
| | Kotzebue | 381 | 446 | 723 | 1,249 | 1,728 | 2,127 | 2,192 | 1,932 | 2,080 | 1,554 | 1,057 | 636 | 16,105 |
| | McGrath | 208 | 338 | 633 | 1,184 | 1,791 | 2,232 | 2,294 | 1,817 | 1,758 | 1,122 | 648 | 258 | 14,283 |
| | Nome | 481 | 496 | 693 | 1,094 | 1,455 | 1,820 | 1,879 | 1,666 | 1,770 | 1,314 | 930 | 573 | 14,171 |
| | St. Paul | 605 | 539 | 612 | 862 | 963 | 1,197 | 1,228 | 1,168 | 1,265 | 1,098 | 936 | 726 | 11,199 |
| | Shemya | 577 | 475 | 501 | 784 | 876 | 1,042 | 1,045 | 958 | 1,011 | 885 | 837 | 696 | 9,687 |
| | Yakutat | 338 | 347 | 474 | 716 | 936 | 1,144 | 1,169 | 1,019 | 1,042 | 840 | 632 | 435 | 9,092 |
| AZ | Flagstaff | 46 | 68 | 201 | 558 | 867 | 1,073 | 1,169 | 991 | 911 | 651 | 437 | 180 | 7,152 |
| | Phoenix | 0 | 0 | 0 | 22 | 234 | 415 | 474 | 328 | 217 | 75 | 0 | 0 | 1,765 |
| | Prescott | 0 | 0 | 27 | 245 | 579 | 797 | 865 | 711 | 605 | 360 | 158 | 15 | 4,362 |
| | Tucson | 0 | 0 | 0 | 25 | 231 | 406 | 471 | 344 | 242 | 75 | 6 | 0 | 1,800 |
| | Winslow | 0 | 0 | 6 | 245 | 711 | 1,008 | 1,054 | 770 | 601 | 291 | 96 | 0 | 4,782 |
| | Yuma | 0 | 0 | 0 | 0 | 148 | 319 | 363 | 228 | 130 | 29 | 0 | 0 | 1,217 |
| AR | Ft. Smith | 0 | 0 | 12 | 127 | 450 | 704 | 781 | 596 | 456 | 144 | 22 | 0 | 3,292 |
| | Little Rock | 0 | 0 | 9 | 127 | 465 | 716 | 756 | 577 | 434 | 126 | 9 | 0 | 3,219 |
| | Texarkana | 0 | 0 | 0 | 78 | 345 | 561 | 626 | 468 | 350 | 105 | 0 | 0 | 2,533 |
| CA | Bakersfield | 0 | 0 | 0 | 37 | 282 | 502 | 546 | 364 | 267 | 105 | 19 | 0 | 2,122 |
| | Bishop | 0 | 0 | 42 | 248 | 576 | 797 | 874 | 666 | 539 | 306 | 143 | 36 | 4,227 |
| | Blue Canyon | 34 | 50 | 120 | 347 | 579 | 766 | 865 | 781 | 791 | 582 | 397 | 195 | 5,507 |
| | Burbank | 0 | 0 | 6 | 43 | 177 | 301 | 366 | 277 | 239 | 138 | 81 | 18 | 1,646 |
| | Eureka | 270 | 257 | 258 | 329 | 414 | 499 | 546 | 470 | 505 | 438 | 372 | 285 | 4,643 |
| | Fresno | 0 | 0 | 0 | 78 | 339 | 558 | 586 | 406 | 319 | 150 | 56 | 0 | 2,492 |
| | Long Beach | 0 | 0 | 12 | 40 | 156 | 288 | 375 | 297 | 267 | 168 | 90 | 18 | 1,711 |
| | Los Angeles | 28 | 22 | 42 | 78 | 180 | 291 | 372 | 302 | 288 | 219 | 158 | 81 | 2,061 |
| | Mt. Shasta | 25 | 34 | 123 | 406 | 696 | 902 | 983 | 784 | 738 | 525 | 347 | 159 | 5,722 |
| | Oakland | 53 | 50 | 45 | 127 | 309 | 481 | 527 | 400 | 353 | 255 | 180 | 90 | 2,870 |
| | Pt. Arguello | 202 | 186 | 162 | 205 | 291 | 400 | 474 | 392 | 403 | 339 | 298 | 243 | 3,595 |
| | Red Bluff | 0 | 0 | 0 | 53 | 318 | 555 | 605 | 428 | 341 | 168 | 47 | 0 | 2,515 |
| | Sacramento | 0 | 0 | 12 | 81 | 363 | 577 | 614 | 442 | 360 | 216 | 102 | 6 | 2,773 |
| | Sandberg | 0 | 0 | 30 | 202 | 480 | 691 | 778 | 661 | 620 | 426 | 264 | 57 | 4,209 |
| | San Diego | 6 | 0 | 15 | 37 | 123 | 251 | 313 | 249 | 202 | 123 | 84 | 36 | 1,439 |
| | San Francisco | 81 | 78 | 60 | 143 | 306 | 462 | 508 | 395 | 363 | 279 | 214 | 126 | 3,015 |
| | Santa Catalina | 16 | 0 | 9 | 50 | 165 | 279 | 353 | 308 | 326 | 249 | 192 | 105 | 2,052 |
| | Santa Maria | 99 | 93 | 96 | 146 | 270 | 391 | 459 | 370 | 363 | 282 | 233 | 165 | 2,967 |
| CO | Alamosa | 65 | 99 | 279 | 639 | 1,065 | 1,420 | 1,476 | 1,162 | 1,020 | 696 | 440 | 168 | 8,529 |
| | Colorado Springs | 9 | 25 | 132 | 456 | 825 | 1,032 | 1,128 | 938 | 893 | 582 | 319 | 84 | 6,423 |
| | Denver | 6 | 9 | 117 | 428 | 819 | 1,035 | 1,132 | 938 | 887 | 558 | 288 | 66 | 6,283 |
| | Grand Junction | 0 | 0 | 30 | 313 | 786 | 1,113 | 1,209 | 907 | 729 | 387 | 146 | 21 | 5,641 |
| | Pueblo | 0 | 0 | 54 | 326 | 750 | 986 | 1,085 | 871 | 772 | 429 | 174 | 15 | 5,462 |
| CT | Bridgeport | 0 | 0 | 66 | 307 | 615 | 986 | 1,079 | 966 | 853 | 510 | 208 | 27 | 5,617 |
| | Hartford | 0 | 6 | 99 | 372 | 711 | 1,119 | 1,209 | 1,061 | 899 | 495 | 177 | 24 | 6,172 |
| | New Haven | 0 | 12 | 87 | 347 | 648 | 1,011 | 1,097 | 991 | 871 | 543 | 245 | 45 | 5,897 |
| DE | Wilmington | 0 | 0 | 51 | 270 | 588 | 927 | 980 | 874 | 735 | 387 | 112 | 6 | 4,930 |
| FL | Apalachicola | 0 | 0 | 0 | 16 | 153 | 319 | 347 | 260 | 180 | 33 | 0 | 0 | 1,308 |
| | Daytona Beach | 0 | 0 | 0 | 0 | 75 | 211 | 248 | 190 | 140 | 15 | 0 | 0 | 879 |
| | Ft. Myers | 0 | 0 | 0 | 0 | 24 | 109 | 146 | 101 | 62 | 0 | 0 | 0 | 442 |
| | Jacksonville | 0 | 0 | 0 | 12 | 144 | 310 | 332 | 246 | 174 | 21 | 0 | 0 | 1,239 |
| | Key West | 0 | 0 | 0 | 0 | 0 | 28 | 40 | 31 | 9 | 0 | 0 | 0 | 108 |
| | Lakeland | 0 | 0 | 0 | 0 | 57 | 164 | 195 | 146 | 99 | 0 | 0 | 0 | 661 |
| | Miami Beach | 0 | 0 | 0 | 0 | 0 | 40 | 56 | 36 | 9 | 0 | 0 | 0 | 141 |
| | Orlando | 0 | 0 | 0 | 0 | 72 | 198 | 220 | 165 | 105 | 6 | 0 | 0 | 766 |
| | Pensacola | 0 | 0 | 0 | 19 | 195 | 353 | 400 | 277 | 183 | 36 | 0 | 0 | 1,463 |
| | Tallahassee | 0 | 0 | 0 | 28 | 198 | 360 | 375 | 286 | 202 | 36 | 0 | 0 | 1,485 |
| | Tampa | 0 | 0 | 0 | 0 | 60 | 171 | 202 | 148 | 102 | 0 | 0 | 0 | 683 |
| | W. Palm Beach | 0 | 0 | 0 | 0 | 6 | 65 | 87 | 64 | 31 | 0 | 0 | 0 | 253 |

Note: See page 735 for footnotes.

**TABLE A6.3**  Normal Total Heating Degree-Days, Base 65°F[a,b] (*Continued*)

| | Location | Jul | Aug | Sep | Oct | Nov | Dec | Jan | Feb | Mar | Apr | May | Jun | Ann |
|---|---|---|---|---|---|---|---|---|---|---|---|---|---|---|
| GA | Athens | 0 | 0 | 12 | 115 | 405 | 632 | 642 | 529 | 431 | 141 | 22 | 0 | 2,929 |
| | Atlanta | 0 | 0 | 18 | 127 | 414 | 626 | 639 | 529 | 437 | 168 | 25 | 0 | 2,983 |
| | Augusta | 0 | 0 | 0 | 78 | 333 | 552 | 549 | 445 | 350 | 90 | 0 | 0 | 2,397 |
| | Columbus | 0 | 0 | 0 | 87 | 333 | 543 | 552 | 434 | 338 | 96 | 0 | 0 | 2,383 |
| | Macon | 0 | 0 | 0 | 71 | 297 | 502 | 505 | 403 | 295 | 63 | 0 | 0 | 2,136 |
| | Rome | 0 | 0 | 24 | 161 | 474 | 701 | 710 | 577 | 468 | 177 | 34 | 0 | 3,326 |
| | Savannah | 0 | 0 | 0 | 47 | 246 | 437 | 437 | 353 | 254 | 45 | 0 | 0 | 1,819 |
| | Thomasville | 0 | 0 | 0 | 25 | 198 | 366 | 394 | 305 | 208 | 33 | 0 | 0 | 1,529 |
| ID | Boise | 0 | 0 | 132 | 415 | 792 | 1,017 | 1,113 | 854 | 722 | 438 | 245 | 81 | 5,809 |
| | Idaho Falls 46W | 16 | 34 | 270 | 623 | 1,056 | 1,370 | 1,538 | 1,249 | 1,085 | 651 | 391 | 192 | 8,475 |
| | Idaho Falls 42NW | 16 | 40 | 282 | 648 | 1,107 | 1,432 | 1,600 | 1,291 | 1,107 | 657 | 388 | 192 | 8,760 |
| | Lewiston | 0 | 0 | 123 | 403 | 756 | 933 | 1,063 | 815 | 694 | 426 | 239 | 90 | 5,542 |
| | Pocatello | 0 | 0 | 172 | 493 | 900 | 1,166 | 1,324 | 1,058 | 905 | 555 | 319 | 141 | 7,033 |
| IL | Cairo | 0 | 0 | 36 | 164 | 513 | 791 | 856 | 680 | 539 | 195 | 47 | 0 | 3,821 |
| | Chicago | 0 | 0 | 81 | 326 | 753 | 1,113 | 1,209 | 1,044 | 890 | 480 | 211 | 48 | 6,155 |
| | Moline | 0 | 9 | 99 | 335 | 774 | 1,181 | 1,314 | 1,100 | 918 | 450 | 189 | 39 | 6,408 |
| | Peoria | 0 | 6 | 87 | 326 | 759 | 1,113 | 1,218 | 1,025 | 849 | 426 | 183 | 33 | 6,025 |
| | Rockford | 6 | 9 | 114 | 400 | 837 | 1,221 | 1,333 | 1,137 | 961 | 516 | 236 | 60 | 6,830 |
| | Springfield | 0 | 0 | 72 | 291 | 696 | 1,023 | 1,135 | 935 | 769 | 354 | 136 | 18 | 5,429 |
| IN | Evansville | 0 | 0 | 66 | 220 | 606 | 896 | 955 | 767 | 620 | 237 | 68 | 0 | 4,435 |
| | Ft. Wayne | 0 | 9 | 105 | 378 | 783 | 1,135 | 1,178 | 1,028 | 890 | 471 | 189 | 39 | 6,205 |
| | Indianapolis | 0 | 0 | 90 | 316 | 723 | 1,051 | 1,113 | 949 | 809 | 432 | 177 | 39 | 5,699 |
| | South Bend | 0 | 6 | 111 | 372 | 777 | 1,125 | 1,221 | 1,070 | 933 | 525 | 239 | 60 | 6,439 |
| IA | Burlington | 0 | 0 | 93 | 322 | 768 | 1,135 | 1,259 | 1,042 | 859 | 426 | 177 | 33 | 6,114 |
| | Des Moines | 0 | 9 | 99 | 363 | 837 | 1,231 | 1,398 | 1,163 | 967 | 489 | 211 | 39 | 6,808 |
| | Dubuque | 12 | 31 | 156 | 450 | 906 | 1,287 | 1,420 | 1,204 | 1,026 | 546 | 260 | 78 | 7,376 |
| | Sioux City | 0 | 9 | 108 | 369 | 867 | 1,240 | 1,435 | 1,198 | 989 | 483 | 214 | 39 | 6,951 |
| | Waterloo | 12 | 19 | 138 | 428 | 909 | 1,296 | 1,460 | 1,221 | 1,023 | 531 | 229 | 54 | 7,320 |
| KS | Concordia | 0 | 0 | 57 | 276 | 705 | 1,023 | 1,163 | 935 | 781 | 372 | 149 | 18 | 5,479 |
| | Dodge City | 0 | 0 | 33 | 251 | 666 | 939 | 1,051 | 840 | 719 | 354 | 124 | 9 | 4,986 |
| | Goodland | 0 | 6 | 81 | 381 | 810 | 1,073 | 1,166 | 955 | 884 | 507 | 236 | 42 | 6,141 |
| | Topeka | 0 | 0 | 57 | 270 | 672 | 980 | 1,122 | 893 | 722 | 330 | 124 | 12 | 5,182 |
| | Wichita | 0 | 0 | 33 | 229 | 618 | 905 | 1,023 | 804 | 645 | 270 | 87 | 6 | 4,620 |
| KY | Covington | 0 | 0 | 75 | 291 | 669 | 983 | 1,035 | 893 | 756 | 390 | 149 | 24 | 5,265 |
| | Lexington | 0 | 0 | 54 | 239 | 609 | 902 | 946 | 818 | 685 | 325 | 105 | 0 | 4,683 |
| | Louisville | 0 | 0 | 54 | 248 | 609 | 890 | 930 | 818 | 682 | 315 | 105 | 9 | 4,660 |
| LA | Alexandria | 0 | 0 | 0 | 56 | 273 | 431 | 471 | 361 | 260 | 69 | 0 | 0 | 1,921 |
| | Baton Rouge | 0 | 0 | 0 | 31 | 216 | 369 | 409 | 294 | 208 | 33 | 0 | 0 | 1,560 |
| | Burrwood | 0 | 0 | 0 | 0 | 96 | 214 | 298 | 218 | 171 | 27 | 0 | 0 | 1,024 |
| | Lake Charles | 0 | 0 | 0 | 19 | 210 | 341 | 381 | 274 | 195 | 39 | 0 | 0 | 1,459 |
| | New Orleans | 0 | 0 | 0 | 19 | 192 | 322 | 363 | 258 | 192 | 39 | 0 | 0 | 1,385 |
| | Shreveport | 0 | 0 | 0 | 47 | 297 | 477 | 552 | 426 | 304 | 81 | 0 | 0 | 2,184 |
| ME | Caribou | 78 | 115 | 336 | 682 | 1,044 | 1,535 | 1,690 | 1,470 | 1,308 | 858 | 468 | 183 | 9,767 |
| | Portland | 12 | 53 | 195 | 508 | 807 | 1,215 | 1,339 | 1,182 | 1,042 | 675 | 372 | 111 | 7,511 |
| MD | Baltimore | 0 | 0 | 48 | 264 | 585 | 905 | 936 | 820 | 679 | 327 | ·90 | 0 | 4,654 |
| | Frederick | 0 | 0 | 66 | 307 | 624 | 955 | 995 | 876 | 741 | 384 | 127 | 12 | 5,087 |
| MA | Blue Hill Obsy | 0 | 22 | 108 | 381 | 690 | 1,085 | 1,178 | 1,053 | 936 | 579 | 267 | 69 | 6,368 |
| | Boston | 0 | 9 | 60 | 316 | 603 | 983 | 1,088 | 972 | 846 | 513 | 208 | 36 | 5,634 |
| | Nantucket | 12 | 22 | 93 | 332 | 573 | 896 | 992 | 941 | 896 | 621 | 384 | 129 | 5,891 |
| | Pittsfield | 25 | 59 | 219 | 524 | 831 | 1,231 | 1,339 | 1,196 | 1,063 | 660 | 326 | 105 | 7,578 |
| | Worcester | 6 | 34 | 147 | 450 | 774 | 1,172 | 1,271 | 1,123 | 998 | 612 | 304 | 78 | 6,969 |
| MI | Alpena | 68 | 105 | 273 | 580 | 912 | 1,268 | 1,404 | 1,299 | 1,218 | 777 | 446 | 156 | 8,506 |
| | Detroit (City) | 0 | 0 | 87 | 360 | 738 | 1,088 | 1,181 | 1,058 | 936 | 522 | 220 | 42 | 6,232 |
| | Escanaba | 59 | 87 | 243 | 539 | 924 | 1,293 | 1,445 | 1,296 | 1,203 | 777 | 456 | 159 | 8,481 |
| | Flint | 16 | 40 | 159 | 465 | 843 | 1,212 | 1,330 | 1,198 | 1,066 | 639 | 319 | 90 | 7,377 |
| | Grand Rapids | 9 | 28 | 135 | 434 | 804 | 1,147 | 1,259 | 1,134 | 1,011 | 579 | 279 | 75 | 6,894 |
| | Lansing | 6 | 22 | 138 | 431 | 813 | 1,163 | 1,262 | 1,142 | 1,011 | 579 | 273 | 69 | 6,909 |
| | Marquette | 59 | 81 | 240 | 527 | 936 | 1,268 | 1,411 | 1,268 | 1,187 | 771 | 468 | 177 | 8,393 |
| | Muskegon | 12 | 28 | 120 | 400 | 762 | 1,088 | 1,209 | 1,100 | 995 | 594 | 310 | 78 | 6,696 |
| | Sault Ste. Marie | 96 | 105 | 279 | 580 | 951 | 1,367 | 1,525 | 1,380 | 1,277 | 810 | 477 | 201 | 9,048 |
| MN | Duluth | 71 | 109 | 330 | 632 | 1,131 | 1,581 | 1,745 | 1,518 | 1,355 | 840 | 490 | 198 | 10,000 |
| | International Falls | 71 | 112 | 363 | 701 | 1,236 | 1,724 | 1,919 | 1,621 | 1,414 | 828 | 443 | 174 | 10,606 |
| | Minneapolis | 22 | 31 | 189 | 505 | 1,014 | 1,454 | 1,631 | 1,380 | 1,166 | 621 | 288 | 81 | 8,382 |
| | Rochester | 25 | 34 | 186 | 474 | 1,005 | 1,438 | 1,593 | 1,366 | 1,150 | 630 | 301 | 93 | 8,295 |
| | St. Cloud | 28 | 47 | 225 | 549 | 1,065 | 1,500 | 1,702 | 1,445 | 1,221 | 666 | 326 | 105 | 8,879 |
| MS | Jackson | 0 | 0 | 0 | 65 | 315 | 502 | 546 | 414 | 310 | 87 | 0 | 0 | 2,239 |
| | Meridian | 0 | 0 | 0 | 81 | 339 | 518 | 543 | 417 | 310 | 81 | 0 | 0 | 2,289 |
| | Vicksburg | 0 | 0 | 0 | 53 | 279 | 462 | 512 | 384 | 282 | 69 | 0 | 0 | 2,041 |

| | Location | Jul | Aug | Sep | Oct | Nov | Dec | Jan | Feb | Mar | Apr | May | Jun | Ann |
|---|---|---|---|---|---|---|---|---|---|---|---|---|---|---|
| MO | Columbia | 0 | 0 | 54 | 251 | 651 | 967 | 1,076 | 874 | 716 | 324 | 121 | 12 | 5,046 |
| | Kansas | 0 | 0 | 39 | 220 | 612 | 905 | 1,032 | 818 | 682 | 294 | 109 | 0 | 4,711 |
| | St. Joseph | 0 | 6 | 60 | 285 | 708 | 1,039 | 1,172 | 949 | 769 | 348 | 133 | 15 | 5,484 |
| | St. Louis | 0 | 0 | 60 | 251 | 627 | 936 | 1,026 | 848 | 704 | 312 | 121 | 15 | 4,900 |
| | Springfield | 0 | 0 | 45 | 223 | 600 | 877 | 973 | 781 | 660 | 291 | 105 | 6 | 4,561 |
| MT | Billings | 6 | 15 | 186 | 487 | 897 | 1,135 | 1,296 | 1,100 | 970 | 570 | 285 | 102 | 7,049 |
| | Glasgow | 31 | 47 | 270 | 608 | 1,104 | 1,466 | 1,711 | 1,439 | 1,187 | 648 | 335 | 150 | 8,996 |
| | Great Falls | 28 | 53 | 258 | 543 | 921 | 1,169 | 1,349 | 1,154 | 1,063 | 642 | 384 | 186 | 7,750 |
| | Havre | 28 | 53 | 306 | 595 | 1,065 | 1,367 | 1,584 | 1,364 | 1,181 | 657 | 338 | 162 | 8,700 |
| | Helena | 31 | 59 | 294 | 601 | 1,002 | 1,265 | 1,438 | 1,170 | 1,042 | 651 | 381 | 195 | 8,129 |
| | Kalispell | 50 | 99 | 321 | 654 | 1,020 | 1,240 | 1,401 | 1,134 | 1,029 | 639 | 397 | 207 | 8,191 |
| | Miles City | 6 | 6 | 174 | 502 | 972 | 1,296 | 1,504 | 1,252 | 1,057 | 579 | 276 | 99 | 7,723 |
| | Missoula | 34 | 74 | 303 | 651 | 1,035 | 1,287 | 1,420 | 1,120 | 970 | 621 | 391 | 219 | 8,125 |
| NE | Grand Island | 0 | 6 | 108 | 381 | 834 | 1,172 | 1,314 | 1,089 | 908 | 462 | 211 | 45 | 6,530 |
| | Lincoln | 0 | 6 | 75 | 301 | 726 | 1,066 | 1,237 | 1,016 | 834 | 402 | 171 | 30 | 5,864 |
| | Norfolk | 9 | 0 | 111 | 397 | 873 | 1,234 | 1,414 | 1,179 | 983 | 498 | 233 | 48 | 6,979 |
| | North Platte | 0 | 6 | 123 | 440 | 885 | 1,166 | 1,271 | 1,039 | 930 | 519 | 248 | 57 | 6,684 |
| | Omaha | 0 | 12 | 105 | 357 | 828 | 1,175 | 1,355 | 1,126 | 939 | 465 | 208 | 42 | 6,612 |
| | Scottsbluff | 0 | 0 | 138 | 459 | 876 | 1,128 | 1,231 | 1,008 | 921 | 552 | 285 | 75 | 6,673 |
| | Valentine | 9 | 12 | 165 | 493 | 942 | 1,237 | 1,395 | 1,176 | 1,045 | 579 | 288 | 84 | 7,425 |
| NV | Elko | 9 | 34 | 225 | 561 | 924 | 1,197 | 1,314 | 1,036 | 911 | 621 | 409 | 192 | 7,433 |
| | Ely | 28 | 43 | 234 | 592 | 939 | 1,184 | 1,308 | 1,075 | 977 | 672 | 456 | 225 | 7,733 |
| | Las Vegas | 0 | 0 | 0 | 78 | 387 | 617 | 688 | 487 | 335 | 111 | 6 | 0 | 2,709 |
| | Reno | 43 | 87 | 204 | 490 | 801 | 1,026 | 1,073 | 823 | 729 | 510 | 357 | 189 | 6,332 |
| | Winnemucca | 0 | 34 | 210 | 536 | 876 | 1,091 | 1,172 | 916 | 837 | 573 | 363 | 153 | 6,761 |
| NH | Concord | 6 | 50 | 177 | 505 | 822 | 1,240 | 1,358 | 1,184 | 1,032 | 636 | 298 | 75 | 7,383 |
| | Mt. Wash. Osby. | 493 | 536 | 720 | 1,057 | 1,341 | 1,742 | 1,820 | 1,663 | 1,652 | 1,260 | 930 | 603 | 13,817 |
| NJ | Atlantic City | 0 | 0 | 39 | 251 | 549 | 880 | 936 | 848 | 741 | 420 | 133 | 15 | 4,812 |
| | Newark | 0 | 0 | 30 | 248 | 573 | 921 | 983 | 876 | 729 | 381 | 118 | 0 | 4,859 |
| | Trenton | 0 | 0 | 57 | 264 | 576 | 924 | 989 | 885 | 753 | 399 | 121 | 12 | 4,980 |
| NM | Albuquerque | 0 | 0 | 12 | 229 | 642 | 868 | 930 | 703 | 595 | 288 | 81 | 0 | 4,348 |
| | Clayton | 0 | 6 | 66 | 310 | 699 | 899 | 986 | 812 | 747 | 429 | 183 | 21 | 5,158 |
| | Raton | 9 | 28 | 126 | 431 | 825 | 1,048 | 1,116 | 904 | 834 | 543 | 301 | 63 | 6,228 |
| | Roswell | 0 | 0 | 18 | 202 | 573 | 806 | 840 | 641 | 481 | 201 | 31 | 0 | 3,793 |
| | Silver City | 0 | 0 | 6 | 183 | 525 | 729 | 791 | 605 | 518 | 261 | 87 | 0 | 3,705 |
| NY | Albany | 0 | 19 | 138 | 440 | 777 | 1,194 | 1,311 | 1,156 | 992 | 564 | 239 | 45 | 6,875 |
| | Binghamton (AP) | 22 | 65 | 201 | 471 | 810 | 1,184 | 1,277 | 1,154 | 1,045 | 645 | 313 | 99 | 7,286 |
| | Binghamton (PO) | 0 | 28 | 141 | 406 | 732 | 1,107 | 1,190 | 1,081 | 949 | 543 | 229 | 45 | 6,451 |
| | Buffalo | 19 | 37 | 141 | 440 | 777 | 1,156 | 1,256 | 1,145 | 1,039 | 645 | 329 | 78 | 7,062 |
| | Central Park | 0 | 0 | 30 | 233 | 540 | 902 | 986 | 885 | 760 | 408 | 118 | 9 | 4,871 |
| | J. F. Kennedy Intl. | 0 | 0 | 36 | 248 | 564 | 933 | 1,029 | 935 | 815 | 480 | 167 | 12 | 5,219 |
| | Laguardia | 0 | 0 | 27 | 223 | 528 | 887 | 973 | 879 | 750 | 414 | 124 | 6 | 4,811 |
| | Rochester | 9 | 31 | 126 | 415 | 747 | 1,125 | 1,234 | 1,123 | 1,014 | 597 | 279 | 48 | 6,748 |
| | Schenectady | 0 | 22 | 123 | 422 | 756 | 1,159 | 1,283 | 1,131 | 970 | 543 | 211 | 30 | 6,650 |
| | Syracuse | 6 | 28 | 132 | 415 | 744 | 1,153 | 1,271 | 1,140 | 1,004 | 570 | 248 | 45 | 6,756 |
| NC | Asheville | 0 | 0 | 48 | 245 | 555 | 775 | 784 | 683 | 592 | 273 | 87 | 0 | 4,042 |
| | Cape Hatteras | 0 | 0 | 0 | 78 | 273 | 521 | 580 | 518 | 440 | 177 | 25 | 0 | 2,612 |
| | Charlotte | 0 | 0 | 6 | 124 | 438 | 691 | 691 | 582 | 481 | 156 | 22 | 0 | 3,191 |
| | Greensboro | 0 | 0 | 33 | 192 | 513 | 778 | 784 | 672 | 552 | 234 | 47 | 0 | 3,805 |
| | Raleigh | 0 | 0 | 21 | 164 | 450 | 716 | 725 | 616 | 487 | 180 | 34 | 0 | 3,393 |
| | Wilmington | 0 | 0 | 0 | 74 | 291 | 521 | 546 | 462 | 357 | 96 | 0 | 0 | 2,347 |
| | Winston Salem | 0 | 0 | 21 | 171 | 483 | 747 | 753 | 652 | 524 | 207 | 37 | 0 | 3,595 |
| ND | Bismarck | 34 | 28 | 222 | 577 | 1,083 | 1,463 | 1,708 | 1,442 | 1,203 | 645 | 329 | 117 | 8,851 |
| | Devils Lake | 40 | 53 | 273 | 642 | 1,191 | 1,634 | 1,872 | 1,579 | 1,345 | 753 | 381 | 138 | 9,901 |
| | Fargo | 28 | 37 | 219 | 574 | 1,107 | 1,569 | 1,789 | 1,520 | 1,262 | 690 | 332 | 99 | 9,226 |
| | Williston | 31 | 43 | 261 | 601 | 1,122 | 1,513 | 1,758 | 1,473 | 1,262 | 681 | 357 | 141 | 9,243 |
| OH | Akron | 0 | 9 | 96 | 381 | 726 | 1,070 | 1,138 | 1,016 | 871 | 489 | 202 | 39 | 6,037 |
| | Cincinnati | 0 | 0 | 54 | 248 | 612 | 921 | 970 | 837 | 701 | 336 | 118 | 9 | 4,806 |
| | Cleveland | 9 | 25 | 105 | 384 | 738 | 1,088 | 1,159 | 1,047 | 918 | 552 | 260 | 66 | 6,351 |
| | Columbus | 0 | 6 | 84 | 347 | 714 | 1,039 | 1,088 | 949 | 809 | 426 | 171 | 27 | 5,660 |
| | Dayton | 0 | 6 | 78 | 310 | 696 | 1,045 | 1,097 | 955 | 809 | 429 | 167 | 30 | 5,622 |
| | Mansfield | 9 | 22 | 114 | 397 | 768 | 1,110 | 1,169 | 1,042 | 924 | 543 | 245 | 60 | 6,403 |
| | Sandusky | 0 | 6 | 66 | 313 | 684 | 1,032 | 1,107 | 991 | 868 | 495 | 198 | 36 | 5,796 |
| | Toledo | 0 | 16 | 117 | 406 | 792 | 1,138 | 1,200 | 1,056 | 924 | 543 | 242 | 60 | 6,494 |
| | Youngstown | 6 | 19 | 120 | 412 | 771 | 1,104 | 1,169 | 1,047 | 921 | 540 | 248 | 60 | 6,417 |
| OK | Oklahoma City | 0 | 0 | 15 | 164 | 498 | 766 | 868 | 664 | 527 | 189 | 34 | 0 | 3,725 |
| | Tulsa | 0 | 0 | 18 | 158 | 522 | 787 | 893 | 683 | 539 | 213 | 47 | 0 | 3,860 |

| | Location | Jul | Aug | Sep | Oct | Nov | Dec | Jan | Feb | Mar | Apr | May | Jun | Ann |
|---|---|---|---|---|---|---|---|---|---|---|---|---|---|---|
| OR | Astoria | 146 | 130 | 210 | 375 | 561 | 679 | 753 | 622 | 636 | 480 | 363 | 231 | 5,186 |
| | Burns | 12 | 37 | 210 | 515 | 867 | 1,113 | 1,246 | 988 | 856 | 570 | 366 | 177 | 6,957 |
| | Eugene | 34 | 34 | 129 | 366 | 585 | 719 | 803 | 627 | 589 | 426 | 279 | 135 | 4,726 |
| | Meacham | 84 | 124 | 288 | 580 | 918 | 1,091 | 1,209 | 1,005 | 983 | 726 | 527 | 339 | 7,874 |
| | Medford | 0 | 0 | 78 | 372 | 678 | 871 | 918 | 697 | 642 | 432 | 242 | 78 | 5,008 |
| | Pendleton | 0 | 0 | 111 | 350 | 711 | 884 | 1,017 | 773 | 617 | 396 | 205 | 63 | 5,127 |
| | Portland | 25 | 28 | 114 | 335 | 597 | 735 | 825 | 644 | 586 | 396 | 245 | 105 | 4,635 |
| | Roseburg | 22 | 16 | 105 | 329 | 567 | 713 | 766 | 608 | 570 | 405 | 267 | 123 | 4,491 |
| | Salem | 37 | 31 | 111 | 338 | 594 | 729 | 822 | 647 | 611 | 417 | 273 | 144 | 4,754 |
| | Sexton Summit | 81 | 81 | 171 | 443 | 666 | 874 | 958 | 809 | 818 | 609 | 465 | 279 | 6,254 |
| PA | Allentown | 0 | 0 | 90 | 353 | 693 | 1,045 | 1,116 | 1,002 | 849 | 471 | 167 | 24 | 5,810 |
| | Erie | 0 | 25 | 102 | 391 | 714 | 1,063 | 1,169 | 1,081 | 973 | 585 | 288 | 60 | 6,451 |
| | Harrisburg | 0 | 0 | 63 | 298 | 648 | 992 | 1,045 | 907 | 766 | 396 | 124 | 12 | 5,251 |
| | Philadelphia | 0 | 0 | 60 | 291 | 621 | 964 | 1,014 | 890 | 744 | 390 | 115 | 12 | 5,101 |
| | Pittsburgh | 0 | 9 | 105 | 375 | 726 | 1,063 | 1,119 | 1,002 | 874 | 480 | 195 | 39 | 5,987 |
| | Reading | 0 | 0 | 54 | 257 | 597 | 939 | 1,001 | 885 | 735 | 372 | 105 | 0 | 4,945 |
| | Scranton | 0 | 19 | 132 | 434 | 762 | 1,104 | 1,156 | 1,028 | 893 | 498 | 195 | 33 | 6,254 |
| | Williamsport | 0 | 9 | 111 | 375 | 717 | 1,073 | 1,122 | 1,002 | 856 | 468 | 177 | 24 | 5,934 |
| RI | Block Is. | 0 | 16 | 78 | 307 | 594 | 902 | 1,020 | 955 | 877 | 612 | 344 | 99 | 5,804 |
| | Providence | 0 | 16 | 96 | 372 | 660 | 1,023 | 1,110 | 988 | 868 | 534 | 236 | 51 | 5,954 |
| SC | Charleston | 0 | 0 | 0 | 59 | 282 | 471 | 487 | 389 | 291 | 54 | 0 | 0 | 2,033 |
| | Columbia | 0 | 0 | 0 | 84 | 345 | 577 | 570 | 470 | 357 | 81 | 0 | 0 | 2,484 |
| SC | Florence | 0 | 0 | 0 | 78 | 315 | 552 | 552 | 459 | 347 | 84 | 0 | 0 | 2,387 |
| | Greenville | 0 | 0 | 0 | 112 | 387 | 636 | 648 | 535 | 434 | 120 | 12 | 0 | 2,884 |
| | Spartanburg | 0 | 0 | 15 | 130 | 417 | 667 | 663 | 560 | 453 | 144 | 25 | 0 | 3,074 |
| SD | Huron | 9 | 12 | 165 | 508 | 1,014 | 1,432 | 1,628 | 1,355 | 1,125 | 600 | 288 | 87 | 8,223 |
| | Rapid City | 22 | 12 | 165 | 481 | 897 | 1,172 | 1,333 | 1,145 | 1,051 | 615 | 326 | 126 | 7,345 |
| | Sioux Falls | 19 | 25 | 168 | 462 | 972 | 1,361 | 1,544 | 1,285 | 1,082 | 573 | 270 | 78 | 7,839 |
| TN | Bristol | 0 | 0 | 51 | 236 | 573 | 828 | 828 | 700 | 598 | 261 | 68 | 0 | 4,143 |
| | Chattanooga | 0 | 0 | 18 | 143 | 468 | 698 | 722 | 577 | 453 | 150 | 25 | 0 | 3,254 |
| | Knoxville | 0 | 0 | 30 | 171 | 489 | 725 | 732 | 613 | 493 | 198 | 43 | 0 | 3,494 |
| | Memphis | 0 | 0 | 18 | 130 | 447 | 698 | 729 | 585 | 456 | 147 | 22 | 0 | 3,232 |
| | Nashville | 0 | 0 | 30 | 158 | 495 | 732 | 778 | 644 | 512 | 189 | 40 | 0 | 3,578 |
| | Oak Ridge (CO) | 0 | 0 | 39 | 192 | 531 | 772 | 778 | 669 | 552 | 228 | 56 | 0 | 3,817 |
| TX | Abilene | 0 | 0 | 0 | 99 | 366 | 586 | 642 | 470 | 347 | 114 | 0 | 0 | 2,624 |
| | Amarillo | 0 | 0 | 18 | 205 | 570 | 797 | 877 | 664 | 546 | 252 | 56 | 0 | 3,985 |
| | Austin | 0 | 0 | 0 | 31 | 225 | 388 | 468 | 325 | 223 | 51 | 0 | 0 | 1,711 |
| | Brownsville | 0 | 0 | 0 | 0 | 66 | 149 | 205 | 106 | 74 | 0 | 0 | 0 | 600 |
| | Corpus Christi | 0 | 0 | 0 | 0 | 120 | 220 | 291 | 174 | 109 | 0 | 0 | 0 | 914 |
| | Dallas | 0 | 0 | 0 | 62 | 321 | 524 | 601 | 440 | 319 | 90 | 6 | 0 | 2,363 |
| | El Paso | 0 | 0 | 0 | 84 | 414 | 648 | 685 | 445 | 319 | 105 | 0 | 0 | 2,700 |
| | Ft. Worth | 0 | 0 | 0 | 65 | 324 | 536 | 614 | 448 | 319 | 99 | 0 | 0 | 2,405 |
| | Galveston | 0 | 0 | 0 | 0 | 138 | 270 | 350 | 258 | 189 | 30 | 0 | 0 | 1,235 |
| | Houston | 0 | 0 | 0 | 6 | 183 | 307 | 384 | 288 | 192 | 36 | 0 | 0 | 1,396 |
| | Laredo | 0 | 0 | 0 | 0 | 105 | 217 | 267 | 134 | 74 | 0 | 0 | 0 | 797 |
| | Lubbock | 0 | 0 | 18 | 174 | 513 | 744 | 800 | 613 | 484 | 201 | 31 | 0 | 3,578 |
| | Midland | 0 | 0 | 0 | 87 | 381 | 592 | 651 | 468 | 322 | 90 | 0 | 0 | 2,591 |
| | Port Arthur | 0 | 0 | 0 | 22 | 207 | 329 | 384 | 274 | 192 | 39 | 0 | 0 | 1,447 |
| | San Angelo | 0 | 0 | 0 | 68 | 318 | 536 | 567 | 412 | 288 | 66 | 0 | 0 | 2,255 |
| | San Antonio | 0 | 0 | 0 | 31 | 207 | 363 | 428 | 286 | 195 | 39 | 0 | 0 | 1,549 |
| | Victoria | 0 | 0 | 0 | 6 | 150 | 270 | 344 | 230 | 152 | 21 | 0 | 0 | 1,173 |
| TX | Waco | 0 | 0 | 0 | 43 | 270 | 456 | 536 | 389 | 270 | 66 | 0 | 0 | 2,030 |
| | Wichita Falls | 0 | 0 | 0 | 99 | 381 | 632 | 698 | 518 | 378 | 120 | 6 | 0 | 2,832 |
| UT | Milford | 0 | 0 | 99 | 443 | 867 | 1,141 | 1,252 | 988 | 822 | 519 | 279 | 87 | 6,497 |
| | Salt Lake City | 0 | 0 | 81 | 419 | 849 | 1,082 | 1,172 | 910 | 763 | 459 | 233 | 84 | 6,052 |
| | Wendover | 0 | 0 | 48 | 372 | 822 | 1,091 | 1,178 | 902 | 729 | 408 | 177 | 51 | 5,778 |
| VT | Burlington | 28 | 65 | 207 | 539 | 891 | 1,349 | 1,513 | 1,333 | 1,187 | 714 | 353 | 90 | 8,269 |
| VA | Cape Henry | 0 | 0 | 0 | 112 | 360 | 645 | 694 | 633 | 536 | 246 | 53 | 0 | 3,279 |
| | Lynchburg | 0 | 0 | 51 | 223 | 540 | 822 | 849 | 731 | 605 | 267 | 78 | 0 | 4,166 |
| | Norfolk | 0 | 0 | 0 | 136 | 408 | 698 | 738 | 655 | 533 | 216 | 37 | 0 | 3,421 |
| | Richmond | 0 | 0 | 36 | 214 | 495 | 784 | 815 | 703 | 546 | 219 | 53 | 0 | 3,865 |
| | Roanoke | 0 | 0 | 51 | 229 | 549 | 825 | 834 | 722 | 614 | 261 | 65 | 0 | 4,150 |
| | Wash. Nat'l. Ap. | 0 | 0 | 33 | 217 | 519 | 834 | 871 | 762 | 626 | 288 | 74 | 0 | 4,224 |
| WA | Olympia | 68 | 71 | 198 | 422 | 636 | 753 | 834 | 675 | 645 | 450 | 307 | 177 | 5,236 |
| | Seattle | 50 | 47 | 129 | 329 | 543 | 657 | 738 | 599 | 577 | 396 | 242 | 117 | 4,424 |
| | Seattle Boeing | 34 | 40 | 147 | 384 | 624 | 763 | 831 | 655 | 608 | 411 | 242 | 99 | 4,838 |
| | Seattle Tacoma | 56 | 62 | 162 | 391 | 633 | 750 | 828 | 678 | 657 | 474 | 295 | 159 | 5,145 |

**TABLE A6.3** Normal Total Heating Degree-Days, Base 65°F[a, b] (*Continued*)

| | Location | Jul | Aug | Sep | Oct | Nov | Dec | Jan | Feb | Mar | Apr | May | Jun | Ann |
|---|---|---|---|---|---|---|---|---|---|---|---|---|---|---|
| WA | Spokane | 9 | 25 | 168 | 493 | 879 | 1,082 | 1,231 | 980 | 834 | 531 | 288 | 135 | 6,655 |
| | Stampede Pass | 273 | 291 | 393 | 701 | 1,008 | 1,178 | 1,287 | 1,075 | 1,085 | 855 | 654 | 483 | 9,283 |
| | Tatoosh Is. | 295 | 279 | 306 | 406 | 534 | 639 | 713 | 613 | 645 | 525 | 431 | 333 | 5,719 |
| | Walla Walla | 0 | 0 | 87 | 310 | 681 | 843 | 986 | 745 | 589 | 342 | 177 | 45 | 4,805 |
| | Yakima | 0 | 12 | 144 | 450 | 828 | 1,039 | 1,163 | 868 | 713 | 435 | 220 | 69 | 5,941 |
| WV | Charleston | 0 | 0 | 63 | 254 | 591 | 865 | 880 | 770 | 648 | 300 | 96 | 9 | 4,476 |
| | Elkins | 9 | 25 | 135 | 400 | 729 | 992 | 1,008 | 896 | 791 | 444 | 198 | 48 | 5,675 |
| | Huntington | 0 | 0 | 63 | 257 | 585 | 856 | 880 | 764 | 636 | 294 | 99 | 12 | 4,446 |
| | Parkersburg | 0 | 0 | 60 | 264 | 606 | 905 | 942 | 826 | 691 | 339 | 115 | 6 | 4,754 |
| WI | Green Bay | 28 | 50 | 174 | 484 | 924 | 1,333 | 1,494 | 1,313 | 1,141 | 654 | 335 | 99 | 8,029 |
| | La Crosse | 12 | 19 | 153 | 437 | 924 | 1,339 | 1,504 | 1,277 | 1,070 | 540 | 245 | 69 | 7,589 |
| | Madison | 25 | 40 | 174 | 474 | 930 | 1,330 | 1,473 | 1,274 | 1,113 | 618 | 310 | 102 | 7,863 |
| | Milwaukee | 43 | 47 | 174 | 471 | 876 | 1,252 | 1,376 | 1,193 | 1,054 | 642 | 372 | 135 | 7,635 |
| WY | Casper | 6 | 16 | 192 | 524 | 942 | 1,169 | 1,290 | 1,084 | 1,020 | 657 | 381 | 129 | 7,410 |
| | Cheyenne | 19 | 31 | 210 | 543 | 924 | 1,101 | 1,228 | 1,056 | 1,011 | 672 | 381 | 102 | 7,278 |
| | Lander | 6 | 19 | 204 | 555 | 1,020 | 1,299 | 1,417 | 1,145 | 1,017 | 654 | 381 | 153 | 7,870 |
| | Sheridan | 25 | 31 | 219 | 539 | 948 | 1,200 | 1,355 | 1,154 | 1,054 | 642 | 366 | 150 | 7,683 |

[a]"Climatic Atlas of the United States," U.S. Government Printing Office, 1968.

[b]One of the most practical of weather statistics is the "heating degree-day." First devised some 60 yr ago, the degree-day system has been in quite general use by the heating industry for more than 40 yr.

Heating degree-days are the number of degrees the daily average temperature is below 65°F. Normally heating is not required in a building when the outdoor average daily temperature is 65°F. Heating degree-days are determined by subtracting the average daily temperatures below 65°F from the base 65°F. A day with an average temperature of 50°F has 15 heating degree-days $(65 - 50 = 15)$ while one with an average temperature of 65°F or higher has none.

Several characteristics make the degree-day figures especially useful. They are cumulative so that the degree-day sum for a period of days represents the total heating load for that period. The relationship between degree-days and fuel consumption is linear, i.e., doubling the degree-days usually doubles the fuel consumption. Comparing normal seasonal degree-days in different locations gives a rough estimate of seasonal fuel consumption. For example, it would require roughly $4\frac{1}{2}$ times as much fuel to heat a building in Chicago, Ill., where the mean annual total heating degree-days are about 6200, than to heat a similar building in New Orleans, La., where the annual total heating degree-days are around 1400. Using degree-days has the advantage that the consumption ratios are fairly constant, i.e., the fuel consumed per 100 degree-days is about the same whether the 100 degree-days occur in only 3 or 4 days or are spread over 7 or 8 days.

**TABLE A6.4** Normal Monthly Average Temperature: Selected Cities[a]

| | Location | Jan | Feb | Mar | Apr | May | Jun | Jul | Aug | Sep | Oct | Nov | Dec | Ann |
|---|---|---|---|---|---|---|---|---|---|---|---|---|---|---|
| AL | Mobile | 53.0 | 55.2 | 60.3 | 67.6 | 75.6 | 81.5 | 82.6 | 82.1 | 77.9 | 69.9 | 58.9 | 54.1 | 68.2 |
| AK | Juneau | 25.1 | 26.8 | 30.4 | 38.0 | 45.6 | 52.3 | 55.3 | 54.1 | 48.9 | 41.6 | 34.3 | 28.4 | 40.1 |
| AZ | Phoenix | 49.7 | 53.5 | 59.0 | 67.2 | 75.0 | 83.6 | 89.8 | 87.5 | 82.8 | 70.7 | 58.1 | 51.6 | 69.0 |
| AR | Little Rock | 40.6 | 44.4 | 51.8 | 62.4 | 70.5 | 78.9 | 81.9 | 81.3 | 74.3 | 63.1 | 49.5 | 41.9 | 61.7 |
| CA | Los Angeles | 54.4 | 55.2 | 57.0 | 59.4 | 62.0 | 64.8 | 69.1 | 69.1 | 68.5 | 64.9 | 61.1 | 56.9 | 61.9 |
| | Sacramento | 45.2 | 49.2 | 53.4 | 58.4 | 64.0 | 70.5 | 75.4 | 74.1 | 71.6 | 63.5 | 52.9 | 46.4 | 60.4 |
| | San Francisco [b] | 50.7 | 53.0 | 54.7 | 55.7 | 57.4 | 59.1 | 58.8 | 59.4 | 62.0 | 61.4 | 57.4 | 52.5 | 56.8 |
| CO | Denver | 28.5 | 31.5 | 36.4 | 46.4 | 56.2 | 66.5 | 72.9 | 71.5 | 63.0 | 51.4 | 37.7 | 31.6 | 49.5 |
| CT | Hartford | 26.0 | 27.1 | 36.0 | 48.5 | 59.9 | 68.7 | 73.4 | 71.2 | 63.3 | 53.0 | 41.3 | 28.9 | 49.8 |
| DE | Wilmington | 33.4 | 33.8 | 41.3 | 52.1 | 62.7 | 71.4 | 76.0 | 74.3 | 67.6 | 56.6 | 45.4 | 35.1 | 54.1 |
| DC | Washington | 36.9 | 37.8 | 44.8 | 55.7 | 65.8 | 74.2 | 78.2 | 76.5 | 69.7 | 59.0 | 47.7 | 38.1 | 57.0 |
| FL | Jacksonville | 55.9 | 57.5 | 62.2 | 68.7 | 75.8 | 80.8 | 82.6 | 82.3 | 79.4 | 71.0 | 61.7 | 56.1 | 69.5 |
| | Miami | 66.9 | 67.9 | 70.5 | 74.2 | 77.6 | 80.8 | 81.8 | 82.3 | 81.3 | 77.8 | 72.4 | 68.1 | 75.1 |
| GA | Atlanta | 44.7 | 46.1 | 51.4 | 60.2 | 69.1 | 76.6 | 78.9 | 78.2 | 73.1 | 62.4 | 51.2 | 44.8 | 61.4 |
| HI | Honolulu | 72.5 | 72.4 | 72.8 | 74.2 | 75.9 | 77.9 | 78.8 | 79.4 | 79.2 | 78.2 | 75.9 | 73.6 | 75.9 |
| ID | Boise | 29.1 | 34.5 | 41.7 | 50.4 | 58.2 | 65.8 | 75.2 | 72.1 | 62.7 | 51.6 | 38.6 | 32.2 | 51.0 |
| IL | Chicago | 26.0 | 27.7 | 36.3 | 49.0 | 60.0 | 70.5 | 75.6 | 74.2 | 66.1 | 55.1 | 39.9 | 29.1 | 50.8 |
| | Peoria | 25.7 | 28.4 | 37.6 | 50.8 | 61.5 | 71.7 | 76.0 | 74.3 | 66.4 | 55.3 | 39.7 | 29.1 | 51.4 |
| IN | Indianapolis | 29.1 | 31.1 | 38.9 | 50.8 | 61.4 | 71.1 | 75.2 | 73.7 | 66.5 | 55.4 | 40.9 | 31.1 | 52.1 |
| IA | Des Moines | 19.9 | 23.4 | 33.8 | 48.7 | 60.6 | 71.0 | 76.3 | 74.1 | 65.4 | 54.2 | 37.1 | 25.3 | 49.2 |
| KS | Wichita | 32.0 | 36.3 | 44.5 | 56.7 | 66.0 | 76.5 | 80.9 | 80.8 | 71.3 | 59.9 | 44.4 | 35.8 | 57.1 |
| KY | Louisville | 35.0 | 35.8 | 43.3 | 54.8 | 64.4 | 73.4 | 77.6 | 76.2 | 69.5 | 57.9 | 44.7 | 36.3 | 55.7 |
| LA | New Orleans | 54.6 | 57.1 | 61.4 | 67.9 | 74.4 | 80.1 | 81.6 | 81.9 | 78.3 | 70.4 | 60.0 | 55.4 | 68.6 |
| ME | Portland | 21.8 | 22.8 | 31.4 | 42.5 | 53.0 | 62.1 | 68.1 | 66.8 | 58.7 | 48.6 | 38.1 | 25.8 | 45.0 |
| MD | Baltimore | 34.8 | 35.7 | 43.1 | 54.2 | 64.4 | 72.5 | 76.8 | 75.0 | 68.1 | 57.0 | 45.5 | 35.8 | 55.2 |
| MA | Boston | 29.9 | 30.3 | 37.7 | 47.9 | 58.8 | 67.8 | 73.7 | 71.7 | 65.3 | 55.0 | 44.9 | 33.3 | 51.4 |
| MI | Detroit | 26.9 | 27.2 | 34.8 | 47.6 | 59.0 | 69.7 | 74.4 | 72.8 | 65.1 | 53.8 | 40.4 | 29.9 | 50.1 |
| | Sault Ste. Marie | 15.8 | 15.7 | 23.8 | 38.0 | 49.6 | 59.0 | 64.6 | 64.0 | 55.8 | 46.3 | 33.3 | 20.9 | 40.6 |
| MN | Duluth | 8.7 | 10.8 | 21.3 | 37.0 | 49.2 | 58.8 | 65.5 | 63.8 | 54.2 | 44.6 | 27.3 | 14.0 | 37.9 |
| | Minneapolis-St. Paul | 12.4 | 15.7 | 27.4 | 44.3 | 57.3 | 66.8 | 72.3 | 70.0 | 60.4 | 48.9 | 31.2 | 18.1 | 43.7 |
| MS | Jackson | 47.9 | 50.5 | 56.5 | 64.9 | 73.1 | 79.8 | 82.3 | 82.0 | 76.5 | 67.0 | 55.5 | 49.4 | 65.5 |
| MO | Kansas City | 31.7 | 35.8 | 43.3 | 55.7 | 65.6 | 75.9 | 81.5 | 79.8 | 71.3 | 60.2 | 44.6 | 35.8 | 56.8 |
| | St. Louis | 31.9 | 34.7 | 42.6 | 54.9 | 64.2 | 74.1 | 78.1 | 76.8 | 69.5 | 58.4 | 44.1 | 34.8 | 55.3 |
| MT | Great Falls | 22.1 | 23.8 | 30.7 | 43.6 | 53.0 | 59.9 | 69.4 | 66.8 | 57.4 | 47.5 | 34.3 | 27.3 | 44.7 |
| NE | Omaha | 22.3 | 26.5 | 36.9 | 51.7 | 63.0 | 73.1 | 78.5 | 76.2 | 66.9 | 55.7 | 38.9 | 28.2 | 51.5 |
| NV | Reno | 30.4 | 35.6 | 41.5 | 48.0 | 53.9 | 60.1 | 67.7 | 65.5 | 58.8 | 49.2 | 38.3 | 31.9 | 48.4 |
| NH | Concord | 21.2 | 22.7 | 31.7 | 43.8 | 55.5 | 64.5 | 69.6 | 67.4 | 59.3 | 48.7 | 37.6 | 25.0 | 45.6 |
| NJ | Atlantic City | 34.8 | 34.7 | 41.1 | 51.0 | 61.3 | 70.0 | 75.1 | 73.7 | 67.2 | 57.2 | 46.7 | 36.6 | 54.1 |
| NM | Albuquerque | 35.0 | 39.9 | 45.8 | 55.7 | 65.1 | 74.9 | 78.5 | 76.2 | 70.0 | 58.0 | 43.6 | 37.0 | 56.6 |
| NY | Albany | 22.7 | 23.7 | 33.0 | 46.2 | 57.9 | 67.3 | 72.1 | 70.0 | 61.6 | 50.8 | 39.1 | 26.5 | 47.6 |
| | Buffalo | 24.5 | 24.1 | 31.5 | 43.5 | 54.8 | 64.8 | 69.8 | 68.4 | 61.4 | 50.8 | 39.1 | 27.7 | 46.7 |
| | New York [b] | 32.2 | 33.4 | 40.5 | 51.4 | 62.4 | 71.4 | 76.8 | 75.1 | 68.5 | 58.3 | 47.0 | 35.9 | 54.5 |
| NC | Charlotte | 42.7 | 44.2 | 50.0 | 60.3 | 69.0 | 77.1 | 79.2 | 78.7 | 72.9 | 62.5 | 50.4 | 42.7 | 60.8 |
| | Raleigh | 41.6 | 43.0 | 49.5 | 59.3 | 67.6 | 75.1 | 77.9 | 76.9 | 71.2 | 60.5 | 50.0 | 41.9 | 59.5 |
| ND | Bismarck | 9.9 | 13.5 | 26.2 | 43.5 | 55.9 | 64.5 | 71.7 | 69.3 | 58.7 | 46.7 | 28.9 | 17.8 | 42.2 |
| OH | Cincinnati [b] | 33.7 | 35.1 | 42.7 | 54.2 | 64.2 | 73.4 | 76.9 | 75.7 | 69.0 | 57.9 | 44.6 | 35.3 | 55.2 |
| | Cleveland | 28.4 | 28.5 | 35.1 | 47.0 | 58.0 | 67.8 | 71.9 | 70.4 | 64.2 | 53.4 | 41.3 | 30.5 | 49.7 |
| | Columbus | 29.9 | 31.1 | 38.9 | 50.8 | 61.5 | 70.8 | 74.8 | 73.2 | 65.9 | 54.2 | 41.2 | 31.5 | 52.0 |
| OK | Oklahoma City | 37.0 | 41.3 | 48.5 | 59.9 | 68.4 | 78.0 | 82.5 | 82.8 | 73.8 | 62.9 | 48.4 | 40.3 | 60.3 |
| OR | Portland | 38.4 | 42.0 | 46.1 | 51.8 | 57.4 | 62.0 | 67.2 | 66.6 | 62.2 | 54.2 | 45.1 | 41.3 | 52.9 |
| PA | Philadelphia | 32.3 | 33.2 | 41.0 | 52.0 | 62.6 | 71.0 | 75.6 | 73.6 | 66.7 | 55.7 | 44.3 | 33.9 | 53.5 |
| | Pittsburgh | 28.9 | 29.2 | 36.8 | 49.0 | 59.8 | 68.4 | 72.1 | 70.8 | 64.2 | 53.1 | 40.8 | 30.7 | 50.3 |
| RI | Providence | 29.2 | 29.7 | 37.0 | 47.2 | 57.5 | 66.2 | 72.1 | 70.5 | 63.2 | 53.2 | 43.0 | 32.0 | 50.1 |
| SC | Columbia | 46.9 | 48.4 | 54.4 | 63.6 | 72.2 | 79.7 | 81.6 | 80.5 | 75.3 | 64.7 | 53.7 | 46.4 | 64.0 |
| SD | Sioux Falls | 15.2 | 19.1 | 30.1 | 45.9 | 58.3 | 68.1 | 74.3 | 71.8 | 61.8 | 50.3 | 32.6 | 21.1 | 45.7 |
| TN | Memphis | 41.5 | 44.1 | 51.1 | 61.4 | 70.3 | 78.5 | 81.3 | 80.5 | 73.9 | 63.1 | 50.1 | 42.5 | 61.5 |
| | Nashville | 39.9 | 42.0 | 49.1 | 59.6 | 68.6 | 77.4 | 80.2 | 79.2 | 72.8 | 61.5 | 48.5 | 41.4 | 60.0 |
| TX | Dallas | 45.9 | 49.5 | 56.1 | 65.0 | 72.9 | 81.3 | 84.9 | 85.0 | 77.9 | 67.8 | 54.9 | 48.1 | 65.8 |
| | El Paso | 42.9 | 49.1 | 54.9 | 63.4 | 71.9 | 81.0 | 81.9 | 80.4 | 74.5 | 64.4 | 51.2 | 44.1 | 63.3 |
| | Houston | 53.6 | 55.8 | 61.3 | 68.5 | 76.0 | 81.6 | 83.0 | 83.2 | 79.2 | 71.4 | 60.8 | 55.7 | 69.2 |
| UT | Salt Lake City | 27.2 | 32.5 | 40.4 | 49.9 | 58.9 | 67.4 | 76.9 | 74.5 | 64.4 | 51.7 | 36.7 | 30.1 | 50.9 |
| VT | Burlington | 16.2 | 17.4 | 26.7 | 41.2 | 53.8 | 64.2 | 69.0 | 66.7 | 58.4 | 47.6 | 35.3 | 21.5 | 43.2 |
| VA | Norfolk | 41.2 | 41.6 | 48.0 | 58.0 | 67.5 | 75.6 | 78.8 | 77.5 | 72.6 | 62.0 | 51.4 | 42.5 | 59.7 |
| | Richmond | 38.7 | 39.9 | 47.7 | 58.1 | 67.0 | 75.1 | 78.1 | 76.0 | 70.2 | 58.7 | 48.5 | 39.7 | 58.1 |
| WA | Seattle-Tacoma | 38.3 | 40.8 | 43.8 | 49.2 | 55.5 | 59.8 | 64.9 | 64.1 | 59.9 | 52.4 | 43.9 | 40.8 | 51.1 |
| | Spokane | 25.3 | 30.0 | 38.1 | 47.3 | 56.2 | 61.9 | 70.5 | 68.0 | 60.9 | 49.1 | 35.7 | 30.1 | 47.8 |
| WV | Charleston | 36.6 | 37.5 | 44.4 | 55.3 | 64.8 | 72.0 | 74.9 | 73.8 | 68.2 | 57.3 | 45.3 | 37.1 | 55.6 |
| WI | Milwaukee | 20.6 | 22.4 | 31.0 | 43.6 | 53.4 | 63.3 | 68.7 | 67.8 | 60.3 | 50.0 | 35.8 | 24.6 | 45.1 |
| WY | Cheyenne | 25.4 | 27.3 | 32.4 | 42.6 | 52.9 | 63.0 | 70.0 | 67.7 | 58.6 | 47.5 | 34.2 | 29.5 | 45.9 |
| PR | San Juan | 74.4 | 74.4 | 75.3 | 76.6 | 78.7 | 80.0 | 80.4 | 80.9 | 80.5 | 80.0 | 78.2 | 76.2 | 78.0 |

[a]In Fahrenheit degrees. Airport data unless noted otherwise. Based on standard 30-yr period, 1931–1960. From "Local Climatological Data," U.S. Department of Commerce, NOAA. For more detailed data see "Climatography of the United States," Publ. 84, U.S. Department of Commerce, NOAA, NCC.

[b]City office data.

**TABLE A6.5** Conductance and Resistance Values for External Air Surfaces[a]

| Wind condition Position of surface | Direction of heat flow | Type of surface | | | | | |
|---|---|---|---|---|---|---|---|
| | | Foil | | Aluminum-coated paper | | Nonreflective building materials | |
| | | Conductance C, Btu/(hr)(ft²)(°F) | Resistance R, 1/[Btu/(hr)(ft²)(°F)] | Conductance C, Btu/(hr)(ft²)(°F) | Resistance R, 1/[Btu/(hr)(ft²)(°F)] | Conductance C, Btu/(hr)(ft²)(°F) | Resistance R, 1/[Btu/(hr)(ft²)(°F)] |
| Still air: | | | | | | | |
| Horizontal | Up | 0.76 | 1.32 | 0.91 | 1.10 | 1.63 | 0.61 |
| 45° slope | Up | 0.73 | 1.37 | 0.88 | 1.14 | 1.60 | 0.62 |
| Vertical | Horizontal | 0.59 | 1.70 | 0.74 | 1.35 | 1.46 | 0.68 |
| 45° slope | Down | 0.45 | 2.22 | 0.60 | 1.67 | 1.32 | 0.76 |
| Horizontal | Down | 0.22 | 4.55 | 0.37 | 2.70 | 1.08 | 0.92 |
| 7.5-mph wind Any position | Any direction (for summer calculations) | ... | ... | ... | ... | 4.00 | 0.25 |
| 15-mph wind Any position | Any direction (for winter calculations) | ... | ... | ... | ... | 6.00 | 0.17 |

[a]Adapted from Johns-Mansville, Denver, Colorado.

**TABLE A6.6** Conductance and Resistance Values for Internal Air Spaces[a]

| Position of air space | Direction of heat flow[b] | Thickness, in. | Temp. cond.[c] | Foil and nonreflective building materials | | Aluminum-coated paper and nonreflective bldg. materials | | Both surfaces nonreflective building materials | |
|---|---|---|---|---|---|---|---|---|---|
| | | | | Conductance C, Btu/(hr)(ft²)(°F) | Resistance R, 1/[Btu/(hr)(ft²)(°F)] | Conductance C, Btu/(hr)(ft²)(°F) | Resistance R, 1/[Btu/(hr)(ft²)(°F)] | Conductance C, Btu/(hr)(ft²)(°F) | Resistance R, 1/[Btu/(hr)(ft²)(°F)] |
| Horizontal | Up | 3/4 | W | 0.45 | 2.23 | 0.59 | 1.71 | 1.15 | 0.87 |
| | | 3/4 | S | 0.44 | 2.26 | 0.61 | 1.63 | 1.32 | 0.76 |
| | | 4 | W | 0.37 | 2.73 | 0.50 | 1.99 | 1.07 | 0.94 |
| | | 4 | S | 0.36 | 2.75 | 0.53 | 1.87 | 1.24 | 0.80 |
| 45° slope | Up | 3/4 | W | 0.36 | 2.78 | 0.50 | 2.02 | 1.06 | 0.94 |
| | | 3/4 | S | 0.36 | 2.81 | 0.53 | 1.90 | 1.24 | 0.81 |
| | | 4 | W | 0.33 | 3.00 | 0.47 | 2.13 | 1.04 | 0.96 |
| | | 4 | S | 0.33 | 3.00 | 0.51 | 1.98 | 1.21 | 0.82 |
| Vertical | Horiz. | 3/4 | W | 0.29 | 3.48 | 0.42 | 2.36 | 0.99 | 1.01 |
| | | 3/4 | S | 0.31 | 3.28 | 0.48 | 2.10 | 1.19 | 0.84 |
| | | 4 | W | 0.29 | 3.45 | 0.43 | 2.34 | 0.99 | 1.01 |
| | | 4 | S | 0.29 | 3.44 | 0.46 | 2.16 | 1.17 | 0.91 |
| 45° slope | Down | 3/4 | W | 0.28 | 3.57 | 0.42 | 2.40 | 0.98 | 1.02 |
| | | 3/4 | S | 0.31 | 3.24 | 0.48 | 2.09 | 1.19 | 0.84 |
| | | 4 | W | 0.23 | 4.41 | 0.36 | 2.75 | 0.93 | 1.08 |
| | | 4 | S | 0.23 | 4.36 | 0.40 | 2.50 | 1.11 | 0.90 |
| Horizontal | Down | 3/4 | W | 0.28 | 3.55 | 0.42 | 2.39 | 0.98 | 1.02 |
| | | 1 1/2 | W | 0.17 | 5.74 | 0.31 | 3.21 | 0.88 | 1.14 |
| | | 4 | W | 0.11 | 8.94 | 0.25 | 4.02 | 0.81 | 1.23 |
| | | 3/4 | S | 0.31 | 3.25 | 0.48 | 2.08 | 1.19 | 0.84 |
| | | 1 1/2 | S | 0.19 | 5.24 | 0.36 | 2.76 | 1.07 | 0.93 |
| | | 4 | S | 0.12 | 8.08 | 0.30 | 3.38 | 1.01 | 0.99 |

[a]Adapted from Johns-Mansville, Denver, Colorado.
[b]Heat flows from hot to cold. For ceiling installation the direction of heat flow would normally be "up" for winter and "down" for summer. In a floor the direction of heat flow would be "down" in winter and "up" in summer. Heat flow in walls would be in a horizontal direction.
[c]W = winter; S = summer.

**TABLE A6.7**  Conductance and Resistance Values for Exterior Siding Materials[a]

| Material | Description | Conductivity $k$, Btu/(hr)(ft²)(°F/in.) | Thickness, in. | Conductance $C$, Btu/(hr)(ft²)(°F) | Resistance $R$, 1/[Btu/(hr)(ft²)(°F)] |
|---|---|---|---|---|---|
| Brick | Common | 5.0 | 4 | 1.25 | 0.80 |
| Brick | Face | 9.0 | 4 | 2.27 | 0.44 |
| Stucco | | 5.0 | 1 | 5.0 | 0.20 |
| Asbestos cement shingles | | | | 4.76 | 0.21 |
| Wood shingles | 16–7½-in. exposure | | | 1.15 | 0.87 |
| Wood shingles | Double 16–12-in. exposure | | | 0.84 | 1.19 |
| Wood shingles | Plus $\frac{5}{16}$ in. Insulated backerboard | | | 0.71 | 1.40 |
| Asbestos cement siding | ¼ in. lapped | | | 4.76 | 0.21 |
| Asphalt roll siding | | | | 6.50 | 0.15 |
| Asphalt insulating siding | | | ½ | 0.69 | 1.46 |
| Wood | Drop siding, 1 × 8 in. | | | 1.27 | 0.79 |
| Wood | Bevel, ½ × 8 in. lapped | | | 1.23 | 0.81 |
| Wood | Bevel, ¾ × 10 in. lapped | | | 0.95 | 1.05 |
| Wood | Plywood, ⅜ in. lapped | | | 1.59 | 0.59 |
| Hardboard | Medium density | 0.73 | ¼ | 2.94 | 0.34 |
| Hardboard | Tempered | 1.00 | ¼ | 4.00 | 0.25 |
| Plywood lap siding | | | ⅜ | 1.79 | 0.56 |
| Plywood flat siding | | | ⅜ | 2.33 | 0.43 |

[a]Adapted from Johns-Mansville, Denver, Colorado.


**TABLE A6.8**  Conductance and Resistance Values for Sheathing and Building Paper[a]

| Material | Description | Conductivity $k$, Btu/(hr)(ft²)(°F/in.) | Thickness, in. | Conductance $C$, Btu/(hr)(ft²)(°F) | Resistance $R$, 1/[Btu/(hr)(ft²)(°F)] |
|---|---|---|---|---|---|
| Gypsum | . . . | 1.11 | ⅜ | 3.10 | 0.32 |
| | | | ½ | 2.25 | 0.45 |
| | | | ⅝ | 1.75 | 0.57 |
| Plywood | . . . | 0.80 | ¼ | 3.20 | 0.31 |
| | | | ⅜ | 2.13 | 0.47 |
| | | | ½ | 1.60 | 0.62 |
| | | | ⅝ | 1.28 | 0.78 |
| | | | ¾ | 1.07 | 0.93 |
| Nail-base sheathing | . . . | 0.44 | ½ | 0.88 | 1.14 |
| Wood sheathing | Fir or pine | 0.80 | ¾ | 1.06 | 0.94 |
| Sheathing paper | Vapor-permeable 2 layers mopped | | | 16.70 | 0.06 |
| Vapor barrier | 15-lb felt | | | 8.35 | 0.12 |
| | Plastic film | | | Negl. | Negl. |

[a]Adapted from Johns-Mansville, Denver, Colorado.

**TABLE A6.9** Conductance and Resistance Values for Masonry Materials[a]

| Material | Description | Conductivity $k$, Btu/(hr)(ft²)(°F/in.) | Thickness, in. | Conductance $C$, Btu/(hr)(ft²)(°F) | Resistance $R$, $1/[Btu/(hr)(ft²)(°F)]$ |
|---|---|---|---|---|---|
| Concrete blocks, three-oval core | Sand and gravel aggregate | ... | 4 | 1.40 | 0.71 |
| | | | 8 | 0.90 | 1.11 |
| | | | 12 | 0.78 | 1.28 |
| | Cinder aggregate | ... | 4 | 0.90 | 1.11 |
| | | | 8 | 0.58 | 1.72 |
| | | | 12 | 0.53 | 1.89 |
| | Lightweight aggregate | ... | 4 | 0.67 | 1.50 |
| | | | 8 | 0.50 | 2.00 |
| | | | 12 | 0.44 | 2.27 |
| Hollow clay tile | 1 cell deep | ... | 4 | 0.90 | 1.11 |
| | 2 cells deep | ... | 8 | 0.54 | 1.85 |
| | 3 cells deep | ... | 12 | 0.40 | 2.50 |
| Gypsum partition tile | 3 x 12 x 30 in. solid | ... | 3 | 0.79 | 1.26 |
| | 3 x 12 x 30 in. 4-cell | ... | 3 | 0.74 | 1.35 |
| | 4 x 12 x 30 in. 3-cell | ... | 4 | 0.60 | 1.67 |
| Cement mortar | | 5.0 | 1 | 5.0 | 0.20 |
| Stucco | | 5.0 | 1 | 5.0 | 0.20 |
| Gypsum | Poured | 1.66 | 1 | 1.66 | 0.60 |
| | Precast | 2.80 | 2 | 1.40 | 0.71 |
| Concrete | Sand and gravel or stone | 12.0 | 1 | 12.0 | 0.08 |
| Lightweight concrete | Perlite or zonolite mixture | | | | |
| | 1:4 mix, 36 lb/ft³ | 0.72–0.75 | 1 | 0.74 | 1.35 |
| | 1:5 mix, 30 lb/ft³ | 0.61–0.72 | 1 | 0.67 | 1.49 |
| | 1:6 mix, 27 lb/ft³ | 0.54–0.61 | 1 | 0.58 | 1.72 |
| | 1:8 mix, 22 lb/ft³ | 0.47–0.54 | 1 | 0.51 | 1.96 |
| Stone | ... | 12.5 | 1 | 12.5 | 0.08 |

[a]Adapted from Johns-Mansville, Denver, Colorado.


**TABLE A6.10** Conductance and Resistance Values for Woods[a]

| Material | Description | Conductivity $k$, Btu/(hr)(ft²)(°F/in.) | Thickness, in. | Conductance $C$, Btu/(hr)(ft²)(°F) | Resistance $R$, $1/[Btu/(hr)(ft²)(°F)]$ |
|---|---|---|---|---|---|
| Maple, oak and similar hardwoods | 45 lb/ft³ | 1.10 | $\frac{3}{4}$ | 1.47 | 0.68 |
| Fir, pine and similar softwoods | 32 lb/ft³ | 0.80 | $\frac{3}{4}$ | 1.06 | 0.94 |
| | | | $1\frac{1}{2}$ | 0.53 | 1.89 |
| | | | $2\frac{1}{2}$ | 0.32 | 3.12 |
| | | | $3\frac{1}{2}$ | 0.23 | 4.35 |

[a]Adapted from Johns-Mansville, Denver, Colorado.

**TABLE A6.11** Conductance and Resistance Values for Wall-Insulation Materials[a]

| Material | Description | Conductivity $k$, Btu/(hr)(ft²)(°F/in.) | Thickness, in. | Conductance $C$, Btu/(hr)(ft²)(°F) | Resistance $R$, 1/[Btu/(hr)(ft²)(°F)] |
|---|---|---|---|---|---|
| Fiber glass roof insulation | . . . | . . . | $\frac{15}{16}$ | 0.27 | 3.70 |
| | | | $1\frac{1}{16}$ | 0.24 | 4.17 |
| | | | $1\frac{5}{16}$ | 0.19 | 5.26 |
| | | | $1\frac{5}{8}$ | 0.15 | 6.67 |
| | | | $1\frac{7}{8}$ | 0.13 | 7.69 |
| | | | $2\frac{1}{4}$ | 0.11 | 9.09 |
| Urethane roof insulation | Thickness includes membrane roofing on both sides | 0.13 | $\frac{4}{5}$ | 0.19 | 5.26 |
| | | | 1 | 0.15 | 6.67 |
| | | | $1\frac{1}{5}$ | 0.12 | 8.33 |
| Styrofoam SM & TG | 2.1 lb/ft³ | 0.19 | $\frac{3}{4}$ | 0.25 | 3.93 |
| | | | 1 | 0.19 | 5.26 |
| | | | $1\frac{1}{2}$ | 0.13 | 7.89 |
| | | | 2 | 0.95 | 10.52 |
| Wood shredded | Cemented in pre-formed slabs | 0.60 | 1 | 0.60 | 1.67 |
| Insulating board | Building and service board, decorative ceiling panels | 0.38 | $\frac{3}{8}$ | 1.01 | 0.99 |
| | | | $\frac{1}{2}$ | 0.76 | 1.32 |
| | | | $\frac{9}{16}$ | 0.68 | 1.48 |
| | | | $\frac{3}{4}$ | 0.51 | 1.98 |
| Thermal, acoustical fiber glass | . . . | 0.39 | $2\frac{3}{4}$ | 0.14 | 7.00 |
| | | 0.36 | 4 | 0.09 | 11.00 |
| | | 0.34 | $6\frac{1}{2}$ | 0.05 | 19.00 |
| Corkboard | 6.4 lb/ft³ | 0.26 | 1 | 0.26 | 3.85 |
| Expanded polystyrene | | | | | |
| Extruded | 1.8 lb/ft³ | 0.25 | 1 | 0.25 | 4.00 |
| Molded beads | 1.0 lb/ft³ | 0.26 | 1 | 0.26 | 3.85 |
| Urethane foam | | | | | |
| Thurane (Dow Chemical) | 1.9 lb/ft³ | 0.17 | $\frac{3}{4}$ | 0.23 | 4.41 |
| | | | $1\frac{1}{2}$ | 0.11 | 8.82 |
| | | | 2 | 0.09 | 11.76 |
| Fiberglas perimeter insulation | . . . | . . . | 1 | 0.23 | 4.30 |
| | | | $1\frac{1}{4}$ | 0.19 | 5.40 |
| Fiberglas form board | . . . | . . . | 1 | 0.25 | 4.00 |

[a]Adapted from Johns-Mansville, Denver, Colorado.

**TABLE A6.12** Conductance and Resistance Values for Roofing Materials[a]

| Material | Description | Conductivity $k$, Btu/(hr)(ft²)(°F/in.) | Thickness, in. | Conductance $C$, Btu/(hr)(ft²)(°F) | Resistance $R$, 1/[Btu/(hr)(ft²)(°F)] |
|---|---|---|---|---|---|
| Asbestos cement shingles | 120 lb/ft³ | . . . | . . . | 4.76 | 0.21 |
| Asphalt shingles | 70 lb/ft³ | . . . | . . . | 2.27 | 0.44 |
| Wood shingles | . . . | . . . | . . . | 1.06 | 0.94 |
| Slate | . . . | . . . | $\frac{1}{2}$ | 20.0 | 0.05 |
| Asphalt roll roofing | 70 lb/ft³ | . . . | . . . | 6.50 | 0.15 |
| Built-up roofing | Smooth or gravel surface | . . . | $\frac{3}{8}$ | 3.00 | 0.33 |
| Sheet metal | . . . | . . . | . . . | Negl. | Negl. |

[a]Adapted from Johns-Mansville, Denver, Colorado.

**TABLE A6.13**  Conductance and Resistance Values for Flooring Materials[a]

| Material | Description | Conductivity $k$, Btu/(hr)(ft²)(°F/in.) | Thickness, in. | Conductance $C$, Btu/(hr)(ft²)(°F) | Resistance $R$, 1/[Btu/(hr)(ft²)(°F)] |
|---|---|---|---|---|---|
| Asphalt, vinyl, rubber, or linoleum tile | ... | ... | ... | 20.0 | 0.05 |
| Cork tile | ... | 0.45 | $\frac{1}{8}$ | 3.60 | 0.28 |
| Terrazzo | ... | 12.5 | 1 | 12.50 | 0.08 |
| Carpet and fibrous pad | ... | ... | ... | 0.48 | 2.08 |
| Carpet and rubber pad | ... | ... | ... | 0.81 | 1.23 |
| Plywood subfloor | ... | 0.80 | $\frac{5}{8}$ | 1.28 | 0.78 |
| Wood subfloor | ... | 0.80 | $\frac{3}{4}$ | 1.06 | 0.94 |
| Wood, hardwood finish | ... | 1.10 | $\frac{3}{4}$ | 1.47 | 0.68 |

[a]Adapted from Johns-Mansville, Denver, Colorado.

**TABLE A6.14**  Conductance and Resistance Values for Interior Finishes[a]

| Material | Description | Conductivity $k$, Btu/(hr)(ft²)(°F/in.) | Thickness, in. | Conductance $C$, Btu/(hr)(ft²)(°F) | Resistance $R$, 1/[Btu/(hr)(ft²)(°F)] |
|---|---|---|---|---|---|
| Gypsum board | | 1.11 | $\frac{3}{8}$ | 3.10 | 0.32 |
| | | | $\frac{1}{2}$ | 2.25 | 0.45 |
| Cement plaster | Sand aggregate | 5.0 | $\frac{1}{2}$ | 10.00 | 0.10 |
| | | | $\frac{3}{4}$ | 6.66 | 0.15 |
| Gypsum plaster | Sand aggregate | 5.6 | $\frac{1}{2}$ | 11.10 | 0.09 |
| | | | $\frac{5}{8}$ | 9.10 | 0.11 |
| Gypsum plaster | Lightweight aggregate | 1.6 | $\frac{1}{2}$ | 3.12 | 0.32 |
| | | | $\frac{5}{8}$ | 2.67 | 0.39 |
| Gypsum plaster on | | | | | |
| Metal lath | Sand aggregate | | $\frac{3}{4}$ | 7.70 | 0.13 |
| Metal lath | Lightweight aggregate | | $\frac{3}{4}$ | 2.13 | 0.47 |
| Gypsum board, $\frac{3}{8}$ in. | Sand aggregate | | $\frac{7}{8}$ | 2.44 | 0.41 |
| Insulating board | | 0.38 | $\frac{1}{2}$ | 0.74 | 1.35 |
| Plywood | | 0.80 | $\frac{3}{8}$ | 2.13 | 0.47 |

[a]Adapted from Johns-Mansville, Denver, Colorado.

**TABLE A6.15**  Conductance and Resistance Values for Glass[a]

| Material | Description | Conductivity $k$, Btu/(hr)(ft²)(°F/in.) | Thickness, in. | Conductance $C$ ($U$), Btu/(hr)(ft²)(°F) | Resistance $R$ (1/$U$), 1/[Btu/(hr)(ft²)(°F)] |
|---|---|---|---|---|---|
| Single-plate | | | | 1.13 | 0.88 |
| Double-plate | Air space, $\frac{3}{16}$ in. | ... | ... | 0.69 | 1.45 |
| Storm windows | Air space, 1–4 in. | ... | ... | 0.56 | 1.78 |
| Solid-wood door | Actual thickness, $1\frac{1}{2}$ in. | ... | ... | 0.49 | 2.04 |
| Storm door, wood and glass | With wood and glass storm door | ... | ... | 0.27 | 3.70 |
| Storm door, metal and glass | With metal and glass storm door | ... | ... | 0.33 | 3.00 |

[a]Adapted from Johns-Mansville, Denver, Colorado.

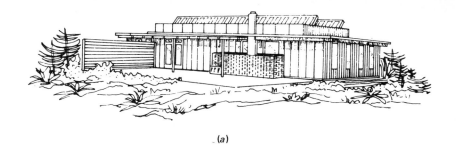

(a)

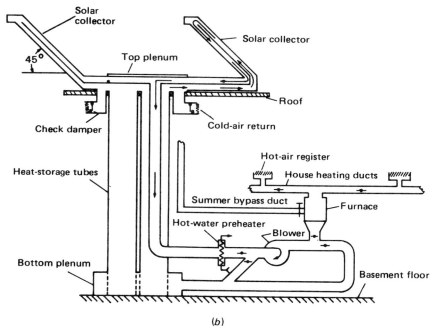

(b)

**FIGURE A6.1** Performance data for a solar-heated dwelling in Denver, Colorado. (a) Denver solar-heated house, (b) Denver solar space-heating system (not to scale). Adapted from Kreider, J. F., and F. Kreith, "Solar Heating and Cooling," revised 1st ed., Hemisphere Publ. Corp., 1977.

Designer: George O. G. Löf
Architect: James Hunter
Building: Residence, single-story [see Fig. A6.1(*a* and *b*)]
Floor area: 2050 ft$^2$
Heat-loss rate: 32,000 Btu/degree-day
Location: Denver, Colorado
Latitude: 40°N
Altitude: 5280 ft
Degree-days: 6283
Collector: Type: Flat plate (overlapping plates)
Fluid: Air
Area: 600 ft$^2$
Position: Mounted on flat roof
Tilt: 45°
Glazing: Half single, half double
Storage: Type: Crushed rock, 2.5-in diameter
Volume: 230 ft$^3$, 11 tons
Container: Two fiber drums
Temperature: 140°F normal maximum
System: Forced air
Hot water: Preheated by thermosiphon air-to-water heat exchanger
Performance: Period: Winter 1959–1960 (5700 degree days)
Average collector efficiency: 34.6 percent
Solar contribution: 25.7 percent
See Table A6.16 for annual performance summary for the winter of 1959–1960.

**FIGURE A6.1** (*Continued*)

**TABLE A6.16** Energy Balance of Denver Solar House Shown in Fig. A6.1[b]

Winter 1959–60: All values in million Btu

| | September (18–30) | October | November | December | January | February | March | April | May | June (1–10) | Total |
|---|---|---|---|---|---|---|---|---|---|---|---|
| 1. Total solar incidence on 45°, 600 ft² collector area | 9.93 | 26.84 | 25.81 | 21.98 | 25.48 | 22.10 | 30.43 | 26.56 | 29.61 | 8.13 | 226.86 |
| 2. Total solar incidence available on 45°, 600-ft² collector area when collection cycles operated | 6.94 | 20.16 | 17.55 | 15.67 | 17.05 | 13.38 | 22.03 | 19.94 | 22.60 | 6.00 | 161.33 |
| 3. Gross collected solar heat | a | a | a | a | 5.99 | 4.33 | 9.08 | 9.16 | 8.86 | 2.40 | a |
| 4. Gross collector efficiency, percent | a | a | a | a | 34.7 | 32.4 | 41.1 | 45.9 | 39.8 | 40.0 | a |
| 5. Useful collected heat | 1.93 | 5.91 | 5.34 | 5.61 | 5.59 | 3.79 | 8.45 | 8.66 | 8.25 | 2.18 | 55.72 |
| 6. Net collector efficiency, percent | 27.9 | 29.4 | 30.3 | 35.8 | 32.7 | 28.3 | 38.3 | 43.5 | 36.5 | 36.4 | 34.6 |
| 7. Solar heat absorbed by storage tubes | 1.12 | 3.04 | 2.77 | 2.79 | 2.64 | 1.92 | 3.99 | 3.68 | 3.01 | 0.50 | 25.46 |
| 8. Storage-tube inventory | −0.03 | 0.07 | −0.002 | −0.008 | 0.017 | −0.023 | −0.008 | 0.07 | 0.17 | −0.05 | a |
| 9. Solar heat absorbed by water pre-heater | 0.11 | 0.35 | 0.30 | 0.32 | 0.27 | 0.21 | 0.51 | 0.72 | 0.87 | 0.29 | 3.95 |
| 10. Heat delivered by natural gas for house heating | 3.19 | 12.26 | 19.65 | 22.16 | 26.90 | 28.10 | 17.45 | 7.09 | 4.78 | 0.25 | 141.83 |
| 11. Heat delivered by natural gas for water heating | 0.67 | 1.49 | 1.79 | 2.01 | 1.84 | 2.68 | 2.60 | 3.23 | 3.24 | 0.88 | 20.43 |
| 12. Total heat load | 5.79 | 19.66 | 26.78 | 29.78 | 34.33 | 34.57 | 28.50 | 18.98 | 16.27 | 3.31 | 217.98 |
| 13. Percent of useful collected heat absorbed by water preheater | 5.7 | 5.91 | 5.63 | 5.7 | 4.84 | 5.55 | 6.04 | 8.2 | 10.56 | 13.4 | 7.09 |
| 14. Percent of total water-heating load supplied by solar energy | 14.1 | 19.0 | 14.4 | 13.75 | 12.80 | 7.26 | 16.4 | 18.25 | 21.20 | 28.40 | 16.25 |
| 15. Percent of house heat load supplied by solar energy (including water preheating but excluding water heating) | 37.5 | 32.3 | 21.4 | 20.2 | 17.2 | 11.87 | 32.6 | 55.3 | 63.4 | 89.6 | 28.20 |
| 16. Percent of house heat load supplied by solar energy (including both water pre-heating and heating) | 37.0 | 30.2 | 19.9 | 18.0 | 16.25 | 10.95 | 29.6 | 45.8 | 50.7 | 65.8 | 25.7 |

[a]Not determined.

[b]From Kreider, J. F., and F. Kreith, "Solar Heating and Cooling," revised 1st ed., Hemisphere Publ. Corp., 1977.

# CLIMATIC AND THERMODYNAMIC DATA FOR COOLING SYSTEMS

**7**
Appendix

**TABLE A7.1** Equivalent Full-Load Cooling Hours for Selected U.S. Cities[a]

| City | | Cooling hours | City | | Cooling hours |
|---|---|---|---|---|---|
| Alabama | Birmingham | 1500 | Minnesota | Minneapolis | 500 |
| Alaska | | n/a | Missouri | St. Louis | 1150 |
| Arizona | Flagstaff | 200 | Montana | Helena | 250 |
| | Phoenix | 2750 | Nebraska | Lincoln | 1000 |
| Arkansas | Little Rock | 2000 | Nevada | Las Vegas | 2350 |
| California | Los Angeles | 550 | | Reno | 650 |
| | Sacramento | 1000 | New Hampshire | Concord | 400 |
| | San Francisco | <50 | New Jersey | Trenton | 450 |
| Colorado | Denver | 650 | New Mexico | Albuquerque | 1150 |
| Connecticut | Hartford | 500 | New York | New York City | 650 |
| Delaware | Wilmington | 600 | North Carolina | Raleigh | 1050 |
| District of Columbia | Washington | 1000 | North Dakota | Bismarck | 450 |
| Florida | Miami | 2400 | Ohio | Columbus | 600 |
| | Jacksonville | 1800 | Oklahoma | Oklahoma City | 1450 |
| Georgia | Atlanta | 1000 | Oregon | Salem | 300 |
| Hawaii | Honolulu | 1350 | Pennsylvania | Philadelphia | 700 |
| Idaho | Boise | 700 | Rhode Island | Providence | 700 |
| Illinois | Chicago | 750 | South Carolina | Columbia | 1350 |
| Indiana | Indianapolis | 750 | South Dakota | Rapid City | 550 |
| Iowa | Des Moines | 600 | Tennessee | Nashville | 1300 |
| Kansas | Topeka | 1150 | Texas | San Antonio | 2000 |
| Kentucky | Louisville | 1050 | Utah | Salt Lake City | 900 |
| Louisiana | New Orleans | 1750 | Vermont | Montpelier | 300 |
| Maine | Bangor | 200 | Virginia | Richmond | 1000 |
| Maryland | Baltimore | 850 | Washington | Seattle | 100 |
| Massachusetts | Boston | 400 | West Virginia | Charleston | 800 |
| Michigan | Detroit | 500 | | | |

[a]Cooling hours over 80° F (rounded to the nearest 50 hr) are taken from Engineering Weather Data, Chapter 6 of Air Force publication AFM88-8. Similar data can also be found in the NAHB "Insulation Manual—Homes/Apartments."

748

**TABLE A7.2**  Properties of Freon-12, $CCl_2F_2$[a]

| | | | Saturated | | | | | Degrees of superheat | | |
| | | | | | | | 25°F | | 50°F | |
| $t$ | $p$ | $v_g$ | $h_f$ | $h_g$ | $s_f$ | $s_g$ | $h$ | $s$ | $h$ | $s$ |
|---|---|---|---|---|---|---|---|---|---|---|
| −40 | 9.32 | 3.911 | 0 | 73.50 | 0 | 0.17517 | 76.81 | 0.18281 | 80.19 | 0.19027 |
| −35 | 10.60 | 3.471 | 1.01 | 74.10 | 0.00235 | 0.17451 | 77.43 | 0.18212 | 80.83 | 0.18952 |
| −30 | 12.02 | 3.088 | 2.03 | 74.70 | 0.00471 | 0.17387 | 78.05 | 0.18147 | 81.47 | 0.18881 |
| −25 | 13.58 | 2.759 | 3.05 | 75.30 | 0.00706 | 0.17328 | 78.66 | 0.18086 | 82.11 | 0.18816 |
| −20 | 15.28 | 2.474 | 4.07 | 75.87 | 0.00940 | 0.17275 | 79.27 | 0.18029 | 82.75 | 0.18755 |
| −15 | 17.16 | 2.224 | 5.09 | 76.46 | 0.01172 | 0.17222 | 79.88 | 0.17974 | 83.38 | 0.18698 |
| −10 | 19.20 | 2.003 | 6.14 | 77.05 | 0.01403 | 0.17175 | 80.49 | 0.17922 | 84.01 | 0.18645 |
| − 5 | 21.44 | 1.808 | 7.20 | 77.64 | 0.01636 | 0.17131 | 81.09 | 0.17874 | 84.64 | 0.18595 |
| 0 | 23.87 | 1.637 | 8.25 | 78.21 | 0.01869 | 0.17091 | 81.69 | 0.17829 | 85.26 | 0.18547 |
| 5 | 26.51 | 1.485 | 9.32 | 78.79 | 0.02097 | 0.17052 | 82.29 | 0.17786 | 85.89 | 0.18502 |
| 10 | 29.35 | 1.351 | 10.39 | 79.36 | 0.02328 | 0.17015 | 82.90 | 0.17747 | 86.51 | 0.18460 |
| 15 | 32.44 | 1.230 | 11.48 | 79.94 | 0.02556 | 0.16980 | 83.50 | 0.17710 | 87.13 | 0.18420 |
| 20 | 35.75 | 1.121 | 12.55 | 80.49 | 0.02783 | 0.16949 | 84.09 | 0.17675 | 87.75 | 0.18382 |
| 25 | 39.32 | 1.025 | 13.66 | 81.06 | 0.03008 | 0.16919 | 84.67 | 0.17643 | 88.37 | 0.18348 |
| 30 | 43.16 | 0.939 | 14.76 | 81.61 | 0.03233 | 0.16887 | 85.25 | 0.17612 | 88.97 | 0.18315 |
| 35 | 47.27 | 0.862 | 15.87 | 82.16 | 0.03458 | 0.16859 | 85.83 | 0.17583 | 89.57 | 0.18285 |
| 40 | 51.68 | 0.792 | 17.00 | 82.71 | 0.03680 | 0.16833 | 86.41 | 0.17554 | 90.16 | 0.18256 |
| 45 | 56.37 | 0.730 | 18.13 | 83.26 | 0.03904 | 0.16808 | 86.98 | 0.17529 | 90.76 | 0.1822 |
| 50 | 61.39 | 0.673 | 19.27 | 83.78 | 0.04126 | 0.16785 | 87.54 | 0.17505 | 91.36 | 0.18203 |
| 55 | 66.73 | 0.622 | 20.41 | 84.31 | 0.04348 | 0.16762 | 88.09 | 0.17481 | 91.94 | 0.18179 |
| 60 | 72.41 | 0.575 | 21.57 | 84.82 | 0.04568 | 0.16741 | 88.64 | 0.17458 | 92.51 | 0.18155 |
| 65 | 78.43 | 0.532 | 22.72 | 85.32 | 0.04789 | 0.16721 | 89.18 | 0.17437 | 93.09 | 0.18134 |
| 70 | 84.82 | 0.493 | 23.90 | 85.82 | 0.05009 | 0.16701 | 89.72 | 0.17417 | 93.67 | 0.18144 |
| 75 | 91.60 | 0.457 | 25.08 | 86.32 | 0.05229 | 0.16681 | 90.25 | 0.17397 | 94.23 | 0.18094 |
| 80 | 98.76 | 0.425 | 26.28 | 86.80 | 0.05446 | 0.16662 | 90.78 | 0.17379 | 94.80 | 0.18075 |
| 85 | 106.3 | 0.395 | 27.48 | 87.28 | 0.05665 | 0.16644 | 91.28 | 0.17361 | 95.33 | 0.18057 |
| 90 | 114.3 | 0.368 | 28.70 | 87.74 | 0.05882 | 0.16624 | 91.77 | 0.17343 | 95.86 | 0.18040 |
| 95 | 122.7 | 0.342 | 29.93 | 88.19 | 0.06100 | 0.16604 | 92.26 | 0.17325 | 96.39 | 0.18022 |
| 100 | 131.6 | 0.319 | 31.16 | 88.62 | 0.06316 | 0.16584 | 92.75 | 0.17308 | 96.92 | 0.18004 |
| 105 | 140.9 | 0.297 | 32.40 | 89.03 | 0.06534 | 0.16564 | 93.21 | 0.17291 | 97.43 | 0.17990 |
| 110 | 150.7 | 0.277 | 33.65 | 89.43 | 0.06749 | 0.16542 | 93.66 | 0.17274 | 97.93 | 0.17976 |
| 115 | 161.0 | 0.258 | 34.90 | 89.80 | 0.06965 | 0.16520 | 94.07 | 0.17254 | 98.40 | 0.17958 |
| 120 | 171.8 | 0.240 | 36.16 | 90.15 | 0.07180 | 0.16495 | 94.47 | 0.17233 | 98.84 | 0.17939 |
| 125 | 183.1 | 0.223 | 37.41 | 90.46 | 0.07395 | 0.16468 | 94.86 | 0.17210 | 99.27 | 0.17920 |
| 130 | 194.9 | 0.208 | 38.69 | 90.76 | 0.07607 | 0.16438 | 95.25 | 0.17186 | 99.71 | 0.17900 |
| 135 | 207.3 | 0.193 | 39.95 | 91.01 | 0.07817 | 0.16404 | 95.64 | 0.17161 | 100.14 | 0.17880 |
| 140 | 220.2 | 0.180 | 41.24 | 91.24 | 0.08024 | 0.16363 | 96.03 | 0.17132 | 100.56 | 0.17859 |

[a]Symbols: $t$, °F; $p$, psia; $v_g$, ft³/lb; $h_f$, Btu/lb; $h_g$, Btu/lb; $s_f$, Btu/lb · °R; $s_g$, Btu/lb · °R; $h$, Btu/lb; $s$, Btu/lb · °R.

# TABLE A7.3  Properties of Ammonia, $NH_3{}^a$

| | | Saturated | | | | | Degrees of superheat | | | |
|---|---|---|---|---|---|---|---|---|---|---|
| | | | | | | | 100°F | | 200°F | |
| $t$ | $p$ | $v_g$ | $h_f$ | $h_g$ | $s_f$ | $s_g$ | $h$ | $s$ | $h$ | $s$ |
| −70 | 3.94 | 61.65 | −31.1 | 584.4 | −0.0771 | 1.5026 | 635.6 | 1.6214 | 686.2 | 1.7151 |
| −65 | 4.69 | 52.34 | −26.0 | 586.6 | −0.0642 | 1.4833 | 638.0 | 1.6058 | 688.7 | 1.6990 |
| −60 | 5.55 | 44.73 | −20.9 | 588.8 | −0.0514 | 1.4747 | 640.3 | 1.5907 | 691.1 | 1.6834 |
| −55 | 6.54 | 38.38 | −15.7 | 591.0 | −0.0381 | 1.4614 | 642.6 | 1.5761 | 693.6 | 1.6683 |
| −50 | 7.67 | 33.08 | −10.5 | 593.2 | −0.0254 | 1.4487 | 644.9 | 1.5620 | 696.1 | 1.6537 |
| −45 | 8.95 | 28.62 | − 5.3 | 595.4 | −0.0128 | 1.4363 | 647.2 | 1.5484 | 698.6 | 1.6395 |
| −40 | 10.41 | 24.86 | 0 | 597.6 | 0 | 1.4242 | 649.4 | 1.5353 | 701.0 | 1.6260 |
| −35 | 12.05 | 21.68 | 5.3 | 599.5 | 0.0126 | 1.4120 | 651.7 | 1.5226 | 703.4 | 1.6129 |
| −30 | 13.90 | 18.97 | 10.7 | 601.4 | 0.0250 | 1.4001 | 653.9 | 1.5103 | 705.9 | 1.6002 |
| −25 | 15.98 | 16.66 | 16.0 | 603.2 | 0.0374 | 1.3886 | 656.1 | 1.4983 | 708.3 | 1.5878 |
| −20 | 18.30 | 14.68 | 21.4 | 605.0 | 0.0497 | 1.3774 | 658.3 | 1.4868 | 710.8 | 1.5759 |
| −15 | 20.88 | 12.97 | 26.7 | 606.7 | 0.0618 | 1.3664 | 660.5 | 1.4756 | 713.2 | 1.5644 |
| −10 | 23.74 | 11.50 | 32.1 | 608.5 | 0.0738 | 1.3558 | 662.6 | 1.4647 | 715.5 | 1.5531 |
| − 5 | 26.92 | 10.23 | 37.5 | 610.1 | 0.0857 | 1.3454 | 664.7 | 1.4541 | 717.9 | 1.5423 |
| 0 | 30.42 | 9.116 | 42.9 | 611.8 | 0.0975 | 1.3352 | 666.8 | 1.4438 | 720.3 | 1.5317 |
| 5 | 34.27 | 8.150 | 48.3 | 613.3 | 0.1092 | 1.3253 | 668.9 | 1.4338 | 722.7 | 1.5214 |
| 10 | 38.51 | 7.304 | 53.8 | 614.9 | 0.1208 | 1.3157 | 670.9 | 1.4241 | 725.0 | 1.5115 |
| 15 | 43.14 | 6.562 | 59.2 | 616.3 | 0.1323 | 1.3062 | 672.9 | 1.4147 | 727.3 | 1.5018 |
| 20 | 48.21 | 5.910 | 64.7 | 617.8 | 0.1437 | 1.2969 | 675.0 | 1.4055 | 729.6 | 1.4924 |
| 25 | 53.73 | 5.334 | 70.2 | 619.1 | 0.1551 | 1.2879 | 677.0 | 1.3966 | 731.9 | 1.4833 |
| 30 | 59.74 | 4.825 | 75.7 | 620.5 | 0.1663 | 1.2790 | 678.9 | 1.3879 | 734.2 | 1.4744 |
| 35 | 66.26 | 4.373 | 81.2 | 621.7 | 0.1775 | 1.2704 | 680.8 | 1.3794 | 736.4 | 1.4658 |
| 40 | 73.32 | 3.971 | 86.8 | 623.0 | 0.1885 | 1.2618 | 682.7 | 1.3711 | 738.6 | 1.4575 |
| 45 | 80.96 | 3.614 | 92.3 | 624.1 | 0.1996 | 1.2535 | 684.5 | 1.3630 | 740.9 | 1.4493 |
| 50 | 89.19 | 3.294 | 97.9 | 625.2 | 0.2105 | 1.2453 | 686.4 | 1.3551 | 743.1 | 1.4413 |
| 55 | 98.06 | 3.008 | 103.5 | 626.3 | 0.2214 | 1.2373 | 688.2 | 1.3474 | 745.2 | 1.4335 |
| 60 | 107.6 | 2.751 | 109.2 | 627.3 | 0.2322 | 1.2294 | 689.9 | 1.3399 | 747.4 | 1.4260 |
| 65 | 117.8 | 2.520 | 114.8 | 628.2 | 0.2430 | 1.2216 | 691.6 | 1.3326 | 749.5 | 1.4186 |
| 70 | 128.8 | 2.312 | 120.5 | 629.1 | 0.2537 | 1.2140 | 693.3 | 1.3254 | 751.6 | 1.4114 |
| 75 | 140.5 | 2.125 | 126.2 | 629.9 | 0.2643 | 1.2065 | 694.9 | 1.3184 | 753.7 | 1.4044 |
| 80 | 153.0 | 1.955 | 132.0 | 630.7 | 0.2749 | 1.1991 | 696.6 | 1.3115 | 755.8 | 1.3976 |
| 85 | 166.4 | 1.801 | 137.8 | 631.4 | 0.2854 | 1.1918 | 698.1 | 1.3048 | 757.9 | 1.3909 |
| 90 | 180.6 | 1.661 | 143.5 | 632.0 | 0.2958 | 1.1846 | 699.7 | 1.2982 | 759.9 | 1.3843 |
| 95 | 195.8 | 1.534 | 149.4 | 632.6 | 0.3062 | 1.1775 | 701.2 | 1.2918 | 761.8 | 1.3779 |
| 100 | 211.9 | 1.419 | 155.2 | 633.0 | 0.3166 | 1.1705 | 702.7 | 1.2855 | 763.8 | 1.3717 |
| 105 | 228.9 | 1.313 | 161.1 | 633.4 | 0.3269 | 1.1635 | 704.2 | 1.2793 | 765.7 | 1.3656 |
| 110 | 247.0 | 1.217 | 167.0 | 633.7 | 0.3372 | 1.1566 | 705.6 | 1.2732 | 767.6 | 1.3596 |
| 115 | 266.2 | 1.128 | 173.0 | 633.9 | 0.3474 | 1.1497 | 706.9 | 1.2672 | 769.5 | 1.3538 |
| 120 | 286.4 | 1.047 | 179.0 | 634.0 | 0.3576 | 1.1427 | 708.2 | 1.2613 | 771.3 | 1.3480 |
| 125 | 307.8 | 0.973 | 185.1 | 634.0 | 0.3679 | 1.1358 | 709.5 | 1.2555 | 773.1 | 1.3423 |

$^a$Symbols: $t$, °F; $p$, psia; $v_g$, ft$^3$/lb; $h_f$, Btu/lb; $h_g$, Btu/lb; $s_f$, Btu/lb · °R; $s_g$, Btu/lb · °R; $h$, Btu/lb; $s$, Btu/lb · °R.

**TABLE A7.4** Psychrometric Table—SI Units. Properties of Moist Air at 101,325 N/m²[a]

| Temperature | | | Properties[b] | | | | | | |
|---|---|---|---|---|---|---|---|---|---|
| C | K | F | $P_s$ | $W_s$ | $V_a$ | $V_s$ | $h_a$ | $h_s$ | $s_s$ |
| −40 | 233.15 | −40 | 12.838 | 0.00007925 | 0.65961 | 0.65968 | −22.35 | −22.16 | −90.659 |
| −30 | 243.15 | −22 | 37.992 | 0.0002344 | 0.68808 | 0.68833 | −12.29 | −11.72 | −46.732 |
| −25 | 248.15 | −13 | 63.248 | 0.0003903 | 0.70232 | 0.70275 | −7.265 | −6.306 | −24.706 |
| −20 | 253.15 | − 4 | 103.19 | 0.0006371 | 0.71649 | 0.71724 | −2.236 | −0.6653 | −2.2194 |
| −15 | 258.15 | + 5 | 165.18 | 0.001020 | 0.73072 | 0.73191 | +2.794 | 5.318 | 21.189 |
| −10 | 263.15 | 14 | 259.72 | 0.001606 | 0.74495 | 0.74683 | 7.823 | 11.81 | 46.104 |
| − 5 | 268.15 | 23 | 401.49 | 0.002485 | 0.75912 | 0.76218 | 12.85 | 19.04 | 73.365 |
| 0 | 273.15 | 32 | 610.80 | 0.003788 | 0.77336 | 0.77788 | 17.88 | 27.35 | 104.14 |
| 5 | 278.15 | 41 | 871.93 | 0.005421 | 0.78759 | 0.79440 | 22.91 | 36.52 | 137.39 |
| 10 | 283.15 | 50 | 1227.2 | 0.007658 | 0.80176 | 0.81163 | 27.94 | 47.23 | 175.54 |
| 15 | 288.15 | 59 | 1704.4 | 0.01069 | 0.81600 | 0.82998 | 32.97 | 59.97 | 220.22 |
| 20 | 293.15 | 68 | 2337.2 | 0.01475 | 0.83017 | 0.84983 | 38.00 | 75.42 | 273.32 |
| 25 | 298.15 | 77 | 3167.0 | 0.02016 | 0.84434 | 0.87162 | 43.03 | 94.38 | 337.39 |
| 30 | 303.15 | 86 | 4242.8 | 0.02731 | 0.85851 | 0.89609 | 48.07 | 117.8 | 415.65 |
| 35 | 308.15 | 95 | 5623.4 | 0.03673 | 0.87274 | 0.92406 | 53.10 | 147.3 | 512.17 |
| 40 | 313.15 | 104 | 7377.6 | 0.04911 | 0.88692 | 0.95665 | 58.14 | 184.5 | 532.31 |
| 45 | 318.15 | 113 | 9584.8 | 0.06536 | 0.90115 | 0.99535 | 63.17 | 232.0 | 783.06 |
| 50 | 323.15 | 122 | 12339 | 0.08678 | 0.91532 | 1.0423 | 68.21 | 293.1 | 975.27 |
| 55 | 328.15 | 131 | 15745 | 0.1152 | 0.92949 | 1.1007 | 73.25 | 372.9 | 1221.5 |
| 60 | 333.15 | 140 | 19925 | 0.1534 | 0.94372 | 1.1748 | 78.29 | 478.5 | 1543.5 |
| 65 | 338.15 | 149 | 25014 | 0.2055 | 0.95790 | 1.2721 | 83.33 | 621.4 | 1973.6 |
| 70 | 343.15 | 158 | 31167 | 0.2788 | 0.97207 | 1.4042 | 88.38 | 820.5 | 2564.8 |
| 75 | 348.15 | 167 | 38554 | 0.3858 | 0.98630 | 1.5924 | 93.42 | 1110 | 3412.8 |
| 80 | 353.15 | 176 | 47365 | 0.5519 | 1.0005 | 1.8791 | 98.47 | 1557 | 4710.9 |
| 85 | 358.15 | 185 | 57809 | 0.8363 | 1.0146 | 2.3632 | 103.5 | 2321 | 6892.6 |
| 90 | 363.15 | 194 | 70112 | 1.416 | 1.0288 | 3.3409 | 108.6 | 3876 | 11281 |

[a]Symbols and units: $P_s$ = pressure of water vapor at saturation, N/m²; $W_s$ = humidity ratio at saturation, mass of water vapor associated with unit mass of dry air; $V_a$ = specific volume of dry air, m³/kg; $V_s$ = specific volume of saturated mixture, m³/kg dry air; $h_a$ = specific enthalpy of dry air, kJ/kg; $h_s$ = specific enthalpy of saturated mixture, kJ/kg dry air; $s_s$ = specific entropy of saturated mixture, J/K · kg dry air. Abstracted from Bolz, R. E., and G. L. Tuve, eds., "CRC Handbook of Tables for Applied Engineering Science," 2nd ed., Chemical Rubber Co., Cleveland, Ohio, 1973.

[b]The $P_s$ column gives the vapor pressure of pure water at temperature intervals of 5°C. For the latest data on vapor pressures at intervals of 0.1°C, from 0 to 100°C, see Wexler, A., and L. Greenspan, Vapor Pressure Equation for Water, *J. Res. Natl. Bur. Stand. Sect. A*, 75(3):213–229, May–June 1971. For very low barometric pressures and wet-bulb temperatures, the values of $h_a$ here are somewhat low; for corrections see the "Handbook of Fundamentals," American Society of Heating, Refrigerating, and Air Conditioning Engineers, 1972.

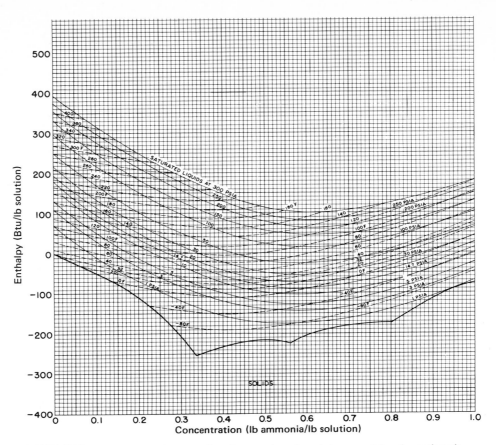

**FIGURE A7.1** Thermodynamic properties of ammonia-water mixtures used for absorption air conditioning.

# GLOSSARY

8 Appendix

**Absorber** Component of a solar collector (generally metallic), the function of which is to collect and retain as much of the radiation from the sun as possible. A heat-transfer fluid flows through the absorber or the conduits attached to the absorber.

**Absorptance** The ratio of absorbed to incident solar radiation. Absorptivity is the property of absorbing radiation possessed by all materials to varying extents.

**Absorption air conditioning** Achieving a cooling effect by an absorption-desorption process without the requirement of large shaft work input; desorption is caused by heat input.

**Air mass** The length of the path through the earth's atmosphere traversed by direct solar radiation, expressed as a multiple of the path length with the sun at zenith (overhead); see Chapter 2.

**Albedo** See reflectance.

**Algae** Unicellular or filamentous plants, usually fast growing and able to live in fresh or sea water.

**Ambient conditions** Conditions of the surroundings.

**Aperture** An opening through which radiation passes prior to absorption in a solar collector.

**ASHRAE** Acronym denoting the American Society of Heating, Refrigerating, and Air Conditioning Engineers. ASHRAE handbooks are sources of basic data on heating and air conditioning.

**Auxiliary system** System that acts as a backup to the solar system during extended periods of extremely cold and/or cloudy weather.

**Azimuth** The angle between the south-north line at a given location and the projection of the earth-sun line in the horizontal plane; see Chapter 2.

**Battery** An electrical energy storage system frequently using reversible chemical reactions.

**Bioconversion** The conversion of solar energy to chemically stored energy

Assembled with assistance from Karen George and Albert Nunez, University of Colorado, Denver, Colorado.

using biological processes. Bioconversion may be used to produce useful materials as well as fuels.

**Black body**  An ideal surface that absorbs all of the radiation incident on it and emits none; alternatively, a body that emits the maximum possible radiation at a given temperature.

**Btu**  British thermal unit; the amount of heat required to raise the temperature of one pound of water (at $4°C$) one degree Fahrenheit.

**C**  Celsius temperature scale wherein water freezes at $0°C$ and boils at $100°C$ at one atmospheric pressure.

**Cadmium sulfide, CdS**  A yellow-orange chemical compound produced from cadmium metal in large quantities; used as a pigment. As a semiconductor, CdS is always $n$-type.

**Capital cost**  Cost of equipment, construction, land, and other items required to construct a facility; different from recurrent operating and maintenance costs.

**Carbon reduction**  Fixation of $CO_2$ as an energy-rich carbohydrate, usually via photosynthesis but possible also by a synthetic process.

**Cellulose**  Fundamental carbon-containing constituent of all plants, composed of long chains of glucose molecules.

**Collector**  Any device gathering the sun's radiation and converting it to a useful energy form.

**Collector efficiency**  The ratio of the energy collected by a solar collector to the radiant energy incident on the collector.

**Collector tilt angle**  Angle at which the collector aperture is slanted up from the horizontal plane.

**Collimated light**  Parallel rays of light; the direct or beam component of solar radiation.

**Comfort zone**  Range of temperatures, humidities, and air flow rates at which the majority of persons are comfortable; varies with activity level.

**Concentration ratio**  Ratio of aperture area of a solar collector to its receiver area.

**Concentrator**  Solar collector that focuses or funnels solar radiation onto a relatively smaller absorber at the focal point or line.

**Conduction**  Heat transfer through matter by exchange of kinetic energy from particle to particle.

**Constant dollars**  Dollars (or other currency) expressed, net of inflation, in terms of today's dollars.

**Convection**  Heat transfer resulting from fluid motion in the presence of a temperature difference.

**Cooling load**  Amount of heat and humidity that must be removed from a building in order to maintain occupant comfort.

**COP**  Coefficient of performance; see Chapters 6 and 7 for various uses of this term.

**Cover plate**  Transparent material used to cover collector-absorber plate so that the solar energy is "trapped" by the greenhouse effect (primarily a convection suppression effect).

**CPC collector**  Compound parabolic concentrator (see Chapter 4).

**Current dollars**  Dollars (or other currency) expressed in future terms that include inflation effects.

**Degree-day**  A convenient measure of monthly or annual heating demands; see Chapter 6.

**Depreciation** Allocation of initial cost of a facility or system to the time period during which it is used.

**Design conditions** Indoor and outdoor temperatures, humidities, and wind speeds used to predict maximum expected heating and cooling loads for a building.

**Diffuse radiation** Scattered radiation from the sun that falls on a plane of stated orientation over a stated period; in the case of an inclined surface, including reflected from the ground as well.

**Direct conversion** Transformation of sunlight to electricity without an intervening thermodynamic cycle.

**Direct radiation** Radiation from the sun received from a narrow solid angle measured from a point on the earth's surface.

**Discounting** Adjusting cash flows for the time value of money; see Chapter 5.

**Discount rate** The opportunity cost of making an investment; see Chapter 5 for detailed discussion.

**Efficiency** Ratio of the measure of a desired effect to the measure of the input causing the effect, both expressed in the same units of measure.

**Electrolysis** Use of an electric current to produce $H_2$ and $O_2$ from water.

**Emittance** The ratio of radiation emitted by a real surface to the radiation emitted by a perfect radiator at the same temperature. Normal emittance is the value measured at $90°$ to the plane of the sample; hemispheric emittance is the total amount emitted in all directions.

**Endergonic reaction** Chemical reaction with absorption of Gibbs free energy.

**Endothermic reaction** Chemical reaction with absorption of heat.

**Energy gap** Range of energies in a semiconductor forbidden to electrons; photons are readily absorbed if their energy exceeds the gap. The electron states below the gap (in energy) constitute the valence band (normally full); above the gap is the conduction band, normally empty, into which valence band electrons can be excited by light. Impurities and atomic displacements produce allowed states in the otherwise forbidden gap.

**Energy plantations** Large areas used for growing plant material for its fuel value; a renewable source of energy-rich fixed carbon produced by photosynthesis.

**Eutectic** Having the lowest fusing points of a mixture of several salts or metals; these substances can be utilized in heat-of-fusion energy storage.

**Evacuated-tube collector** A collector manufactured from specially coated concentric glass tubes with an evacuated space between the outer two tubes.

**Evaporator** A heat exchanger in which a fluid undergoes a liquid-to-vapor phase change.

**Exergonic reaction** Chemical reaction with evolution of Gibbs free energy.

**Exothermic reaction** Chemical reaction with evolution of heat.

**Fan coil** A heat exchanger used to transfer heat from a liquid to air; used in space heating or cooling.

**Flat-plate collector** A collector of diffuse and beam solar radiation, with five basic component parts:

1. Glazing to reduce heat loss from the illuminated surface of the absorber.
2. Tubes for directing the heat-transfer fluid from the inlet to the outlet.
3. Absorber plate to which tubes are attached in a manner producing a good thermal bond.
4. Insulation to minimize heat loss from the nonilluminated side of the absorber plate.

5. Housing surrounding the foregoing components and keeping them free of dust and moisture.

**Fossil fuels** Fuels derived from the remains of carbonaceous flora and fauna; include petroleum, natural gas, coal, oil shale, and tar sands. Fossil fuels all had their origin millions of years ago when ancient plants trapped solar energy and were buried under layers of sediment. About 28 million tons of carbon form new fossil sediments every year, but current consumption of fossil fuels is about 6 billion tons of carbon per year.

**Fuel cell** Device using chemical reactions to produce electricity directly.

**Glazing** Glass, plastic, or other transparent covering of a collector-absorber surface.

**Greenhouse effect** A heat-transfer effect wherein heat loss from surfaces is controlled by suppressing the convection loss; frequently incorrectly attributed to suppression of radiation from an enclosure.

**Gross area** The total frontal area of a collector, including framing and structural supports.

**Heat exchanger** Device used to transfer heat between two fluid streams without mixing them.

**Heating load** The amount of heat and humidity that must be added to a building to maintain occupant comfort.

**Heat of fusion** The quantity of heat that must be added (removed) to melt (freeze) a unit mass of a substance at a constant pressure.

**Heat of vaporization** The quantity of heat required to vaporize a unit mass of liquid at a constant pressure.

**Heat pipe** A passive heat exchanger employing principles of evaporation and condensation to transfer heat at high levels of effectiveness.

**Heat pump** Device that transfers heat from a relatively low-temperature reservoir to one at a relatively higher temperature by the input of shaft work.

**Heliochronometry** Telling of time by means of the sun.

**Heliodon** A solar angle simulator useful in conducting shading assessments on buildings or solar collector arrays.

**Heliostat** Device to direct sunlight toward a fixed target.

**Heliothermal process** Process that uses the sun's radiation to produce heat.

**Heliotropic** An adjective referring to devices or plants that track the sun's apparent motion across the sky.

**Hole** Vacant electron state in valence band; behaves like a positively charged electron.

**Hydronic** Trade term for liquid-based space-heating system.

**Incident angle** The angle between the sun's rays and a line normal to the irradiated surface. In the case of tubular collectors, the angle between the normal to the tube centerline and the sun's rays.

**Infiltration** Uncontrolled air leakage into a building through cracks and pores.

**Infrared radiation** Thermal radiation at wavelengths longer than those of the red end of the visible spectrum.

**Interest rate** Cost of borrowing money; see discount rate.

**Irradiation; irradiance** The radiant energy falling per unit area on a plane surface per unit time, normally stated in watts per square meter (Btu per square foot per hour).

**ISES** Acronym for the International Solar Energy Society. A worldwide society composed of professionals, tinkerers, and enthusiasts interested in the utilization of solar energy.

**Joule** Energy unit equal to one newton-meter.

**Kinetic energy** Energy possessed by virtue of an object's motion.

**Knot** Measure of wind speed; equal to one nautical mile per hour.

**Langley** Unit of solar radiation intensity equivalent to 1.0 gram-calorie per square centimeter.

**Latent heat** Internal energy change required to effect a phase change (q.v.).

**Latitude** The angular distance north (+) or south (−) of the equator, measured in degrees.

**Load factor** The ratio of the average power output of an energy system to the rated output.

**Local solar time, LST** System of astronomical time in which the sun always crosses the true north-south meridian at 12 noon. This system of time differs from local time according to longitude, time zone, and equation of time.

**Micrometer** Unit of measure equal to one millionth of a meter; one micron.

**$n$-type; $N$-type** Semiconductor doped with impurities so as to have free electrons in the conduction band.

**Net area** Area of a solar collector aperture through which radiation may pass; gross area net of all opaque area.

**Normal** Perpendicular, in the geometric sense.

**Ocean thermal conversion** The use of temperature differentials between warm layers and cold layers of the ocean to drive a low-pressure turbine connected to a generator to produce electricity.

**Ocean thermal gradient** Differences in the temperature of layers of the ocean potentially useful for running a heat engine.

**Ohmic contact** Areas comprising the electrical contacts for solar cells for connection to external circuitry.

**Open circuit voltage** Photovoltage developed on an open circuit, which is the maximum available at a given irradiance.

**$p$-type** Semiconductor doped with impurities so as to have vacancies (holes) in the valence band.

**Parabola** Locus of a point moving in a plane so that its distances from a fixed point (focus) and a fixed straight line (directrix) are equal; equation in Cartesian coordinates $y = x^2/4f$ where $f$ is the focal length.

**Pascal** Unit of pressure equal to one newton per square meter.

**Passive systems** Systems using the sun's energy without mechanical systems; achieved by properly placed storage masses, overhangs, and fenestration, use of vegetation, proper building materials, and energy-conserving design.

**Payback period** Length of time required to recover the investment of a project by benefits accruing from the investment; payback period is an incomplete economic index; see Chapter 5.

**Peak watt** Unit used for the performance rating of solar electric power systems. A system rated at one peak watt will deliver one watt at the specified working voltage under peak solar irradiation.

**Phase change** Change of state from solid to liquid, solid to vapor, etc.

**Photochemistry** The branch of chemistry concerned with the study of chemical reactions initiated by interactions of light with chemical matter.

**Photon** A quantum of electromagnetic radiation; its energy $E_\nu$ is related to the frequency $\nu$ of the radiation by the equation $E_\nu = h\nu$, where $h$ is Planck's constant.

**Photosynthesis** The process by which the sun's radiation causes water, carbon dioxide, and nutrients to react in plants in the presence of chlorophyll to produce oxygen.

**Photovoltaic effect** The generation of an electromotive force when radiant energy falls on the boundary between certain dissimilar substances in close contact.

*pn* **junction** Junction of dissimilar semiconductor materials, where electrons move from one type to another, under specific conditions.

**Potential energy** Energy stored in matter because of its position or because of the arrangement of its parts; e.g., water stored behind a dam.

**Power** The rate at which work is done.

**Power coefficient** The ratio of the power extracted by a wind machine to the power available in the wind stream.

**Power density** The amount of power per unit of cross-sectional area of a wind stream.

**Power tower** A tower placed so that the reflected direct radiation from heliostat mirrors can be focused onto a receiver at its top. Heat exchange takes place at the top of the tower.

**Present value** The value of a future cash flow discounted to the present; based on the premise that a dollar today is worth more than a dollar received in the future by virtue of the amount of interest (or return) it earns.

**Pyranometer** A solar radiometer that measures total *diffuse* and *direct* radiation.

**Pyrheliometer** An instrument used to measure the sun's *direct* radiation.

**Pyrolysis** Decomposition of organic material into chemical constituents by the action of heat.

**Quad, Q** Unit of energy equivalent to $10^{15}$ Btu (occasionally defined as $10^{18}$ Btu).

**Quantum** A discrete unit of energy. A quantum of light is called a photon.

**Quantum yield** Ratio of the number of photon-induced reactions occurring to the total number of absorbed photons (or less commonly, to the total number of incident photons).

**Radiometer** Instrument to measure the intensity of any kind of radiation.

**Rankine cycle** A closed heat engine cycle using various components, including a working fluid pumped under pressure to a boiler where heat is added; an expander (turbine) where work is generated; and a condenser used to reject low-grade heat to the environment.

**Reflectance** The ratio of radiation reflected from a surface to that incident on the surface. Reflectivity is the property of reflecting radiation possessed by all materials to varying extents; called the albedo in atmospheric references.

**Retrofit** The installation of solar energy systems in already existing structures.

**Selective surface** A surface that absorbs more or less energy than it emits; from the original term *wavelength selective*.

**Semiconductor** A material whose atomic structure is such that it has an electrical conductivity intermediate between an insulator, such as glass, and a metal, such as copper. Its conductivity can be greatly influenced by the addition of minute amounts of certain elements called doping agents. Depending on the agent used, the doped semiconductor will have either an excess (*n*-type) or a deficiency (*p*-type) of energy-conducting electrons.

**Shroud** A structure around a propeller used to concentrate a wind stream.

**Silicon cells** Photovoltaic cells made principally of silicon, a semiconductor.

**Solar-altitude angle** The angle between the line joining the center of the solar disc to the point of observation at any given instant and the horizontal plane through that point of observation.

**Solar cell** See photovoltaic effect.

**Solar constant** The intensity of solar radiation beyond the earth's atmosphere, at the average earth-sun distance, on a surface perpendicular to the sun's rays. The value for the solar constant is 1353 $W/m^2$, 1.940 $cal/cm^2 \cdot min$, or 429.2 $Btu/ft^2 \cdot hr$ ($\pm$ 1.6 percent).

**Solar distillation** Process in which the sun's energy is utilized for the purification of brackish or poor quality water. The greenhouse effect is utilized to trap heat to evaporate the liquid. The vapor so formed then condenses on the cover plate and can be collected for use.

**Solar energy** That energy, in the form of radiation, emitted from the sun and generated by means of a fusion reaction within the sun.

**Solar furnace** A solar device used to obtain high temperatures by focusing the sun's rays onto a small receiver. The largest solar furnace now in existence is located at Odeillo in the French Pyrenees and is capable of obtaining temperatures of nearly $6000°F$.

**Solar ponds** Ponds of stratified water that collect and retain heat. Convection, normally present in ponds, is suppressed by imposing a stable density gradient of dissolved salts.

**Solar radiation** Radiant energy received from the sun both directly as a beam component and diffusely by scattering from the sky and reflection from the ground.

**Solar rights** The right of a person who uses a solar energy device not to have sunlight blocked by another structure.

**Solar-thermal electric conversion** The conversion of solar energy to thermal energy, which powers turboelectric generators.

**Specific heat** Amount of heat required to raise the temperature of a unit mass a unit amount.

**Spectral energy distribution** A curve showing the variation of spectral irradiance with wavelength.

**Spectral irradiance** The monochromatic irradiance of a surface per unit bandwidth at a particular wavelength. Units often in watts per square meter per nanometer bandwidth.

**Spectral reflectance** Ratio of the energy reflected from a plane surface in a given defined waveband to the energy incident in that waveband.

**Specular reflection** Mirrorlike reflection in which incident and reflected angles are equal.

**SRTA** Stationary reflector, tracking absorber concentrating collector: a spherical dish with movable absorber rod that tracks the movement of the sun.

**Suntime** Time measured on a basis of the sun's virtual motion; see local solar time.

**Thermal conductivity** The amount of heat that can be transferred by conduction through a material of unit area and thickness per unit temperature difference.

**Thermocouple** A thermoelectrical device consisting of two dissimilar wires with their ends connected together. A small voltage is generated when two junctions are at different temperatures; if one junction is kept at a reference temperature, the voltage generated in the other is a measure of the temperature of the other junction above the reference.

**Thermopile**   A large number of thermocouples connected in series.

**Thermostat**   Temperature sensor used to monitor the temperature in a space; provides a signal to a building's thermal control system.

**Thermosyphon**   The convective circulation of fluid occurring in a closed system wherein less dense, warm fluid rises, displaced by denser, cooler fluid in the same fluid loop.

**Ton of refrigeration**   The removal of heat at the rate of 12,000 Btu/hr; derived from melting a ton (2000 lb) of ice, requiring $2000 \times 144$ Btu over a 24-hr period.

**Transmittance**   The ratio of the radiant energy transmitted by a given material to the radiant energy incident on a surface of that material; depends on the angle of incidence.

**Turbidity**   Atmospheric haze resulting from aerosols and particulates.

$U$   Overall heat-loss coefficient; frequently called the $U$ value.

**Ultraviolet radiation**   Radiation having wavelengths shorter than those of X-rays.

**Watt**   Energy rate of one joule per second.

**Wind energy**   Kinetic energy resulting from air motion caused by the interaction of earth's rotation and solar-driven atmospheric currents.

**Work**   Energy transfer causing a force to move through a distance with measure equal to the dot product of the force and displacement vectors.

# NOMENCLATURE

| | |
|---|---|
| $A$ | Area, $m^2$ $(ft^2)$ |
| $a_s$ | Solar-azimuth angle (positive east of south), deg |
| $a_w$ | Wall-azimuth angle (positive east of south), deg |
| $\bar{B}$ | Monthly averaged, daily beam radiation, $kW \cdot hr/m^2 \cdot day$ $(Btu/day \cdot ft^2)$ |
| $B$ | Daily total beam radiation, $kW \cdot hr/m^2 \cdot day$ $(Btu/day \cdot ft^2)$ |
| $C$ | Cost, $ |
| $C_D$ | Drag coefficient |
| $C_{se}$ | Average annual cost of delivered solar energy or energy savings by conservation, $/gJ ($/MMBtu) |
| $c$ | Speed of light, m/sec (ft/sec) |
| $C_p$ | heat capacity |
| $c_p$ | Specific heat at constant pressure, $kJ/kg \cdot K$ $(Btu/lb_m \cdot {}^\circ F)$ |
| $c_v$ | Specific heat at constant volume, $kJ/kg \cdot K$ $(Btu/lb_m \cdot {}^\circ F)$ |
| $CC$ | Cloud cover in tenth of sky covered; e.g., for a clear sky $CC = 0$, for a sky fully covered with clouds $CC = 10$ |
| $COP$ | Coefficient of performance of heat pump or refrigeration system |
| $CPC$ | Compound parabolic concentrator |
| $CR$ | Concentration ratio |
| $CRF$ | Capital-recovery factor |
| $\bar{D}$ | Monthly averaged, daily diffuse (scattered) radiation, $kW \cdot hr/m^2 \cdot day$ $(Btu/day \cdot ft^2)$ |
| $D$ | Daily total diffuse (scattered) radiation, $kW \cdot hr/m^2 \cdot day$ $(Btu/day \cdot ft^2)$; diameter, m (ft) |
| $D_H$ | Hydraulic diameter, m (ft) |
| $E$ | Energy |
| $E_{b\lambda}$ | Spectral emissive power of a black body at $\lambda$, $W/m^2 \cdot \mu m$ $(Btu/hr \cdot ft^2 \cdot \mu m)$ |
| $e$ | Specific internal energy; eccentricity of earth orbit |
| $F$ | Fin efficiency; force |
| $F'$ | Plate efficiency of a flat-plate collector |

The symbols in this nomenclature are used in several chapters. A few additional symbols, used only in one chapter, are defined in the text.

761

$F''$      Flow factor of a flat-plate collector

$F_{ij}$      Radiation shape factor between surfaces $i$ and $j$

$F_R$      Heat-removal factor of a flat-plate collector

$f$      Fanning friction factor; frequency, $s^{-1}$; focal length

$f_s$      Fraction of energy demand delivered by solar system

$G$      Mass flow per unit area ($= \rho V$), kg/m$^2$ · hr (lb/ft$^2$ · hr)

$Gr_L$      Grashof number based on length dimension $L$

$g$      Gravitational acceleration, m/sec$^2$ (ft/sec$^2$)

$g_c$      Inertia proportional factor occurring in relation: force = mass × acceleration/$g_c$

$\bar{H}$      Monthly averaged, daily total horizontal radiation, kW · hr/m$^2$ · day (Btu/day · ft$^2$)

$H$      Daily total horizontal radiation, kW · hr/m$^2$ · day (Btu/day · ft$^2$)

$h$      Specific enthalpy, kJ/kg (Btu/lb); Planck's constant; altitude

$h_c$      Convection heat-transfer coefficient between a surface and a fluid, W/m$^2$ · K (Btu/hr · ft$^2$ · °F)

$h_s$      Local solar-hour angle (measured from local solar noon, 1 hr = 15°)

$h_{ss}, h_{sr}$      Hour angle between sunset (or sunrise) and local solar noon

$I$      Insolation, defined as the instantaneous or hourly solar radiation on a surface, W/m$^2$ (Btu/hr · ft$^2$) with appropriate subscripts to denote beam, diffuse, reflected, horizontal, tilted, etc.; current, A

$I_0$      Solar constant, W/m$^2$ (Btu/hr · ft$^2$)

$i$      Interest rate, percent; incidence angle, rad (deg)

$J$      Mechanical equivalent of heat; flux

$K$      Extinction coefficient, m$^{-1}$ (ft$^{-1}$)

$\bar{K}_T$      Ratio of monthly averaged, total horizontal radiation on a terrestrial surface to that on the corresponding extraterrestrial surface, i.e., the monthly clearness index

$K_T$      Daily ratio of horizontal total radiation on a terrestrial surface to that on the corresponding extraterrestrial surface, i.e., the daily clearness index

$k$      Thermal conductivity, W/m · K (Btu/hr · ft · °F); Boltzmann's constant

$k_T$      Instantaneous or hourly ratio of horizontal total radiation on a terrestrial surface to that on the corresponding extraterrestrial surface, i.e., the instantaneous or hourly clearness index (sometimes called percent of possible radiation)

$L$      Length, m (ft); latitude, rad (deg); thermal load or demand

$l$      Length, m (ft)

$m$      Mass, kg (lb$_m$); air mass, i.e., the distance through which radiation travels from the outer edge of the earth's atmosphere to a recovery point on the earth divided by the distance radiation travels to a point on the equator at sea level when the sun is overhead

$\dot{m}$      Mass flow rate, kg/sec (lb$_m$/hr)

$N$      Avogadro's number

$n$      Index of refraction; number index; angular velocity, $s^{-1}$; polytropic exponent in $pv^n$ = const

$NTU$      Number of transfer units (dimensionless)

$Nu_L$      Nusselt number based on length dimension $L$

$P$      Power, W (hp)

$p$      Pressure, Pa = N/m$^2$ (lb/in$^2$)

$PP$      Percent of possible sunshine

| | |
|---|---|
| Pr | Prandtl number |
| $PV$ | Present value |
| $PWF$ | Present-worth factor |
| $Q$ | Quantity of energy or heat, kJ or kW $\cdot$ hr (Btu) |
| $q$ | Rate of heat flow, W (Btu/hr) |
| $R$ | Thermal resistance, K $\cdot$ m$^2$/W ($^{\circ}$F $\cdot$ ft$^2$ $\cdot$ hr/Btu); tilt factor; gas constant in $pv = RT$ |
| $r$ | Radius, m (ft) |
| Ra | Rayleigh number |
| Re$_L$ | Reynolds number based on length dimension $L$ |
| Ri | Richardson number |
| $S$ | Surface area, m$^2$ (ft$^2$) |
| $s$ | Specific entropy, kJ/kg $\cdot$ K (Btu/lb $\cdot$ $^{\circ}$R) |
| $T$ | Temperature, K or $^{\circ}$C ($^{\circ}$F or $^{\circ}$R); tax or tax rate |
| $t$ | Time, sec (hr); thickness, m (ft) |
| $U$ | Overall heat-transfer coefficient, W/m$^2$ $\cdot$ K (Btu/hr $\cdot$ ft$^2$ $\cdot$ $^{\circ}$F) |
| $u$ | Specific internal energy, kJ/kg (Btu/lb$_m$) |
| $V$ | Volume, m$^3$ (ft$^3$); voltage, V |
| $v$ | Velocity, m/sec (ft/sec); specific volume, m$^3$/kg (ft$^3$/lb$_m$) |
| $W$ | Humidity ratio, kg water/kg dry air (lb water/lb dry air); work, kW $\cdot$ hr (ft $\cdot$ lb$_f$) |
| $w$ | Width, m (ft) |
| $X$ | Quality, mass fraction of vapor in a two-phase mixture |
| $z$ | Zenith angle, deg; altitude above mean sea level, m (ft) |
| $\alpha$ | Absorptance; solar-altitude angle |
| $\beta$ | Volumetric thermal expansion coefficient $[\beta = (1/v)\partial v/\partial T = -(1/\rho)\partial \rho/\partial T]$; collector tilt angle from horizontal plane |
| $\gamma$ | Solar profile angle; specific heat ratio |
| $\delta$ | boundary-layer thickness, m (ft); declination; optical intercept factor |
| $\Delta$ | Difference |
| $\epsilon$ | Emittance |
| $\varepsilon$ | Heat-exchanger effectiveness |
| $\eta$ | Efficiency; effectiveness |
| $\theta, \phi$ | Angles |
| $\lambda$ | Wavelength; mean free path, m (ft) |
| $\mu$ | Dynamic viscosity; mean value |
| $\nu$ | Frequency; kinematic viscosity |
| $\rho$ | Reflectance; density |
| $\sigma$ | Stefan-Boltzmann constant; standard deviation |
| $\tau$ | Transmittance; shear stress |
| $\omega$ | Solid angle |
| $\dot{\omega}$ | Angular velocity |

## Subscripts

| | |
|---|---|
| $a$ | Air |
| ab | Absorbent |
| amb | Ambient conditions |
| ann | Annual |
| atm | Atmospheric |

| | |
|---|---|
| aux | Auxiliary |
| $b$ | Beam; particle bed |
| $C$ | Carnot cycle |
| $c$ | Collector; convection |
| $D_H$ | Diameter (hydraulic) |
| $d$ | Diffuse (scattered); delivery conditions |
| day | Day |
| $dn$ | Direct normal |
| eff | Effective |
| exp | Expansion |
| $f$ | Fluid; fin; fuel |
| $g$ | Glass |
| $h$ | Horizontal surface |
| $hp$ | Heat pump |
| $hw$ | Hot water |
| $hx$ | Heat exchanger |
| $i$ | Incident; infiltration |
| in | Inlet or inside |
| $ir$ | Infrared |
| $k$ | Conduction |
| $L$ | Length dimension |
| $l$ | Loss |
| $m$ | Air mass; monthly totals; maintenance |
| max | Maximum |
| min | Minimum |
| $n$ | Normal |
| $o$ | Extraterrestrial; optical |
| opt | Optimum |
| out | Outlet or outside |
| $R$ | Rankine cycle |
| $r$ | Reflected; roof |
| rad | Radiation |
| ref | Refrigerant |
| $s$ | Solar; surface |
| sky | Sky |
| $ss$ | Sunset |
| $sr$ | Sunrise |
| $T$ | Total |
| $t$ | Tilted surface; top |
| terr | Terrestrial |
| $u$ | Useful |
| $v$ | Ventilation |
| $w$ | Wall; water |
| $y$ | Yearly |
| $z$ | Zenith |
| $\infty$ | Sink or environmental conditions |
| 0 | Reference or standard |
| $\parallel$ | Parallel |
| $\perp$ | Perpendicular |

# BIBLIOGRAPHY

AIA Research Corporation, "Solar Dwelling Design Concepts," U.S. Government Printing Office, Washington, D.C., 1976. A book with excellent graphics illustrating many features of the solar schematic design process; for the generalist, planner, and architect.

American Society of Heating, Refrigerating, and Air Conditioning Engineers, *ASHRAE Journal*, New York. Published monthly, the *ASHRAE Journal* frequently includes articles on the application of solar energy in the space-conditioning industry and on the topic of solar component testing and standards.

American Society of Mechanical Engineers, *Transactions of the ASME*, New York. Published quarterly, the ASME *Journal of Heat Transfer* and *Journal of Engineering for Power* contain technical articles on solar energy conversion, collection, and storage.

Association for Applied Solar Energy, Stanford Research Institute, University of Arizona, "World Symposium on Applied Solar Energy," Stanford Research Institute, Palo Alto, 1956. Papers presented at the AFASE meeting held in Phoenix on November 1–5, 1955; proceedings of a gathering of world authorities on thermal, photochemical, and photoelectrical applications of solar energy. (Available from Johnson Reprint Corporation, New York.)

Beach, C., and E. Fordyce, "Proceedings of the 1977 Annual Meeting American Section of the International Solar Energy Society," American Section ISES, Winnipeg, 1977. Three volumes of papers, presented at the Florida meeting of the AS–ISES, that cover a broad range of technologies and socioeconomic topics. Abstracts of poster sessions are included.

Beckman, W. A., S. A. Klein, and J. A. Duffie, "Solar Heating Design," John Wiley & Sons, 1977. Describes in great detail the *f*-chart method applied to air- and liquid-based solar-heating systems with application to economic optimization in the United States.

Bockris, J. O'M., "Energy: The Solar-Hydrogen Alternative," The Architectural Press, Ltd., London, 1976. Presents a highly optimistic future energy scenario based on solar collection with hydrogen storage and distribution.

Carpenter, E. F., ed., "Transactions of the Conference on the Use of Solar Energy—The Scientific Basis," University of Arizona, Stanford Research

Institute, Association for Applied Solar Energy, Tucson, 1958. Basic collection, in five volumes, of papers on solar radiation, thermal applications, photochemistry, and solar electricity from the conference held in Tucson, October 31–November 1, 1955.

Commoner, B., "The Poverty of Power," Alfred A. Knopf, New York, 1976. Contains a chapter on solar energy as viewed from a second law perspective.

Daniels, F., "Direct Use of the Sun's Energy," Yale University Press, New Haven, 1964. A complete description of all direct methods of solar energy use for the generalist; an excellent introduction to the field.

Daniels, F., and J. Duffie, eds., "Solar Energy Research," University of Wisconsin, Madison, 1955. A collection of 38 presentations to a conference of solar researchers held at the University of Wisconsin on September 12–14, 1953; major topics include solar radiation, solar heating, solar power, solar distillation, and photosynthesis. (Available from University Microfilms, Ann Arbor.)

Deutsche Gesellschaft für Sonnenenergie, "Grundlagen der Solar-Technik I," D. Ges. für Sonn., P.O. 200604, D-8000, München, 2 DRG. Contains meteorological data for Europe, and technical information on flat-plate collectors, corrosion, and storage for applications in Europe.

deWinter, F., ed., Workshop Proceedings—Solar Cooling for Buildings, *NSF Rept.* NSF-RA-N-74-063, 1974 (*USGPO* 3800-00189). Papers and discussion for the first NSF solar-cooling workshop held in Los Angeles, February 6–8, 1974.

deWinter, F., ed., The Use of Solar Energy for the Cooling of Buildings, *ERDA Rept.* SAN/1122-76/2, 1976. Papers of the second solar-cooling workshop held at UCLA on August 4–6, 1975.

Duffie, J. A., and W. A. Beckman, "Solar Energy Thermal Processes," John Wiley & Sons, New York, 1974. A thorough technical treatise on solar radiation, collector methods (flat-plate analysis), and modeling by mathematical simulation; emphasis is on building heating and cooling.

General Electric, Solar Heating and Cooling of Buildings, Phase O, *NTIS Repts.* PB-235431–PB-235435, 1974. Massive study of the capture potential of solar energy in the United States; appendixes contain massive amounts of data, albeit not always well cataloged.

Hovel, Harold J., "Solar Cells," Volume 11 of the series "Semiconductors and Semimetals," Academic Press, New York, 1975.

International Solar Energy Society, American Section and Solar Energy Society of Canada, "Sharing the Sun, Solar Technology in the Seventies," ISES, Winnipeg, 1976. Complete papers in 10 volumes presented at the ISES American Section meeting in Winnipeg, Manitoba, August 15–20, 1976.

International Solar Energy Society, United Kingdom Section, "Solar Energy—A U.K. Assessment," UK–ISES, The Royal Institution, London, 1976. A technology assessment of the potential of solar energy for the United Kingdom. Considers all forms of solar energy conversion and suggests research priorities for the United Kingdom.

Jensen, J. S., "Applied Solar Energy Research," Association for Applied Solar Energy, Phoenix, 1959. An annotated bibliography of 2916 solar writings published prior to 1960.

Jordan, R. C., ed., "Applications of Solar Energy for Heating and Cooling Buildings," ASHRAE, New York, 1977. An update of an earlier reference with new information on design and modeling and on air systems for space heating.

Kreider, J. F., and F. Kreith, "Solar Heating and Cooling," revised 1st ed., Hemisphere Publ. Corp., Washington, D.C., 1977. A practical introductory handbook for designers of solar-heating and solar-cooling systems for buildings; extensive tabulated reference data.

Messel, H., and S. T. Butler, eds., "Solar Energy," Pergamon Press, New York, 1975. A collection of invited articles on solar energy; topics include resource magnitude, bioconversion, thermal conversion, and direct conversion.

Sargent, S., ed., Proceedings of the Workshop on Solar Collectors for Heating and Cooling of Buildings, *NSF Rept.* NSF-RA-N-75-019, 1975. Papers presented at the solar collector workshop held in New York City on November 21–23, 1974; numerous flat-plate and concentrating designs commercially available or under development are described.

Szokolay, S. V., "Solar Energy and Building," John Wiley & Sons, New York, 1975. Views solar energy from the architectural perspective; includes detailed descriptions of more than two dozen solar-heated and solar-cooled buildings.

TRW, Inc., Solar Heating and Cooling of Buildings, Phase O, *NTIS Repts.* PB-235422–PB235424, 1974. (See entry for General Electric.)

United Nations, "Proceedings of the United Nations Conference on New Sources of Energy (1961)," United Nations, New York, 1964. A basic reference containing full papers presented at the conference in Rome, August 21–31, 1961; vols. 4, 5, and 6: "Solar Energy"; vol. 7: "Wind Energy."

U.S. Energy Research and Development Administration, Solar Energy—A Bibliography, *NTIS Rept.* TID-3351-RIPI, 1976. Annotated bibliography of 9732 solar works published prior to March 1976; updated periodically.

U.S. Energy Research and Development Administration, Passive Solar Heating and Cooling Conference and Workshop Proceedings, *Los Alamos Rept.* LA-6637-C, 1977. Three hundred and fifty pages of information useful to the passive system enthusiast and designer. The majority of the material is qualitative with little of use to the engineer requiring quantitative design criteria.

Westinghouse, Inc., Solar Heating and Cooling of Buildings, Phase O, *NTIS Repts.* PB-235426–PB-235429, 1974. (See entry for General Electric.)

Zarem, A. M., and D. D. Erway, eds., "Introduction to the Utilization of Solar Energy," McGraw-Hill Book Co., New York, 1963. A collection of papers on all practical solar energy applications by preeminent investigators of the era; a useful reference today. (Available from University Microfilms, Ann Arbor.)

# AUTHOR INDEX

# SUBJECT INDEX